International Conference on Automation, Computing and Communication (ICACC-2021)

ITM Web of Conferences Volume 40 (2021)

Navi Mumbai, India
14 - 15 July 2021

Editors:

Mukesh D. Patil
Vishwesh A. Vyawahare

ISBN: 978-1-7138-3468-7

Printed from e-media with permission by:

Curran Associates, Inc.
57 Morehouse Lane
Red Hook, NY 12571

CURRAN ASSOCIATES INC.
proceedings.com

Some format issues inherent in the e-media version may also appear in this print version.

No changes have been made to the content of these proceedings. There may be changes to pagination and minor adjustments for aesthetics.

Printed with permission by Curran Associates, Inc. (2022)

For additional information, please contact EDP Sciences – Web of Conferences at the address below.

EDP Sciences – Web of Conferences
17, Avenue du Hoggar
Parc d'Activité de Courtabœuf
BP 112
F-91944 Les Ulis Cedex A
France

Phone: +33 (0) 1 69 18 75 75
Fax: +33 (0) 1 69 28 84 91

contact-edps@webofconferences.org

Additional copies of this publication are available from:

Curran Associates, Inc.
57 Morehouse Lane
Red Hook, NY 12571 USA
Phone: 845-758-0400
Fax: 845-758-2633
Email: curran@proceedings.com
Web: www.proceedings.com

TABLE OF CONTENTS

Preface

The second International Conference on Automation, Computing and Communication (ICACC-2021) was organized by Ramrao Adik Institute of Technology (which is under the ambit of D. Y. Patil deemed to be University), Nerul, Navi Mumbai, India, between July 14-15, 2021. It was technically co-sponsored by the IEEE Bombay Section and IETE Mumbai Chapter. Due to the COVID 19 pandemic, the conference was conducted online.

ICACC-2021 was organized with a motivation to provide a platform for researchers working in the areas of automation, computing and communication to showcase their research and to create a strong community which thrive on the frontier of technology.

Like our previous year's conference, ICACC-2020, this year too we received overwhelming response for our call for papers. This year we received a total of 145 submissions. All the papers were first checked for plagiarism and then were sent to external reviewers for review for the originality and novelty reported. After a very meticulous and strict review, only 65 papers (amounting to 45%) were selected for publication in the conference proceedings.

ICACC is gaining popularity among the researchers in India and abroad and it is a matter of prestige for the authors to get their paper included in ICACC proceedings. We received papers from researchers across the country.

The accepted submissions report original and novel results in various fields like control systems, signal and image processing, AI and ML applications, intelligent systems, PV cell technology, Deep Learning, Blockchain, virtual reality and many more. The keynote speakers and authors presented their work online. The sessions were interactive and brainstorming.

We would like to take this opportunity to express our deep sense of gratitude towards our management, especially Hon. Dr D Y Patil (Founder President), Hon. Dr Vijay Patil (President), Hon. Mrs Shivani Patil (Managing Trustee) and Dr. Shirish Patil, Hon. Vice-Chancellor DYPU for the encouragement and support. We put on record our sincere thanks to the members of international programme committee and technical programme committee, keynote speakers, reviewers, session chairs, session coordinators, authors and participants for making this conference a grand success. Thanks are due to our colleagues in the institute for supporting us. The tireless efforts and meticulous planning by the organizing team to make this event successful deserve special appreciation. We would also like to acknowledge the cooperation and support by the teams at Easychair conference management system and MATEC, Web of Conferences.

We sincerely hope that research community will be truly benefitted by the ICACC-2021 contributions.

We look forward to welcome all in ICACC-2022.

Prof Mukesh D. Patil, Conference Chair, ICACC-2021
Prof Vishwesh A. Vyawahare, Convener, ICACC-2021
July 2021.

Statement of Peer review

In submitting conference proceedings to *Web of Conferences*, the editors of the proceedings certify to the Publisher that

1. They adhere to its **Policy on Publishing Integrity** in order to safeguard good scientific practice in publishing.
2. All articles have been subjected to peer review administered by the proceedings editors.
3. Reviews have been conducted by expert referees, who have been requested to provide unbiased and constructive comments aimed, whenever possible, at improving the work.
4. Proceedings editors have taken all reasonable steps to ensure the quality of the materials they publish and their decision to accept or reject a paper for publication has been based only on the merits of the work and the relevance to the journal.

Title, date and place of the conference

International Conference on Automation, Computing and Communication, July 14-15, 2021
Ramrao Adik Institute of Technology, Nerul, Navi Mumbai, India

Proceedings editor(s):

Dr. Mukesh D Patil,
Dr. Vishwesh A Vyawahare

Date and editor's signature

July 30, 2021

ITM Web of Conferences **40**, 01001 (2021)
ICACC-2021
https://doi.org/10.1051/itmconf/20214001001

Performance Comparison of Various controllers on Semi-Active Vehicle Suspension System

Sarvesh Walavalkar[1,*], *Viraj* Tandel[1,**], *Rahul Sunil* Thakur[1,***], *V.V Pramod* Kumar[1,****], and *Supriya* Bhuran[1,†]

[1]Ramrao Adik Institute of Technology, Nerul, Navi Mumbai 400706, Maharashtra, India

Abstract. The value of a self-tuning adaptive semi-active control scheme for automotive suspension systems is discussed in this paper. The current vehicle suspension system uses fixed-coefficient springs and dampers. The ability of vehicle suspension systems to provide good road handling and improve passenger comfort is usually valued. Passive suspension allows you to choose between these two options. Semi-Active suspension(SAS), on the other hand, can provide both road handling and comfort by manipulating the suspension force actuators directly. The semi-active suspension system for a quarter car model is compared to passive and various controllers such as Proportional-Integral, Proportional-Integral-Derivative, Internal model control (IMC)-PID, IMC-PID with filter, FUZZY, and Adaptive-network-based fuzzy inference system(ANFIS) in this analysis. This research could be relevant in the future for designing better car suspension adjustments to eliminate vertical jerks and rolling motion experienced by the vehicle body on bumps and humps.

1 Introduction

1.1 Motivation and background

Suspension systems are vital for vehicles' health and their life expectancies, and also for rider's health, safety, and comfort. The chore of the vehicle suspension system is to ensure ride comfort by alienating unwanted disturbances in various conditions The current passive suspension systems limit the motion of the Vehicle's body and wheel by limiting their relative velocities to a rate that fails to provide the required ride comfort. So, developing and designing an advanced controller for semi-active suspension systems will reinforce the existing semi-active suspension systems and provide better resistance against the vertical acceleration caused by road disturbances. The underlying motive for conducting a comparative study of various controllers' responses to speed bumps is because speed breakers are a common sight in India and also an indicator of disruption in road transport and safety.

1.2 Contribution and paper structure

In this paper, a control technique for linear time-invariant based semi-active suspension for a quarter car model are presented. The study focuses on bumps and humps, which are specified by the Indian government in section IRC 99-1988 of the Indian Road Congress and customary design approved by different manufacturers all over India. Each of the bumps and humps considered has a distinct shape and height. According to safety standards, the Indian government has contemplated a certain highest arrival speed in which the heave motion would be minimum and considered to be safe for high-value cargo vehicles and ambulances. The article suggested does not take into account the speed component, but rather the response of the model to displacement caused by road disturbances.This study could be beneficial to automobile makers by offering a simple method of developing suspension systems for cars that deliver outstanding dampening response on government-approved speed bumps.

2 Literature Review

Many valuable contributions have recently been published in the literature for the design of semi-active suspension systems, with each publication focused on distinct characteristics such as invariable road disturbances, design framework and so on. Each paper has a distinct ideology to tackle the heave motion, and the intrusion created due to road profile. The passive system is considered in many research as of open-loop type and pairing it with any controller improves the transient response significantly. The advantage gained by this approach notably reduces peak overshoot, settling time, and other time-domain parameters[4]. Another approach to this problem is by implementing higher accuracy controller like Model predictive controller cascaded with the secondary controller, the output of semi-active system is much better than passive in invariable road disturbances by improving vertical chassis acceleration and peak chassis acceleration at bump points[6].

Logic-based controllers like fuzzy are proved to provide the best response in such random inference sys-

*e-mail: sarvesh.walavalkar@gmail.com
**e-mail: virajtan60@gmail.com
***e-mail: thakurrahul62690@gmail.com
****e-mail: pramodvepada99@gmail.com
†e-mail: supriya.bhuran@rait.ac.in

tems. When fuzzy is optimized with the right optimization method, such as particle swarm optimization, the maximum peak overshoot and stabilization time are reduced [8]. Advanced research in this area has revealed that semi-active systems can be nonlinear and that by taking nonlinear factors into account, a fuzzy controller may be implemented on a nonlinear model to reduce heave, rolling, and pitch motion acting on the chassis[5].

A certain approach toward developing a cheap model of SAS is by using magnetorheological dampers(MR) much research is focused on developing a stable model of MR dampers and increasing its overall response. Considering a Bouc–Wen model of the MR damper can pivot the direction of the research to non linearities present in the model.Bouc–Wen model provides more insight on acceleration acting on various sections of the body. With a proper state space model a more accurate model can be designed and various parameters such as acceleration,velocities can be derived and reduced accordingly[12]. Complex controllers like fuzzy and sliding mode control can also be implemented on such systems. One of the approaches toward developing is to pair complex controllers with neural networking[1]. This strategy proves useful for annihilating a certain type of disturbance from the system. Variable universal fuzzy control, Fuzzy neural network and particle swarm optimization (VUFC-FNN0-PSO) are more effective to remove random road vibrations than passive. Particle swarm optimization provides backward integration to training and learning for the fuzzy neural network (FNN) which can improve the response further[13].

A new perspective towards this anomaly is by constructing a skyhook model. These models improve the researchers attitude towards ride comfort qualities. These methods improve the comfort quality by 24% than traditional systems without sacrificing the handling qualities[2][3]. This method is only useful for random noise generated by the road. Skyhook control algorithm maintains the degradation of handling performance within the range of 10% which improves the system significantly than the standard models[14]. Linear Parameter Varying (LPV), and Frequency Estimation-Based (FEB) are some of the approaches to increase handling and road grip of the tires. Comfort improvements during the LPV approach are 36% greater in pitch reduction than the passive control strategies and 24% better in chassis control. Both controllers reduce the tire compression by 30% and reduce heave motion by 40% than any standard suspension system[15].

3 Proposed Algorithms

The study's approach is to use mechanical equations of the system to construct a proper and reliable model of the passive suspension system. Now, this passive model is converted into a linearized control system model to get the transfer function. Using this transfer function, the stability analysis of the model can be obtained. The transfer function provided the stability analysis and the value of damping coefficient (). The next step is to turn the open loop passive model into a closed loop model with a controller and a feedback element gain that can be cascaded. The reaction of the suspension system connected with the controller can be compared to the passive model response by picking a correct controller each time. If the controller's response is worse than the passive system's, it can be fine-tuned or paired with a different optimization strategy. The process continues until the desired response is achieved.

4 Methodology

A closed-loop system with feedback elements and external road disturbance can be simplified with the given block diagram in Fig. 1.

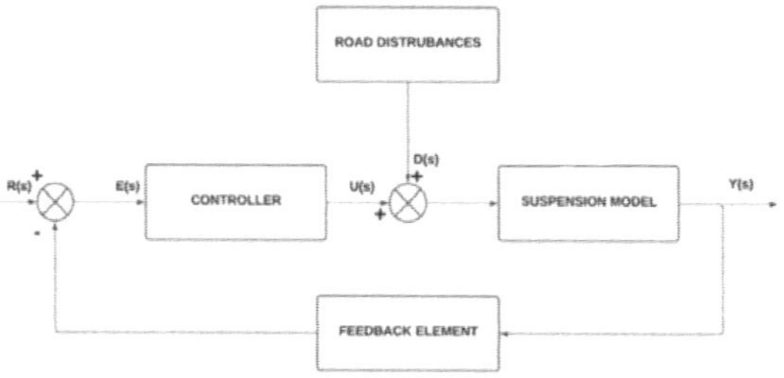

Figure 1. Block Diagram for closed loop system

Vertical motions and disturbances will be detected by sensors positioned on the vehicle's body. These values are sent to the controller, which will take corrective action to bring the system back into balance. The damping coefficient will be controlled directly by the controller by manipulating current values into system. Fig. 2. demonstrates the dynamic model for the quarter car system.

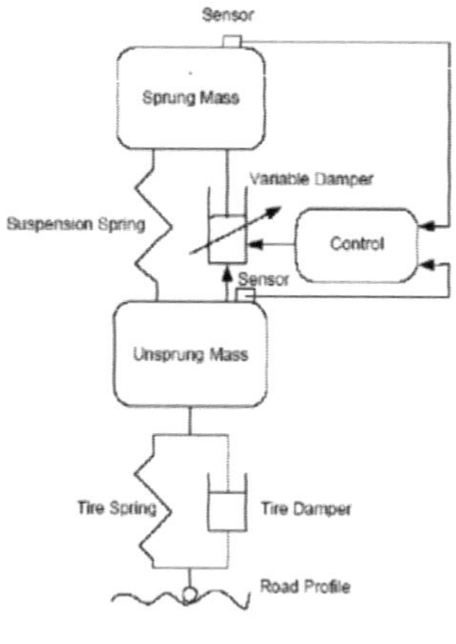

Figure 2. Dynamic model for Quarter car

4.1 Passive Suspension System

The characteristics of springs and dampers are fixed in the passive suspension system. The designer of the suspension determines these characteristics concerning for design goals and the desired application. Consider M_2, M_1, and W to be wheel mass, vehicle mass, and tire contact point to the road respectively. An unrealistic assumption that the axles move in the same way is made considering the vertical movement of the car body and axles so that they are carried together and the rotational movement of the car body can be disregarded. Since the road is flat, we can consider the car body and the axle can move autonomously, x_1 and x_2 are reference variables which are position variables, the position of the car body (mass M_1), and of axle (mass M_2) under the impact of gravity of the normal road level. The tires are modeled as rigid with a high stiffness that is combined in parallel with a damper for energy dissipation through the rubber. A linear spring and a linear damper connecting the axle to the car body are used to model the main suspension systems, which include coil springs and shock absorbers. Let x_1 be the difference between the usual road level and the point of contact between the tire and the road. The profile of the road and the location of the car decide x_1. The tire spring force (in an upward direction) associated with the tire's deviation from the standard road level is low, while the corresponding frictional force is $b2W$. The state-space equations can be obtained by applying Newton's second and third law to Fig. 3.

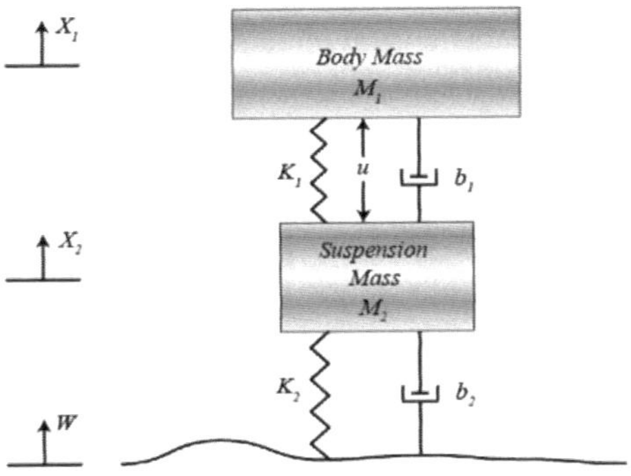

Figure 3. Semi - Active Suspension for Quarter car Model [4]

The equation of motion would be as follows if Newton's second law is applied to the two masses and Newton's third law is applied to the interaction between the two masses [4]:

$$M_1\ddot{x}_1 + b_1(\dot{x}_1 - \dot{x}_2) + k_1(x_1 - x_2) = 0 \quad (1)$$

$$M_2\ddot{x}_2 + b_1(\dot{x}_2 - \dot{x}_1) + k_1(x_2 - x_1) + b_2\ddot{x}_2 + k_2x_2 = b_2\dot{W} + kW \quad (2)$$

$$M_1S^2x_1(s) + b_1S(x_1(s) - x_2(s)) + k_1(x_1(s) - x_2(s)) = 0 \quad (3)$$

$$M_2S^2x_2(s) + b_1S(x_2(s) - x_1(s)) + k_1(x_2(s) - x_1(s)) + 2S\,x^2(s) + k_2x^2(s) = b_2S\,W(s) + k_2W(s) \quad (4)$$

$$M_1\ddot{x}_1 + \widetilde{b}_1(x_1 - x_2) + k_1(x_1 - x_2) = 0 \quad (5)$$

$$M_2\ddot{x}_2 + \widetilde{b}_1(\dot{x}_2 - \dot{x}_1) + k_1(x_2 - x_1) + b_2\dot{x}_2 + k_2x_2 = b_2\dot{W} + k_2W \quad (6)$$

$$M_1S^2x_1(s) + \widetilde{b}_1S(x_1(s) - x_2(s)) + k_1(x_1(s) - x_2(s)) = 0 \quad (7)$$

$$M_2S^2x_2(s) + \widetilde{b}_1S(x_2(s) - x_1(s)) + k_1(x_2(s) - x_1(s)) + b_2S\,x^2(s) + k_2x^2(s) = b_2S\,W(s) + k_2W(s) \quad (8)$$

To obtain the transfer function, We have considered a linearized open-loop model while maintaining its linear invariant characteristics.The obtained Fourth order Transfer Function is

$$\frac{422.1s + 8442}{s^4 + 44.93s^3 + 555.5s^2 + 1327s + 8442} \quad (9)$$

The parameters considered are M_1=241.5 kg, M_2=41.5 kg, K_1=6000 N/m, K_2=14000 N/m, b_1=300 N-s/m, b_2=1500 N-s/m

4.2 Semi-Active Suspension with PID for Quarter Car Model

Only the damping speeds, an embedded controller, and a series of sensors are handled by external power in the Semi-active suspension system. Based on a control strategy, the controller computes the amount of damping and automatically adjusts the damper to achieve the damping. The proportionality constants Kp, Ki, and Kd are determined by the method of trial and error.

When tuning controllers, deciding which controller to use is crucial. There are four basic control combinations: proportional (P) only, proportional plus integral (PI), proportional plus derivative (PD), and proportional plus integral plus derivative (PID). For comparative analysis, we have used PI and PID controllers.

K_i=35.2368, K_d=23.6584, and K_p=50.3652 are the integral, derivative, and proportionality constants, respectively[4].

4.3 Semi – Active Suspension with IMC based PID (with and without filter) for Quarter Car Model

The IMC-based PID method provides the same feedback output as IMC for stable processes with no time delay. The IMC-based PID procedure will not provide the same output as IMC for stable processes with a time delay since a time delay is introduced in Padé's deadtime approximation. Calculation of dead time from the graph(θp), and compared with standard equations and calculated gains and selected lambda from the given dead time observed with formulas. For IMC-based PID tuning, we have reduced the 4th order transfer function to 2nd order transfer function. To obtain 2nd-order transfer function we have observed dead time(θ) from the graph. The selection of λ and θ is essential[3].
Second order Transfer Function =

$$\frac{1}{0.0557s^2 + 0.0704s + 1} \quad (10)$$

Standard equation =

$$\frac{Kp}{(\tau)s^2 + 2(\zeta)(\tau)s + 1} \tag{11}$$

K_i=65.9198, K_d=45.6972, K_p=55.3254, and N=196.2526 are the integral, derivative, proportionality constants, and Filter coefficient respectively.

4.4 Semi–Active Suspension with Fuzzy controller for quarter car model

A mathematical model of the system to be controlled is needed for classical control. However, since it is based on fuzzy logic which is shown in Table 1., it does not require a mathematical model. a structure based on rules, when applied to ill-defined and dynamic systems, FLC has an advantage over classical controllers[9].

The steps for developing a Fuzzy Logic-based suspension system are as follows:

- Fuzzification: the sharp inputs are converted into fuzzy outputs.
- Rule design: the truth values of the fuzzy output are computed,
- Computation: the requisite control function is computed, and
- Defuzzification: the fuzzy output is transformed into crisp values[1].

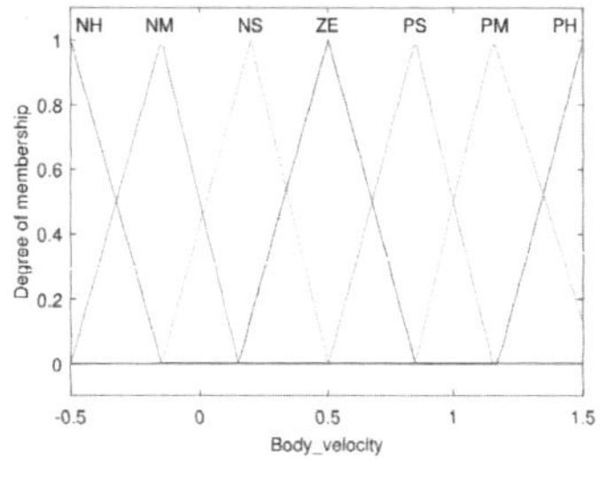

Figure 4. Membership Function of i/p variable "Body Velocity"

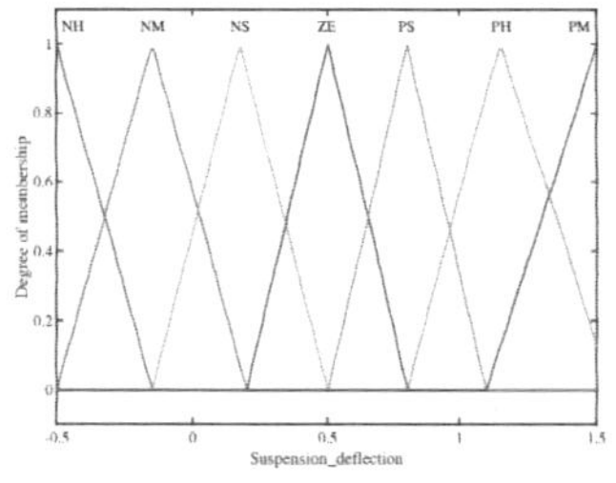

Figure 5. Membership Function of i/p variable "Suspension Deflection"

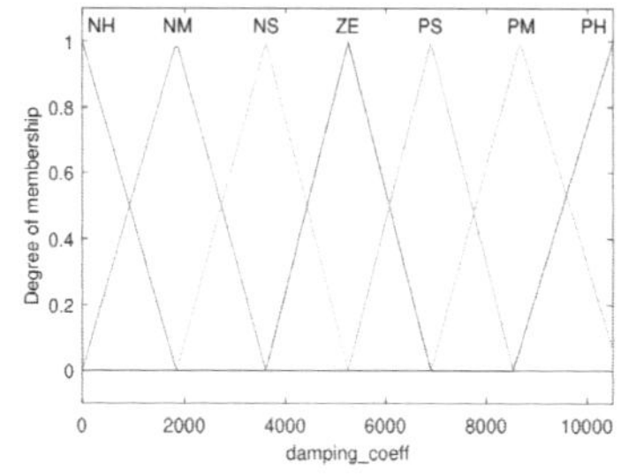

Figure 6. Membership Function of o/p variable "Damping Coefficient"

A membership function is a fuzzy set with pre defined range of 0 to 1 as shown in Fig 4,5 and 6, Fig. 7. explains the output surface of the proposed fuzzy system.

	PL	PM	PL	ZE	NS	NM	NL
NL	PM	PS	ZE	NS	NS	NS	NM
NM	PM	PM	PS	NS	NM	NM	NM
NS	PL	PL	PM	NS	NM	NM	NM
ZE	PL	PL	PM	ZE	NM	NL	NL
PS	PL	PM	PM	PS	NM	NL	NL
PM	PM	PM	PM	PS	NS	NM	NM
PL	PM	PS	PS	PS	ZE	NS	NM

Table 1. Fuzzy Rule Table [7]

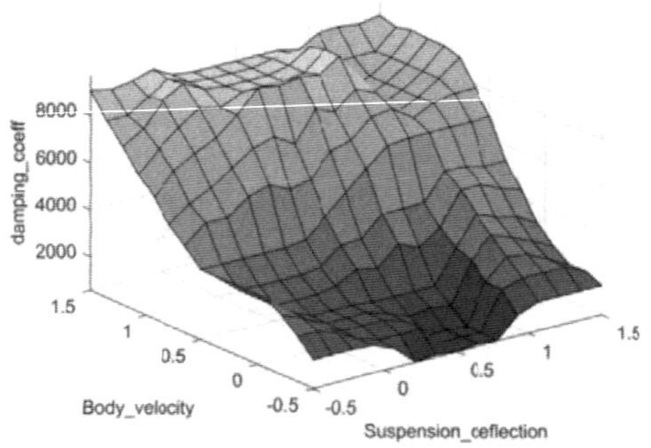

Figure 7. Surface view of Fuzzy i/p vs o/p functions

4.5 Semi-Active Suspension with ANFIS for quarter car model

The ANFIS controller combines the benefits of a fuzzy controller and a neural network. Fuzzy logic demonstrates computational effectiveness in integrating human expertise related to the control system being modelled, while neural networks are highly competitive in dealing with mathematical data.

As a result, the ANFIS controller, when used in conjunction with a neural network and a fuzzy controller, has a higher capacity due to its ability to learn from mathematical results[10]. Based on the selected Membership fuctions shapes and input-output pairs, the structure and fuzzy if-then rules of FIS in the ANFIS controller are automatically constructed based on the neural network learning algorithm.[2] A membership function is a fuzzy set with pre-defined range of 0 to 1 as shown in Fig 8 and 9, Fig. 10. demonstrates the Surface view of ANFIS for i/p vs o/p functions.

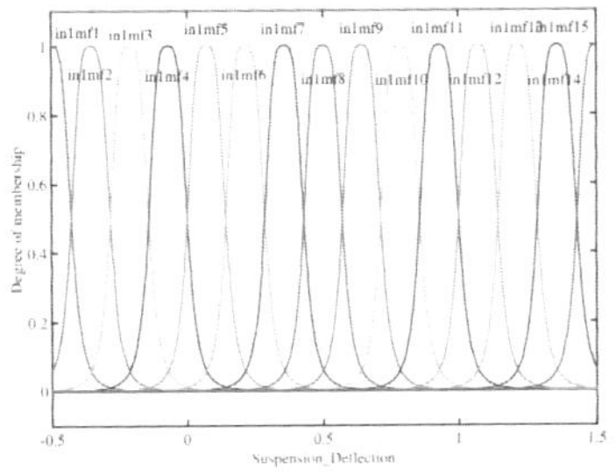

Figure 8. Membership Function of i/p variable "Suspension Deflection"

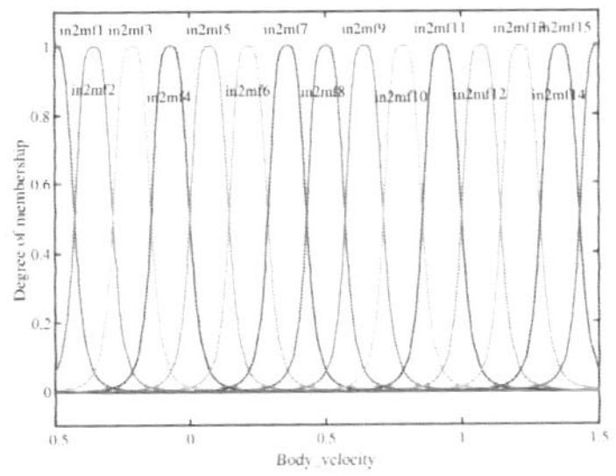

Figure 9. Membership Function of i/p variable "Body Velocity"

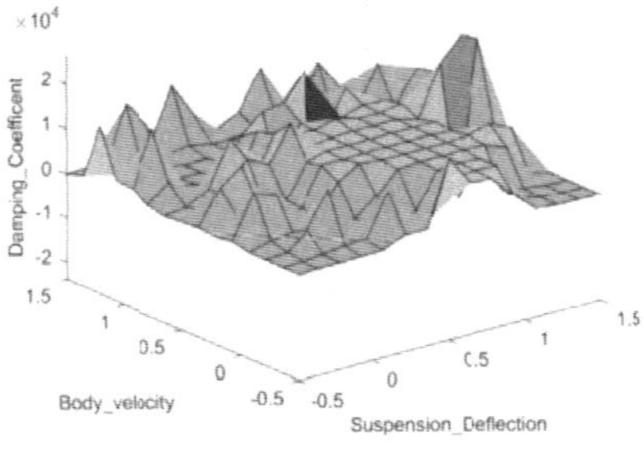

Figure 10. Surface view of ANFIS for i/p vs o/p functions

A neuro-fuzzy system is a fuzzy system that determines its parameters (fuzzy sets and fuzzy rules) by processing data samples using a learning algorithm derived from neural network theory [11].

5 Results & Discussion

Ride comfort is influenced by several elements while driving over a speed bump including vertical spring action, the oscillation that occurs during the settling process, the time required to settle after oscillatory behavior, and the maximum heave motion are typically associated with a smooth ride. The relatively low-frequency bounce and rebound motions of the suspension system are a role in the ride quality, which is typically linked with the vehicle's response to bumps. After a bump, a vehicle's undamped suspension (without variable shocks) will suffer a sequence of oscillations that will cycle according to the system's natural frequency.

Measuring and quantifying ride comfort not only assists development teams in satisfying essential standards and laws, but also provides the necessary insight to troubleshoot, comprehend, and improve the vehicle's noise and vibration comfort.

The handling characteristics of an automobile are concentrated on the characteristics of the dampers. They are the final link in the driver's chain of output directives, and they regulate the input of forces and disturbances from the road. The effect of the road on the vehicle, as well as the effectiveness of the output forces that control vehicle stability and cornering characteristics, are all influenced by damper properties.

The system of springs, mass, tire, and linkages that control how tires move and react to disturbances and control inputs manages the tire's basic characteristics. When a speed breaker is encountered, the oscillatory behavior that results might impair handling quality and could represent a threat if arrived at a specific speed.

The vehicle models created in MATLAB (Figures 4 and 5) are utilized to investigate the performance of the various controllers on disturbances in a realistic manner. These models simulate the behavior of a vehicle with continuous damping control. The results will be a form of linear displacement of the system after the disturbance has acted on the system such simulations have been carried out using the following benchmark speed breakers employed in industrial tests:

5.1 Circular speed bump

The performance of car suspension systems is put to the test with circular wave bumps. The test are simulated in such a way that pitch movements are successfully stimulated because the length of the hole is bigger than the distance between the front and rear axles (Gillespie, 1992). Furthermore, bounce performances may be assessed based on the amplitude characteristics of such road bumps.

The road disturbance/profile is taken according to Indian standards which are mentioned in Indian congress section IRC 99-198 as shown in Fig 11 and 13, Where 'a and b' denotes their bump amplitudes respectively, and their width are denoted as 'c and d'. The frequencies of these bumps are 0.1HZ and 1 HZ. Fig 12 and 14 demonstrates the response of various controllers.

$$W(x) = a1[u(t-5) - u(t-10)]sin(0.2\pi t) \quad 5 \leq t \leq 10 \quad (12)$$

Where,(a=0.1m (road bump height) and c=5m (width))[4]

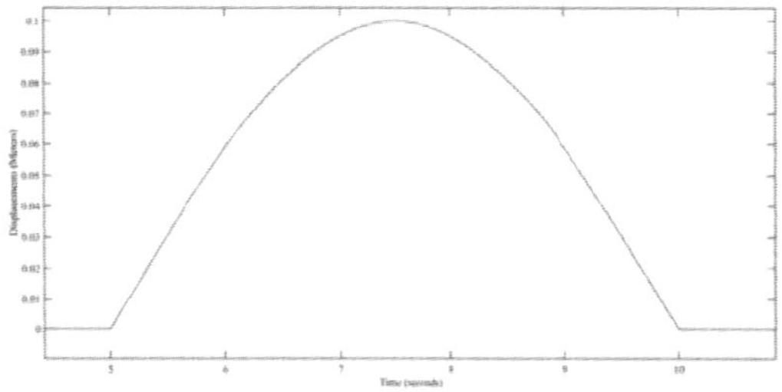

Figure 11. Road profile 1a

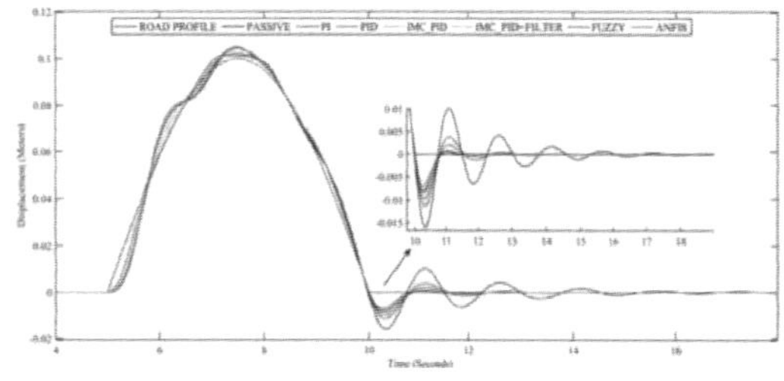

Figure 12. Response against Road profile 1a

$$W(x) = a1[u(t-30) - u(t-30.5)]sin(2\pi t) \quad 30 \leq t \leq 30.5 \quad (13)$$

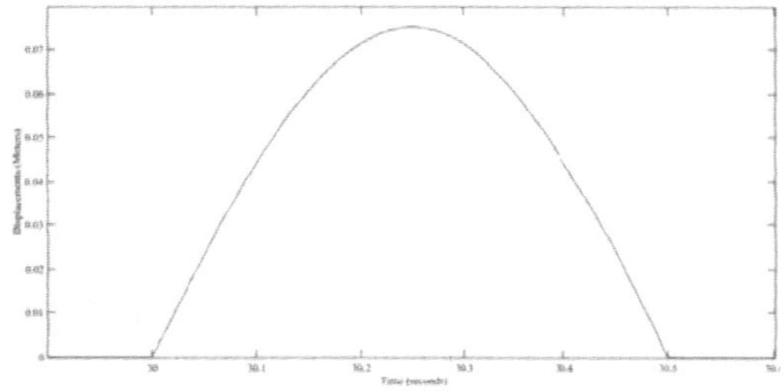

Figure 13. Road profile 1b

Where b=0.75m (road bump height), d=0.5m(width)

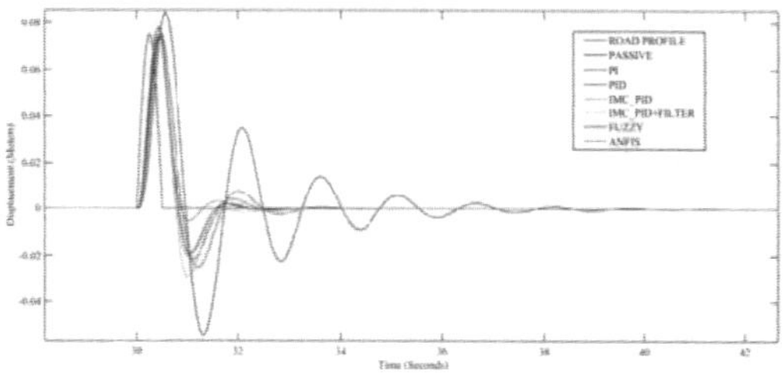

Figure 14. Response against Road profile 1b

5.2 Sinusoidal speed bumps

The sinusoidal curve is simulated with the help of the gaussian curve equation. For the desired signal curve, the values of sigma and c are considered accordingly. The height and width of the bump are denoted by x and y respectively. The road profile 2 (Fig 15) can be obtained from Equation 14 and Fig. 16. shows the response of various controllers against road profile 2.

$$W(x) = f(x;\sigma,c) = e^{-\frac{(x-c)^2}{2\sigma^2}} \quad (14)$$

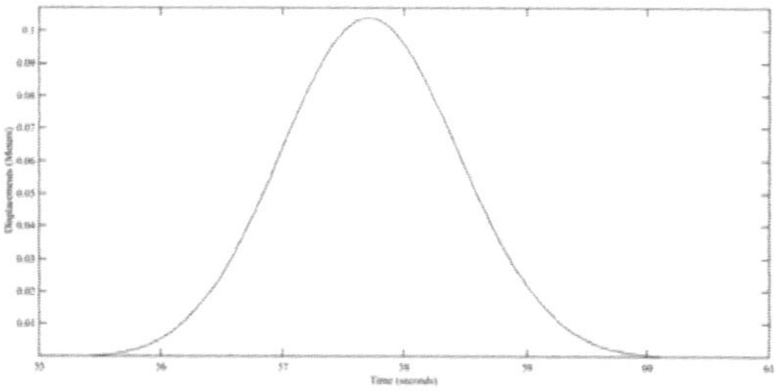

Figure 15. Road profile 2

Where, (x=0.1m(road bump height), y=5m(width))

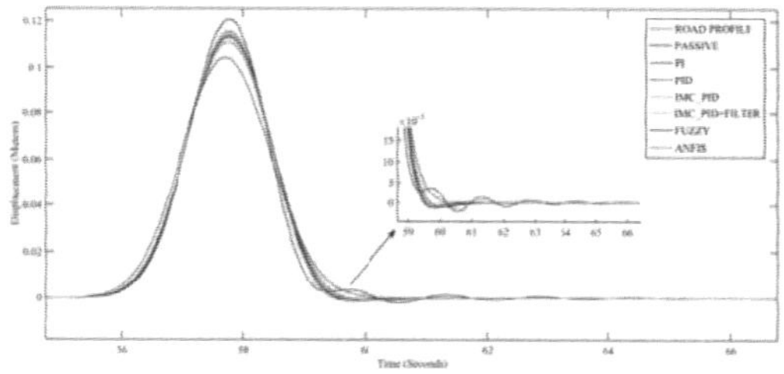

Figure 16. Response against Road profile 2

5.3 Trapezoidal speed bumps

The distinctive design of the trapezoidal speed bump assists drivers in reducing vehicle speed without compromising vehicle handling.The shape of the breaker is distinctive

https://doi.org/10.1051/itmconf/20214001001

and can be visible from far distance providing users ample amount of time to reduce and control the speed of the vehicle. The features such as height and width are taken accordingly as vendors specified design.

The height and width of the bump is denoted by p and q respectively. The road profile 3 (fig 17) can be obtained using equation 15 and Fig. 18. demonstrates the response of various controllers against the road profile 3

$$f(x) = \begin{cases} r(t-75) & 75 \le t \le 76.2 \\ u(t-76.2) & 76.2 \le t \le 78.8 \\ -r(t-78.8) & 78.8 \le t \le 80 \\ 0 & otherwise \end{cases} \quad (15)$$

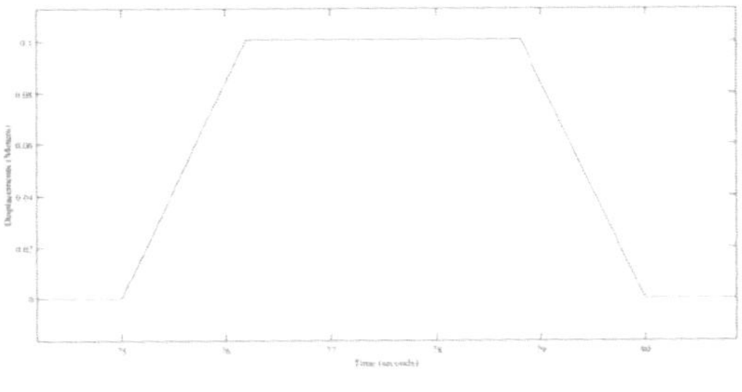

Figure 17. Road profile 3

Where, (p=0.1m (road bump height), q=5m(width))

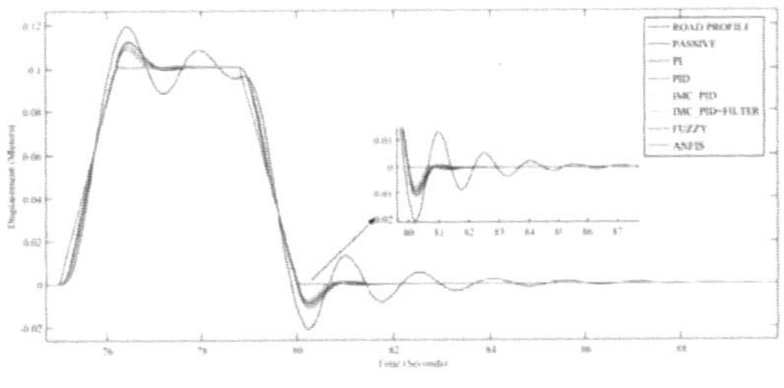

Figure 18. Response against Road profile 3

5.4 Overall speed bump

The design of the semi active suspension and its linear displacement caused due to bumps are focused individually to provide a more detailed view of the performance. The final perspective on the causes and effects is summed up by contemplating a road profile with these disturbances cascading to one another.

This outcome can be obtained in simulation by adding all of the responses together to obtain a final model of the road profile(Fig. 19).

W(x)= summation of all signal in equation

Adding equations (12) +(13) +(14) +(15)

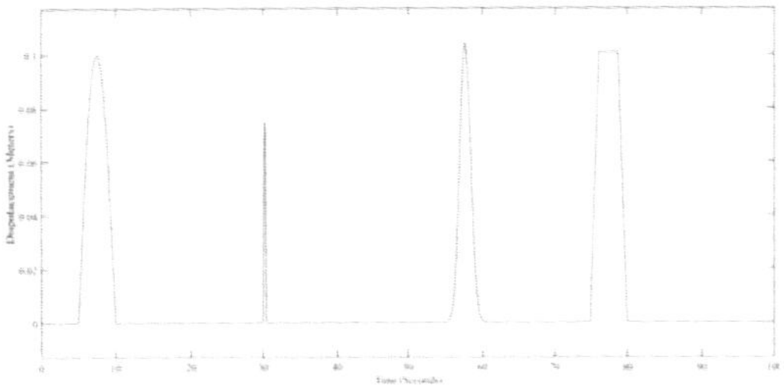

Figure 19. Road profile Overall

Fig. 20. is the response of various controllers against summation of road profiles

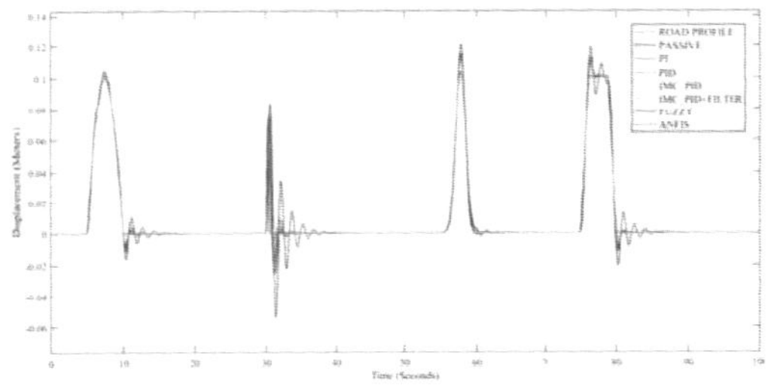

Figure 20. Response against Overall Road profile

The performance of the quarter car Semi-active suspension system is compared with the passive, PID controller, IMC based PID, IMC based PID with Filter, Fuzzy, and ANFIS controllers under various road profiles in this segment.

The immediate value of the controller output is defined as the control effort. It could be stated as an integral part of the power of a control signal. The controller effort gives you an idea of how different controllers compare the RMS value of the controller efforts of various controllers is shown in Table 2 throughout the simulation.

Controller	Controller Efforts
Passive	2.55e-5
PI	3230.8172
PID	4810.9117
IMC-PID	7619.6825
IMC-PID + Filter	8910.2395
FUZZY	2445.9078
ANFIS	3007.8301

Table 2. Controllers Efforts

Controller	ISE	ITAE	ITSE	IAE	Error parameter
ANFIS	0.0004325	2.384	0.01669	0.05286	Suspension on deflection
	0.08216	39.35	3.662	0.8668	Body Velocity
FUZZY	0.00041	2.50	0.01677	0.05512	Suspension on deflection
	0.08715	40.16	3.84	0.9005	Body Velocity
IMC-PID	787.7	4492	3.888e+04	94.53	Body Displacement
IMC-PID+Filter	786.6	4506	3.881e+04	94.93	Body Displacement
PI	803.2	4554	3.979e+04	96.43	Body Displacement
PID	793.2	4515	3.925e+04	95.2	Body Displacement

Table 3. Performance Index

Parameter	PI	PID	IMC-PID	IMC-PID with filter	Passive	FUZZY	ANFIS
RiseTime(ms)	264.249	177.935	96.0865	95.1098	103.4889	48.9998	12.8251
Settling Time(s)	3.1248	2.256	2.084	1.840	8.564	1.642	1.4473
Overshoot	65.7	50.3148	60.6402	72.026	69.806	34.5382	45.6255

Table 4. comparison table

6 Conclusion

This study compares the performance of damping control schemes such as PI, PID, IMC-PID, IMC-PID with Filter, FUZZY, and ANFIS for the quarter-car model. Simulations on the semi-active vehicle suspension system using MATLAB/Simulink is used to evaluate the performance of each semi-active control in terms of vertical displacement and suspension travel. Performance indices are also used to assess the performance of controllers. This research could be relevant in the future for designing real-time optimal nonlinear state feedback car suspension adjustment to improve ride comfort and stability.

References

[1] J. Wang and L. Sun, "Design of semi-active air suspension system based on backstepping sliding mode control," 2017 Chinese Automation Congress (CAC), 2017, pp. 4378-4382, doi: 10.1109/CAC.2017.8243550.

[2] E. Enders, P. Karle, G. Bonelli, D. Killian and G. Burkhard, "Modal Vertical Vehicle Dynamics Control for Semi-Active and Active Suspension Systems," 2020 Fifteenth International Conference on Ecological Vehicles and Renewable Energies (EVER), 2020, pp. 1-9, doi: 10.1109/EVER48776.2020.9242985.

[3] J. Wu, H. Zhou, Z. Liu and M. Gu, "Ride Comfort Optimization via Speed Planning and Preview Semi-Active Suspension Control for Autonomous Vehicles on Uneven Roads," in IEEE Transactions on Vehicular Technology, vol. 69, no. 8, pp. 8343-8355, Aug. 2020, doi: 10.1109/TVT.2020.2996681.

[4] Gowda, Dr.Dankan and Chakrasali, Sadashiva. (2014). Comparative Analysis of Passive and Semi-Active Suspension System for Quarter Car Model using PID Controller.

[5] Qazi, Abroon & Khan, Dr. Afzal & Khan, Muhammad & Noor, Sahar. (2013). A Parametric Study on Performance of Semi-active Suspension System with Variable Damping Coefficient Limit. AASRI Procedia. 4. 154-159. 10.1016/j.aasri.2013.10.024.

[6] Karthik Murali Madhavan Rathai. Experimental Implementation of Model Predictive Control Scheme for Control of Semi-active Suspension System. ScienceDirect, 2019.

[7] Wen, Jinyu & Liu, Ju & Long, Yao & Yao, Wei. (2016). Solution to short-term frequency response of wind farms by using energy storage systems. IET Renewable Power Generation. 10. 669-678. 10.1049/iet-rpg.2015.0164

[8] Abroon Jamal Qazi Afzal Khan, M. Tahir Khan, Sahar Noor. A Parametric Study on Performance of Semi-Active SuspensionSystem with Variable Damping Coefficient Limit ,2014.

[9] Kazima, Sosthene and Josee, Musabyimana and Hui, Xiong. (2018). Fuzzy Logic Controller for Semi Active Suspension Based on Magneto-Rheological Damper. International Journal of Automotive Engineering and Technologies. Volume 7. 10.18245/ijaet.438045.

[10] Hung, N.C., Nhung, N.T.B., Vu, L.T.Y., Khai, V.Q., Son, T.A., Thanh, T.Q., 2020. Apply a Fuzzy Algorithm to Control an Active Suspension in a Quarter Car by Matlab's Simulink. AMM 902, 23–32. https://doi.org/10.4028/www.scientific.net/amm.902.23

[11] Abellan-Nebot, Jose. (2010). A review of artificial intelligent approaches applied to part accuracy prediction. International Journal of Machining and Machinability of Materials. 8. 10.1504/IJMMM.2010.034486

[12] Yao, G. Z. et al. "MR damper and its application for semi-active control of vehicle suspension system." Mechatronics 12 (2002): 963-973.

[13] Pang, Hui Liu, Fan Xu, Zeren. (2018). Variable universe fuzzy control for vehicle semi-active suspension system with MR damper combining fuzzy neural network and particle swarm optimization. Neurocomputing. 306. 10.1016/j.neucom.2018.04.055.

[14] Durieux, Olivier Abramov, Sergey Mannan, Samjid. (2009). Semi-active suspension system simulation using Simulink. International Journal of Engineering Systems Modelling and Simulation. 1. 101-114. 10.1504/IJESMS.2009.027573.

[15] Juan C. Tudon-Martınez, Rubén Morales-Menéndez, Ricardo Ramirez-Mendoza, Olivier Sename, Luc Dugard. A Comparison between a Model-free and Model-based Controller of an Automotive Semi-active Suspension System : Independent Wheel-stations. IFAC Joint conference SSSC - 5th Symposium on System Structure and Control, Feb 2013, Grenoble, France. pp.869-874,

Design of cost-efficient Glucose meter

Sanjana Rao[1], Sattwik Mohanty[2], Chirag Amarnani[3], Kausar Fakir[4], Sandip Dhobale[5*], Trupti Agarkar[6*]

Ramrao Adik Institute of Technology, Nerul, Navi Mumbai

sanjana.raviprasad99@gmail.com[1], mohanty.sattwik123@gmail.com[2], chiragamarnani@gmail.com[3], kausar.fakir@rait.ac.in[4], dhobale.sandip@gmail.com[5]*, trupti.agarkar@rait.ac.in[6*]

*Corresponding author

Abstract- **A blood glucose metre is a small, portable device that measures blood glucose levels. To avoid long-term complications from diabetes, careful blood glucose control is required. New blood glucose metres have a small size, a large memory capacity, blood glucose manipulation techniques, and computer-based data processing capabilities.**

The Glucose Detection System is implemented in tandem with an improved Op-based Potentiostat and an automatic test system. This project's goal is to investigate an improved OP-based Three-Electrode Potentiostat that can be used in an Electrochemical Glucose Biosensor device to achieve comparable measurement accuracy. This Project enables us to make the Glucometer Cost-Efficient, Affordable and Easy to Use. They hold a lot of promise in the field of bio-detection on the go systems for use in health care and biomedicine.

Keywords: Potentiostat, Glucometer, affordable, easy to use, health- care.

1.INTRODUCTION

A glucose metre is also available as a "glucometer," is a medical device used to determine the estimated glucose concentration in the blood. It could also be a piece of sensing elements dipping in a substance and its measurement against a glucose map. It is an important part of home blood glucose monitoring (HBGM) for people suffering from diabetes or hypoglycemia. A single drop of blood is deposited on a disposable test strip by pricking the skin with a lancet, which the metre reads and measures blood glucose levels.
Clark and Lyons proposed the enzymatic electrode of glucose oxidase in Cincinnati, Ohio, in 1962. Guilbault and Lubrano employed the amperometric method in the electrochemical in 1973 to feel the amount of hydrogen peroxide (H2O2) in order to determine the glucose concentration. The use of a metal surface reaction-based electrode without an enzyme membrane to determine glucose concentration is defined. A new material is needed. The different glucose monitors are compared and contrasted. The amperometric technology is the most widely used, however, it is more difficult to design circuits for and is not appropriate for long-term measurements.

It's impossible to construct an Amperometric with a high impedance amplifier. The transducer calculated the flow of current between a biochemically sensitive and a reference electrode, which is proportional to the concentration of electro-active analytic in the solution. The glucose can be detected using a two-electrode (RE and WE) potentiostat due to concentration polarization and the higher degree of voltage at the RE terminal. The offset voltage was provided by the induced current flow through the RE electrode, which increased the measurement error. The CE electrode is used in the Potentiostat, a three-electrode bio-sensor that has fast sensing, high performance, and high reliability. The CE electrode creates a current loop, while the RE electrode creates a voltage direction, reducing the over-voltage effect in RE and reducing measuring error.

The oxygen electrode-based (O_2-based) sensor that generates the oxidation current and the hydrogen peroxide electrode-based sensor (H_2O_2-based) sensor, which induces the reduction current, are the most commonly electrochemical glucose biosensors with electron propagation mediators were used. The Potentiostat is a device that amplifies and reads current in a bio-signal detection system. Aerometric electrochemical sensor with three electrodes, a current metre, and a control amplifier make up the basic setup. A current proportional to the concentration of substrate is detected as an electrochemical reaction on the electrode.

The Potentiostat are a big category's electronic interface of Amperometric sensors with the ability to track many biologically analytics that are critical for the environment. In system-on-chip designs, potentiostat and interface circuits are integrated in CMOS chips to improve functionality while reducing device size.

The primary goals of a glucometer are to avoid excessive insulin secretion, excessive food intake, diabetic patient analysis and prescribed treatment, and to determine the average blood glucose value. Furthermore, the glucometers currently on the market are quite expensive and out of reach for many people. As a result, the cost-effective Glucometer is both affordable and accurate. [1-8]

 https://doi.org/10.1051/itmconf/20214001002

2. SYSTEM DESIGN

An enzymatic reaction and a detector are the two most important components of glucose meters. The enzyme part of a glucose meter is usually packed in a single-use strip or reaction cuvette in a dehydrated state. The glucose in the the patient's blood sample is rehydrated and interacts because of the enzymes to create a detectable substance. A few metres produce hydrogen peroxide or an intermediary that is capable of reacting with a dye, causing a colour shift proportional to glucose solution concentration. Other metres include enzymes in a biosensor, which generates an electron, which the metre detects. Current glucose meters use three different enzymatic reactions: glucose oxidase, glucose dehydrogenase, and hexokinase. Each enzyme has its own set of benefits and drawbacks.

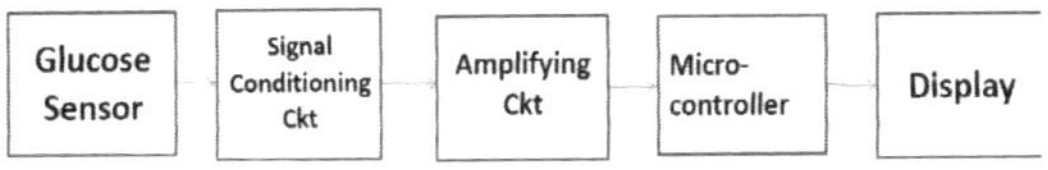

The System Design includes two parts, namely Sensor Design and Software Design.

2.1 Sensor Design and Glucose detection:

The below figure shows the hardware representation of the Glucometer. The Sensing Film is used to sense the amount of glucose present in the blood. The Sensing Film consists of components such as the Glucose Oxidase enzyme(GOx) immobilized on conducting polymer (CP) film deposited on intedigitated PCS board which catalyses the oxidation of glucose to hydrogen peroxide nd D-glucono-delta-lactone,

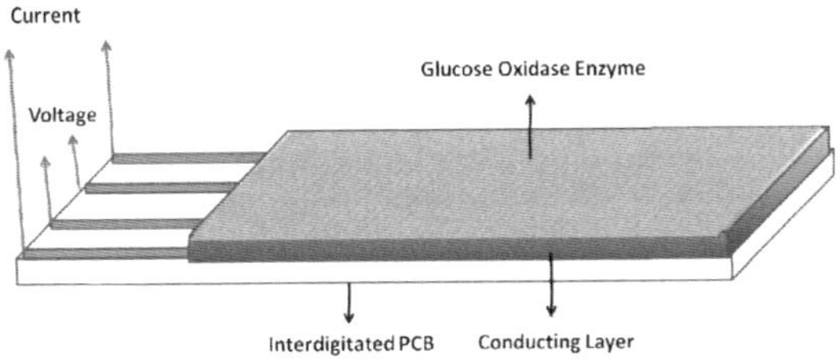

Glucose detection has been done with different glucose concentration from normal to higher levels (100 mg/dl to 270 mg/dl). Different glucose concentrations has been prepared in standard phosphate buffer solutions and I-V responses are recorded for different glucose concentration mentioned above.

The reactions taking place are as shown:

- Surface Reaction

$\frac{1}{2} O_2 \rightarrow O$ (adsorbed)
O (adsorbed) + 2e- $\rightarrow O^{2-}$ (adsorbed).

- Enzyme Reaction

- Interface Reaction

2p+ + $2O^{2-}$ (adsorbed) $\rightarrow O_2$ + 2e- $\rightarrow$ CP (reduction).
Here, SnO2 is responsible to decrease resistance.
$H_2O_2 + 2O^{-} \rightarrow H_2O + 2O_2$ + 2e-
[6-8].

2.2 Software Design

The overall representation of how it would look like on the breadboard after connecting it to the Arduino UNO is as shown below: -

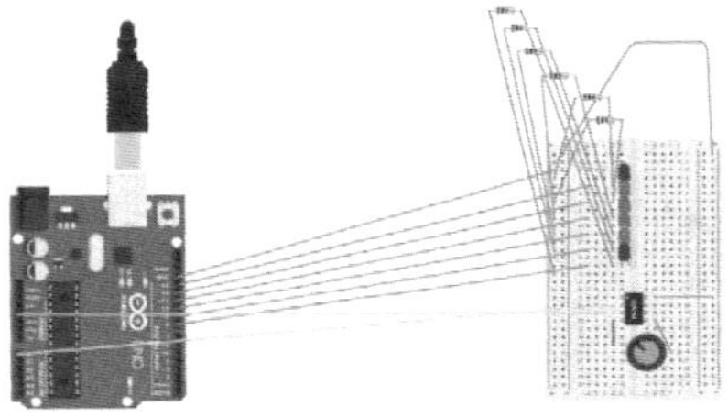

3. GLUCOSE LEVEL STAGES

The test for checking blood glucose can be done when one is:

- Fasting
- After Eating
- 2-3 Hrs After Eating

Glucose level according to voltage is as shown below:

Voltage	Glucose Level
Under 2mV	Very Low
Between 2.8mV to 6.5mV	Low
Between 7mV to 11mV	Normal
Between 12mV to 18mV	High
Greater than 18mV	Very High

 https://doi.org/10.1051/itmconf/20214001002

4. SIGNAL CONDITIIONING CIRCUIT

A basic glucometer consists of a signal conditioning circuit. Signal conditioning is a data acquisition procedure that is carried out with the aid of a signal conditioner instrument. This device transforms one form of electrical or mechanical signal (input-signal) into a different type of signal (output-signal) (output-signal).

A signal conditioning circuit's functions are as follows: -

i. Signal Conversion- A signal conditioner's primary function is to obtain a signal and it should be converted to a higher electrical signal level. Signal conversion is frequently used in industrial applications that use a variety of sensors to perform measurements.

ii. Linearization- Certain signal conditioners can perform linearization when the signals produced by a sensor do not have a straight-line relationship with the physical measurement. This is the process of interpreting a signal generated by software, which is commonly used for thermocouple signals.

iii. Amplifying- The process of increasing the signal in preparation for processing or digitization is known as signal amplification. Signal amplification can be accomplished in two ways: by increasing the signal-to-noise ratio or increasing the resolution of the input signal.

iv. Filtering- The frequency spectrum of the signal is filtered to include only valid data and exclude noise. Filters can be made with both passive and active components, as well as a digital algorithm.

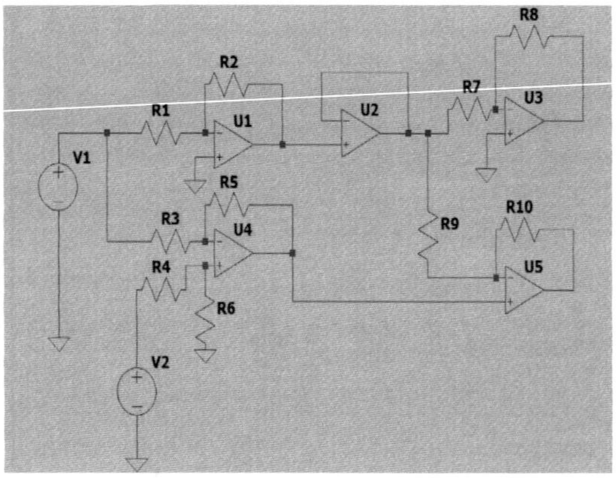

The above figure shows the signal conditioning circuit for the glucometer.

5. WORKING OF A GLUCOMETER

Diabetics find it difficult to keep track of their blood sugar levels, both metaphorically and literally. They prick their finger several times a day obtaining a blood droplet, which they then apply to a plastic strip that is inserted

The pancreas is usually in charge of monitoring blood sugar levels which are maintained by secreting glucagon and insulin to keep them between 100 and 200 milligrammes per deciliter of blood. However, glucose testing is a do-it-yourself proposition diabetics, either because their pancreas does not function properly or because their body is unable to process the hormones secreted by the pancreas. And it's a big one. Blood sugar levels indicate whether it is time to inject a few units of insulin — or to eat a life-saving snack.

This is where the glucometer comes into play. Glucometers today use test strips containing glucose oxidase, an enzyme that reacts to glucose in blood droplets, as well as an interface to an electrode inside the metre. When the strip is inserted into the metre, the flux of the glucose reaction generates an electrical signal. The glucometer has been calibrated such that the number displayed in its digital display equates to the current intensity: the greater the number, the more glucose throughout the sample. Routine glucometer tests are an essential part of a patient's course of treatment, however some new systems fall far short of providing an accurate picture of real-time glucose fluctuations. [9-14]

6. RESULTS

The below images show the stages at which the LEDs blink after the multi-meter has been stopped at a certain voltage value. They indicate if the level of blood glucose is high, low, or regular according to the table shown in section 3.

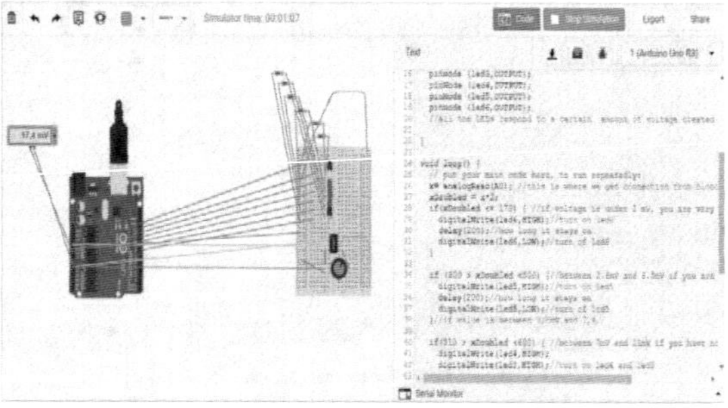

Fig: When Glucose level is high, the LED 2 (yellow) blinks

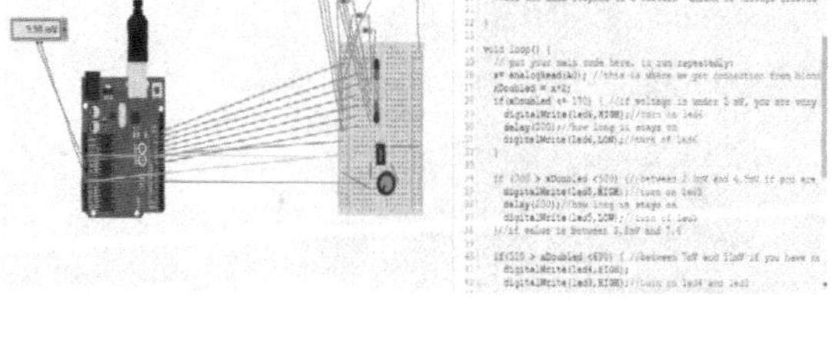

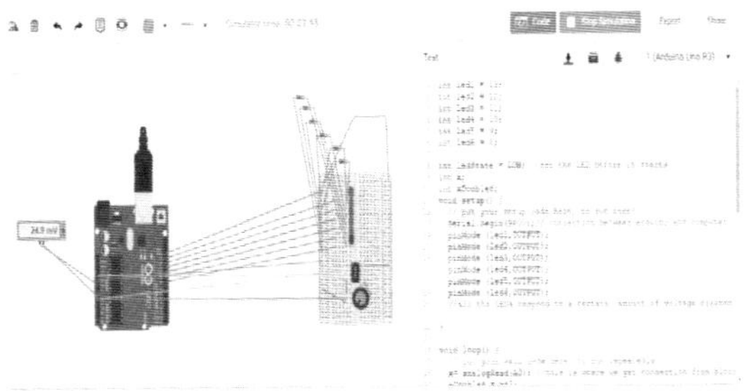

Fig: When Glucose level is very high, the LED 1 (red) blinks.

7. ADVANTAGES

There are many benefits of having your own blood sugar testing kit at home. These include:

The convenience of use - One of the biggest pros of a glucometer is that it offers a patient the facility of testing their blood sugar levels from the comfort of their home without needing to always go to a doctor This is especially great for patients who need to check their blood sugar levels daily. If the weather is bad — such as heavy snow — a patient need not suffer in the dark but can keep track of their blood sugar levels and collect the data or even send these across via apps to their doctor.

Possibility to make medical decisions- A glucometer gives you an accurate reading of your blood sugar level that enables you to make crucial decisions about your diet or medication. For instance, your sugar levels show as low, you might want to consume a quick snack before working out to avoid feeling sick or unconscious. Additionally, a sudden onset of high blood sugar is a warning sign that something is not right and you need medical help. This can actually make the difference between life and death.

Accuracy of reading- Based on the latest technology, modern glucometers have highly sensitive strips that are based on certain chemicals to detect every component of blood accurately. So, you can get near-to-laboratory accuracy from the comfort of your home or even while traveling.

Suitable for seniors - While everyone with diabetes should have a glucometer, the benefits are especially notable for seniors. This is because seniors often find it hard to leave their homes and head to a clinic each time that they need their blood glucose checked. Now, with glucometers, they can check their blood sugar readings at home and also share these reports with their doctors online. This ensures the good health of the patient and helps the doctors too, by easing their workload in laboratories.

Cost-effective - Finally, a glucometer is a one-time purchase. This definitely helps save money on laboratory tests for blood sugar levels. Plus, you also save money that you would spend on your commute. You do not even have to physically visit your doctor to across online. So, overall, this is a worthwhile medical investment that is a must-have for a diabetes patient.

8. LIMITATIONS.

While there are many advantages, there are a few disadvantages too. They are given below: -

It hurts- Using an invasive glucometer hurts! Pain is present regardless of how small and thin the needle is. While some people become accustomed to it and thus tolerate it, it can be problematic for others. Measuring blood glucose levels is routine for diabetics, so incorporating this pain into your routine becomes difficult. This is especially true for children and adolescents with type 1 diabetes. Even mild discomfort is difficult for young kids to tolerate.

It may result in Non-Compliance - As stated earlier, using an intrusive glucometer could result in pain. Patients may develop a mental barrier as a result of the pain that inhibits them from having to check their sugar levels on a daily basis. This noncompliance can eventually affect their blood glucose control, causing hypoglycemia and hyperglycemia. Noncompliance is risky because it raises the risk of developing diabetes.

It has the potential to cause infection - The skin barrier that protects against infection is breached when you prick your finger. Diabetics are more susceptible to infection, so even the smallest gap can allow bacteria to enter and cause infection. Because diabetics have a low immune system, any infection can be dangerous. Continuous pricking at the same location can also cause skin changes. The skin at that place may become hard, making it hard to prick and acquire a blood sample.

9. CONCLUSION

Our main aim is to Provide a Blood Glucometer that will be determined by a variety of factors, including the patient's requirements, Easy to Utilize and Affordable.

The enhanced OP-based Potentiostat within the bio-detection structure uses the glucose concentration is measured using an electrochemical enzyme-immobilized test strip. The electrode responds to the current in a proportional manner to the concentration of glucose activity. The polarity of VOX governs the role of oxidation and reduction. The difference value of Rf alters the measurement spectrum. It also shows Device parameters.

Cost effective glucometer is designed with signal-conditioning unit and micro controller, it shows potential compatibility for glucose detection. The results obtained from the sensing kit shows good response for range glucose concentrations present in healthy person as well diabetic patient.

9. REFERENCES

[1] L. Clark and C. Lyons, "Electrode Systems for Continuous Monitoring in Cardiovascular Surgery," Ann. N. Y. Acad. Sci, vol. 102, 1962.

[2] Lai, J.L., Wu, H.N., Chang, H.H. and Chen, R.J., 2011, June. Design a portable bio-sensing system for glucose measurement. In 2011 International Conference on Complex, Intelligent, and Software Intensive Systems (pp. 71-76). IEEE.

[3] Saha, Shumit and Sarker, Nayan and Hira, Avijit. (2014). Design and Implementation of a Low Cost Blood Glucose Meter with High Accuracy. 1st International Conference on Electrical Engineering and Information and Communication Technology, ICEEICT 2014. 10.1109/ICEEICT.2014.6919050.

[2] R. F. B. Turner, D. J. Harrison, and H. P. Baltes, "A CMOS Potentiostat for Amperometric Chemical Sensors," IEEE Journal of Solid-State Circuits, vol. 22, pp. 473-478, 1987.

[3] C.W. Liao. "Development of Micro Electrochemical Electrode Array Platform for Blood Sugar and Cholesterol Measurement," Master Thesis, National Taiwan University, Graduate Institute of Biomedical Engineering, Taiwan, 2005.

[4] W. Gopel and P. Heiduschka, "Interface Analysis in Biosensor design," Biosensors & bioelectronics, Vol. 10, pp. 853-8838, 1995.

[5] J. Bard and L. R. Faulkner, Electrochemical Methods: Fundamentals and Applications, 2nd Edition, Wiley, New York, 2001.

[6] S. Dhobale, P. Joshee, G. Deore, S. L. Laware, *and* S. N. Kale "Nanostructured glucose-oxidase immobilized SnO2SnO2 thin films for glucose sensing", Applied Physics Letters 98, 073704 (2011)

[7] Sejin park, Hankil Boo, and taek Dong Chung, "Electrochemical Nonenzymatic Glucose sensors," Analytica Chemica Acta, Vol. 556, pp. 46- 57, 2006.

[8] Carlos Eduardo Ferrante do Amaral, Benhard Wolf, " Current Development in Non-invasive Glucose Monitoring," Medical Engineering & Physics, Vol. 30, pp. 541-549, 2008.

[9] Audrey Sassolas, L. Blum, B.D. Leca-Bouvier, "Polymeric Luminol on Pre-treated Screen-printed Electrodes for the Design of Performant Reagentless bio-Sensor," Sensors and Actuators B: Chemical, Vol. 139, pp.214 – 221, 2009.

[10] Martin, S.M.,Gebara, F.H., Strong, T.D., Brown, R.B., "A Low-voltage, Chemical Sensor Interface for Systems-on-chip: the Fully-differential Potentiostat,". Proceedings of the 2004 International Symposium on Circuits and Systems, ISCAS '04, Vol.4, 23-26, Page(s):IV - 892-5, May 2004.

[11] N. A. Latha, B. R. Murthy, U. Sunitha , " Design And Development Of A Microcontroller Based System For The Measurement Of Blood Glucose," International Journal of Engineering Research and Applications (IJERA), Vol. 2, No. 5, pp.1440-1444, Sep. 2012.

[12] M. Bularzik, D. Price, M. Rivera; "Accessible Blood Glucose Monitor," University of Connecticut, Connecticut, Biomedical Engineering Senior Design, 2006.

[13] B. Feldman, "Electrochemical Blood Glucose Test Strips for People with Diabetis," Diabetes Technol. Ther., 2003.

[14] R. Graham, "Self-Monitoring of Blood Glucose: Considerations for IntensiveDiabetes Management," A CME/CPE/CNE/CCM–Certified Supplement Tech. Rep., 2005.

Condition Monitoring and Predictive Maintenance of Process Equipments

Manthan Deshmukh[1,*], *Rohan* Dumbre[1,**] *Shubham* Anekar[1,***] *Heramb* Kulkarni[1,****], and *Sushant* Pawar[1,†]

[1]Department of Instrumentation Engineering, Ramrao Adik Institute of Technology, Navi Mumbai

Abstract. Industry 4.0 the proclaimed fourth industrial revolution is unfolding at the moment. It is characterized by interconnectedness and vast amounts of available information. Industrial production has evolved enormously over the last centuries due to modern instruments. Hence issue of the instrument failure is very paramount in any industry. Even if one machine fails it halts the whole production. Overall, it may cost us with more man-hours, project delay, process latency and all this sums up as a huge loss. The life of the instruments should be taken care by continuously monitoring its health. Any faulty or unnatural disturbance in usage of the instrument may lead to its failure. Every instrument needs proper maintenance, even with the slight negligence towards the anomaly it may lead to instrument failure. In, predictive maintenance historic data is utilized and analyzed with the help of advance analytics and modelling techniques using Machine learning, moreover we can predict failures and can schedule the maintenance beforehand and predict failure in advance. With the help of relevant sensor dataset, we can estimate the remaining runtime of the instruments. This maintenance approach helps to lower the costs which are incurred due to system shut downs. It also ease the scheduling and maintenance activities.In this work, three different industrial case studies are considered like shell and tube type heat exchanger, plate type heat exchanger, and semiconductor manufacturing process.Here the predictive maintenance is carried out for heat exchanger by utilizing the concept of multi linear regression and time series analysis. For the semiconductor manufacturing dataset, support vector machine algorithm is implemented to find out the good and bad quality of semiconductor production slots.

1 Introduction

In the environment of Industry 4.0, proper cost allocation and time management is the primary concern to lead in this competitive Industry [10]. Equipment failure or process shutdown will cost an industry a heavy fortune. To avoid this, proper timely maintenance is especially important. In early days before the digitalization preventive maintenance technique was used, were-in equipment was kept running in the process until failure. This was not a cost saving and time efficient method [11]. Predicting failure well in advance with the help of advance analytics can significantly outperform traditional maintenance approaches. In contrast to preventive maintenance, predictive maintenance can be especially useful as it increases productivity, reduces breakdown and lowers maintenance cost. In process industries, a lot of sensor data is recorded, and it plays an important role in predictive analysis. Using this recorded data and machine learning algorithms like Linear/Multi Linear Regression [1], Logistic Regression [2], Time-Series Forecasting [3], Cluster Analysis, etc we can build a model which will help us with monitoring the health of the equipment and forecasting the future results well in advance and apparently helping us to schedule a maintenance beforehand. These data have a lot of hidden information which can be an asset to the company. Analysing these data can help to know the behaviour of the equipment. Also, various conditions can be analysed in order to distinguish between operating and abnormal conditions. The data can also be helpful for analysing the process behaviour and can be used to make the process more efficient. Hence data can be the driver for the betterment for the industry and for the future growth.

1.1 Motivation and Background

Everyday a lot of data is generated in every field which varies from business to business. In case of instruments a lot of sensor data is generated through various field instruments. This data can be made to work using predictive modelling techniques [16]. Though traditional techniques i.e., preventive and reactive maintenance are widely used in many industries. A lot of advancements are made, and continuous research is going in improving the quality of data for optimum results. Hence it marks a new way in maintenance approach, better manage inventory, eliminate unplanned downtime and maximize equipment lifetime.

In 1990's Predictive Maintenance came into the picture, still most of the industries, were unaware due to lack of digitization. But now-a-days predictive maintenance is emerging, and various industries are shifting towards it. Predictive maintenance has evolved from the first method that is visual inspection to digitized methods using the

*e-mail: manthandeshmukh100@gmail.com
**e-mail: dumbre1999@gmail.com
***e-mail: anekarshubham50@gmail.com
****e-mail: Kulkarni.heramb@outlook.com
†e-mail: sushant.pawar@rait.ac.in

statistical and analytical techniques based on trend recognition and machine learning [7]. Predictive maintenance mainly focuses on three aspects: 1) Downtime Reduction – Lesser the downtime due to equipment failures, more the productivity. 2) Cost saving – Equipment failure is prevented, by scheduling the maintenance while gradually leads to low equipment expenditure 3) Energy efficiency – Preventing intermittent and unscheduled downtime resulting in maximizing the productivity leading to proper energy management.

1.2 Related Research

Maintenance is an important activity in various process equipment. It plays a significant role in process industries with respect to production, cost, safety. Hence the maintenance was carried out through various methods since olden days. In early days of production technology breakdown maintenance was used and was reactive in nature [5]. Equipment was kept running until its failure occurred and then replaced this is called as reactive maintenance. As it can lead to serious consequences and is an issue of safety. Due to which it led to time-based maintenance, known as preventive maintenance [8]. In this maintenance of equipment has been schedule before its failure but it may happen that the equipment can work for more period than its maintenance schedule this led to maintenance cost, unnecessary utilization of manpower and equipment life may also get affected. This led to the evolution of predictive maintenance, in which maintenance is schedule well before its failure with the help of historical data and continuous monitoring of equipment to detect and verify the defects [14]. Only when a defect is detected, the maintenance is scheduled. Till the early 90s, spreadsheets were used to obtain approvals for condition-based maintenance programs [13]. Now a days, this is no longer the case. The obvious benefit of predictive maintenance is that it maximizes runtime, reducing maintenance cost, operations, long shutdowns, increases plant safety [18]. Repairs can be carried out just before a breakdown. That represents a major advantage, knowledge shows that unplanned maintenance work led to high downtime cost. Predictive maintenance improves repair time [17]. Since the specific equipment problems are known in advance, maintenance work can be scheduled accordingly. This makes the maintenance work faster and smoother. As necessary action is taken well before failure, there is no secondary damage, thus reduces repair time [12]. As every industry is shifting towards digitalization, it gives an instant boost to productivity helping projects to move faster and manufacturers to hit more aggressive deadlines. So, with this process industries have also marked their way. "Data is the new Oil" and hence if used in a proper way can drive the growth of the industries benefiting cost, safety, life of instruments, sustainability, and future [6]. According to studies it has shown that proper scheduled maintenance reduces costs up to 25maintenance costs in addition it has increased production in less amount of time [15].

1.3 Contribution and Paper Structure

Predictive analytics is an important concept when it comes to industry 4.0. In order to incorporate in process industries a primary step is the acquisition of large amount of sensor data. Along with data analytics there is also need of various Machine Learning algorithms for prediction [9]. In this paper 1) Predictive maintenance of Heat exchanger is done using Linear/Multi Linear Regression: It is generally used because it gives linear relationship between dependent and independent variables. Fouling factor which is dependent variable in heat exchanger is predicted using independent variables from the dataset 2) Auto Regression Integrated Moving Average (ARIMA) model: This model is used for Time Series Forecasting i.e.to predict the failure for future days based on the historical data. Fouling resistance of heat exchanger for future days was predicted using ARIMA model. 3) Semiconductor industry case study which manufactures equipment wafers, the response of these manufactured wafer was classified using Support Vector Machine (SVM): This algorithm is generally used for regression and classification problems. To classify the targeted class based on the labelled data. In this case the targeted response i.e. the manufactured wafer whether it is ''Good'' or ''Bad'' was classified using SVM. 4) Principal Component Analysis (PCA): It is a dimensionality reduction algorithm used to reduce the higher dimensional data into lower without losing the features. The features of the higher dimensional data is reflected in lower dimensions which can be further processed for modelling. In semiconductor case the data was having a higher dimension which was lowered to two dimensions without losing the features and further the targeted response is classified by Support Vector Machine.

Paper Structure is as follows: Section 1 represents the introduction, overall motivation and background of the idea and related research of the same. Actual problem and its brief solution is discussed in section 2. Section 3 describes the various algorithms used to achieve the required results. . Simulations and Results are shown in section 4 which elaborates three case studies and their visualizations. Conclusion and Future Scope talks about the end result and future scope of the predictive maintenance in section 5.

2 Problem Statement

In conventional method of maintenance i.e., Preventive Maintenance unplanned downtimes resulting in the reduction of output, which gradually impacting the direct loss of profit. In, 21st century, plants are increasingly turning towards machine learning (ML) to recognize patterns in sensor data. As a result of shifting towards digitalization, it benefits the business as maintenance activity is predicted well in advance. Hence it reduces the cost of the maintenance. Predictive maintenance is also useful against unnecessary shutdown due to unplanned maintenance. It also increases the safety of the workers because continuous monitoring alerts in advance for any faulty condition and we can stop the workers from coming in exposure to

https://doi.org/10.1051/itmconf/20214001003

any hazard. Now-a-days lot of accidents at various process plants are taking place which occurs due to human mistake or due to unplanned maintenance. Traditional methods of monitoring and maintenance did not comment upon idea of lifespan of the equipment, where in contrast Predictive Maintenance can give a predicted picture over remaining useful life and with planned maintenance life of an equipment can be increased with preventing the downtimes. Proper behavioural monitoring of instruments and predicting their remaining useful life in advance can protect the plants from hazards.

- Monitoring the health of equipment:- The main aim is to continuously monitor the health of the instruments i.e., whether it is functioning healthy or not. If any fault occurs in the machine or an equipment immediately it is identified by the system and gets notified to the controller in the control room. This can be identified with the help of analysing the historic data and setting the range or the threshold range, where disrupting these ranges will describe the unhealthy situations.
- To detect anomalies, present in the dataset:- When the raw sensor data is obtained it contains a lot of anomalies or noise. These should be detected, and data should be cleaned in order to use it for future predictions and visualizations.
- Predictive analysis of the equipment:- Every process instrument or machine needs a maintenance for its smooth functioning. But also, if it is carried out anonymously without proper schedule it may lead to decrease the life of machine and lead to maintenance cost. Hence using predictive analysis, we can predict in advance and plan the maintenance well in advance.
- Comment upon the remaining useful life of equipment:- The prediction analysis is done using a lot of sensor data, it also includes the timestamp. Hence using the time and the output to predict, we can use time series forecasting algorithm in order to predict the future output i.e., Remaining Useful Life well in advance and hence accordingly predict the maintenance activity.

3 METHODOLOGY

The basic idea of our project is to build predictive maintenance model of process instruments and to predict on the remaining useful life (RUL) using different Machine Learning algorithms. In various process industries various instruments carry out their task efficiently and hence the maintenance of this instruments is especially important for their smooth working and for the protection of assets. As these instruments are a part of different loop processes it has a lot of sensors which are mounted for accurate measurements of different variables. These data are recorded periodically on various timeframes and the same can be used for the predictive maintenance. The data can be processed using different data analytics techniques. The raw data is firstly processed, and noise and pure data is filtered. After we must find the condition indicators to bifurcate between healthy and faulty conditions. The next step is to build the machine learning model supervised learning algorithms. The model is trained using training dataset, 70 percent of total data is used for training the machine learning model and remaining 30 percent data is used for testing. Based on the training model the testing data or the future predictions can be done. The various types of regression algorithms, classification algorithms are considered in this work such as Linear Regression, Multiple Linear Regression, Support Vector Machine (SVM), Principal Component Analysis (PCA), Auto Regression Integrated Moving Average (ARIMA) and Decision Tree.

- Linear Regression are basically of two types simple and among which simple Linear regression is a technique for predetermining the value of one variable dependent on the value of another. The variable to be estimated is known as the dependent variable. It is a basic linear methodology for predetermining a response Y based on a solitary variable X. It has linear or straight relationship among X and Y. So linear Regression is the way toward discovering a line that best fits the data points accessible on the plot, so we can utilize it to anticipate yield esteems for inputs that are absent in the data set we have, with the conviction that those yields would fall on the line. And Multiple regression is a type of linear regression that is expansion of simple linear regression. It is utilized to forecast the value of a dependent term depending on two or more independent terms or variables. Mathematically represented way of the relationship is shown ahead, this regression model is written as $Y = C + m_1X_1 + m_2X_2 + ...m_nX_n + E$ where 'X_n' is the variable is used to represent the nth predeterminer and 'm_n' compute the dependence between that variable and the response. 'm_n' is interpreted as the mean effect on Y as a single unit increase in 'X_n', holding all other predeterminers fixed.

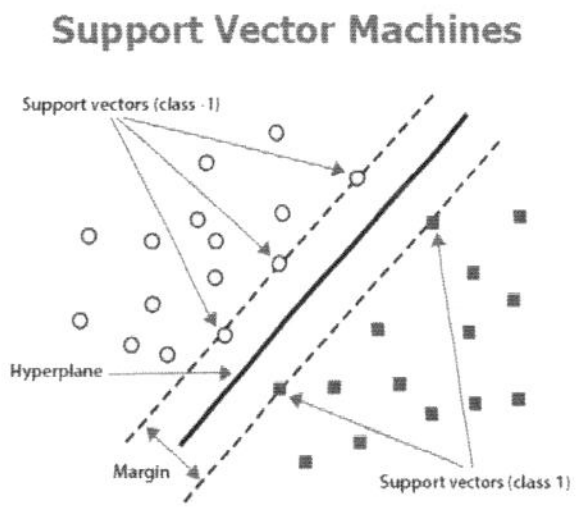

Figure 1. Support Vector Machine representation

- Support Vector Machine:- SVM is a sort of machine learning algorithm that can be utilized to solve classification or regression problems as shown in Figure 1. It transforms the data using a technique known as the kernel trick and then finds an ideal boundary between the possible outputs depending on these transformations. It consists of a hyperplane which divides the classes for which classification is to be done[4]. Support vectors are nothing but the data points that are nearer to the hyperplane which impact the position and direction of hyperplane. As per the SVM calculation we discover the points nearest to the line from both classes. These focuses are called support vectors. Presently, we process the distance between the support vectors and the line. This distance is known as the margin. We will probably augment the margin. The hyperplane for which the margin is greatest is the ideal.

- Principal Component Analysis: - Principal component analysis (PCA) means the method by which its components are computed, and the proper utilisation of these components in understanding the data. It is a dimensionality-reduction process which is often used to shrink the dimensionality of big data sets, by converting a large set of variables into a simple and small one that still contains most of the information in the huge set as shown in Figure 2. PCA is a non-supervised learning approach, as it includes only a set of features $X_1, X2, ..., X_p$, and no other related response Y. PCA is used to produce derived variables for supervised learning approach. It even works as a tool for data visualization. It converts the data in lower dimensions from higher dimensions. Theory here is that every '*n*' observation lies in m-dimension, but not every dimension is similarly interesting. Decreasing the quantity of variables of a data set obviously reduces the accuracy, but

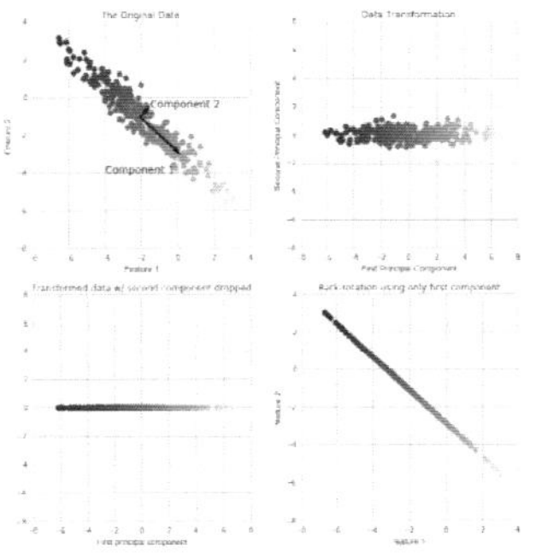

Figure 2. Principal Component Analysis

what can we do here is, in dimensionality reduction is we can adjust little accuracy with simplicity. Since small data sets are easy to explore and visualize and do analysing data much easy and fast with machine learning algorithms without external variables to process as shown in Figure 2. PCA asks for a small quantity of dimensions that are interesting like possible, here the theory of interesting is to level by the number that the observations differ along each dimension. Every dimension discovered by PCA is linear combination of the p features.

- Time Series Forecasting:-ARIMA stands for Auto Regressive Integrated Moving Average and it is one of the simplest and most powerful machine learning algorithms for forecasting time series. This is the function of combining Auto Regression and Moving Average. ARIMA is a easiest stochastic time series model, forecasting future time points can be done if use it to train and then forecast. ARIMA can detect complex dependencies as it takes terms and observations of delayed terms. These models depend on regressing a variable on previous values. Auto regression is a time series model that predetermines the value of the upcoming time step by using measurements from past time steps as input to the regression equation. In other words, it predicts t by doing regression in the previous time stage t-1. Error terms of past time points are used to predetermine future and present point's results. Moving average (MA) get rid of non-determinism or non-uniform movements from a time series. The property Q depicts Moving Mean (Average) in ARIMA. It is expressed as MA(x) where x depicts past observations that are utilized to calculate present observation. Moving average models have a definite window and weights are related to the time. This means that the MA models are extra responsive to present event and are more reactive. MA is an abbreviation for moving average, which is also known as the rolling mean. Fundamentally, we are considering the simple average in each time frame and dividing it by the total number of time frames taken. Whereas Integration denotes the number of times the variations are taken. If a trend is there, then the time series is taken as non-stationary and depicts seasonality. Integrated is a quality which reduces seasonality from a time series. ARIMA models have differencing degree which removes seasonality.

4 SIMULATIONS AND RESULT

4.1 Case Study 1

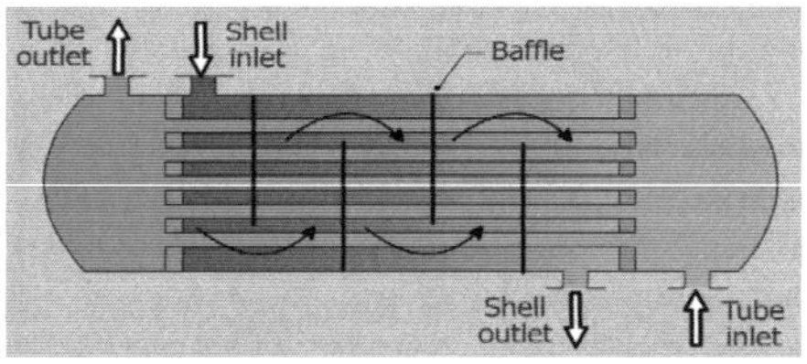

Figure 3. Shell and Tube Heat Exchanger

Shell and Tube type Heat Exchangers are popular types of heat exchanger because of design flexibility and wide range of temperatures and pressures. It consists of number of tubes mounted inside a shell as shown in the Figure 3. Two fluids or gases can exchange heat, one flows through the tubes and another from the outside of the tubes. The heat exchangers are generally manufactured according Tubular Exchanger Manufacturers Association (TEMA). As fluids passes through the shell and tube side during their regular operations, after a specific period of time the dirt or unwanted substances from the fluids gets settled on the surfaces. This deposition of unwanted material is called as Fouling and the theoretical resistance which represents the fouling is called as Fouling Factor. Hence maintenance is an important activity in heat exchangers to prevent fouling. This maintenance can be carried out using predictive maintenance. The first step for any predictive modelling is a data. A sensor data set of heat exchanger consisting of five months data was obtained. Now once we receive data, it has to be classified whether it is structured or unstructured, continuous or logistic (Binary). Further the dataset is checked for null values, if any than they are eliminated. The dataset should now be classified for dependent and independent variable. Now, our data is ready for visualization. Data can be visualized using various tools like excel, tableau or python. The trend component in the data visualized in our case study, i.e. fouling factor is the dependent variable which is to be predicted. Then using multi linear regression machine learning algorithm the fouling factor is predicted using independent variables by training the model and testing. In this case study regression is used because the variable to be predicted i.e., fouling factor is completely dependent on independent variables. Table 1 depicts the Primary Variables i.e. measured variables and Secondary Variables i.e. calculated variables. The Independent variables in our case where Log Mean Temperature difference (LMTD), U-Transfer Rate and Q which is amount of heat transfer which are responsible for predicting fouling. Fig. 4. and Figure 5. depicts the relation between the U Transfer Rate and Cummulative Flow Tonnes per day with Fouling Resistance. The multi linear regression uses simple mathematical equation of $Y = mX + C$. Here Y is our dependent variable and X is independent variable. Using this model, the accuracy of the prediction came near about 97 percent and as shown in Figure 9. The visualization of test vs predicted data is shown in Figure 6. Also, for the same case study the fouling factor for future days was predicted using Time Series Forecasting as shown in Figure 7. and Figure 8. This time series forecast was predicted using Auto Regressive and Integrated Moving Average (ARIMA) model. Hence with the help of predictive maintenance using sensor data the fouling factor was predicted in order to get the trends and schedule the maintenance activity in advance. Also using ARIMA model the useful life was predicted till which the heat exchanger can continue its operations within the limits.

Table 1. Variable List

Primary Variables		Secondary Variables	
Hot Temp In	Hot Temp Out	Hot Temp Increase	Q
Cold Temp In	Cold Fluid Temp Out	Cold Temp Decrease	Hot-Out Cold In
Cold Flow Rate	Hot Flow Rate	Fouling Resistance	Hot-In Cold-Out
Furnance Inlet Temp		U Transfer Rate	CMTD
		LMTD	

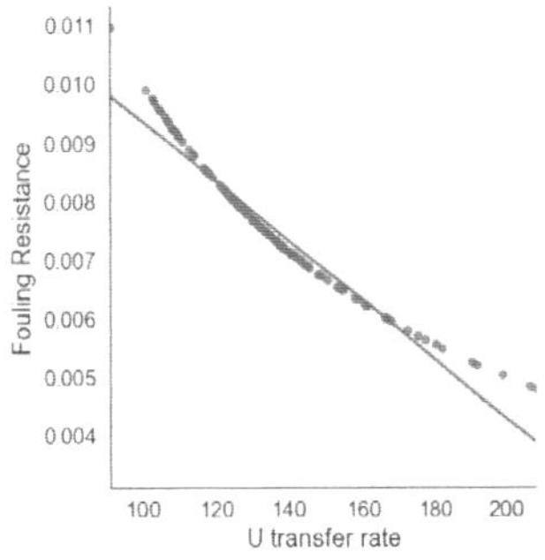

Figure 4. U Transfer Rate vs Fouling Resistance

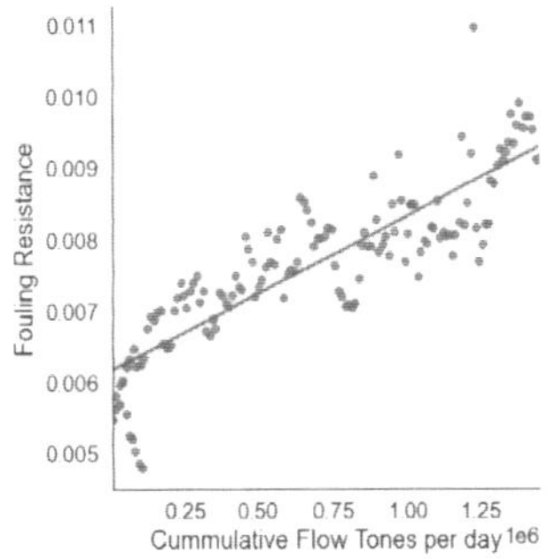

Figure 5. Cummulative Flow Tonnes Per Day vs Fouling Resistance

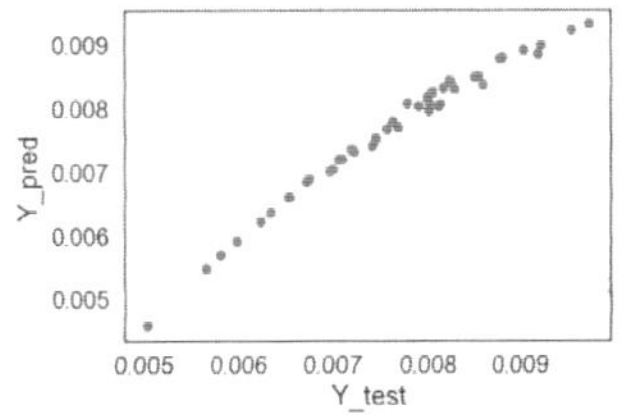

Figure 6. Fouling factor prediction of shell and tube heat exchanger

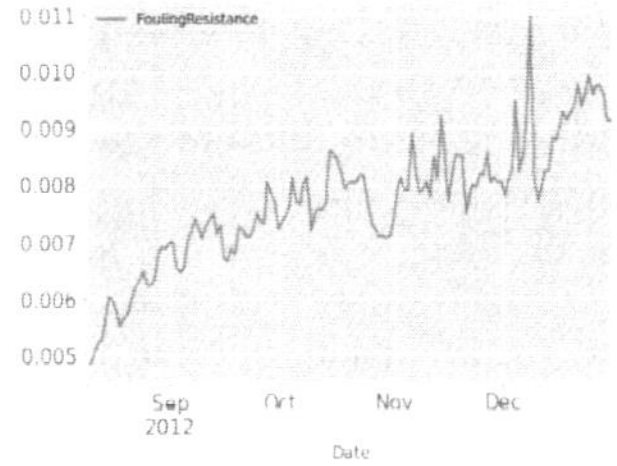

Figure 7. Working months vs Fouling Resistance

4.2 Case Study 2

Plate type heat exchanger is a specialized design to transfer heat between medium and low-pressure fluids. It is

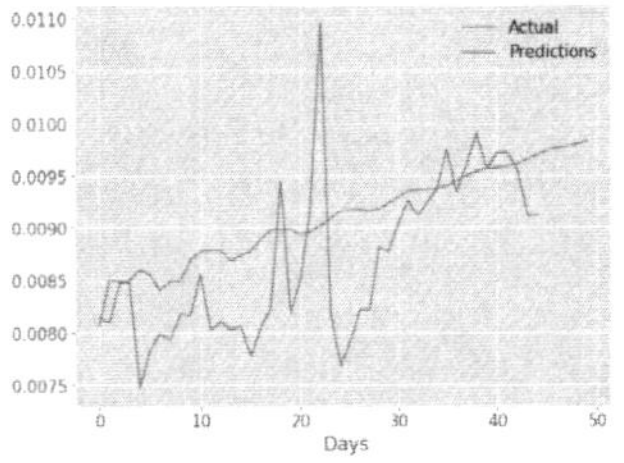

Figure 8. Actual vs Predicted fouling values

```
Intercept: 0.013079680465101827
                                  Coefficient
U transfer rate                  -4.296929e-05
Cummulative Flow Tones per day    4.609254e-10
R2 Score is: 0.9741388398089696
```

Figure 9. R2 Score for shell and tube heat exchanger

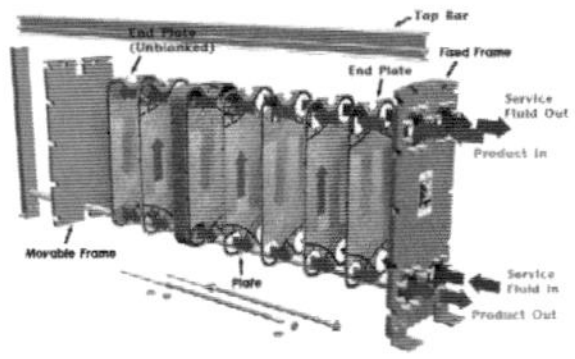

Figure 10. Plate Heat Exchanger

found to be most efficient of all heat exchangers. Plate type heat exchanger consist of a series of parallel plates as shown in the Figure 10. that are placed one above the other to allow the formation of series channels for fluids to flow between them. These heat exchangers are mainly used in oil, gas and chemical industries, food industries because they are exceptionally durable and ideal for transferring corrosive fluids or fluids which have high temperature. Fluid flows through plate type heat exchanger for transfer of heat for further processes during this process after some period there is accumulation of dirt inside or outside the tubes which forms small coating that adds resistance to heat transfer know as fouling factor. So, maintenance is an important activity to avoid fouling resistance in heat exchanger. This can be done using predictive maintenance. Initial requirement is data with which we can predict the future outcome. Now once we receive data, it must be classified whether it is structured or unstructured, linear, or logistic (Binary). Further the dataset is checked for null values, if any than they are eliminated. The dataset should now be classified for dependent and independent variable. our data is ready for visualization. Data can be visualized using various tools like excel, tableau or python with this Data Trend is visualized. Then using multi linear regression machine learning algorithm the fouling factor is predicted using independent variable by training and testing model. Fouling factor is completely dependent on independent variable so there is requirement to use regression. Independent variables in this case study where Log Mean Temperature difference (LMTD), U-Transfer Rate and Q which is amount of heat transfer which are responsible for predicting fouling. Table 2 depicts the primary Variables i.e. measured variables and secondary Variables i.e. calculated variables. Linear Regression equation is used to show the dependence of one variable on another variable. As fouling resistance is dependent variable in both case studies 1 and 2, all the visual representations of this case study are similar to that of Case Study-1 i.e. Linear in nature. This model gave us accuracy and predicted value as 82 percent as shown in figure 11. Hence using predictive modelling, we can schedule maintenance of equipment well before its failure.

Table 2. Variable List

Primary Variables		Secondary Variables	
Thickness	Fluid 1	Q	**Qr**
Temp1 In	Fluid 2	Ae	LMTD
Temp1 Out	dP1	V1 In	V1 Out
Temp2 In	dP2	m1	m2
Temp2 Out	Area	Theta 1	Theta 2
		U	F
		A	M
		Me	NP

```
Intercept: 7.238990730530886e-06
       Coefficient
Lmtd  9.247570e-05
U     9.149892e-10
Q    -1.470825e-07
Qr    2.328605e-08
Me    6.552722e-06
M    -5.752454e-06
R2 Score is: 0.8235290396306056
```

Figure 11. R2 Score for plate heat exchanger

4.3 Case Study 3

This set of data is created as per the semiconductor industry and contains sensor accounts from high exactness and innovative creation gear. Essentially, the semiconductor creation comprises of many cycle steps performing physical and substance procedure on so called wafers, for example cuts dependent on semiconductor material. Normally, lots of wafers are totalled into purported loads of size 25, which consistently go through similar tasks in the creation chain. This informational index was created as per the semiconductor business and contains sensor accounts from high-exactness and cutting-edge creation gear. Essentially, the semiconductor creation comprises of many cycle steps performing physical also, substance procedure on purported wafers, for example cuts based on semiconductor material Regularly, lots of wafers are collected into supposed bunches, which consistently passthrough similar tasks in the creation chain. In this chain, every process component is equipped with different sensors, for recording various parameters such as gas flow, temperature, voltages etc. resulting in the recording of the data at each step of the process. Using a time frame in each one of the sensor data recorded and found important for the process by the experts, key numbers (KNs) among the dependent sensor data are recorded. The key numbers are used to interrupt in case of miss direction for keeping the whole process balanced. After the creation, every gadget on the wafer is tried in the most cautious route coming about in purported wafer test information. In a few cases, dubious examples happen in the wafer test data potentially prompting disappointment. For this situation, the underlying driver must be found in the creation chain. So, the key numbers were provided for the purpose. And the purpose was to find the main cause by finding the correlations in between the key numbers and the wafer test data. Henceforth during Exploratory Information Analysis (EDA) it was tracked down that out of 50 KNs as it were some KNs were showing varieties. Hence forth these sensors were utilized to anticipate the class of disappointment whether 'Good' or on the other hand 'Bad' as visualized in Figure 12. and Figure 13. There is a threshold value specified i.e., 0.75. If response goes beyond it then the classification will be bad otherwise good. If the response is 'Good' it will be used, otherwise it will be discarded. The expect action calculations utilized for this situation study was Support Vector Machine (SVM) and Principal Component Analysis (PCA). SVM is a type of supervised machine learning algorithm that can be used to solve classification or regression problems. Principal Component Analysis (PCA) is an unsupervised, nonparametric statistical technique basically used for dimensionality reduction in machine learning. The precision drew close to around 80 percent from Figure 14.

Hence by using previous data we can train the model for the "Good" and "Bad" response and therefore it can be used in real time on the field with proper testing to automatically classify the faulty wafers in the industry.

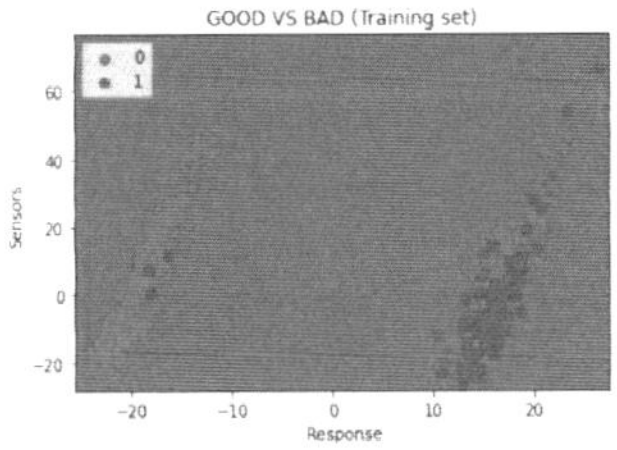

Figure 12. Training set of Good vs Bad data

5 CONCLUSION AND FUTURE SCOPE

The methods undertaken in this work are industry intensive. They help us obtain fairly accurate inferences about the state of the system. It is possible that instrument of measurement improve with time but not the method of performing analysis, diagnosis and repair, as they have remained more or less the same over many years. The

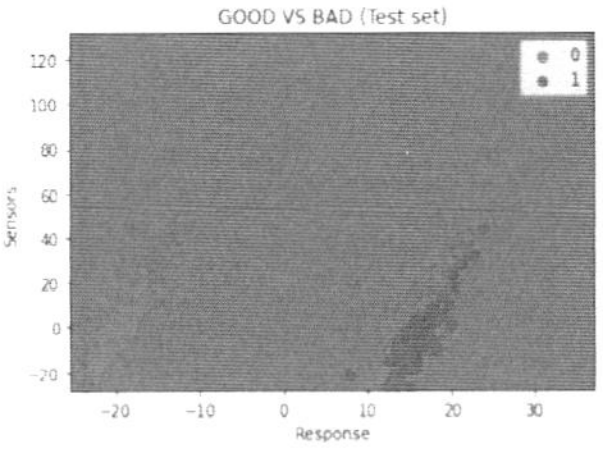

Figure 13. Testing set of Good vs Bad data

```
from sklearn.metrics import confusion_matrix
cm = confusion_matrix(Y_test,Y_pred)
accuracy = float(cm.diagonal().sum())/len(Y_test)
print("\nAccuracy Of SVM For The Given Dataset : ", accuracy)

Accuracy Of SVM For The Given Dataset :  0.8
```

Figure 14. Accuracy of svm algorithm for semiconductor manufacturing process

methodology and technology is an exact science, but engineering isn't hence quality of work depends upon the skill and experience of the performer of monitoring and maintenance operation. More conformity in quality of reading as the data transmission is automated. Hence through this work, we predictied the fouling resistance of heat exchanger. The model was trained using machine learning and hence the fouling resistance is predicted using testing dataset which came out with prediction accuracy of approximately 97 percent. The prediction was done using R-squared technique which is calculated by the difference between the testing dataset and predicted value.The value should be very much nearer to 1. The more the R score nearer to 1 the more accurate will be the predicted value. Also done the time series forecasting using ARIMA model to predict the Remaining Useful Life(RUL) of process equipment. In the manufacturing wafer industry the faulty sensors are identified and the class is predicted using SVM and PCA. Hence there are tremendous opportunities in variety of industries with the use of huge amount of data to study the behaviour of the process instruments and for the predictive analytics and monitoring. Proper domain knowledge and analytical skills will support the digital transformations taking place and lead to a better future. The introduction of reliable cloud platforms used in condition monitoring, the development trend towards predictive maintenance, and strong demand from new applications are expected to drive the growth of the system condition monitoring.

References

[1] P. Roback, J. Legler, Review of Multiple Linear Regression, in *Beyond Multiple Linear Regression* (Chapman and Hall/CRC, 2021), pp. 1–38, https://doi.org/10.1201%2F9780429066665-1

[2] A. Gelman, J. Hill, A. Vehtari, Logistic regression, *Regression and other stories* (Cambridge University Press, 2020)

[3] D. Shaub, Fast and accurate yearly time series forecasting with forecast combinations, International Journal of Forecasting **36**, 116 (2020)

[4] J. Cervantes, F. Garcia-Lamont, L. Rodríguez-Mazahua, A. Lopez, A comprehensive survey on support vector machine classification: Applications, challenges and trends, Neurocomputing **408**, 189 (2020)

[5] L. Silvestri, A. Forcina, V. Introna, A. Santolamazza, V. Cesarotti, Maintenance transformation through Industry 4.0 technologies: A systematic literature review , Computers in Industry **123**, 103335 (2020)

[6] H. Truong-Ba, M.E. Cholette, P. Borghesani, L. Ma, G. Kent, Condition-based inspection policies for boiler heat exchangers, European Journal of Operational Research **291**, 232 (2021)

[7] T. Zonta, C.A. da Costa, R. da Rosa Righi, M.J. de Lima, E.S. da Trindade, G.P. Li, Predictive maintenance in the Industry 4.0: A systematic literature review, Computers & Industrial Engineering p. 106889 (2020)

[8] A. Jimenez-Cortadi, I. Irigoien, F. Boto, B. Sierra, G. Rodriguez, Predictive maintenance on the machining process and machine tool, Applied Sciences **10**, 224 (2020)

[9] T.P. Carvalho, F.A. Soares, R. Vita, R.d.P. Francisco, J.P. Basto, S.G. Alcalá, A systematic literature review of machine learning methods applied to predictive maintenance, Computers & Industrial Engineering **137**, 106024 (2019)

[10] M. Compare, P. Baraldi, E. Zio, Challenges to IoT-enabled predictive maintenance for industry 4.0, IEEE Internet of Things Journal **7**, 4585 (2019)

[11] S.M. Lee, D. Lee, Y.S. Kim, The quality management ecosystem for predictive maintenance in the Industry 4.0 era, International Journal of Quality Innovation **5**, 1 (2019)

[12] V. Radhakrishnan, M. Ramasamy, H. Zabiri, V. Do Thanh, N. Tahir, H. Mukhtar, M. Hamdi, N. Ramli, Heat exchanger fouling model and preventive maintenance scheduling tool, Applied Thermal Engineering **27**, 2791 (2007)

[13] H.M. Hashemian, State-of-the-art predictive maintenance techniques, IEEE Transactions on Instrumentation and measurement **60**, 226 (2010)

[14] S. Selcuk, Predictive maintenance, its implementation and latest trends, Proceedings of the Institution of Mechanical Engineers, Part B: Journal of Engineering Manufacture **231**, 1670 (2017)

[15] Z. Zhou, B. Yao, W. Xu, L. Wang, Condition monitoring towards energy-efficient manufacturing: a review, The International Journal of Advanced Manufacturing Technology **91**, 3395 (2017)

[16] R. Sahal, J.G. Breslin, M.I. Ali, Big data and stream processing platforms for Industry 4.0 requirements mapping for a predictive maintenance use case, Journal of manufacturing systems **54**, 138 (2020)

[17] G. James, D. Witten, T. Hastie, R. Tibshirani, *An introduction to statistical learning*, Vol. 112 (Springer, 2013)

[18] Position Paper- Deloitte Analytics Institute, https://www2.deloitte.com/content/dam/Deloitte /de/Documents/deloitte-analytics/ (2017)

Plus Shaped Chipless RFID Tag

Vinay Ganesh Bhagure[1,*] , and *Somdotta* Roy Choudhury[1]

[1]R.A.I.T., Department of Electronics and Telecommunication Engineering, Nerul, Navi-Mumbai, India

Abstract. Chip-less radio frequency identification (RFID) technology is capable for detecting, identification, and chasing for the future Internet of Things (IoT) systems and applications. The chipless RFID technology gives us the impression as a rising substitute of the traditional RFID tags and barcodes by encoding data from tags employing the radio frequency waves. The paper presents the design as well as simulation of chip-less Plus shaped RFID tags. The structure is composed of Rogers RT/Duroid 5880 by means of a substrate having the physical dimensions 40mm x 40mm over which eight resonators are placed. This tag is compact, and the resonators are closely placed. The tag is designed using Ansys Electronics Desktop.

1 Introduction

Radio Frequency Identification (RFID) is a radio technology used to interconnect between two short-range objects namely, a reader and a tag. The main features of this technology are tracking, identification, and finding of objects i.e., animal's persons, logistics, etc, but more generally applications ranging from supply chain management to livestock management. RFID and its applications seem straightforward and easy. But in reality, RFID systems has foundations in many areas like system design, software development, RF and microwave theory, antenna theory, circuit theory, receiver design, integrated circuit design,encryption and network engineering, etc. The fact that RFID offers advantages over barcode systems, the cost of traditional RFID's is excessively high.

Since the presence of silicon chip has increased the prize of RFID tag, the research has been giving rise to chipless RFID technology [14][15][16]. Therefore, in chip-less RFID tags, the tag contains a printed pattern containing ID code. By using low-cost printing techniques, the prize of the tags can be therefore reduced. The aim of this paper is to present a plus shape multi resonator assembly arranged in chipless RFID tag.

2 Design of plus shape

A super slim double layer absorber utilizing resistively stacked frequency selective surface is introduced to give broad absorption bandwidth. Every layer has four strips of metal. The combination of metal strips and resistors on each layer are imprinted on the substrates. The structure accomplishes a broad absorption bandwidth in full-wave electromagnetic simulator.

The proposed design of structure as depicted in Fig.1, the substrate is composed of FR-4/Rogers RT Rogers RT/duroid 5880. On the substrate eight metal resonators are imprinted on the substrate. Relative permittivity of substrate FR-4 is 4.4, Rogers RT/duroid 5880 is 2.2 and dielectric loss tangent for both the substrate combinations is 0.02.

Fig.2 represents the top view of design. The geometric dimensions of substrate is: L_g= 40mm, W_g= 40mm and Δ= 0.3mm. Where L_g, W_g and Δ represents length, width and thickness of the substrate.

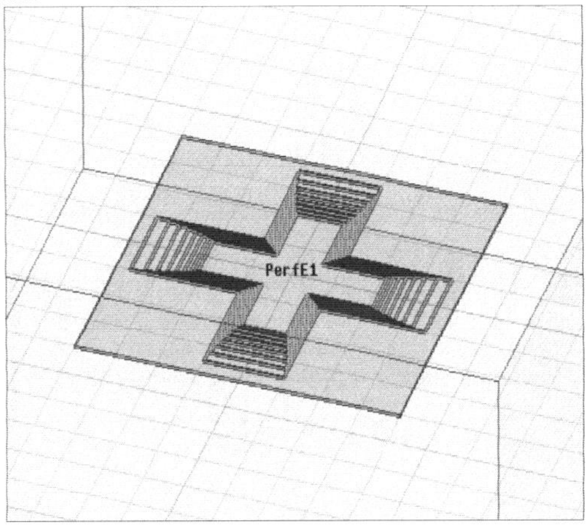

Fig. 1. 3D view of structure

https://doi.org/10.1051/itmconf/20214001004

2.1 Design Properties

The key stages for summarizing the design of the chip-less tag are given as follows:

- Determination of the length of the ith resonator in vertical and horizontal directions i.e., i= 1 ,2 ,3 , 4...., n, according to the formulas [1]:

$$L_i^v = \frac{c}{2f_i^v}\sqrt{\frac{2}{1+\varepsilon_r}} \quad (1)$$

$$L_i^H = \frac{c}{2f_i^H}\sqrt{\frac{2}{1+\varepsilon_r}} \quad (2)$$

Where, f_i^V and f_i^H are the ith resonator's resonant frequency in the horizontal and vertical directions, ε_r and c are the relative permittivity and speed of light of the substrate.

- The design parameters are considered as per the guidelines given in the base paper which is been selected [1].
- With the help of formulas given below, length of vertical and horizontal resonators i.e., L_i^V and L_i^H can be achieved by substituting the frequency equations into length of vertical and horizontal resonators,

$$f_i^H = 5GHz + (i-1) * 0.5\ GHz \quad (3)$$

$$f_i^V = 5.05GHz + (i-1) * 0.5\ GHz \quad (4)$$

In this tag, there is a spacing between two resonators for attaining difference of 50 Mhz to accomplish various coding for horizontal and vertical resonators.

- The design parameters chosen for the chipless tag is,

Table 1. Selected design parameters

Parameter	Lg	Wg	L_i^H	L_i^V	W	Δ
Value(mm)	40	40	18.25	18.07	0.3	0.3

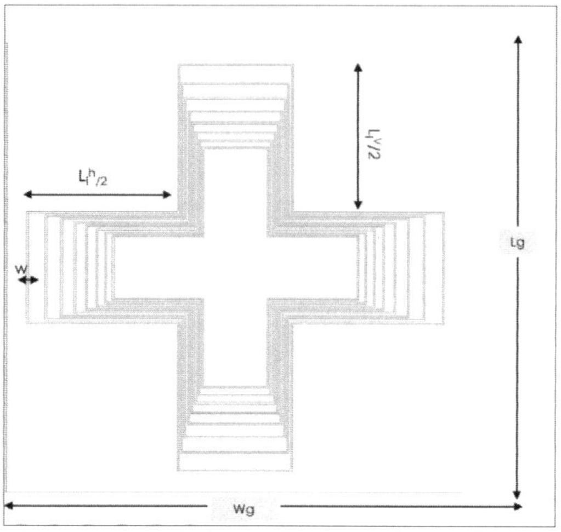

Fig. 2. Top view of design

3 Process, simulation and analysis of chipless RFID tag

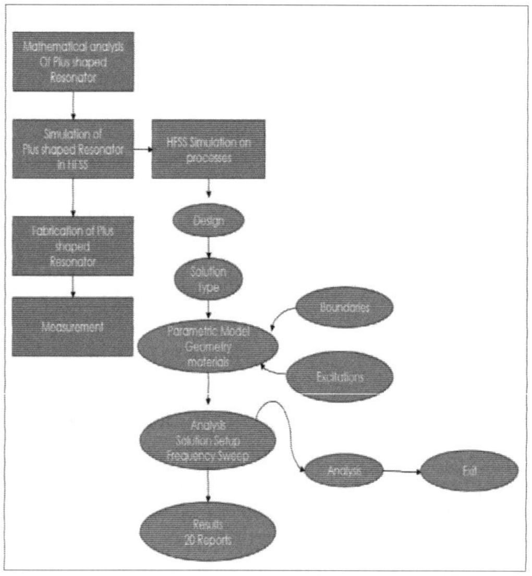

Fig. 3. Design Process

Figure 3 shows the various processes that were followed to achieve the aim of this thesis.

4 Simulation using two resonators

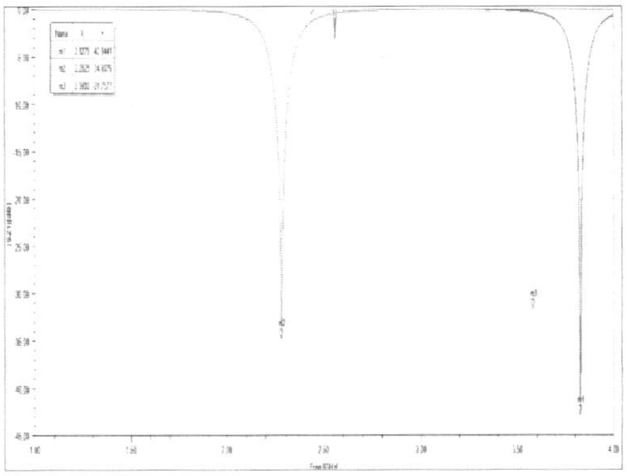

Fig. 4. Simulation result of S-parameters of 1st and 2nd resonator

The maximum depth observed in |S21| depends on the size of the plus shape and length & therefore in the plus shape with a length of 36.5 mm (L_1^H), maximum depth of -34.8 dB is observed at resonating frequency of 2.28 GHz, while in the smaller plus size plus shape with length of 33.18 mm (L_2^H), slightly less depth of -31.75 dB is observed at resonating frequency of 3.58 GHz.

When these two plus shaped resonators are kept together on the substrate depth of return loss of -42.94 dB is observed in the |S21|, which is greater than single resonator of lengths mentioned above when placed alone on the substrate shown in figure 4.

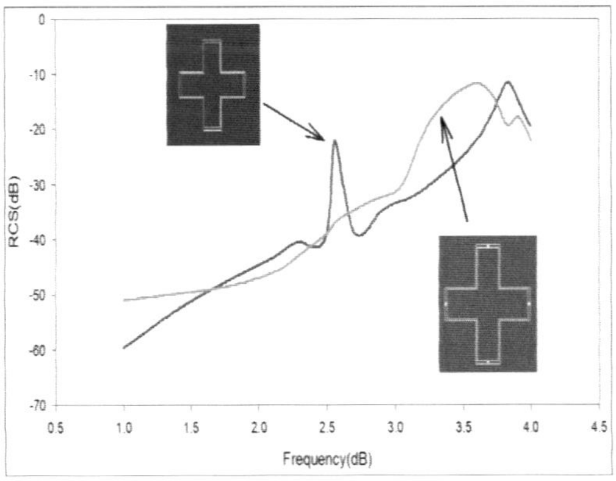

Fig. 5. RCS of 1st and 2nd resonator, open and short circuit

As we can see in the figure 5, due to the interaction between the adjacent rings (when the two rings are joined at all 4 sides of the plus shape), the spike is suppressed due to the contact and only single spike is observed in the monostatic RCS plot. On the other hand, when the two rings are independent (adjacent plus rings are not attached), two spikes can be seen in the RCS plot [11].

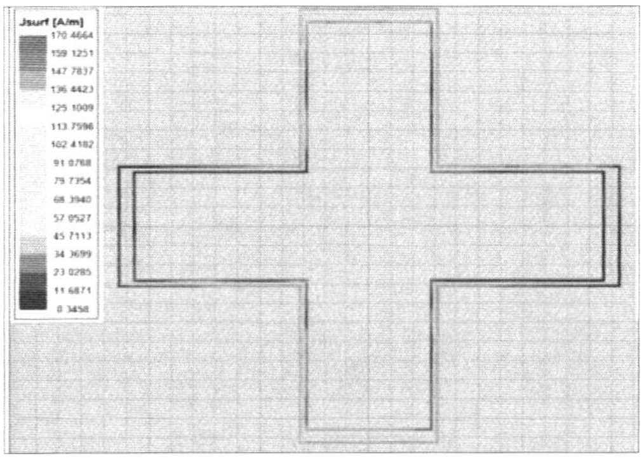

Fig. 6(a). Surface current of two resonators with open circuit

In order to input ID code to tag, adjacent rings should be attached as the above methodology is applied. First the reference tag is introduced in which all the plus shaped rings are in open circuit and we get the result as logic 1 up to 8 notches. In the second step, when we need a different coding, according to the reference tag we have to print short circuits in the assembly to suppress the resonance between two rings. Thus if, we apply short circuits between all neighbouring rings, we will get the logic 0.

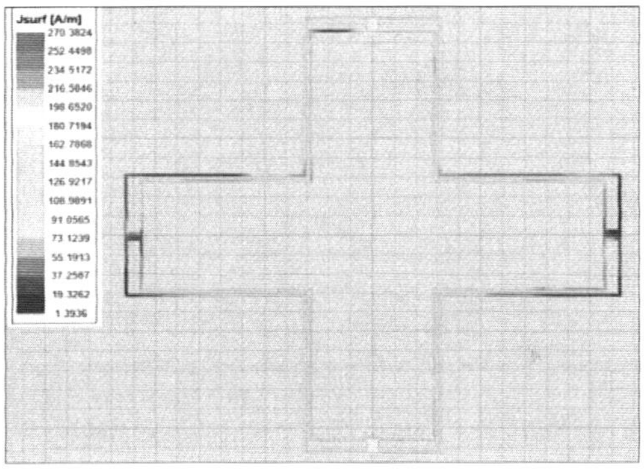

Fig. 6(b) Surface current of two resonators with closed circuit

By addition of set of short circuit's between two adjacent plus shaped, circulation of the surface currents is shown in figure 6(a) and 6(b).

These short circuits when applied on all four sides of the plus assembly. The standing waves (J_{in} and J_{out}) have a similar phase and therefore, the RCS produced from these sources is beneficially affected in free space, and subsequently, the resonance of the lower resonator is absent shown in figure 5.

5 Simulation of 8 resonators using Rogers RT/duroid 5880 substrate

 https://doi.org/10.1051/itmconf/20214001004

According to the figures 7(a) and 7(b) it's been seen that 8-bits are encoded according to 8 resonances. The occurrence of single resonance will let a particular frequency to pass in result the signal being at that frequency when again transmitted to reader. The logic "1" can be encoded based on the resonance levels. The absence of the resonance, which can be done by removing the particular resonator from figure 3 or sorting the two resonators as mentioned above, will attenuate due to the absence of stopband. This would represent logic "0". In the figure 7(b), it's been seen that the tag for the 8-resonator tag, the first 3 resonance phase jumps were expected to give return loss above 10 db. But due to less return loss, the first 3 resonators will not resonate and give Logic "0".

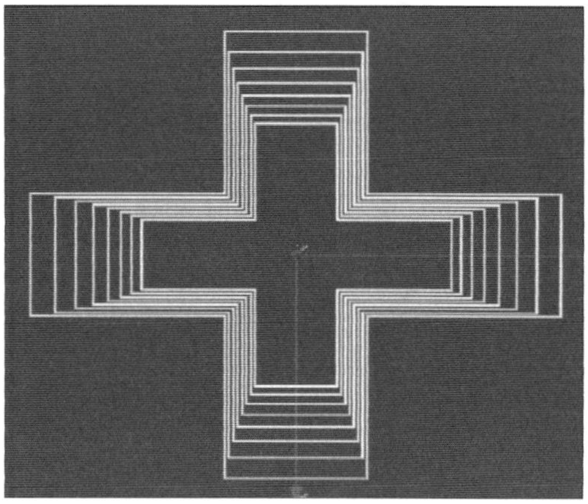

Fig. 7(a) Substrate with 8 resonators

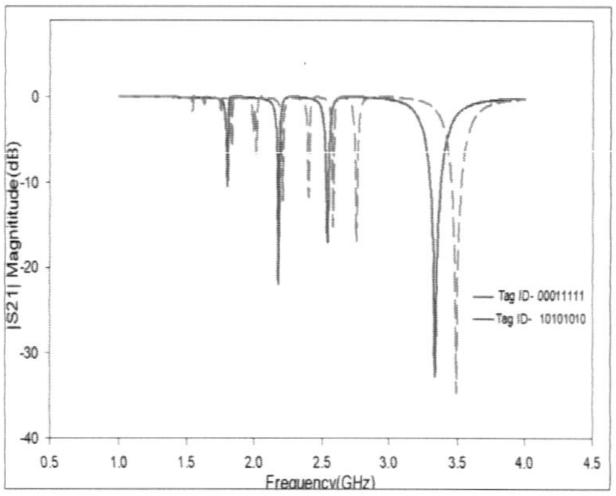

Fig. 7(b) |S21| with different sets of coding

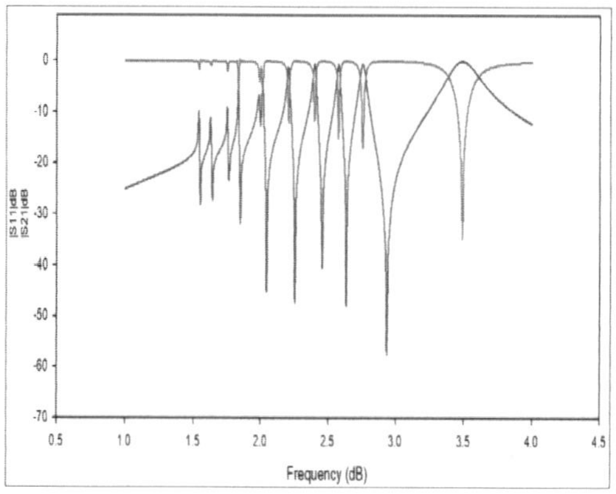

Fig. 7(c) Simulated S-parameters of 8 resonators

As mentioned above this chip-less RFID works in the frequency range of (1-4) GHz, in the figure 7(c) and figure 7(d) the simulated values of S parameters (S11 and S21) and the radiation pattern are shown.

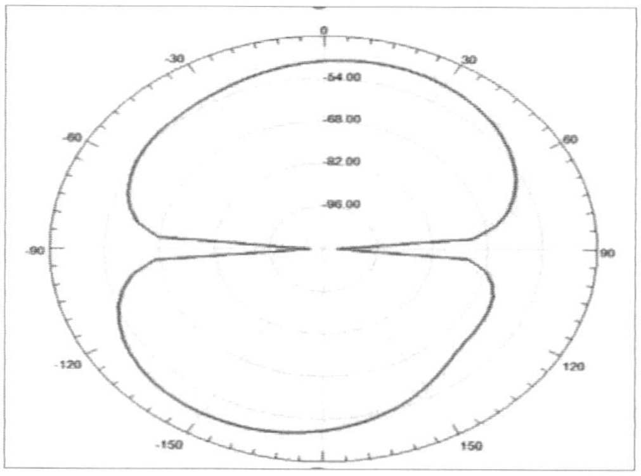

Fig. 7(d) Simulated polar plot

In order to see performance of plus shaped RFID tag, the RCS of tag is calculated at 5.5 GHz shown in figure 8(a) and 8(b). Tags used in measurnment setup include following ID's- 11111111 and 10101010.

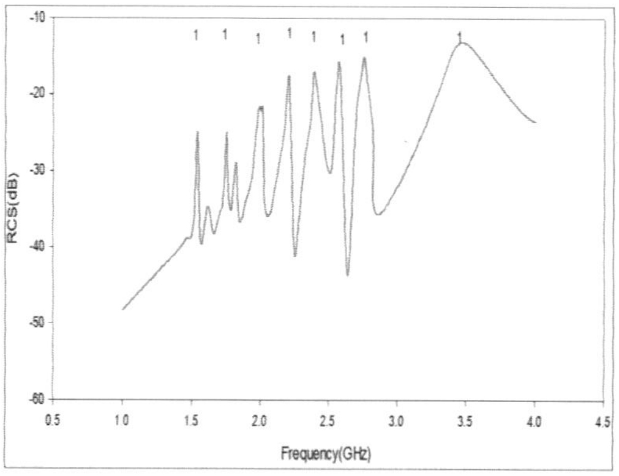

Fig. 8(a) Simulated results of RCS parameters of 8 resonators

For all the 8 resonators combined the simulated results are shown in figure 8(a), its seen that there are 8 different resonant peaks at resonant frequency 5.5 GHz that give us 8-bit output in RCS. On the other hand, only four resonators are taken, which all four resonators are equally spaced. Therefore, the encoded output of 4-bits is obtained which is shown in the figure 8(b) which gives is the output ID-10101010[12][13].

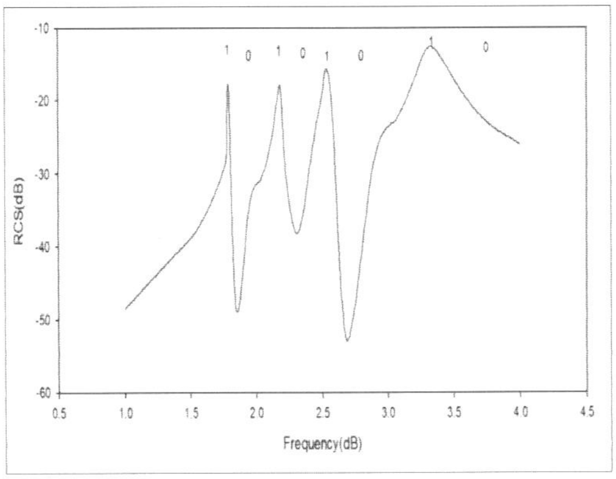

Fig. 8(b) Simulated results of RCS parameters of 4 resonators

6 Simulation of 8 resonators using FR-4 Substrate

As mentioned above the chip-less tag has no ground plate included, it consists of substrate and few resonators as shown above. The study and simulation have been done by changing the substrate to FR-4. The solutions of Rogers RT-duroid 5880 substrate have been kept for reference of performance against the FR-4 substrate. The FR-4 substrate (relative permittivity=4.4 and dielectric loss tangent= 0.02) is cheaper than Rogers RT-duroid 5880.

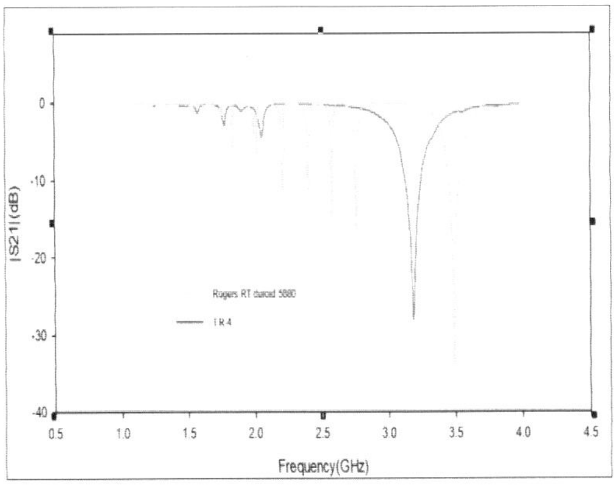

Fig. 9 Result of difference between simulation using two different substrates

As the simulation had been done with substrate as FR-4, it was observed in the return loss in the |S21| graph as shown in figure 9, the return loss using the FR-4 as a substrate is very low and only 1-bit of encoded output is obtained.

On the other hand, in as compared to FR-4 substrate, 6-bit resonant peaks are obtained and the highest return loss obtained is -37dB. The encoded output of the tag with Rogers RT-duroid substrate is 00111111. The 1st three resonators return loss is less than -10 dB, therefore only 6 resonators resonate at 5.5 GHz operating frequency.

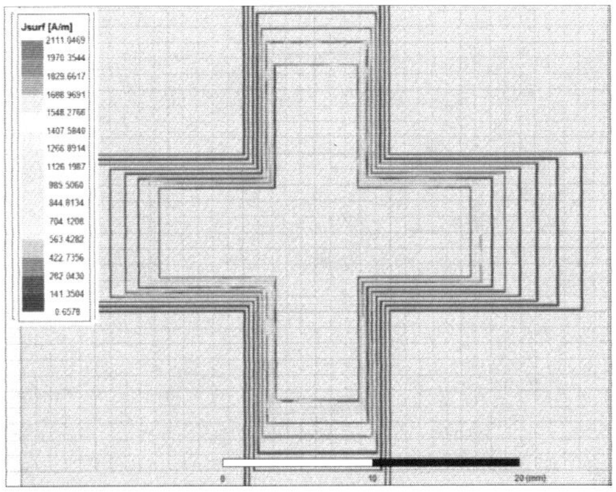

Fig. 10 Simulated result of surface current (Jsurf)

With the help of surface current distribution, the proposed structure can be explained. Here the surface current has been analyzed at the resonating frequencies 5.5 GHz. It demonstrates that the surface current is minimum in the outer resonators while the surface current is more in the inner resonators. As observed in the above figure 10 , surface current distribution is maximum in 6th,7th, and 8th resonators where bits of green, orange and red patches can be seen on those resonators[2][3].

The Plus Shaped Chip-less tag can be fabricated using lithographic techniques [8][9][10] and inkjet printing [4][5][6][7] which reduces production costs of the tag and yield high efficiency.

7 Conclusion

A multi-resonator structure for plus shaped chip-less RFID tags is presented. The proposed structure consists of 8 resonators. The designed chip-less RFID has uniform radiation pattern. The simulated RCS plot and measured S21 plot are similar for the proposed tag which indicates that the L and S bands of 1 GHz to 4 GHz can be effectively used. For generating new codes, the design of RFID tag can be modified, simply by varying the resonator's lengths by an amount equal to the slots spacing. The key performance parameters are enhanced with many simulations having resonant notches of 33 dB depth. The operation of the proposed structure can be explained with the help of surface current distribution in operational frequency of 5.5GHz. The proposed tag can encode 6 bits. The proposed chip-less RFID tag can be incorporated in many industrial and IoT applications.

References

1. Issa, Khaled. (2017). A Novel L-Shape Ultra Wideband Chipless Radio-Frequency Identification Tag. International Journal of Antennas and Propagation.
2. P. P. Sahu, D. Prasad Mishra, T. K. Das and S. Kumar Behera, "*Design of a Chipless RFID Tag for 2.4 GHz and 5.8 GHz ISM Band Applications*," 2020 IEEE International Students' Conference on Electrical,Electronics and Computer Science (SCEECS), 2020, pp. 1-4, doi: 10.1109/SCEECS48394.2020.108.
3. O. Rance, R. Siragusa, P. Lemaître-Auger and E. Perret, "Toward RCS Magnitude Level Coding for Chipless RFID," in IEEE Transactions on Microwave Theory and Techniques, vol. 64, no. 7, pp. 2315-2325, July 2016, doi: 10.1109/TMTT.2016.2562625.
4. H. Jang, W. Lim, K. Oh, S. Moon and J. Yu, "Design of Low-Cost Chipless System Using Printable Chipless Tag With Electromagnetic Code," in IEEE Microwave and Wireless Components Letters, vol. 20, no. 11, pp. 640-642, Nov. 2010, doi: 10.1109/LMWC.2010.2073692.
5. Shao, Botao & Chen, Qiang & Liu, Ran & Zheng, Li - Rong. (2012). Design of fully printable and configurable chipless RFID tag on flexible substrate. Microwave and Optical Technology Letters. 54. 10.1002/mop.26499.
6. B. Shao, Y. Amin, Q. Chen, R. Liu and L. Zheng, "Directly Printed Packaging-Paper-Based Chipless RFID Tag With Coplanar LC Resonator," in IEEE Antennas and Wireless Propagation Letters, vol. 12, pp. 325-328, 2013, doi: 10.1109/LAWP.2013.2247556.
7. A. Vena, A. A. Babar, L. Sydänheimo, M. M. Tentzeris and L. Ukkonen, "A Novel Near-Transparent ASK-Reconfigurable Inkjet-Printed Chipless RFID Tag," in *IEEE Antennas and Wireless Propagation Letters*, vol. 12, pp. 753-756, 2013, doi: 10.1109/LAWP.2013.2270932.
8. S. Preradovic, S. M. Roy and N. C. Karmakar, "RFID System Based on Fully Printable Chipless Tag for Paper-/Plastic-ltem Tagging," in *IEEE Antennas and Propagation Magazine,* vol. **53,** no. **5**, pp. 15-32, Oct. 2011, doi: 10.1109/MAP.2011.6138421.
9. A. Vena, E. Perret and S. Tedjni, "A Depolarizing Chipless RFID Tag for Robust Detection and Its FCC Compliant UWB Reading System," in *IEEE Transactions on Microwave Theory and Techniques,* vol. **61**, no. **8**, pp. 2982-2994, Aug. 2013, doi: 10.1109/TMTT.2013.2267748.
10. V. P. Plessky and L. M. Reindl, "Review on SAW RFID tags," in IEEE Transactions on Ultrasonics, Ferroelectrics, and Frequency Control, vol. 57, no. 3, pp. 654-668, March 2010, doi: 10.1109/TUFFC.2010.1462.
11. Betancourt, D. et al. "*Bending and Folding Effect Study of Flexible Fully Printed and Late-Stage Codified Octagonal Chipless RFID Tags.*" *IEEE Transactions on Antennas and Propagation* 64 (2016): 2815-2823.
12. S. Preradovic, I. Balbin, N. C. Karmakar and G. Swiegers, "*A Novel Chipless RFID System Based on Planar Multiresonators for Barcode Replacement,*" 2008 IEEE International Conference on RFID, 2008, pp. 289-296, doi: 10.1109/RFID.2008.4519383.
13. Xu, Lei and Huang, Kama. (2015). *Design of Compact Trapezoidal Bow-Tie Chipless RFID Tag. International Journal of Antennas and Propagation.* 2015. 1-7. 10.1155/2015/502938.
14. Karmakar N.C., Kalansuriya P., Azim R.E., Koswatta R. *Chipless Radio Frequency Identification Reader Signal Processing*. John Wiley & Sons, Inc.; Hoboken, NJ, USA: 2016.
15. Hunt V.D., Puglia A., Puglia M. *RFID: A Guide to Radio Frequency Identification.* Wiley-Interscience; Hoboken, NJ, USA: 2007.
16. Karmakar N.C., Amin E.M., Saha J.K. *Chipless RFID Reader Architecture.* John Wiley & Sons, Inc.; Hoboken, NJ, USA: 2013.

 https://doi.org/10.1051/itmconf/20214001005

Vehicle Owner Recognition and Speed Estimation through LPD using Deep learning (VORSELD)

Mudit Shrivastava[1] , *Rahul Jadhav*[2], *Pranjal Singhal*[3] *and Savita R. Bhosale*[4,*]

[1]Ramrao Adik Institute of Technology, Electronics Engineering Department, India
[2]Ramrao Adik Institute of Technology, Electronics Engineering Department, India
[3]Ramrao Adik Institute of Technology, Electronics Engineering Department, India
[4]Ramrao Adik Institute of Technology, Electronics Engineering Department, India

Abstract. As name characterizes understanding of a number plate accordingly, from past decades the use vehicles expanded rapidly, taking into account of this such a majority number of issues like overseeing and controlling trafficantc keeping watch on autos and managing parking area zones to overcome this tag recognizer programming is required. The proposed work aims to detect speed of a moving vehicle through its license plate. It will fetch vehicle owner details with the help of CNN model. In this project the main focus is to detect a moving car whenever it crosses dynamic markings. It uses Tensor-flow with an SSD object detection model to detect cars and from the detection in each frame the license plate gets detected and each vehicle can be tracked across a video and can be checked if it crossed the markings made in program itself and hence speed of that vehicle can be calculated. The detected License plate will be forwarded to trained model where PyTesseract is used, which will convert image to text.

1 Introduction

The huge reconciliation of data advancements, under various parts of the advanced world, has prompted the treatment of vehicles as reasonable assets in data frameworks. Since an automation data framework has no importance with no information, there is a need to change vehicle data among the real world and the data system.This can be accomplished by human specialists or by unique clever hardware that will permit recognizable proof of vehicles by their number plates in genuine conditions.

All smart equipment, mentioned is based on the system of detection and recognition of the number plates of vehicles.

The arrangement of vehicle number plate identification and acknowledgment is utilized to identify the plates. At that point make the acknowledgment of the plate that is to extract the content from a video outline which will straightforwardly be shipped off site for web scratching out the owner details, additionally to assess the speed of vehicle through the vehicle positions at various edges.All that gratitude to the computation modules. In India vehicle and its owner identification is an important and challenging problem because of large number ion plate and character of registered vehicles. Vehicle license plate recognition for vehicle and its owner identification becomes necessary and the need of hour in case of suspected vehicles and their owners who break traffic rules and are involved in other illegal activities[1-2].

There are multiple Vehicle License Plate Recognition systems but they are only restricted to the vehicle license plate recognition and the challenging task is the retrieval of vehicle and owner details because of unavailability of a publicdatabase.

The system should be capable of identifying, localizing, detecting, and transporting extracted text as REQUESTS and hence displaying vehicle and owner details.

This design incorporates a vehicle detection which is followed by tracking speed through frames and then recognition of license plate.

The final goal is to establish and implement the full working CNN Model with adequate accuracy.

Using a set of 100 frames as training data set, clarified the arrangement of pictures by drawing the boundary box over the number plates to send it for the training stage[3-4].

* Corresponding author: savita.bhosale@rait.ac.in

2 Literature survey

In the past authors focused on different approaches to automatic number plate recognition considering the different factors like image size, lightning conditions, character segmentation using Probabilistic Neural Network but the system is limited to vehicle licenseplate detection and an extension for vehicle owner identification.

Many authors introduced vehicle identification by capturing the vehicle image followed by license plate extraction and optical character recognition. Recognized vehicle license plate number is then compared with existing database records retrieving details like vehicle owner, place of registration. The identification is dependent on the database coverage limited to a certain data entries.

Authors were also suggested new algorithm for license plate detection involving sobel vertical edge detection, thresholding and bounding box analysis for vehicle plate extraction. For character segmentation the author uses projection analysis and connected component analysis. This system is also restricted to License plate detection.

Many of the authors were discussed about the use of template matching for optical character recognition. Template Matching used a database of templates (A-Z a-z 0-9), upon matching the segmented characters, character with maximum correlated template were matched.

The authors were discussed four methods digitization of image, edge detection, separation of characters and template. Authors used morphological operations for plate detection and connected component analysis for character segmentation.

The authors explored mid-level filtering using a 3x3 kernel for noise reduction and histogram equalizer for enhancing the contrast of the image. Edges were detected using mathematical morphological transform.

The authors discussed on using Artificial Neural Network and K Nearest Neighbor Classification algorithm for the classification and recognition of characters respectively with the highest accuracy rate of up to 87%.

3 Proposed Methodology

3.1. Training CNN Model

In spite of the plentiful data gathered by cameras encompassing us, little of it is handled and perceived by machines. This task takes utilization of genuine chronicles from vehicle front-camera in endeavour to foresee vehicle movement properties. Such movement identification assignments are valuable in investigating recordings pre-recorded by gadgets like scramble cams, CCTV giving information on at-the-time vehicle status, helping other downstream errands, for example, driver expectation surmising when joined with consequences of other PC vision undertakings like item discovery, and can be handily reached out to other moving articles. In spite of the plentiful data gathered by cameras encompassing us, little of it is handled and perceived by machines. This task takes utilization of genuine

$$T_s(l,t) = T_g(l,t) \quad (1)$$

$$T_s(l,t) = T_g(l,t)\, T_b(x \rightarrow -\beta, t) = 0 \quad (2)$$

Use italics for variables (u) and bold (**u**) for vectors. chronicles from vehicle front-camera in endeavour to foresee vehicle movement properties.

In light of the anticipated qualities, a last state is figured that solidifies the vehicle's movement in four classifications: Still, Forward, Turning Left, and Turning Right. Forward speed fv decides the vehicle's "quietness", where vehicle is anticipated as moving "Forward" when fv > 2m/s. For any moving vehicle, its preparing point is additionally classified dependent on precise speed av, where vehicle is anticipated as "Turning Left" when av > 2 ◦ , and "Turning Right" when av < −2 ◦ .

Table 1. Best Overall Train, Validation and test report

Train	Validation	Test
4.522	3.509	6.85
4.556	2.854	7.04
4.591	3.538	7.22

Choosing the appropriate thresholds for the consolidated results is a rather subjective process. The informational collection (Data-Set) doesn't give such sorts of truth information, and therefore, we envision the anticipated qualities on the video outlines for natural comprehension of the outcomes, and pick the edges dependent on perceptions on the approval set. The imagined results generally match the misfortune we see from the quantitative outcomes.

The Fig. 1 and 2 show two model representations on the testing outlines, where the red boxes address the info object veil and text box under the edges show next to each other correlation of the properties of interest, just as the last united status. Several observations are made from these visualizations.

The significant goal of this work is to follow the moving (vehicles) out and about. The different ideas of profound learning and PC vision have been used for this reason. Track by location system was applied for constant vehicle following. YOLOv3 object identification framework was utilized to identify the vehicles and the ideas of Deep SORT calculation was applied for following.

By changing the re-recognizable proof model of the first Deep SORT framework and preparing the organization on the vehicle data-set created without any preparation, the proposed framework improves the following presentation by lessening the quantity of personality switches. With the utilization of all the more remarkable equipment, the exhibition of the framework can be improved.

https://doi.org/10.1051/itmconf/20214001005

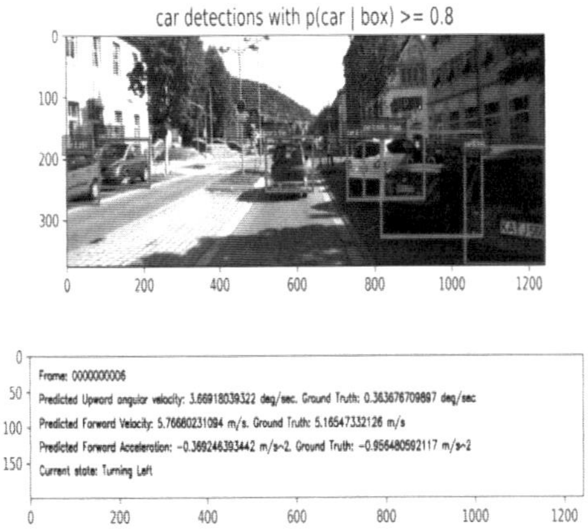

Fig 1 : Visual Results

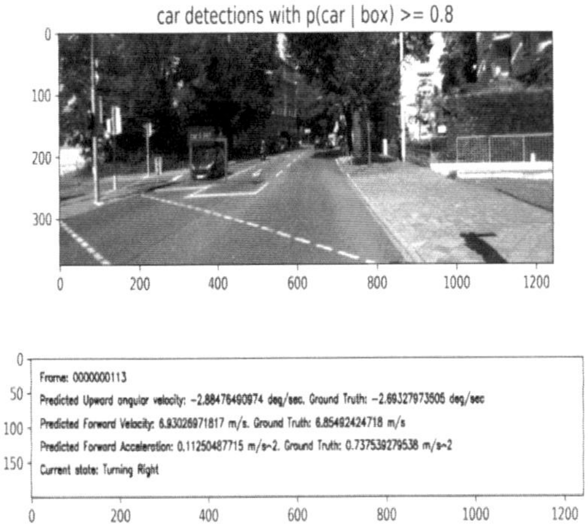

Fig 2 : Visual Results - 2

Lopsided Error Distribution Among every one of the three forecasts, rakish speed have all the earmarks of being significantly more mistake inclined contrast with the others. This perception could be brought about by the way that we right now train a solitary model to anticipate every single intrigued property and rakish speed's commitment to the general misfortune is nearly little because of the little extent in sweep.

Skewed Status Prediction Fig. 2 is an illustration of the model dishonestly accepting that the vehicle is making a left turn. Indeed, we have noticed the model is overall slanted towards accepting that the vehicle is making a left turn. This perception could be because of the way that all video chronicles in the informational index are from right-hand streets, where pixels on the privilege are relied upon to have higher optical streams, and they are nearer to the vehicle and move quicker. Fortuitously, this conduct is like that of a left-turning vehicle, where objects on the right-hand side are relied upon to have more sensational optical stream change. Truth be told, we improve results when the item covers are remembered for the information channels, and hope to see more clear contrasts with extended dataset and more events of turning.

3.1.1 Training Phase

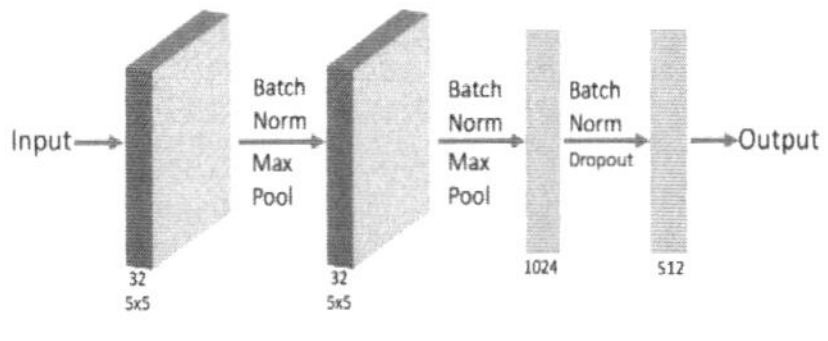

Fig. 3: Architecture of 2-layer baseline CNN

The architecture of 2-layer baseline CNN is illustrated in Fig. 3.
The optical stream calculation use the OpenCV execution of Gunnar Farneback's two-outline movement assessment calculation dependent on polynomial extension, where quadratic polynomials are utilized to in exact neighbourhoods of successive casings and gauge the dislodging fields between them. Thick optical stream vectors of every pixel address the heading and greatness of the pixel's development starting with one casing then onto the next, and are consequently expected to correspond firmly with the speed of the camera mounted on the vehicle.

The article recognition measure utilizes the FasterRCNN calculation carried out in Tensorflow , which is made out of a profound completely convolutional RPN that proposes districts of applicant objects, trailed by a FastRCNN with VGG-16 identifier that performs picture order on top and yields class softmax probabilities and per-class jumping box counterbalances. Consideration instruments are utilized to direct the identifier toward the suitable proposed districts. In light of the item recognition results, we develop twofold article covers for the class vehicle.

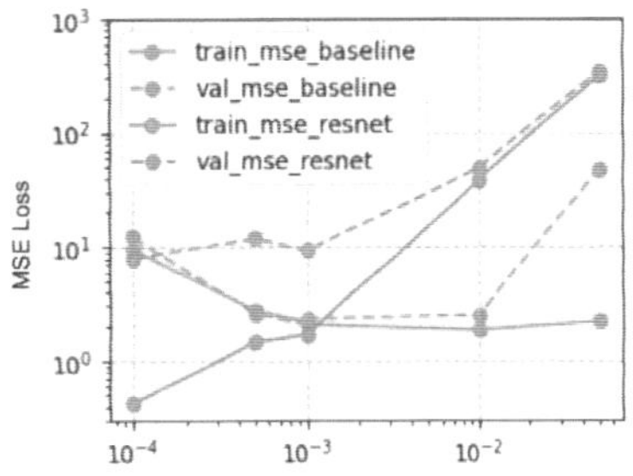

Fig. 4: Initial Learning Rate tuning results

Fig.4 shows the after effects of starting learning rate tuning on pictures down sampled to 100x300. True to form, both the gauge CNN and ResNet-17 don't perform well when the underlying learning rate is excessively enormous, for example more prominent than 10–2 . The AlexNet and pre-prepared AlexNet model (not appeared) show comparable conduct to the benchmark CNN and ResNet-17, separately, so we pick 0.0001 for ResNet-17 and pretrained AlexNet, and 0.001 for the standard and AlexNet models, with the perception that the best learning rate for ResNet is lower than that of different models.

Given our information dat–a(Frames from Video), we carry out three diverse CNN designs to yield the anticipated forward speed, speed increase, and precise speed. Figure 4 shows the benchmark 2-layer CNN design, which comprises of 2x conv-relu-cluster standard max pooling layers, 2x relative relubatch standard dropout layers.

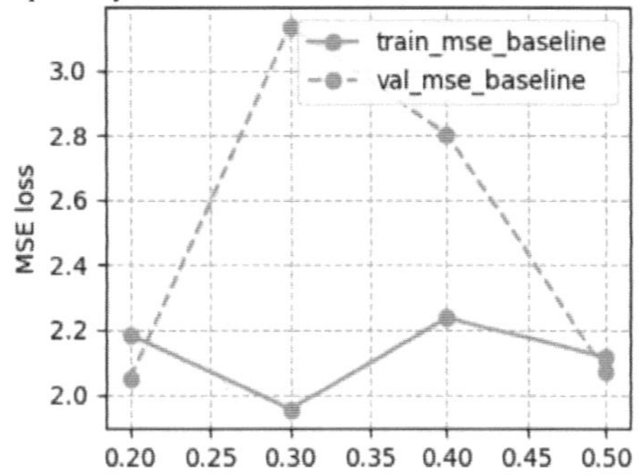

Fig. 5: Dropout Rate Tuning Results

Fig. 5 shows the consequences of dropout rate tuning on pictures down-examined to 100x300. Generally speaking, dropout rate from 0.2 to 0.5 doesn't have as huge an impact on the train and approval Minimum Mean Square Error (MSE's) as the learning rate. This can be clarified by the way that every one of the info pictures are as of now found the middle value of and down-examined, which basically adds a regularization impact on the model. A dropout worth of 0.2, the worth that accomplish the best approval MSE in this analysis, is picked for the excess tests with this mode.

The primary goal of this segment is to recognize over speed vehicles, utilizing Deep Learning and Machine Learning Algorithms. After procurement of arrangement of pictures from the video, trucks are identified utilizing Haar Cascade Classifier. The model for the classifier is prepared utilizing loads of positive and negative pictures to make a XML document. This is trailed by finding the vehicles and assessing their velocities with the assistance of their individual areas, Pixels Per Meter)(PPM) and fps (outlines each second). Presently, the edited pictures of the distinguished trucks are sent for License Plate location.

The Connected Component Analysis (CCA) aids Number Plate discovery and Characters Segmentation. The SVC model is prepared utilizing characters pictures (20X20) and to build the precision, 4 cross crease approval (Machine Learning) is likewise done. This model guides in perceiving the sectioned characters. After acknowledgment, the determined speed of the trucks is taken care of into a dominate sheet alongside their tag numbers. These trucks are likewise appointed a few IDs to produce an arranged data set.

Fig. 6 : Detecting Car Speed under selected area

Detecting car speed under selected area is shown in Fig.6.This venture utilizes YOLOv3 for Vehicle recognition and SORT(Simple Online and Realtime Tracker) for vehicle.

This venture carries out the accompanying undertakings in the task:

1. Vehicle counting
2. Lane Division
3. Lane change detection
4. Speed estimation
5. Dumps all these details into a CSV file

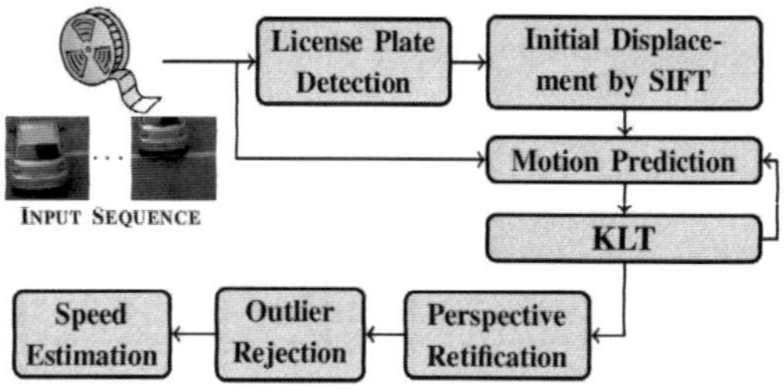

Fig. 7: Overview of the proposed system of Speed Detection

Deciding vehicle speed is a significant assignment for metropolitan traffic reconnaissance. The data might be utilized not exclusively to give fines when drivers surpass speed limits, yet in addition to take care of frameworks, for example, traffic regulators.

The framework utilizes text identification to find the tags of moving vehicles, which are then used to choose stable highlights for following.

The followed highlights are then separated and amended for point of view mutilation. Vehicle speed is assessed by looking at the direction of the followed highlights to known true measures. In tests performed on recordings caught under genuine activity conditions, our framework achieved an exactness of 0.87 and a review of 0.92 for tag identification. Vehicle speeds were assessed with a normal blunder of 0.59 km/h, remaining inside the +2/ - 3 km/h limit, controlled by administrative experts in a few nations, in more than 75% of the cases[5].

In this work, a novel framework is depicted for assessing vehicle speed from recordings caught in metropolitan roadways. The framework's pipeline is appeared in Fig. 7. Initial, a content finder is utilized to identify the tags of passing vehicles. Stable highlights inside the identified areas are then followed, utilizing a blend of the SIFT and Kanade-Lucas-Tomas (KLT) calculations. Subsequent to sifting through irregularities, the vehicle speed is assessed by contrasting element directions with known genuine measures, which permit us to amend the point of view bending and acquire a meter-per-pixel connection. As far as anyone is concerned, our framework is quick to gauge vehicle speed by following corner and locale highlights separated from the tag.

To assess our framework, we utilized recordings caught under genuine activity conditions related with ground truth information got by an inductive circle indicator. Our framework accomplished 0.87 exactness and a 0.92 review for tag identification. Vehicle speed was assessed with normal mistake of 0.59 km/h, remaining within the +2/ - 3 km/h limit dictated by administrative experts in a few nations, in more than 75% of the cases[6-7].

4 Implementation of LPD Model

a) License Plate Detection

Tag discovery is essential to our framework's exhibition. For this errand we use SNOOPERTEXT, openly accessible at, a cutting edge calculation for recognizing text in metropolitan scenes (like structure numbers, bulletins, and so on) A few boundaries of the indicator were chosen to improve its presentation for identifying tags. As demonstrated in Fig. 2, the SNOOPERTEXT indicator comprises of four principle modules: picture division, character sifting, character gathering, and text area filtering. Input image is shown in Fig. 8 .

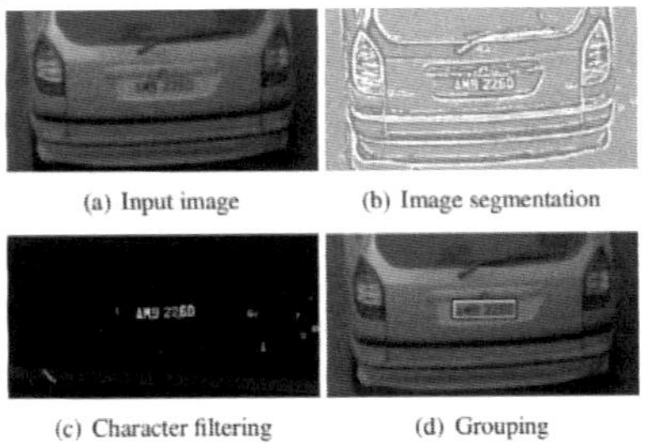

Fig. 8: License plate detection by SNOOPERTEXT

The division calculation utilized by SNOOPERTEXT is an altered variant of switch planning, an administrator for neighborhood contrast improvement and limit that utilizes the nearby closer view and foundation levels Fig. 8 (b). To discover splendid content on dim foundation, the division is rehashed on the negative (pixel-wise supplemented) picture. This subsequent stage can be disregarded if the tags consistently have dull characters on splendid foundation (that shifts from one country to another). This can impressively diminish the calculation exertion, and furthermore facilitate keeping away from bogus recognition.

The sectioned forefront locales are separated dependent on basic mathematical measures (territory, width and stature) and named character/non-character, in view of shape descriptors prepared on a dataset of divided letters Fig. 8 (c). The excess areas are then gathered to frame the up-and-comer tag locales Fig. 8 (d).

Every one of these means are acted in a multi-scale style, to effectively deal with various character sizes, to smother insignificant detail and to keep away from the utilization of excessively huge pieces in the division. The contingent upon the camera position and video outline size, needn't bother with in excess of 2 pyramid levels to identify tags went against to the 5 levels required in a free setting situation. In the last advance, competitor text districts are approved by a double book/non-text locale classifier that oddballs applicant areas that don't seem to contain a solitary line of text. This classifier utilizes the T-HOG descriptor, which depends on the multi-cell Histogram of Oriented Gradients (HOG).

b) Feature Extraction and Tracking

A vehicle's tag is identified in a casing, implemented framework separates highlights from the tag locale, and tracks these highlights across outlines. The tag area is utilized just a single time for every vehicle, to decide the arrangement of highlights to be followed. To concentrate and track highlights, we join the Kanade-Lucas-Tomas (KLT) and the Scale-Invariant Feature Transform (SIFT) calculations.

A "great" highlight is an area with extreme focus variety in both x and y headings, like finished areas or corners. Let [Ix Iy] = ∇ I = [∂I/∂x ∂I/∂y] be the picture subsidiaries in the x and y headings of I and Z a 2 × 2 grid given by

$$Z = \sum_{W} \begin{bmatrix} I_x^{\,2} & I_x I_y \\ I_x I_y & I_y^{\,2} \end{bmatrix} \quad (1)$$

Where W is a window with $n \times n$ pixels centered on some pixel within the license plate region. The region covered by the window is selected if both eigen values of Z are above a given threshold (set as 1 in our system).

Selected highlights followed with sub-pixel exactness we have utilized the pyramidal KLT calculation. Leave I and J alone two progressive video outlines. The KLT calculation takes I, J and a bunch of n layout areas T{1,2,...,n} ∈ I with n × n pixels covered by a little picture window W that contains the highlights to be looked for. For each chose format Ti focused at u = (x, y), it returns a changed position d = (x ' , y ') on J to such an extent that the local W of u + d in J is generally like Ti . That is, it discovers the dislodging d which limits Σw [J (u + d) I(u)]2 . The greatest pixel dislodging d that the pyramidal KLT can deal with is given by d = (2 ℓ +1 - 1)δ, where is the quantity of pyramid levels and δ is the pixel movement permitted by rudimentary optical stream calculation (of the request for one pixel). For ℓ = 3 the most extreme relocation this can be generat around 15 pixels. Notwithstanding, contingent upon the vehicle speed and the video outline rate, this can be deficient. To dodge this issue, we need an underlying assessment of the vehicle speed. This assessment is gotten by figuring SIFT highlights inside the primary event of the tag locale and coordinating with them in the following casing.

The normal area dislodging originate by the SIFT coordinating is utilized as a speculated position for the KLT calculation. That is, record a harsh forecast of the vehicle removal d in the following edge by utilizing the SIFT, this dislodging is additionally refined with sub-pixel exactness by the KLT calculation, to track down the best uprooting d. The SIFT highlights are utilized distinctly for the underlying assessment, when the tag is first recognized for the leftover edges, the expectation is figured from the normal locale uprooting found by KLT, since we as of now have a coarse assessment of the vehicle speed.

In this progression, the contrast between current casing and foundation outline is determined. Since all pixels in the primary edge is considered as foundation, so All pixels in the current casing ought to be inspected with the principal edge to decide if it's have a place with foundation or item dependent on edge activity. Limit esteem is figured from the outright contrasts between the current picture and foundation.

c) Outlier Rejection (Motion Blur)

The movement of each termplate locale Ti in a sequential pair of casings, we remove a movement vector di = (x, y). To dispose of movement vectors that compare to bungles, for example exceptions, the mean and standard deviation of the relocations in the x and y tomahawks, disposing of movement vectors outside the three-sigma deviation toward any path. This method is rehashed until the standard deviation in both x and y tomahawks become more modest than 0.5 pixel. The cycle is exemplified in Fig. 9.

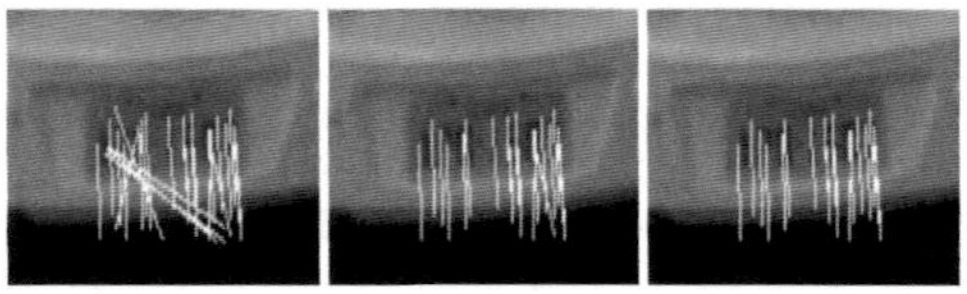

(a) (b) (c)

Fig. 9: Outlier Rejection: a) Motion Vectors
b) Outliers elimination c) Final Result

After procurement of various continuous edges, optical stream calculation is performed on these arrangements of pictures. For this reason above technique is utilized. Areas with high detail, for example, tag edges and front barbecue are chosen for following. Subsequent to performing optical stream, highlight determination is acted in two stages. In the initial step, a sifting activity is applied to wipe out superfluous movement vectors. Since the entirety of the pieces of a vehicle is moving together, movement vectors of these parts must be predictable. With this data identified, exceptions are disposed of. At the subsequent stage, sizes of movement vectors are prepared to eliminate superfluous movement other than the vehicle of interest. Optical stream data is consolidated to produce a solitary movement vector showing the bearing of stream. A model is given in Fig. 10.

Fig. 10: Resultant motion (Red arrow) and feature locations (Blue dot)

d) Background Update

Because of basic circumstances like presence of clamor, light changes and fleeting/perpetual varieties in foundation, the foundation model should be refreshed in each new casing of the picture grouping to adjust it after some time.

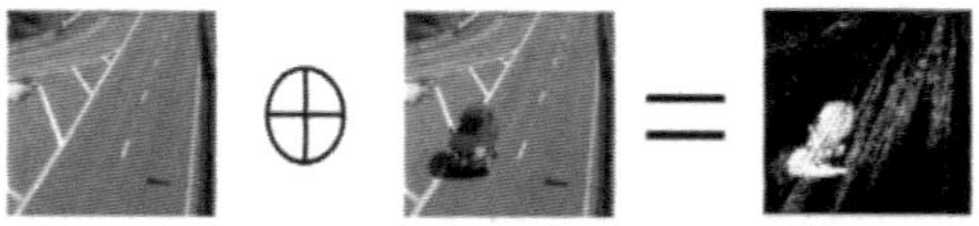

Fig. 11: Background Image Subtraction using XOR

We utilized XOR capacity to distinguish changes between two pictures, since pixels which didn't change yield 0 and pixels which changed outcome in 1. In Fig. 11, we can see the SRBI and the video outline picture contains a moved article, though the fixed parts nearly vanished in the XOR result.

Because of the impacts of clamor, we can in any case see a few pixels in the fixed parts.

A noise cancellation channel is expected to improve the applicant frontal area cover dependent on data got from foundation picture deduction, as we dispenses with commotion that don't compare to real moving items, by applying middle channel, as demonstrated in Fig 12.

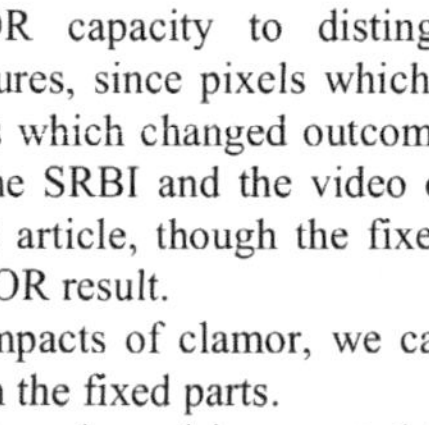

Fig. 12:Foreground Mask (Median Filter)

The yield of the last cycle is a frontal area cover which addresses by a paired picture where foundation pixels are named as 0 and any closer view objects are named as 1. At that point we apply this cover to the video casing to extricate the moving items.

At long last we applied an information approval measure by utilizing a profound learning calculation to order the distinguished article as a vehicle. We prepared an information base comprised of 500 vehicle pictures by a Convolutional Neural Network (CNN). At that point we separated the highlights of the up-and-comer object by a CNN, after that we thought about the highlights removed of that item to the vehicles pre-prepared CNN, at that point the best comparability is considered to discover a match. To quantify the likeness, two completely associated neural organization layers are used[9-10].

e) Color Segmentation

After foundation Subtraction, shading division is acted in the proposed calculation. The point of this progression is to isolate highlights dependent on colors from casings of the picture succession. In shading pictures, tint, immersion and power esteem are valuable for deciding limits. Shading division method utilized in this paper depends on changing over picture shading space into HSV model. Along these lines, shading can be ordered in scopes of tint and determined relying upon immersion and worth. At the point when the vehicle is distinguished and arranged accurately it is included as an approved article in vehicle counter data set. At that point the vehicle approval is checked in the event that it is inside the functioning hours to be gone through the entrance.

f) Vehicle Speed Estimation

Vehicle speed assessment begins from the subsequent casing, after the recognition of the tag, and depends on the highlights followed by the KLT calculation. For each new edge, each followed highlight (with the exception of the anomalies) will bring about a movement vector
di = (x, y) in pixels per outline. To change the movement vector over to a speed vector in meters each second we need to decide a connection between the pixel movement in the picture and the movement in reality.

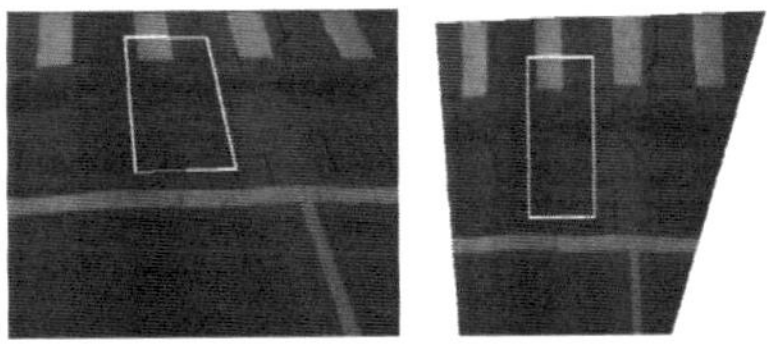

(a) (b)

Fig.13: Image improvement: Original image with perspective result; Improved Image

Image Rectification is shown in Fig.13. At that point, for each adjoining pair of edges, a redressed movement vector R = (xr, yr) is determined by the normal of all ri . The standard of R is given by ||R||=Sqrt(x2 + y2), and genuine dislodging in meters by D = ||R||/M, where D is the relocation in meters. At long last, the vehicle speed s in km/h is processed by s = 3.6 * (D/T) , where T is the casing span right away. Note that amendment is performed uniquely for the movement vectors the highlights are as yet followed in the first information pictures.

Fig. 14 : Vehicle Speed Detected

The primary goal of this venture is to distinguish over speed vehicles, utilizing Deep Learning and Machine Learning Algorithms. After obtaining of arrangement of pictures from the video, trucks are identified utilizing Haar Cascade Classifier. The model for the classifier is prepared utilizing heaps of positive and negative pictures to make a XML record. This is trailed by finding the vehicles and assessing their velocities with the assistance of their separate areas, Pixels Per Meter (PPM) and fps (outlines each second).

Presently, the trimmed pictures of the recognized trucks are sent for License Plate location. The Component Analysis (CCA) Connected aids Number Plate discovery and Characters Segmentation. The SVC model is prepared utilizing characters pictures (20X20) and to build the precision, 4 cross overlay approval (Machine Learning) is likewise done. This model guides in perceiving the fragmented characters. After acknowledgment, the determined speed of the trucks is taken care of into a dominate sheet alongside their tag numbers. These trucks are likewise appointed a few IDs to create an arranged data set. Estimating speeds through different frames is shown in Fig. 15.

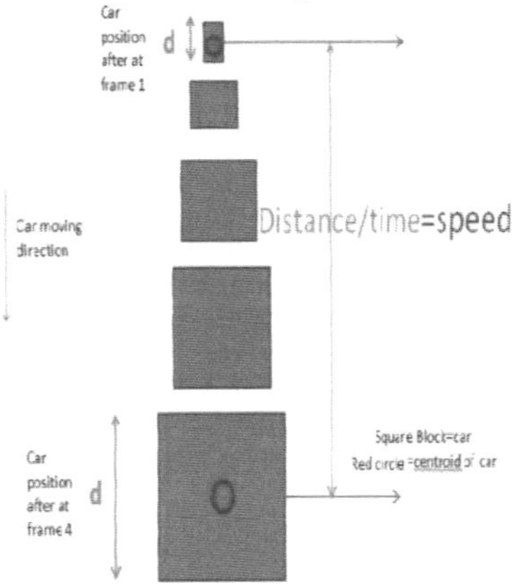

Fig. 15 : Estimating Speeds through Different Frames

f) Experimental Evaluation

A dataset gathered with seventy five vehicle groupings from a metropolitan street path with related real speed.

The dataset was gained from a video, with outline goal of 768×480 pixels and 31.25 edges. The real speeds were gotten from a high accuracy speed meter inductive circle locator, appropriately aligned and endorsed by a public metrology organization.

5 Result

License Plate detection performance

The tag recognition execution measured as far as exactness (P) the extent of distinguished articles that were surely tags and review (R) the extent of tags that were identified. As just the principal event of each tag is utilized by implemented framework to concentrate and track includes, every actions consider just the edge in which the main discovery happened.

Leave r alone the rectangular locale distinguished as a tag, and s be the genuine tag area in the picture. A genuine positive (T P) is tallied if A(r ∩ s)/A(r s) > 0.7, where A(t) is the space of the littlest square shape encasing a set t. Else, we have a bogus positive (F P). A bogus negative F N alludes to a missed tag and a genuine negative T N alludes to a vehicle without tag. From these markers, we have P = T P/(T P + F P) and R = T P/(T P + F N). The general accuracy and review of our framework, considering all the vehicle successions, were P = 0.87 and R = 0.92 individually (for T P = 60, F P = 9, F N = 5 and T N = 1). The F-measure (the symphonious mean of P and R), given by F = 2/(1/P + 1/R), was F = 0.90. A few examples of tags found by character recognition are appeared in Fig. 16.

Fig 16 : License plates samples through character detection

The speed performance was calculated by comparing the estimated speed, returned by proposed system, with the real speed as shown in Fig. 17. The average error, for the complete dataset, was 0.59 km/h and a standard deviation of 1.63 km/h.

Speed Estimation Performance

The speed performance was computed by comparing the estimated speed, returned by our system, with the real speed shown in Fig. 17. The average error, for the whole dataset, was 0.59 km/h with a standard deviation of 1.63 km/h.

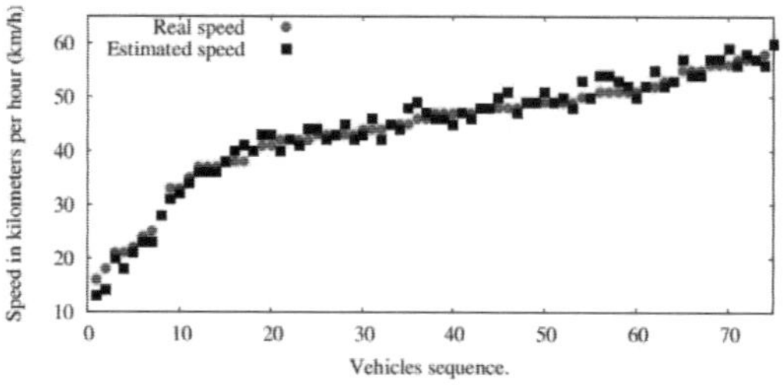

Fig. 17: Vehicle speed performance: estimated speed and the real speed

Conclusion

This project detects speed of a moving vehicle through its license plate and with the outcome as license plate it will fetch vehicle owner details with the help of CNN model (Machine Learning). In this project a moving car whenever it crosses dynamic markings, it is detected . It uses Tensor-flow with an SSD object detection model to detect cars and from the detection in each frame the license plate gets detected and each vehicle can be tracked across a video and can be checked if it crossed the markings made in program itself and hence speed of that vehicle can be calculated. The detected License plate will be forwarded to trained model where it uses PyTesseract, which converts Image to Text.

References

1. Muhammad Tahir Qadri, Muhammad Asif, "Automatic Number Plate Recognition System for Vehicle Identification Using Optical Character Recognition" IEEE 2009 International Conference on Educational Technology and Computer Singapore, (2009).
2. Satadal Saha, Subhadip Basu, Mita Nasipuri,DipakKumar Basu, "License Plate Localization fromVehicle Images: An Edge Based Multi-stage Approach", International Journal of Recent Trends in Engineering, **1**, 1(2009).
3. Chirag Patel, Dipti Shah, Atul Patel, "Automatic Recognition System: A survey" International Journal of Computer Applications (0975 – 8887) **69**,9(2013).
4. Ronak Patel, Narendra Patel, Keyur Brahmbhatt, "Automatic License Plate Recognition" IJCSMC, **2**, 4 (285– 294), (2013).
5. Sushama Bailmare H, A.B Gadicha, "A Review Paper on Vehicle Number Plate Recognition (VNPR) using Improved Character Segmentation Method" International Journal of Scientific and Research Publications, **3**, 12, (2013).
6. Suprokash Dey, Amitava Choudhury,Joydeep Mukherjee, "An Efficient Technique to Recognize License Plate using Morphological Edge Detection and Character Matching Algorithm" International Journal of Computer Applications (0975-8887) **101**,. 15(2014).

7. Divya K.N, Ajit Danti, "Recognition of vehicle number plates and retrieval of vehicle owner's registration details" International Journal of Innovative Research in Technology &Science(IJIRT), **2**,3(2014).
8. Aniruddh Puranic, Deepak K.T, Umadevi V, "Vehicle Number Plate Recognition System: A Literature Review and Implementation using Template Matching" International Journal of Computer Applications (0975 – 8887)**134** ,1(2016).
9. Ishraq Haider Chowdhury, Afsana Abida, Nur-E-Zannat, Naresh Singh Chauhan "Universal Vehicle Identification and Classification System (UVICS) Using Image Processing" International Journal of Pure and Applied Mathematics, **118,**18(2018).
10. Monika Arora, AnubhaJain, Shubham Rustagi, Tushar Yadav, "Automatic Number Plate Recognition System Using Optical Character Recognition", International Journal of Scientific Research in Computer Science, Engineering and Information Technology (IJSRCSEIT), ISSN : 2456-3307, **5**, 2, (986-992),(2019).

ITM Web of Conferences **40**, 01006 (2021) https://doi.org/10.1051/itmconf/20214001006
ICACC-2021

Real Time Weather Monitoring System Using Iot

Puja Sharma[1] *and Shiva Prakash*[2]
Department of Information Technology and Computer Application
Madan Mohan Malaviya University of Technology, Gorakhpur

Abstract: In Today's World, knowing live environmental condition is one of the biggest issues because there is an IoT of hurdles arrives when live environmental condition is measured. The proposed system will remove this problem since it monitors real-time weather conditions. In this proposed work we will monitor the live weather's parameter of the Gorakhpur Region. The proposed system will work on the client-server architecture model using IoT. The system is organized in Two-tier Architecture. Our proposed system contains a various sensor which will monitor the temperature of the region, humidity, Rain value and pressure of the system. The sensor captured data and send it to the node MCU controller. Arduino ide is used to upload the sensed data. The serial monitor has worked as a gateway between the sensor and the cloud. The data is pushed by the sensor on a serial monitor. The serial monitors an IP address. The HTTP protocol is used to view the data on the webserver. This paper displays the data on the webserver and monitor the real-time data of weather using environmental parameter or sensor. Using a webserver, everyone can monitor the weather's condition from anywhere without depending on any application or website. The data is available publicly. With the help of this proposed system, we measure the weather condition of the Gorakhpur Region. After getting results from the various sensor, it is observed that our proposed model achieves better results in comparison with the standard weather parameter.

1 Introduction

Monitoring weather's condition plays an extensive role in every person's life. The impact of the environment's condition causes numerous challenges in various fields like agriculture, industry, constructions as well as more other fields. But the measured impact occurs mostly in agriculture and industry. As we all know, agriculture perform a crucial role in India's Economy. Almost One-fourth of India's economy is acquired from Agriculture. Over the recent year, smart agriculture was a debated topic in the world. In IoT [13,16], the smart word indicates that the use of minimum parameters produces a better result. It reduces the use of land, water, time as well and the uses of the new technology and science for the enhancement of crops. The security is one of the major issues [17,18] in IoT network, lot of security techniques are available but still many several possibilities to enhance the existing security.

In agriculture, Before the yield, the Farming process consists of several phases and in that weather plays the most extensive role. In Gorakhpur Region, the situation of rain mostly happens. It is situated near the border of India "Nepal" and Nepal is a hilly area due to which the rainy situation occurs mostly. Due to this situation, the farming problem occurs due to heavy rainfall. In this situation, an indication of weather's condition is an important aspect before sowing or reaping the crops. Hence in this situation, the monitoring of weather's condition [14,15] would help farmers with the help of a weather monitoring system. S It will help the farmer before reaping and sowing. In Gorakhpur Region, the total geographical area of this region is around 3488.8 square km and cultivable land is 26428 hectares, In which the percentage of irrigated area is 76.5 %. The main crop of this region is Paddy. The total area is around 152655 hectares. The amount of water if became high then it will be a problem for farmers. Because in the sowing of paddy it needs water sufficiently but if the quantity of water will grow then it should be a problem for farmer. So, to remove this problem, if the weather's condition is known then it will be beneficial for them. The main objective of our system is to provide an indication of whether that would be beneficial for the farmers before sowing crops or reaping crops. In this paper, Section-II shows the literature review. Development of weather monitoring system is done in section III. Section IV describes the hardware parts of our system. Section V portrays the analysis of our experimental result in real-time. The next part will be the summary or conclusion parts of this research.

2 Literature Survey-

The author in [1], proposed a robust and affordable Automatic weather station. In this paper, the author elaborates how the weather prediction system is becoming a crucial challenge in every Weather extreme event that causes an adverse effect of the system on lives and property as well. Hence the accuracy of weather data is being one of the critical challenges to enhance the

*Puja Sharma: shravi.pujasharma012@gmail.com

weather prediction skills and build up the resilience to effect of detrimental weather report condition. The author describes that Uganda and various other developing countries have looked challenges in developing timely & accurate weather data due to scarce weathers observation. The scarce weather monitoring is a part of the high cost of developing automatic weather situations. The restricted funding is available to national meteorological services of the respective countries. In this proposed system the author firstly takes care of the problems and then applies them. The author proposed an Automatic weather monitoring Station based on a wireless sensor network. The planning of the author is to develop three generations of Automatic weather stations or AWS prototypes. In this research, the author evaluates the 1st-generation AWS prototype to improve the 2nd generation depending upon the need and generation. The author provides a suggestion to improve the non-functional requirement such a power consumption, data accuracy, reliability, and data transmission in order to have an Automatic Weather Station. The non-functional requirement collapsed with cost reduction in order to produce a robust and affordable Automatic Weather Station (AWS) Therefore the proposed work, like developing countries like Uganda will be able to acquire the AWS in suitable quantities. So that it can improve the weather forecasting

The author in [2], presents an IoT-based weather monitoring system. In this research, the environmental parameter can be retrieved through sensors. The author uses a different sensor to scale the various parameter like humidity, temperature, pressure, rain value & the LDR sensor is used. The system also calculates the dew point value from the temperature prototype. The temperature sensor can be used to measure the value of the particular area, room, or any place. With the help of the LDR sensor, the light intensity can be used as described by the author. The author in this used an additional functionality of the weather monitoring as SMS alert system based on the exceed the value of the sensing parameters as temperature, humidity, pressure, light intensity, and rain value. The author also adds an email and tweet post alerting system. The author in this system uses node MCU 8266, and various sensors.

In this paper [3], the author represents a low-cost live weather monitoring system using OLED display, in which the author displays the various fields where the IoT has produced innovative things in the system. The author described A new revolutionary system. Which measures the real-time Weather's condition. The monitoring weather situation is very much helpful for everyone either for farmer or industry or daily working people or for school as well. So, the author by developing a live weather monitoring system reduced the difficulty level for farmers and industry as well. In this paper, the author uses an OLED display that will display the weather conditions and In the proposed model, the author uses an ESP8266-EX microcontroller-based WeMos D1 board executed on Arduino, that retrieved the data from the cloud. WeMos D1 is a wifi module that is developed on ESP- 8266EX microcontroller. It has a 4MB flash memory. It one of the Excellent which is programmed with node MCU and Arduino ide. In this paper Author uses only two gadgets to measure the weather conditions i.e., Wemos and OLED, After the connection, it will store the data on the cloud for storing data a thingspeak website is used to display the data regarding weather. The system displays the data on OLED and thing speak cloud. The author's aim is to obtain live information on weather conditions on OLED display.

The author in [4], proposed a system that monitors and predicts the weather condition by which anyone can plan for our day-to-day life. This activity became helpful in every field either in agriculture or industry. So as to achieve monitoring and predicting weather info, the author uses 2 stages of the weather management system. In which they amalgamated the information from the sensors, bus mobility, and deep learning technology is used to allow a weather reporting system in stations and buses in real-time. Forecasting of weather is achieved through the friction model. Depending upon the sensing measurement from vehicles like buses, the work incorporates the strength of local information processing. The author talks about in stage-I, sensing of weather's condition, multilayer perception model and long-term memory are trained and then it will verify using temperature data, humidity, air pressure of test environment. In Stage-II, the training is applied to learn the time series of weather information. To get accurate data or not, to check the system performance, the author comparing the predicted weather data and actually obtained data from the environment Protection Administrator and central Baeuro of Taichung observation system that calculate the prediction of accuracy. The author finally talks about the proposed system has reliable performance on monitoring of weather. And this model also proposed a one-day weather forecast or prediction via the training model. So finally, the author demonstrates that this system presents a real-time weather monitoring and prediction system using bus information management. The author represents 4 basic components 1- Information management. 2- Interactive bus stop 3- Machine learning predictive model 4- weather information platform. In this, information shown via dynamic chart.

The author [5], implement an IoT-based weather monitoring system, in this research paper, the author describes that how with the help of IoT technology, the weather can be monitored. And which provide the info of climate-changing conditions. With the help of this project, people can be aware of the climate condition changes. It gives an accurate and efficient output and the algorithm as the swarm is used to implement for further improving the accuracy. So, in this project, the author aims to make a weather monitoring with the help of IoT. In this project, the hardware and software are used which makes it easy to implement. In the project, the author uses a different sensor to collect the information of the climate and stored it in the cloud. For this storage, the website www.thingspeak.com is commonly used for Internet of things projects. And from the cloud storage space, it extracts the whole weather data and uploads it to the android mobile application using an API key.

Tools which detect the rain drops, is called rain sensor. Once the plague reveals the raindrops on the strips and the voltage is considered from that. And there is no short circuit condition that occurs because water is a bad conductor and the sensor acts like variable resistance. Once a measurement of voltage is completed then the circuit takes the output. The voltage is measure through a potentiometer and to transfer analog signal to digital signal the system applies LM393 Comparator. The LED turns on when the power supply system is observed and there have no water drops on the sensor then the digital output is high. And the sensor also dictates when the sensor has moisture on the sensor plate the digital output is actively low. For measuring the humidity three sensors can be used as humidity sensing components, TC thermistors, and An IC at the back of sensors. For humidity measurement humidity component is used it has two electrodes. Node MCU is used as a controller. In IoT projects, the Node MCU is basically used. This programmed using Arduino IDE. The scripting language LUA is used for programming. NodeMCU runs on the ESP8266 WIFI module. This Android STUDIO is used it is developed by google developers. The aim of this application is to make processing easier.

Android studio is used to implement a common application to show the streaming of data from thing space. With this smartphone application, the weather condition can be monitored. Now after designing the application, it will display temperature, humidity, heat, and raindrops and is stored on the cloud. The author provided an API Key to bring in our data from the thingspeak website to the android device.

This paper [6] presents a paper, i.e. Design and implementation of weather monitoring and controlling system. In this project, the author emphasized that how the weather monitoring system works. According to the author, weather monitoring system plays a major in our daily life. Hence the collection of information of temporary changes of environment situation is crucial. In the industrial field, there have certain hazards available. In the earlier systems, the industry is totally dependent upon mechanical and electromechanical instruments. And it has some drawbacks of that system such as poor rigidity, human intervention is required and associated parallax errors, durability requires. Hence, to remove this drawback a remote monitoring system is needed and the principal goal of this system to produce a standalone modular weather monitoring system with the in-built facility of remote monitoring and communicating to capture & send the metrological parameter. Therefore, it is very much important to monitor the weather. In this research paper, the author implements an embedded system and plot a weather reporting and monitoring system which enables weather monitoring system. This type of system consists of a couple of sensors as gas, temperature, rain, humidity sensor, etc monitored. LPC1768 microcontroller (ARM9) is used as a core system as all the sensors are connected with this sensor and collected the information regarding weather as temperature, humidity, pressure, etc. After collecting all the data send it to the LABVIEW by applying serial communication. This module kept the details information in the excel sheet and that can get SMS in the mobile with the help of the Global System for Mobile communication model. The dense circuitry is developed around the LPC1768(ARM) microcontroller is used in this system. The program like embedded C using the IDE is used in this project. JTAG is used to loading programs into a microcontroller. The LABVIEW is a platform that develops the environment for a visual programming language from National Instrument. The name of the graphical is "G". Data transferring from hardware could be received in LABVIEW using a serial communication system. The reading shows the different data with the indication like metering etc. LABVIEW result from view on EXCEL FILE and locate the file into the folder and updating of the value can be displayed in the excel file. Hence, in this way, this system monitors the environmental parameter. The embedded controlled sensor has been demonstrated reliable themselves.

In paper [7], discusses the implementation of a weather monitoring system using a WFI module. The weather monitoring system performs the process through the connection of Wi-Fi. The system contains a transmitter section and receiving section. The transmitting section contains the Wi-Fi module, controller, various sensors. The receiving section contains the web server, router, website, etc. the working of the system is defined by the author as, multiple sensors such as sensor, pressure, humidity sensor reads the environmental temperature, pressure, humidity respectively. The rain gauge sensor uses the tipping bucket technique. In this technique, measure the no. of times, a calibrated bucket is exhausted from the rainwater. And in that water is filled to calibrated mark, and hence the emptied one tip is implied and multiplied by 0.02mm and that's the final amount of rain. The wind gauge is applied to sense the wind speed and wind direction. The signal conditioning is mandatory for getting the exact wind velocity value. Signal conditioning contains an optocoupler in the system and the output is shown to the Schmitt trigger, which is a bi-stable circuit. In this circuit when the input raises across a certain threshold, the output increases to a continuously increases maximum and decreases to the early 0 when the voltage input value falls down another threshold. Each sensor's output is accepted by the D1 controller and handled in a compulsory format. Those values are transmitted using Wi-Fi to the main server. All the values of sensed data are maintained on the main server and routed and show continuously on the website. The refreshing rate of 10 seconds. The database is created using the MySQL query language. A PHP scripting language is used to display the information on the website.

In this project [8], the author proposed, "A LOW-COST ZIGBEE BASED AUTOMATIC WIRELESS WEATHER STATION WITH GUI AND WEB HOSTING FACILITY". The author focuses on the current climate condition and conceptualized a model, in which a ZigBee wireless link is used. Because seeing the today's condition stirring the quality of human life and also affect the increasing population activity. In the current situation, the metrological parameter data collection is necessary rather than limited, and keep used of the comprehensively monitoring system due to cost constraint and availability of human is just a drawback in the system. In this research paper, the author suggested and implemented a low-cost hardware module based on the Arduino Uno Board. With this board, it measures the metrological data, it includes the typical temperature, sensor, atmosphere, wind speed, wind direction and sends all the information to the GUI system which is running on a laptop through Zigbee wireless link. The GUI is a convenient and standalone application for the Windows operating system. It is implemented in the LABVIEW. The LABVIEW application graphically displays the current situation and logs the data that is accessible by MS excel. The system has a facility to host the same information on the webpage. This system has a mathematical model that is efficient in creating and generating a short-time local alert on weather parameters.

The system provides a real-time effect. The main objective of this model to detect the condition mapping, and weather forecasting most necessarily, it warns the people of its devastating effects. This would be very much helpful for landing the aircraft, cloud burst, healthcare effect, tornado, tsunami, navigational, and the shipborne effect.

In the paper [9], The author proposed an "An Intelligent weathering system using the internet of thing" research in which, with the help of machine learning algorithm and IoT technology, climate condition is measure and predicts the next upcoming condition. The author presents the system that is a very much cost-effective and efficient method for controlling the weather condition and sends it to the cloud so that the value is fetched and available on the internet and can be seen anywhere by connecting through the internet. There is a significant role of temperature, humidity, pressure in the system. also have a different area used in the system such as agriculture, logistics, etc. weather monitoring, and forecasting is important for the advancement of these industries. IoT technology used in the formation and development of this research, that is an advanced & efficient approach for connecting the sensor to the cloud that stored the real-time data and connects the whole world of things through a network. The thing might be anything as the sensor, electronic gadget, and automotive electronic equipment as well. The whole system deals with the monitoring and controlling the environmental condition like temperature, smoke, gas, wind, pressure relative humidity level, and many other gases with sensor transfer the data or information to the cloud platform and store the data on it. A machine learning prediction algorithm is used in this system. Machine learning is a branch of AI (Artificial Intelligent) which deals with analysing and predicting the given dataset. The collected data is analysed. The real-time data sent through the sensor can be reachable through the internet in the entire world. The methodology which is used by the author as node MCU microcontroller is used to gather the information from discrete sensor which is basically on the code dumped in Nodemcu, send the data to the specific cloud as Thingspeak is used for displaying both the data with web view and also application view. Each sensor is connected with nodeMCU. It transmits the sensed data from the cloud and data analysis is performed and shows the result in the resultant view on the website. Data is fetched from the cloud in the CSV file format. And after that for further processing, it includes the data from the cloud in R Studio, and then data is processed using various algorithms.

This paper [10], presents the "Embedded weather station with Remote wireless control". In this paper, the author elaborates on the importance of a weather monitoring system. The author describes that how the weather monitoring system works in now a day. It is very important to know the climate condition before going outside for any specific work. This paper describes what is the field where weather monitoring is important. The field like agriculture, military, entertainment, industries, etc. there is so various solution for monitoring the weather parameter. It describes the three solutions which are a solution for monitoring the weather. The first one, using a static weather station. The second one is based on the wireless sensor network. and the 3rd one is, using a low dimension weather station. In this research paper, the author represents a weather station made up of temperature, luminosity, pressure, humidity, sensor implanted with a microcontroller based on the board. The weather station is restrained with an SMS service of mobile phones.

The author [11] conceptualized the weather monitoring concept using Artificial intelligence. According to the author, weather forecasting is a statical method instead of using a binary decision concept. The intention of the author is to implement an intelligent weather prediction system. Since the system has a mandatory tool and this tool used to measure the maximum and minimum temperature, rainfall for a sampled period of the day and then analysed. The prediction is done on the basis of available data, which is achieved using machine learning Algorithms. The prediction and analysis are based on the linear regression which predicts the upcoming day's weather condition with better accuracy. The accuracy is achieved more than 90% is achieved with machine learning techniques. The recent research strikes that with the use of machine learning techniques the system achieved better accuracy and better performance. The machine learning technique is a branch of AI. It has proven that it is a robust technology in analysing and predicting. Machine learning performs an important role in various field as Industrial, agriculture, logistics, healthcare and etc. And where weather forecasting is an important area. In this paper, the result is verified with a multiclass classification algorithm using a logistic regression algorithm with ANN i.e., Artificial neural network. This paper also talks about the disadvantages of using ANN (artificial neural network), multi-classification. Because they did not provide any value to the nearest of the day, not the probability of the day that how the day is going to be. In this system, the support vector machine learning techniques can be used in prediction and it works on large data and provides the best result. But in this classification, there has presented some redundant data. And that must be escaped. The only power is to be taken by the system that gives the best result is to update the data set in order. to reflect the accurate result. It performs well when the data set is in huge amounts. And provided over a seventh of the data is to be forecasted. This module assists in predicting and monitoring the weather condition with a better rate of accuracy in an effective manner.

In [12], the author talks about the real-time accessible system. The author proposed a method in this paper using IoT technology. In this research, the system deals with monitoring the weather and climate condition changes as UV radiation, pressure, temperature, humidity, and even carbon monoxide levels present in the air. It uses different sensors to collect the info and sent it to the web page to view and sensed data presents in the visual view. If the info is upload on the web server, then it can be accessed from anywhere in the world. The project consists of a mobile application that sends a notification for an alert system to provide warnings to people about drastic and sudden changes in weather. To forecasting the most complex weather, it cannot be done using a sensor only but we need an API that analyses the data is collected by the sensor and predict the accurate result. This API can be used to access the data anywhere, anytime with ease. And it also stores the data for further use. This system requires less maintenance because it uses fewer parts and has a compact design. The component doesn't consume more power .and it can also powered by a solar panel. The author said that if we compare it to another device that is available in the market is cheaper and more cost-effective. This type of project used in the metrological department, aviation, weather station, marine industries, etc.

3 Development of Weather monitoring System using Multi Sensors

The aim of this paper is to develop a Weather monitoring system which will observe the weather's situation and monitor live environmental parameter continuously through wi-fi. The weather monitoring system makes use of various sensors to measure the parameters.

Table 1. Multiple modes in weather monitoring System

Mode 1	If mode 1 presents, DHT11 i.e., temperature and pressure, if it is on, then it captures the temperature value and represents it on the webpage
Mode 2	If mode 2 presents, BMP 180 i.e., it shows the barometric pressure of the environment.
Mode 3	If mode 3 present, the Rain sensor module i.e., it shows the rain value of the environment.

Table 2. Grouped sensor nodes under various modes.

Sensor node-1	Sensor node-2	Sensor node-3
DHT11 (Temperature and pressure sensor)	BMP 180 (Barometric pressure)	Raindrop sensor (Rain sensor)

A novel weather monitoring system using IoT is proposed. In this paper, the author presents 3 different sensors organized in different forms of bundles in the system. The model describes how the sensor is connected to the microcontroller node MCU. Sensors are connected with the node mcu8266 in the architecture. The system is arranged in this manner that data will automatically fetch from sensors and will upload while connecting with wi-fi and displays the data on the webpages. The proposed model uses three sensors meant to acquire temperature, pressure, humidity, raindrops which are shown with mode1, mode2, and mode3 in above table 1. The modes are described as Temperature and humidity (MODE-1), Barometric pressure (MODE-2), and Raindrop sensor (MODE-3) represents in table 2 as well. By consolidating table 1 and table 2 shows as-Mode 1 indicates as Sensor node 1 which contains 1 DHT11sensor for measuring the temperature and humidity of the environment. Mode 2, indicates sensor 2 which contains BMP 180 sensor for measuring the barometric pressure from the environment. And the last one is Mode 3, which indicates sensor 3 that contains a raindrop sensor, its name suggest it indicate, it is used to measure the raindrops from the environment. The proposed system is an effective weather monitoring system with less no. of sensors. It is reliable and also free and available publicly. we can use multiple sensors for weather monitoring like wind speed, wind direction, PIR sensor, and many more. But in our proposed model we are using three sensors to check the working of the system and reliability of the system. The proposed model measures the accurate value of the environmental condition

4 Hardware Illustration of the proposed system

Fig 1 represents the architecture diagram. With the help of this architecture, Every sensor is connected with node MCU pins. The power supply is given to the node MCU by USB which is connected with pc.

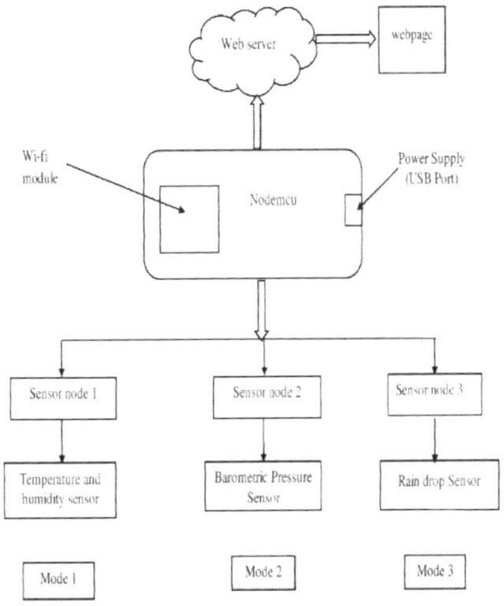

Fig 1.: The proposed model architecture of weather monitoring system.

4.1 Sensors, Gateway, and Data logging

4.1.1 Gateway or data collector: In our proposed system, nodeMCU works as a gateway of the model. From the connection of the system, data is sent to the local network and that will display on the web page of our pc.

4.1.2 Mode 1 Sensor (Temperature and humidity Sensor): DHT11 sensor allows us to measure both to temperature and humidity of our system. This sensor operates at 5V and utilizes a current of 2.5 mA. The sensor can consider temperature values from 0 degrees to 50 degrees. The humidity measures from 20%-90%. With an accuracy of +_ 1 degree and +-1 % as well. It uses 3 to 5V power and input/output.

- It has 2.5 mA max current usage during conversion.
- It is good for 20-80% humidity with an accuracy of 5%
- It is just good for)-50 degree Celsius.
- It has 4 pins with 0.1 spacing between them

4.1.3 Mode 2 Sensors (Barometric pressure)
BMP 180 is a digital barometric pressure sensor that measures the pressure of the environment. It follows BMP085 and brings many enhancements.

- It has ultra-low power consumption down to 3micro ampere
- It operates on 3 to 5V dc voltage
- The pressure-sensing rate of BMP 180 is 300-1100 hPa (9000-500m above the sea level)
- The operational range of BMP is -40 degrees to +85 degrees Celsius.
- This chip uses I2C 7- bit address 0X77.

4.1.4 Mode 3 Sensors (Raindrop sensor)

- The raindrop sensor is used to measure the detection of water beyond humidity sensor detects.
- Rain sensor detects water on its sensor board of raindrop sensor.
- The board act as a variable that will change from 100K Ohm.

5 Circuit Diagram

The circuit diagram of our proposed system is given below. The diagram represents the connection of the sensor and how the connection will be done.

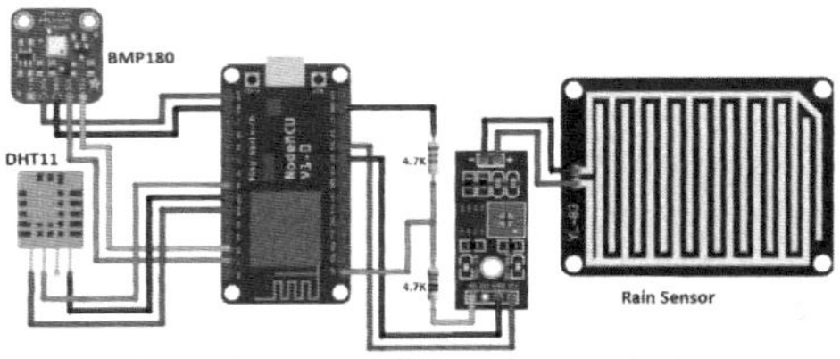

Fig.2. Circuit diagram of the weather monitoring system

The Dht11 sensor, BMP 180 sensor, rain sensor, all are connected with the node MCU pins and the power supply is done by USB cable to connect the hardware to the system. The prototype model is represented in the above images. All the connections should be done in the same manner then will get a proper result. The below tables show the pin connection for each sensor.

Table 3. Pin configuration between node MCU and DHT1

VCC	3V3
DATA	D5
GND	GND

Table 4. Pin Configuration between node MCU and BMP sensor

SDA	D2
SCL	D1
GND	GND
VIN	3V3

Table 5. Pin Configuration between node MCU and Rain sensor:

A0	GND (including with 4.7k register)

https://doi.org/10.1051/itmconf/20214001006

GND	GND
VCC	3V3

6 Flow chart of the proposed model

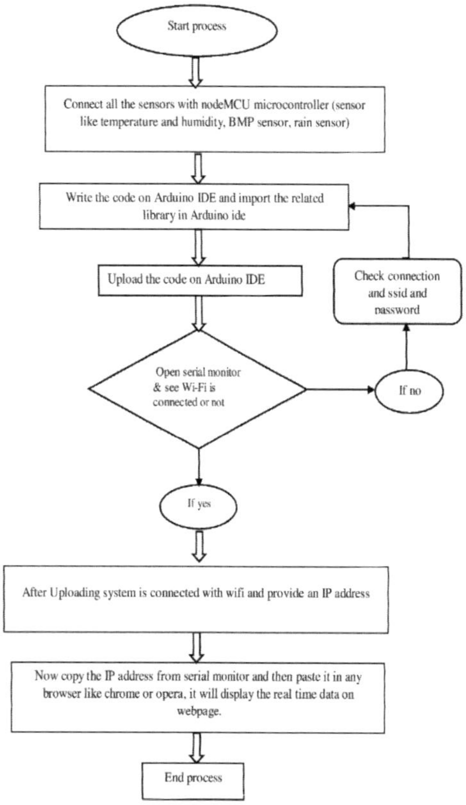

Fig.3. Flow chart of the proposed model

The flowchart describes the functioning of our proposed system. This proposed system is developed as a weather monitoring system for the Gorakhpur region. This system is comprising of only three sensors and display the result on the webpage using HTTP.

7 Working Process

After making all the connections as shown in fig.2. The power supply is provided to node MCU through a USB cable. Which is connected with our personal computer. We have written two code modules. The first one is to connect to the different sensors in Arduino code and the second code module is for the web page which is written in java script.

After providing connection to hardware. Then upload the code which we have written. But before uploading code, it needs to verify. So, Open Arduino ide then write code there and click on verify and wait for almost 1-2 min to complete verification and then choose an option to upload the code then it compiles successful and done uploading successfully. After that open the serial monitor. There, it shows that system is connected with Wi-Fi. When hardware connection is established, HTTP request starts processing, and then it will show an IP address on the serial monitor. Then copy the IP address and paste it on a web browser like chrome, internet explorer, opera, etc. You can select any browser and it will show the corresponding temperature value, humidity, pressure, and rain value. The result is showing below

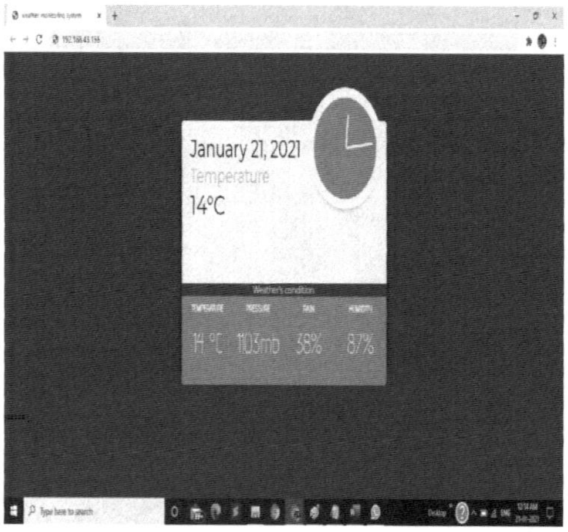

Fig.4. Final Result another web page

Hence, the above fig.4 shows the result on the webpage contains different parameters information from the weather. As the weather of Gorakhpur region is very cold and rain possibility is also very high. The temperature, pressure, and rain sensor have a different variation available.

8 Experimental Results and Analysis

The function of "weather monitoring system using IOT", the agenda of this system to design a system which keeps track of weather condition and simultaneously measuring environmental variables through the internet of things. Fig shows a prototype development for a weather monitoring system with some parameters. The developed weather monitoring system monitor different parameters like humidity, pressure rain value, etc. After making the connection as per in fig.2 the value from the temperature and humidity sensor, pressure sensor, and rain sensor value is observed on the webserver. Here we are storing observed data on the web server through HTTP requests on the webpage.

Table 6.: Experimental reading obtained at Kushinagar place (near Gorakhpur)

Experimental result Date- 21 January 2021						
S. N.	Time	Weather	Temperature value	Humidity value	Pressure value	Rain value
1.	9:00am	Fog	14 °C	87%	1016mbar	38 %
2.	12:00 pm	Cloudy	13 °C	78%	1018mbar	32%
3.	2:00pm	Partially sunny	14 °C	72%	1019mbar	34%
4.	5:00pm	Fog	15 °C	73%	1018mbar	39%

In the above table, we have taken the weather information of Kushinagar nearby Gorakhpur. We have gathered sensor data on 21 January 2021 at different time slots. The sensed value is shown in table 6.

Table 7: Experimental reading obtained at Gorakhpur (MMMUT college)

Experimental result Date-25 Jan 2021						
S. N.	Time	Weather	Temperature value	Humidity value	Pressure value	Rain value
1.	10am	Fog	12 °C	88%	1018mbar	30 %
2.	11am	Fog	14°C	78%	1019mbar	32%
3.	12:30pm	Cloudy	16 °C	70%	1133mbar	30%
4.	2pm	Partially sunny	15 °C	70%	101mbar	28%

Here, we have sensed the weather's information for MMMUT college. In this experiment, we measured the temperature, humidity, rain, and pressure value dated 25 January 2021. The above table 7 represents the sensed value with our proposed model for MMMUT location.

Table 8: Experimental reading obtained at Gida Gorakhpur

Experimental result Date-25 Jan 2021						
S. N.	Time	weather	Temperature value	Humidity value	Pressure value	Rain value
1.	8:30am	Cloudy	11 °C	78%	1019mbar	29%
2.	11:00am	Cloudy	12°C	76%	1020mbar	26%
3.	2:30pm	Partially sunny	16 °C	66%	1016mbar	30%
4.	4:30pm	Sunny	15 °C	72%	1015mbar	28%

In table 8 we have measured the environment parameter with the help of different sensors for Gida Location Gorakhpur.

This experiment is developed using a prototype model of the environment. This has been developed within the Gorakhpur region in three different locations. The locations are Kushinagar, MMMUT college Gorakhpur and Gida Gorakhpur. Primarily we have tested in the Kushinagar region in the month of January and dated 21 Jan 2021. Secondly in MMMUT college Gorakhpur and then Gida Gorakhpur location. We have placed temperature and humidity sensors, pressure sensor, and rain sensor and start detecting the environment's parameters. When the reading obtained from sensors, we have tabulated all the readings captured through sensors according to their location, date, and time for analysis. The sensor's analysis is shown below.

8.1Temperature

As shown in table 6 we have first sensed temperature value at the Kushinagar location. On 21 January 2021, at time 9:00 am, we measure the minimum temperature as 7 and the maximum temperature is 21degree and found the average temperature is 14 degrees and the real-time temperature is 13 degrees. To analyse more, we have taken a temperature reading through the sensor at the different time slots. At 12pm, the average temperature or relative temperature is 13 degrees and the weather is cloudy. On the same day, at 2:00 pm, the average temperature value is 14degree and next time instance the average temperature value is 15 degrees. Hence at all the time, we have compared with real-time temperature value we found that temperature value is approximately accurate at Kushinagar place.

Like, Kushinagar place, On 25 January 2021, we have also measured temperature value at different time instance of MMMUT College Gorakhpur Location is shown in table 7 In the morning, at 10:00 am, we measured average temperature value is 12 degree and real-time temperature value is 12degree. At 11:am, the average temperature value captured value is 12degree and the real-time temperature is 13degree. Another time interval is 12:30 pm, captured average temperature value is 16degree but the real-time temperature is 15degree and at 2:00 pm, the observed average temperature value

is 15degree, and the available real-time temperature is 15degree. Hence in this way, we can say that at the MMMUT College Gorakhpur location the sensed temperature value is extremely nearer to the real-time temperature.
Same as the previous location, we have measured temperature value at Gida Location at multiple time instance on 26 January 2021 which is shown in table 8. Now we measure the temperature value at 8:30 am, the average temperature value is 11degree and the real-time temperature is 12degree. At 11:00 am, the relative captured temperature value is 12degree and the real-time temperature is 13degree. At 2:30 pm, the observed average temperature is 16degree and the actual temperature is 16degree and the last time instance, the sensed average temperature data is 15degree and the real-time observed temp is 15 degrees.
Hence with the above result, we can say that the observed result through our proposed model is sensed temperature data is closed by real-time data.

8.2 Humidity

As we have measured the temperature value of a particular location, in the same manner, we have measured the humidity value. Firstly, measure the humidity value of Kushinagar then MMMUT College Gorakhpur, and then Gida Gorakhpur location. In Kushinagar, the sensed humidity is 87 % and the humidity value in real-time is 89% at 9:00 am. Next, at 12:00pm, the average humidity percentage is 78% and the real-time humidity value is 77%. At 2:00 pm, the average humidity value in percentage is 72% and in real-time, the humidity value is 72%. And last time reading which we have measured i.e., at 5:00 pm, humidity value is 73% and real-time observed data is 74%. So, after taking the reading for Kushinagar, we have analysed that humidity value is almost accurate when compared to the real-time observed data through the internet.
Likewise, for MMMUT College and Gida Gorakhpur, the humidity value is sensed through our proposed model, dated on 25 Jan and 26 Jan 2021. We have seen that observed and real-time percentage value of humidity gives nearly accurate value. We have sensed reading for both locations with date and time is shown in table 7 **a**nd table 8.

8.3 Pressure

Pressure value of Kushinagar, MMMUT College Gorakhpur and Gida Gorakhpur is observed and is referred to in table 6, table and table. The first reading has been taken for the Kushinagar location and tabulated in table 6**.** The first reading is taken at 9:00 am and the average value of pressure is 1016mbar (millibar pressure unit) and the real-time pressure value is 1016mbar. Next, at 12:00pm, the average pressure value is 1018 mbar and the average pressure value in real-time is 1018mbr. Next at 2:00 pm, the pressure value on average is 1019mbar and the real pressure value is 1017mbar. And at 5:00 pm, the average value of pressure is 1018 and the real-time pressure value is 1018mbr. Hence in this way, we have observed and compare with real-time pressure and we have seen that the pressure value is very much similar to real-time obtained data for the Kushinagar location.
In the same manner, we have sensed pressure value for MMMUT College and Gida Gorakhpur location and we found that observed pressure is nearer to the real-time pressure value. The sensed pressure data is tabulated in table 7 and table 8.

8.4 Rain

Rain value is observed in percentage form. Like all the above parameters is measured, the rain sensor value is also measured. In the same way, the Rain value is measured and compared with real-time data and found that like other parameters it is also very much accurate. So first we measure for Kushinagar location and the date is 21 January 2021. The average rain value is 38% at 9:00 pm and the real value is 35%. Again, we measured at 12:00pm, the average rain value is 32% and the instantaneous value is 33%. We measure again at another time i.e., 2:00 pm then the rain value is 34% and at the same time, the real-time value is 33%. At 5:00 pm, the rain value is 39% and the rain value in real-time is 38%. These readings are obtained at the different time slots. On the basis of the above rain value data, we can say that it is almost similar to the real value. And we can say the possibility of rain is less.
Hence like the above location we have also analysed that the value of rain for MMMUT College and Gida location is also the same and almost gives accurate values of rain value. For these two locations, the readings are available in table 7 and table 8.
Hence with this experiment analysis, we can take an idea for our future planning. This will especially help in industry, agriculture, and various field. So, if we provide this facility to the farmer to do an experiment with this system then they will collect data with these sensors and make analysis and perform their action according to the gathered information of weather.

9 Conclusion

Devices that monitor weather's parameters with minimum cost in our proposed system. The proposed system works on the client-side architecture model. The proposed approach observed various environmental information using multiple sensors. The system which is designed has used less sensor than the existing model. The main aim of our proposed model is to make the system cost-effective, affordable. So that everyone can use it freely. In our proposed system, capturing multiple data from multiple sensors and send all the data to the webpage by HTTP request protocol on the webserver. Here, the proposed system performs working in the region of Gorakhpur. In this region, we have sensed the value for three locations i.e. Kushinagar, MMMUT College, and Gida Gorakhpur. Then we have arranged all the values in table form with respect to their data and time and done analysis. The proposed model is not only collecting data but also making decisions on the basis of observed data. It is very much helpful for farmers because it collects environmental data making an interpretation for the former. But it is also very important

https://doi.org/10.1051/itmconf/20214001006

for making another decision such as or industry work purpose, transportation. The accuracy of the proposed model nearly accurate with real data

Reference

[1] Mary Nsabagwaa, Maximus Byamukamab, Emmanuel Kondelaa, "Towards a robust and affordable Automatic Weather Station ", journal homepage: www.elsevier.com/locate/deveng.

[2] Ravi Kishore Kodali and Snehashish Mandal "IoT Based Weather Station" 2016 International Conference on Control, Instrumentation, Communication and Computational Technologies (ICCICCT) 978-1-5090-5240-0/16/$31.00, IEEE, (2016)

[3] Ravi Kishore Kodali and Archana Sahu "An IoT based Weather Information Prototype Using WeMos" 2016 2nd International Conference on Contemporary Computing and Informatics (ic3i), 978-1-5090-5256-1/16/$31.00, IEEE, (2016)

[4] Zi-Qi Huang, Ying-Chih Chen and Chih-Yu Wen, "Real-Time Weather Monitoring and Prediction Using City Buses and Machine Learning", Vols. 3 to 21 Published 10 September (2020)

[5] Kavya Ladi, A V S N Manoj, G V N Deepak, "IOT Based Weather Reporting System to Find Dynamic Climatic Parameters", International Conference on Energy, Communication, Data Analytics and Soft Computing (ICECDS-2017)

[6] P. Susmitha, G. Sowmyabala "Design and Implementation of Weather Monitoring and Controlling System", International Journal of Computer Applications (0975 – 8887) Volume 97– No.3, (July 2014)

[7] Tanmay Parasharl, Shobhit Gahlot2, Akash Godbole3, Y.B. Thakare4 "Weather Monitoring System Using Wi-Fi", (IJSR) ISSN (Online): 2319-7064 Index Copernicus Value (2015): 78.96, 2015

[8] Nutant Sabharwal, Rajesh Kumar, Abhishek Thakur, Jitender Sharma "A LOW-COST ZIGBEE BASED AUTOMATIC WIRELESS WEATHER STATION WITH GUI AND WEB HOSTING FACILITY" e-ISSN: 1694-2310 | p-ISSN: 1694-2426, Vol. 1, Spl. Issue 2 (May 2014)

[9] M. Prasanna, M. Iyapparaja, M. Vinothkumar, B Ramamurthy, S.S. Manivannan," An Intelligent Weather Monitoring System using Internet of Things", International Journal of Recent Technology and Engineering (IJRTE) ISSN: 2277-3878, Volume-8 Issue-4, November (2019)

[10] Mircea Popa and Catalin Iapa "Embedded Weather Station with Remote Wireless Control", 19th Telecommunications forum TELFOR 2011 Serbia, Belgrade, November 22-24, 2011, 978-1-4577-1500-6/11/$26.00, IEEE, 2011

[11] T.R.V. Anandharajan G. Abhishek Hariharan, K.K. Vignajeth, R. Jijendiran4kushmita "Weather Monitoring Using Artificial Intelligence" 2016 International Conference on Computational Intelligence and Networks. 2375-5822/16 $31.00, IEEE DOI 10.1109/CINE.2016.26, (2016)

[12] Yashaswi Rahul, Rimsha Afreen, Divya Kamini "Smart weather monitoring and real-time alert system using IoT" International Research Journal of Engineering and Technology (IRJET) Volume: 05 Issue: 10, Oct(2018)

[13] Ravi Sharma, Shiva Prakash and Pankaj Kumar, "Methodology, Applications and Challenges of WSN-IoT", IEEE International Conference of Electrical and Electronics Engineering (ICE3—2020), jointly organized by Madan Mohan Malaviya University of Technology Gorakhpur and North Dakota State University, Fargo, USA, Pid197, Feb. 14-15, (2020)

[14] Garima Verma, Shiva Prakash, "A Comparative Study Based on Different Energy Saving Mechanisms Based on Green Internet of Things (GIoT)", IEEE 8thInternational Conference on Reliability, Infocom Technology and Optimization (ICRITO-2020), IEEE Conference Record Number 48877, Amity University, Noida, India, Pid631, June 4-05, (2020)

[15] Ferdin Joe John Joseph "IoT Based Weather Monitoring System for Effective Analytics "International Journal of Engineering and Advanced Technology (IJEAT) ISSN: 2249 – 8958, Volume-8 Issue-4, April (2019)

[16] Raj Kumar, Shiva Prakash, "Performance & Parametric Analysis of IoT's Motes with Different Network Topologies", International Conference on Electrical and Electronics Engineering, jointly organized by School of Engineering The University of Malaya, Kuala Lumpur, Malaysia & Centre of Excellence- Power Engineering and Clean Energy Integration, Galgotias University, India in the collaboration with NPTI, Pid 175, January 2nd -3rd, (2021)

[17] Garima Verma, Shiva Prakash, "Emerging Security Threats, Countermeasures, Issues and Future Aspects on Internet of Things (IoT): A Systematic Literature Review", 2nd International Conference on Future Learning Aspects of Mechanical Engineering (FLAME 2020), Organized by Department of Mechanical Engineering, Amity School of Engineering & Technology. Amity University, Noida, Uttar Pradesh, India, Pid 231, August 5th -7th(2020)

[18] Saijshree Srivastava, Shiva Prakash, "An Analysis of Various IoT Security Techniques: A Review", IEEE 8thInternational Conference on Reliability, Infocom Technology and Optimization (ICRITO-2020), IEEE Conference Record Number 48877, Amity University, Noida, India IA, Pid360, June 4-05(2020)

[19] Garima Verma and Shiva Prakash, "Internet of Things for Healthcare: Research Challenges and Future Prospects", Springer, 1st International Conference on Advanced Communication and Computational Technology (ICACCT), NIT Kurukshetra, Kurukshetra, Haryana, India, Pid 75, December 6-7, (2019)

https://doi.org/10.1051/itmconf/20214001007

Design and Assessment of Electric Vehicle Performance Parameters based on Drive Cycle

Binsy Joseph[1,*] and *Dr Deepak Vishnu* Bhoir[1,**]

[1]Department of Electronics Engineering, Fr Conceicao Rodrigues College of Engineering, Bandra(W), Mumbai-50, India

Abstract. Electric vehicle plays a significant role, in the future transportation across the world. EV has the potential to reduce air pollution and emission of Greenhouse gasses significantly compared to the existing fossil-fuel-based vehicles. Even though substantial progress can be expected in the area of embarked energy storage technologies, charging infrastructure, customer acceptance of Electric Vehicles is still limited due to the problems of Driving range anxiety and long battery charging time. We can solve most of these problems with the infrastructure development ,optimum sizing and design of the vehicle components and extensive study on vehicle dynamics under various real-time driving conditions. This research focuses on the Matlab software based co-simulation of Electric Vehicle system, including the battery pack and motor, to predict the vehicle performance parameters like driving range, efficiency, power requirement, and energy characteristics under different driving scenarios. The vehicle's acceleration performance, energy consumption, and efficiency are determined by simulation and verified analytically. Using ADVISOR software the fuel economies and tail pipe emission for various vehicle models are determined by simulation and results are compared with Hybrid Electric vehicle models.

1 Introduction

In recent years, Electric Vehicle (EVs) have gained renewed interest in research and the automotive sector. Electric vehicle presents a tangible solution to reduce environmental pollution levels worldwide. Electric Vehicles and Hybrid Electric Vehicles have been regarded as the most promising alternative to Internal Combustion Engine (ICE) vehicle. Governments promote EV technology by providing subsidies and deploying public charging infrastructure in major city Highways and public parking places.Despite the fact that considerable advancement can be expected in embarked energy storage technologies, charging infrastructure, due to the problems of Driving range anxiety and battery charging speed limitation etc. To increase the customer acceptance of Electric vehicle, researchers and automotive companies are consistently working on modified, reliable and affordable vehicle models depending on the country wise requirements.

Electric Vehicles (EV's) batteries are charged from the grid. The charging process takes more time than refuelling of internal combustion engine vehicles. This problem significantly reduces the acceptance of EV's. Today's EV's are equipped with Li-ion batteries supporting a Fast Charging current[1]. Most modern EV's can accept a charging current of magnitude xC where x is the multiplication factor and C is the capacity of the battery[2].

The development of EV's starts changing the way that people travel. One of the major issue hindering the development of EV usage is range anxiety[3]. In this paper, we address the driving range estimation of an Electric vehicle, permitting drivers to beat range anxiety. This approach has the method of simplicity. This paper focuses on the analysis of vehicle energy consumption with the help of powertrain simulation and driving behaviour analyses.

When designing battery packs, it is important to be able to simulate the operation of the pack before actual fabrication. This helps to ensure that the pack will be able to meet all performance requirements before a large investment.

In order to understand the energy consumption of an actual Electric vehicle during real-world operating conditions and to evaluate the performance of various coupled Drive Train components before the actual fabrication and testing, manufacturers normally use sophisticated vehicle Co-simulation software. ADVISOR software is one commonly used software for Vehicle Simulation, but it has limited flexibility to simulate various design parameters of battery pack, motor specifications and vehicle parameters. Many researchers over the years studied different Optimization and simulation method to study vehicle performance under real-time driving conditions. A drive cycle is a time series of vehicle speed pattern created to simulate real world driving conditions. Driving cycles are utilized for assessing vehicle on-street energy utilization and vehicle emissions[4] . Jony J Eckert et al. studied Electric vehicle drive train optimization using a genetic algorithm to improve vehicle range and performance .Stephan Günther et al. investigated the influence of driving cycle on the design optimisation of permanent magnet syn-

*e-mail: binsy_joseph@fragnel.edu.in
**e-mail: bhoir@fragnel.edu.in

chronous motor drive for electric vehicle application [5]. Today's vehicle design mostly based on powertrain design and control , with no information on future driving plans[6]. Petru Palcu et al presented modelling and simulation results for Whole-Day Driving Prediction(WDDP vehicles on the real-world logged driving cycles of different drivers. Sanghpriya H. Kamble et al studied real world driving cycle for Indian city drive. They developed an urban driving cycle for estimating vehicular emissions and fuel consumption using micro-trips extracted from real-world data [7].

2 Vehicle Dynamics and Powertrain Component Modelling

This section discuss the modelling of complete EV drive train components and the basics of vehicle dynamics. The main difference between electric car and their conventional counterparts is in the drivetrain. Conventional ICE vehicle (Internal combustion Engine) has a multi speed gear box where as EV has a single speed transmission. In EV models, a single fixed -ratio transmission is used, consist of single reduction gear. Several Electric car models like Tesla Model S, Chevrolet Bolt, Nissan Leaf etc. use a single speed transmission coupled with electric motor which reduces emissions and eliminate the use of clutch. As per the speed, torque is varied automatically and vehicle operate at high efficiency.

2.1 Electric Vehicle Power Train

The powertrain of a Electric Vehicle (EV) consists of an electric drive system with a battery as an energy source and driven by an efficient electric motor, normally an Induction Motor. The chemical energy stored in the battery pack is used to drive the motor with the help of a Power Electronic Converter, accompanied by a control system. The control system controls the frequency and magnitude of the output of the Inverter depending on the driver's commands which are communicated through sensors connected to acceleration and brake pedal.

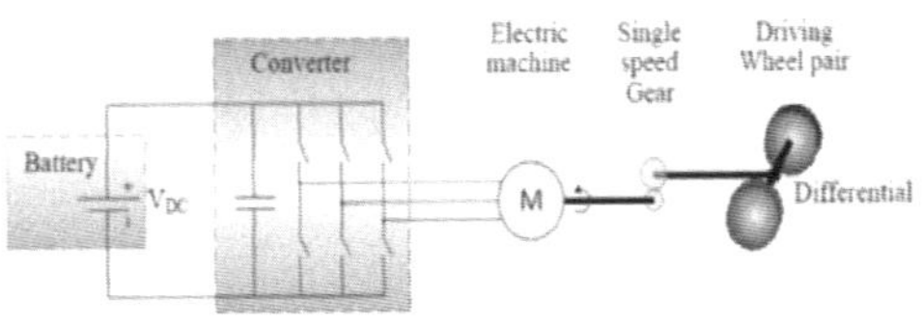

Figure 1. Schematic sketch of BEV powertrain

Electrical machines are normally designed with higher speed levels typically to reduce the physical volume of the machines. Gears are used for speed reduction at the wheels. To give the left and right wheels to have somewhat different speeds during turning, there is additionally a need for differential between the wheels.

2.2 Vehicle Dynamics

Vehicle dynamics describes how a vehicle moves forward affected by frictional powers between the tyre and the road.

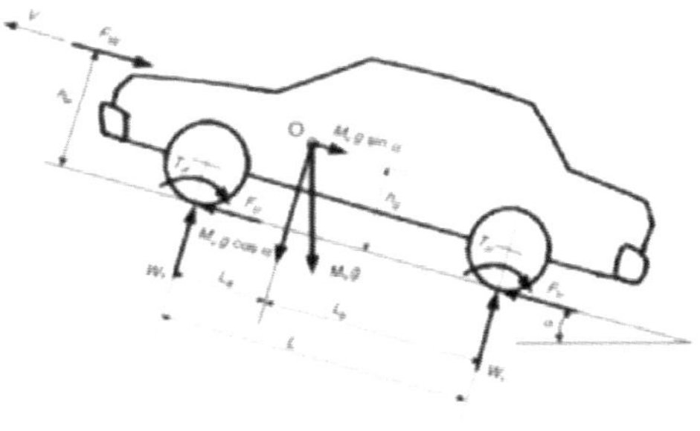

Figure 2. Forces acting on vehicle [9]

Understanding vehicle dynamics is very important while analyzing the impact of power train on vehicle performance to determine whether to accelerate or average energy consumption for a driven distance.

As per Newton's second law, vehicle acceleration can be written as

$$ma = m\frac{dv}{dt} = F_{tractive}(t) - F_{resistive}(t) \tag{1}$$

'm' is the equivalent mass of the vehicle in Kg, $a(m/s^2)$ is the acceleration, $F_{tractive}(t)$ (Nm) is the sum of all the tractive forces and $F_{resistive}(t)$ is the total resistive forces. The tractive force is applied from the powertrain through the gear, differential and the wheel shaft to the contact region between the wheels and the road.

Aerodynamic drag is the major resistive force acting on the vehicle while moving. Because of the complex shape of automobiles and complex nature of fluid dynamics, it is difficult to model accurate analytical model even using CFD(Computational Fluid Dynamics)software. Mathematically aerodynamic drag force is given by

$$F_a = \frac{1}{2}\rho_a C_d A_f (V_{vehicle} - V_{wind})^2 \tag{2}$$

$\rho_a(kg/m^3)$ is air density, C_d is aerodynamic drag coefficient, $A_f(m^2)$ cross-sectional area of the vehicle, $V_{vehicle}(m/s)$ vehicle speed, and $V_{wind}(m/s)$ is wind speed moving in the direction of the vehicle. The rolling resistance force, F_r acting on a vehicle in the longitudinal direction, is

$$F_r = C_r mg cos(\alpha) \tag{3}$$

where m is vehicle mass, g is gravity constant, $\alpha(rad)$ is road inclination angle.

The grad force, F_g and wheel force F_{wheel} acting on a vehicle is

$$F_g = mg sin(\alpha) \tag{4}$$

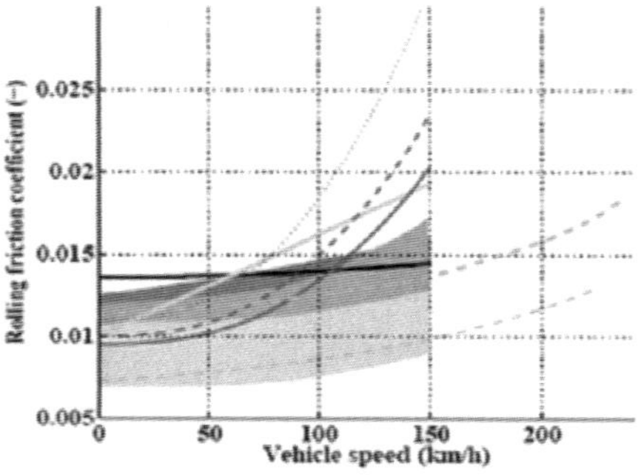

Figure 3. Rolling Frictional coefficient Vs Vehicle speed[10].

$$F_{wheel}(t) = F_{acc}(t) + F_a(t) + F_r(t) + F_g(t) \quad (5)$$

$F_{acc}(Nm)$ is the force required to accelerate the vehicle mass at a certain magnitude of acceleration.

2.3 Differential

All vehicles need a type of differential. It is imperative when turning a vehicle. To turn securely, the external wheels should turn quicker than the internal wheels, provided by differential.

2.4 Wheel Power and Energy

The wheel power P_{wheel} is determined by the tractive force (F_{wheel}) and the vehicle speed ($V_{vehicle}$) as

$$P_{wheel}(t) = F_{wheel}(t).V_{vehicle}(t) \quad (6)$$

The total energy at the wheel can be found from the time integral of the power as

$$E_{wheel}(t) = \int P_{wheel}(t)dt \quad (7)$$

2.5 Modelling of Electric Motor

An electric motor converts stored electric energy in the battery into mechanical energy to move the vehicle. It can also act as generator, gathering the mechanical power from the transmission to recharge the battery.Torque-power region is divided 2 stages: into constant torque, constant power and high speed region. Electric vehicles, do not require a multi-gear transmission. It start from zero speed and do not any consume any fuel. Thus contributes to a better efficiency without polluting emissions.

2.5.1 Modelling of Permanent Magnet Synchronous Machine

A Permanent Magnet Synchronous Machine (PMSM) consist of a rotor with permanent magnets and a wound stator.

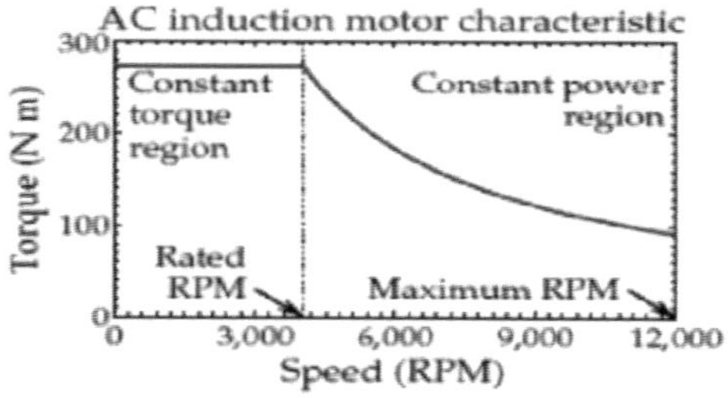

Figure 4. Torque-speed characteristics of Induction motor.

Equivalent circuit model: The commonly used model is the equivalent dynamic dq-model of a PMSM which is shown in Figure 5, where dq denote the rotor frame of reference or synchronous coordinates. Direct axis (d-axis) and Quadrature axis (q-axis) 90^0 displaced.

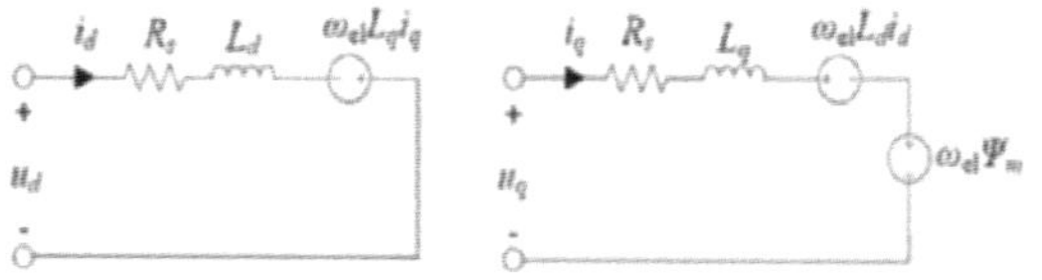

Figure 5. Equivalent circuit model of a PMSM machine

The stator voltage equations as functions of the dq axis stator currents (i_d and i_q) are;

$$U_d = L_d\frac{di_d}{dt} + R_s i_d - \omega_{el}L_q i_q \quad (8)$$

$$U_q = L_q\frac{di_q}{dt} + R_s i_q + \omega_{el}L_d i_{qd} + \omega_{el}L\phi_m \quad (9)$$

where Rs is stator winding resistance, ω_{el} is angular speed ($\omega_{el} = n_p W_r$) , W_r is rotor angular speed, and n_p is number of pole pairs, L_d and L_q are dq-axis winding inductance, and ϕ_m is flux linkage related to the permanent magnet.

3 Methodology

A robust methodology is followed for developing a simple, cost-effective and general-purpose Electric vehicle simulation model using Matlab software. The results are verified with Standard vehicle simulation model using ADVISOR software.

The steps involved in this methodology are a selection of Test Electric Vehicles model with its specification, collection of drive data(speed Vs Time) under which Electric vehicle performance is evaluated using a computer programme, Experimentation, Test Data generation and Drive Cycle Validation, Analysis of performance of EV and Assessment efficiency of all components in the Electric drive drain including Battery pack, Power Converter and Electric Motor.

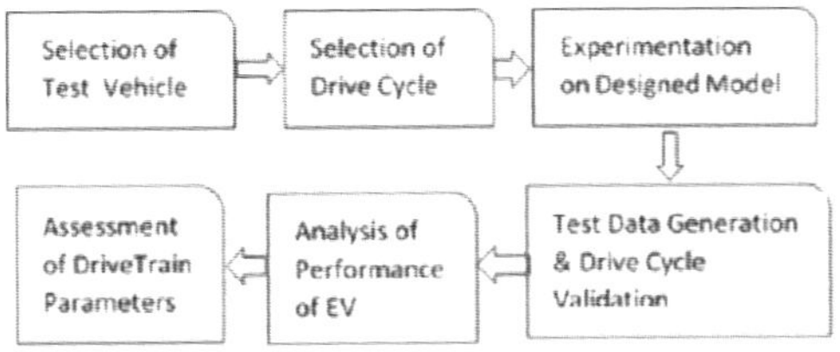

Figure 6. Block diagram of the methodology

To simulate an EV, an accurate description of vehicle and knowledge of the task the vehicle is required to accomplish are needed.The vehicle description includes all factors needed for dynamic simulation,including drivetrain characteristics ,battery-pack characteristics, motor and inverter characteristics .The vehicle driving patteren is simulated with drive-cycle profile(desired speed Vs time).The urban dynamometer driving schedule (UDDS) profile is used to analyse city driving and highway fuel efficiency test (HW-FET) to predict EV performance in highway driving. Simulation approach is to compute on a sample by sample basis,the desired acceleration as per selected drive cycle profile,to match the desired speed profile and estimate achievable torque and power values by electric vehicle.The desired toque and power values are restricted by the specification of the motor chosen for the vehicle.

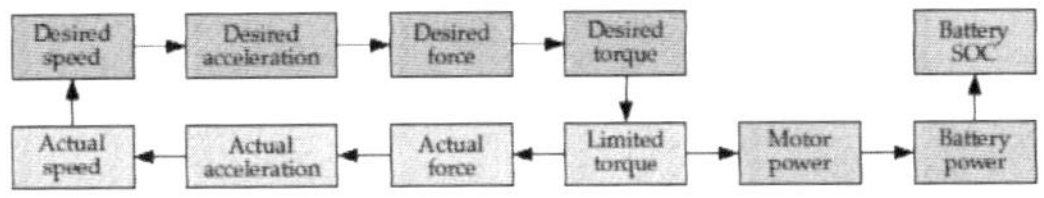

Figure 7. Simulation strategy

The battery power computed based on motor power and battery state of charge (SOC) is estimated and projected driving range of the electric vehicle is extrapolated from the remain battery SOC data.

To determine vehicle range based on battery capacity, battery pack must be simulated with the vehicle and load.Instantaneous power required by the motor is given by

$$MotorPower = 2\pi * \frac{Motorspeed * Torque}{60} \qquad (10)$$

Battery Power is calculated as Battery Power = overhead power + motor Power * $\eta_{drivetrain}$

where $\eta_{drivetrain}$ is the efficiency of drivetrain

$$Batterycurrent = \frac{batterypower}{batteryvoltage} \qquad (11)$$

Battery SOC at k^{th} instant is updated as

$$BatterySOC_k = [BatterySOC_{k-1}] - \frac{BatteryCurrent * \Delta t}{BatteryCapacity * 3600} \qquad (12)$$

Driving range of the vehicle is calculated from SOC and total distance of simulated drive cycle ($TotDistance_{dc}$) as

$$range = TotDistance_{dc} * \frac{(SOC_{max} - SOC_{min})}{(SOC_{initial} - SOC_{end})} \qquad (13)$$

4 Simulation Result and Discussion

For the purpose of verification, the values of a conceptualized EV obtained from the system using the parameters either specified by the manufacturer or calculated using the formulas mentioned above are compared to the values of a vehicle of the same make and model. This helps in proving the effectiveness of the system and also proves the relevancy of the system with the current automotive industry standards. The vehicle which is being used as a test benchmark is the General Motors Chevy Volt operating in pure electric mode. Vehicle performance is simulated for City Drive cycle (UDDS) and Highway drive cycle (HWFET).

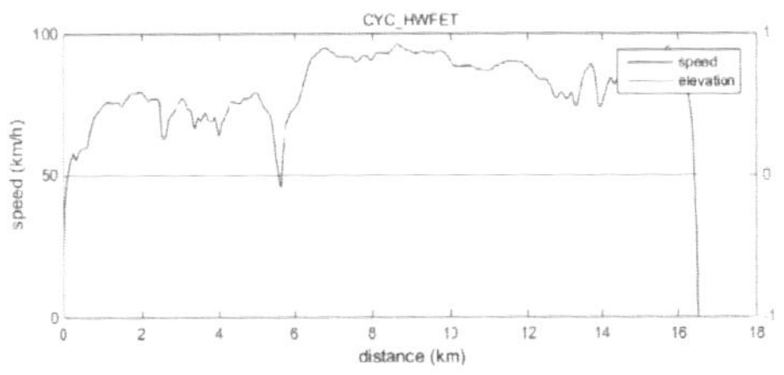

Figure 8. Highway drive cycle (HWFET)

ADVISOR software is used to compare the performance of conventional Gasoline powered vehicle and Hybrid vehicle. Using this software various vehicle models including conventional and Hybrid Electric vehicle(HEV) are simulated over city and Highway drive cycle and results are tabulated.

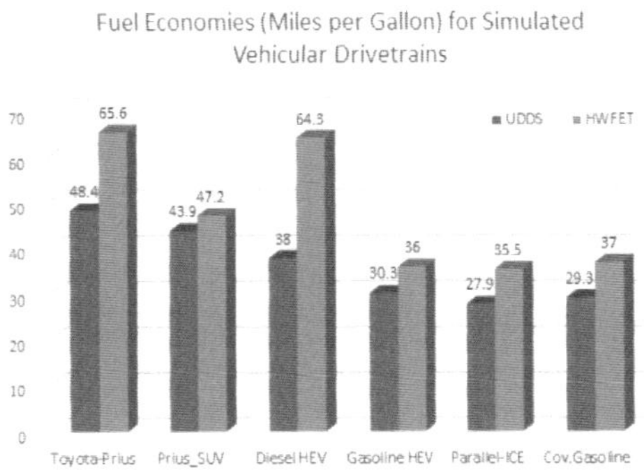

Figure 9. A Fuel Economies (mpg) obtained for various vehicle.

Fuel economies in Miles per Gallon for Hybrid drivetrain vehicle models (HEVs)are much better than conventional counterparts as shown in Fig.9.

Different vehicle models are simulated over UDDS and HWFET drive cycle.Simulated vehicle's fuel consumption and emission results are shown in Fig.10.

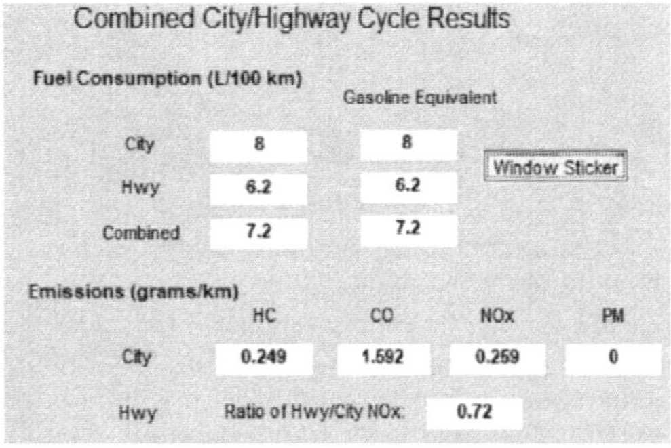

Figure 10. Fuel consumption and Emission result

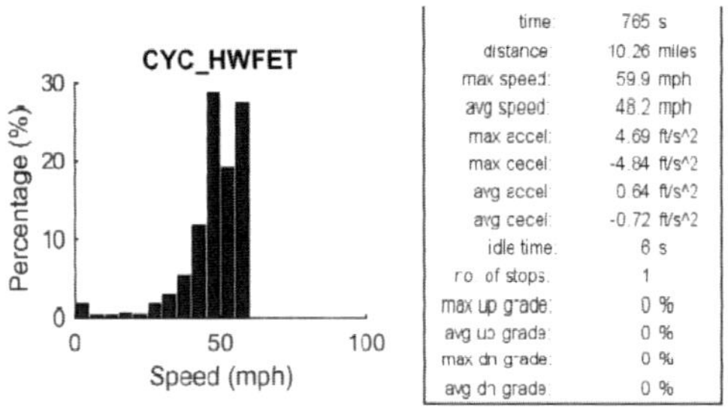

Figure 11. HWFET drive cycle ,EV speed Profile

For simulated vehicle model ,the difference between requested and obtained vehicle speed(km/h) profile under the UDDS drive cycle is shown in Fig.12. Simulated Electric Vehicle Speed profile is plotted in Fig. 11.

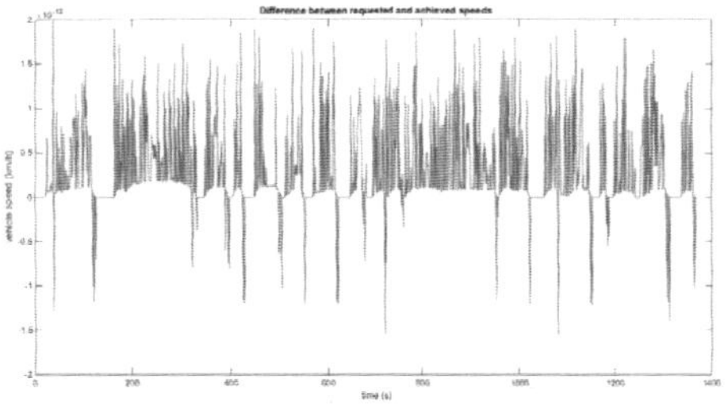

Figure 12. Difference between requested and achieved speeds

Fig.12 shows the difference between requested and achievable vehicle speed under selected driving profile. Vehicle performance is evaluated using simulation and Torque-Speed characteristics is plotted in Fig.13.

EV models,General Motors Chevy Volt operating in pure electric mode,is simulated using proposed algorithm using Matlab software. Vehicle specifications used for

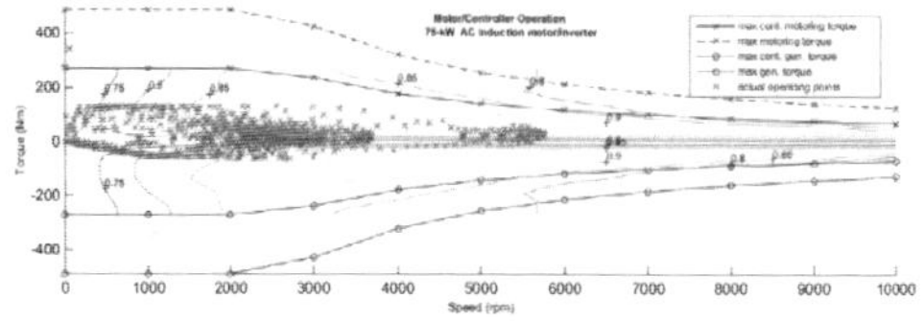

Figure 13. Motor Speed -Torque Characteristics

Table 1. Vehicle Specification:

SrNo	Description	Value
1	Weight	1500 Kg
	Aerodynamic Frontal Area	$1.84m^2$
	Drag coefficient Cd	0.22
	Constant Overhead power	200 W
	Wheel rolling radius	0.35m
	Rolling coefficient Cr	0.0111
	Gear ratio	12:1
	Gear Inertia	$0.05kg/m^2$
	Gear efficiency in percent	97
2	Battery nominal voltage	3.8 V
	Cell capacity	15 Ah
	Battery module	3P8S
	Battery pack	12 modules
	Pack nominal Voltage	365 V
3	Inverter efficiency	90
	Motor max torque	275Nm
	Rated speed RPM	4000 rpm
	Max speed	12,000 rpm
	Motor efficiency	95

simulation are tabulated in Table 1. All vehicle parameters are stored as structure and passed to Matlab function for each specific task. System is modelled and Software program is written for each specific module using the mathematical model and formulas discussed.The vehicle performance parameters are analysed and plotted.

5 Conclusion

This project focuses on the Matlab based simulation of Electric Vehicle system to predict the vehicle performance parameters like driving range, efficiency, power requirement, and battery current demand under different driving scenarios. This work also discuss the advantages of converting an conventional vehicle to an EV and also we can protect the environment by reducing the usage of fuel. The absence or reduction of tailpipe emissions results in improving the air quality and removal of greenhouse gas emissions. The additional benefit of using electricity is that there is a greener option to charge from renewable power thus benefiting the planet. The system also calculates the capacity of the battery which will be needed to run the vehicle for a said amount of distance, thus, providing the manufacturer the ideal battery capacity. In this paper, we address the strategy for assessing the driving range of an Electric vehicle, permitting drivers to defeat range tension.

References

[1] Christen, Daniel and Jauch, Felix and Biela, Jurgen, *Ultra-Fast Charging Station for Electric Vehicles with integrated split Grid Storage,17th European Conference on Power Electronics and Applications,* EPE'15, ECCE-Europe, 2015, 1-11

[2] Bertoluzzo M, Buja G, Pede G.*Design considerations for fast AC battery chargers.* (World Electric Vehicle Journal, 2013) 147-154.

[3] B. Joseph and D.V. Bhoir*Design and Simulation of Wireless Power Transfer for Electric Vehicle,* (2019 Int. Conf. Adv. Comput. Commun. Control, 2019) 1-5.

[4] S. K. Pathak, Y. Singh, V. Sood, and S. A. Channiwala*Drive Cycle Development for Electrical Three Wheelers,* (SAE Tech. Pap., vol. 2017-March, 2017)

[5] S. Günther, S. Ulbrich, and W. Hofmann,*Driving Cycle-Based Design Optimization of Interior Permanent Magnet Synchronous Motor Drives for Electric Vehicle Application,* (2014) pp.25-30

[6] P. Palcu and J. Bauman*Whole-Day Driving Prediction Control Strategy: Analysis on Real-World Drive Cycles,* (IEEE Trans. Transp. Electrif., vol. 4, no. 1, 2017) pp.172-183

[7] S. H. Kamble, T. V. Mathew, and G. K. Sharma,*Development of real-world driving cycle: Case study of Pune, India,* (Transp. Res. Part D Transp. Environ., vol. 14, no. 2, 2009) pp. 132–140

[8] C. Daniel Pires Rodrigues, J. Humberto Oliveira Seabra Co-Advisor, and J. António dos Santos Almacinha,*Design of a high-speed transmission for an electric vehicle.*

[9] A. E. Mehrdad Ehsani, Yimin Gao,*Modern Electric ,Hybrid Electric and Fuel cell vehicles - Fundamentals,Theory and DesignBook.*1390

[10] R. Carlos Daniel Pires,*Design of a high-speed transmission for an electric vehicle,* (2018) pp. 309

Sixth Sense Jacket

Prabhat S. Pandey [a]
prabhatpandey_@outlook.com

Shreya S. Chakraborty [a]
shreya.04chakraborty@gmail.com

Jay S. Gaikwad [a]
jaygaikwad121@gmail.com

Divya K. Shah [a]
divya.shah@rait.ac.in

[a] *Department of Electronics Engineering, Ramrao Adik Institute of Technology, Nerul, Navi* Mumbai,400 706,India

***Abstract*—In a time when the entire world is going through a pandemic which can be curbed or substantially avoided and reduced by maintaining distance, it is crucial for every individual to follow the social distancing norms. However, it can be a challenging and daunting task to ensure safe distance all the time. Thus a Sixth Sense Jacket is a great way to keep the user mindful and alert of their surrounding without actually being alert all the time. It can not only alert the user but can also keep daily essentials such as keys and cards clean using dry sanitization. The jacket comes with a bunch of special features such as Pulse Oximeter which can be used to check vital health stats such as heart rate and blood saturation (SpO_2) levels. The prime purpose of the Sixth Sense Jacket is to keep the user passively mindful of it's surrounding all the time and thus can be used in any field where it is crucial to stay alert in a given surrounding. The various domains this jacket can be used are women safety, battlefield and to add an extra sense to visually impaired people. The jacket can be a useful means to pre-empt any given danger that might befall the user and hence the name "Sixth Sense" Jacket.**

***Index Terms*—Social Distancing, PIR Sensor, Raspberry Pi Zero, Dry sanitization, Pulse Oximeter, MAX30100, Sixth Sense**

I. Introduction

In the recent times, all have realized the importance of social distancing while in public or in crowded places. So a safety jacket would help the person wearing it maintain social distancing more effectively with the help of some electronics devices and components. The implementation of this jacket or the Sixth Sense Jacket will be based on the Raspberry Pi Zero single board computer which is a small, lightweight device and is highly compatible with our application. The jacket comes with additional features such as Ultraviolet LEDs in the outer pocket of the jacket. UV light is an effective method for dry sanitization and can be effectively used for sanitization of day to day object. Along with this a Pulse Oximeter Sensor which would deliver an output on a LED display is attached on the sleeve of the Jacket.

The main objective behind the designing of the sixth sense jacket is to allow people go out into the world because sometimes it is absolutely inevitable and one has to go out of the safety of their homes. In such a situation, this jacket will enable the user to stay mindful of his or her surrounding and at the same time keep a check of their health parameters by measuring and displaying various health vitals such as heart rate and Oxygen saturation levels. This way the user can be rest assured that he or she is taking all the necessary precautions amidst the pandemic.

II. Existing Techniques

Development Paper of a Sensing Module for Standing and Moving Human Body Use the Shutter and PIR Sensor to improve the sensor and body sensor module using Shutter and pyroelectric infrared (PIR) sensor [7]. Test results show that the hearing distance is 7.0 meters hearing angles around 110 degrees at room temperature. In these cases, the transmission rate is 100% and the power distribution of the module is rated as 100 m.

Human Movement Detection and Identification Using Pyroelectric Infrared Sensors [8]. With three PIR-based modules in the monitoring hall; one roof module; two modules on opposite walls facing each other. Data were collected from eight subjects when traveling in three different, three intervals and three speed levels. Two types of feature sets were used: a raw data set and a reduced feature set made for size and height. The findings show that with the setting of raw data taken from one PIR sensor in three modules, more than 92% accuracy in distinguishing index. And in addition to 94% accuracy in classification, speed and distance identification using a reduced feature set extracted from two PIR sensors for each of the three modules.

Modifying UV blocking effect using polyester fabric [2]. The influence of conventional thermal and alkaline dyeing processes on fabric formation is being investigated. A spectrophotometer with a connecting sphere is used to measure the UV transfer of polyester fabrics according to AATCC tests. The slight difference in the weight of the fabrics after the blinding processes already affects the UV transfer of the fabrics. Orange and blue dyed fabrics show a high enough UPF, while pale red fabric does not achieve such values. Double-dyed fabrics with both fabrics after being treated with a UV absorber reach high UPF values.

Low Cost Pulse Oximetry Design Based on the Raspberry Pi. Pulse oximetry is a non-invasive method of monitoring oxygen saturation [1]. Oxygen intake (SpO_2)

is a percentage of oxygen-containing hemoglobin. Pulse Oximetry operates on the condition that hemoglobin with dissolved oxygen (HbO_2) captures light at different wavelengths from hemoglobin that does not bind oxygen (Hb). This method uses different wave- lengths of red light (660 nm) and infrared light (940 nm) carried by the light sensor after passing through the veins and bones at the end of the index finger [8].

III. Block Diagram

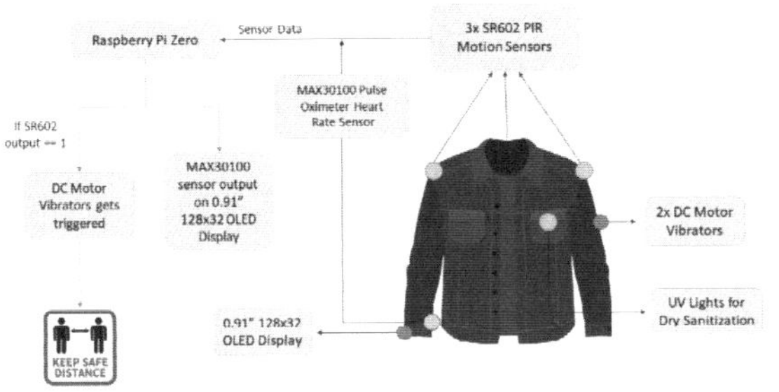

Fig. 1. Block Diagram

The Sixth Sense Jacket incorporates various features for achieving its purpose, depicted using the block diagram(fig.1.).

Distancing Mechanism: The Sixth Sense Jacket has three SR602 PIR Sensors. One sensor each on the left and the right shoulder of the jacket and the third one on the back, below the collar of the jacket. The output of the PIR sensor will be reflected on the DC Motor Vibrators. Four such DC motor vibrators are used in this project, two vibrators each on the right and left arm of the jacket.

Health Stats Mechanism: One MAX30100 Pulse Oximeter Heart Rate Sensor is being used along with an 0.91" OLED display for displaying the data accumulated by the PO Sensor. The MAX30100 and OLED Display will be mounted on a wide wrist band, with the display on the upper end of band visible to the user.

Dry Sanitization: The UV Light which will be used for dry sanitization will be in the outer pocket of the jacket, it will be controlled manually using a switch. For safety purpose, the pockets will be coated with a thick layer of dyed polyester fabric to prevent the UV Light's effect on the user.

IV. Working

By referring the circuit diagram (fig. 2), the circuitry of the project can be understood. The sensors are setup when the Raspberry Pi initializes. The SR602 gives conditional output. If the sensor gets a reading which is less than 5 meters then the output will be triggered and the DC motor will start vibrating. There are three PIR sensors to be mounted on the left shoulder, the right shoulder and the backside of the jacket

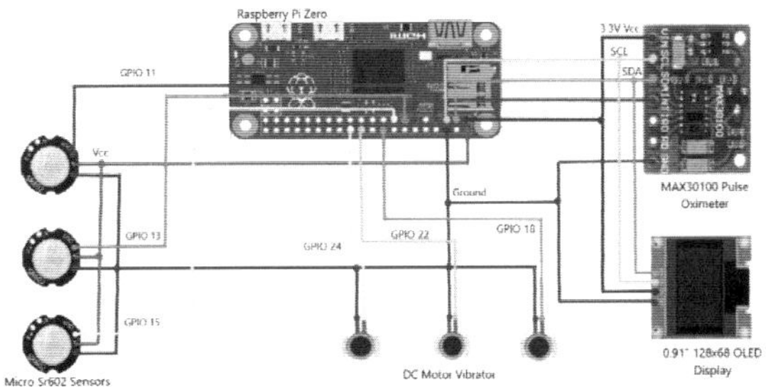

Fig. 2. Circuit Diagram

while there are two DC motor vibrators mounted on each arm of the jacket, all of them are connected on separate GPIO pins on the RPi. So if the sensor on the right side of the jacket gets an input within the range then the two DC motors on the right arm of the jacket will start vibrating thereby alerting the user, similarly with the left arm sensor triggers the motors on the left arm. In the situation when someone is approaching from the back, all the four DC motor vibrators will be triggered simultaneously alerting the user.

For the MAX30100 Pulse Oximeter sensor no GPIO pins needs to be allotted but the sensor needs to be connected to the SDA (Serial Data Line) and SCL (Serial Clock Line) of the Raspberry Pi along with the SDA and SCL of the OLED display. So as the PO sensor starts reading the pulse reading and the oxygen saturation of the user, the output will be received on the OLED display in real-time. The PO sensor will be fixed on the wrist of the user using a band, while the OLED Display will be mounted right on the opposite side of the band. For the purpose of power consumption a 2 pin switch can be allotted to turn this peripheral off. Proper flow of all the process of the project can be clearly understood by referring the flow-diagram (fig.3.).

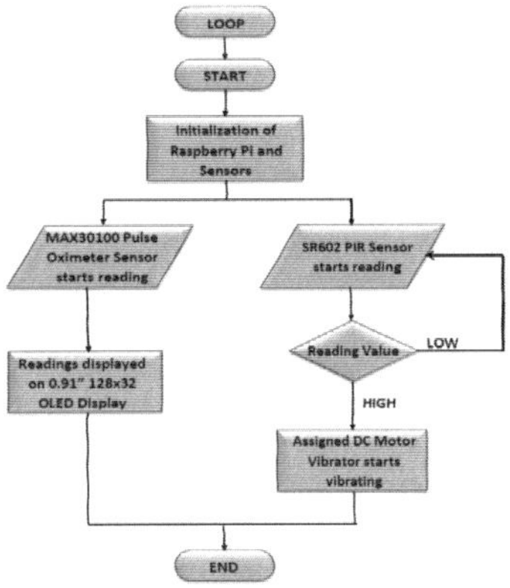

Fig. 3. Flow-diagram

V. FLOWCHART

The flowchart (fig. 4) represents the working of the SR602 PIR Sensor. It shows the condition of the three sensors which will be on the left shoulder, right shoulder and the back of the user. All the sensors will run in a loop, meaning the PIR sensor will continuously fetch the data within the range of less than 200cm in this case (cm<=200). Depending on the location where the jacket is to be used, the unit can be modified by very minor changes in the code.

The 'readCentimeter' parameter is being used here for detecting the distance, so as soon the condition becomes true, the DC Motor Vibrators assigned to each sensor gets triggered thereby alerting the user. The sensor on the left shoulder and the right shoulder will be assigned two DC motor vibrators each whereas for producing the output for the sensor on the back of the jacket, all the four DC motor vibrators are assigned. LOOP3 below represents the sensor on the backside of the jacket whereas LOOP1 and LOOP2 depicts the sensor on either of the left or right shoulder.

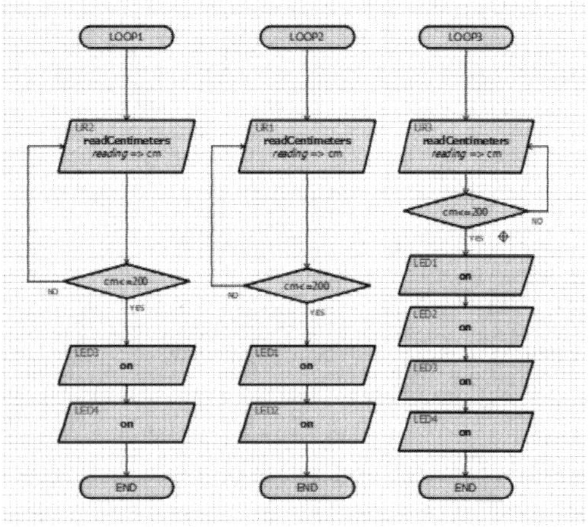

Fig. 4. Flowchart

VI. RESULTS

The integration of the components results in desired output .i.e. the SR602 sensors giving accurate output if a human comes within the range (<=5 meters), the range can be adjusted by some minor changes in the code or by mounting a potentiometer on the sensor itself that will adjust its sensitivity. The output of PIR sensor triggers the assigned DC Motor Vibrator which vibrates thereby alerting the user. The SR602 has considerably less power consumption of 20uA and also has an accuracy of over 94% [8]. For getting a depth of power requirement of the jacket, the current and voltage ratings of all the devices can be referred from the datasheet (fig. 5.).

Ratings / Components	Voltage	Current	Pieces	Total Consumption
SR602 PIR Motion Sensors	3.3V	0.02mA	3	0.06mA
MAX30100 PO Heart Rate Sensor	5V	1.2mA	1	1.2mA
0.91" OLED Display	5V	2mA	1	2mA
DC Motor Vibrators	3V	80mA	4	320mA
Raspberry Pi Zero	3.3V	120mA	1	120mA
Net Charge Consumption (Ideal state)				≈450 mA

Fig. 5. Datasheet

The MAX30100 Pulse Oximeter Sensor apart from the 5V Vcc and GND simply occupies the SDA and SCL pins. of the Raspberry Pi which is then shorted with the SDA and SCL terminal of the 0.91" OLED Display. This minimal connection makes the function of heart rate module easy and less complex additionally reducing the strain on Raspberry Pi.

Calculation:

Battery Capacity (mAh): 10000
Net Charge Consumption (mA): 450

Formula: $$\text{Charge/discharge time (in Hours)} = \frac{\text{Amp-hour rating}}{\text{Continuous current (in Amps)}}$$

Here, Amp-hour rating of the battery = 10000 mAh
Continuous current (charge consumption) = 450mA

Hence, Discharge time = 22 hours 13 mins

Fig. 6. Calculation

After analysis and calculation (fig. 6) it can be noticed that an 10,000mAh power bank is capable enough to fulfil the power requirements of the Sixth Sense Jacket for 22+ hours in ideal conditions, considering that all the sensors and the output device will be functioning constantly. The UV Light is not taken into consideration as ideally it may be used for a very short span. The charge consumption can be further lowered by using high end sensors and output devices which may increase the expense of the jacket.

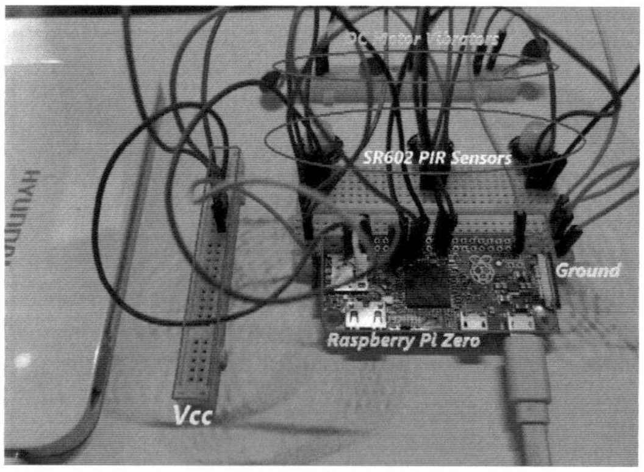

Fig.7. Hardware Assembly

Proteus Design Suite is being used for the simulation (fig. 8.) of the SR602. As the library of the PIR sensor is unavailable for Raspberry Pi in Proteus, Grove Ultrasonic sensors are used for demonstrating the distancing mechanism of the jacket. When one of the sensor, representing the SR602 of the left or shoulder gets an reading of distance less than 200cm, the 2 LED lights glow up which represents the two assigned DC motor vibrators.

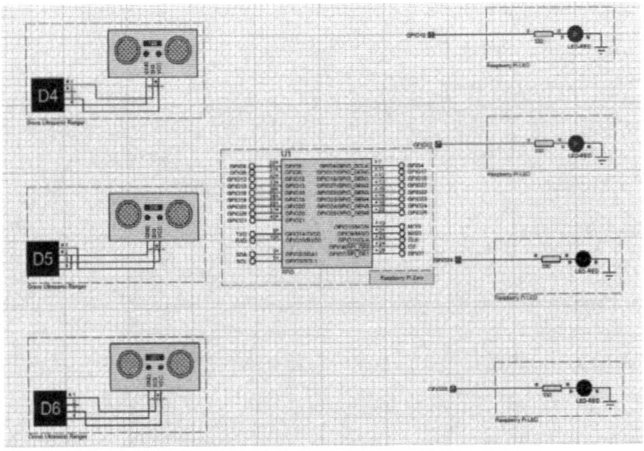

Fig. 8. Simulated Results

VII. Conclusion

The Sixth Sense Jacket is everything one can need in today's situation. The concept of the jacket is not very complex which makes the jacket very basic yet user friendly and affective when it comes to daily use. The sensors SR602 and MAX30100 are really cost effective and compact. In total there are 8 I/O devices yet the charge consumption of the devices are low enough to comfortably make if through the day. While this jacket may have few manageable drawbacks, the purpose and implementation can make this project notable. The biggest feature of the Sixth Sense Jacket is that it's not limited to a specific domain or task and the utility of the jacket can be ever-increasing.

VIII. FUTURE SCOPE

The Sixth Sense Jacket can have immense scope in present and future as well. The range of this jacket is vivid.

The number of PIR sensors can be increased enhancing the sensitivity of the jacket. On top of that a dedicated alerting mechanism can be created within a group of people wearing the Sixth Sense Jacket in a small geographical scale. An additional SOS feature can prove very useful for the military application of this jacket.

An IFTT based approach for alerting the user through phone alerts may prove beneficial by significantly reducing the number of output devices on the jacket thereby making space for more sensors that can be implemented on the same discharge time margin. The jacket if integrated with microphone and a compact speaker can be used for speak-to-translate feature in real-time. This feature will be an heavily IoT based approach and the audio device can simply be the audio device that will be in user by the user during the entire process.

Additionally it can be used for visually impaired people, women safety, construction sites and may more places where one needs to be alert.

IX. References

1. S. Bakhri, E. Rosiana and RC Saputra. (2020). *"Design of low cost pulse oximetry based on Raspberry Pi"*. J. Phys.: Conf.Ser.1501012003.

2. Marija Gorensek and Franci Sluga. (June, 2014) "Modifying the UV blocking effects of polyester fabrics". *Textile Research Journal*, https://doi.org/10.1177/004051750407400601.

3. Yun J, and Sang-Shin Lee. 2014. *"Human Movement Detection and Identification Using Pyroelectric Infrared Sensors" Sensors 14*, no. 5: 8057-8081. https://doi.org/10.3390/s140508057

4. P. Zappi, E. Farella and L. Benini, "Tracking Motion Direction and Distance With Pyroelectric IR Sensors," in *IEEE Sensors Journal, vol. 10, no. 9, pp. 1486-1494*, Sept. 2010, doi: 10.1109/JSEN.2009.2039792.

5. Jian-Shuen Fang, Qi Hao, David J. Brady, Mohan Shankar, Bob D. Guenther, Nikos P. Pitsianis, and Ken Y. Hsu, "Path-dependent human identification using a pyroelectric infrared sensor and Fresnel lens arrays," *Opt. Express 14, 609-624* (2006)

6. Mohan Shankar, John B. Burchett, Qi Hao, Bob D. Guenther, and David J. Brady "Human-tracking systems using pyroelectric infrared detectors," *Optical Engineering 45(10), 106401* (1 October 2006). https://doi.org/10.1117/1.2360948

7. Ronnie O. Serfa Juan, Jin Su Kim, Yui Hwan Sa, Hi Seok Kim and Hyeong Woo Cha. "Development of a sensor module for standing and moving human body using a shutter and PIR Sensor".*International Journal of Multimedia and Ubiquitous Engineering Vol.11,No.7(2016)*,pp.47-56. http://dx.doi.org/10.14257/ijmue.2016.11.7.05

8. E. Susana and H. Tjahjadi, "Handheld pulse oximeter based on single board computer raspberry Pi B +" 2017 *15th International Conference on Quality in Research (QiR) : International Symposium on Electrical and Computer Engineering*, 2017, pp.141-145, doi:10.1109/QIR.2017.8168470.

Food Reviews Classification using multi-label convolutional neural network text classifier

Krutuja S Lasne[1], *Sejal S Nandrekar*[2], *Ashraf A Khan*[3], *Tushar Ghorpade*[4]
Department of Computer Engineering
Ramrao Adik Institute of Technology
Nerul, Navi Mumbai, India.
[1]krutuja.lasne999@gmail.com, [2]sejalnandrekar@gmail.com, [3]aashrafakhan@gmail.com, [4]tushar.ghorpade@rait.ac.in

Abstract—Most commercial websites, such as Amazon, encourage users to leave reviews of the goods and services they get after purchasing them. For certain consumers, this analysis is critical when determining whether or not to buy a product. Understanding the consequences of feedback and correctly classifying their utility may therefore be an advantageous method for websites. The classification results can also be used as a review and recommendation program for ongoing success. Nowadays people visit several restaurants on various occasions. They get confused most of the time after having a look at the food menu. Based on the ratings and reviews of the dish it becomes easier for them to decide the dish they wish to order. However they unable to read each review of the previous customers. So to overcome this issue, we have proposed NLP (Natural Language Processing) technique and Spacy CNN (Convolutional Neural Network) pipeline system which will classify all the reviews in a single rating. Each review is labelled with a reviewer's score indicating the sentiment of the reviewers. Our task is to predict a reviewer's score on a scale of 0 or 1. Where 1 indicates the users like the dish while 0 indicates that the reviewers were not satisfied with the dish.
Index Terms—Summarization, Sentiment, Review Recommendation, Machine Learning, CNN model.

1. Introduction

People visit several restaurants on various occasions. They get confused most of the time after having a look at the food menu. Based on the ratings and reviews of the dish it becomes easier for them to decide the dish they wish to order. But it is impossible for them to read each and every review of the previous customers [6],[9]. This proposal aims to create an automatic text-based classification model that can forecast the usefulness of Zomato, Swiggy and Reddit feedback accurately. The challenge at hand is to perform a binary classification using a mixture of text-based functionality and machine learning algorithms. The binary classes will be described as "like" and "dislike," with "1" denoting "like" and "0" denoting "dislike". Text encoding, tokenization, translating uppercase to lowercase, lemmatization, and other text-based features were used to create this model [8].

Further paper breakdowns are as follows, section 2 describes the Literature Survey, section 3 describes the problem definition, section 4 describes the Proposed System and evaluation parameters for implementation, section 5 concludes the paper along with section 6 which shows the future work.

2. Literature Survey

This section describes the methodology adopted for the literature review. This paper represents an exploration of the contributions that have already been made in the academic field.

In this paper author Oman Somantri et. al. [1] proposed a hybrid method called PSO-IG means Particle Swarm Optimization and Information Gain. Author have used four different algorithm as K-NN, Support Vector Machine, Naive Bayes and Decision Tree. The main purpose of this hybrid model is to classify food reviews using sentiment analysis approach of culinary food which will be very helpful to culinary shopkeepers and seekers.

In this paper author Sindhu B Hegde et. al. [2] suggest a system of recommendations feedback from customers in forecasting star ratings for food classification for restaurant business. Positive, unfavourable, moderate, and divisive user feedback are grouped into four groups. SVM grouping is used to determine training data in accordance with divisions. The proposed framework is used two main tasks four course meal and sentiment food classification.

The author Hemalatha S et. al. [3] have proposed feature based sentiment analysis for restaurants. They have used ambiance, price, service and food quality features for classifying reviews as neutral, negative and positive. As part of the Yelp Database Challenge, highlights the introduction of machine learning algorithms that process textual and mathematical data presented in the Yelp database. The aim was to categorise business feedback so that they could be sorted consistently from negative to positive. Naive Bayes, Multinomial Naive Bayes, Bernoulli Naive Bayes, Logistic Regression, and Linear SVC are the algorithms used (Support Vector Clustering).

A data analysis or statement on the TripAdvisor's web page about products in restaurants in Surabaya is used as a source in this paper [4]. The author Masrur Adnan et. al.used Web Harvy software to gather the results. The NLTK (Natural Language Tool Kit) library was used to process the data in order to promote the Python programming language. Wrapping events, removing marks, token-ups, altering sentences, suspending deletions, and stopping are all moves in the progression process. Using the Decision Tree, it figures out how many times each word appears in the text.

Author Jiayu Wu and Tianshu Ji [5] tested whether recursive neural network is effective with insufficient tree labelling or not. They proposed Recursive Neural Network for multiple sentences, to handle multiple sentences at once. However the accuracy was challenging.

Present Systems are totally random review based if we rely on star based review and does not contribute towards understanding the salient features of the dish. Star based system is totally depend on the mood of the user. If the consumer is angry it will result in decrease of average rating on the food. The consumer way change it not depending upon the environment and may lead to rating the dish thing of other parameters like waiter behaviour, environment of restaurant etc. which affect the average rating of dish.

3. Problem Definition

This proposal aims to create an automatic text-based classification model that can forecast the usefulness of Zomato, Swiggy, and Reddit feedback accurately. The challenge at hand is to perform a binary classification using a mixture of text-based functionality and machine learning algorithms. The binary classes will be described as "like" and "dislike," with "1" denoting "like" and "0" denoting "dislike". Text encoding, tokenization, translating uppercase to lowercase, lemmatization, and other text-based features were used to create this model.

4. Proposed Methodology

A. Proposed Work :

This proposal is aimed at the development of an application system through which the users enter the food review which gets converted into tokens and is passed through all methods which have been implemented in "Fig. 2"

The primary focus of the system is to understand the user vocabulary review and rate the dish automatically without relying on star-based system.

The proposed methodology is quite simple. The food or dish is to be rated and it is important to understand the vocabulary of the review. Most rating systems are inefficient to perform this task. In the new approach, the author is using spacy and food review dataset from Reddit.

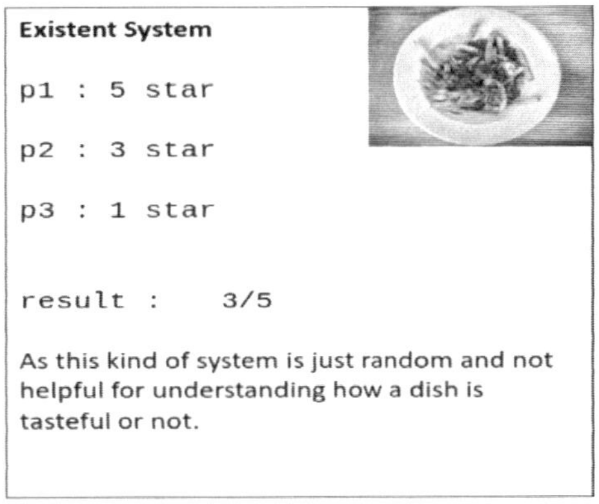

Fig. 1 Existing System

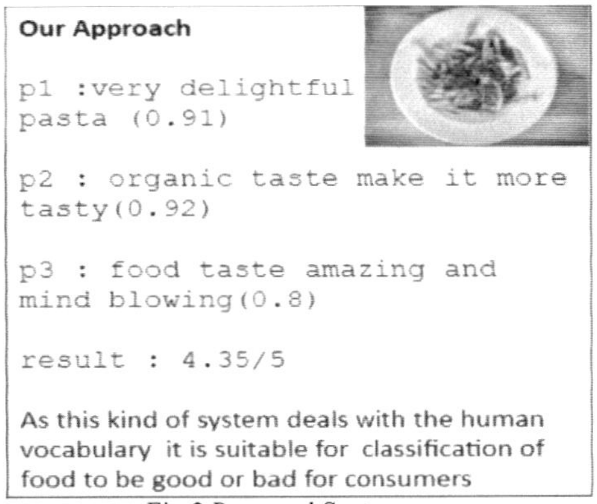

Fig.2 Proposed System

 https://doi.org/10.1051/itmconf/20214001009

In this system, we learned how Spacy works and understood how to use this technology to understand the user review to predict the rating of the dish. As we are using spacy library for advanced NLP we are considering all parameters of the consumer review which may lead to an optimal solution.

B. Implementation Details:

The model is trained on a heavy datasets and spacy library in python is used to develop this system. This model is trained to understand a review from its core like understanding noun-adjective, adverb, verb, etc. So by using this module it will be able to achieve the score while classification of the review.

Data is taken from Kaggle and also some scraped data from an online food review website is used. As this proposal is based on supervised learning the reviews were labelled. By pre-processing the user review and passing the review through spacy filter, tokens will be created to understand grammar, nouns from the sentences.

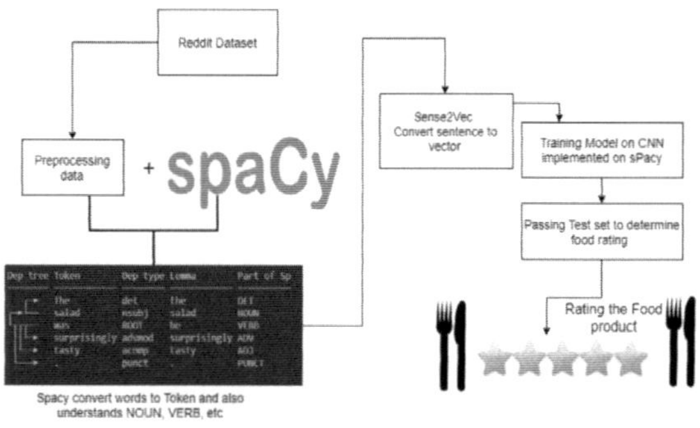

Fig. 3 System Design

As this machine can understand sentences so the tokens that are generated will be converted to vector using sense2vec. This will create vector for of every words and it understand every word and word related to another word.

For Eg: doc = spacy token (“peanut butter”)

[((‘peanut butter’, ‘NOUN’), 1.0), ((‘peanut butter’, ‘NOUN’), 0.9349336624145508), ((‘almond butter’, ‘NOUN’), 0.9219189882278442), ((‘apple slices’, ‘NOUN’), 0.9212321043014526)]

As this all food belong to same category they have same value. The data is trained in the further step. As spacy has it’s own training algorithm called TextCategorizer it helps to train the model and store it in file so that it can be exported and can be stored for future use.

C. Word vectors and similarity:

Sentence to vector is a program that converts words into sentences. The aim is to create a model that is superior to the word to a vector model. The theory behind sense to vector is quite simple. If the problem is that duck in the sense of waterfowl and duck in the sense of crouch are two different concepts, the easy solution is to have just two entries, duck as noun and duck as a verb. Trask et al, (2015) presented a series of tests that demonstrated the concept’s viability. It assigns sections of speech tags to phrases, such as verb, noun, and adjective, which are then used to provide meaning.

- Please [VERB] book my ticket.
- Read the [NOUN] book.

Reddit has a lot of discussions about food, we can acquire strong similarity vectors for food goods.

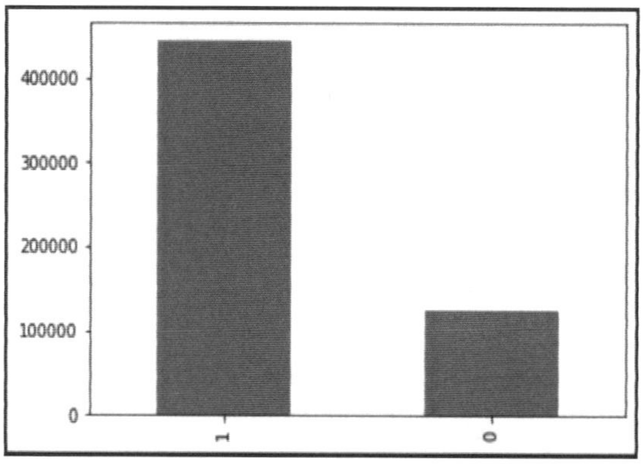

Fig. 4 Histogram plot on data set

```
Loading food reviews data...
Using 30000 examples (24000 training, 6000 evaluation)
```

Fig. 5 Number of data for training and testing

D. Spacy Text Categorizer:

We have trained a multi-label convolutional neural network text classifier on our food evaluations using spacy's newest text categorizer component.

Spacy provides a categorization paradigm that includes a number of labels that are not mutually exclusive. The model setup can be changed, but by default, the Text Categorizer class uses a convolutional neural network to assign position-sensitive vectors to each word in the document. To stop exchanging weights with other pipeline elements, the Text Categorizer uses its own CNN model. The text sensor is summarised after max and mean pooling, and a multilayer perceptron is used to predict an output vector of length no class, before elementwise logistic activation. The value of each output neuron represents the chance of any class being present.

5. Results And Analysis

5.1 Dataset:

Fig. 4 describes graph of fine cuisine reviews from Amazon are included in this collection. All 500,000 reviews up to October 2012 are included in the data, which covers more than a decade. In addition, reviews include product and use statistics, ratings, and a plain- language summary. Ratings from all of Amazon's other categories are also included.

Data includes:

Since October 1999 to October 2012, the reviews are taken. The number '1' is considered as a Positive review and number '0' is considered as a negative review. The total number of reviews are 30000. Out of that 24000 reviews used for training and 6000 reviews used for validation.

5.2 Performance Measure:

As SpaCy is a powerful tool to detect sentiment analysis on the review it will understand all pros and cons of the review depending on the trained dataset. As Food is the primary need of a consumer we should understand what kind of food is most likely to have by a consumer and trending food needs to have a better ranking algorithm. As per the study of existing system from few reputed papers authors have uses different algorithms to do sentiment analysis [7] of review but complex algorithm takes more time to implement and take more time to predict as Spacy has an encorewebsm library spacy gets a plus to understand noun,pronoun,adjective etc which is very help to understand the review and rate the review.

LOSS	P	R	F
327.039	0.836	0.851	0.843
200.148	0.852	0.864	0.858
165.254	0.855	0.869	0.862
138.109	0.858	0.867	0.862
123.523	0.866	0.864	0.865
112.838	0.865	0.875	0.870
103.065	0.866	0.872	0.869
94.570	0.870	0.865	0.867
89.310	0.869	0.865	0.867
82.493	0.864	0.870	0.867

Fig. 6. Training data for 10 epochs

The following table 1 and Fig. 7 shows the comparison of different models with evaluation parameters are precision, recall and f-score.

Table 1: Comparison between various models.

Evaluation Parameters	Proposed Model using Spacy CNN	Naive Bayes	RNNMS	Decision Tree
Precision	86.40 %	69.00 %	75.00 %	48.00 %
Recall	87.00 %	69.00 %	75.00 %	36.80 %
F-score	86.70 %	69.00 %	74.00 %	41.40 %

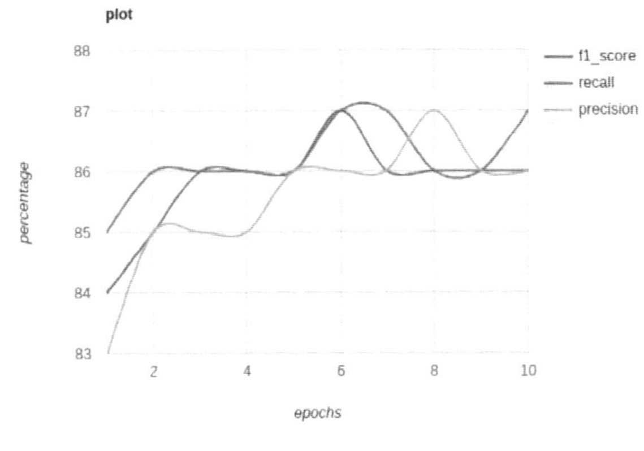

Fig. 7. Graph plot

The above graph plot is for Recall, Precision and F-score. It shows percentage for 10 epochs.

6. Conclusion

On the basis of the ratings and reviews of the dish, it becomes easier for customers to decide the dish they wish to order. This kind of system will help many online restaurant/online food markets to create a satisfying review process for the users and with this kind of mechanism the Hotel/food industries can create a food ranking page automatically. By reviews, the online food company will understand what user likes and what user doesn't like independent on where the users are located. We also can create a hybrid model where we will consider both the rating system and review rating system in the future. In Conclusion, the rating system don't work independently to provide how the product is useful to the consumer but with the help of review classification, we can generate rating with the help of users review.

References

1. O. Somantri, D. A. Kurnia, D. Sudrajat, N. Rahaningsih, O. Nurdiawan and L. Perdana Wanti, "A Hybrid Method Based on Particle Swarm Optimization for Restaurant Culinary Food Reviews," Fourth Inter-national Conference on Informatics and Computing (ICIC), Semarang, Indonesia, 2019, pp. 1-5, doi: 10.1109/ICIC47613.2019.8985842.(2019)

2. S. B. Hegde, S. Satyappanavar and S. Setty, "Sentiment based Food Classsification for Restaurant Business," International Conference on Advances in Computing, Communications and Informatics (ICACCI), Bangalore, 2018, pp. 1455-1462, doi: 10.1109/ICACCI.2018.8554794.(2018)

3. H. S. and R. Ramathmika, "Sentiment Analysis of Yelp Reviews by Machine Learning," International Conference on Intelligent Computing and Control Systems (ICCS), Madurai, India, 2019, pp. 700-704, doi: 10.1109/ICCS45141.2019.9065812.(2019)

4. M. Adnan, R. Sarno and K. R. Sungkono, "Sentiment Analysis of Restaurant Review with Classification Approach in the Decision Tree-J48 Algorithm," International Seminar on Application for Tech-nology of Information and Communication (iSemantic), Semarang, Indonesia, 2019, pp. 121-126, doi: 10.1109/ISEMANTIC.2019.8884282.(2019)

5. Jiayu Wu, Tianshu Ji "Deep Learning for Amazon Food Review Sentiment Analysis"

6. Yinyang Sasikala P, L.Mary Immaculate Sheela, "Sentiment Analysis of Online Food Reviews using Customer Ratings," Volume 119 No. 15 2018, 3509-3514 ISSN: 1314-3395 (on-line version)(2018)

7. Feiran Y. M. Aye and S. S. Aung, "Enhanced Sentiment Classifica-tion for Informal Myanmar Text of Restaurant Reviews," IEEE 16th International Conference on Software Engineering Re-search, Management and Applications (SERA), 2018, pp. 31-36, doi: 10.1109/SERA.2018.8477231.(2018)

8. K. Zahoor, N. Z. Bawany and S. Hamid, "Sentiment Analysis and Classification of Restaurant Reviews using Machine Learning", 21st International Arab Conference on Information Technology (ACIT), 2020, pp. 1-6, doi: 10.1109/ACIT50332.2020.9300098.(2020)

9. S. M. Asiful Huda, M. M. Shoikot, M. A. Hossain and I. J. Ila, "An Effective Machine Learning Approach for Sentiment Analysis on Popular Restaurant Reviews in Bangladesh,"1st International Conference on Artificial Intelligence and Data Sciences (AiDAS), 2019, pp. 170-173, doi: 10.1109/AiDAS47888.2019.8970976.(2019)

10. Tsung- A. Bhatt, A. Patel, H. Chheda, K. Gawande, — "Amazon Review Classification and Sentiment Analysis" ,IJCSIT, Vol. 6 (6) , 5107-5110(2015)

EEG Signal Enhancement Using OWA Filter

Soham Yadav[1], *Jeevika* Pawar[1], *Girish* Patil[1], and *Shivangi* Agarwal[1]

[1]Department of Electronics Engineering, Ramrao Adik Institute of Technology, Nerul, Navi Mumbai, 400706, India

Abstract. Biomedical signal monitoring and recording are an integral part of medical diagnosis and treatment control mechanisms. For this, enhanced signals with appropriate peak preservation are required. The OWA (Ordered Weighted Aggregation) Filter used in this paper helps in non-linear signal filtering and preservation of peaks for accurate medical diagnosis. Weights are an important aspect of the OWA filter, the Gaussian method and the KDE (Kernel Density Estimation) function are used to obtain a precise output which helps in filtering the signal. This filter is further compared with another non-linear filter that is the median filter to understand the compatibility and the preciseness of the filter in a much deeper sense.

OWA | filter | peak | kernel density estimation | probability density | EPD (Estimated Probability Density)

1 Introduction

When it comes to signal filtering, noise reduction is a crucial aspect. Filtering biomedical signals should be done with extreme caution. The need for effective noise reduction must be balanced against the signal's need for as little filtration as possible. Filtering wave-forms like the electroencephalogram signal (EEG) or the electrocardiograph signals (ECG) should be given special attention. In this paper we consider the EEG signals for signal filtering and enhancement. This paper reflects on the refinement of signals. Noise filtering, smoothing, and other signal improvement methods are also used to improve the analytical signals' signal-to-noise ratio. In order to remove unwanted noise from the signal, a careful selection of the experimental conditions and a detailed review of the sources of noise should be developed as the first approach.[1] [2]

There are times where for instance, smoothing of the analytical signal is needed to evaluate or validate the presence of a peak in a noisy baseline. Any peak narrowing or deconvolution techniques often involve simultaneous filtering of the noise. Filtering can aid in the visual identification of the signal's significant characteristics. For such signal filtering and enhancement, we use a novel nonlinear filter called the OWA filter. OWA filter is better known for its precise measurements and preservation of peak signals, through this paper we venture the field of signal enhancement using the OWA filter.[3] [4] [5]

2 Literature Survey

OWA operators were first introduced by Ronald R. Yager in 1988[6] and have been subjected to numerous research studies later on. Since its appearance, it has been used in a wide variety of different applications that include decision making, neural networks, database systems, fuzzy logic controllers, expert systems, learning systems, market research, linguistic quantified propositions and mathematical programming [7]. It is a parameterised class of mean type aggregation operators [9]. If the weights in OWA operator are determined, then special aggregation operators can be obtained, such as arithmetic average, max and min operators, etc. An application of OWA operator to suppress noise in ECG signal processing is presented in Pander (2017)[9]. Along with this, a new approach to robust filtering with OWA aggregation operator is proposed. The process of using OWA operators in signal filtering has come to light.

3 OWA Filter

Filters are used to remove unwanted frequencies from the signal. But the normal subtraction of the noise or the unwanted frequencies from the original signal might lead to the loss of some important data of the signal required. This loss can compromise the accuracy of the signal, which may affect the results. OWA (Ordered weighted Aggregation) filter is used to overcome this data loss, by processing the weighted average of the ordered data from the signal in a moving window. The process of OWA filter is as follows, the 1 st step is to "rearrange the signal data" in the descending order, step 2 is finding the weights of the signal data (Eq 5) and the step 3 is to find the mean type aggregation by calculating the weights and the data points [6].

$$Yi = \sum_{i=-N}^{N} WiXi \tag{1}$$

where. Xi denotes the discrete input signal and Yi denotes the filtered output, Wi denotes the weights calculated, and i=(1.....M).[7] [9] The window length "M" will be asked from the user and the window will iterate over the set of data points in the signal.

$$OWA(Y1......Yn) = \sum_{i=-N}^{N} WiXi \quad (2)$$

The "M" denotes the window length where M = (2N+1) and i is the range of the elements in a window, where i = (1......M). The operation of OWA is parameterised by the choice of weights. Assuming, if the weight wi = 1/M, where i = (1....M), [7] [9]

$$OWA(Y1......Yn) = \frac{1}{M}\sum_{i=-N}^{N} Xi \quad (3)$$

It turns into a simple averaging filter. The main operation of the OWA depends on the aggregation of the signal based on the weights of its data points. If the summation of all the weights of the signal elements is;

$$\sum_{i=-N}^{N} Wi = 1 \quad (4)$$

i=(1...M),for the appropriate working of the OWA filter.

3.1 Flowchart

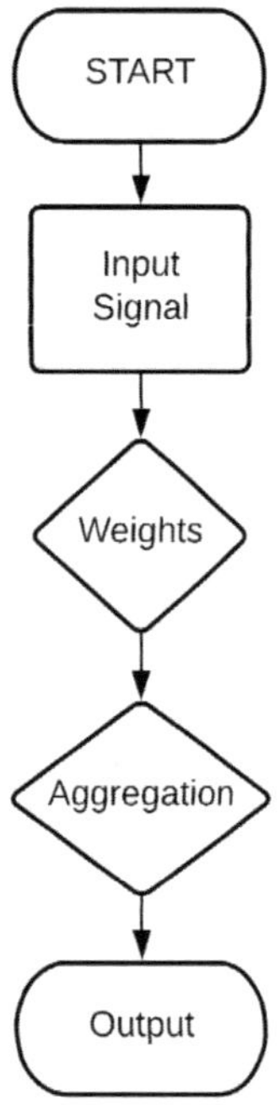

Figure 1. Flowchart of OWA Filter Process

- Any signal can be used as an input, here we have used an EEG signal as an input.
- Finding weights is the most important part, weights are found based on the importance or the value of the data points.
- Then the data points are aggregated based on their weights and output is generated.

4 Weights

All the data points are allotted with weights which gives the summation as one. There are many ways for finding weights of the data points, one such method is:

- Kernel Density Estimation:
  Kernel Density Estimation (KDE), is introduced to estimate the probability density function (Eq 6,7). It effectively captures the complex data distribution characteristic of input values without using complicated classification methods .

$$Wj = \frac{P(Vj)}{\sum_{l=1}^{n} P(Vl)} \quad (5)$$

5 Kernel Density Estimation

Determining weights of the input data is one of the most important part of the OWA filter. The process of finding weights is very confusing and one of the most searched topics. Many different processes have been suggested among which Kernel Density gives a precise value for determination of weights. In other processes, the weights are found by using common ways, neglecting the complex data and without considering their distribution characteristics, kernel density over comes these draw backs. Since there is no practical demonstration of use of KDE for biomedical signal processing, this paper demonstrates how KDE can be Used practically.

EPD function is an essential part of KDE. This mathematical tool can be used to find the probability density which helps to describe the distribution of input values.

$$P(v) = \frac{1}{nh}\sum_{i=1}^{n} K(\frac{v - vi}{h}) \quad (6)$$

where p(v) denotes the EPD function of random variable(v), n is the input value, h is the smoothing parameter also known as bandwidth, "K" is the kernel [9]. The accuracy of EPD depends on the smoothing factor and the kernel. Using the EPD and kernel we can find weights of the input signal.

$$Wj = \frac{P(Vj)}{\sum_{l=1}^{n} P(Vl)} = \frac{\frac{1}{nh}\sum_{i=1}^{n} K(\frac{v-vi}{h})}{\sum_{i=1}^{n}(\frac{1}{nh}\sum_{i=1}^{n} K(\frac{v-vi}{h}))} \quad (7)$$

where j denotes the maximum element of an array of data, Wj is the maximum weight where j=(1....M), similarly Vj is the maximum input data and p denotes the probability density. l = (1...n) and Vl is (V1......Vn). The summation of all the weights; i.e. $\sum_{i=-N}^{N} Wi = 1$.

6 SNR and FFT

6.1 SNR

SNR calculations may either be easy or complicated, and it depends on the devices in question and obtainable knowledge. If the SNR measurements are in dB, then you can deduct the noise amount from the specified signal: SNR = S - N. The distinction within the numbers equals the SNR. as an example, considering a radio wave with a strength of -10 sound unit and a noise signal of -50 sound unit. Therefore, -10 - (-50) = 40 dB.

As shown explicitly earlier, shrewd SNR may be concerned, as well. So, for complicated calculations, divide the worth of the specified signal by the quantity of the noise, taking logarithm of the result into consideration, i.e., log (S / N). After this, if the signal strength measurements are in watts (power), then multiply by twenty. However, if the units are in voltage, then may be multiplied by ten. Furthermore, for power, SNR = 20log(S / N) and for voltage, SNR = 10log(S / N). Also, the ensuing calculation is that the SNR is in dB. as an example, measured noise worth (N) is 2V, and signal (S) is 300mV. The SNR is 10log(0.3 / 0.000002) or around 62 dB.

6.2 FFT

FFT of Fourier analysis converts a signal from original domain of that signal into frequency domain or it converts the frequency domain into time domain.

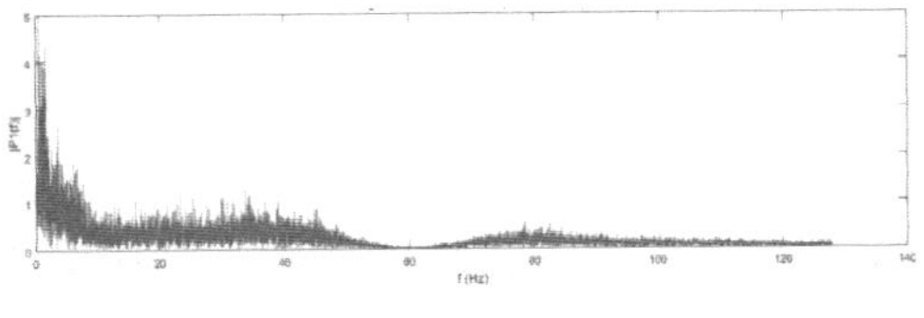

Figure 2. FFT of Input Signal

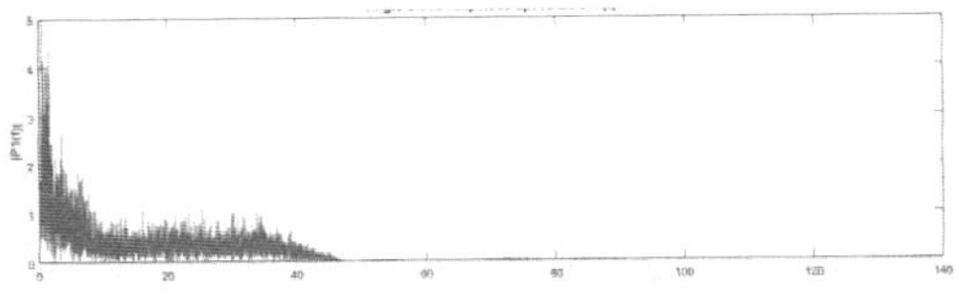

Figure 3. FFT of Output Signal

7 Median Filter

It is a non-linear filter and is simple in comparison to other non-linear filters and hence widely used. The process of median filter is as follows, Step 1: is to arrange the number in numerical order, in step 2: the input data is counted and in step 3 in case 1 if the count is odd, the input data is divided by 2, where as in case 2 if the count is even then the average of the input data and the next higher data is calculated.

$$Case1 : \frac{Xi}{2} = Yi(absolutevalue) \tag{8}$$

$$Case2 : \frac{Xi + (Xi + 1)}{2} = Yi \tag{9}$$

where Xi is the data points and Yi is the filtered output, i=(1....n)

8 Software

MATLAB (an abbreviation of "matrix laboratory") is a multi-paradigm programming language and a numeric computing environment developed by MathWorks. MATLAB allows matrix manipulations, plotting of functions and data, implementation of algorithms, creating user interfaces, and interfacing with programs written in other languages. It is used to write special-purpose language that can be used to write moderate-size programs that can solve problems involving the manipulation of numbers. MATLAB is used in a wide variety of domains from the natural sciences, to engineering, to finance, and beyond, and it is heavily used in industry. In this paper MATLAB has been used for data implementation and simulation of the OWA filter. An OWA program was created using the general OWA formulas and implementation of the KDE to find out weights which were then used in filtering of the signal, finding the SNR and also for Peak Preservation. The use of MATLAB makes it easy to generate complex outputs, compare multiple filters, their features, etc.

9 Comparison

Sr.No.	Parameters	OWA Filter	Median Filter
1	SNR	25.586db	22.213db
2	Peaks	0.05960139583	0.4412551886

Table 1. Comparison of OWA and Median Filters

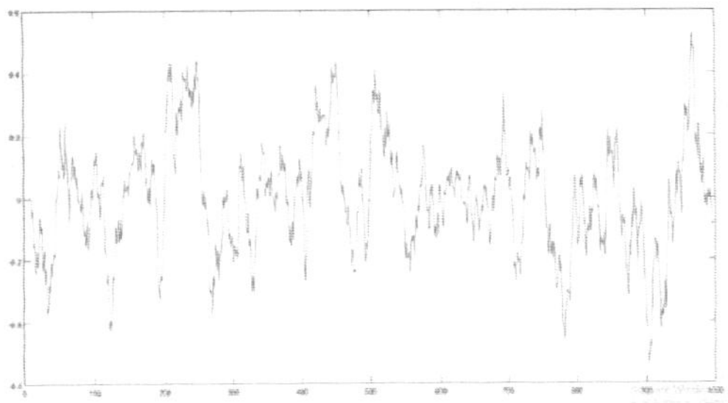

Figure 4. OWA Filtered Output

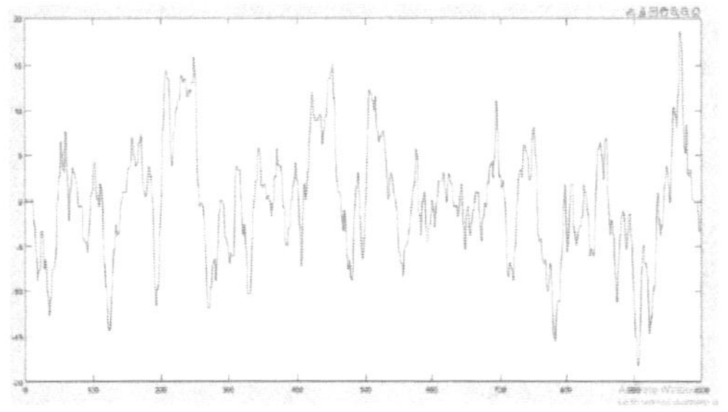

Figure 5. Median Filtered Output

10 Results and Analysis

From the signal to noise ratios of different EEG signals we can figure out that the OWA filter is capable of filtering signal while maintaining peaks without losing any data.

Sr no.	Chb Types	OWA SNR	Median SNR
1	Chb0101	24.628 db	22.765 db
2	Chb0102	23.071 db	18.453 db
3	Chb0103	19.641 db	17.098 db
4	Chb0201	24.367 db	21.876 db
5	Chb0202	21.351 db	21.021 db
6	Chb0203	22.439 db	20.854 db
7	Chb0301	16.432 db	15.583 db
8	Chb0302	18.986 db	18.637 db
9	Chb0303	19.942 db	15.734 db
10	Chb0401	21.895 db	17.872 db
11	Chb0402	18.678 db	14.093 db
12	Cbh0403	26.554 db	24.476 db
13	Chb0501	19.734 db	15.089 db
14	Chb0502	17.318 db	16.347 db
15	Chb0503	17.648 db	14.987 db

Table 2. Signal to Noise Ratio

Sr no.	Chb Types	OWA Peaks	Median Peaks
1	Chb0101	0.0615	0.0453
2	Chb0102	0.0642	0.0519
3	Chb0103	0.0665	0.0521
4	Chb0201	0.0672	0.0485
5	Chb0202	0.0691	0.0213
6	Chb0203	0.0750	0.0691
7	chb0301	0.0623	0.0563
8	Chb0302	0.0894	0.0474
9	Chb0303	0.0973	0.0703
10	Chb0401	0.0566	0.0274
11	Chb0402	0.0611	0.0469
12	Cbh0403	0.0644	0.0295
13	Chb0501	0.0776	0.0603
14	Chb0502	0.0631	0.0576
15	Chb0503	0.0576	0.0289

Table 3. Peaks Preservation of OWA Filters

The design and the algorithm used for the designing of the OWA filter specifies that the peak preservation of this filter is the most eminent and distinguished feature. The peak preservation is expected to preserve the important data which denotes the presence of SSPE (Subacute Sclerosing Panencephalitis). The presence of SSPE is only and only detected because of its peaks and specific time intervals and OWA helps in capturing the peaks efficiently. In preserving the peaks, even a minor variance in the output can turn fatal at times. For such instances, OWA can be put to use as it has the highest peak preservation. The output of this filter was studied by extracting the peak data points from the input as well from the output and were cross verified using a MATLAB function "Goodness of fit". The output of this function is in the table given below.

https://doi.org/10.1051/itmconf/20214001010

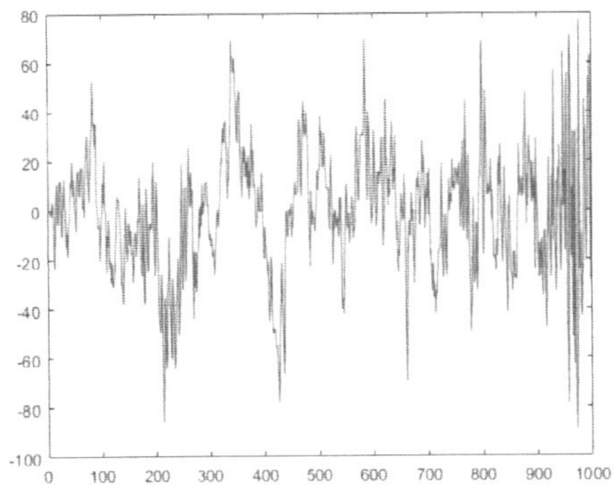

Figure 6. chb01-01 OWA Output

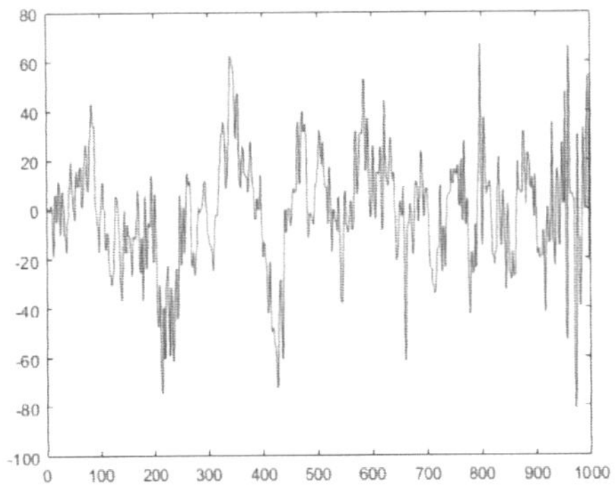

Figure 7. chb01-01 Median Filter Output

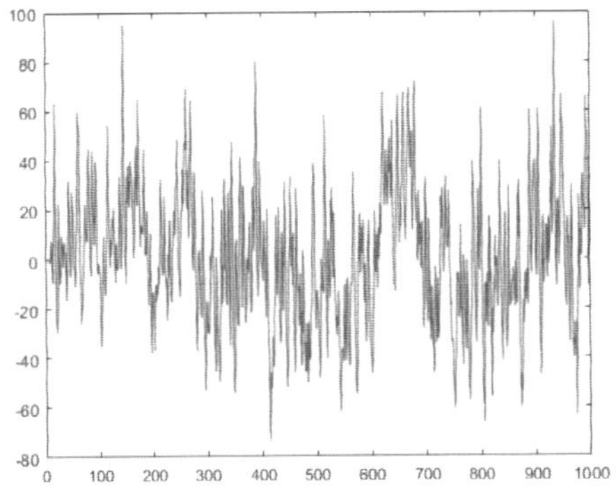

Figure 8. chb02-02 OWA Output

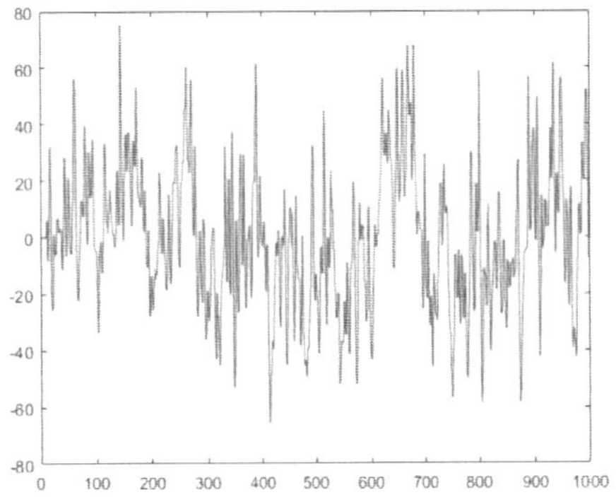

Figure 9. chb02-02 Median Filter Output

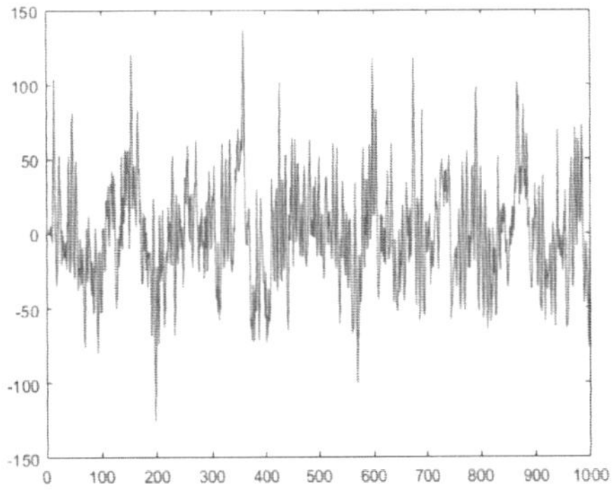

Figure 10. chb02-03 OWA Output

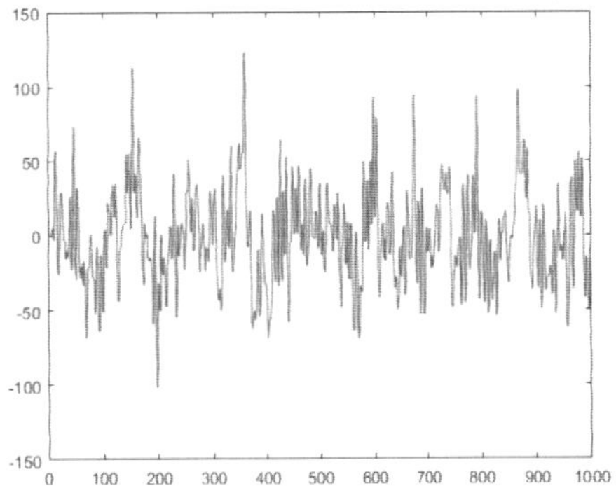

Figure 11. chb02-03 Median Filter Output

https://doi.org/10.1051/itmconf/20214001010

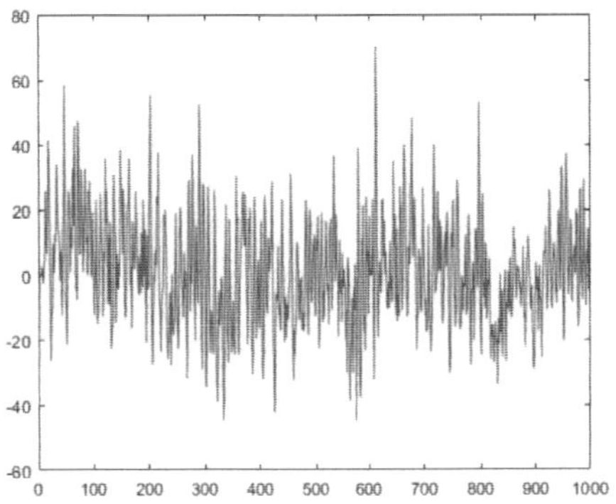

Figure 12. chb03-03 OWA Output

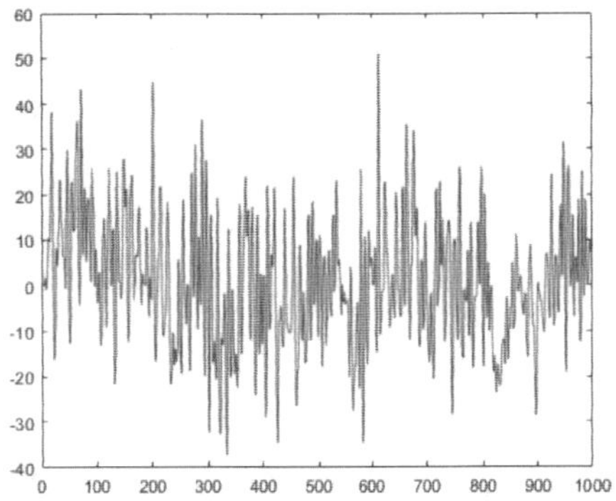

Figure 13. chb03-03 Median Filter Output

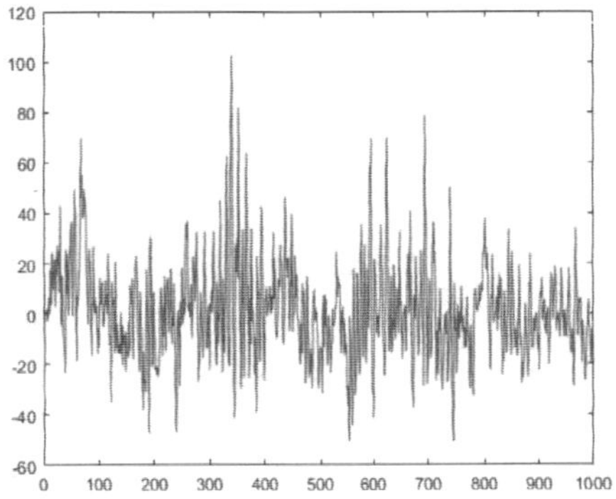

Figure 14. chb04-01 OWA Output

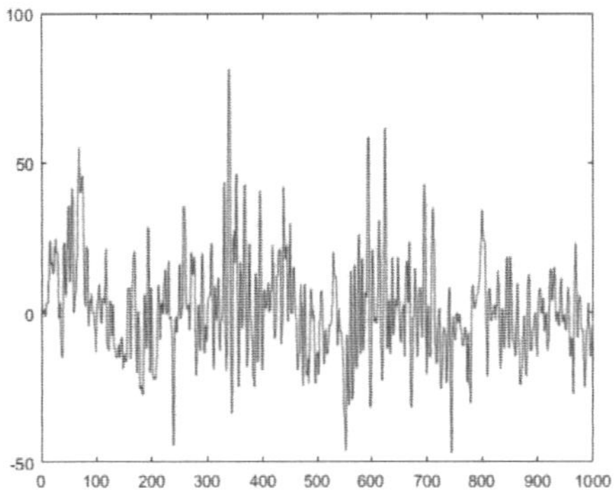

Figure 15. chb04-01 Median Filter Output

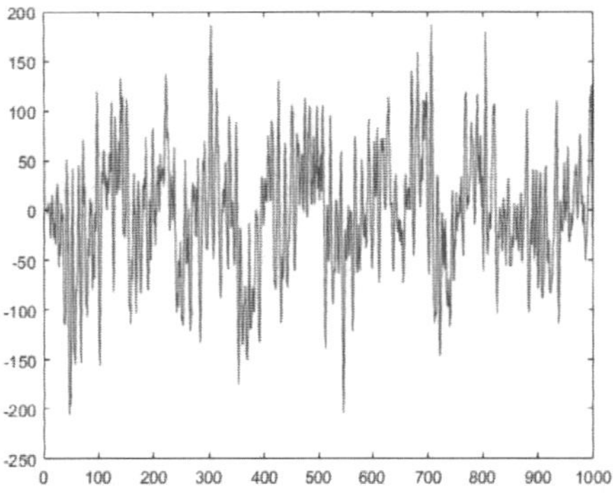

Figure 16. chb05-02 OWA Output

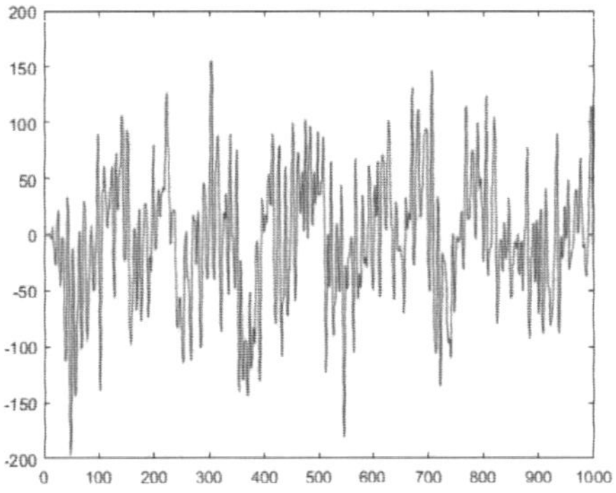

Figure 17. chb05-02 Median Filter Output

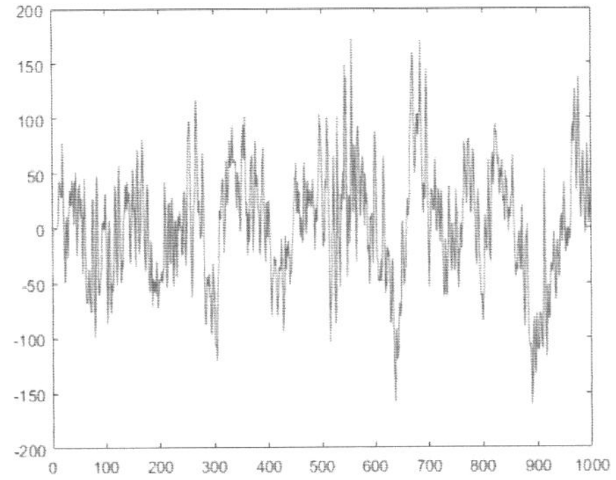

Figure 18. chb05-03 OWA Output

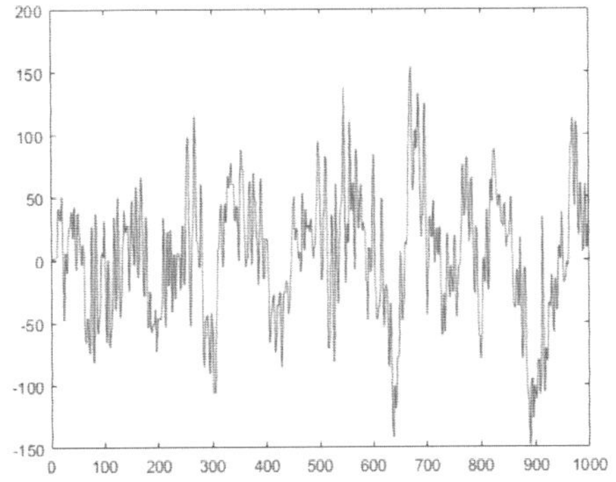

Figure 19. chb05-03 Median Filter Output

10.1 Result and discussion

This paper lights a torch on differentiating between the SSPE signals and normal epilepsy signals. The OWA filter is considered more accurate for biomedical signal processes. The signal is applied as an input to the filter and the weights are assigned to the data based on their value. The data is then aggregated on the basis of their weights and an output is obtained. While considering the window length for the signal, the window should be small and odd in number to increase the accuracy of the output filter. The appropriate frequency for this given filter is 60Hz and the sampling frequency is 260. The SNR is obtained by constructing a formula which provides the ratio of signal to noise.

11 Conclusion

EEG signals should be filtered in real time without losing their original form. To ensure this, OWA filter was used and implemented and compared with a median filter to obtain the output signal as close to its original form as possible. It is shown here OWA is better non-linear filter compared to other non-linear filters by comparing their SNR values, peak preservation and other aspect of the signal.

References

[1] Schaefer, P.A. and Mitchell, H.B., 1999. A generalized OWA operator. *International Journal of Intelligent Systems*, 14(2), pp.123-143.

[2] Mianabadi, H., Sheikhmohammady, M., Mostert, E. and Van de Giesen, N., 2014. Application of the ordered weighted averaging (OWA) method to the Caspian Sea conflict. *Stochastic environmental research and risk assessment*, 28(6), pp.1359-1372

[3] Torra, V., 1997. The weighted OWA operator. *International Journal of Intelligent Systems*, 12(2), pp.153-166.

[4] Xu, Z., 2005. An overview of methods for determining OWA weights. *International journal of intelligent systems*, 20(8), pp.843-865.

[5] Lin, M., Xu, W., Lin, Z. and Chen, R., 2020. Determine OWA operator weights using kernel density estimation. *Economic Research-Ekonomska Istraživanja*, 33(1), pp.1441-1464.

[6] Yager, R.R., 1988. On ordered weighted averaging aggregation operators in multicriteria decisionmaking. *IEEE Transactions on systems, Man, and Cybernetics*, 18(1), pp.183-190.

[7] Pander, T., 2019. EEG signal improvement with cascaded filter based on OWA operator. *ignal, Image and Video Processing*, 13(6), pp.1165-1171.

[8] Pander, T. and Przybyla, T., 2018, June. ECG signal enhancement with serial cascade OWA filter. In*2018 25th International Conference" Mixed Design of Integrated Circuits and System"(MIXDES)*, (pp. 366-370). IEEE.

[9] Pander, T. and Wróbel, J., 2017, June. OWA aggregation operator in robust filtering. In *IEEE 2017 MIXDES-24th International Conference" Mixed Design of Integrated Circuits and Systems*, (pp. 532-536). IEEE.

A Hybrid Approach for Landmark Recognition using Deep Local Features and Residual Network-50

Nishant Nimbare, Parth Shah, Shail Shah, Ramchandra Mangrulkar*

Computer Engineering Department, Dwarkadas J Sanghvi College of Engineering, Mumbai, India.

Abstract. As smartphones and mobile data become universal in modern society, the opportunities to interact with the real world would grow tremendously. Latest Technologies such as Oculus Rift and Google Glass attempt to bridge the gap between the virtual and the material. With advancements in computing speed and image recognition, the idea of augmented reality (AR) becomes more tangible. However, the sheer complexity of image processing and feature recognition is an area of concern for AR. A successful AR system must distinguish among many landmarks and identify or classify the existence of new landmarks. AR algorithms naturally lend themselves to using deep learning because of the adaptability required to various factors. This paper aims to develop and refine a deep learning algorithm that can distinguish landmarks from images using a google landmark database of known landmarks. Instance-level recognition is universally used in areas of Landmark recognition and is also the upcoming research area. Instance-level recognition is the brain behind Landmark recognition. As in Landmarks, the goal is to seek an instance of a common group instead of a group, requiring new deep learning techniques. In this paper, three different VGG16, Inceptionv3, and ResNet50 models are trained using the transfer learning technique and a Pure Convolutional Neural Network (CNN) model is also trained from scratch. This paper proposes a modified version of the ResNet50 model to increase the accuracy and performance of the models used. The revised version of Resnet50 contains an additional Deep Local Features (DeLF) processing layer before generating the final output.

Keywords. Landmark Recognition, DeLF, ResNet50, Transfer learning, CNN, Inceptionv3, VGG16, Instance-Level Recognition

1 Introduction

Deep learning approaches have outperformed previous advanced machine learning strategies in many areas, including video, audio, imaging, medical, social, and sensor. Object recognition has gained tremendous interest by engineers and scientists in artificial intelligence and computer vision. Deep learning enables models with multiple computing layers to learn and represent data at multiple levels of abstraction, simulating how the brain perceives and understands multimodal information and indirectly capturing complex constructs of large-scale data. Among various methods developed in object recognition, Convolutional Neural Network (CNN) is of interest for this project. CNNs (Convolutional Neural Networks) are influenced by the structure of the human visual system and hence outperform conventional machine learning approaches in computer vision and pattern recognition. CNNs are used in a variety of applications, including self-driving cars, segmentation, object identification, facial recognition, and many more. [1].

This paper presents a deep learning model based on CNN. A technology that can accurately predict landmark labels directly from image pixels can broadly benefit applications in various areas such as photo management, maps, aviation, and satellite images.

Image processing and classification is a fascinating and interesting problem. This branch of Computer Science is developing exponentially, given the industry's current rise of computational power. As Image Processing has complex matrix-based calculations, the time to train the model depends on processing power. The availability of storage to masses has made image processing fast and inexpensive [2].

Anything from mobility pattern to stop signs in a self-driving car to cancers in a brain cell can be predicted using image recognition. Convolutional Neural Nets (CNNs) are one of the most widely used image recognition models. This paper presents one such Image Processing application of the real-world, Landmark Recognition.People often fly to foreign countries or abstract places, take several photographs, and then discover a month later that they have no idea what they were standing in front of. People would be able to remember exactly what it was that made the picture so memorable if they are able to train models that can identify landmarks. Analysing popular landmarks is also appealing because it is one of the basic building blocks for making AI smarter. Recognizing landmarks in sequences of images is a challenging problem for several

* Corresponding author: shail.shah822@gmail.com

https://doi.org/10.1051/itmconf/20214002001

reasons. To begin with, the appearance of any particular landmark varies significantly from one observation to the next. In addition to variation due to various aspects, illumination change, external clutter, hardware of the device capturing and changing geometry of the imaging devices are other factors affecting the variability of the observed landmarks. Due to given problems, normal object detection is not applied for Landmark Recognition, instance-level learning is used for such types of problems [3].

Instance-level recognition (ILR) is a method in image recognition or computer vision, where the goal is to find a specific instance of an object. For example, instead of labeling an image as "water-fall" it identifies the image as Niagara waterfall. ILR distinguishes the instance of the object class from the object class. There are many problem statements that require to find a specific instance from a collection, like artwork, landmarks, products and can also be useful in visual searching, shopping, copyrighting etc. Instance-level recognition will unravel the true potential of deep learning technologies for semantic image classification/retrieval for eCommerce, travel, media & entertainment, agriculture, etc. [4].

2 Literature Review

Authors Asia Kausar, Mohsin Sharif, Jinhyuck Park, Dong Ryeol Shin, published a paper with title "Pure-CNN: A Framework for Fruit Images Classification" have implemented a system to classify fruits based on PCNN. Unlike a traditional CNN where the last few layers are fully connected layers, they used a Global Average Pooling (GAP) after Convolutional and Pooling layers, thus using a Pure-CNN (PCNN). GAP is used to reduce the number of parameters and prevent the model from overfitting. GAP behaves like a Max pooling layer but performs more reduction compared to Max pooling. Finally, the GAP layer was attached to the last SoftMax layer for prediction. ReLU was used as an activation function, and the whole architecture was 7-layers deep. They were able to achieve 98.8% accuracy. Authors make a case that fully connected (FC) dense layers at the end of a usual CNN model are computationally expensive and tend to overfit quickly. Also, layers of CNN behave as object detectors, but this ability vanishes on using FC layers for classification. There are two suggested solutions for this. First is the use of dropout layers, wherein dropout layers are introduced. In the Dropout concept, some connections between the layers are randomly dropped off during training. The second way to reduced overfitting due to FC layers is to replace them with a Global average pooling (GAP) layer. Paper compares a traditional CNN model, with and without dropout layers, to a pure-CNN (PCNN) model, consisting only of convolution layers and a GAP layer on top. The pure-CNN model was found to better at classifying fruit images. Dataset used was Kaggle's fruit-360 dataset with 81 classes of fruits. The model was trained over 40,000 iterations with a batch of 50 images. Images were RGB 100 x 100. Global Average Pooling (GAP) was used to minimize the number of parameters and protect the model from overfitting. It reduces filter size by merely averaging the entire function map. GAP, although similar to the max-pooling layer, performs a more drastic method of dimension reduction. Here dimension [h, w, d] is reduced to dimension size [1, 1, d].

Authors Andrei Boiarov, Eduard Tyantov, in published paper titled "Large Scale Landmark Recognition via Deep Metric Learning", have stated metric-based learning of the model to train on the dataset of the Landmarks. Since there isn't a lot of data about a certain landmark, the model ought to have a high accuracy and a low false-positive score. In this paper, the authors have used a deep neural network to train the model. This paper proposes an algorithm to be used while training the model. It also mentions the cleaning of the dataset required in order to apply the metric learning. The paper has also provided overall architecture, learning process, and feature extraction by measured approach. The authors used a Convolutional Neural Network approach to train their landmark recognition model. They divided the CNN into three parts – the main network, the embedding layers, and the classification layer. They trained their model to extract as many characteristics as possible using the fine-tuning method, which is widely used in computer vision. As Landmarks and Scenes are similar with respect to the number of features thus, the authors first trained their model on the scene's dataset. The authors also removed the last fully connected layer and instead added a fully connected layer along with batch normalization. It is also a fast method and has been tested with other methods. The method is like the state-of-the-art method but is scalable and is faster in deployment. The concept has been applied and scaled up for use in image recognition on user images.

Authors Ilja Kuzborskij; Fabio Maria Carlucci; Barbara Caputo, in published paper titled "When Naïve Bayes Nearest Neighbors Meet Convolutional Neural Networks", mentions that since it cannot use CNN activations as input, NBNN has lost traction since introduction of CNN, and it cannot be used as the final layer of CNN architecture. NBNN cannot handle huge datasets as it is not scalable. In paper [7], the authors introduce a structure that solves these issues while also reintroducing NBNN into the scene. Firstly, CNN activation is extracted from local patches at multiple scaling levels. Then, "a scalable version of Naïve Bayes Non-Linear Learning (NBNL)" [7] is addressed, which exploits the learning power of local SVM. This paper provides a method to use CNN activation features and NBNN-based classifiers together. The key elements are:

1. Obtaining CNN activations from local patches at different scales

2. a scalable NBNN based algorithm that makes use of the learning power of locally linear SVMs [7].

Authors: Mahbub Hussain, Jordan Bird, and Diego Faria, in their published paper titled "A Study on CNN Transfer Learning for Image Classification" have used Convolutional Neural Network for image classification. Image classification has been a core problem in the computer vision field with a significant variety of applications. Machine Learning has gained a lot of attention when dealing with image classification. This paper has proposed a CNN architecture model, InceptionV3, to find out its accuracy and efficiency when dealing with new datasets via Transfer Learning. This model is trained on human-face, cars, and animals. Higher accuracy was achieved when InceptionV3 was used on the above-classified images.

Authors: Hyeonwoo Noh, Andre Araujo, Jack Sim, Tobias Weyand, Bohyung Han, in their published paper titled "Large-Scale Image Retrieval with Attentive Deep Local Features", have suggested a local attribute descriptor appropriate for large-scale image retrieval. It is based on Convolutional Neural Network (CNN). This system generates dependable confidence scores in order to exclude false positives.

The four main blocks used are: -

I. dense localized feature extraction,

II. selection of keypoint,

III. dimensionality reduction,

IV. indexing and retrieval.[9]

The dense features are extracted by giving the image to Fully Convolutional Network (FCN). Then, the features are localized based on their respective fields, which can be computed considering the configuration of pooling and convolutional layers of FCN. Hence, Local descriptors indirectly learn representations that are more applicable to a topic. It can be deduced from the results that although DELF shows great empirical results improving the accuracy nontrivially when combined with Deep Image Retrieval (DIR), it does not show optimum results on its own. Author mentions that "This indicates that DELF has capability to encode complementary information that is not available in global feature descriptors" [9].

3 Methodology

3.1 Transfer Learning

Transfer learning is a machine learning technique where a model trained on one task is re-purposed for a second related task. It is an optimization that improves performance and allows rapid progress when modeling the second task [10].

Transfer learning is the most used deep learning alternative since deep learning models need significant resources to train as well as huge and difficult datasets in which deep learning models are trained.

Transfer learning is effective when features learned by the model in the first task are generic. This deep learning transfer learning technique is known as an inductive transfer. The selection of possible models for the task, is done on the basis of its previous learning so that the model can fit the new task by making small changes.

Usual Transfer Learning Workflow: -

1. Selecting Base Model: A pre-trained base model is selected from a range of models. Many research institutions publish models on the internet that have been conditioned on vast and difficult datasets and are part of a collection of candidate models from which to select.
2. Reuse Model: The pre-trained model will then be used to construct a model for the next task of concern. Depending on the modeling process, this can include any portions of the model.
3. Tune Model: The base model may not be used for the new task in its raw or original state and thus the model needs to be tuned by adding some layers according to the problem statement.

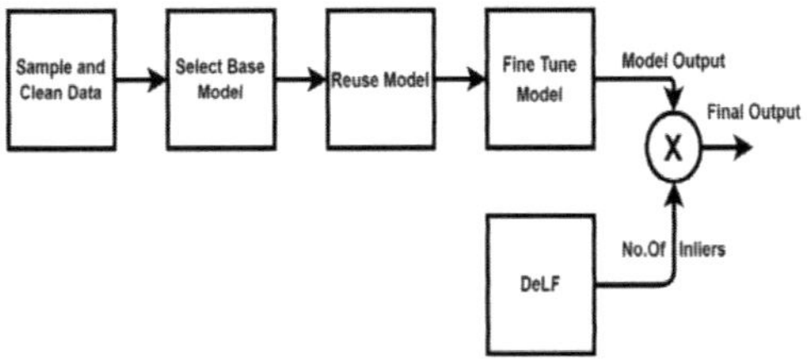

Fig. 1. Workflow

Fig. 1 shows the basic workflow of the system used for the training of the model.

3.2 Optimizing the tuned transfer learning model

After generating the fined tuned model, by transfer learning, of various CNN models (InceptionV3, VGG16, and RESNET50), a maximum accuracy of 90.4 on our "Test" dataset was achieved on the RESNET50 model.

On analyzing the Test dataset, it was found that some structural similarities, color luminance, and shortage of training images were causing errors in the prediction of the model. Thus, to tackle the problems mentioned earlier, DeLF, a Convolutional Neural Network that identifies semantically local features, was used.

The pre-trained DeLF module can be used for image retrieval as a drop-in substitute for other keypoint detectors and descriptors. It describes each noteworthy point in each image with 40-dimensional vectors known as the feature descriptor.

So, to optimize the desired result from the CNN model, a new layer was added in front of our Tuned model, which took the output from both the predicted RESNET values and multiplied that by the top 3 prediction matches with the input image using DeLF.

The maximum multiplied result determined the prediction of the model.

This inclusion of DeLF increased our model's accuracy from 90.4 to 95.6.

3.3 Dataset

This paper uses the Google landmark recognition challenge dataset. The dataset contains around 15k classes and 1.2 M images. Due to limitations of computing resources, a sample of 100 classes containing between 10 and 50 images was taken [11].

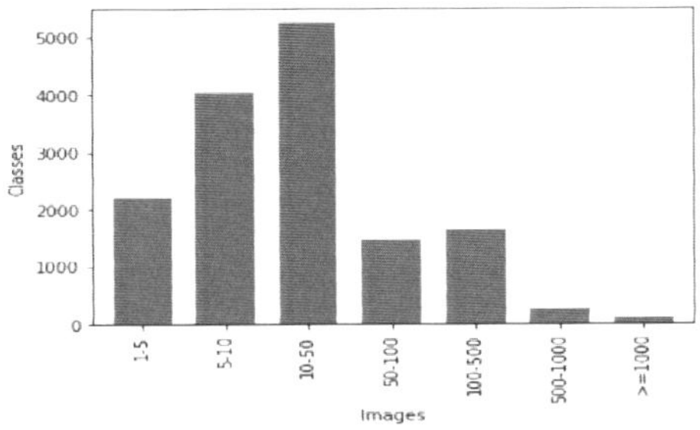

Fig. 2. Distribution of images per class

Fig. 2 shows the distribution of number of images on different class in the dataset.

Total Images (over 100 classes) = 4100
Training split = 75% [3128]
Validation split = 20% [764]
Test split = 5% [208]

3.4 Selecting the base model

The base model was selected from the following three models: VGG 16, InceptionsV3, Resnet 50.

The ImageNet dataset was used to train all the models mentioned above. The ImageNet dataset is a massive archive of images that have been annotated by humans. It was created by academics for computer vision science. The dataset contains slightly more than 14 million images, slightly more than 21 thousand categories or classes (synsets), and slightly more than 1 million images with bounding box annotations. (e.g., boxes around identified objects in the images).

3.5 Architectural Design

For all three of the above models, the following architecture was prepared.

Input Images are 224 x 244 with RGB channels.

Input shape 224 x 224 x 3

Fig. 3 shows the basic architecture of the models used for the paper. The base model is fine-tuned and trained with the dataset to produce accurate results.

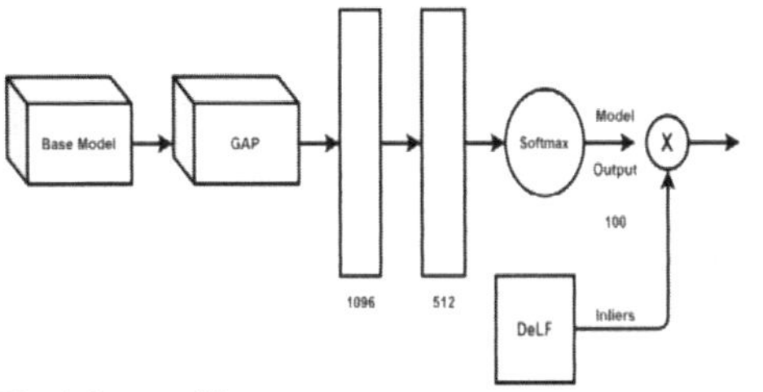

Fig. 3. Base architecture

1. VGG 16:

VGG 16 was proposed by Karen Simonyan and Andrew Zisserman of the Visual Geometry Group Lab of Oxford University in 2014. VGG-16 was one of the best performing architecture in the ILSVRC challenge 2014.

VGG 16 has a 16-layer architecture: 13 convolution layers, two fully connected dense layers, one softmax layer, as shown in Fig. 4.

Number of parameters: ~ 138 million
Output shape: 7 x 7 x 512

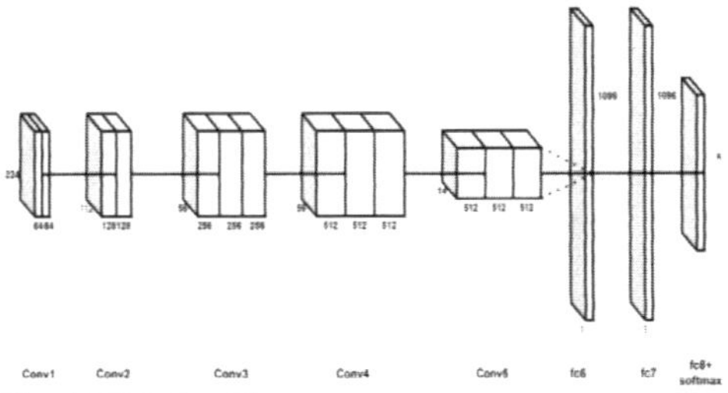

Fig. 4. VGG16 architectures

2. Inception V3:

Inception v3 is a popular image recognition model that has been shown to achieve greater than 78.1 percent accuracy on the ImageNet dataset. The model is the result of several concepts generated over time by various researchers.

Inception-v3 is a 48 layers deep convolutional neural network, as shown in Fig. 5.

Number of parameters: ~ 24 million
Output shape: 5 x 5 x 2048

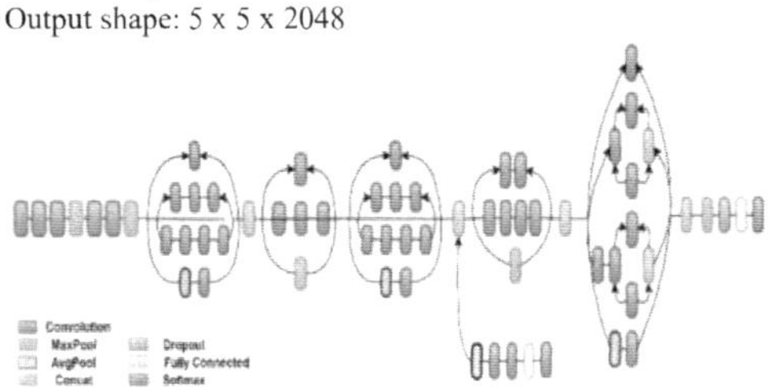

Fig. 5. InceptionV3 architecture

 https://doi.org/10.1051/itmconf/20214002001

3. ResNet50:

ResNet-50 is a smaller version of ResNet 152. Residual Network is a classic neural network used as a backbone for many computer-vision tasks. In 2015, this model was the winner of the ImageNet challenge.

Number of parameters: ~ 22 million
Output shape: 7 x 7 x 2048

Fig. 6 shows the ResNet50 architecture used to train the model.

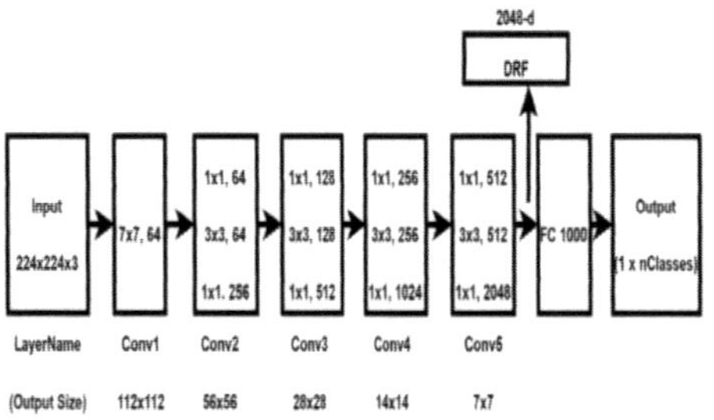

Fig. 6. ResNet50 architecture

4. Pure CNN :

Unlike in traditional Transfer Learning apporaches in Convolution Neural Networks, Pure CNN, as the name suggests, does not contain any Fully Connected layer on the top. On Top of the base model (ResNet 50), the model consisted of another conv2D layer and GAP layer and finally a softmax layer, as shown in Fig. 7.

Total no. of parameters: ~ 23.7 million
Trainable parameters: 105,000

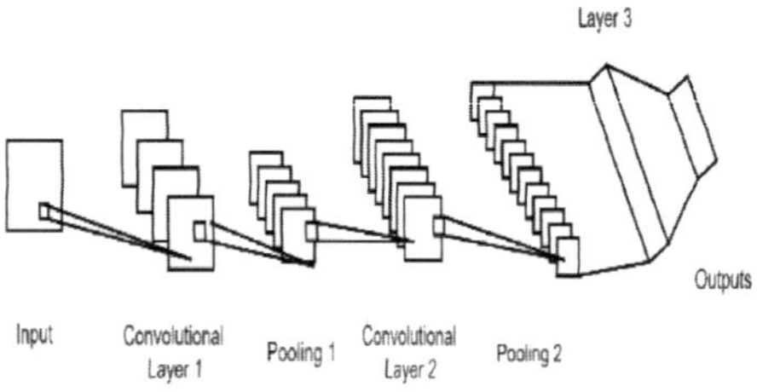

Fig. 7. Pure CNN architecture

5. ResNet50 + DeLF :

This model is an extension of ResNet 50 model used, to increase the accuracy of the earlier trained model. A new DeLF module, produces number of inliers matched between test and an image of classes of three predictions from the ResNet model, as shown in Fig. 8. The number of inliers produced by the DeLF module multiplied by the probability of the class produced by the ResNet model, gives the final output.

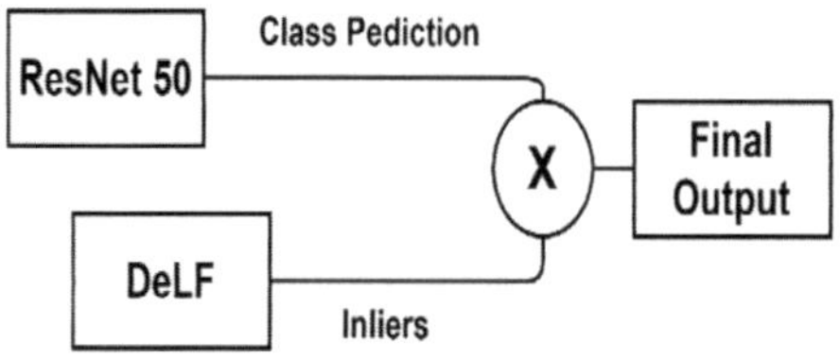

Fig. 8. ResNet50 + DeLF architecture

3.6 Results

Table 1 shows the comparison of various models used in this paper. It is seen from the table that the testing accuracy of most of the models is comparble, but that of ResNet 50 + DeLF is the greatest.

Table 1. Results

Model	Training Accuracy (%)	Testing Accuracy (%)
VGG 16	99.8	89.1
Inception V3	27.3	18.6
ResNet 50	100	90.4
Pure CNN	97.55	89.67
ResNet 50 + DeLF	100	95.67

Fig. 9 shows the comparison of all the models with their training and testing accuracy, on the dataset.

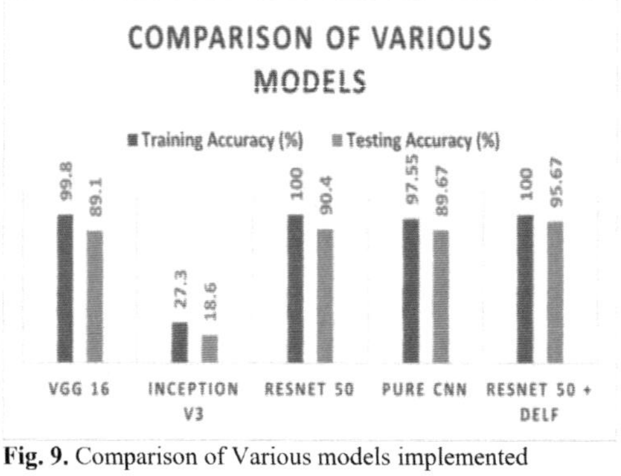

Fig. 9. Comparison of Various models implemented

4 Conclusion

A Landmark Recognition model was implemented, using transfer learning techniques with three different models, namely VGG16, InceptionV3, ResNet50. The accuracy of the models is not high and can be improved using further fine-tuning techniques.

VGG 16 was one of the two pure models which performed well, though slightly behind ResNet it should good empirical results, even with certain manual tests. Overall, it performed extremely well considering it had the shortest training time amongst others. InceptionV3 turned out to be the worst performing base model. It showed a clear case of under-fitting, which was shocking considering it is a complex model. ResNet 50 was by far the most accurate simple model, it showed the best empirical results. Considering this it was the chosen as the base model for the next two models (PCNN and DeLF).Pure CNN model was based on ResNet (base model). Though it initially looked promising, the PCNN model quickly started to show signs of overfitting. It was deduced that the lack of a Fully connected layer was the cause. Combination of ResNet and DeLF outperformed all other approaches discussed above. But addition of the DeLF stage highly increased the recognition processing time, which would be a hindering factor in testing large datasets.

The model was also hosted on a web app made using basic HTML, CSS, and Javascript webpage with python flask framework at server side.

References

1. Athanasios Voulodimos, Nikolaos Doulamis, Anastasios Doulamis, Eftychios Protopapadakis, "Deep Learning for Computer Vision: A Brief Review", *Computational Intelligence and Neuroscience*, vol. **2018**
2. A Data Geek's Guide to Recognize Landmarks https://medium.com/@abhinaya08/google-landmark-recognition-274aab3c71ae
3. Visual Learning for Landmark Recognition http://www.cs.cmu.edu/~takeuchi/iuw97/iuw97.html
4. Instance-level Recognition https://towardsdatascience.com/instance-level-recognition-6afa229e2151
5. A. Kausar, M. Sharif, J. Park, and D. R. Shin, "Pure-CNN: A Framework for Fruit Images Classification," 2018 International Conference on Computational Science and Computational Intelligence (CSCI), Las Vegas, NV, USA, 2018,**pp.404-408.**
6. Andrei Boiarov and Eduard Tyantov. 2019. Large Scale Landmark Recognition via Deep Metric Learning. In Proceedings of the 28th ACM International Conference on Information and Knowledge Management Association for Computing Machinery, New York, NY, USA, 169–178.
7. I. Kuzborskij, F. M. Carlucci and B. Caputo, "When Naïve Bayes Nearest Neighbors Meet Convolutional Neural Networks," 2016 IEEE Conference on Computer Vision and Pattern Recognition (CVPR), Las Vegas, NV, 2016, pp. 2100-2109
8. Mahbub Hussain, Jordan Bird, and Diego Faria, "A Study on CNN Transfer Learning for Image Classification" at School of Engineering Aston University,Birmingham.
9. Hyeonwoo Noh, Andre Araujo, Jack Sim, Tobias Weyand, and Bohyung Han, "Large-Scale Image Retrieval with Attentive Deep Local Features"
10. A Comprehensive Hands-on Guide to Transfer Learning with Real-World Applications in Deep Learning. https://towardsdatascience.com/a-comprehensive-hands-on-guide-to-transfer-learning-with-real-world-applications-in-deep-learning-212bf3b2f27a
11. Google-Landmarks Dataset https://www.kaggle.com/google/google-landmarks-dataset

Assessment of Water Quality using Fuzzy-AHP and TOPSIS

Asmita Sonavane[1, *], *Devyani* Narkhede[2], *Shruti* Pawar[3] and *Tabassum* Maktum[4]
[1,2,3,4]Ramrao Adik Institute of Technology, Nerul, Navi Mumbai, Maharashtra, India.

Abstract. The quality of water available to people has deteriorated because of the vast increase in global industries. The overall quality of water has reduced due to the high use of fertilizers in farms and chemicals in sectors such as mining and construction. The quality of water has a direct impact on the health of humans and hence it is necessary to evaluate the quality of drinking water. The quality of water is dependent on various substances like pH, ammonia, iron, arsenic etc. present in water. If these parameters are available in an appropriate amount in water then only the water is considered as of good quality. The problem of assessing water quality is usually considered as multi-criteria decision problem. In this paper an approach to evaluate quality of water using Fuzzy Analytical Hierarchical Process (Fuzzy-AHP) and Technique for Order Preference by Similarity to Ideal Solution (TOPSIS) is proposed. The parameters which are considered for evaluation include Arsenic, Ammonia , Iron , Nitrate , Dissolved Oxygen and Ph. The relative importance of these parameters is utilized while applying the Fuzzy-AHP process and weights for each parameter are computed. The various datasets containing values for considered parameters are collected and the weights are used to assign quality levels to these water samples. The paper also gives the performance analysis of the proposed method in terms of accuracy. The accuracy is measured in terms of how many water samples are assigned correct quality level and the results show that the proposed method has better accuracy as compared to traditional AHP and Fuzzy- AHP method.

Keywords- *Water Quality, AHP, Fuzzy-AHP, Standard Error Calculation, GWQI, MCDM, TOPSIS, BWM TOPSIS*

1 Introduction

Water covers 70 percent of our planet, and it's simple to assume that it'll continually be quite sufficient. However, freshwater that's used for drinking purposes, bathing, and irrigating farm fields is rare. World's 3 percent of water is fresh whereas the 2/3 percent of it is frozen glaciers and the left part is unprocurable for usage relating to human health, due to lack of sanitation as well as the dearth of potable water. Natural Phenomena like Erosion and other industrial activities like accidents and urban sprawl cause Water Pollution [1]. Whereas, they also embody threats through exposure to pathogens from chemical toxicants as well as organic occurrences. For example, the irrigation by contaminated water or bioaccumulation of ototoxic chemicals by aquatic animals throughout swimming in contaminated water. [2]. Therefore, in Life and Trade the highest priority is the interception of pollution.

A system that would check the accuracy of necessary elements in water is required so water quality is analyzed. The project proposes a technique to identify water quality on basis of various parameters, as, arsenic, ammonia, iron, nitrate, dissolved oxygen and Ph.

Potable water with low levels of arsenic over an extended period is associated with polygenic disorder and raised risk of cancers of organs in excretory and urinary system specifically. It additionally causes skin issues, such as development of corns, lesions and also discoloration of skin [3]. So, the acceptable range of water is below 0.010 ppm [4].

Ammonia is toxic to aquatic animals at concentration less than 1ppm in water as it leads to toxic build up in internal tissues and blood, and potentially death [5]. Human beings are least affected by ammonia in potable water, but if water with more than 1 ppm ammonia is ingested for a longer period then it may damage the internal organ system [6].

Iron is not hazardous to health, but it is considered a secondary or aesthetic contaminant because iron can carry bacteria in drinking water. High concentration of iron in

* Corresponding author: asmitass0403@gmail.com

water gives a yellow, reddish orange, or brownish look. Iron causes your water to taste metallic as well as affect the taste of foods cooked and also blackens vegetables and absorbs the metallic taste [7]. Therefore, the acceptable range for iron in water is less than 0.3 mg/l[8].

The Nitrate cannot be tasted, smelled, or seen in water. Consuming too much nitrate can affect the blood which carries oxygen and can be especially harmful for babies. If there is an excess level of nitrates, plants and algae will grow excessively whereas if we drink water that is high in nitrates it will intervene the ability of RBC's to transport oxygen[9].The acceptable range for nitrate in water is less than or equal to 10 ppm[10].

Dissolved oxygen is the most important parameter of water quality as most aquatic organisms need oxygen to survive and grow. When dissolved oxygen becomes too low aquatic organisms cannot survive. Also, it affects human health [11]. Therefore, the acceptable range required is between 6.5 to 8 mg/l [12].

Ph is important in body to keep balance as well as to synchronize metabolic processes. Here as, a diet with high pH causes weight gain and the immune reactions also become slow whereas an alkaline diet causes an inability to make sync with key nutrients in body. Therefore, water with more free hydrogen ions in water indicates it is acidic, whereas more free hydroxyl ions in water indicates it is basic [13]. pH is measured on a scale of 0 – 14 and the acceptable range is 6.5 to 8.5[14].

In this paper a design and development of a system for monitoring of the water quality is presented. The proposed system will help people in water redundant areas to detect the quality of water. Water resources in remote villages get contaminated due to numerous reasons, thus the proposed system will help to tell if particular water is drinkable or not.

The major contributions of the paper are as follows:

1. To monitor the quality of water on the basis of different parameters such as Arsenic, Ammonia, Iron, Nitrate, Dissolved Oxygen and Ph.
2. To find relative importance among the considered parameters by applying AHP, Fuzzy-AHP and TOPSIS Method.
3. To assign different quality levels to the considered water samples.
4. To compare performance of the proposed method with different datasets of water samples.
5. To achieve the optimal solution for water quality analysis using TOPSIS.

This research paper is organized as follows:

Section 1 is Introduction which gives the techniques and information of the various parameters to identify water quality. Section 2 is Related Work which gives the survey of existing systems to tell how people all over the world are trying to solve the problem of scarcity of clean drinking water by developing various real time water quality monitoring systems. Section 3 demonstrates the Proposed System in which the system design of the proposed method is presented. Finally, the Conclusion and Future Work is expounded in Section 5.

2 Related Work

In An Giang the Northeast and the Southeast regions have bad quality water due to high Arsenic amount in water caused in monsoon season due to sediment deposition. Whereas, it also has bad ground water quality due to the combined effect of human and natural activities [2,15]. In areas, wherever groundwater is extracted for irrigation it had high values of GWQI links with high Arsenic content. Hence, Fuzzy AHP was successfully applied based upon GWQI for assessing ground water quality in An Giang.

The eight wells in An Giang have improved their groundwater quality from 2009 to 2018 within the monsoonal season as collated to other seasons. As this is because during wet season there is a decrease in deposition of small solid particles near the flood plains which lessens the river discharge. Therefore, there is a reduction in the movement of pollutants in water by filling of unused wells for them to be groundwater aquifers. There is no sufficient evidence to show that there is a relation between Arsenic and Sediment Deposition at present.

The Vietnamese Mekong Delta (VMD) of An Giang the secondary canals have a high amount of nutrients and that's why people rely on it as the sources underneath the land function are the most important supplies not only for domestic usages but partially for irrigation as well and it also release the rice intensifications of untreated agricultural efflux. Hence, the Water Quality mostly exceeds the Vietnamese standards for domestic water system.

Water bodies portray an important role in the construction business, day-to-day life, farming irrigation and fire protection in the ecological surroundings. China's rapid development in economy has resulted in increased water pollution which indirectly resulted in forming a compounded water scarcity problem thus becoming an obstacle for the country's growth and modernism. Hence, a case study was conducted on six regions in North China and further the integrated Fuzzy BWM-TOPSIS method was applied.

The unsuitable use Groundwater may be a significant issue as the rapidly increasing population and also the expansion of social economy has incremented the utilization of water resources [15]. Due to which there is a disparity between accessible water and demanded water in few areas. Simultaneously, Groundwater Pollution is also there due to the obvious downturn of groundwater levels, Therefore, the water resources also do not satisfy the basic water-use standards due to which there is also decrement in the ecosystem services.

The Qing 'an Country is being focused for a case study which aspires to progress an improved Fuzzy AHP method to allot water rights to corporations. Also, an evaluation index system has been accepted for the assigning of initial water rights and has also subsumed physical geography, economic, and ecological standard. This system gets categorized into four parts [15]. Here, Fuzzy AHP method helps to determine the order of priority of the evaluation index and the total weight of initial water rights for different corporations. Assigned initial water rights got ranked in the descending order of their total weights coefficient according to the results observed.

3 Proposed Work

In the proposed system the six parameters are considered for evaluating water quality such as, Arsenic, Ammonia, Iron, Nitrate, Dissolved Oxygen, pH. The proposed architecture contains a generation of pairwise comparison matrix, weight calculation, score generation and assigning quality levels. The system involves weight calculation using MCDM methods namely Fuzzy AHP and AHP. Then the TOPSIS algorithm is applied to assign quality levels such as excellent, good, bad or unsuitable.

In 1981, Hwang and Yoon firstly proposed TOPSIS whereas in 1992, Chen and Hwang developed it.The TOPSIS method calculates the distance between the best scheme and the worst scheme in the continuous time series of samples and use the relative degree of the ideal solution as the standard of comprehensive evaluation. TOPSIS detects the optimal solution as well as the distance of worst solution for sorting.Whereas for being optimal the evaluation object should be close to the optimal solution and as distant as possible compared to worst solution otherwise it cannot be optimal.

Therefore, Technique for Order Preference by Similarity to Ideal Solution (TOPSIS) is at present one of the most approved MCDM methods and also used in various application areas such as manufacturing, industries as well as government sectors [16].

The pairwise comparison matrix is computed to decide priority among the considered parameters. This matrix is further utilized by the AHP and Fuzzy-AHP algorithm and weight for each parameter is computed. Finally the score is generated for the weighted parameters using the weighted sum method. Then the quality levels are assigned to the samples respectively.

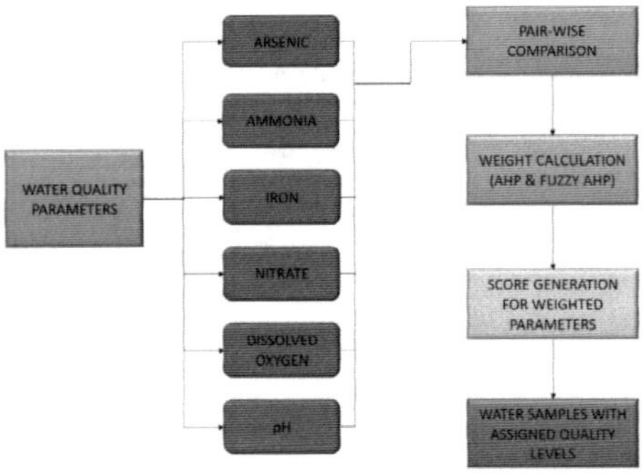

Fig. 1. System Design of AHP & Fuzzy AHP

The entire process is explained in following steps:

1 Develop pair-wise comparison matrix for n parameters as M_{nxn} with the help of scale of relative importance.

2 Conversion of crisp numeric values to Fuzzy numbers associated with triangular membership function.
If crisp value is n then fuzzification is done by the formula (n-1, n, n+1) except at 1 and 9 where the Fuzzy numbers associated will be (1,1,1) and (9,9,9). If crisp value is 1/n then Fuzzy number is of the form($\frac{1}{n+1}, \frac{1}{n}, \frac{1}{n-1}$).

3 Calculate the Row-wise Multiplications of all the Fuzzy numbers, where the calculated number is (x,y,z).
Formulae for the multiplication and addition of Fuzzy numbers:

$$(l_1, m_1, u_1) \otimes (l_2, m_2, u_2) = (l_1 * l_2, m_1 * m_2, u_1 * u_2)$$
$$(l_1, m_1, u_1) \oplus (l_2, m_2, u_2) = (l_1 + l_2, m_1 + m_2, u_1 + u_2) \quad (1)$$

4 Calculation of Fuzzy geometric mean value r_i, where r_i for each attribute in matrix M is r_i = (pow (x, 1/n), pow (y, 1/n), pow (z, 1/n))

5 Calculation of the Fuzzy weights (w_i)
Where,

$$w_i = r_i \otimes (r_1 \oplus r_2 \; .... \oplus r_n)^{-1} \quad (2)$$

6 Defuzzify the Fuzzy weights to get the crisp weights (W_i) by using the Defuzzification method 'Centre Of Area'(COA).

$$W_i = \frac{a+b+c}{3} \quad (3)$$

Where,(a,b,c) is the Fuzzy weight(w_i).

7 Generate the score as

$$\text{Score} = \sum_{i=1}^{n} W_i * p_i \quad (4)$$

Where,
W_i=Normalized weight
p_i=Normalized Parameter value

The weight parameters are calculated using above steps 1 to 7, now following process is used to assign Quality Levels to water sample.

8 The Normalized Matrix is calculated using,

$$r_{ij} = \frac{X_{ij}}{\sqrt{\sum_{i=1}^{n} X_{ij}^2}} \quad (5)$$

Where,
X_{IJ} = Parameter value

9 The Weighted Normalized Matrix is calculated as

$$v_{ij} = r_{ij} * W_j \quad (6)$$

Where,
i = 1, 2, ...n; j = 1, 2, ..., m
Then the Calculation of the Ideal best v_j^+ and Ideal worst values v_j^- using v_{ij} is done.

10 Next is the calculation of the Euclidian distance from the ideal best S_i^+ and the ideal worst S_i^- .

$$S_i^+ = \left[\sum_{j=1}^{m}\left(v_{ij} - v_j^+\right)^2\right]^{0.5}$$

$$S_i^- = \left[\sum_{j=1}^{m}\left(v_{ij} - v_j^-\right)^2\right]^{0.5} \qquad (7)$$

11 After this, calculate the performance Score P_i,

$$P_i = \frac{S_i^-}{S_i^+ + S_i^-} \qquad (8)$$

12 If $P_i >= 0.8$ then $level_i$ = Excellent
Else if $0.6 < P_i >= 0.8$ then $level_i$ = Good
Else if $0.3 <= P_i <= 0.6$ then $level_i$ = Bad
Else $level_i$ = Unsuitable

4 Results & Discussion

In this paper the method is implemented to assign quality levels for different water samples. The prototype of proposed method is implemented in java with a minimum hardware configuration of core i3 processor and 4GB RAM. The different datasets with varying numbers of water samples are considered. These samples include the amount of Arsenic, Ammonia, Iron, Nitrate, Dissolved Oxygen, pH present in water. The five different datasets with 100, 200, 300, 400 and 500 samples are considered. The values of all these parameters are collected from [17-20]. The value of each parameter is normalized to range [0-1] by applying min-max normalization method. The proposed method assigns a quality level as excellent, good, bad, unsuitable to water samples. In order to assign quality levels to water samples the weights must be assigned to the considered parameters. The weights are calculated using AHP and fuzzy- AHP algorithms. The weighted sum method is utilized to assign a final score to the water sample. The method to assign quality level to water samples by applying TOPSIS method is also implemented. If the final score is found to be greater than or equal to 0.8, the water quality is considered to be excellent, if it is found to be in between 0.6 and 0.8 then quality of water is good. If the final score is between 0.3 and 0.6 then the level assigned will be bad and if it is below 0.3 the water quality level will be unsuitable.

In Figure 2 the comparison of Standard Error while assigning quality levels for water is presented, the three methods such as Fuzzy AHP, TOPSIS with Fuzzy AHP weights are considered for this comparison. The accuracy of each method is measured in terms of Standard Error. The smaller Standard Error value indicates more accuracy. It can be observed from the Figure 2 that the error is more for Fuzzy AHP method as compared to TOPSIS method for all datasets. The error for TOPSIS with AHP and Fuzzy AHP weights are almost similar. For increased number of Samples also, the Standard Error value for TOPSIS with Fuzzy AHP weights remains smaller than the Fuzzy AHP method. Thus, the proposed method with TOPSIS and Fuzzy AHP weights assigns quality levels to water samples more accurately and hence it is an appropriate method for calculating quality levels.

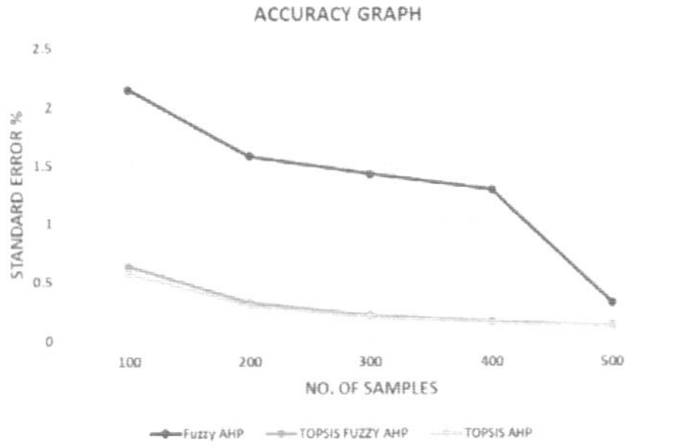

Fig. 2. Comparison of Standard Error with respect to No. of Samples

In Figure 3 the comparison of Mean Execution Time is presented for two methods such as TOPSIS with Fuzzy AHP weights and standard Fuzzy AHP method. The different datasets with varying numbers of water samples are considered. The mean execution time of each method is measured in terms of time in seconds which indicates time required for execution. The mean value of ten executions of each algorithm is considered for comparison. Hence, we can conclude that the execution speed of both algorithms is almost similar for all the datasets.

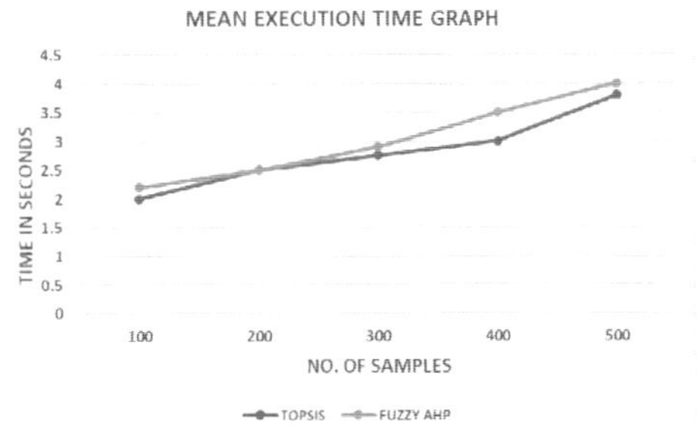

Fig. 3. Comparison of Mean Execution Time

In the Figure 4 the comparison of Standard Error by varying weights of parameters is presented for the TOPSIS with Fuzzy AHP weights method. The accuracy of TOPSIS Fuzzy AHP method is measured in terms of Standard Error. The original Fuzzy AHP weights and two sets of random weights are considered. It can be observed that TOPSIS with Fuzzy AHP weights gives the minimum Standard Error % and is the most effective way for Water Quality Assessment. Hence, the weights calculated using Fuzzy AHP are more appropriate than the random considered weights.

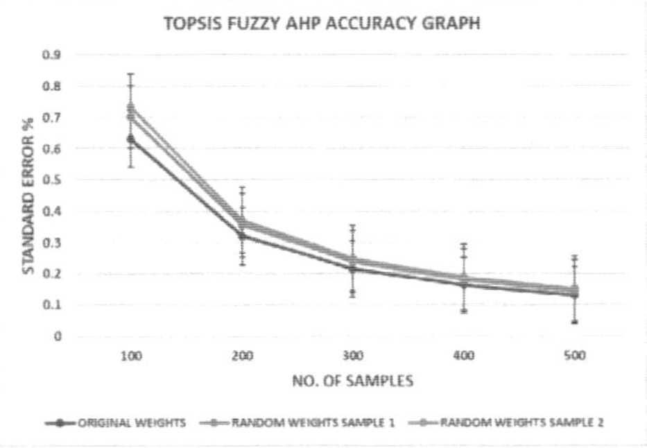

Fig. 4. Comparison of Standard Error by varying weights of Parameters

The Figure 5 shows the comparison of final score assigned for different water samples belonging to various quality levels. The results of all four methods i.e., TOPSIS with Fuzzy AHP weights, TOPSIS with AHP weights, standard Fuzzy AHP and manual assignment is considered for this comparison. The Figure 5 indicates that the score assigned by standard Fuzzy AHP method are quite overestimated as compared to other three methods. Also, the scores of manual assignment method are quite similar to TOPSIS with fuzzy AHP weights approach and hence it can be concluded that this method computes the scores in appropriate manner.

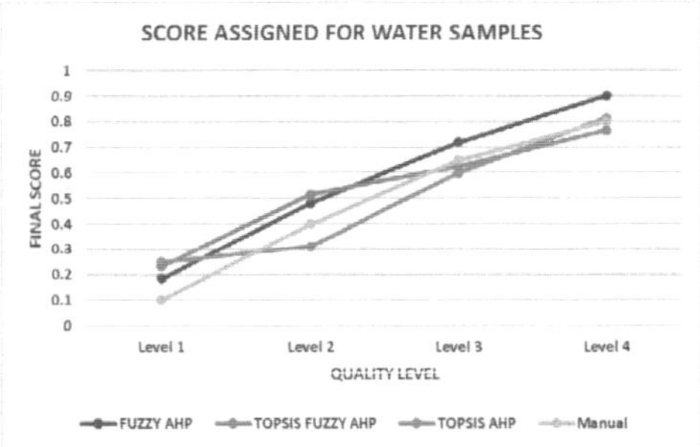

Fig. 5. Comparison of final score assigned for different water samples belonging to various quality levels

5 Conclusion

The system to monitor the quality of water on the basis of different parameters is proposed in this paper. The considered water parameters are Arsenic, Ammonia, Iron, Nitrate, Dissolved Oxygen & Ph. The proposed system applies a Fuzzy- AHP and AHP method to compute weights for these parameters. The priorities of these parameters are taken into consideration while calculating weights. The proposed system also applies the TOPSIS algorithm to assign quality levels to the water samples and the weights computed by Fuzzy-AHP are used during this process. The results show that the TOPSIS with Fuzzy-AHP weights approach is more accurate for assigning quality levels to water samples. The accuracy is majored in terms of standard error and standard error is less for TOPSIS method as compared to only Fuzzy-AHP method. The performance analysis of TOPSIS with random weights is also carried out. Here also the TOPSIS with Fuzzy AHP weights outperforms the method with random weights. Thus, the generated weights are appropriate and give better performance while evaluating quality levels to water samples. In future the proposed method can be improved by using any machine learning technique and considering other parameters of water.

References

1. H. Minh, R. Avtar, P. Kumar, Dat Q. Tran, T.Van, H. Behera, M. Kurasaki, *Groundwater Quality Assessment Using Fuzzy-AHP in An Giang Province of Vietnam* (2019)
2. Y. Tu, K. Chen, H.Wang, Z. Li , *Regional Water Resources Security Evaluation Based on a Hybrid Fuzzy BWM-TOPSIS Method* (2020)
3. *Arsenic in Drinking Water* (2019) https://www.health.state.mn.us/communities/environment/water/contaminants/arsenic.html#HeatlhEffects
4. *Arsenic In Well Water*(2019) https://www.michigan.gov/documents/deq/deq-wd-gws-wcu-arsenicwellwater_270592_7.pdf
5. Oregon Department of Human Services, *Health Effects Information* (2020) https://www.oregon.gov/oha/PH/HealthyEnvironments/DrinkingWater/Monitoring/Documents/health/ammonia.pdf
6. Insights Editor, *Spike in ammonia levels in Yamuna* (2020) https://www.insightsonindia.com/2020/11/03/spike-in-ammonia-levels-in-yamuna/
7. Passaic Bergen Water Softening, *Passaic Bergen Water Softening* (2017) https://www.pbwatersoftening.com/5-harmful-effects-high-levels-iron-water/
8. *Iron in Drinking Water* (2017) https://dnr.wi.gov/files/pdf/pubs/dg/dg0035.pdf
9. *Nitrate in Drinking Water* (2019) https://www.health.state.mn.us/communities/environment/water/contaminants/nitrate.html#HealthEffects
10. *Frequently Asked Questions About Nitrates and Drinking Water* (2018) https://www.cdc.gov/healthywater/emergency/dwa-comm-toolbox/before/tools/FAQ-nitrates.docx
11. *Vital Signs:The Five Basic Water Quality Parameters* (2018) https://www.arroyoseco.org/wqparameters.htm
12. *Dissolved Oxygen (DO)*(2016) https://www.enr.gov.nt.ca/sites/enr/files/dissolved_oxygen.pdf#:~:text=Healthy%20water%20should%20generally%20have,between%20about%2080%2D120%20%25
13. Eldorado Natural Spring Water, *Why Is the pH Of Your Water So Darn Important?* (2016)

https://www.eldoradosprings.com/blog/water-and-health-0

14. Dr J. Cotruvo, Mr J.K. Fawell, Ms M. Giddings, Mr P. Jackson, Prof. Y. Magara, Dr E. Ohanian, *pH in Drinking-water* (2007)

15. J. Gao, H. He, Q. An, J. Xie, Y. Cui, X. Xie, *An Improved Fuzzy Analytic Hierarchy Process for the Allocation of Water Rights to Industries in Northeast China* (2020)

16. Z. Chen, H. Zhang, M. Liao, *Integration Multi-Model to Evaluate the Impact of Surface Water Quality on City Sustainability: A Case from Maanshan City in China* (2019)

17. Md. M. Haque, M. Alam, L.Carpenter-Boggs, *Sustainability of Water Quality and Ecology*(2017)

18. G. Banerjee, *Water Quality Value Of Ground Water Treatment Scheme* (2016)

19. J. Salman, *Assessment Of Water Quality Of Hilla River For Drinking Water Purpose By Canadian Index (CCME WQI)* (2015)

20. M. Rabbi, J. Hossen, M. D. Sarwar, P. Roy, S. Shaheed, M. Hasan, *Investigation of Waste Water Quality Parameters Discharged from Textile Manufacturing Industries of Bangladesh* (2018)

 https://doi.org/10.1051/itmconf/20214002003

Cost-Effective Disinfection Robot Using UVC Radiation

Mansi Dhikle, Vinaya Dharne, Pankaja Gaikar, Kausar Fakir
Department of Electronics and Telecommunication, Ramrao Adik Institute of Technology, Navi Mumbai, India
mansidhikle@gmail.com, vinayadharne20@gmail.com, piyagaikar1999@gmail.com, kausarmf@gmail.com

***Abstract*—Sanitization with human efforts is not an easy task. Chances of contracting infections increases which leads to additional spread of bacteria. Currently, normal cleaning robots are used in most of the places but looking at the current situation the sanitization techniques need to be improved. The robot uses radiation of UV rays to kill the microrganisms. It gives a live video streaming of its surrounding using a Wi-fi based camera. With the help of Bluetooth module and android mobile, we can control the movement of the robot inside the room without being physically present. It is built with PIC Microcontroller and Ultraviolet-C (UVC) Sanitization LED. UV-C has bandwidth range of 200-280nm and is most powerful when it comes to killing pathogens in the room. This allows us to sterilise the room effectively. By killing the germs, the UV light restricts their multiplication by destroying their reproductive system. Thus use of this robot lowers the threat of infection, cost of traditional cleaning and sterilisation and increases security in medical facilities. Thus, we are trying to implement a more efficient way of sanitization by building a Low cost UV sanitization Robot which can be used in small clinics and for household purpose.**

***Index Terms*—UVC Sanitization LED, PIC Micrcontroller, Bluetooth module**

I. Introduction

The Covid-19 pandemic due to the novel coronavirus SARS-CoV-2 has led to more than 2.5 million deaths and has infected more than 103 million people worldwide, which makes us rethink how organizations and societies can work with minimum or no physical contact. The economic and social loss caused by the epidemic are horrific: Millions of people at risk of falling into extreme poverty. Adherence to safety and work to create and promising them decent work and protection of employees rights in all industries is important. The actions aimed at saving lives and livelihoods should include expanding social protection with regard to the provision of universal health care and income support to the most respected.

Every day health workers are fighting to help control the spread of virus as it has infected many lives all over the world. Robots were also used to treat and provide support for isolated patients. The World Health Organization recommends physical isolation from nearby people world to prevent the transfer of Covid-19 public standards.

Cleaning and disinfection are very critical steps in stopping the spread of contagious diseases inside closed doors . However, due to a shortage of manpower, the persistent cleaning of indoor spaces has become a critical issue. Also, involving manpower carries a high threat of infection while working for a long time in such areas.

Sanitization in the high exposure areas of the virus, that are hospitals and medical wards, is indeed challenging and requires very high measures to be taken. But in spite of all these high-end measures taken, there is always a risk associated with it.[7]

Even though the virus is relatively fragile, it continues to hold on to both porous and non-porous surfaces for hours, This leads to increase in the chance of spreading and risk of infection. The lipids that coat the genetic cloth of the virus is easy to disrupt, and therefore sanitising surfaces can certainly destroy the virus.

Thus, mobile robots are considered as a feasible option to the issues associated with conventional cleaning and disinfecting techniques because of their validated capacity to assist humans in various software regions together with hospitals, aged homes, and industries.

The objective of this project is to create a robotic controlled vehicle which is designed to sanitize using UV light and control the operation through cell phone. The robot provides live streaming of the surrounding area.With easy algorithm and program, the robot is able to cover big ground areas[5]. Here, the use of robots reduces human contact with the bacteria.Use of this robotic, lowers the risk of contamination, cost of traditional cleansing and disinfection and most importantly acquires self belief and safety in medical facilities.

II. Literature Review

Robots have wide use in Health Care Areas. These bots use drifting machines mounted with ipads to produce doctor tele-presence,drones for transportation of important or different medical instrumentation. Socially conscious bots utilized in the supervision of the older or physically/mentally restricted individual, and industrial robots like those want to sterilize patient areas or offer delivery.

A few bots are presently within the procedure of evaluation and development. There is a lot to recognize about the usage of robots in nearby settings. The comfort of robots can rely upon many variables, such a person, culture, a particular software, or a business.

To clear up the clinic disinfection disadvantage, many style requirements had to be enforced. There's a of lot of design requirements for a robotic disinfector and more if the robot is to be operated in a clinic.

A few analysis papers associated with medical reports are been studied and they show the influence on the management of good health professional.

The paper by Pacharawam Chanprakon[1] of "An Ultra-violet sterilization mechanism for disinfection" has developed a ultraviolet radiation mechanism for sterilization in an operative or a patient room. Their ultraviolet radiation larva has three 19.3 W of ultraviolet radiation LEDs mounted on the radiation bot platform covering 360° direction.

The paper by Himadri Nath Saha[2] of "Iot Based Garbage Monitoring and Clearance Alert System" proposed IoT based smart garbage monitoring and clearance alert system in which RGB lights attached to the bin indicates the garbage level of the bin. An android app was also developed to send alerts from the microcontroller to the system.

The paper by Noriyuki Yagi[3] of "Sterilization Using 365 Nm Uv-Led" has proposed Sterilization using 365nm UV-LED. This paper surveys the sterilization results of UV-led and proves that it is able to destroy the harmful bacteria . This paper suggests that UV-led is smaller and brighter than low-pressure mercury lamp and can be used for the motive of disinfection effectively.

The paper by T.Sivaranjani[4] of "Design Of a Portable Assistive Device for Disinfection Of Floor Using Bluetooth Technology" has proposed a multilevel planning for cleaning tasks to be performed by lowcost floor cleaners. This paper helps efficient ground cleaning device with sweeping, suction and mopping Operations. In addition to the cleaning, a UV Germicidal lamp has been used to kill the micro organisms present on the floor.

The paper by G Tuangzhi Dai[5] of "Design on measurement and control system of cleaning robot based on sensor array detection" has proposed use of grid scanning algorithm based on electric map realize floor coverage task, and designed a synthesis detection system based on sensor array finding method technology according to algorithm characteristics, experimental outcomes for obstacle detection by static finding indicate that the designed detection device improves cleaning robotic's surroundings notion and route search capability significantly.

The paper by Sahil Parmar[6] of "A Design Of Hybrid Floor UV-C Sterilizer Robot" has developed a Hybrid Floor UV-C sterilizer Robot, a fluid less device which can be used to perform the disinfection using UVC light and has a hidden vaccum device cleaner and a built in UPS system. The robot is cost-effective and user friendly.

The paper by Apeksha Wadibhasme[7] of "Sanitization Robot" has designed a wise medical assistant mechanism by exploring numerous contactless device technologies and is compact for economical management and induce a real time environment identification technology for its movement in an exceedingly clinic area.

The paper by Thomas Rubaek[8] of "Evaluation Of The Uv-Disinfection Robot" has developed a UV-Disinfection robot to decrease the spread of diseases and Hospital Acquired Infections (HAIs). This robotic is used to disinfect locations in health facility and other environments. This robot is designed as an addition to the existing cleaning cycle. The test outcomes of the designed Robot are that the robotic is able to damage extraordinary types of micro organisms.

III. Proposed System

The paper being presented here is a cost effective, portable UV Sanitization Robot. It is capable of sterilising the room or small objects without being physically present in the room. The major outcome of this robot is to support healthcare workers by reducing their exposure to bacteria as much as possible.

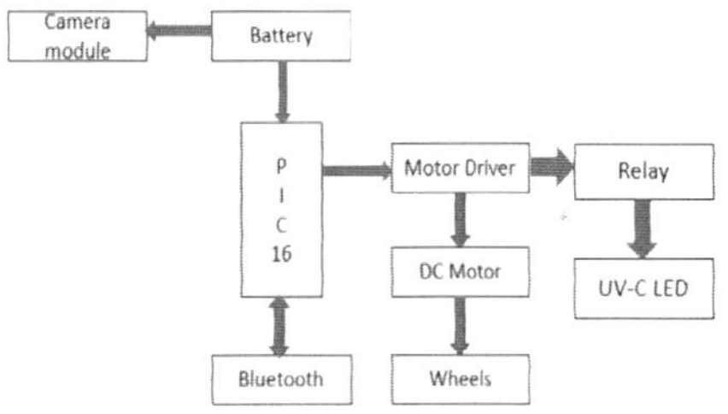

Fig. 1. Block Diagram of Proposed System

The Fig 1 shows the basic block diagram of the UV Sanitization Robot. This project uses Bluetooth decoder module and an andriod app to control the motion of the robot as well as to switch the UVC leds on and off.

The PIC microcontroller performs main actions like receiving data from the bluetooth module through serial communication and controls the motors based on the received

https://doi.org/10.1051/itmconf/20214002003

data. The HC-05 Bluetooth module is interfaced with PIC microcontroller through serial communication.

IV. Hardware Description

A. *PIC 16F877A Micrcontroller:*

The microcontroller plays a significant role in controlling the autonomous device. The main advantage of PIC is that it has a flash memory technology. An EEPROM is additionally present that allows the storage of data like transmitter codes and receiver frequencies and a few different needed data permanently. The cost of the controller is less and its use is simple. They are reliable and out of whack of PIC proportion is incredibly less. And performance of the PIC is incredibly quick due to exploitation computer architecture. Power consumption is also less compared to different microcontrollers. The price of this controller is less and its hosting is also easy [4]. It has wide range of applications and can be used in areas of microprocessor applications and timer functions
1)Instructions set - 35.
2) Operational frequency- 20MHz
3) Operating Voltage range -4.2V to 5.5V.

B. *HC-05 Bluetooth Decoder:*

It provides ASCII code output. This receiver enables wireless serial data transmission. It has a distance of 10 meters. Android phone and Bluetooth decoder attached on robot is used to control the robot The decoder is always connected to the Robot and the cell phone is used to control Robot movement.
The bluetooth decoder installed in the robot should be connected with Android phone. Every cell phone key is assigned a specific task. After connection, if the button is pressed, the corresponding button code is transmitted to the bluetooth decoder. With the help of Bluetooth decoder, the microcontroller will process the received code.
The android app used is HC-05 App which is available on any android mobile. Each key of the mobile is assigned a particular job.

C. *L293D Motor Driver:*

It is an electronic device that allows voltage to be applied across a load in both direction. It is used to convert 5V to 12V which is required to drive the motor and has an operational voltage of 5V. It is an electronic circuit which enables a voltage to be applied across a load in either direction.

D. *DC Motors:*

The motor uses power to provide energy, usually through magnetic field and current carrying coils. The DC motor generates power from the DC power provided by motor by converting DC to AC inside motor, constant magnetic field and rotating magnets.

E. *UVC LED:*

UVC leds are used as they are powerful and effetive in destroying viruses and other bacterial substances present in air and moisture [2]. This LED uses ultraviolet germicidal irradiation for killing the virus. Its operational voltage is 12V and wavelength is of 270 nm. The radiation kills the RNA in the virus and thus reduces the chance of transmission. From decades, the radiation are efficiently used for limiting the spread of bacteria. So UV-C lights are also known as "germicidal" lights.

F. *Power Supply:*

The battery is an integral part of the transmission module. The battery is used to produce 5V and 12V for DC supply ie 12/500 mA volts. Battery has output voltage of 12V and current rating = 500mA

G. *Voltage Regulator:*

Two different power regulators 7805 and 7812 are used behind the capacitor filter to produce a permanent DC power supply of 5 volts and 12 volts. Both are three pin ICs namely input, ground and output.

H. *V380- Wifi based Camera:*

The camera is used for the live video streaming of the surrounding. It can be remotely controlled from a smartphone. The camera's resolution of 720 Pixel helps to obtain a clear view of the surrounding and rotates in 360 degree to give a direct view of every corner of the room. IR LEDs with ICR present in the cameras offers clear night vision for viewing and recording.

I. *Relay:*

Relays are electromagnetic switches which help to convert small electrical cuurents into very large currents and is used to provide isolation between the UV LED and the microcontroller.

V. Working

- The robot is provided with an android control which uses bluetooth communication.
- The android app can be used to control the device i.e. to move forward, right, left or reverse and to change the status of UV LEDs.
- The HC-05 Bluetooth module receives the input signal and sends it to the microcontroller.
- The microcontroller then executes these commands and sends it to the motor driver IC to drive the motor.
- The controlling of the UV Led's is done through the motor driver IC.
- The output pins of the motor driver ic is connected to the relay and then to the UV LEDs.

VI. Softwares Used

- Eagle PCB design Software- EAGLE stands for Easily Applicable Graphical Layout Editor. It provides accurate solutions of pcb design and hence was used for designing of pcb.

- KEIL Microcontroller Software - This software was used to program the PIC microcontroller IC for movement of the robot. The programming was done in Assembly Language.

VII. Flowchart

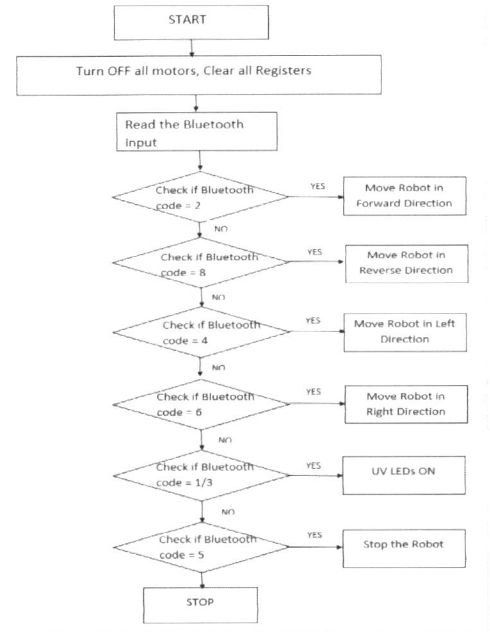

Fig. 2. Flowchart of the Proposed System

VIII. Algorithm

Once the robot is switched ON, the connection is established between android and Bluetooth decoder. The microcontroller will check the input of Bluetooth decoder. If the input of the decoder is 2, robot will move in forward direction. If Bluetooth decoder input is 8, robot moves in reverse. For 4 in left and 6 in right direction. Apart from movement of robot, the microcontroller also receives signal from Bluetooth decoder for switching ON/OFF of UVC LEDs. If the input of the decoder is 1 or 3, it will switch on the LEDs. If microcontroller receives 5 as an input, it will switch off the LEDs and stop the robot.

IX. Result and Discussion

The robot is suitable for any type of floor and is operated by Bluetooth technology. The motors consumes less power, so it will be energy saving and cost savings too. The harmful micro-organisms present on the floor is destroyed by Ultra-violet LED. The UV LED destroys the DNA/RNA of the microorganisms using the powerful UV rays. This is an eco-friendly method as UV light does not leave any residue. It is very essential to design the device in way that UV-C light will not affect the humans or its rays will not come in the eyes of users. Here, we have developed the design that follow the guidelines of using UV-C light and it will not affect the humans as it is very dangerous to direct exposure [6].

Fig. 3. Working model of UV Sanitization Robot

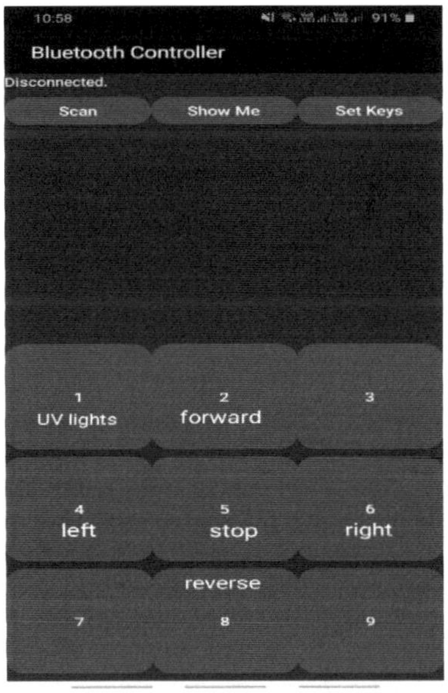

Fig. 4. Control Interface

X. Conclusion

The creation of clinical robotics has appreciably increased the safety and standard of health control systems due to Healthcare digitization.Thus, we have developed a sanitization device to perform disinfection of objects, floor etc. Ultraviolet sanitization is the latest disinfection technology with increasing importance. It can be used in clinics and other commercial purposes to improve the way of life. The main benefit of this project is to support health workers. The future scope for this project can be making this robot autonomously driven so that even less human intervention is required and making the robot base more resilient to rough terrain for smoother operation.

References

[1] Pacharawam Chanprakon, Tapparat Sae-Oung, Treesukon Treebupachatsakul "An Ultra-violet sterilization robot for disinfection" IEEE Xplore, August 19.

[2] Himadri Nath Saha, Sourav Gon, Annesha Nayak, Samabrita kundu, Sumandrita Moitra , "IoT Based Garbage Monitoring and Clearance Alert System" 2018 IEEE 9th Annual Information Technology, Electronics and Mobile Communication Conference (IEMCON) Pages: 204 – 208.

[3] Noriyuki Yagi, Mirei Mori, Akiko Hamamoto, Masayuki Nakano, Masatake Akutagawa, "Sterilization Using 365 Nm Uv-Led", 29th Annual International Conference of the IEEE EMBS Cité Internationale, 2007. Karthick.T, Ravikumar.A, Selvakumar.L, Viknesh.T, Parthiban.B Gopinath.A. "Simple Autonomous Cleaner Robot". International Journal of Pure and Applied Mathematics, Volume 118 No. 20 2018.

[4] Sivaranjani, Mayalvizhi, Swathy, Venkatesan, Janarth, "DESIGN OF A PORTABLE ASSISTIVE DEVICE FOR DISINFECTION OF FLOOR USING BLUETOOTH TECHNOLOGY". International Journal of Science,Engineering and Technology Research(IJSETR), Volume 4,Issue-5,May 2015.

[5] G Tuangzhi Dai and Tiequn Chen,"Design on measurement and control system of cleaning robot based on sensor array detetion",In IEEE International conference on control automation Guangzhou, 30 to June 1, 2015.

[6] Sahil Parmar, Hardik Bhatt, "A DESIGN OF HYBRID FLOOR UV-C STERILIZER ROBOT",International Research Journal of Engineering and Technology (IRJET), Gujarat, India.

[7] Apeksha Wadibhasme, Yedhubooshan M M, Kaushik Moolya, Shireen Farhath, Dipti Darade, Sumana Hati, "SANITIZATION ROBOT".International Research Journal of Engineering and Technology (IRJET),Volume 7, Issue 8, Aug 2020.

[8] Thomas Rubaek, Merima Cikotic, Simon Falden, "Evaluation Of The Uv-Disinfection Robot", 2016.

https://doi.org/10.1051/itmconf/20214002004

Design and Implementation of Drone in Healthcare Applications

Pratik Khandagale[,*], *Dharmaraj* Eranjikal[,**], *Siddhi* Parab[,***], and *Bhawana* Garg[,****]

[1]Ramrao Adik Institute Of Technology, Nerul, Navi Mumbai

Abstract. A drone, in simple words, is a type of flying robot that can be remotely controlled by a human operator. It is in technology terms also known as Unmanned Aerial Vehicle (UAV). With many new advancements in drone technology, it is expected that drones may expand further to include diagnostic capabilities, namely in the imaging field. The drone presented in this paper will also be used for the delivery of medical supplies. With a lock system implemented on the container attached, it ensures the safety of the supplies till they reach the rightful person. The drone could be used by online pharmaceuticals, which are booming as people are gaining interest in them. The designed system is an internet-based drone using the principles of IoT.

1 Introduction

Nowadays, the use of drones is increasing in multiple fields such as military, photography, and delivery purposes. Also, in the medical field, it is observed that a drone is been used to deliver medicines, vaccines to distant places where roadway delivery isn't possible. This has helped a lot in the fast and safe delivery of medical supplies. In this pandemic, the use of a drone has increased because of the contact-less delivery, which has hugely impacted society.
Drones used for commercial purposes become more popular when Amazon announced it intends to use drones for delivering packages to their customers. This is a very fascinating and life-changing idea with many intended and unintended consequences [1].
Drones have already been used to deliver small aid packages to places affected by major disasters, including the 2010 earthquake in Haiti, the 2012 hurricane called Superstorm Sandy that affected the northeastern United States, Canada, and the Caribbean, the 2015 category 5 cyclone named Pam that struck the islands of Vanuatu, and the 2015 earthquake (Gorkha) in Nepal. Supplies like medicine, defibrillators, blood samples, and vaccines were being delivered [2], [3], [4].
In rural areas where the mode of roadways delivery isn't possible because of poor road infrastructure and where poor weather the vehicles will not be able to transport the medicines and vaccines. Hence, people might be left without essential medical supplies. So drones can easily navigate through airways and above blocked roads to deliver essential supplies safely and quickly. The drone will be not affected by the road's infrastructure and it will transport the medicines to the people who are living in such rural areas in no time as compared to other vehicles. Drone delivery is a fast, cost-effective strategy to ensure rural patients don't need to go without.
During the medicines transit process, the packages are vulnerable to getting lost or stolen. At times, packages are damaged during the process of delivery. This can cause considerable damage if the package carries some medicines or vaccines in fragile containers. Newly emerging online pharmacies require a faster and secure way of medical transportation. Since, the traditional process is land based, the there is a high risk for medicines getting stolen or smuggled. This may also lead to black marketing of the medicines which are in high demand.
Considering the pandemic scenario caused due to COVID-19, the Remdesivir injections and other in-demand antibiotics and anti-body medications required for treatment of patients were falling short in supplies to the patients which were in need. Also, these medicines were sold in unauthorized ways at a much higher rate than the Maximum Retail Price marked on them. For such conditions, a check on supplies and inventory management by a centralized is needed. Also, delivery to the person in need must be ensured.
Patients being treated for diseases like Diabetes, Asthma, cardiac-related diseases, etc. which require a daily dose of medicines and cannot be skipped, at times run out of supplies. In such situations, an urgent delivery of medicines is required[5].

1. Reduces response time during emergency.
2. Serves as an excellent medi-aid.
3. Medkit box attached to transport medicines.
4. Locking mechanism ensures safety of medicines.
5. Prevents medicine contamination.

The next section covers the block diagram of the drone with the locking mechanism used for the container. The fourth section gives technical insights into all the hardware and software used for the drone. In section 5, the

*e-mail: pkhandagale02@gmail.com
**e-mail: mandareranjikal@gmail.com
***e-mail: siddhiparabs00@gmail.com
****e-mail: garg.bhawana@rait.ac.in

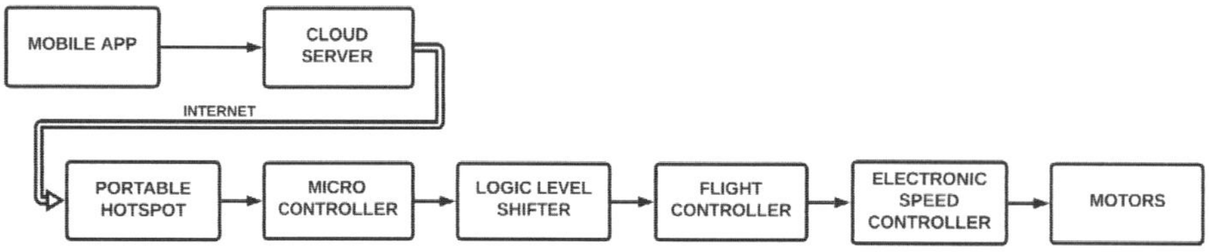

Figure 1. Block Diagram of Drone

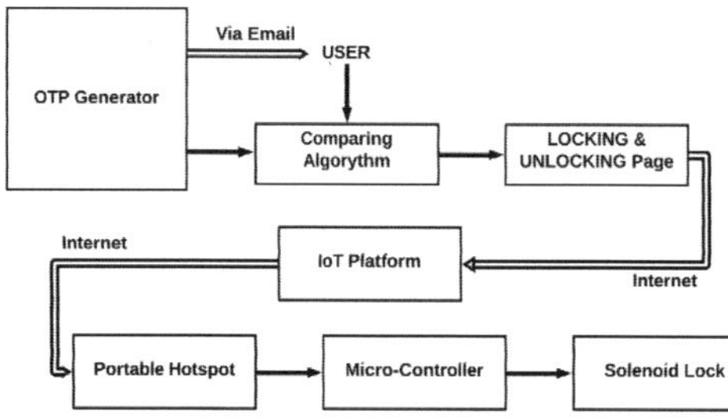

Figure 2. Block Diagram of Locking Mechanism

complete design and the components implementation are reviewed. The sixth section will consist of the snaps taken of the constructed drone. The final section consists of the concluding remarks. The references are also mentioned at the end.

2 Proposed Work

The system consists of Flight Controller, ESP8266 microcontroller, Electronic Speed Controller (ESC), and four Brushless DC motors (BLDC) assembled on a quadcopter frame. The system is installed with an onboard portable hotspot device to provide internet connectivity through WiFi. Logic Level Shifters are used for interfacing 3.3V logic with 5V logic.

The connectivity between the drone and the remote controller routes via an Internet of Things platform. The remote controller is replaced by a mobile application that holds buttons and a joystick to simulate and produces values similar to a Radio Frequency based remote controller.

The process of controlling the drone starts with a mobile-based remote controller. Primarily, the parameters required for controlling the flight dynamics of the drone are generated via the user interface on the remote controller app. The generated parameter values are uploaded to a specific channel of the server provided by the Internet of Things platform.

The parameter values uploaded to the server are available to be read by the drone. The drone is provided with an onboard portable hotspot. The portable hotspot produces a WiFi zone around the drone which also propagates along with the drone. The portable hotspot enables the drone to connect with the internet to read the parameter values produced and upload by the mobile application.

To read the parameter values, an ESP8266 microcontroller is employed. ESP8266 having an onboard WiFi module that connects with the portable hotspot device to gain access to the internet. Once the ESP8266 is online, it can start reading data from the Internet of Things server. The ESP also serves as the receiver for the drone. On reading the parameter values which were previously uploaded by the mobile application, the ESP microcontroller starts generating PWM (Pulse Width Modulation) signal to represent the parameter values. These generated PWM signals are similar to the PWM signals that would be generated by a traditional receiver. Generation of PWM signals represents the success of the transfer of data from mobile (remote controller) to the drone which is traditionally done by a radio transmitter and receiver [6].

The PWM signals generated by the ESP8266 are fed to respective channels of the Flight Controller of the drone. Hence, the Flight Controller can be controlled by the remote controller on the mobile application.

The Flight Controller controls the flight dynamics of the drone. It drives the Electronic Speed Controllers (ESC). The ESCs acts as an interface between the Flight Controllers and the motors. The ECS regulates the speed of the Brushless DC Motor (BLDC Motor) by the PWM signals fed by the ESP8266. The BLDC motors are fitted with propellers to generate lift. Thus, the drone is controlled by the remote mobile application via the internet [7].

FOR LOCKING MECHANISM:

The locking mechanism is controlled by a desktop application. The locking vessel consists of an ES8266 microcontroller which operates a solenoid lock or a servo-motor in a combination with a door latch.

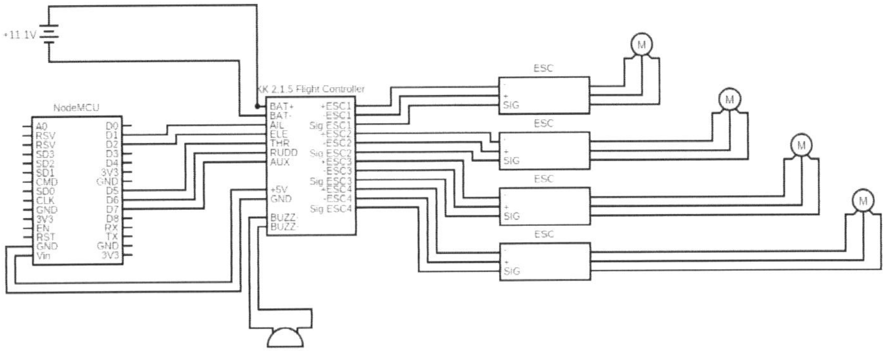

Figure 3. Circuit Diagram for Drone

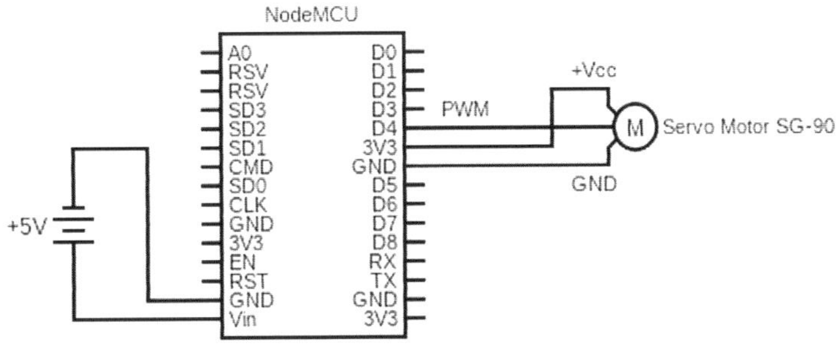

Figure 4. Circuit Diagram for Locking Mechanism

The desktop application generates a One Time Password (OTP) which is sent to the user's email address for verification. It then compares the OTP entered by the user with the OTP which was previously generated by the application. The desktop application establishes a connection with an Internet of Things (IoT) platform. The Lock and Unlock functions send a secret code that represents lock and unlock order. The generated code is sent to the server allotted by the IoT platform. The onboard ESP8266 reads the code uploaded by the application and performs the desired operation on locking or unlocking the container box in accordance with the code sent by the application. The micro-controller controls a solenoid lock.

In this paper, a drone is been proposed that can be used in the medical field.
When there is an urgent need of medicines to a patient who can't go out themselves and bring medicines, or when due to some reason the patient forgets to buy medicines, this drone will deliver the medicines in a box to the patient's location.
Chances of the medicines getting contaminated will be less and the patient will be satisfied [8].

3 Technical Insights

3.1 Hardware Implemented

Frame: The frame of a quadcopter is the main structure, or the skeleton upon which the rest of components will be attached.
Flight Controller: The Flight Controller (aka "FC") is the brain of a quadcopter, it has sensors on the board so it can understand how the craft is moving.
ESP8266: It is a microcontroller which helps us to connect with the wifi, or send or receive data. With the respective of the project, the ESP is employed as a receiver. It produces PWM signals similar to the signals produced by the receiver output for corresponding values sent by the transmitter.
Electronic Speed Controller: An ESC is a device that interprets signals from the flight controller, and translates those signals into phased electrical pulses to determine the speed of a brushless motor.
Motors: The motors are the main drain of battery power on your drone.
The propellers: The propeller converts the motion into lifting power. Because of the special shape of the blades, the

air pressure is uneven on two sides while they are in motion, what creates lifting power. The principle can be easily modelled by Newton's third motion law and Bernoulli's principle [9].
Battery: LiPo batteries are the power sources of the quadcopters. LiPo is used because of the high energy density and high discharge rate.
Logic Level Shifter : It is a bi-directional circuit which converts 3.3V logic to 5V logic or vice-versa.
JioFi Portable Hotspot: It is a device which creats the wifi signal. It is used to connect our project to the internet and the cloud-based server.
Microcontroller: It will be used to manipulate the data and controlling the lock.
Servo-motor SG90: It is used for activating and deactivating lock.
Enclosure: It's an object for holding or transporting something.

3.2 Software Implemented

Blynk App: Blynk is a new platform that allows you to quickly build interfaces for controlling and monitoring your hardware projects from your iOS and Android device.
Arduino IDE: The Arduino Integrated Development Environment - or Arduino Software (IDE) - contains a text editor for writing code, a message area, a text console, a toolbar with buttons for common functions and a series of menus. It connects to the Arduino and Genuino hardware to upload programs and communicate with them.

4 Design and Implementation Details

4.1 Design Methodology

2200 mAh Lipo Battery for the main power supply is used for the flight controller of the drone and KK2.15 Flight controller as the brain of the drone. Also, the controller's ESC1, ESC2, ESC3, and ESC4 ports are used to give power supply to the motors. NodeMCU is also employed to operate the drone via WiFi. The D1 port is connected AIL port, D2 is connected to the Elevator port. Then, D5 was connected to the Throttle. D6 was connected to the rudder and D7 connected to the Auxiliary pin of the Flight Controller. Vin was connected to a 5V supply while both the ground pins are also connected. The buzzer is also connected to the Flight Controller for signaling. The motors are also connected to the propellers and the power supply was given by the ESC. The Servo Motor has 3 pins, Vcc, ground, and PWM. We connected Servo Motor's ground pin to NodeMCU's ground pin and PWM pin to the D4 pin. D4 pin is allocated to generate a PWM signal equivalent to the desired angle for locking and unlocking the box. The VCC for the Servo motor is provided through the +3.3 V on-board supply of NodeMCU. Arms of the Servo motors are connected to the handle of the latch using a metal frame that acts as a lever. The Blynk app serves as an interface to operate the microcontroller which in turn controls the locking mechanism.

Figure 5. The implemented Drone with container attached (1)

Figure 6. The implemented Drone with container attached (2)

4.2 Selection Criteria for Components:

To cut down the cost, an 11.1V 2200mAh Li-po Battery was employed. Also, NodeMCU was used as a transmitter rather than an Arduino because it had Wifi-module and it is less costly than Arduino. A 30V ESC battery to power supply the motors were utilized. Also, a Flight Controller KK2.1.5 was used because it has an onboard gyroscope and display for configuration setup. Also, it is inexpensive as compared to other flight controllers in the market. A stand for the drone to lift it from the ground for ground clearance was attached. The plastic box was used because it's lightweight and durable. Also, it can be carried easily by our drone.

5 Results and Discussion

As planned, the drone will deliver the required pharmaceutical products to the patient's location.
Just like this, the drone can be used by doctors to transfer medicines to a region where roadways are not good for transport.
While making this drone, major issues arose during connecting the NodeMCU to the flight controller. Also when the drone was taking off, it was quite unstable and because of that, the drone was tilted towards one side. So to overcome the issue, the flight controller was programmed again and reduced the previous values. After these changes, the drone was able to successfully take off. The drone can carry around 1 kilogram including the medkit box containing the medicines. The payload in the Medkit box is considered to be 250 gms for one flight and it will transport the medicines in roughly around 50 minutes. Considering the Medkit box size and structure, it will be able to carry around 750 to 800 grams of medicines in it. As the payload is increased, the time will also increase simultaneously.
Approach time can be reduced and a patient's life will be saved.

Table 1. Payload associated with Time

Payload(gm)	Time(mins)
250	50
350	80
450	110
550	140
650	170
750	200

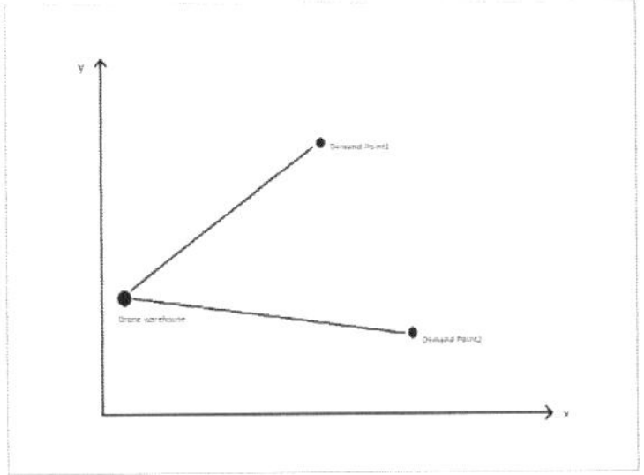

Figure 7. Location of the warehouse and demand zone

In rural areas, a warehouse will be set up to operate, charge and maintain the drone. In case of emergency, one can contact the warehouse near the area and the operator can put on the medical supplies in the Medkit box, and then he can fly the drone to the demand locations. The drone can also cover multiple demand locations in a single flight.

6 Acknowledgements

With great pleasure, we avail this opportunity to express our profound gratitude and deep regards to our Head Of Department of Instrumentation Engineering, RAIT, Dr. Sharad P. Jadhav, for his generous support and constant encouragement. It helped us to work in the right direction to attain the desired objective.

7 Conclusion

Drones can provide beneficial and humanitarian applications, especially related to healthcare. Furthermore, extreme weather and city congestion can be an obstacle to emergency medical delivery. Consequently, drone healthcare delivery to inaccessible locations is likely to become more present in the near future, in any disaster the medical supplies and basic needs supply is mostly transported using vehicles, by this people will not be able to receive proper help and will have to wait for transportation systems to arrive in their area. So, drones will be a great help in such disastrous situations in supplying medicines and other medical supplies to the people. This paper discusses some innovative applications of drones in healthcare. Healthcare costs are increasing day by day in both developed and developing countries. Since time is of the essence in an emergency, a faster response would prevent medical trauma and potentially save lives. Our models could be used by drone companies, delivery companies, healthcare organizations, humanitarian organizations, or governments. In some cases, governments could partner with drone companies, such as Pharmeasy and Netmeds. Drone healthcare delivery is a new field with many opportunities for impactful future research.
Considering the drone's future scope, a GPS will be added to track the location of the drone and to recover the drone from a lost location in case of any system or power failure. Multiple Medkit boxes could be added in order to deliver medical supplies to multiple demand points in a single flight. High powered batteries could be installed later in order to receive the long flight duration.

References

[1] Amukele, Timothy, Current state of drones in healthcare: challenges and Opportunities **2**, 296–298 (2019)

[2] Balasingam, Manohari, Drones in medicine—the rise of the machines **71**, e12989(2017)

[3] Scott, Judy and Scott, Carlton, Drone delivery models for healthcare (2017)

[4] MStalberg, Malachi, Designing and Developing a Commercial Delivery Drone **1**, (2020)

[5] Bhatt, Kunj and Pourmand, Ali and Sikka, Neal, Targeted applications of unmanned aerial vehicles (drones) in telemedicine **24**, 833–838 (2018)

[6] Mesquita, João and Guimarães, Diana and Pereira, Carlos and Santos, Frederico and Almeida, Luís, Assessing the ESP8266 WiFi module for the Internet of Things **1**, 784–791 (2018)

[7] Cheng, Zhuoqun and West, Richard and Einstein, Craig, End-to-end analysis and design of a drone flight controller **37**, 2404–2415 (2018)

[8] Zailani, Mohamed Afiq Hidayat and Sabudin, Raja Zahratul Azma Raja and Rahman, Rahana Abdul and Saiboon, Ismail Mohd and Ismail, Aniza and Mahdy, Zaleha Abdullah, Drone for medical products transportation in maternal healthcare: A systematic review and framework for future research, **99**, 36(2020)

[9] Dalai, Rama Prasanna and Sabyasachi, Sidharth, Future of Drone Technology in India **1673**, 064X

[10] A. Josephin Arockia Dhivya, J. Premkumar, Quadcopter based technology for an emergency healthcare, 1–3 (2017)

Analysis of GPU Computation of Parabolic, Bessel, Wright and Riemann Zeta Functions

Ashish A. Jadhav[1], Abhijeet D. Kalamkar[2], Pritish A.Gaikwad[3], Vishwesh Vyawahare[4] and Navin Singhaniya[5]
Department of Electronics Engineering,
Ramrao Adik Institute of Technology,
Nerul, Navi Mumbai.
Email : 23jadhavashish@gmail.com[1], abhijeetkalamkar03@gmail.com[2], pritish.gaikwad05@gmail.com[3], vishwesh.vyawahare@rait.ac.in[4] and navin.singhaniya@rait.ac.in[5]

Abstract. This paper deals with GPU computing of special mathematical functions that are used in Fractional Calculus. The graphics processing unit (GPU) has grown to be an integral part of nowadays's mainstream computing structures. The special mathematical functions are an integral part of Fractional Calculus. This paper deals with a novel parallel approach for computing special mathematical functions used in Fractional Calculus. NVIDIA's GPU hardware is used to speed up the parallel algorithm. A comparison of the sequential code, vectorized code and GPU code is performed. We have successfully reduced the computation time of special mathematical functions using the parallel computing capabilities of GPU.

Keywords: GPU computing, Parallel computing, Fractional Calculus, Parabolic function, Bessel function, Wright function, Riemann Zeta function.

1. Introduction

Fractional calculus (FC) is widely used in the field of engineering because of its ability to model real-world systems with more precision [1]. In recent years the use of this technology has grown exponentially these days as a growing community of researchers has come together to revitalize the field with the help of advanced computing technology [3]. In most cases, this leads to certain differential equations whose solutions are given by Special Mathematical Function. These are non-elementary functions that are found to be useful in some applications and satisfy some special properties. Computation of such functions has proven to be a resource and time intensive task. To tackle this problem, coprocessors such as GPU and parallel algorithms have been used by researchers in the past [4]. The general purpose use of GPUs for computing is commonly known as GPGPU (General Purpose Computing on Graphics Processing Unit) [3]. This is achieved by Parallel Computing. In Parallel Computing, a large number of computing nodes or CPU cores are used to carry out logical and arithmetic operations from an algorithm in parallel fashion. This increases efficiency and reduces compute time.

It has been found that, for the special functions of mathematics are also very useful for finding solutions to the initial-and / or regional issues governed by partial differential equations, and fractional partial differential equations. These special functions are having wide application in different areas of mathematics. Some commonly used special functions are Wright function, Bessel function, Riemann Zeta function, Parabolic Cylinder function, etc.

In this paper we have done computation of different special functions using hardware Section 2 deals with the Introduction GPU. Section 3 will go through the Definition of Special Mathematical Function which we have done the analysis. In section 4 deals with the Implementation and the design part. In section 5 we have taken the results of the functions. And lastly in Section 6 we have concluded.

2. Graphics Processing Unit (GPU)

The Graphics Processing Unit (GPU) is a many-core device used for parallel computing applications. Initially intended only for graphics rendering purposes, now have found its way into scientific computing as well. The CPU has been designed for low latency to be able to switch between operations quickly. In the past, CPU designers opted towards making the multicore design for overcoming the ludicrously high power requirements that came with increasing clock speed [7]. Because graphics rendering is a simultaneous task, GPU designers chose the many-core technique early on. GPUs are designed for high throughput, allowing users to perform as many operations as possible. There is an ample amount of infrastructure on the chip to get low latency on the CPU, such as caches to ensure that a vast amount of data is readily accessible. There is a significant amount of control flow silicon. Counter to that, the balance has shifted in the GPU, where more arithmetic and logic units are needed. It is designed for compute-intensive, highly parallel computations and hence is specialized for them.
The GPUs use stream processing architecture that is well suited for compute-intensive parallel tasks [6]. Coupled with the current spike in interest in similar research endeavours, collectively known as GPGPU computing, GPUs have become a viable and efficient tool for implementing compute-intensive algorithms [8].

MATLAB is an abbreviation for Matrix Laboratory. It is one of the most widely used proprietary software and programming language by scientists and engineers for numerical computation, due to its excellent graphical capabilities, it has become a well-known tool for data analysis. Researchers also have been known to be using MATLAB scripts and toolboxes to construct immensely complicated systems [7]. The most crucial part of MATLAB is its Toolboxes. MATLAB Toolboxes are software packages/libraries tailored for a specific domain or application. One such toolbox was used in this study, The Parallel Computing Toolbox (PCT), which is the backbone of parallel computing in MATLAB [6]. We will be using the GPU enabled functions from PCT for a single GPU system, a detailed explanation of GPU focused part of PCT is in [3]. At this time, PCT is only available for NVIDIA CUDA enabled GPUs only.

This following table is GPU device specifications used during computing the function.

GPU Specification :

PARAMETER	SPECIFICATION
Speed	536@1.46-1.59 GHz 192 Bit @12000 MHz
Architecture	Turing
Core Speed	1455-1590(Boost) MHz
Memory Speed	12000 MHz
Memory Type	GDDRG6
Max.Amount of Memory	6144 MB
DirectX	DirectX 12.1
Transistor Count	6600 Millon
Cuda Cores	1536

Table 1: NVIDIA GTX 1660Ti GPU Specifications

CPU Specification :

PARAMETER	SPECIFICATION
Processor Name	Intel@CoreTMi7-9750H Processor
Lithography	14nm
No. of Cores	6
No.of Threads	12
Processor Based Frequency	2.6 GHz
Maz Turbo Frequency	4.50 GHz
Cache	12 MB

Table 2: CPU Intel@CoreTMi7-9750H Processor Specifications.

3. Special Mathematical Functions Used in Fractional Calculus

Special functions implemented in this paper are Wright function, Bessel function, Riemann Zeta function, Parabolic Cylinder function. The Wright function plays an important part in the theory of the partial differential equations of the fractional order that are used for modeling of most phenomena including. The anomalous diffusion processes or in the theory of the complex systems. Bessel functions are used to solve in 3D wave equation at a given frequency. The solution is generally a sum of spherical bessels functions that gives the acoustic pressure at a given location of the 3D space. The Riemann zeta function used in analytic number theory and has applications in applied statistics, physics, probability theory. We have used numerical method of this functions and coded that using MATLAB.

3.1. Parabolic Function:

The parabolic cylinder functions (PCFs), also known as Weber parabolic cylinder functions. It is a class of functions sometimes called Weber functions.

$$y_1(x) = e^{-x^2/4} F_1\left(\frac{1}{2}a + \frac{1}{4}; \frac{1}{2}; \frac{1}{2}x^2\right) \quad (1)$$

$$y_2(x) = xe^{-x^2/4} F_1\left(\frac{1}{2}a \frac{3}{2}; \frac{3}{2}; \frac{1}{2}x^2\right) \quad (2)$$

The above two equations 1 and 2 are the even and odd equation respectively. Where, 1F1(a:b:z) is a confluent hypergeometric function of the first kind. We have checked the results for the values of N = 200, and parameter A = 1, 4, 6, and having step size *-10 : 1e-3 : 10, -10 : 1e-4 : 10, -10 : 1e-5 : 10.*

3.2. Bessel Function:

Bessel's function was first formulated by mathematician Daniel Bernoulli. Bessel's function is a solution of differential equation which is known as Bessel equation

$$x^2\frac{d^2y}{dx^2} + x\frac{dy}{dx} + (x^2 - n^2)y = 0 \quad (3)$$

For an arbitrary complicated range α, the order of the bessel function. Although α and −α produce the equal differential equation, it is conventional to define different bessel capabilities for those two values in this sort of manner that the bessel features are by and large easy capabilities of α. Bessel function are mainly important for most of the problems of wave propogation and in static potentials and in cylindrical co-ordinate systems. It is can be used to define the function by its series expansion around x = 0

$$J_\alpha(x) = \sum_{m=0}^{\infty} \frac{(-1)^m}{m!\,\Gamma(m+\alpha+1)}(x/2)^{2m+\alpha} \quad (4)$$

Here, Γ(z) is the gamma function. Using the above equation 4 we have analysed the function and obtained the results. We have checked the results for the values of N=200, parameter Alpha = 0.9, 1.5, 1.9, and have plot against step size *0 : 1e-3 : 3, 0 : 1e-4 : 3, 0 : 1e-5 : 3.*

3.3. Wright Function:

The Wright function is defined by

$$\emptyset = (\alpha, \beta; z) = \sum_{k=0}^{\infty} \frac{z^k}{k!\,\Gamma(\alpha k + \beta)} \quad (5)$$

Where, Γ(z) is the gamma function. It became added for the primary time via E.M. Wright in [8] in connection together with his investigations inside the asymptotic principle of partitions. The Wright function characteristic belongs to the magnificence of the hypergeometric capabilities; it may be reduced to the recognized special function for some values of the parameters ρ and β because of the relation. The Wright function along with the Mittag-Leffler function plays a prominent role in the theory of the partial differential equations of the fractional order. Using the above equation 5 we have analysed the function and and computed results for the values N=200, and have plot against step size *0 : 1e-2 : 3, 0 : 1e-3 : 3, 0 : 1e-4 : 3.*

3.4. Riemann Zeta Function :

Riemann expanded the definition of Euler ξ (s) function in all complex numbers (except for a simple pole in s = 1 and one remainder). Euler's product description for this function is still valid if the actual s is more than one. To help understand the values of some of the more complex numbers, Riemann found the performance efficiency of the Riemann zeta function:

$$\xi\,(s) = \sum_{n=1}^{\infty} \frac{1}{n^s} \quad (6)$$

Using the above equation 6 we have analysed the function and results for the values of N=200, and have plot against step size: *1e6, 1e4, 1e5.*

4. Design and Implementation

While changing over MATLAB code to run on the GPU, it is best to begin with a correct and optimized MATLAB code. The code should provide the required results accurately. For CPU programming, vectorization is considered to be a great practice, whereas for GPU, vectorization is essential for achieving high execution. Subsequently it is essential to vectorize the code by replacing looped scalar operations with MATLAB matrix and vector operations. Profiling gives accurate information about the time taken by each line or section of the code. The Run & Time include of MATLAB is an viable tool which profiles the code while executing it. The section of code that the profiler appears taking the maximum execution time on the CPU are the ones that we need to focus on. Writing the GPU code utilizing the functions in MATLAB which are overloaded to run on GPU is very useful. The variables or data required for execution on GPU device is to begin with transferred to device memory using gpuArray() function. The capacities with these variables as arguments are specifically executed on GPU. After the handling on GPU, the results are replicated back utilizing gather() function.

Following algorithm is used for converting the MATLAB code for GPU execution:

1. Start with the code that gives correct results.
2. Make the vectorization code of the function.
3. Run the code using Run & Time Function.
4. Analyze and Convert the section that takes more time to execute on GPU.
5. Transfer the variables used in the section described in step #4 to GPU memory using GPU Array function.
6. After GPU execution, transfer the output variables back to CPU memory using gather function.
7. After receiving the gathered data from GPU, rest of the execution is performed on CPU.

To calculate the speed up use the following equation

$$\text{SPEEDUP TIME} = \frac{\text{CPU EXECUTION TIME}}{\text{GPU EXECUTION TIME}}$$

5. Result and Analysis

First we have written sequential code using the definition of functions. After checking the correctness of the code we have profiled the sequential code using MATLAB profiler. Then we have observed the section of code which consume more time for executions, that section we have vectorized using Parallel Computing toolbox of MATLAB. Output of vectorized code is also benchmark with sequential code for verifying correctness of code. Finally we have converted the vectorized code to GPU code by transferring data to GPU memory using inbuilt

 https://doi.org/10.1051/itmconf/20214002005

function of MATLAB like gpuArray(). Fig 1, Fig 2, Fig 3 and Fig 4 shows the output benchmarking plots of different functions. From the benchmarking plots it is clear that output obtained from vectorized code is same as the output obtained from GPU code.

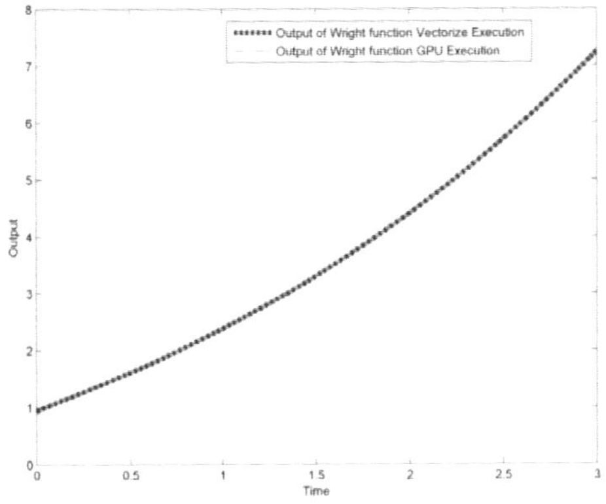

Fig. 1. Output Benchmarking plots of Wright function.

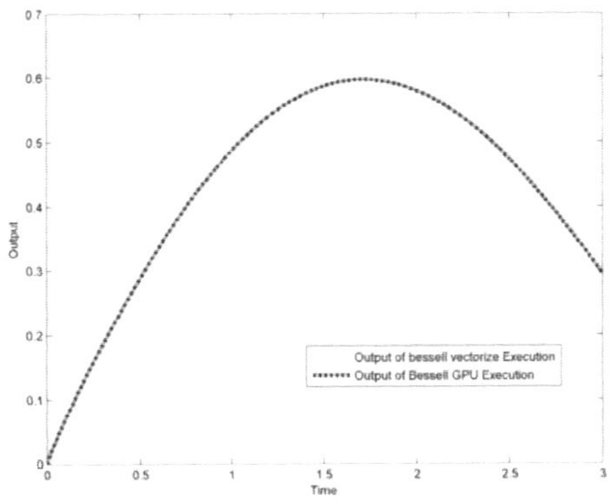

Fig. 2. Output Benchmarking plots of Bessell function.

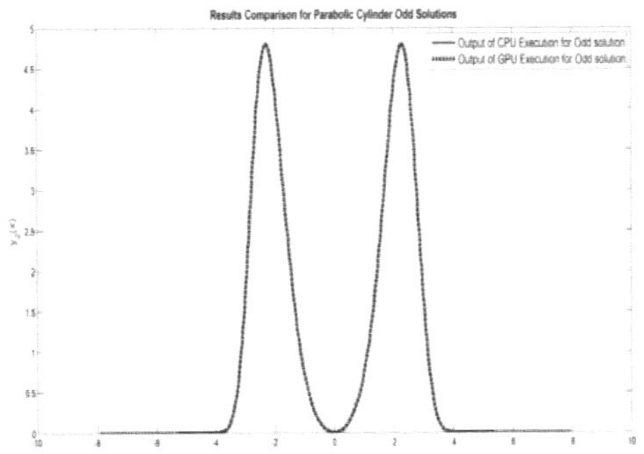

Fig. 3. Output Benchmarking plots of Odd Parabolic Cylinder function.

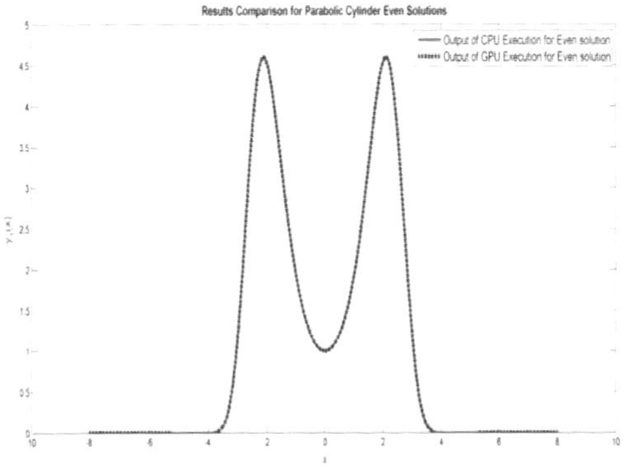

Fig. 4. Output Benchmarking plots of Even Parabolic Cylinder function.

The speed up obtained are in the range of 2x to 180x for systems with NVIDIA GTX 1660Ti GPU. From the results it is observed that speedup depends on the amount of calculation operation like multiplication, division, power, number of iterations of an operation and amount of data. We can further extend this work by writing more optimized MATLAB and using more powerful platforms like CUDA.

In the Results tables T_S=Sequential Time for execution of code, T_V=Vectorize Time for execution of code, T_G= GPU Time for execution of code. We have observed the speed up with respect to T_S=Sequential Time for execution of code verses T_V=Vectorize Time for execution of code (T_S/T_V), T_V=Vectorize Time for execution of code verses T_G= GPU Time for execution of code (TV/TG) and T_S=Sequential Time for execution of code verses T_G= GPU Time for execution of code (T_S/T_G). For the Wright function T_S/T_V speedup is approximately 152x, T_V/T_G speedup is approximately 1.2x, and T_S/T_G speedup is approximately 181x. In the table of Bessel function T_S/T_V speedup is approximately 76x, T_V/T_G speedup is approximately 1.2x, and T_S/T_G speedup is approximately 79x. . In the result table of Riemann Zeta function T_S/T_V speedup is approximately 3x, T_V/T_G speedup is approximately 1.4x, and T_S/T_G speedup is approximately 5x. In the table of Even and Odd Parabolic cylindrical function, T_V/T_G speedup is approximately 1.2x. From the results we have observed that speedup increases as the data increases like for example speedup increases as we change step size from stepsize of 1e-2 to 1e-4. Time taken for Sequential code is more as compare to vectorize code and GPU code, from all the three codes GPU code is fastest and very optimized.

WRIGHT FUNCTION							
Parameters		Seq uen tial	Vecto rize	GPU Time	Speedup		
ALPHA ,BETA	Step Size	T_S	T_V	T_G	T_S/T _V	T_V/ T_G	T_S/T _G
0.6	1.00E -02	0.34 83	0.004	0.002	79.15	1.56	124.0 69
0.8	1.00E -02	0.39 98	0.003	0.002	120.4	1.16	140.6 643
0.9	1.00E -02	0.27 34	0.005	0.004	52.57	1.13	59.43
0.6	1.00E -03	3.96	0.019	0.018	208.7	1.00	210.3 0
0.8	1.00E -03	4.01	0.022	0.020	178.8 4	1.08	194.7 3
0.9	1.00E -03	3.59 4	0.017	0.014	206.0 6	1.17	242.8 3
0.6	1.00E -04	27.5 3	0.164	0.135	167.7 7	1.21	203.8 1
0.8	1.00E -04	36.2	0.175	0.140	206.9 2	1.25	258.7 2
0.9	1.00E -04	26.8	0.18	0.135	149.2	1.33	199.0 3
		Total Average Speed			1.52E +02	1.214	181.5 1

Table 3. : SpeedupTime for Wright Function

EVEN PARABOLIC FUNCTION				
Parameters		Vectorize	GPU Time	Speedup
A	Step Size	T_V	T_G	T_V/T_G
A=1	1e-3	0.129	0.0777	1.6641
A=4	1e-3	0.0938	0.0781	1.2002
A=6	1e-3	0.1074	0.0762	1.4094
A=1	1e-4	0.7124	0.682	1.0446
A=4	1e-4	0.7651	0.6838	1.119
A=6	1e-4	0.7292	0.6829	1.0678
A=1	1E-5	OUT OF MEMORY		
	TOTAL AVERAGE SPEED			1.25085

Table 4. : Speedup Time for Even Parabolic Function

RIEMANN ZETA FUNCTION							
Parameters		Sequ ential	Vect orize	GPU Time	Speedup		
S	Step Size	T_S	T_V	T_G	T_S/ T_V	T_V/ T_G	T_S/ T_G
2	1.00E +06	0.03	0.01	0.0052	3.25	2.27	7.384
3	1.00E +06	0.13	2.93E- 02	0.03	1.06	1.06	4.64
4	1.00E +06	0.12	0.03	0.0239	3.88	1.39	5.389
2	1.00E +05	0.002	0.00	0.0011	1.93	1.36	2.636
3	1.00E +05	0.01	0.01	0.0093	1.98	1.08	2.129 0
4	1.00E +05	0.017 7	0.003 6	0.0031	4.916	1.161	5.709 6
2	9.00E +05	0.04	0.01	0.0043	3.31	2.63	8.70
3	9.00E +05	0.11	0.03	0.0247	4.05	1.13	4.574 8
4	9.00E +05	0.10	2.97E- 02	0.0222	3.65	1.34	4.887
		Total Average Speed			3.11	1.49	5.116

Table 5. : SpeedupTime for Remann Zeta Function

BESSEL FUNCTION							
Parameters		Sequ ential	Vect orize	GPu Tim e	Speedup		
Alpha	Step Size	T_S	T_V	T_G	T_S/ T_V	T_V/ T_G	T_S/T _G
0.9	1e-3	2.701	0.03	0.02	90.05	1.111	100.0
1.5	1e-3	3.291	0.029	0.02	112.7 2	1.057	119.2
1.9	1e-3	2.268	0.029	0.02	78.23	1.043	81.61
0.9	1e-2	0.428	0.006	0.00	62.97	1.283	80.79
1.5	1e-2	0.367	0.006	0.00	58.31	1.166	68.03
1.9	1e-2	0.260	0.004	0.00	54.35	1.066	57.97
0.9	1e-5	350.99	OUT OF MEMORY				
		Total Average Speed			76.11	1.121	79.14

Table 6. : Speedup Time for Bessel Function

ODD PARABOLIC FUNCTION				
Parameters		Vectorize	GPU Time	Speedup
A	Step Size	T_V	T_G	T_V/T_G
A=1	1e-3	0.0848	0.072	1.1786
A=4	1e-3	0.0886	0.072	1.2314
A=6	1e-3	0.0862	0.0736	1.1705
A=1	1e-4	0.7974	0.6784	1.1754
A=4	1e-4	0.7456	0.6768	1.1017
A=6	1e-4	0.7354	0.6767	1.0869
A=1	1E-5	OUT OF MEMORY		
	TOTAL AVERAGE SPEED			1.1574166 67

Table 7. : Speedup Time for Odd Parabolic Function

6. Conclusion

The graphical processing unit is visually and visibly changing the path of standard cause computing. In this work, different special mathematical functions are computed using Graphical processing unit (GPU). While computing special function we have observed acceleration because of multiple processing cores of graphics processor. Special codes for these functions were developed. Then using Parallel Computing Toolbox of MATLAB vectorized version of these codes are developed and computed using GPU hardware. Thus we can observe that the special mathematical functions can be successfully implemented on GPU.

References

1. Kiryakova, Virginia. "The special functions of fractional calculus as generalized fractional calculus operators of some basic functions." Computers mathematics with applications 59, no. 3 (2010): 1128-1141.
2. Gil, Amparo, Javier Segura, and Nico M. Temme. Numerical methods for special functions. Society for Industrial and Applied Mathematics, 2007.
3. Patil, Parag, Navin Singhaniya, Chaitanya Jage, Vishwesh A. Vyawahare, Mukesh D. Patil, and P. S. V. Nataraj. "GPU Computing of Special Mathematical Functions used in Fractional Calculus." Frontiers in Fractional Calculus 199 (2018).
4. Keluskar, Yugesh C., Megha M. Navada, Chaitanya S. Jage, and Navin G. Singhaniya. "Implementation of Airy function using Graphics Processing Unit (GPU)." In ITM Web of Conferences, vol. 32, p. 03052. EDP Sciences, 2020.
5. Choy, Ron, and Alan Edelman. "Parallel MATLAB: Doing it right." Proceedings of the IEEE 93, no. 2 (2005): 331-341.
6. Sharma, Gaurav, and Jos Martin. "MATLAB®: a language for parallel computing." International Journal of Parallel Programming 37, no. 1 (2009): 3-36.
7. Diener, Joseph E., and Atef Z. Elsherbeni. "FDTD Acceleration using MATLAB Parallel Computing Toolbox and GPU." Applied Computational Electromagnetics Society Journal 32, no. 4 (2017).
8. Owens, John D., Mike Houston, David Luebke, Simon Green, John E. Stone, and James C. Phillips. "GPU computing." Proceedings of the IEEE 96, no. 5 (2008): 879-899.
9. Banks, Nicola E. "Insights from the parallel implementation of efficient algorithms for the fractional calculus." (2015).
10. Singhaniya, Navin G., Mukesh D. Patil, and Vishwesh A. Vyawahare. "Implementation of special mathematical functions for fractional calculus using DSP processor." In 2015 International Conference on Information Processing (ICIP), pp. 811-816. IEEE, 2015.
11. "Parallel Computing", Wolfram Mathworld. [Online].Available:http://mathworld.wolfram.com/ParallelComputing.html
12. "Best Practices for MATLAB GPU Coding", Cornell University Center for Advanced Computing [Online].Available:https://www.cac.cornell.edu/matlab
13. N.E.Banks,"Insights from the parallel implementation of efficient algorithms for the fractional calculus", University of Chester, United Kingdom, 2015.

Android-based object recognition application for visually impaired

Akilesh Salunkhe[1], Manthan Raut[1], Shayantan Santra[1] and Mrs. Sumedha Bhagwat[2]

[1] Department of Information Technology, Ramrao Adik Institute of Technology, Nerul, India
[2] Mentor from Department of Information Technology, Ramrao Adik Institute of Technology, Nerul, India

Abstract - Detecting objects in real-time and converting them into an audio output was a challenging task. Recent advancement in computer vision has allowed the development of various real-time object detection applications. This paper describes a simple android app that would help the visually impaired people in understanding their surroundings. The information about the surrounding environment was captured through a phone's camera where real-time object recognition through tensorflow's object detection API was done. The detected objects were then converted into an audio output by using android's text-to-speech library. Tensorflow lite made the offline processing of complex algorithms simple. The overall accuracy of the proposed system was found to be approximately 90%.

Keywords - Tensorflow, TensorFlow Lite, Machine learning, Computer vision, Object recognition, and Android

1. Introduction

Out of all the human senses, eyesight is the most important sense. It allows a person to analyse and understand their surrounding environment. At least more than 285 million people are facing eyesight challenges or are visually impaires as per the data gathered from WHO. Eyesight issues can cause disturbance in the daily activities of a person. Identifying objects in day-to-day lives, reading text, crossing a road are a few examples of such problems. The proposed system is a simple android app based object detection application named "digital eyes" to help the visually impaires. This application tries to replicate the human eye with the help of a smartphone camera using object detection. The normal life of the people can be improved by using modern computer vision techniques. Object detection is one of the methods of computer vision which have had many broader applications over recent years. Object detection technology [3] uses the contrasting features intensity, edge, and shape to recognize the object from the input image. The advancement in object detection algorithms has enabled us to incorporate complex algorithms into an android application. The SSD algorithm and the trained tensorflow models are used for object detection in our android application. Image processing techniques are now currently being used in object detection domain for various applications. [4], which are used for social and many other applications. The purpose of this project is object detection for the visually impaired by using speech feedback and extracting features from live camera feeds. The application is easy to use and it is equipped with speech synthesizing so that the detected object is communicated to the blind people as voice output. In section ii, describes a comparative study of several object detection methods and their statistics in tabular format. Section iii describes the system description along with the technology stack used in the system. In sections iv and v, this paper explains about the advantages and disadvantages of the proposed system in this paper. In section vi it explains the proposed system technically giving an idea about the important components in the system followed by a conclusion and future scope.

2. Related work of different algorithms

Table no. 1. Related work of different algorithms [13]

Mode Type	Latency of model	(Accuracy)mAP	FPS	Real-Time
R-CNN	High	~60	<1	NO
Fast R- CNN	Medium	~70	<1	NO
Faster R-CNN	Medium	~70	7	NO
YOLO	Low	~60	46	YES
SSD	Low	~74	59	YES

The table above shows the comparison of the various models regarding latency, average accuracy (map), images per second (FPS), and whether they could be used in real-time or non real-time applications. Analysis of the outcome of a system or algorithm was based on certain parameters. The most common parameters were efficiency, time, resources, accuracy, etc. that were undertaken in almost all analyses. On applying the general parameters over the R-CNN method of object detection, the results showed that it was much faster than the old methods based on the classification methods. Instead of a huge number of regions, RCNN used the selected search to retrieve only 2000 regions per frame. So the feature extraction would run over only 2000 regions. A new version of R-CNN called fast R-CNN was far better than R-CNN as it did not transmit 2000 regional proposals to CNN each time. Instead, the CNN operation was carried out once per frame. The implementation of the new method was similar to the previous methods but instead of a selective search algorithm, an independent network was used to anticipate the proposed regions. A new method You Only Look Once (YOLO) was proposed for the recognition of the objects. As the above methods used the suggested regions to identify the objects in the image, it never considers the entire image. Regions with a high likelihood of having objects were instead passed through the system for object detection. But in YOLO there was only a single convolutional network and the entire image was analysed by that network [13]. The SSD was very close to R-CNN in terms of accuracy. This made the SSD the best algorithm that balances speed and accuracy. Due to this, the SSD algo was used broadly in object based detection systems.

3. System description

3.1 Object Detection

Object detection is a computer technology related to computer vision and image processing that deals with detecting the presence of objects with a limitative box and types or classes of objects located in an image in digital images and videos [5]. By Using object detection, visually impaired users can understand their surrounding environment without any challenges and remain independent of others

Input: a picture with one or several objects, like a photograph.

Output: one or more limiting boxes (e.g., defined by a dot, width and height) and a class label for each limiting box.

3.2 Tensorflow

Tensorflow is an open-source software library framework, which was used to implement object detection and recognition. This consists of a pre-trained object detection model, which uses an SSD algorithm to detect objects more efficiently and accurately. This method of object detection uses the COCO mobile net SSD v1 model, which also consists of datasets of 80 object categories, which are commonly found around us.

3.3 Android Studio

Android SDK was being used to make the android application, which can be easily used by visually impaired users for detecting objects and understanding their surrounding environment. The application's front end and backend are implemented using this platform. This platform provides all the libraries and packages required for implementing this system.

3.4 Mobile device-based object recognition

With the ever-increasing advance in smartphone technology, many have tried to implement identification of objects on smartphones [14]. Thanks to smartphones, applications adapted to the blind can be made user-friendly, portable and widely available, eliminating the need for special equipment to do the processing. However, because of a mobile phone's limited processing power, some such applications rely on a client-server architecture [19]. One such well-known application is google goggles, which requires an internet connection and cannot add new images to the application systems for visually impaired users that utilize only local processing bases on the calculation resources of a smartphone like in this document an application was developed for android that performs all the processes locally giving the result in the form of auditory feedback [5]. This implementation uses the functionalities of the SIFT. However, the work proposes in this document is not dedicates to real-time processing. Commercial object recognition applications are available to blind individuals. Looktel developed two purpose-based apps with a particular focus on the visually impaired and they are: Looktel recognizer and Looktel money reader, which was designed for IOS-enabled devices. Both applications perform in real time and do not require an internet connection to operate. Looktel recognition [17] works by pronouncing object names when they are paired to a database that is normally pre-built for the user by a blind person.

4. Proposed system

The system was being implemented on an android app that detects diverse objects in live video feed along with a real-time text reader. In proposed system an object recognition android app was developed used google's tensorflow object detection API model which implemented using SSD algorithm and real-time text reader feature which was using google's TTS engine and google played services mobile vision API which describes the used of text recognizer class to detect text from a real-time video feed. SSD algorithm based object detection model was used for real-time and offline object detection.

4.1. System Outline

The system uses a mobile phone to capture incoming data in live video feed. The application gives two options to the user for detecting objects and reading the text. The camera of the application is automatically accessible and it begins to capture the surrounding objects and texts. Data is sent to the TensorFlow object detection model for processing and later it identifies the class of the objects detected and returns the output as spoken feedback. In the case of reading text, it uses Google play services mobile vision API which consists of TextRecognizer class to detect texts from real time video feed and sends it to google's TTS (text-to-speech) engine for converting text to speech and thus reads out the text detected by the phone's camera.

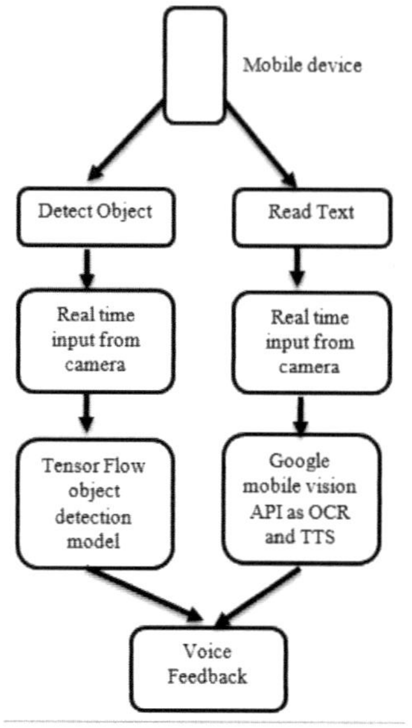

Fig no.1 Flowchart of the proposed system

4.2. Implementation

The system was developed by integrating various technologies, which are mentioned below. Android SDK was used for developing the application because it is the official Integrated development environment (IDE) designed specifically for developing Android applications. [8]. The Android framework supports capturing images and video through the android.hardware.camera2 API or camera Intent. It is a package used for capturing real-time video for object detection and reading text.

Tensorflow library is used for implementing object detection models inside the android application. It provides high performance numerical computing. It has a flexible architecture, making it easy to deploy the calculation through a variety of all possible platforms[9]. SSD-Mobile-Net-COCO model was being used for real time processing. The SSD architecture is a single convolution network that learns to predict bounding box locations and predicts the detected object in the form of limitation boxes. The system uses two object detector modules and real time text reader.

4.2.1 Object Detection: The app is using the SSD-MobileNet-COCO model detecting objects. It utilizes only one neural network for the entire input image. The network model then separates the real time input img into various different regions and predicts the objects using a quadrilateral surrounding the object along with its probability score [12].

4.2.2. Text Reader: The text reader is using Google mobile vision API for detecting texts in real time just like an OCR and then converts the text into speech by using Googles TTS library which was available in android SDK while making application.Using this a user can easily read the menu cards in restaurants, texts on objects(medicines, food,etc.), hotel room no., or even read a paper document, etc.

TextRecognizer: This object processes the images and determines the text contained therein. Once initialized, it can be used to detect text in all picture types.
Reading text feature was implemented using Google Text-to-Speech, which speaks the detected text and acknowledged objects.

4.3. Dataset

In this project, the Common-Object-in-Context (COCO) dataset was used for training the model i.e. SSD MobileNet model, which was able to recognize 81 different categories [22].

5. Analysis of system

In this project, Tensorflow's Object detection model was used which used an SSD algorithm in the backend, and it was able to work by balancing between accuracy and speed. This model successfully detects approximately 81 objects. This model has 74.3 mAP (Mean Average Precision) value, which is highest among the models targeted for real-time processing. After implementing this project, it was expected for a speech feedback for the object which was being detected. But the same object was getting called out multiple times as it got detected. But it will be undesirable to speak out the same object name even if the detection result is the same. Also, it was undesirable if two object names spoken are overlapping or very closely that the user would not be able to distinguish. To solve this problem, if one object was getting detected in the first frame and was speaking out. Then the program will not speak out its class for next five seconds, even if it gets detected. By this the problem of detecting a single object multiple times was being solved.
Upon testing on several objects we found that the results may vary sometimes and the accuracy for detecting objects depends on several parameters. In order to improve the accuracy the model needs to be trained by using objects under different scenarios and test cases. Different cases may include under different light, distance, state of object, direction of object.,etc.
Following are some of the results which show the prediction value of detecting an object accurately. Thus giving us an idea about the accurate performance from the model while detecting objects correctly. Possible Objects can be detected at a time but only objects which have precision value higher than fixed threshold value will be told to visually impaired users using voice output feedback. Multiple objects can even be accurately detected at a single time.

Read text example
(a)

Detecting teddy bear
(b)

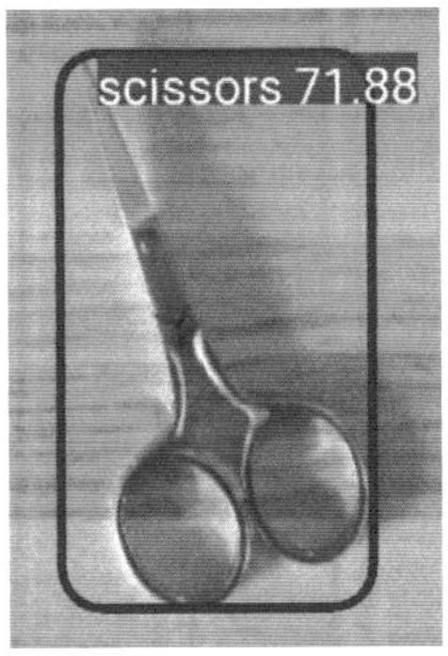

Detecting scissors
(c)

Detecting bottle
(d)

Detecting TV
(e)

Detecting a Laptop
(f)

Table no. 2. Confidence results from object detection of some objects under different cases.

Objects	Normal	Dark	Shadowed object	Long dist.
Mobile	95%	81%	78%	66%
Person	93%	78%	71%	73%
Dog	89%	77%	-	69%
Orange	92%	82%	67%	65%

Laptop	96%	88%	75%	69%
car	91%	-	-	74%

6. Conclusion

In this paper, a model, which was using the SSD algorithm, was made use for creating an application for object detection, which uses TensorFlow object detection API for working offline, and giving maximum accuracy as possible. An object detection API was used for the purpose of detecting objects. The future work includes further enhancing the efficiency of the model by training a large number of images, working on live stream image capturing and recognition, and training the model a higher number of steps for better results. This system's voice synthesis provides convenience features for the visually impaired. Tensorflow lite module was used to create a mobile compatible object recognition model for easy use by visually impaired users. The Android application can be further improved on its stability and functionality.

7. Future scope

For security reasons, wired serial communications were used instead of a wireless server. If the information is linked to a server, it could be leaked onto the Internet. Since the information in question contains a lot of privacy and camera-based observations, such leaks could create critical security issues for users. However, a wired connection can secure the information by keeping it offline [13]. Continuous research is expected to solve server security problems, eliminate blind spots in observations by connecting Internet of Things (IOT) cameras to a secure network, and increase precision in object recognition [18]. This study can be used widely to provide the blind with privacy and convenience in everyday life. With the addition of a face recognition feature, the application can be trained to store information on people closely associated with the person, which would help them to distinguish between peers and outsiders.

References

[1] D. Choi, and M. Kim, "Trends on Object Detection Techniques Based on Deep Learning," Electronics and Telecommunications Trends, Vol. **33**, No. 4, pp. 23-32, Aug. 2018, Article
[2] World Vision Disability Data, WHO, https://www.who.int/blindness/publications/globaldata/en/.
[3] Tom M. Mitchell "Machine Learning". McGraw Hill Education 2017.
[4] Raphael C. Gonzalez & Richard E. Woods "Digital Image Processing," Pearson 2018..
[5] Salman TOSUN, Eni KARARSLAN "RealTime Object Detection App for Visually Impaied:Third Eye". IEEE Conferences 2018.
[6] "Jayshre Punsare, Adit Gaikwad, Vaishnavi Ankarm, Priyanka Kare & Sikha Sharma "Real time Text Reader" International Journal of Computer Applications 182(34):42-45, December 2018
[7] Anirban sarkar, ayush goyal, david hicks and saikat hazra "Android Application Development: A Brief Overview of Android Platforms and Evolution of Security Systems" 2019 Third International conference on I-SMAC.
[8] Tensorflow,https://www.tensorflow.org/ https://www.tensorflow.org/lite
[9] Joseph Redmon, Santosh Divvala, Ross Girshick, Ali Farhadi "(YOLO) You Only Look Once: Unified, Live Object Detection". Cornell University, Jun 2015
[10] W. Liu et al., "SSD: Single Shot MultiBox Detector," European Conference on Computer Vision, Vol. **9905**, pp. 21-37, Sept. 2016, Article (CrossRef Link)
[11] Prince Kumar, Vaibhav Garg, Pavan Somvanshi and Pathanjali C "A Comparative Study of Object Detection Algorithms in A Scene" in IJERT published on 20-05-2019, ISSN (Online) 2278-0181.
[12] A.Quattoni, and A.Torralba. Recognizing Indoor Scenes. IEEE Conference on Computer Vision and Pattern Recognition (CVPR), 2009.
[13] K. Jung, "Object recognition on mobile devices," in Consumer Electronics - Berlin (ICCE-Berlin), 2012 IEEE International Conference on, 2012, pp. 258–262.
[14] R. Girshick, J. Donahue, T. Darrell, and J. Malik, "Object Detection Techniques: Detection and comparison," Proceedings of IEEE Computer Society's conference on computer vision and pattern recognition, 11 2013.
[15] R. Girshick, "Fast R-CNN," in 2015 IEEE International Conference on Computer Vision (ICCV), Dec 2015, pp. 1440–1448.
[16] "Looktel." [Online]. Available: www.looktel.com [Accessed: August 2013] Sarkale, Shah, Chaudhary, & Nagarhalli.
[17] "A novel machine learning approach for object detection and recognition.," 2018 Second International Conference on Creative Communication and Information Technology.
[18] K. Matusiak, P. Skulimowski and P. Strumiłło, "Object detection in a mobile phone app for blind users," Sixth International. Conference on Human System Interactions (HSI), Sopot.
[19] Live object detection using TensorFlow Detection Model. [Available: https://towarddatascience.com/live-object-detection-using-tensorflow-detection-model-e7fd20421d5d]
[20] Microsoft "COCO Dataset", http://cocodataset.org

HealthDesk- A One Stop Health Application Using Machine Learning Techniques

Preet Soni
Computer Engineering, Dwarkadas J. Sanghvi College of Engineering

Yash Gandhi
Computer Engineering, Dwarkadas J. Sanghvi College of Engineering

Vruddhi Mehta
Computer Engineering, Dwarkadas J. Sanghvi College of Engineering

Dr. Ramchandra S. Mangrulkar, Associate Professor, Computer Engineering, Dwarkadas J. Sanghvi College of Engineering

ABSTRACT - HealthDesk is a mobile solution for all health-related queries. The application is for all the users of the healthcare system. It covers all the major domains in the hierarchy. People have now started digitizing in this domain too. People have shown great trust in this. This helps us to provide the best. Despite this, individuals often face problems ordering their medicines online. There are high chances of people taking medications without a prescription. So, this application recommends top medicines that are similar to the one being ordered to avoid drug-drug interactions. In the scenario of patient emergency many applications for patient health monitoring and appointment scheduling have been developed. However, in the wake of an emergency, people tend to blank out or are unaware of nearby emergency services. So, the application has a feature that enables users to search the most nearby doctor and provide the doctor with the user's current location. The user can be provided with first aid immediately so that he doesn't succumb to death. The healthcare system has frequent updates. The doctors must remain at par with the updates. However, doctors find it strenuous to sit by and search. The app provides them with relevant news according to their preferences. Concluding, this app covers the most important stakeholders of the healthcare system.

Keyword - NLP, DNR, Natural language tool kit, Support vector machine, Machine Learning, SMILES, RDKit.

1. INTRODUCTION

In this project the user can call for an emergency whenever needed, the patient who has the application also to help the patient in classifying and suggesting them the best results by making a comparison of data based on x features. The project is categorized into 3 phases. Nowadays, with advancement in technology, there is an increase in number of websites, and mobile applications that are developed to solve real-life problems people were once facing. The scope for the development of this android application is focusing on the healthcare domain, which has three purposes. One is to help people order correct drugs. With users able to buy drugs from the platform, we make a judicious use of data of which drug the patient has ordered. The reason of choosing this is due to increased medical costs, high consultation fees, fear of being put through unnecessary medical tests, many people tend to turn to self-medication. With regard to this they may intake drugs at irregular intervals without realizing it. Increased intakes can cause adverse effects. For example, higher doses of sleeping pills can result in dire consequences. In order to curb the high intake of drug if any, whenever users buy a drug through the app, it alerts the user if the medicines are interacting using Machine learning model based on Clustering technique. The tag line of this application for public users is "Find. Plan. Search." Also, if the user is in any kind of emergency like accident, cardiac arrest, shock, etc., the app will

immediately search for a doctor in your shortest distance and provide the doctor the user's current location. Healthcare is an ever-changing field of practice with advances in medicine, expanded evidence sources, new treatment options, and changing governmental regulations and models of care. Lifelong learning is not an option in healthcare, it's required by healthcare workers to remain relevant and up to date with such information and continue providing safe, effective patient care. Healthcare professionals are busy round the clock with an increasing amount of patients seeking care. For this reason, developing an app in order to provide the relevant information to healthcare professionals within very less amount of time and in a structured manner is necessary. The app offers immediate up-to-date medical information and reference material at the touch of the finger that is continuously updated.

2. LITRATURE SURVEY

A. Drug Name Recognition:

Approaches and Resources Authors: Shengyu Liu, Buzhou Tang, Qingcai Chen, Xiaolong Wang With increased development in the field of Computer Science, there is an increase in the amount of medical information available. This information can be obtained from various sources like medical journals, medical patent applications, clinical trial documents, medical literature, electronic medical records, etc. In order to exploit the information available, it is necessary to extract valuable information and process it for relevant purposes. Drugs, being a basic entity for medical evaluations become an important target here to be extracted effectively. This Paper mentions an approach termed Drug Name Recognition (DNR), which seeks to recognize drug mentions in unstructured medical texts and classify them into predefined categories, to

help for gaining further insights. Thus, would apply the above approach, which would help us in achieving the objective of providing relevant information to doctors.
The paper mentions basic steps to be followed viz Gather Unannotated Texts from Various Sources, Pre-processing of the text, Selecting Appropriate DNR approach, and finally Post Processing. The main aim would be to select and implement the DNR approach according to the use case. The different approaches cited in the paper include Dictionary based approach, Rule-based approach, Machine Learning approach, and hybrid approach. As of now the Machine Learning Based approach seems to be effective. The process that is mentioned is that this method formalizes DNR as a classification problem or a sequence labelling problem. The tokens generated in pre-processing steps are each designated with a class label following the BIO label technique. The four types of drugs (group, nonhuman, brand, drug) are defined. An instance of Bio tagging results of a sentence from the DDI Extraction is demonstrated below. It consists of a sentence and the BIO tag generated represented as BIO: Sentence: PXR ligands include a wide variety of pharmaceutical agents, such as antiepileptic drug, Taxol, rifampicin, and human immunodeficiency virus protease inhibitors such as ritonavir and saquinavir. BIO: PXR-ligands\O include\O a\O wide\O variety\O of\O pharmaceutical\O agentst\O .\such\O as \O antiepileptic \B-group drugs \ I-group ,\O taxol\B-brand \O rifampicin\Bdrug\O and\O human\ B-group immunodeficiency\l-group virus \lgroup protease\l-group inhibitors\I-group such\O ritonavir\B-drug and \O saquinavir \B-drug .\O. This tagging shows us to extract the drug names and so would use it for further implementation to achieve the objective.[1]

B. A Counter Medication Classifier Using Machine Learning and Computer Vision Techniques.

Authors: Eduardo Ceh-Varela , Gandhi HernándezChan , Marisol Villanueva-Escalante, Giner AlorHernández , José Self-medication is one of the major health concerns worldwide and the World Health Organization (WHO) has emphasized correctly
investigating and controlling it. Self-medication or self-prescription can be defined as the action of obtaining and consuming one or more medicines without the advice of a physician nor prescription for treatment. This paper presents the risks about self-medication and proposes a machine learning approach to reduce the risks associated. This paper has the main motive to present a tool that helps in better decision making as to which counter medications to buy. This is done by classifying drugs into different classes/clusters and then providing relevant info like class, description, use of the drug. The authors have used K-means Algorithms to generate clusters. The K factor was determined by comparing two methods, Rule of Thumb and the Elbow Method. Finally, after averaging the K value from both methods, it came out to be 147. Further, the SVM algorithm, a supervised ML algorithm is employed to generate the classification model. The SVM classifier is proposed of achieving an accuracy of 97%. Though the paper implements this along with image processing, the objective was to note the underlying concept of the classification approach used, to aid in this task.[2]

C. ANDROID APPLICATION FOR EMERGENCY MEDICAL ASSISTANCE (DOCTORS NEARBY)

Authors: Dhanesh Sharma, Priyanka Dubey, Navin Singh, Prof. Sulachana Devi The time-lapse in medical help to accident victims is a concern of increasing urgency in India and other countries. To know the locations of victims would be an important function in this case. Google Map API can be used to fetch the location. Objective is to develop a mobile application with integration of this API and other added features. According to statistics , 92.59 % of the total Smartphones sold in 2019 in India were the Smartphones that run on Android Operating System. This paper provides the implementation of an android application in order to provide a solution to the concerned issue . Searching for a nearby doctor during an emergency is one of several mentioned implementations in the application. In case of emergency when the user presses the "emergency button" present in the application, a list of nearby doctors, using the latitude and longitude of the present location and destination location is fetched from the internet using Google Maps API and the distance is calculated. The doctor nearest to the emergency spot, with respect to the minimum distance calculated, is notified about the patient's location, and the shortest route to the patient's spot is displayed to the doctor's screen. This involves implementing different modules. Modules cited in the paper include Login Module: The Signup and Login pages with database connectivity. Emergency Module: The module that covers the functionality of the emergency button which is a major component of the application. Google API module: The display of maps, markers for locations, calculation of routes is handled by this module. Database Module. The User credentials, doctor credentials, which doctor is currently assigned to which patient and all such information is needed to be effectively stored.[3]

D. LeMeNo: Personalised News Using Machine Learning

Authors: D. Khandelwal, D. Shanbhag, A. Shriyan, R. Thorve and Y. Borse The paper demonstrates an application "LeMeNo" that provides user personalized new recommendations. In the day and age, people are not able to keep up with the latest news. With an abundance of news flashing on screens, it is essential to read only the relevant data which appeals to the individual. The need to say "Let Me Know" inspired the author. Based on the user's preferences and engagements in the articles, the news feed is customized. The author presents a literature survey of various topics needed for the recommendation system. He later proposes a model for the same.

The author has further described his implementation steps in detail. Firstly, the author gathers the information using the NLTK library. Keywords are extracted from the titles and descriptions of articles and stored. Then, he classifies them into different categories and subcategories using Naïve Bayesian & SVM algorithms. The application asks for preferences at the user signup. Each article can be liked or disliked. Further recommendations are made, based on this and his engagements. On liking or disliking a particular news article, the system identifies all the keywords associated with that news and creates a vector of these keywords. The author obtains a dataset for a user, which

https://doi.org/10.1051/itmconf/20214003002

consists of tags and scores. The score is added by 0.5 for every like. Similarly, subtracted by 0.5 for disliking a particular tag. The author uses this data to match the scores with other users and design a new recommendation based on it. Lastly, the author has created an event timeline to keep the users aware of future events. There is an imperative need to create a solution that can direct users to the relevant articles based on his preferences.[4]

3. RESEARCH GAP

The system design has certain drawbacks. Firstly, despite trying to bring the project to the highest accuracy possible. The project can be worked open in the future. With healthcare and medication into the picture, multiple things are to be taken into consideration. We have tried to calculate similarities based on the structure of the drugs. There are various other types of measures to be considered like phenotypic, therapeutic, genomic, etc. Secondly, the role of belief towards IT healthcare. Despite alerting the user on self-medication, if the user is a person who resists technology in terms of reliability, then the application lacks support. The application will not do any good for the user. They may not resist the technology but the effect of it in their lives. Social influence also takes a makes a curve. Another drawback is the use of healthdesk in remote locations. In areas such as villages with limited doctors available already, usage of the app will not make a difference in assistance. During emergencies, it will be difficult to find a doctor nearby.

4. PROPOSED ARCHITECTURE WITH MODULAR DESCRIPTION

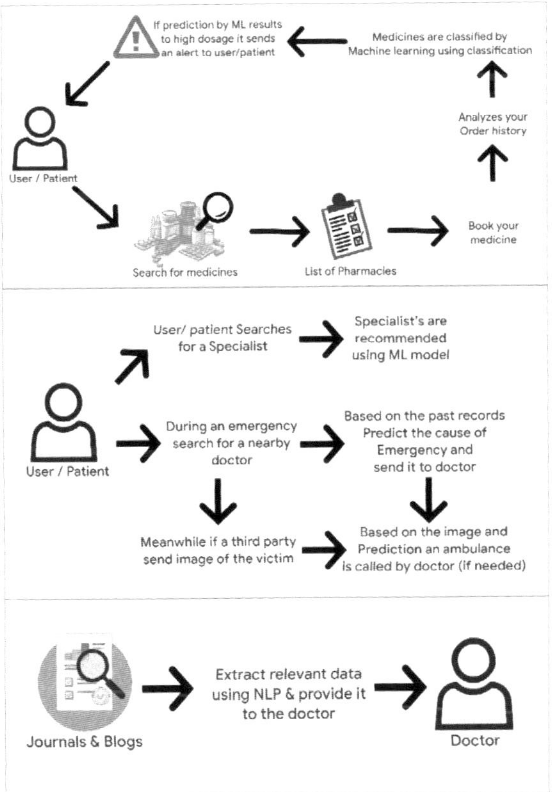

Figure 4.1. Proposed Architecture

The figure 4.1 above is the proposed architecture of HealthDesk. An android application that searches for a doctor during an emergency within the nearest radius from the point of the accident. From a database of hundreds of registered doctors on the application, the nearest doctor is assigned for the patient's assistance, providing the doctor with all the details of the patients through the application including location, medical history, etc. Thus, the doctor can immediately reach out to the patient and initial medication is provided. To curb the issue of high drug intake, an application is developed which keeps a check on the drugs purchased, and alerts the patient if the drug is a Schedule H drug and the patient is unaware of it, or if the drug he purchased from one pharmacy would have a side effect with a drug purchased a few days back from another pharmacy.[5] This is implemented using Machine Learning and applying appropriate learning models on clusters which are different groups of medicines based on similar effects on consumption. This helps prediction to be more dynamic. With the evolvement of information technology, medical documents are available progressively, which contains a magnanimous amount of medical data, such as medical entities and relations between them. It is crucial to extract valuable information from them, to take leverage of medical texts. Drugs being one type of basic medical elements, also need to be recognized. Drug names that appear in unstructured medical texts can be classified into predefined categories. This can be done by a technique termed DNR (Domain Name Recognition), which is a fundamental task of medical information extraction and is a key component of many medical relation extraction systems and applications (e.g., information management, drug discovery, information tracking, information retrieval, clinical decision support, and drug development).This project uses NLP with the abovementioned concept of DNR and thus provide relevant updates to the doctors.[6][7].

5. EXPERIMENTATION AND RESULTS

Indians prefer self-medication. They may go to the doctor once, and if they get a similar type of problem, for example, fungal infections, they tend to assume and take the same medicine. Consumption of self-medication can prove to be fatal. In the case of adults, most of them consume multiple drugs, which is called polypharmacy. They might interact and cause unknown problems. Hence the application can guide such people by providing the top medicines which interact with what they want to order. So, there are various ways to find the similarity. The application uses structural similarity. The drug name generates as smiles data, and then we calculate the fingerprint. The project makes use of the Tanimoto coefficient, like cosine similarity, to find the drug similarity.

https://doi.org/10.1051/itmconf/20214003002

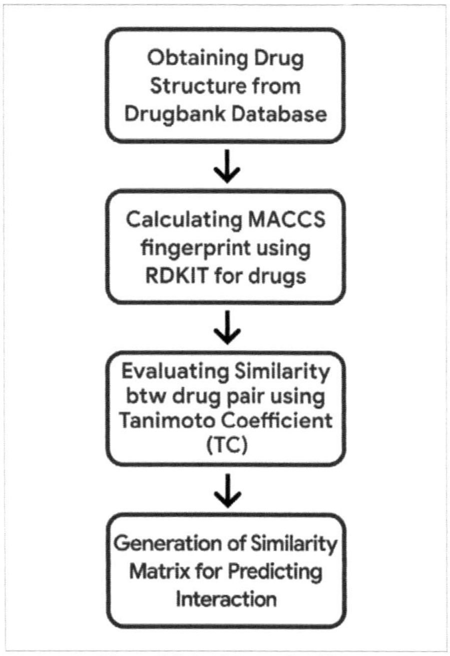

Figure 5.1. Implementation Flow

The model has taken into consideration the Tanimoto coefficient formula for the calculations as in figure 5.2:

$$\mathbf{T = N_{A\&B} / N_A + N_B - N_{A\&B}}$$

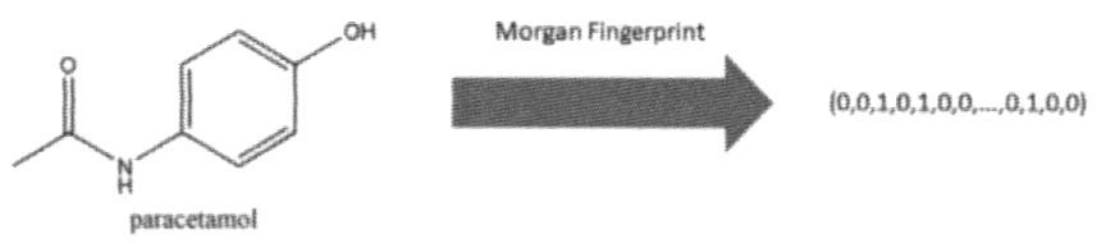

Figure 5.2. Fingerprint

- ***NA represents the number of "on" features (bits) in structure A.***
- ***NB represents the number of "on" features (bits) in structure B.***
- ***NA&B represents the number of "on" features (bits) common to both fingerprints A and B.***

The molecular structure is denoted as strings of special characters using SMILES. The similarity between compounds can be computed, since each molecule is represented as a string.

Example: ***Aspirin SMILES: CC(=O)OC1=CC=CC=C1C(O)=O***

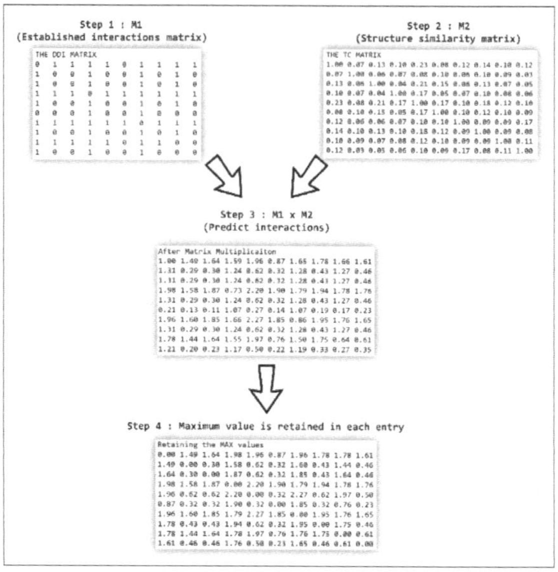

Figure 5.3. Flowchart for DDI

The flow chart in the figure 5.3 above represents establishing a DDI prediction model through a combination. In step one, the M1 is computed by the interaction of approved drugs in the Drug Bank. In step two, M2 is created by the TC values of structural similarity. "1" stands for the interaction which has occurred between two drugs. "0" stands for the opposite situation. Next, multiply the matrix and retain the maximum value in each entry. In step 4, M is formed by retaining the maximum values. Since the interaction of the drug with itself is not considered, set the values in the diagonal of the matrices to "0".[5] Figure 5.4 shows the Confusion Matrix for a TC cut-off value of 0.17 whereas figure 5.5 shows the Confusion Matrix for a TC cut-off value of 0.22.

```
Accuracy: 0.68
Mis-Classification: 0.32
Sensitivity: 0.63
Specificity: 0.74
Precision: 0.74
f_1 Score: 0.68
```

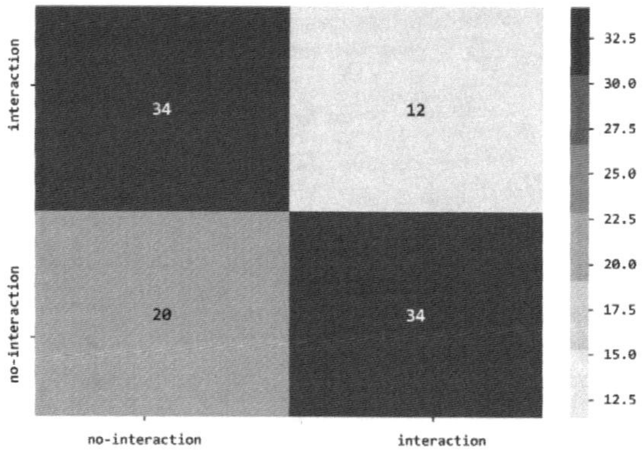

Figure 5.4. Confusion Matrix for threshold value 0.17

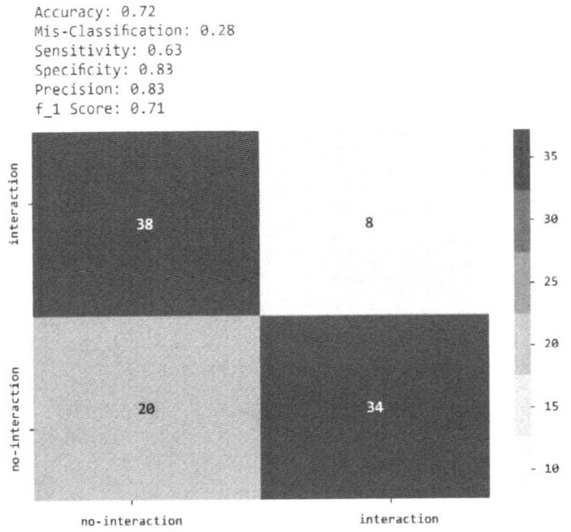

Figure 5.5. Confusion Matrix for threshold value 0.22

According to the results ,the performance measures are higher in case when a higher TC cut-off value was taken . An accuracy of 0.68 and 0.72 for the two cut-offs. Since TC value is the measure of the extent of similarity , this model justifies that Drugs with higher degree of similarity would have higher chances of interactions between them

Parameters\ TC	***0.22***	***0.17***
Accuracy	***0.72***	***0.68***
Mis-Classification	***0.28***	***0.32***
Sensitivity	***0.63***	***0.63***
Specificity	***0.83***	***0.74***
Precision	***0.83***	***0.74***
F-1 Score	***0.71***	***0.68***

Table 5.1. Comparison for different threshold values.

6. CONCLUSION

The project highlighted a Health based application that would help a user/patient to call for an emergency whenever needed. The developed application is a Machine learning-based Health application. The application can be used for calling an emergency through the mobile phone. The user can also search for a particular medicine and a list of pharmacies will be displayed from where the users can directly book the medicine. The application predicts top drugs which interact with the drug being ordered. The last one is designed especially for doctors where the model will extract relevant data using NLP and provide useful information to the doctors through the application.

7. REFERENCES

1. Liu, Shengyu & Tang, Buzhou & Chen, Qingcai & Wang, Xiaolong. (2015), "Drug Name Recognition: Approaches and Resources." Information. 6. 790-810. doi : 10.3390/info6040790.
2. Ceh varela, Eduardo & Hernández-Chan, Gandhi & Villanueva-Escalante, Marisol & Alor-Hernández, Giner & Sánchez-Cervantes, José. (2017). A Counter Medication Classifier Using Machine Learning and Computer Vision Techniques. Research in Computing Science. 132. 9-22. 10.13053/rcs-132-1-1.
3. Dhanesh Sharma, Priyanka Dubey, Navin Singh, Prof. Sulachana Devi, " ANDROID APPLICATION FOR EMERGENCY MEDICAL ASSISTANCE (DOCTORS NEARBY) ".International Journal of Scientific & Engineering Research, February 2018, ISSN 2229-5518. Web Address : https://www.ijser.org/researchpaper/android- application-for-emergency-medical-assistance-doctors-nearby.pdf
4. D. Khandelwal, D. Shanbhag, A. Shriyan, R. Thorve and Y. Borse, "LeMeNo: Personalised News Using Machine Learning," 2018 Fourth International Conference on Computing Communication Control and Automation (ICCUBEA), Pune, India, 2018, pp. 1-4, doi: 10.1109/ICCUBEA.2018.8697560.
5. Huang, De-Shuang & Bevilacqua, Vitoantonio & Premaratne, Prashan. (2016). Intelligent Computing Theories and Application: 12th International Conference, ICIC 2016, Lanzhou, China, August 2-5, 2016, Proceedings, Part I. 10.1007/978-3-319-42291-6.
6. Santiago Vilar, Rave Harpaz, Eugenio Uriarte, Lourdes Santana, Raul Rabadan, Carol Friedman, Drug—drug interaction through molecular structure similarity analysis, Journal of the American Medical Informatics Association, Volume 19, Issue 6, November 2012, Pages 1066–1074, https://doi.org/10.1136/amiajnl-2012-000935
7. Vilar S, Uriarte E, Santana L, et al. Similarity-based modeling in large-scale prediction of drug-drug interactions. Nat Protoc. 2014;9(9):2147-2163. doi:10.1038/nprot.2014.151

Fake News Detection using Machine Learning

Prasad Kulkarni[1,*], *Suyash* Karwande[1,**], *Rhucha* Keskar[1,***], *Prashant*Kale[1,****], and *Sumitra* Iyer[1]

[1]*Electronics and Telecommunication Department, SIES, Graduate School of Technology, Navi Mumbai, India.*

Abstract. Everyone depends upon various online resources for news in this modern age, where the internet is pervasive. As the use of social media platforms such as Facebook, Twitter, and others has increased, news spreads quickly among millions of users in a short time. The consequences of Fake news are far-reaching, from swaying election outcomes in favor of certain candidates to creating biased opinions. WhatsApp, Instagram, and many other social media platforms are the main source for spreading fake news. This work provides a solution by introducing a fake news detection model using machine learning. This model requires prerequisite data extracted from various news websites. Web scraping technique is used for data extraction which is further used to create datasets. The data is classified into two major categories which are true dataset and false dataset. Classifiers used for the classification of data are Random Forest, Logistic Regression, Decision Tree, KNN and Gradient Booster. Based on the output received the data is classified either as true or false data. Based on that, the user can find out whether the given news is fake or not on the webserver.

1 Introduction

The term 'Fake news' refers to the news content that is false, misleading, or fabricated, in which the facts, sources, or quoted statements of the news content are unverified. Fake news has existed in the form of gossip, rumor, and misinformation throughout human history [1]. To increase its effectiveness this Fake news is spread throughout social media. Along with the billions of people using social media, there are also robots, or simply bots, residing within. These Bots help to propagate fake news faster and boost up its popularity on social media.

Fake news detection is used to avoid rumors from spreading across various platforms, such as social media and messaging platforms. The impetus for this work is to avoid the spread of Fake-news which can even lead to worse activities. There has been a rise in the news lately about lynchings and riots that result in mass deaths; fake news detection aims to detect these and stop similar activities, thereby protecting society from these unwelcome violent acts [3]. The proposed system helps to find the authenticity of the news. The news given by the user is classified as true or false based on the data collected using Web Scraping. This task uses five various classification models, including Random Forest, Logistic Regression, Decision Tree, KNN, and Gradient Booster. To improve prediction accuracy, a mixture of these models is tested.

Further, the paper is structured as follows: section 2 takes a glance at previous work done in fake news detection. In the next section, data extraction, pre-processing and classifiers are discussed. Section 4 depicts the classifier accuracies and related results. Finally, In section 5 concluding remarks are mentioned.

2 Related Work

There have been quite several initiatives taken to achieve fake news detection. In [3], Mykhailo Granik et. al. showed a simple approach for the fake news detection system using a naive Bayes classifier model. This was implemented as a software system and then tested against a dataset of Facebook news or the posts on Facebook. The news was gathered from three Facebook pages, as well as three large mainstream political news pages (Politico, ABC News, CNN). They were able to achieve an accuracy of around 74 percent. Classification accuracy for false news is a little worse. This could have been caused by the skewness of the dataset, only 4.9 of it is fake news.

The author uses different ideas for processing the text dataset such as TF-IDF, Count Vectors, and Word Embedding [5]. Further, the author implements the comparison on various classification models which includes SVM, Recurrent Neural Network model, Logistic Regression (LR), and Naïve Bayes Method. Based on the comparison the author has examined the scores like recall and precision etc of the various models.

An overview of qualitative data cleaning with error repairing and error detection approaches is discussed in [6]. Cleaning of data techniques was focused on the errors like duplication, inconsistency, and missing values were dealt with. It also described a statistical perspective on qualitative data cleaning with the help of Machine Learning techniques.

*e-mail: prasadpkulkarni99@gmail.com
**e-mail: suyashkarwande20@gmail.com
***e-mail: rhucha.keskar17@siesgst.ac.in
****e-mail: prashant.kale17@siesgst.ac.in

In [4] Avinash Shakya et. al. used aggregators in their study Smart System for Fake News Detection to see news from various sources in a single convenient location. Checking RSS Feeds regularly, extracting articles from various news sites, and gathering information are all part of the basic methodology. The proposed plan is a mixture of Naive Bayes classifiers, SVM, and semantic investigation due to the multi-dimensional nature of fake news. The proposed plan is entirely based on Artificial Intelligence approaches, which are essential to precise order between the genuine and the fake. The three-section strategy combines Machine Learning calculations, which are subdivided into managed learning procedures, with traditional language preparation techniques [4].

In [7], a variety of topics of web scraping, starting with a simple introduction and a brief review of various web scraping software and applications. The process of web scraping, as well as the numerous sorts of web scraping techniques, before closing with web scraping's pros and downsides, as well as a full discussion of the numerous fields in which it can be employed have been discussed. Open Data, Big Data, Business Intelligence, aggregators and comparators, development of new applications and mashups, and so on are just a few of the possibilities available with this data.

The researchers in [8] proposed to focus on different feature engineering methods for generating feature vectors, such as count vector, TF-IDF, and word embedding. Seven distinct ML classification algorithms are trained to categorize news as false or real, and the top one is chosen based on accuracy, F1 Score, recall, and precision.

3 Methodology

Any Machine Learning model primarily requires a set of data to train or test model. To extract vast volumes of data from websites and save it in table format to a local file or a database, we used web data extraction popularly known as the web scraping technique. The methodology used is by collecting all of the data retrieved from multiple sources using the vivid characteristics of the web crawler 'Scrapy' and python scripts and then analyzing it according to the requirements. The python-based web crawler 'Scrapy' may also assist us in retrieving the desired result, as we analyze the process with specific code and provide the necessary URL for the iteration to scrape the data from the source URL [7]. Figure 1 represents the workflow.

Further, the collected data is separated into two groups: A training set and a testing set. Train/Test is a method to measure the accuracy of the model. The general idea is to train an algorithm on a huge number of manually examined web pages [1].

Raw content needed certain data pre-processing before it could be fed into the simulations. Data Preprocessing is a technique for data exploration that converts original data into a suitable form. Actual data (real-life data) is often inaccurate and therefore could not be sent over the design with that information. This may cause some mistakes. So while we send over a system, we have to pre-process data. After reading the dataset we use some preprocessing functions like tokenizing, stemming, etc. The material and information were taken from websites that were thought to be involved with fake news. Before using a machine learning system, text must be translated into numbers. The predictive algorithm takes documents as input and generates a class label as output for document classification. For the algorithm to accept the texts as input, they must be transformed into fixed-length vectors of numbers. After parsing the text, a procedure known as tokenization is used, in which particular terms are deleted. With the help of the feature selection and extraction method, we are manually selecting relevant features which will contribute most to the prediction variable and to increase the accuracy of the model. For feature selection, techniques like bag-of-words and n-grams and then TF-IDF weighting from sci-kit learn python libraries were used. The Bag of Words model is a basic and effective machine learning model for parsing text texts. The model ignores all word order information and simply looks at how many times the words appear in the document. This model can be implemented in two different ways.

- TF-IDF Vectorizer.
- Countvectorizer.

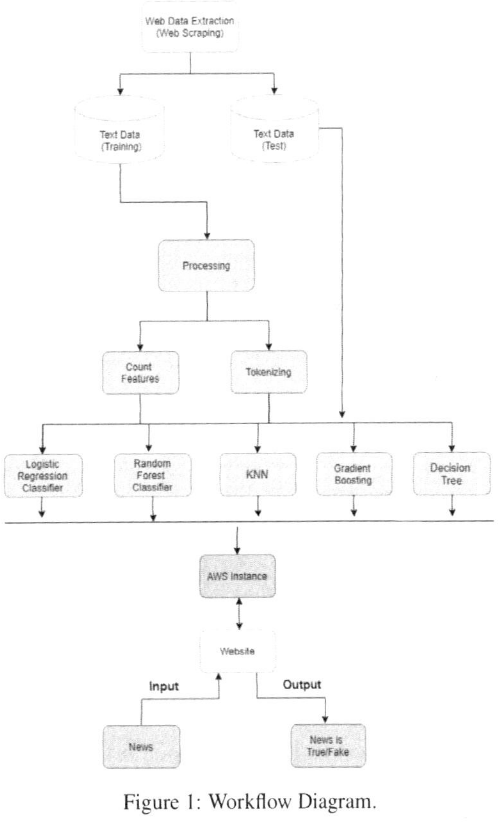

Figure 1: Workflow Diagram.

The Count Vectorizer creates an encoded vector that comprises the full vocabulary's length as well as the frequency with which each word appears in the document. A popular approach is to use the TF-IDF algorithm to calculate word frequencies (Term Frequency Inverse Document). Each document is then assigned a TF - IDF score.

TF-IDF uses word frequency to identify words that are more important (occur more frequently) in a document. The TF-IDF Vectorizer converts documents into tokens, learns vocabulary, inverses document frequency weightings, and allows you to cipher new documents [2]. When compared to a non-vectorized implementation, vectorization in this technique can significantly speed up the calculation process.

We are using different classifiers for predicting fake news. A classifier will help us in ordering data automatically or categorizes data into one or more of a set of 'Classes'. Different classifiers will use the extracted features. All of the extracted features will be used by the classifiers. The classifiers employed in the Machine Learning model are as follows

3.1 Logistic Regression

The advantages of logistic regression include probability modeling, the capacity to depend on features, and the flexibility to update the model. However, for higher accuracy, logistic regression requires a big data set, but Naive Bayes may function with small datasets as well.

3.2 Decision Trees

A decision tree is made up of decision nodes that start at the top and work their way down. Dependent characteristics, no need for linear class separation, fast management of outliers, and intuitive decision tree interpretation are all advantages of employing a decision tree. When there are a significant number of sparse features, however, a decision tree will overfit and perform poorly on the testing data [2].

3.3 Random Forest Classifier

Many decision trees are built by the random forest algorithm. Utilizing a subset of features, each decision tree is created. Each decision tree produces one class and eventually bootstraps the votes to obtain better accuracy from the Random Forest technique. A tree-shaped pattern is used to describe the plan of action in a decision tree. At any node, a decision will be made.

3.4 KNN Classifier

The KNN method assumes that the new case/data and existing cases are similar and places the new case in the category that is most similar to the existing categories. This means that new data can be quickly sorted into a suitable category using the KNN algorithm. The KNN algorithm can be used for both regression and classification, though it is more commonly used for classification problems.

3.5 Gradient Boosting Classifier

Gradient boosting creates an ensemble of weak prediction models, usually decision trees, as a prediction model. The resulting technique is called gradient boosted trees when a decision tree is a weak learner, and it usually outperforms random forest. It constructs the model in the same stage-by-stage manner as other boosting approaches, but it broadens the scope by allowing optimization of any differentiable loss function.

We will compare the score and examine the confusion matrix after we have fitted the model. When all of the classifiers have been fitted, we will dump those models and vectorization models which will be used while connecting the model with the webserver.

Next, to connect the webpage to the model we chose flask as a web development framework, which is deployed on Amazon Web Services (AWS) instance. The input received from the website will be given to the AWS-EC2 instance. This instance has all required files of machine learning models, flask files, and the front-end web server. Once the required models are dumped from the machine learning model we will further use only those dumped models in the flask file. For the front-end of the website, we used HTML, CSS, and Bootstrap which will help us to provide easy access to the webserver. The output received will be again displayed to the webserver. This deployment will help us to provide access to the webserver on any internet-enabled devices. Since it uses responsive code, the fluidity of the frontend makes it platform-independent. Whenever we run the instance we will be getting a unique public DNS which will provide us direct access to the webserver. This public DNS will keep changing whenever we run the instance.

4 Implementation and Results

For implementation and better result purpose we created a dataset a CSV file we made it by using web scraping method we scraped the news articles for authentic news web sites and created a dataset having more than 20,000 news here below is the screenshot of the result of making the dataset using web scraping[9] [10].

Further, a Jupyter Notebook was created to implement the ML program. We have used Logistic Regression, Gradient Boosting, k-nearest neighbors, Decision Tree, and Random Forest. After TF-IDF vectorization and cleaning the data we train and test them according to these classifiers we got the accuracy for Logistic Regression as 85.04%, Decision Tree as 78.11%, Gradient Boosting as 77.44%, Random Forest as 84.50%, KN Neighbors Classifier as 80.20%

The columns of Table 1 are defined as follows

- Classifiers: Models that are used to train and test data are known as classifiers.
- Accuracy: How often a data point is correctly classified by the algorithm.
- Precision: The number of accurately predicted positive observations divided by the total number of predicted positive observations.

Classifiers	Accuracy	Precision	Recall	F1-Score
Logistic Regression	85	84	87	85
Decision Tree	78	79	79	79
Gradient Boosting	77	70	84	76
Random Forest	84	83	86	84
KN Neighbors	80	75	85	79

Table 1: Tabular comparison of the classifiers (in %)

- Recall: The percentage of accurately anticipated positive observations to the total number of observations in a class is known as recall.
- F1 Score: The F1 Score is the weighted average of Precision and Recall.

Confusion Matrix provides the values needed to calculate the F1 score, Recall, and Precision. It is a table that demonstrates how well a classification model (or a classifier) performs on a set of test data with known True values [8].

Figure 2: User-Interface.

These prediction values are calculated for all classifiers and then the average of all these prediction values will be taken as the final percentage. Based on these values we are setting a range that will help to find the percentage of the truthness of the news. The webserver we created is displaying the result followed by the news. To catch the attention of users we used some emoticons as a symbol to show the results more effectively. The created web server is platform-independent. It means that all of the arrangements of the webserver will be independent of devices. Bootstrap helped us to make our web server device-independent. The final output is in the form of a message which is different according to different percentage values. It is displayed as shown in figure 2.

5 Conclusion

In this paper, we looked at a computerized model for verifying news extracted from social media, which provides expository demonstrations for recognizing fake news. Following the demonstration that even the most basic algorithms in domains such as AI and Machine Learning can produce a reasonable result on such a critical issue as the spread of fake news around the world. As a result, the findings of this investigation suggest that systems like this could be very useful and effective in dealing with this critical issue. Web scraping is also a key part of this paper as the scraped data will be based on real-time news and will be more reliable than the ready-made datasets available all over the internet. It is an efficient and fast process and also it is very easy to maintain. The dataset used in this study is expected to be used in arrangements that use machine learning-based statistical calculations, for example, Logistic Regression (LR), Decision Tree, Gradient Booster, Random Forest, and KN Neighbours. In the future, the prototype's efficiency and accuracy can be improved, as well as the proposed model's user interface.

References

[1] Burkhardt, Joanna M. Combating fake news in the digital age. Vol. 53, no. 8. American Library Association, 2017.

[2] de Lima Salge, Carolina Alves, and Nicholas Berente. "Is that social bot behaving unethically?." Communications of the ACM 60, no. 9 (2017): 29-31.

[3] Granik, Mykhailo, and Volodymyr Mesyura. "Fake news detection using naive Bayes classifier." In 2017 IEEE First Ukraine Conference on Electrical and Computer Engineering (UKRCON), pp. 900-903. IEEE, 2017.

[4] Jain, Anjali, Avinash Shakya, Harsh Khatter, and Amit Kumar Gupta. "A smart System for Fake News Detection Using Machine Learning." In 2019 International Conference on Issues and Challenges in Intelligent Computing Techniques (ICICT), vol. 1, pp. 1-4. IEEE, 2019.

[5] Mahir, Ehesas Mia, Saima Akhter, and Mohammad Rezwanul Huq. "Detecting fake news using machine learning and deep learning algorithms." In 2019 7th International Conference on Smart Computing Communications (ICSCC), pp. 1-5. IEEE, 2019.

[6] Yalçın, Mehmet Adil, Niklas Elmqvist, and Benjamin B. Bederson. "Keshif: Rapid and expressive tabular data exploration for novices." IEEE transactions on visualization and computer graphics 24, no. 8 (2017): 2339-2352.

[7] Singrodia, Vidhi, Anirban Mitra, and Subrata Paul. "A Review on Web Scrapping and its Applications." In 2019 International Conference on Computer Communication and Informatics (ICCCI), pp. 1-6. IEEE, 2019.

[8] Smitha, N., and R. Bharath. "Performance Comparison of Machine Learning Classifiers for Fake News Detection." In 2020 Second International Conference on Inventive Research in Computing Applications (ICIRCA), pp. 696-700. IEEE, 2020.

[9] Thomas, David Mathew, and Sandeep Mathur. "Data analysis by web scraping using python." In 2019 3rd International conference on Electronics, Communica-

https://doi.org/10.1051/itmconf/20214003003

tion and Aerospace Technology (ICECA), pp. 450-454. IEEE, 2019.

[10] Diouf, Rabiyatou, Edouard Ngor Sarr, Ousmane Sall, Babiga Birregah, Mamadou Bousso, and Sény Ndiaye Mbaye. "Web Scraping: State-of-the-Art and Areas of Application." In 2019 IEEE International Conference on Big Data (Big Data), pp. 6040-6042. IEEE, 2019.

https://doi.org/10.1051/itmconf/20214003004

Indian Sign Language Recognition using Convolutional Neural Network

Rachana Patil[1,*], *Vivek* Patil[1,**], *Abhishek* Bahuguna[1,***], and *Mr. Gaurav* Datkhile[1,****]

[1]Ramrao Adik Institute Of Technology,Nerul, India

Abstract. Communicating with the person having hearing disability is always a major challenge. The work presented in paper is an exertion(extension) towards examining the difficulties in classification of characters in Indian Sign Language(ISL). Sign language is not enough for communication of people with hearing ability or people with speech disability. The gestures made by the people with disability gets mixed or disordered for someone who has never learnt this language. Communication should be in both ways. In this paper, we introduce a Sign Language recognition using Indian Sign Language.The user must be able to capture images of hand gestures using a web camera in this analysis, and the system must predict and show the name of the captured image. The captured image undergoes series of processing steps which include various Computer vision techniques such as the conversion to gray-scale, dilation and mask operation. Convolutional Neural Network (CNN) is used to train our model and identify the pictures. Our model has achieved accuracy about 95%

Keywords: Convolutional Neural Network(CNN), Hand Gesture, Deaf people, Sign Language, Sign language Recognition(SLR), ISL(Indian Sign Language).

1 Introduction

One of the most important requirements for social survival is communication. Deaf and dumb peoples communicate with one another using sign language, but it is difficult for non-deaf and dumb people to understand them. While much study has been done on the recognition of American sign language, Indian sign language varies greatly from American sign language. ISL communicates with two hands (20 out of 26), while ASL communicates with a single hand. Because of the overlapping of hands when using both hands, features are often obscured. Furthermore, a lack of datasets, combined with the fact that sign language varies depending on location, has limited ISL gesture detection efforts. This paper aims to take the first step in using Indian sign language to bridge the communication gap between normal people and deaf people. The extension of this project to words and common phrases will not only make it easier for deaf and dumb people to communicate with the outside world, but it may also help in the development of autonomous systems for understanding and assisting them.

The aim of this paper is to use the corresponding gesture to recognise alphabets in Indian Sign Language. The identification of gestures and sign languages is a well-studied subject in American Sign Language, but it has received little attention in Indian Sign Language. We want to solve this issue, but instead of using high-end technologies like gloves or the Kinect, we want to recognise gestures from photographs (which can be accessed from a webcam), and then use computer vision and machine learning techniques to extract specific features and classify them.

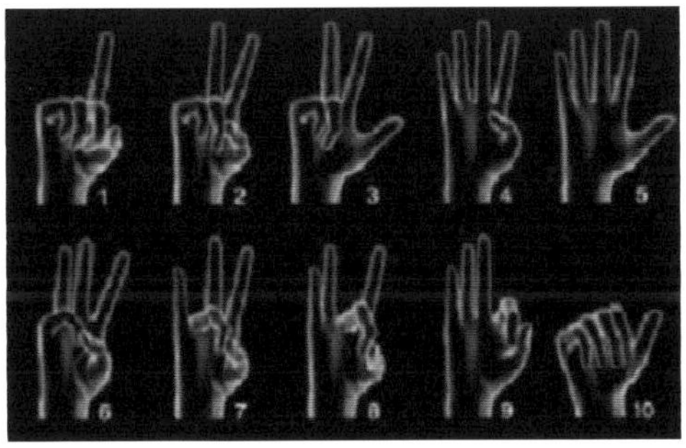

Figure 1. Indian Sign Language Alphabets.

1.1 Problem statement

Understanding the precise meaning of deaf and dumb people's symbolic gestures and converting it into understandable language(Text).

2 Literature Survey

A analysis of the literature for proposed framework reveals that many attempts have been made to tackle sign recognition in videos and images using various methods and algorithms.

*e-mail: rachanakp22@gmail.com
**e-mail: vivek.a.patil202@gmail.com
***e-mail: abhigolu1229@gmail.com
****e-mail: gaurav.datkhile@rait.ac.in

In Jing-hao Sun[1] The human hand was separated from the complex context, and the CamShift algorithm was used to detect real-time hand gestures. Then, using a convolutional neural network, the region of hand movements that was observed in real time is recognised, resulting in the identification of 10 common digits. The proposed system has dataset of total 1600 pictures for training dataset, 4000 hand gesture, 400 images for each type. This experiment shows accuracy about 98.3 percent. Hasan[2] used scaled normalisation to recognise gestures using brightness factor matching. With a black background, thresholding techniques are used for segmenting the input images . At the X and Y axis origins, the coordinates of any segmented image are shifted to match the centroid of the hand unit. and the image's centre mass is determined. Using a boundary histogram, Wysoski et al[3]. provided rotation invariant postures. The input image was captured with a camera, a filter for skin colour detection was applied, and then a clustering procedure was used to find the border line of each category in the pooling image using a standard contour tracking algorithm. Grids were created from the picture, and the boundaries were normalised. Geethu Nath and Arun C.S. [6] developed an ASL symbol recognition system based on the ARM CORTEX A8 processor. The machine recognises numbers using the Jarvis algorithm and alphabets using the template matching algorithm.Using Principal Component Analysis (PCA) and various distance classifiers, Kumud Tripathi [7] developed a framework for recognising continuous ISL gestures. The features from the keyframes are extracted from the own data set using Orientation Histogram and provided as input to the device.Noor Tubaiz [8] proposed using the Modified k-Nearest Neighbor (MKNN) approach to classify sequential data. Data gloves are used to detect hand movements.To supplement the raw data, window-based statistical features are calculated from previous raw feature vectors and future raw feature vectors. To recognise terms in ISL, the proposed framework was developed using novel techniques based on existing systems (ISL).Describe an approach for a continuous sign language recognition method (B. Bauer et al.). It is a framework that depends on continuous hidden Markov models images (HMM). It employs German sign language (GSL). Feature vectors that represent manual signs are fed into the device. [9]

3 Proposed Methodology

3.1 Image Aquistion:

It is the action of extracting an image from a source, typically a hardware-based source, for process of image processing. WebCamera is the hardware-based source in our project. It is the first step in the workflow sequence because no processing can be done without an image. The picture that is obtained has not been processed in any way.

3.2 Segmentation:

The method of separating objects or signs from the context of a captured image is known as segmentation. Con-

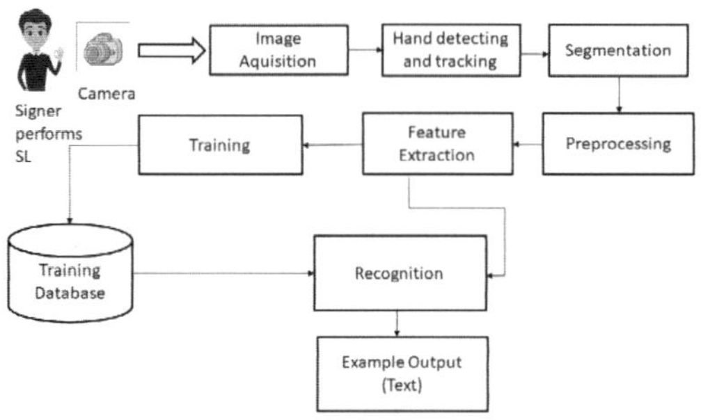

Figure 2. Proposed Methodology.

text subtracting, skin-color detection, and edge detection are all used in the segmentation process. The motion and location of the hand must be detected and segmented in order to recognise gestures.fig:threshold

3.3 Features Extraction:

Predefined features such as form, contour, geometrical feature (position, angle, distance, etc.), colour feature, histogram, and others are extracted from the preprocessed images and used later for sign classification or recognition. Feature extraction is a step in the dimensionality reduction process that divides and organises a large collection of raw data.reduced to smaller, easier-to-manage classes As a result, processing would be simpler. The fact that these massive data sets have a large number of variables is the most important feature. To process these variables, a large amount of computational power is needed. As a result, function extraction aids in the extraction of the best feature from large data sets by selecting and combining variables into functions.reducing the size of the data These features are simple to use while still accurately and uniquely describing the actual data collection.

3.4 Preprocessing:

Each picture frame is preprocessed to eliminate noise using a variety of filters including erosion, dilation, and Gaussian smoothing, among others. The size of an image is reduced when a colour image is transformed to grayscale. A common method for reducing the amount of data to be processed is to convert an image to grey scale.The phases of preprocessing are as follows:

3.4.1 Morphological Transform (Morphological Transform) :

Morphological operations use a structuring feature on an input image to create a similar-sized output image. It compares the corresponding pixel in the input image with its neighbours to determine the value of each pixel in the output image.There are two different kinds of morphological transformations Erosion and Dilation.

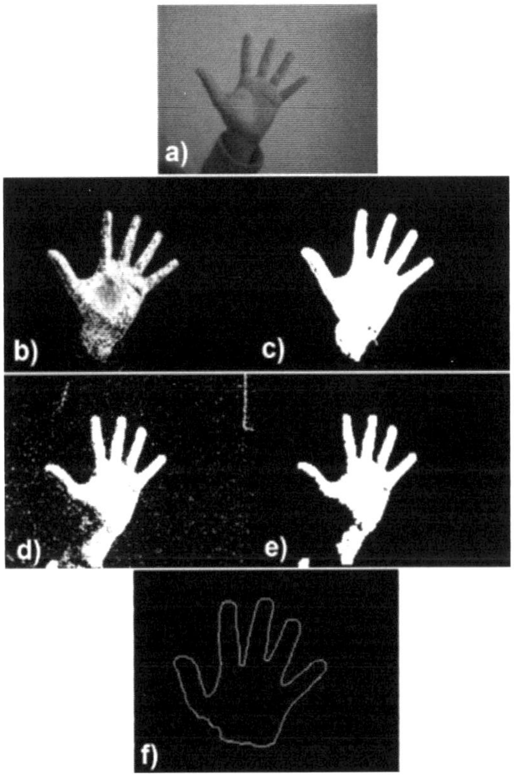

Figure 3. (a) original image with hand; (b) image of hand after skin color detection; (c) after morphological operations and binarization; (d) image of hand after background extraction ; (e) after binarization and morphological operations; (f) hand's contour made of image c and image e concatenation

i. Dilation: The maximum value of all pixels in the neighbourhood is the value of the output pixel. A pixel in a binary image is set to 1 if all of its neighbours have the value 1 Morphological dilation increases the visibility of artefacts and fills in small gaps.

ii. Erosion: The o/p pixel's value is the minimum of all pixels in the neighbourhood. A pixel in a binary image is set to 0 if all of its neighbours have the value 0.small artefacts are eroded away by morphological erosion, leaving behind substantial objects.

3.4.2 Blurring:

Adding a low-pass filter to an image is an example of blurring. The word "low-pass filter" refers to eliminating noise from an image while leaving the rest of the image intact in computer vision. A blur is a simple operation that must be completed before other tasks such as edge detection.

3.4.3 Thresholding:

Thresholding is a form of image segmentation in which the pixels of an image are changed to make it easier to interpret the image. Thresholding is the process of converting a colour or grayscale image into a binary image, which is simply black and white. We most commonly use thresholding to pick areas of interest in a picture while ignoring the sections we are not concerned with.

3.4.4 Recognition:

We'll use classifiers in this case. Classifiers are the methods or algorithms that are used to interpret the signals. Popular classifiers that identify or understand sign language include the Hidden Markov Model (HMM), K-Nearest Neighbor classifiers, Support Vector Machine (SVM), Artificial Neural Network (ANN), and Principle Component Analysis (PCA), among others. However, in this project, the classifier will be CNN. Because of its high precision, CNNs are used for image classification and recognition. The CNN uses a hierarchical model that builds a network, similar to a funnel, and then outputs a fully-connected layer in which all neurons are connected to each other and the output is processed.

3.4.5 Text output:

Understanding human behaviour and identifying various postures and body movements, as well as translating them into text.

4 Proposed Algorithm

4.1 Creating the sign language recognition dataset:

Any frame that detects a hand within the ROI (region of interest) generated can be transferred to a directory that contains a pair of directories, train and take a look at, every containing 10 folders containing images captured mistreatment the produce gesture knowledge.py perform. Now, to create the dataset, we have a tendency to use OpenCV to urge the live cam feed and create a ROI, that is solely the portion of the frame wherever we have a tendency to want to find the hand for the gestures. For differentiating between the background we have a tendency to calculate the accumulated weighted avg for the background then cipher this from the frames that contain some object ahead of the background which will be distinguished as foreground.This can be accomplished by computing the accumulated weight for specific frames and the context's accumulated average.After we've the accumulated average for the background,we tend to subtract it from each frame that we read after sixty frames to seek out any object that covers the background.

4.2 Calculate threshold value:

We now measure the threshold value for each frame and evaluate the contours using cv.findContours.The max contours i.e. object's most outermost contours is returned using function segment. We may decide whether there is

https://doi.org/10.1051/itmconf/20214003004

some foreground object identified in the ROI using the contours, in other words, whether there is a hand in the ROI. When model detects contours, it starts saving the picture of the ROI in the train and test sets for the letter or number we're looking for (or a hand is present in the ROI). The dataset for 1 is generated in the preceding example, and the ROI's thresholded image is displayed in the following window.In the train dataset, we save 600 images for each number to be detected, and in the test dataset, we generate 80 images for each number.

4.3 CNN Layer:

To classify the static images in our first dataset, we used a Convolutional Neural Network, or CNN, model. While building neural network our main objective was to define input layer. Each of the 784 pixels in a 28x28 image is represented by a grayscale value ranging from 0 (black) to 1 (white) (white). We translate the data into a computer-readable format by converting each image to a series of numbers. The hidden layers of the neural network will process the input layer once it has been prepared. Our neural network's architecture can be seen in the fig:blk2 The first hidden layer is made up of a number of nodes, each of which receives a weighted total of the 784 input values. The inputs is then passed through an activation function named as rectified linear unit, or ReLU. The ReLU will produce 0 when the input is negative, but will not alter the input otherwise, as seen in the graph above. The ReLU's outputs will be used as inputs for the network's next hidden layer.

Table 1. CNN Layers

Layer (type)	Output shape	Param#
conv2d(Conv2D)	(None, 62, 62, 32)	896
max_pooling2d (MaxPooling2D)	(None, 31, 31, 32)	0
conv2d_1 (Conv2D)	(None, 31, 31, 64)	18496
max_pooling2d_1 (MaxPooling2D)	(None, 15, 15, 64)	0
conv2d_2(Conv2D)	(None, 13, 13, 128)	73856
max_pooling2d_2 (MaxPooling2D)	(None, 6, 6, 128)	0
flatten (Flatten)	(None, 4608)	0
dense (Dense)	(None, 64)	294976
dense_1 (Dense)	(None, 128)	8320
dropout_1 (Dropout)	(None, 128)	0
dense_2 (Dense)	(None, 128)	16512
dropout_2 (Dropout)	(None, 128)	0
dense_3 (Dense)	(None, 12	1548

4.4 Training CNN:

We now train a CNN on the newly generated data collection.To begin, we'll use keras ImageDataGenerator, which lets us use the flow from directory function to load the train and test set data, with the names of the number folders serving as the class names for the images loaded.Reduced LR on plateau and early training callbacks are used, all of which are based on the validation dataset failure. The validation dataset is used to measure the accuracy and loss after each epoch, and if the validation loss is not decreasing, the model's LR(Learning Rate) is reduced using ReduceLR to prevent the model from overshooting the loss minima. we also use the early stopping algorithm to stop the training if the validation accuracy continues to decrease for some epochs. The two optimization algorithms used are SGD i.e. stochastic gradient descent which means the weights are changed at any training instance and Adam i.e. a combination of Adagrad and RMSProp. We discovered that the model SGD had higher accuracies. As can be seen, we achieved 100 training accuracy and an 81 percent validation accuracy while training.

Epoch No.	Loss	Accuracy	Val_Loss	Val_Acc
1	12.71	0.18	0.81	0.8
2	1.44	0.53	0.30	0.96
3	0.84	0.71	0.15	0.99
4	0.56	0.81	0.02	1.0
5	0.38	0.86	0.01	1.0
6	0.30	0.88	0.0059	1.0
7	0.25	0.89	0.00	1.0
8	0.27	0.90	0.00	1.0
9	0.16	0.94	0.00	1.0
10	0.15	0.95	0.00	1.0

4.5 Predicting the gesture

We generate a bounding box for detecting the ROI and measuring the cumulative average, just as we did when creating the dataset. This is done in order to recognize a foreground entity. Now we look for the maximum contour, and if one is found, it means a hand has been identified, so the ROI's threshold is used as a test picture. Using keras.models.load_ Model we load the previously saved model and then feed the threshold picture of the ROI containing the hand to the model for prediction as an input.

5 Result:

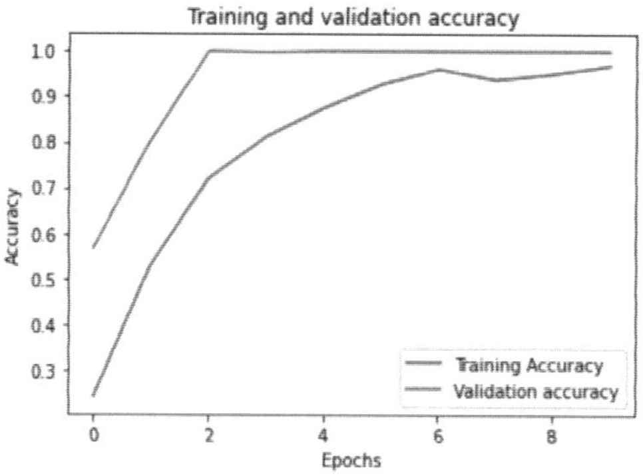

Figure 4. Outputs from the first hidden layer.

When we are training the model, accuracy and loss in model for validation data could be variating with various cases. Normally with every increasing epoch, loss should be going lower and exactness should be going higher.But with validation loss(keras validation loss) and validation

accuracy, numerous cases can be conceivable like underneath:
1)Validation loss starts increasing, validation accuracy starts diminishing. This implies model is cramming values not learning.
2)Validation loss starts increasing, validation accuracy likewise will increase. This could be instance of overfitting or diverse probability values in situations where softmax is being used in output layer.
3)Validation loss starts decreasing, validation accuracy starts increasing. This is additionally fine as that implies model fabricated is learning and dealing fine. After testing our model we have aquired the following results,we have plotted the graph of accuracy and loss with respect to epochs.

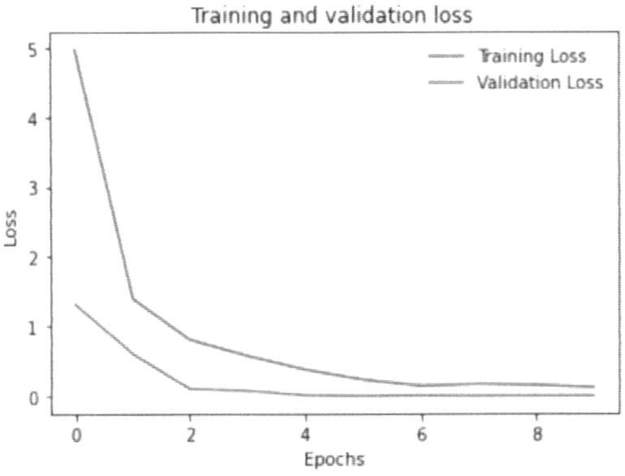

Figure 5. Training and validation loss

Fig 6 shows overall accuracy evolution of model. In which it has been seen that validation loss is decreasing and validation accuracy is increasing noticeably.

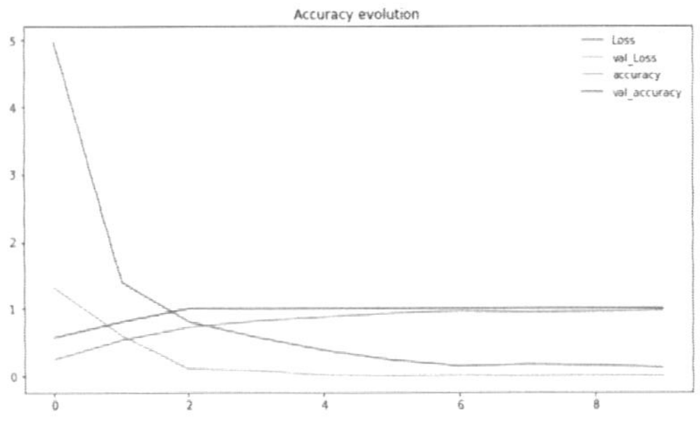

Figure 6. Accuracy Evolution

6 Conclusion and Future Work:

The Sign Language Recognition (SLR) system is a method for recognising a collection of formed signs and translating them into text or speech with the appropriate context. The significance of gesture recognition can be seen in the development of effective human-machine interactions. We attempted to build a model using a Convolutional Neural Network in this project. This results in a validation accuracy of about 95%
The Image Processing section of future work should be enhanced so that the system can interact in both directions, i.e. it should be capable of translating normal language to sign language and vice versa.

References

[1] Jing-Hao Sun,Ting-Ting Ji, Shu-Bin Zhang, Jia-Kui Yang, Guang-Rong Ji "Research on the Hand Gesture Recognition Based on Deep Learning",07 February 2019

[2] Mokhtar M. Hasan, Pramoud K. Misra, (2011). "Brightness Factor Matching For Gesture Recognition System Using Scaled Normalization", International Journal of Computer Science Information Technology (IJCSIT), Vol. 3(2).

[3] Simei G. Wysoski, Marcus V. Lamar, Susumu Kuroyanagi, Akira Iwata, (2002). "A Rotation Invariant Approach On Static-Gesture Recognition Using Boundary Histograms And Neural International Journal of Artificial Intelligence Applications (IJAIA), Vol.3, No.4, July 2012

[4] Stergiopoulou, N. Papamarkos. (2009). "Hand gesture recognition using a neural network shape fitting technique," Elsevier Engineering Applications of Artificial Intelligence, vol. 22(8), pp. 1141– 1158, doi: 10.1016/j.engappai.2009.03.008

[5] V. S. Kulkarni, S.D.Lokhande, (2010) "Appearance Based Recognition of American Sign Language Using Gesture Segmentation", International Journal on Computer Science and Engineering (IJCSE), Vol. 2(3), pp. 560-565.

[6] Geethu G Nath and Arun C S, "Real Time Sign Language Interpreter," 2017 International Conference on Electrical, Instrumentation, and Communication Engineering (ICEICE2017)

[7] Kumud Tripathi, Neha Baranwal and G. C. Nandi, "Continuous Indian Sign Language Gesture Recognition and Sentence Formation", Eleventh International MultiConference on Information Processing-2015 (IMCIP-2015), Procedia Computer Science 54 (2015) 523 – 531

[8] Noor Tubaiz, Tamer Shanableh, and Khaled Assaleh, "Glove-Based Continuous Arabic Sign Language Recognition in User-Dependent Mode," IEEE Transactions on Human-Machine Systems, Vol. 45, NO. 4, August 2015.

[9] B. Bauer,H. Hienz "Relevant features for video-based continuous sign language recognition", IEEE International Conference on Automatic Face and Gesture Recognition, 2002.

[10] Pigou, L., Dieleman, S., Kindermans, P.-J., Schrauwen, B. (2014). Sign Language Recognition Using Convolutional Neural Networks.

Identification of Lost Children using Face Aging with Conditional GAN

Amisha Pupala[1,3,*], *Samruddhi* Mokal[2,**], *Neha* Pandit[3,***], and *Smita* Bharne[4,****]

[1]Ramrao Adik Institute of Technology, Navi Mumbai, India

Abstract. Face recognition technology is a big area that consists of the many features in it but it also comes with some of the factors which affect this technology, one of the factors is Face aging which makes face recognition more difficult. As in India, a large number of children go missing every year. Also just using a photograph is not enough for the process to proceed smoothly and it results in a huge percentage of the missing child cases remain untraced. This paper presents a novel use of face recognition with face aging to overcome the limitation of existing systems. The proposed system has a portal where the public can upload an image of a suspected child and also have a feature where searching for any lost child is possible. The proposed system has mainly concentrated on an Age Conditional generative adversarial network (C-GAN) algorithm for face aging and the FaceNet algorithm for face feature extraction and face recognition.

1 Introduction

Children represent a significant percentage of the total population of India. Unfortunately, many children go missing every year because of numerous reasons like kidnapping, trafficked children and lost children, run-away children, and abduction. According To the National Crime Record Bureau 2019 report, the number of missing children has increased by 8.9 percent since 2019 giving a total of 73,137 children reported missing last year. Many NGOs claim that the actual number of missing children cases is much more than the reported number. A child lost in one location can be found in some other location. Even if the child is found it gets difficult to recognize him/her. Face recognition is thus the most promising biometric technology for recognizing and finding missing children. But if the child is found after many years from the date when he/she is lost the state-of-art face recognition systems fall short to identify children from the photograph because a child's face undergoes many temporal variations due to aging like a change in skin texture, change in the shape of the face, color, weight, facial hair growth, etc. Hence, as the time difference between a true mate image and probe image gets larger, the performance of the face recognition algorithm decreases, and thus the search gets harder. Thus the primary goal of the proposed system is to generate an age-progressed image of the child rather than enhancing the face recognition performance. Thus this system enhances the ability of face matching algorithms which will be used to identify and locate children who are lost at a young age by aging face features to reunite them back with their families. Developed system will be helpful to the police and higher authorities for tracking down missing children quickly.

*e-mail: amishapupala20@gmail.com
**e-mail: samruddhimokal10@gmail.com
***e-mail: panditneha124@gmail.com
****e-mail: smita.bharne@rait.ac.in

2 Literature Survey

In 2018, S. Chandran publish a paper [1] in which authors have used CNN and VGG face technologies for the identification of missing children. The faces of children are detected and features are extracted by using CNN. CNN is an important tool that works appropriately with image data. Blocks of Relu pooling layers, convolutional, and fully connected layers are composed in CNNs or ConvNets. For classification, authors have used the SVM Classifier. But these technologies slow the process of classification and extraction of faces from images.

In 2020, a Face-to-face matching method is based on IPE-based blind computing which is used to restrict access to photos [2]. Using Multitask FaceNet Model, the system can safely run and provides privacy for both the unsure and missing children parts. To provide a more effective face recognition module, the output of images with encryption schemes needs to be improved.

In [3], VGG16, ResNet50, and Mobile-FaceNet these three Convolutional neural network algorithms were used. From which Mobile-FaceNet has the smallest model size and VGG16 has the largest model in size.

In [4] authors have used DeepId and modified Inception structures which contain different types such as DeepId2, DeepId3 modules which give the best performance with higher accuracy when there is very little difference between a deep module and a wide module. however, the mutual information goes down with additional branches such as DeepId3 in a wide module.

Face aging with Age-cGAN to generate artificial images within different age categories [5]. The face recognition method used by this system is the Identity preserving latent vector optimization method used to conserve the original photo's identity while face aging which can be improved by combining Pixelwize and Identity-Preserving methods into one singular method.

3 Proposed Work

It is known that the performance of the face recognition algorithm is compromised due to aging. It is still a problem to consider the aging parameter under consideration for face recognition as most of the algorithms fail to detect the face of the child after a few years. A child's face undergoes various changes including facial hair, weight, the texture of the skin, shape, and size of the face, etc. Several studies were performed to analyze the effects of face aging on the performance of the face recognition algorithms. Two conclusions can be drawn from these studies:

1. The performance of traditional face recognition algorithms decreases with an increase in time-lapse.

2. Performance of face recognition and detection algorithms decreases more in the case of younger individuals as compared to older individuals.

Hence it is important to consider the age progression of the face for enhancing the performance of the face recognition algorithm, especially when enrolled at a young age. A. In particular, our contributions are as follows:

- A portal for finding missing children that compare the image of the missing child with the images of the already registered missing children.
- To take into consideration many kinds of variations in face images which are taken under uncontrolled conditions such as change of expression, illumination, pose, change in lighting condition, presence of noise, and blurred image.
- Using Age Conditional Generative Adversarial Network to reconstruct synthetic images of the required. Using methodology like Facenet and C-Gan the process of recognition gets easier which will help the authorities to find the missing children quickly and solve many cases which have been untraceable because of the above issues.

Thus the proposed system presents a novel deep learning methodology to find the lost children using the photo of the children currently available using face recognition. Whenever anyone finds any suspicious child on the road he/she can click the photograph of the child and upload it into the common portal with landmarks and remarks. The public has to upload the photograph of the missing child into the portal. This photo will automatically be compared with the already saved images in the repository. If no matching record is found then the image will be age-progressed and again the age-progressed image will be compared with the registered photographs. For this, a deep learning model Aging cGAN is used. And the face recognition is done using FaceNet. Thus the proposed system outperforms the earlier methods in face recognition. As mentioned earlier, the face aging model is based on cGAN, a model for the synthesis and generation of human faces of the required age category. Hence the purpose of cGAN network is to generate the image of a child at different ages based on the input image and target age. The structure of the model consists of three parts: encoder, age condition, and decoder. The encoder is used for feature extraction of the input face image. Proposed system is using three groups of convolution layers and nine residual blocks. Each convolution layer consists of a ReLU non-linearity layer, spatial batch normalization layer stride, and a convolution layer. After this, it will give a 56 feature map of the image. Since there is a huge range of images of a large number of age groups in the dataset system is dividing the image into y number of groups which is represented as a one-hot age vector. Each block represents a specific age group. The task of the decoder is to convert the face to generate an image of the required age group based on the features.

4 Proposed Methodology

A. Age C-GAN Algorithm Conditional Generative Adversarial Network is conditioned on extra information and is based on the idea of Vanilla GAN which allows to control the output of the generator network. This system have to give extra information y as an additional layer to the generator. As compared to vanilla GAN, In vanilla GANs no one can control the category of the retrieved images as it can learn only one category, however, in CGAN can generate images of the specific category using condition y. y can be any data depending on the application for which the model for like integer data or class label has been built. Thus CGAN can be used to generate models with different categories and conditions. The Face Aging C-GAN has four networks: Encoder: An encoder is a deep convolutional neural network used to generate the latent vector of images. It is used to learn the inverse mapping of input image and age condition with the latent vector Z.

- It generates a latent vector of 100 dimensions of the input face images of dimension(64,64,3). There are 2 dense layers and 4 convolution blocks.
- All the convolution blocks except the first layer have a convolution layer which is followed by a batch normalization layer and then an activation function.

Generator Network: It is a deep convolutional network that takes a face image and condition vector and tries to generate realistic images of the face.

- It is a CNN having upsampling, convolutional, and dense layers which take a condition vector which is additional information added to the network which is age, and a latent vector to generate a realistic image of a dimension of (64, 64, 3). The condition vector is the additional information that is provided to the network. For the Age- cGAN, this will be the age.

Discriminator Network: It is CNN that is used to discriminate between the real image and the fake image. It contains several convolutional blocks which contain a batch normalization layer, convolutional layer, and an activation function.

https://doi.org/10.1051/itmconf/20214003005

FaceNet: It is a face recognition model that learns the difference between the generated image x' and the original image x (input image). It is used to compare the person's identity in the image. The proposed system is using a ResNet2

For FaceNet Inception ResNet, ResNet-50, or pre-trained Inception can be used. This system using the ResNet2 model without fully connected layers. FaceNet is used to identify the person's identity in the given input image. The extracted embeddings for the original input image and the reconstructed image can be found by calculating the Euclidean distance between the embeddings.

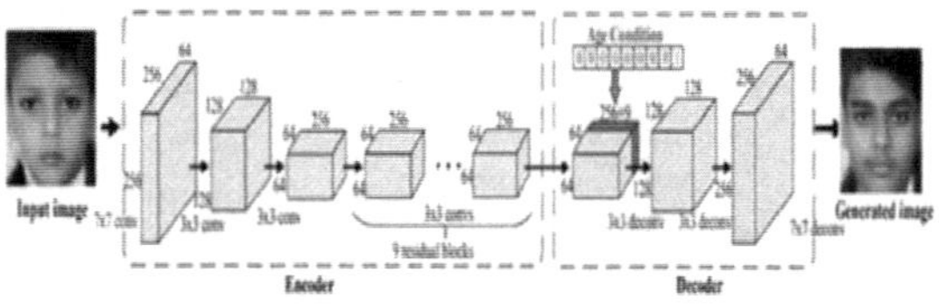

Figure 1. Layers used in Aging cGAN

Stages of Age-cGAN:

1. Age-cGAN has four networks and it is trained in three stages: Conditional GAN training: In this stage,The System trained the generator network which generates the blurred images of the face after training and the discriminator network.
2. Initial latent vector approximation: In this stage,The System trained the encoder network to approximate the latent vector using both the generated images and the real images which generates the latent vector from learned distribution.
3. Latent vector optimization: In this stage,The System optimized both the encoder and the generator network.

4.1 FaceNet

This System uses FaceNet for feature extraction and face recognition.The System first passed all the images through Multi-Task Cascaded Convolutional Neural Network (MTCNN) for finding and extracting faces from the images. Then system used FaceNet for extracting high-quality features from children's faces which been used to train a face identification system. It uses Convolution layers to learn embeddings from the face directly which are used for face verification and face recognition. These embeddings are of 128 dimensions and insert them into feature space such that the squared distance between faces of the same person is less whereas the squared distance between images to two different people is large.

5 Dataset

A. Dataset for Face Recognition

This system used two separate datasets for the training of

Figure 2. Model architecture of FaceNet.

face recognition and the Age Progression model. For face recognition, dataset been created by collecting images of children from various sources. In a proposed system there is a directory structure where each child to be recognized has a dedicated directory with their images saved in it. The photos in the directory provide a range of orientation, sizes, lighting conditions. If an image contains multiple faces then the face having the highest probability will be considered for further steps. The system uses this dataset as the basis for the classifier which is trained on a training dataset only.

B. IMDB Wiki-Crop Dataset

For training the Age progression model,The System used the IMDB WIKI-crop dataset which is the largest publicly available dataset of human faces. It contains about 400 thousand images where each image is labeled in the format gender, age, the name with all the metadata information like dob, the year when the photo was taken, gender, face location, face, and secondary face score, etc in the .mat file. The dataset contains images of all the age categories from 0 to 100 years of age which is not required for the model. So the system have to filtered the dataset and considered the images of children in the age group from 0-15 years of age and have deleted the rest of the images. The dataset contains images with huge variations in facial expression, illumination, resolution, occlusion, pose, etc. The dataset also provides the corresponding landmarks. This dataset can be used for a variety of tasks including age estimation, age progression, landmark localization, progression/ regression, face detection, etc.

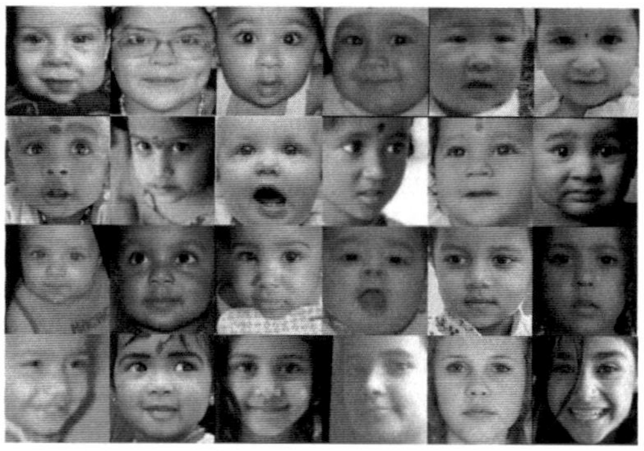

Figure 3. Wiki Sample Dataset

6 System Design

The proposed system consists of a portal where the public can store the details with a photograph of the missing

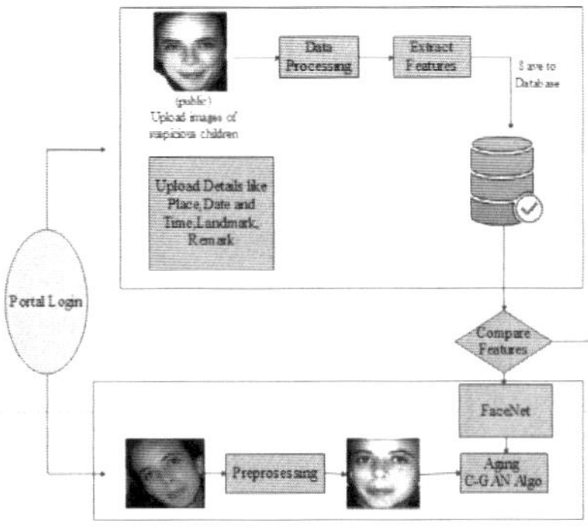

Figure 4. System design for lost child idetification using FaceNet and C-GAN.

child. The public here refers to the parent of the missing child or the police officer. By using the photograph the public can find the matching children from the already stored data. The system will prompt the most matching cases after applying a face recognition algorithm to the uploaded image. Once the matching is done then the concerned person will get the details of the child. The proposed model is divided into various phases. First, whenever the public finds any missing or suspicious child they can upload the photograph of the child with the details like name of the child, age, place where the child was found, date and time when the child was found and remarks like what clothes the child was wearing when he was spotted or any information the child might have told to the person which can be helpful to find the child. This uploaded image goes through an image pre-processing phase. The input to this is the raw image that is uploaded by the public and the output of this phase is an aligned frontal face image. The main purpose of this step is to align all the face images based on eye coordinates such that all the images are aligned with the same standard size.This system used a Multi-task Cascaded Convolutional Networks (MTCNN) model which is a framework used for face alignment and face detection which consists of three stages of convolutional networks that can detect landmark location such as eyes, nose, and mouth and can recognize face very efficiently. The preprocessing step includes removing noise from the image, straightening the image, and detecting and cropping the face from the whole image. The extracted face features will be stored in the database for further processing. To search for a missing child parent of the child needs to upload a photograph of the child then this image will go through the face recognition algorithm after the image pre-processing and feature extraction step.If any matching photo is found in the database then the image of the child will be displayed on the screen.

If no matching image is found then the uploaded image goes through the Age CGAN algorithm. The GAN algorithm is used to generate a new age-progressed image with the same statistics as the training images. As the child may be spotted after many years, the face of the child changes from the last known photograph taken before the child had

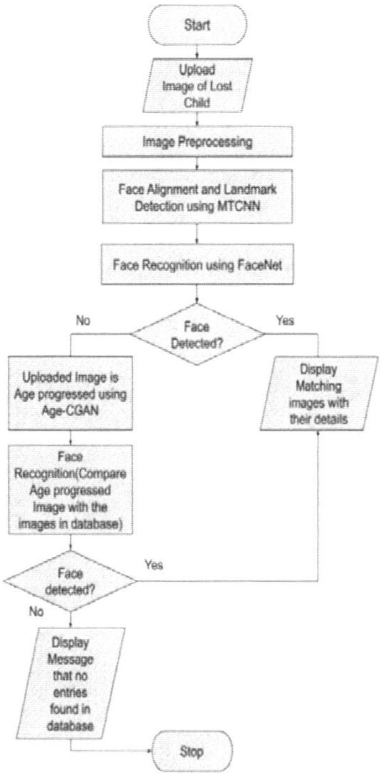

Figure 5. Proposed Flow of the System.

gone missing. So this model will be used to synthesize the child's face to produce an age-progressed facial image of the child. So basically this model will be used to synthesize child faces to the required age category. This face synthesized and age-progressed image goes through the FaceNet algorithm for feature extraction and face recognition. The main distinction between FaceNet and other face recognition algorithms is that FaceNet does not use any bottleneck layer for verification and recognition tasks but rather learns the mapping directly from images and creates embedding. The FaceNet is used to automatically compare the age-progressed photo of the missing child with the already present images in the database and if the photo matches with any of the photos present in the database then it will be displayed on the screen along with details like name, location and time where the child was found.

7 Result and Analysis

The performance of the face recognition algorithm was evaluated on a child dataset before and after age progressing the child images. The Age cGAN model was trained on the IMDB Wiki- Crop dataset. The IMDB Wiki-Crop Dataset contains more than 400k images with system The system filtered that dataset and have just used images from the age group 0-15 years to train the model.

Based on their experiments, the following conclusions are derived:

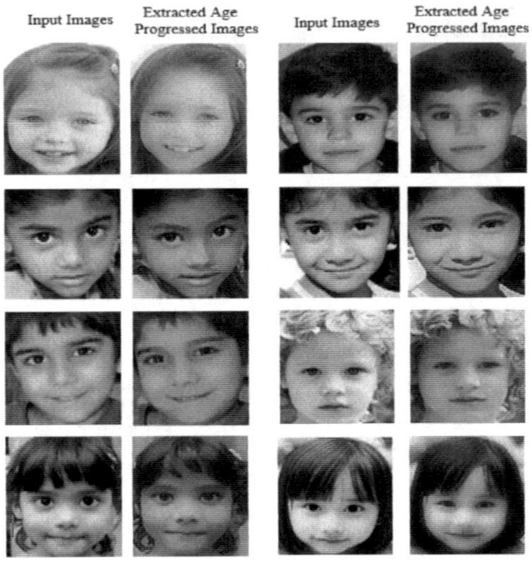

Figure 6. Retrieved Age progressed images of the proposed system

- The accuracy of the face recognition algorithm with the images of the same age category is train Accuracy: 87.38% and test Accuracy: 81.07%.
- The Accuracy of face recognition algorithm with images of different age categories: 44.76%.
- The Accuracy of the FaceNet algorithm with images of different age categories after age progression images using Age-cGAN algorithm: 47.32%.

Based on the experiments, the following conclusions are derived:

- The model correctly recognizes the misaligned faces and images clicked in an uncontrolled environment like the change of expression, illumination, pose, change in lighting condition, and presence of noise.
- The average decrease in the performance of face recognition is more than 40% due to age variation in the faces of the children
- Intra-person variations like expression and pose are degrading the performance of the system is similar to face aging. Also, large self-occlusion, blur, and misaligned faces are slightly degrading the performance of the model.

1) Limitations:

- Training of the deep learning model which requires a large dataset remains a challenge when training with a small dataset hence accuracy is compromised.
- The performance of the face recognition model heavily depends on the accuracy of age-progressed images which are obtained.
- Due to the long duration of code execution and heaviness of the model proposed system have trained the model for 300 epochs due to which the performance of the model is compromised.

2) Suggested Solutions:

- As face aging depends on various other As face aging depends on various other factors other than aging, it is required to consider all the other factors like having a single face aging database with correct demographic details and a lot of images of the child of various age groups is the need of the hour for better accuracy of the face aging model.
- The model age progresses the images of the child to just one age group, this can further be improved in the future by using a bigger dataset and categorizing the images in the dataset in various required age groups.
- The Age progression model should be trained for at least 500 epochs for getting better accuracy for age-progressed images.

Table 1 Identification accuracy for previous work with and without the proposed feature aging module.

Paper	Method Used	Dataset	Without Aging Module	With Aging Module
FaceNet: A unified embedding for face recognition and clustering[6]	FaceNet	CFA	38.16	55.30
CosFace: Large margin cosine loss for deep face recognition[7]	CosFace	ITWCC	60.72	66.12
A Review of Face Recognition against Longitudinal Child Faces[8]	Cognitec	ITWCC	-	41.1
Li et al	MFDA	FG-NET	-	47.50
Face aging conditional generative adversarial networks[9]	C-GAN	MNIST	-	47.5

Table 2 Model Accuracy

Model	Age Category	Accuracy
Face Recognition	Same	81.07%
Face Recognition	Different	44.76%
Age Progressed	Different	47.32%

8 Conclusion

In this work, a proposed system is used for an age-invariant face recognition system for finding missing children. The system reduces the manual work of scanning

 https://doi.org/10.1051/itmconf/20214003005

Figure 7. Welcome page of a missing child finding portal.

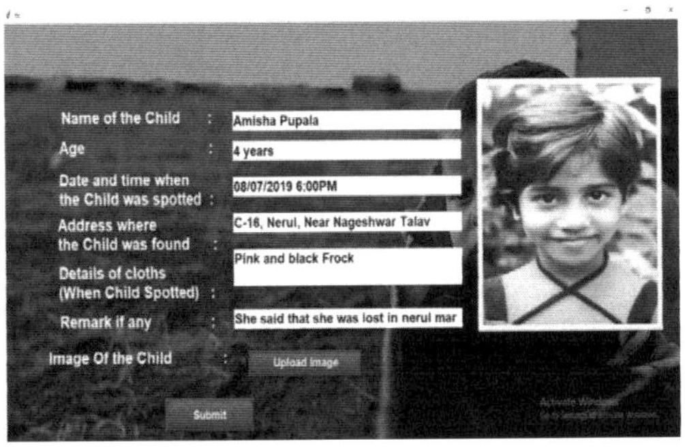

Figure 8. Page where the public can upload details of a suspicious or lost child.

Figure 9. Page to search for a lost child using his/her photograph.

Figure 10. Details of the child found in the database

through all the photographs of the lost children to find a match. The System has tried to solve the problem of age variation by age progressing the images of the child using the Aging cGAN deep learning algorithm. The model improves the performance of the FaceNet model for face recognition from 44.76% to 47.32%. These results suggest that age progression of the faces of the children before face recognition enhances the ability of the system to identify the lost children. This will be beneficial for solving cases of untraced missing children and thus will help to reduce the increasing number of missing child cases.

References

[1] S. Chandran, Pournami Balakrishnan, Byju Rajasekharan, Deepak N Nishakumari, K Devanand, P M Sasi, P. (2018). "Missing Child Identification System Using Deep Learning and Multiclass SVM ".

[2] 2. Xin Jin, Shiming Ge, Chenggen Song, Xiaodong Li, Jicheng Lei, Chuanqiang Wu, and Haoyang Yu "Double-Blinded Finder: A Two- SidePrivacy-Preserving Approach for Finding Missing Children"(2020).

[3] Shun Lei Myat Oo, Aung Nway Oo University of Information Technology, Yangon, Myanmar " Child Face Recognition with Deep Learning"(2019).

[4] Chang Shu School of Communication and Information Engineering, University of Electronic Science and Technology of China Chengdu, China "Optimizing deep neural network structure for face recognition" (2017).

[5] Grigory Antipov, Moez Baccouche, Jean-Luc Dugelay "Face Aging with Conditional Generative Adversarial Networks" (2017) arXiv:1702.01983v2.

[6] Florian Schroff, Dmitry Kalenichenko, and James Philbin. Facenet: A unified embedding for face recognition and clustering. In CVPR, pages 815–823, 2015.

[7] Hao Wang, Yitong Wang, Zheng Zhou, Xing Ji, Dihong Gong, Jingchao Zhou, Zhifeng Li, and Wei Liu. Cosface: Large margin cosine loss for deep face recognition. In CVPR, 2018

[8] Karl Ricanek Jr., Ph.D. Senior Member IEEE, Shivani Bhardwaj, Michael Sodomsky A Review of Face Recognition against Longitudinal Child Faces

[9] FACE AGING WITH CONDITIONAL GENERATIVE ADVERSARIAL NETWORKS GrigoryAntipov, MoezBaccouche, Jean-Luc Dugelay

https://doi.org/10.1051/itmconf/20214003006

Speech Emotion Recognition using Time Distributed CNN and LSTM

Beenaa Salian[1,*], *Omkar* Narvade[1,**] *Rujuta* Tambewagh[1,***], and *Smita* Bharne[1,****]

[1]Ramrao Adik Institute of Technology, Navi Mumbai, India

Abstract. Speech has several distinguishing characteristic features which has remained a state-of-the-art tool for extracting valuable information from audio samples. Our aim is to develop a emotion recognition system using these speech features, which would be able to accurately and efficiently recognize emotions through audio analysis. In this article, we have employed a hybrid neural network comprising four blocks of time distributed convolutional layers followed by a layer of Long Short Term Memory to achieve the same.The audio samples for the speech dataset are collectively assembled from RAVDESS, TESS and SAVEE audio datasets and are further augmented by injecting noise. Mel Spectrograms are computed from audio samples and are used to train the neural network. We have been able to achieve a testing accuracy of about 89.26%.

1 Introduction

As humans, our thoughts are best articulated using speech. Therefore, in this increasingly technologically-driven world, the next step forward would be extending this understanding to machines. Although Speech Emotion Recognition (SER) has been around for almost a decade, it has regained attention due to recent developments in this field (for eg. Voice-based virtual assistants like Siri, Alexa, etc and automated help-center assistance, self-driving cars, etc). Although there is a rising demand for voice-controlled technologies, The recognition of emotion from speech is the main challenge in human-machine interaction. Despite significant progress in speech recognition, we are still a long way from determining underlying emotions from the speaker's audio signals since the machine does not understand the speaker's emotional state. Because speech is the simplest and most effective means of communication for humans, a computer must be able to grasp the user's mood in this more technologically-driven world. We'll look at which aspects of speech are the most effective in discriminating various emotions. An improved speech emotion identification model must be able to recognise seven primary emotions: anger, disgust, happiness, sorrow, neutral, surprise, and fear. We must develop a robust deep learning model which can accurately and efficiently classify emotions from speech alone.
There are various hybrid model implementations in the SER domain. The most commonly used hybrid model is the CNN-LSTM model. Where CNN is used for learning local correlations and LSTM is used to learn long-term dependen- cies from the learned local features. In [1] 13 MFCC (Mel Frequency Cepstral Coefficient) with 13 velocities and 13 acceleration components are used as features and a 1D CNN and LSTM model are used for classification. In this paper, the EMODB dataset is used.To compute MFCCs, Discrete Cosine Transform(DCT) is applied. DCT is needed to decorrelate filter bank coefficients. It is a linear operation and therefore discards any non-linear useful information present in the speech signal. In [4] three different features, MFCCs, magnitude spectrogram, and log-mel spectrogram, are compared with several architectures, such as CNN, BLSTM, and CNN-LSTM, to determine which architecture and feature combination is best for speech emotion recognition. All of the models were tested on two different datasets. – EMODB and IEMOCAP where 4 emotions are classified where the length of audio files is kept 3 seconds. The design is shown to perform effectively with Log-Mel Spectrograms when combined with CNN+LSTM architecture in this article. The aim of [3] is to find the relation between the duration of speech length and the recognition rate of emotions. In this paper, analysis is performed using a CNN model having two convolution layers where magnitude spectrograms are taken as features. Performance for the system is analyzed using speech sequences of length from 0.25s to 1.5s. It is observed as the length of the speech signal increases, the accuracy of the system increases.

2 Proposed Methodology

2.1 Speech Corpus

The data sets used here cover people of different ages – ranging from young to old, covers both the genders, and people having different accents.

2.1.1 *Toronto emotional speech set (TESS)*

200 special words are spoken using the phrase "Say the word - " by two actresses. The two actresses are aged 26

*e-mail: beenaasalian09@gmail.com
**e-mail: nomkar99@gmail.com
***e-mail: rujuta.tambewagh@gmail.com
****e-mail: smita.bharne@rait.ac.in

https://doi.org/10.1051/itmconf/20214003006

and 64 years. The audios ae recorded on set potraying one of the seven emotions

2.1.2 Ryerson Audio-Visual Database of Emotional Speech and Song (RAVDESS)

RAVDESS consists of gender balanced audio samples from males and females (12 females and 12 males). The audio samples contain seven different emotions with two levels of intensity - high and low.

2.1.3 Surrey Audio-Visual Expressed Emotion(SAVEE)

The audio samples in the dataset are recorded with four native english speakers. They are identified as DC, JE, JK, KL. This results in a total of 120 utterances per speaker in which they 7 emotions are covered.

2.1.4 Augmentation

The prediction accuracy of any deep learning model is largely dependent on the amount and the diversity of data available during training. A common method to increase the diversity of your dataset, is to augment your data artificially. To generate syntactic data for audio, we can apply noise injection. We have used the numpy library to add noise to our existing dataset.Table 1 shows the final dataset after augmentation.

Table 1. Final Dataset after Augmentation

Total Dataset After Augmentation								
Code	0	1	2	3	4	5	6	Total
Org	652	652	655	652	652	652	808	4723
Aug	652	652	655	652	652	652	808	4723
Total	1304	1304	1310	1304	1304	1304	1616	9446

$$Emocode\ 0 : Happy, 1 : Sad, 2 : Angry, 3 : Angry,$$

$$4 : Fear, 5 : Surprise, 6 : Neutral$$

2.2 Feature Extraction

There is quite a lot of varied information present in speech signals, most of which doesn't aid us in our objective of recognizing emotions from audio files. Hence, we extract the relevant information from these speech signals and provide these as the feature vector input to our classifier. We have used Mel Spectrograms as our feature to be extracted which would be further used to train our classifier. The steps for feature extraction can be seen in figure 1.

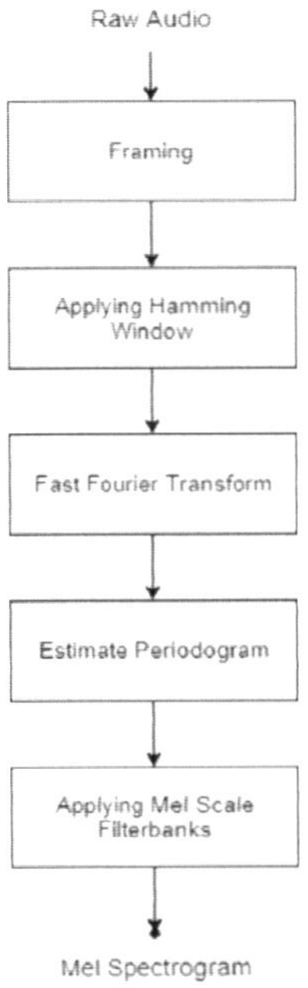

Figure 1. Feature Extraction Steps

2.2.1 Preprocessing

Every audio sample in our dataset is first sampled at 21,500 Hz with an offset of 0.5 seconds and according to the calculations, the average length of audio samples in our dataset is around 3 seconds. Further, Z-distribution, also known as standard normal distribution is performed on the samples and then, we adjust these audio files to be approximately 3 seconds long, i.e 650,000 samples per audio. Adjustment is made by truncating the files, if they have a length greater than average or by padding it with zeroes if it is lesser than the same.

2.2.2 Time Window Framing

We assume that the properties in a non-stationary speech signal signal remain constant over a very short period of time. These short time periods are termed as 'frames', and these frames are long enough to contain vital characteristics and short enough for it to be considered stationary. Here, window size is 23ms with an overlapping of 50frames.Framing begins with the first N = 256 samples, the second frames starts with a hop of M = 128 samples and overlaps the first frame by N M , and this is repeated for the entire signal. This overlap smoothenes the transitions between the frames.

2.2.3 Applying Hamming Window

The signal is further passed through a Hamming window to smoothen out the signal and makes sure that the neighbouring window-ends match. Also, it leads to a better signal clarity and helps in reducing the spectral leakages The

https://doi.org/10.1051/itmconf/20214003006

equation to calculate the same is given by (1)

$$w(k) = 0.54 - 0.46cos(\frac{2\pi k}{N-1}) \quad (1)$$

where $w(k)$ is the window function and $0 \leq k \leq N-1$

2.2.4 Fast Fourier Transform

Fast Fourier Transform takes a sequence of discrete signal amplitudes as input, and converts it into it's frequency con- stituents and is given by (2). We perform an N-point FFT on every frame in the signal to calculate the overall frequency spectrum. This is also termed as Short Term Fourier Transform (STFT).

$$X_k = \sum_{n=0}^{N-1} X_k e^{\frac{-i2\pi}{N}kn}$$

$$k = 0, 1, 2, 3, ..., N-1 \; and \; N = 512 \quad (2)$$

2.2.5 Periodogram Estimate

To compute a periodogram estimation, we square the absolute value of the result derived from the complex fourier transform operation. This estimation helps us in identifying which frequencies are present in every frame extracted from the audio sample. The speech frame's periodogram-based power spectral estimate is given by

$$P_i(k) = \frac{1}{N} \mid S_i(k) \mid^2 \quad (3)$$

2.2.6 Applying Mel-Scale Filterbanks

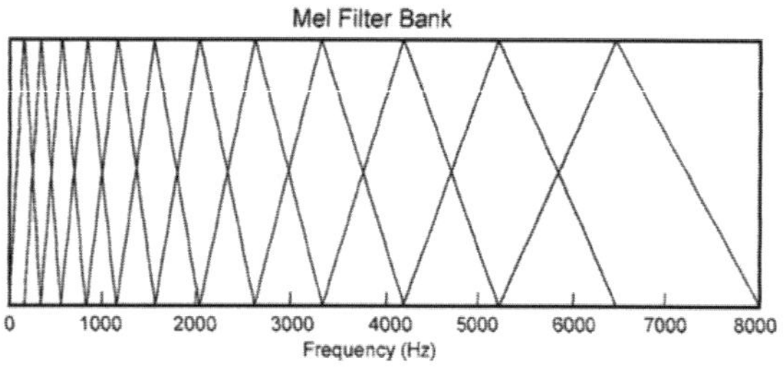

Figure 2. Mel Filterbank

Mel scale is a non-linear transformation and is based on the perception of the audio sample by the human auditory system. To compute the Mel Spectrogram, we just need to apply the mel-spaced filterbanks as seen in figure 2, which is a set of overlapping triangular filters.Starting of one filter overlaps with the centre of the previous one whereas the ending part overlaps with the centre of the succeeding one and so on. It has a response of 1 at the centre and decreases as we move to the ends, where it is 0. The filters are closely spaced and are narrow at the lower frequencies, As the frequency increases, the filters get wider and they turn less discriminative or less sensitive to the variations in the frequency. Mel-Scale tells us how to space the filterbanks on the scale and gives an estimation on how wide to make them as the frequency increases. For converting f Hertz to m mels, we apply

$$Mel(m) = 2595 \times \log_{10}(1 + \frac{f}{700}) \quad (4)$$

3 Classifier

As seen in Figure 3, a hybrid neural network model consisting of Time Distributed CNN followed by a LSTM layer has been proposed. CNN has proven to be a breakthrough in terms of performance related to image recognition and various computer vision tasks. Long Short Term Memory has proven to be very useful in the case of analyzing sequential data. Therefore, it would help the model to learn both short-term and long-term feature dependencies by using the two in succession. CNN and LSTM, can hence, take advantage of the strengths of both networks.

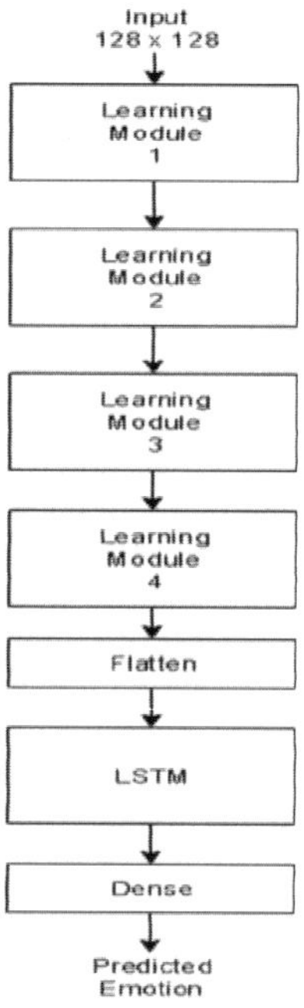

Figure 3. Classifier Model

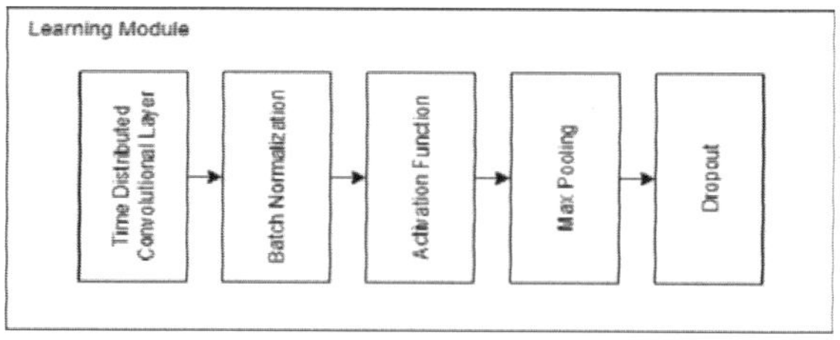

Figure 4. Learning Module

3.1 Time Distributed Convolution Layers

The main idea of the time distributed convolutional layers is applying a rolling window over our input feature, i.e Mel spectrogram. Hence, we get a sequence of images, and the sequence of these images is provided as an input to the first layer in our neural network. As seen in figure 4, this part of the model is subdivided into four Learning Modules (LM). Figure 4 shows that each of these LMs constitutes a time distributed convolutional layer, batch normalization, activation function, max pooling layer, and a dropout layer. Exponential Linear Unit (ELU) is used as the activation function in these four LMs followed by a Max Pooling layer. This is used to lower the size of the image's feature maps, which helps to minimise the number of trainable parameters even more. The dropout regularization helps to avoid overfitting by randomly dropping out a few neurons from the neural network layer.

3.2 Long Short Term Memory

Speech signal is time-varying, and also spectrograms primarily have a time component, so it is well worth trying to explore these temporal properties of speech audio. Long Short Term Memory has proven to be very useful in the case of analyzing sequential data. Therefore, it could help to identify and extract the global temporal features from the mel spectrogram. An LSTM layer having 256 nodes is added as a learning layer in the model followed by a dense fully connected layer having 'softmax' as its activation function.

4 Experiments

The compiled dataset is split into train and test sets, where 80% of the total is used for training the model and the remaining 20% is evaluating the model's performance. Hence, we train with 7557 audio samples and test on 1889 audio samples. After applying a rolling window on the mel spectrogram, a sequence of 6 overlapping images is generated for each audio file, and this is provided as input to our first Learning Module (LM). As seen in figure [3] are four LM's in our model and each one consists of a time distributed convolutional layer, batch normalization layer, an activation function layer, dropout layer, and lastly the max pooling layer. The convolutional layers have a kernel size of 3×3. The layers in the first two blocks have 64 feature maps, and the subsequent two have 128 feature maps. The activation function used in all four blocks is Exponential Linear Unit as this function tends to converge faster and provide better accuracies. After the four LM blocks, the resultant output is flattened and provided as an input an LSTM layer, followed by a fully connected layer with a softmax activation function which provides an output that maps to the predicted emotion for every audio sample. The optimizer for this model is Stochastic Gradient Descent (SGD), with a learning rate of 0.01. We applied early stopping having patience as 15, with respect to maximizing the validation accuracy.

5 Results

The model is trained for a set of 100 epochs and compiled with categorical cross-entropy as the loss function. By using 'Model Checkpoint', we monitored and saved the weights that provide maximum value for validation accuracy. Our best model saved through this provides an accuracy of 90.64% on the training set and 89.26% on the testing set.

5.1 Performance Matrix

Table 2. Performance of the model

	Evaluation Metrics		
Emotions	**Precision**	**Recall**	**F1-score**
Neutral	92	90	91
Happy	82	87	84
Sad	92	81	86
Angry	90	92	91
Fear	91	92	91
Disgust	84	96	90
Surprise	95	87	91

Table 2 summarises our model's performance in terms of various performance metrics. Precision is the ability of our model to check the correctly predicted positives from all the predicted positives and is given by Eq. (5). Our model has the highest precision of 95% for surprise emotion and lowest for happy emotion with precision of 82%.

$$Precision = \frac{TruePositive}{TruePositive + FalsePositive} \quad (5)$$

Recall measures the model ability to check the correct positive from all the existing positives in the test dataset and is given by Eq. (6). We can see from table 2 that our model has the best recall score for disgust emotion of 96%and the worst recall score of 81% for sad emotion.

$$Recall = \frac{TruePositive}{TruePositive + FalseNegative} \quad (6)$$

The F1 score is the harmonic mean of precision and recall, which is used to measure an emotion's overall performance. Eq. (7). The F1 score is The best F1 score is for surprise, neutral, angry, fear emotions which is 91% and the worst F1 score is 84% for happy emotion.

$$F1 - score = \frac{2 * Recall * Precision}{Recall + Precision} \quad (7)$$

Model accuracy score is the most intuitive performance measure, and it is simply a ratio of correctly predicted observation to the total observations. Our model has an accuracy of 90.64% on training set and 89.26% on validation data set.

https://doi.org/10.1051/itmconf/20214003006

Table 3. Comparison with Existing System

Parameters	Proposed Model	System [4] pt. 1
Architecture	Time Distributed CNN + LSTM	CNN
Dataset	RAVDESS + SAVEE + TESS	EmoDB
Features Used	Log Mel-Scale Filterbanks	Mel Spectrogram
Dataset Distribution	7720 and 1450	271 and 68
Accuracy	**89.26 %**	**78.16 %**

Parameters	Proposed Model	System [4] pt. 2
Architecture	Time Distributed CNN + LSTM	Bi-LSTM
Dataset	RAVDESS + SAVEE + TESS	Iemocap
Features Used	Log Mel-Scale Filterbanks	Mel Frequency Cepstral Coefficient
Dataset Distribution	7720 and 1450	4424 and 1107
Accuracy	**89.26 %**	**46.21 %**

In the above tables, we have compared our system to two pre-exiting systems presented in [4], based on various parameters.

5.2 Loss and Accuracy Curves

A learning curve is a diagnostic tool that shows the performance of the model over the period of time. We have in the plotted the loss curve and accuracy curve for the training of our model. We can see that Figure 5 shows a steep decline in training loss for the first 20 epochs and then we notice steady decline of training loss till 100th epoch and Figure 6 shows a sharp increase in the accuracy for the first 20 epochs and then we see that there is a steady increase of the accuracy till 100th epoch.

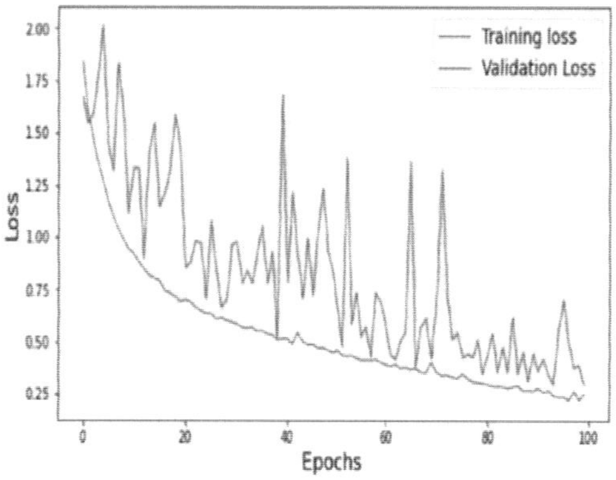

Figure 5. Loss Curves

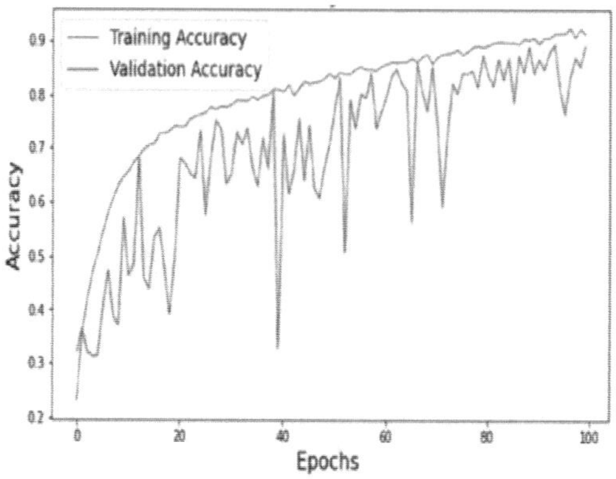

Figure 6. Accuracy Curves

5.3 Confusion Matrix

Confusion matrix helps us evaluate the performance of our neural network, when it makes predictions on the test data, and it helps us to analyse how accurate our recognition model is with respect to specific emotions. From the confusion matrix seen in figure 7, we can conclude that:

1. The model performs exceptionally well, while predicting fear and disgust emotions as these emotions have a result of 251/274 and 249/259 respectively.

2. On the other hand emotions like happy and sad have a result of 220/254 and 208/258,hence there is scope of improvement in these emotions.

3. The model often predicts a sad emotion as neutral emotion. By coming up with a solution for this problem, we can improve the performance of the system to a great extent.

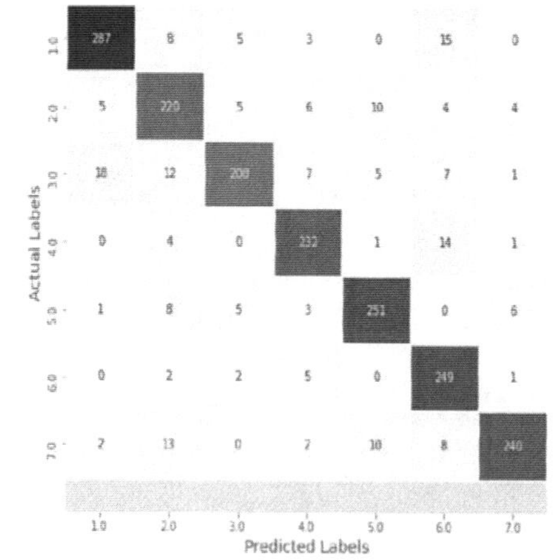

Figure 7. Confusion Matrix

6 Conclusion

This research presents a hybrid neural network strategy for detecting underlying emotions in audio samples. In terms of performance, a Time Distributed CNN + LSTM model trained on a large gender-balanced dataset of speakers with various accents and nationalities outperforms other models. Neutral, Angry, Fear, Disgust, and Surprise all had testing accuracy of 90% or higher. This model's performance could be improved even further through targeted training, by focusing on emotions such as happy and sad, which have accuracy rates of around 84% and 86%, respectively.

References

[1] S. Basu, J. Chakraborty and M. Aftabuddin, "Emotion recognition from speech using convolutional neural network with recurrent neural network architecture," 2017 2nd International Conference on Communication and Electronics Systems (ICCES), 2017, pp. 333-336, doi: 10.1109/CESYS.2017.8321292.

[2] J. Umamaheswari and A. Akila, "An Enhanced Human Speech Emotion Recognition Using Hybrid of PRNN and KNN," 2019 International Conference on Machine Learning, Big Data, Cloud and Parallel Computing (COMITCon), 2019, pp. 177-183, doi: 10.1109/COMITCon.2019.8862221.

[3] B. Puterka and J. Kacur, "Time Window Analysis for Automatic Speech Emotion Recognition," 2018 International Symposium ELMAR, Zadar, 2018, pp. 143-146. doi: 10.23919/ELMAR.2018.85.

[4] S. K. Pandey, H. S. Shekhawat and S. R. M. Prasanna, "Deep Learning Techniques for Speech Emotion Recognition: A Review," 2019 29th International Conference Radioelektronika (RADIOELEKTRONIKA), 2019, pp. 1-6, doi: 10.1109/RADIOELEK.2019.8733432.

[5] A. B. Abdul Qayyum, A. Arefeen and C. Shahnaz, "Convolutional Neural Network (CNN) Based Speech-Emotion Recognition," 2019 IEEE International Conference on Signal Processing, Information, Communication Systems (SPICSCON), 2019, pp. 122-125, doi: 10.1109/SPICSCON48833.2019.9065172.

[6] M. S. Likitha, S. R. R. Gupta, K. Hasitha and A. U. Raju, "Speech based human emotion recognition using MFCC," 2017 International Conference on Wireless Communications, Signal Processing and Networking (WiSPNET), 2017, pp. 2257-2260, doi: 10.1109/WiSPNET.2017.8300161.

[7] B. Puterka, J. Kacur and J. Pavlovicova, "Windowing for Speech Emotion Recognition," 2019 International Symposium ELMAR, 2019, pp. 147-150, doi: 10.1109/ELMAR.2019.8918885.

[8] M. K. Pichora-Fuller and K. Dupuis, "Toronto emotional speech set (TESS)," 2010

[9] Livingstone SR, Russo FA (2018) The Ryerson Audio-Visual Database of Emotional Speech and Song (RAVDESS): A dynamic, multimodal set of facial and vocal expressions in North American English. PLoS ONE 13(5): e0196391. https://doi.org/10.1371/journal.pone.0196391

[10] B. Vlasenko, B. Schuller, A. Wendemuth, and G. Rigoll, "Combining frame and turn-levelinformation for robust recognition of emotions within speech,"Proceedings of Interspeech,pp. 2249–2252, 01 2007.

 https://doi.org/10.1051/itmconf/20214003007

Comparative Analysis of Machine Learning Algorithms for Heart Disease Prediction

Ruby Hasan[1,*]

[1]Department of Computer Science and Engineering, Jamia Hamdard, New Delhi, India

Abstract. In the last few years, cardiovascular diseases have emerged as one of the most common causes of deaths worldwide. The lifestyle changes, eating habits, working cultures etc, has significantly contributed to this alarming issue across the globe including the developed, underdeveloped and developing nations. Early detection of the initial signs of cardiovascular diseases and the continuous medical supervision can help in reducing rising number of patients and eventually the mortality rate. However with limited medical facilities and specialist doctors, it is difficult to continuously monitor the patients and provide consultations. Technological interventions are required to facilitate the patient monitoring and treatment. The healthcare data generated through various medical procedures and continuous patient monitoring can be utilized to develop efficient prediction models for cardiovascular diseases. The early prognosis of cardiovascular illnesses can aid in making decisions on life-style changes in high hazard sufferers and in turn lessen the complications, which may be an outstanding milestone inside the field of medicine. This paper studies some of the most widely used machine learning algorithms for heart disease prediction by using the medical data and historical information. The various techniques are discussed and a comparative analysis of the same is presented. This report compares five common strategies for predicting the chance of heart attack that have been published in the literature. KNN, Decision Tree, Gaussian Naive Bayes, Logistic Regression, and Random Forest are some of the approaches used. Further, the paper also highlights the advantages and disadvantages of using the various techniques for developing the prediction models.

Keywords- *Machine Learning, Heart disease prediction, Logistic regression, Decision Tree, Random Forest, Gaussian Naïve Bayes, KNN, Cross-Validation.*

1. INTRODUCTION

Coronary illness has caused serious concerns over the last decade. One of the most significant difficulties in coronary illness is recognizing the symptoms and the correct diagnosis of the illness. The early procedures were not enough productive and efficient in predicting the heart disease [1]. There are different clinical instruments accessible on the lookout for anticipating coronary illness. Such instruments have two significant issues in them, the first is that they are a lot costly and second one is that they are not efficient to correctly compute the coronary illness in human. As indicated by most recent review directed by WHO, the clinical experts are ready to effectively anticipated just 67% of coronary illness [2], so there is a tremendous extent of exploration in the space of predicting coronary illness in humans. There are numerous kinds of coronary illness, and everyone has its own side effects and treatment. For a few, way of life changes and medication can have a gigantic effect in improving individual wellbeing. For others, they may require a medical procedure to make it work [3]. Due to the heart diseases there is a lot of impact on population worldwide. GBD (Global Burden of Disease) 2019 is a worldwide cooperative examination study that gauges illness trouble for each country on the planet. The investigation is a continuous exertion, refreshed every year, and is intended to consider predictable correlation after some time from 1990 to 2019, by age and sex, and across areas. The examination produces standard ep id em io lo g i cal e s t im a tes l ik e f r equ en cy, pervasiveness, and demise rates just as synopsis proportions of wellbeing, like DALYs. DALYs address the amount of long periods of everyday routine lost rashly and years experienced with incapacity; can be assessed from life tables, evaluations of commonness, and handicap loads; and might be communicated as checks or rates [4]. Headway in software engineering has brought tremendous openings in various territories. Clinical Science is one of the fields where the instrument of software engineering can be utilized. Clinical science additionally utilized a portion of the major accessible devices in software engineering. In a decade computerized reasoning has acquired its second in view of progression in calculation power.

Machine Learning is one such instrument which is broadly used in various spaces since it doesn't need extraordinary calculation for various dataset. Reprogrammable limits of AI bring a ton of solidarity and opens new entry ways of chances for region like clinical science. In clinical science coronary illness is one of the significant difficulties; in light of the fact that a great deal of boundaries and detail is include for precisely predicating this infection. AI could be a superior decision for accomplishing high exactness for predicating coronary illness as well as another infections since this change instrument uses include vector and its different information types under different condition for predicating the coronary illness, calculations like Innocent Bayes, Choice Tree, KNN, Neural Organization, are utilized to predicate hazard of heart sicknesses every calculation has its claim to fame, for example, Innocent Bayes utilized likelihood for predicating coronary illness, though choice tree is utilized to give arranged report to the coronary illness, while the Neural Organization gives freedoms to limit the blunder in predication of coronary illness. Every one of these procedures are utilizing old patient record for getting predication about new tolerant. This predication framework for coronary illness assists specialists with anticipating coronary illness in the beginning phase of infection bringing about saving large number of life [5].

https://doi.org/10.1051/itmconf/20214003007

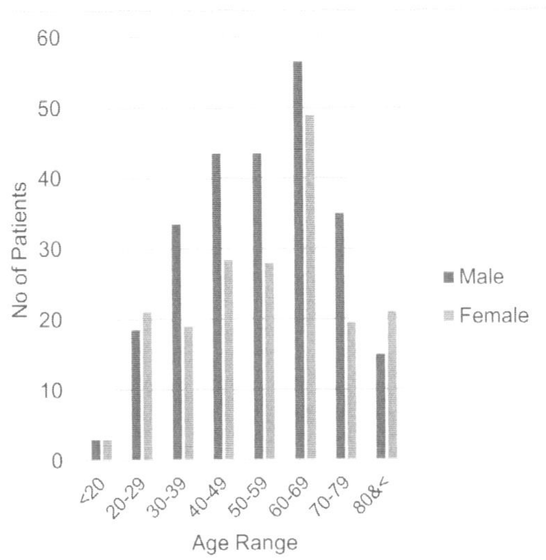

Fig1. Analysis of heart disease with respect to age group (2009),(https://www.researchgate.net/figure/Bar-chart-of-age- years-sex-distribution-of-patients-with-heart-failure_fig1_26831468.)

In scientific centers, records mining strategies and system getting to know algorithms are crucial within the evaluation of records. The strategies and algorithms can be carried out directly to a dataset to create models or draw crucial conclusions and inferences. Commonplace attributes used for coronary heart disease are Age, sex, Fasting Blood stress, Chest pain type, Resting ECG(test that measures the electrical hobby of the heart), number of foremost vessels colored with the aid of fluoroscopy, Thrust Blood strain (excessive blood pressure), Serum cholesterol (decide the chance for growing heart ailment), Thalach (most heart fee executed), ST despair (finding on an electrocardiogram, hint in the ST phase is abnormally low under the baseline), painloc (chest ache vicinity (substernal=1, otherwise=0)), Fasting blood sugar, Exang (workout included angina), smoke, high blood pressure, meals conduct, weight, peak and obesity[6]. Table 1 summarizes the most common types of the coronary heart disorder.

Table 1. Different types of heart disease [7]

Arrhythmia	If the coronary heart rhythm is abnormal, too sluggish, or too rapid, it is a sign that something is incorrect.
Cardiac arrest	Heart function, consciousness, and breathing all stop working unexpectedly.
Congestive heart failure	Chronic heart failure takes place whilst the coronary heart does not pump blood as successfully as it can.
Congenital coronary heart disease	An abnormality of the heart that appears before start.
Coronary artery disease	The main blood vessels of the heart can be damaged, and any disease that affects the blood vessels can be fatal.
Excessive Blood stress	It has a disorder in which the blood stress in opposition to the artery walls is excessive.
Peripheral artery disease	The circulatory disorder is characterized by narrowed blood vessels that restrict blood flow in the limbs.
Stroke	Interruption of blood deliver occur damage to the mind.

1.1 PAPER ORGANIZATION

The paper is divided into five sections. Section two presents an overview of the past researches and studies conducted on the on machine learning based heart disease prediction. Section three provides information about the widely used machine learning algorithms used for prediction especially for heart attacks and other heart diseases. The analysis of the approaches used and their results is presented in section four. Section five presents the conclusion.

2. RELATED WORK

With increasing research in the field of healthcare along with advanced machine learning, various experiments and researches were carried out in the last few years that provides significant information about the potential of modern day technologies in the healthcare sector. Marjia Sultana et.al, endorse coronary heart ailment prediction the usage of KStar, J48, SMO and Bayes internet and Multilayer perceptron using WEKA software [8]. Relying on the performance of various thing SMO (89% of accuracy) and Bayes internet (87% of accuracy) generate most beneficial performance than KStar, Multilayer perceptron and J48 techniques the use of k- fold go validation. These algorithms has not been able to generate satisfactory results of performance. If one can improve the performance of accuracy then it can help in better decision making to diagnosis heart disease. S. Musfiq Ali et.al, research has been conducted using Cleveland dataset for heart diseases which involves 303 instances and used 10-fold Cross Validation, mentioning 13 attributes, putting 4 different algorithms, they resulted that Gaussian Naïve Bayes and Random Forest has given the maximum accuracy of 91.2% [9]. The usage of the same dataset of Framingham, Massachusetts, the experiments had been executed the usage of 4 fashions and had been skilled and tested with most accuracy ok- neighbors Classifier: 87%, support Vector Classifier: 83%, selection Tree Classifier: 79% and Random wooded area Classifier: 84% [10]. To improve accuracy and analysis to coronary heart disease, Abdullah and Rajalaxmi introduced a DM model using the RF Classifier (CHD) [11]. Within the studies, more than one CHD incidents have been tested, which includes angina and acute myocardial infraction (AMI) and skip graft surgical treatment. The study showed that in predictive mode for CHD, an ensemble approach to the

RF classification algorithm can be very effective .Lafta et.al, based on the groundbreaking time series prediction algorithm, developed an intelligent prediction method at the University of Southern Queensland [12]. The scheme offered decision support for medical practitioners based on the medical details of each patient. By improving the capacity of the algorithm, the method has the potential to improve accuracy and efficiency. A total of 303 records with 76 medical attributes were used for the prediction model by Sonam Nikhar et al. for the prediction model, only 19 attributes were used for Naive Bayes Classifier, Decision Tree Classifier algorithm. They also studied that, compared to Naïve Bayes Classifier, Decision Tree has better accuracy [13]. For the Heart Disease Prediction Survey, Ravindra Yadav et.al, used a Machine Learning methodology with Decision Tree algorithm, Naïve Bayes, Neural Network, Deep Learning and SVM. CART, ID3, CYT, C%.0 and J48 are used to generatethe outcome of the decision tree [14]. Devansh Shah et.al, used the dataset which includes 303 instances and 76 attributes, of these14 attributes were used for supervised learningalgorithm which were Decision Tree, Naïve Bayes, K- NN and Random forest [15]. The results portray that the highest accuracy has achieved by K-NN.Archana Singh et.al, used dataset of 14 attributes for the models to trained and tested to get the maximum accuracy Linear Regression: 78%, decision Tree: 79%, Support Vector Machine: 83% and K-NN : 87%. The results showed that the maximum accuracy has with K- NN [16]. PushkalaV et.al, possess dataset of 75 attributes and 303 instances, out of which only 14 attributes are used for Decision Tree, Random Forest, SVM, K- Nearest Neighbors Classifier, Gaussian Naïve Bayes. The Naïve Bayes gives the highest accuracy of 91% [4]. Jaymin Patel et.al, this research paper compares extraordinary algorithms of class in search of better performance in coronary heart disease prognosis using WEKA [17]. Mustafa Jan, Salman Nasir et.al, in this observe, ensemble mastering combines five classifiers version approaches which includes assist vector system, synthetic neural network, naïve Bayesian, regression analysis and random forests to expect and diagnose the recurrence of cardiovascular disease. The dataset was taken from VCI data repository, 14 attributes were taken to train and test the models. The highest accuracy is 98.17% for the random forest algorithm [18]. Jianping Li,Asif Khan et.al, In this paper the classification algorithms were based on SVM, Logistic Regression, Artificial neural networks, KNN, Naïve Bayes and Decision tree. The model achieved accuracy of 92.37%, as compared to previous models [19]. Senthil Kumar Mohan et.al the primary aim of this paper was to find sufficient features by applying machine learning techniques like decision trees, language model, SVM, random forests, naïve Bayes, neural networks, KNN. The proposed hybrid HRFLM method became used for combining the characteristics of random forests and linear technique. This model achieved the accuracy of 88.4% [20]. Prediction and evaluation of the prevalence of heart sickness the use of data Mining strategies turned into counseled by Chala Beyene et al. the key intention is to are expecting the occurrence of heart sickness in an effort to make an early computerized analysis of the sickness with a short end result. In healthcare establishments with specialists who lack enjoy and ability, the cautioned method is also crucial. It uses a ramification of medical traits, which includes blood sugar and heart price, as well as age and sex, to determine whether or not someone has heart disease. WEKA software is used to compute dataset analyses [21]. Kavitha B S, M.Siddapa carried out a survey for the prediction of heart disease.In this study it was found that most of the data was taken from Cleveland repository various machine learning classifiers are used to build heart disease prediction model. Based on the survey, the RF algorithms showed highest accuracy as compared to other models [22]. N. Komal, G. Sarika et.al .In this the various machine learning techniques were used for the prediction of cardio vascular disease.The proposed model showed that random forests having highest accuracy achieved of 85.71% as compared to other classifier techniques [23]. Aditi Gavhane et.al In this paper the model was built for the prediction of heart disease using a machine learning algorithm MLP (Multilayer Perceptron) which provide prediction results that gives the state of a user leading CAD to its users [24].

Table 2. Performance Analysis Chart based on the Reviewed work

Authors	Techniques used	Accuracy
Sonam Nikhar et.al[13]	Naïve Bayes Decision tree	78%
Archana Singh et.al[16]	KNN Decision Tree	87% 79%
Pushkala V et.al[4]	Naïve Bayes	91%
Mustafa et.al[18]	Random Forest	98.17%
Senthil Kumar Mohan et.al[20]	Random Forest, Linear Method (HRFLM)	88.4%
Thankgod Obasi et.al[28]	Random Forest Logistic Regression Naïve Bayes	92.44% 59.7% 61.96%

S.Kathiresan in this article, supervised algorithms of machine learning are used to forecast heart disease. The learning algorithm is implemented on a variety of parameters in a real-time four-stage cloud system. The productivity is estimated to be over ten times higher, and cloud technology for forecasting and diagnosing heart disease has been integrated [5]. Himanshu Sharma, M A Rizvi .This paper summarizes current state-of-the-art strategies and available tools for disease prediction. Deep learning, a new branch of artificial intelligence, has shown promise in other fields, such as high-accuracy medical diagnosis [25].

Coronary illness is perhaps the main sources of mortality on the planet today. Expectation of cardiovascular contamination is a primary check inside the area of scientific records research. AI (ML) has been established to be powerful in supporting with settling on alternatives and expectancies from the great quantity of data created by the hospital therapy enterprise. We've additionally seen ML tactics being utilized in late enhancements in numerous spaces of the web of things (IoT). Unique examinations provide only a short investigate foreseeing heart infection with ML techniques. From the above noted paper recognizing the preparing of crude hospital therapy records of coronary heart records will help in the drawn out saving of human lives and early region of irregularities in coronary heart situations. AI techniques were applied in this work to deal with crude facts and deliver every other and novel insight in the direction of coronary infection. Coronary infection expectation is trying out also, crucial in the medical discipline. Anyways, the mortality fee may be appreciably managed if the infection is distinguished at the start phases and guard measures are embraced as earlier than long as might be anticipated. Table 2 presents the performance analysis of various machine learning algorithms along with their accuracies used for prediction of heart diseases.

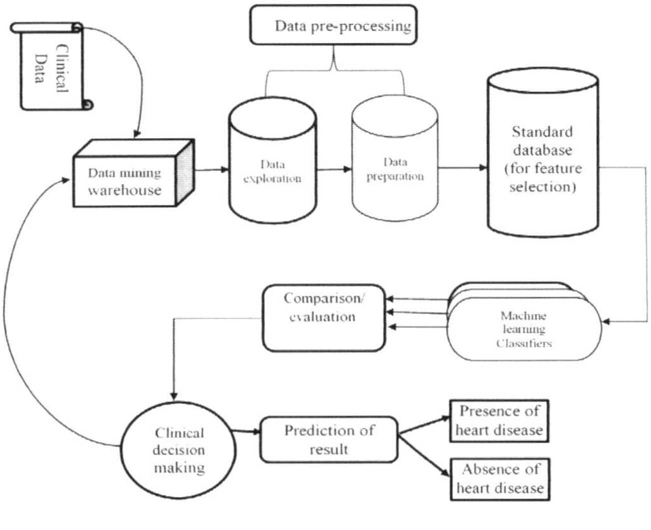

Fig 2. Data mining methodology and heart disease prediction process [18]

3. METHODS FOR PREDICTION

Figure 2 shows the process of predicting heart diseases using data mining and machine learning techniques. The process of data collection, preprocessing, classification and prediction for predicting of results, whether the person has presence or absence of heart disease is highlighted. Numerous classification techniques, such as KNN, Decision Tree, Gaussian naïve Bayes, Random Forest, and Logistic Regression, are studied in this paper to determine the best suited machine learning algorithm for accuracy, as well as many feature selection strategies, such as backward elimination and recursive feature elimination are also reviewed.

3.1 Logistic Regression

Logistic Regression is a supervised classification algorithm. It's a probabilistic analysis algorithm that predicts outcomes. By estimating probabilities using the underlying logistic equation, it assists in measuring the relationship between the dependent variable (TenyearCHD) and one or more independent variables (risk factors) (sigmoid function).
The following logistic function is used in the logistic regression algorithm:

$$p=1\backslash(1+e^{\wedge}-x) \qquad (1)$$

The logistic coefficients for each instance x1, x2, x3..... xn will be b0, b1, b2..... bn during the training stage. Stochastic gradient descent is used to estimate and update the coefficient values.

$$\text{Values} = b0x0+b1x2+.......bnxn \qquad (2)$$

$$b=b+l*(y-p)*(1-p)*p*x \qquad (3)$$

Each training case's output value is represented by y, and all of the coefficients are initially set to 0. For b0's biassed input, l is the learning rate, and x is always 1. It changes the coefficient values at the training level until it predicts the correct performance.

Logistic Regression relies highly on the proper presentation of data. So, to make the model more powerful, important features from the available data set are selected using backward elimination and recursive elimination techniques [26].

3.2 Backward Elimination Method

Only the features that have a major impact on the goal variable should be chosen when designing a machine learning model. The first step in the backward elimination process for feature selection is to choose a significance level or P-value. We chose a 5% significance rating, or a P-value of 0.05, for our model. The function with the highest P-value is marked, and it is excluded from the dataset if its P-value is greater than the significance level. The model is fit to a new dataset, and the process is repeated until all remaining features in the dataset have a significance level less than the significance level. Male, age, cigsPerDay, prevalent Stroke, diabetes, and sysBP were chosen as significant factors in this model after using the backward elimination algorithm [26].

3.3 Recursive Feature Elimination Using Cross Validation

The RFECV algorithm is a greedy optimization algorithm that seeks out the best performing function subset. Recursive Feature Elimination (RFE) suits a model several times, removing the weakest feature each time until the desired number of features is reached. RFE scores various feature subsets and selects the highest scoring range of features, which is RFECV, using the optimal number of features. The key disadvantage of this algorithm is that it can be costly to implement. As a result, it is preferable to reduce the number of features in advance. Correlated features can be removed before

RFECV because they provide the same details. The correlation matrix is plotted to solve this, and the correlated features are omitted. The arguments for instance of RFECV are:

- estimator - model instance (RandomForestClassifier)
- step - number of features removed on each iteration
- cv – Cross-Validation (Stratified Fold)
- scoring – scoring metric (accuracy)

Once RFECV is run and execution is finished, the features that are least important can be extracted and dropped from the dataset. Top 10 features ranked by the RFECV technique in our model listed below from least importance to highest importance [26].

- prevalent Stroke
- diabetes
- BPMeds
- current Smoker
- prevalentHyp
- male
- cigsPerDay
- heart rate
- glucose
- diaBP

3.4 Decision Trees

A decision tree is a tree-like structure or a flowchart-like structure that is used as a decision-support tool in both classification and regression problems to help build automated predictive models. It's a non-parametric supervised learning algorithm. It creates a model that predicts the cost of a target variable via mastering simple selection regulations inferred from facts capabilities. The decision tree model is generated by putting the best attribute of the dataset at the root of the tree. There are several subsets in the training package. Each subset contains information for the same attribute. The algorithm is repeated until all of the leaf nodes have been discovered, i.e., the desired result has been discovered. There are almost no hyper-parameters that need to be set. The decision tree is associated with decision rules and conditions along the path form and if else rule to allow the algorithm to decide on the basis of the situation, where the outcome is the contents of the leaf node. As the algorithm becomes more complex in order to reduce the error of the training set, the test set's error will increase.

This over fitting problem will occur frequently when building a decision tree model. The use of a decision tree aids in the visualization of logic. It generates all possible decision outcomes. The decision tree algorithm's main goal is to prioritize the attribute that can provide the highest level of accuracy. It creates a model to predict the variable "num" using the inferred dataset [4].

3.5 Random Forests

Random Forest produces a forest in an ad hoc manner. The ensemble of decision tree classifiers is used to build the forest. Bootstrap Aggregation or Bagging, which combines the predictions of multiple decision trees and brings them together to construct a forest to make more precise predictions, is the foundation of the random forest algorithm. To fit a sub-sample of the dataset, it employs a decision tree classifier. It's a supervised learning process, and the algorithm is similar to classification and regression trees (CART). They respond to the data gathered during the training. The CART algorithm bagging generates a number of random sub-samples from the dataset, each of which is used to train the CART model, and the average prediction from each model is then determined using the test results. Low bias and high variance characterize these trees. The planning for this process will take a long time, but the problem of overfitting will not arise. CART's problem is that they are self-centered. Random Forest redesigned the CART algorithm in such a way that the sub-trees are known. The resulting prediction from these sub-trees has a lower correlation. Out-Of-Bag samples are the samples left over from the bootstrap samples taken from the training results. The estimated accuracy is determined by the output of each model on its left out samples. It is possible to selectively select the features needed for the prediction process from the dataset, a process known as feature significance. The random forest can be built without the less important feature that contributes less to the prediction process [4].

3.6 Gaussian Naïve Bayes

The Naive Bayes classifier is built using the Bayes Theorem. By assuming a Gaussian distribution, it can be extended to real-valued attributes. A version of Naive Bayes is Gaussian Naive Bayes. It's a supervised classification learning algorithm for binary and multi-class classification problems. This algorithm is also known as Idiot Bayes because the probabilities of hypothesis estimation are tractable. There is no need for optimum coefficient fitting because data training is easy. When using this classifier to make a prediction, the main assumption is that the attributes of the data set are independent. The probabilities in each class, the mean and standard deviations for each input variable in each class are determined [4].

$$P(X/Y) = [P(Y/X)\times P(X)]/P(Y) \quad (4)$$

P(X/Y) is the posterior probability, P(X) is the class prior probability, P(Y) is the predictor prior probability, P(Y/X) is the likelihood, probability of predictor.

Nave Bayes is a non-linear, difficult data categorization algorithm that is simple, straightforward to implement, and efficient. However, because it is reliant on assumption and class conditional independence, there is a loss of accuracy [15].

3.7 KNN (K- Nearest Neighbor)

The K nearest neighbor algorithm is an example-based learning algorithm that is widely used in real-life scenarios. The K Nearest Neighbors algorithm can be used to solve both classification and regression problems. The K Nearest Neighbor algorithm is another name for lazy learning. The K Nearest Neighbor algorithm involves preprocessing the dataset, training the model, and testing the model. Cleaning and removing erroneous and outlier values from a dataset is usually part of the preprocessing phase. In the algorithmic process, this is the most important step. It's also crucial to check the dataset's accuracy before running algorithmic tests on it. It is important to pay attention to missing values and outliers. The K Nearest Neighbor uses the curve to plot the latest test results. The K element is the number of neighbors that the classification takes into account. In most cases, the K value should be a single digit number. After the test data point is plotted, the distance formula is used to calculate the K closest neighbors. [4].

Table 3. Comparisons of Machine Learning Techniques [27]

Technique	Primary Problem	Predictors	Raw Implementation	Interpretability	Regression Also	Normalization
KNN	Multiclass or binary	Numeric	Easy	Good	No	Required
Logistic regression	Binary	Numeric	Easy	Good	No	No
Naïve Bayes	Multiclass or binary	Categorical	Medium	Good	No	Required
Decision Tree	Multiclass or binary	Numeric or Categorical	Difficult	Good	Yes	No
Random forest	Multiclass or binary	Numeric or Categorical	Difficult	Good	Yes	No

4. CONCLUSION

Machine learning based solutions are widely used in healthcare sector for analyzing patients' data, predicting diseases and suggesting possible treatments. With a number of machine learning techniques available today, it is important to identify the most efficient and accurate technique especially in critical domains like healthcare. A comparative analysis of the various Machine learning algorithms used in the heart disease prediction is presented. Five of the most widely used techniques, KNN, Logistic regression, Naive Bayes, Decision tree and random forest are discussed and compared to identify the best suited classifier for heart disease prediction. Many previous researches and studies related to heart disease prediction were identified and analyzed. Based on the identified researches and studies a performance analysis of various machine learning algorithms along with their accuracies used for prediction of heart diseases was done as presented in table 2. The findings shows that in most cases machine learning based approaches have shown significant potential to transform the healthcare sector and improve the entire process of disease predictions and suggesting treatments.

REFERENCES

1. M. A. Jabbar, P. Chandra, and B. L. Deekshatulu, "Prediction of risk score for heart disease using associative classification and hybrid feature subset selection," Int. Conf. Intell. Syst. Des. Appl. ISDA,pp. **628–634**, (2012).
2. V. Kirubha and S. M. Priya, "Survey on Data Mining Algorithms in Disease Prediction," vol. **38, no. 3, pp. 124–128**, (2016).
3. https://www.webmd.com/heart-disease/heart-disease-types-causes-symptoms (23 October 2020)
4. PushkalaV, A. T., & Angayarkanni, S. A. (2019). Comparative Study of Heart Disease Prediction Using Machine Learning Algorithms. *International Journal of Innovations in Engineering and Technology (IJIET) http://dx. doi. org/10.21172/ijiet. 124, 10*,**2319-1058.**
5. http://ictactjournals.in/paper/IJDSML_Vol_2_Issue_2_Paper_7_153_156.p df (25 October 2020)
6. T.Mythili, Dev Mukherji, Nikita Padaila and Abhiram Naidu, "A Heart Disease Prediction Model using SVM- Decision Trees- Logistic Regression (SDL)", *International Journal of Computer Applications*, **vol. 68**, (16 April 2013).
7. https://www.medicalnewstoday.com/articles/257484.php.(28 October 2020)
8. A. H. M. S. U. Marjia Sultana, "Analysis of Data Mining Techniques for Heart Disease Prediction," (2018).
9. M. I. K. A. I. S. Musfiq Ali, "Heart Disease using Machine Learning Algorithms.
10. K. Bhanot, "towarddatascience.com," (13 Feb 2019 .) [Online]. Available: https:/ / towardsdatascience.com/predicting-presence- of - heart-diseases-using-machine - learning-36f00f3edb2c. (2 November 2020).
11. Abdullah AS, Rajalaxmi RR. A data mining model for predicting the coronary heart disease using random forest classifier. International C o n f e r e n c e o n R e c e n t T r e n d s i n Computational Methods, Communication and Controls (ICON3C 2012) Proceedings published in *International Journal of Computer Applications® (IJCA); (2012).*
12. Lafta R, YanLi, Tseng VS. An Intelligent Recommender System based on Short Term Risk Prediction for Heart Disease patients. IEEE/WIC/ ACM *International Conference on Web Intelligence and Intelligent Agent Technology. Singapore: IEEE; (2015).*
13. SonamNikhar, A.M. Karandikar, "Prediction of Heart Disease Using Machine Learning Algorithms", International Journal of Advanced Engineering, Management and

Science (IJAEMS) [Vol-2, Issue-6, June- 2 0 1 6] *I n f o g a i n P u b l i c a t i o n* (Infogainpublication.com) ISSN: **2454-1311**, (2016).
14. h t t p s : / / w w w . i j c r t . o r g / p a p e r s / IJCRT1807425.pdf (13 November 2020)
15. Devansh Shah, Samir Patel, Santosh Kumar Bharti, "Heart Disease Prediction using Machine Learning Techniques", *Springer Nature Singapore Pte Ltd (2020).*
16. Archana Singh, Rakesh Kumar, "Heart Disease Prediction using Machine Learning Algorithms", *International Conference on E l e c t r i c a l & E l e c t r o n i c s Engineering(ICE3-2020)* : **452-457**.
17. Patel, J., TejalUpadhyay, D., & Patel, S. (2015). Heart disease prediction using machine learning and data mining technique. *Heart Disease*, *7*(1), **129-137.**
18. Jan, M., Awan, A. A., Khalid, M. S., & Nisar,S. (2018). Ensemble approach for developing a smart heart disease prediction system using classification algorithms. *Research Reports in Clinical Cardiology*, *9*, **33-45**.
19. Li, J. P., Haq, A. U., Din, S. U., Khan, J.,Khan, A., & Saboor, A. (2020). Heart Disease Identification Method Using Machine Learning Classification in E-Healthcare. *IEEE Access*, *8*, **107562-107582**.
20. Mohan, S., Thirumalai, C., & Srivastava, G. (2019). Effective heart disease prediction using hybrid machine learning techniques. *IEEE Access*, *7*, **81542-81554**.
21. Mr. Chala Beyene, Prof. Pooja Kamat, "Survey on Prediction and Analysis the Occurrence of H e a r t D i s e a s e U s i n g D a t a M i n i n g Techniques", *International Journal of Pureand Applied Mathematics*, (2018).
22. Kavitha, B. S., & Siddappa, M. A Survey on Machine Learning Techniques to Predict Heartt Disease.
23. Kumar, N. K., Sindhu, G. S., Prashanthi, D. K., & Sulthana, A. S. (2020, March). Analysis and prediction of cardio vascular disease using machine learning classifiers. In *2020 6th International Conference on Advanced Computing and Communication Systems (ICACCS)* **(pp. 15-21).** *IEEE.*
24. Gavhane, A., Kokkula, G., Pandya, I., & Devadkar, K. (2018, March). Prediction of heart disease using machine learning. In *2018 Second International Conference on Electronics, Communication and Aerospace Technology (ICECA)* **(pp. 1275-1278).** IEEE.
25. Sharma, H., & Rizvi, M. A. (2017). Prediction of heart disease using machine learning algorithms: A survey. *International Journal on Recent and Innovation Trendsin Computing and Communication*, **5(8), 99-104**.
26. https:// www. academia. edu/ 42249626 / Mini_Project_Report_On_Heart_Disease_Prediction_(27 November 2020)
27. https://jamesmccaffrey.files.wordpress.com/ 2 0 1 8 / 1 0 / ml_classification_algorithms_comparison.jpg(16 December 2020)
28. O b a s i , T. , & O m a i r S h a f i q , M .(2019). Towards comparing and using Machine Learning techniques for detecting and predicting Heart Attack and Diseases. *2019 IEEE International C o n f e r e n c e o n B i g D a t a (B i g D a t a) .*

*Coressponding author: ruby.hasan92@gmail.com

 https://doi.org/10.1051/itmconf/20214003008

Voice Feature Extraction for Gender and Emotion Recognition

Dr. Madhu M. Nashipudimath[1,*], *Pooja* Pillai[1], *Anupama* Subramanian[1], *Vani* Nair[1], and *Sarah* Khalife[1]

[1]Department Of Computer Engineering, Pillai College of Engineering, New Panvel – 410 206

Abstract. Voice recognition plays a key function in spoken communication that facilitates identifying the emotions of a person that reflects within the voice. Gender classification through speech is a popular Human Computer Interaction (HCI) method on account that determining gender through computer is hard. This led to the development of a model for "Voice feature extraction for Emotion and Gender Recognition". The speech signal consists of semantic information, speaker information (gender, age, emotional state), accompanied by noise. Females and males have specific vocal traits because of their acoustical and perceptual variations along with a variety of emotions which bring their own specific perceptions. In order to explore this area, feature extraction requires pre-processing of data, which is necessary for increasing the accuracy. The proposed model follows steps such as data extraction, pre-processing using Voice Activity Detector(VAD), feature extraction using Mel-Frequency Cepstral Coefficient(MFCC), feature reduction by Principal Component Analysis(PCA) and Support Vector Machine (SVM) classifier. The proposed combination of techniques produced better results which can be useful in healthcare sector, virtual assistants, security purposes and other fields related to Human Machine Interaction domain.

Keywords – Human Computer Interaction(HCI), Voice Feature Extraction, Gender Recognition, Emotion Recognition, Voice Activity Detector(VAD), Mel-Frequency Cepstrum (MFC), Mel-Frequency Cepstral Coefficient(MFCC), Principal Component Analysis(PCA), Support Vector Machine (SVM).

1 Introduction

The rapid growth of technology and increasing human demand for automation has made voice recognition systems one of the most desired software programs in various devices. Speech recognition technology converts speech into text and lets users control digital devices by speaking instead of using conventional tools such as keystrokes, buttons, keyboards etc. Some examples of such software are Google voice, digital assistants, car blue-tooth etc. Speech signals contain large amounts of information. Two such pieces of information are gender and emotion which can be distinguished relatively more easily by humans than by computers. At present, the research on voice recognition mainly focuses on the identification of single information, which is not enough to understand the true meaning of speech. Here we intend to use voice feature extraction to identify the gender and emotion of the person using SVM classifier, PCA and MFCC.

2 Literature Survey

Speech is the most fundamental way of communication. The main purpose of this project is to determine which classifiers and pre-processors will be most effective at recognising the speaker's gender and emotional state from their speech. Convolutional Neural Networks (CNN), Support Vector Machines (SVM), Long Short Term Memory (LSTM), Deep Neural Network (DNN), and Recurrent Neural Networks (RNN) have all been employed in recent years. The review of such techniques is as follows.

2.1 Pre-processing voice signals

Speech signals from predefined datasets [1] cannot be directly passed into the feature extraction module. The input signals are first pre-processed using Voice Activity Detector (VAD) to identify active frames [12] and filter out silent frames that carry no information.The benefit of employing VAD is that, even if there is a significant pause at the start or end of an utterance, the classifier's actions will not be influenced.

2.2 Feature Extraction

For the purpose of gender and emotion recognition, it is critical to extract relevant features from speech signals In this step the processed speech signals are transformed into a concise but logical representation which is more distinct and reliable than the actual signal. Jiang et al [5] proposed feature extraction module, a heterogeneous unification module, and a fusion network module for a hybrid neural network.

The short term power spectrum of sound is described by Mel-Frequency Cepstrum (MFC) [4], on the basis of a

*e-mail: madhu.nashipudi@yahoo.in

linear cosine transform to log power spectrum with a non-linear Mel scale of frequency. By converting the conventional frequency to Mel Scale, MFCC [11] accounts for human perception for sensitivity at acceptable frequencies. MFC is easy to implement and hence has become a widely used method for speech recognition. The accuracy of the system decreases if the sound samples used have low emotional intensity. It is observed that accuracy is increased when multiple datasets are involved.

Linear Prediction Coding (LPC) [16] approximates speech samples as a linear combination of past samples. Then, over a finite interval, a unique set of predictor coefficients can be calculated by minimising the total of the squared differences between the real speech samples and the linearly predicted samples. An automatic vowel classification system can be presented based on LPC and neural networks. Where traditional linear prediction suffers aliased auto-correlation coefficients LPC gives a very accurate estimate of speech parameters and is comparatively efficient for computation. At the same time, the performance of LPC degrades on the presence of noise in audio signals. Predictive coding [18] determines the best suitable representation of the speech using minimum mean-squared technique.

Linear Predictive Cepstral Coefficients (LPCC) [7] gives a stable representation of the input speech signal in compressed form as compared to LPC. LPCC are derived from the fourier transform of the log magnitude spectrum of LPC. The input signal is analyzed by approximating the frequency bands by removing their effects from the signal and approximates the intensity and frequency of the remaining signal. With the help of Discrete Wavelet Transform (DWT), time domain and frequency domain information of the signal can be fetched. DWT [2] decreases the quantity of signals required to recognize the emotions. Different feature extraction techniques like MFCC [10], pitch, energy, Zero Crossing Rate (ZCR) and DWT are used to extract maximum information of the speech signal and achieve better accuracy with less processing time[2].

2.3 Feature Selection

There are many features in a speech signal but not all of them are needed for implementation of the proposed system. Feature selection [1] is required to extract features from audio signals for selection of principal components, as well as to remove redundant and unused information. Xavier et al [15] used prosodic feature parameters like pitch contour, utterance timing and energy contour. Principal Component Analysis (PCA)[17] is used to find the principal components out of all available features. PCA is a statistical tool [13] which is used to convert a set of observations of correlated variables to a set of values of linearly uncorrelated variables . It also reduces the processing time since a large set of information requires more processing time.

2.4 Classification

Emotion and gender recognition is a supervised learning problem. Each pattern used for the training of the classifier carries the correct emotion/gender class label. The most popular approaches for classifications include Bayesian learning[6], the Linear Discriminant Analysis (LDA), the Support Vector Machine (SVM) [1, 3] which is used as an extension of LDA with a high-dimensional feature space, the multi-layer Neural Network (NN) [14], and the Hidden Markov model (HMM) which captures temporal state transitions. The intensity of emotion [3] fluctuates on a voice from a low to a high level of emotion.. Hossain et al [9] used Cepstral Coefficient (CC) as voice feature and a fixed valued k-means clustered method for feature classification where value of k is determined by the number of emotional events that are evaluated in human physiology.SVM[7, 8, 11] is the most widely used classifier due to its efficiency in classifying high dimensional data where the number of features is greater than number of observations. SVMs have a major benefit over Artificial Neural Network(ANN) that, unlike ANNs, the solution to an SVM is global and exclusive. SVM has a straightforward geometric interpretation and produces a sparse solution.

3 Proposed System Model

Detecting users' emotion and gender accurately, from his/her voice input, requires complex algorithms and intricate deep learning models. To overcome this, we pre-processed the input data meticulously, followed by classification with the help of SVM, which resulted in precise results without the need of complicated Neural Networks. The proposed system shown in Fig. 1 is trained using pre-defined datasets. More emphasis is given to pre-processing of input voice signals contrary to most of the existing systems.

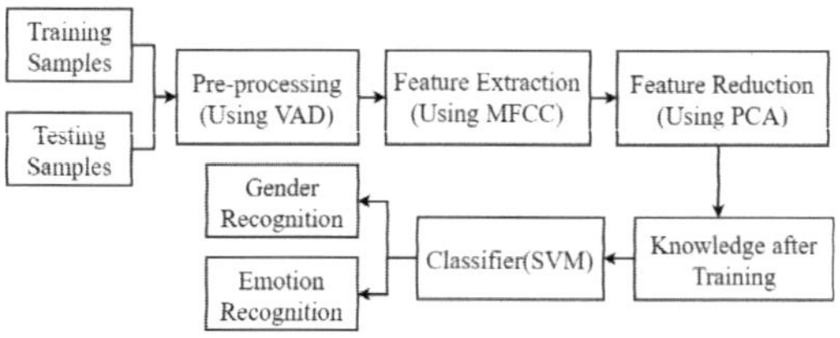

Figure 1. Proposed Model Architecture.

The signals are first pre-processed using Voice Activity Detection (VAD) which is used to determine whether the input signals contain speech or not. The next step is feature extraction. Various acoustic features are extracted using Mel- frequency Cepstral Coefficients (MFCC). The next step is to reduce the number of features. This is done using Principal Component Analysis (PCA). The set of correlated features are transformed into a new set of uncorrelated features called as principal components. Based on the data available after performing the pre-processing steps on the input signal, Support Vector Machine (SVM) classifier is trained to get accurate results.

In the following section, the methodology of Data Selection, Data Pre-processing, Feature Extraction and Data

 https://doi.org/10.1051/itmconf/20214003008

Classification to obtain the classification results from the system are discussed in detail.

3.1 Data Selection

The training of the proposed model is employed using RAVDESS [2, 11] (Ryerson Audio-Visual Database of Emotional Speech and Song) dataset. The database of size 24.8 GB contains a total of 7356 files consisting of voice samples of 24 professional actors ,12 female and 12 male, vocalizing 2 lexically-matched statements. These are in a neutral North American accent. Speech consists of 8 emotions like happy, sad, angry, surprise, calm, disgust and fearful expressions. With additional neutral expression, every expression is produced at 2 levels of emotional intensity- normal and strong.

3.2 Data Pre-processing using Voice Activity Detection (VAD)

The input signals are pre-processed using Voice Activity Detection (VAD). It is a technique used to detect the presence or absence of speech [12] in an input signal. The unvoiced portions in a signal are removed. It is a binary decision i.e. the output can either be 0 or 1. 0 represents absence of speech whereas 1 represents presence of speech. The function $y = VAD(x)$ is used to express VAD algorithm , wherein the desired target output is:

$$y* := 0, xisnotspeech; = 1, xisspeech \quad (1)$$

VAD can also be determined as a probability that an input signal contains speech or not. This is called Speech Presence Probability (SPP). The SPP is expressed as the probability which is always in the range 0 to 1. The main objective of calculating SPP is to determine whether the input signal contains speech or not. The SPP is calculated and compared with a threshold value. If the probability is less than the threshold value, then speech is absent in the input signal. A possible definition for the VAD is then

$$VAD(X) := \{0, SPP(X) < \theta; 1, SPP(X) \geq \theta\} \quad (2)$$

where θ is a scalar threshold.

3.3 Feature Extraction using Mel-Frequency Cepstral Coefficient (MFCC) and Principal Component Analysis (PCA)

3.3.1 Mel-Frequency Cepstral Coefficient (MFCC)

One of the most popular audio feature extraction methods is the Mel-Frequency Cepstral Coefficient (MFCC) [10]. It is a technique which takes voice samples as inputs and determines coefficients unique to a particular sample after processing. It provides enough frequency channels to analyze the audio. MFCC involves: framing and windowing, applying the DFT, Mel frequency warping, computing the log of the magnitude, and then applying the inverse DFT.

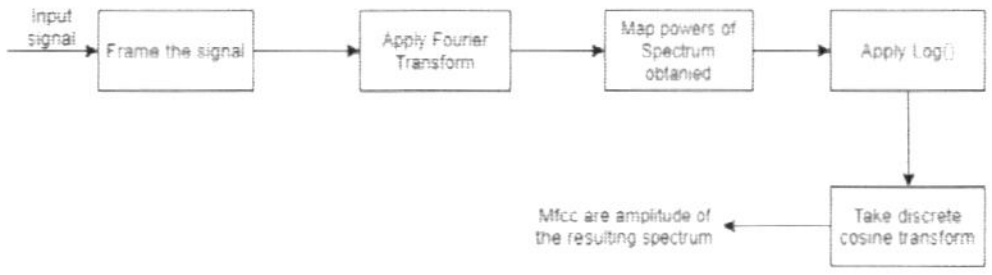

Figure 2. Flowchart of MFCC Feature Extraction process

The flow of extracting the MFCC features is shown in Fig 2.

Acoustic features are extracted using MFCC from the output of VAD i.e. the pre-processed speech signals. The Mel scale compares the perceived frequency of a sound to the measured frequency. It scales the frequency to best match what the human auditory system can hear. The following formula is used to convert a frequency measured in Hertz (f) to the Mel scale

$$Mel(f) = 2595log(1 + f/100) \quad (3)$$

3.3.2 Principal Component Analysis

Principal Component Analysis (PCA)[17] is applied to the output of MFCC to reduce the size of the signals. It is a statistical technique which is used to reduce the dimension of the data which is then plotted with lesser dimension compared to original data. It is used for dimensionality reduction of the extracted features without any loss of information. One set of variables is transformed into another smaller set using PCA. The newly created variable is difficult to interpret. In many implementations, PCA is used to produce information on true dimensionality of the data set. Consider there are X variables in the data set, among those all X variables will not represent the needed information. PCA transforms a set of correlated variables into a new set of uncorrelated variables called principal components. This reduces the amount of time required to train the classifier as well as the memory space required. Fig 3 illustrates the steps involved in PCA.

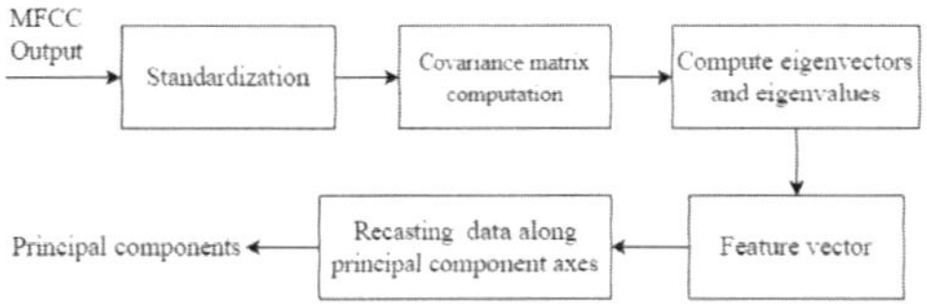

Figure 3. Steps involved in PCA

3.4 Classification using Support Vector Machine (SVM)

The classification model of gender and emotion recognition proposed here is based on the SVM classifier. SVM

[7, 8, 11] uses binary or multi-class classification. It uses hyper planes in the feature space of high dimensions. This helps to differentiate values based on a particular specification. SVM classification is identical to supervised learning which includes feature extraction to generate desired outputs. The major advantage of SVM is that it is very effortless to train. It is capable of scaling high dimensional data better than neural networks[3].

Based on the knowledge available after pre-processing the speech signals, SVM classifier is used for classification and pattern recognition. Thus, pre-defined datasets and integrated algorithms will be used to classify the gender and emotion of the speech signal. All the pre-processing steps are applied to the speech signal. This pre-processed signal along with the knowledge obtained after training is given as input to the SVM Classifier.

4 Experimental Evaluation

4.1 Data Sources

The dataset used here is Ryerson Audio-Visual Database of Emotional Speech and Song (RAVDESS). It consists of 7356 files available in three modality formats Audio-only, Audio-Video and Video-only. Audio-only files are present in two forms- Speech and Song. For the implementation of this project, audio speech files are used. The audio speech database contains 1440 files recorded by 24 professional actors (12 female, 12 male), vocalizing two lexically-matched statements in a neutral North American accent. It includes various emotions such as calm, happy, sad, angry, fearful, surprise, and disgust to name a few, wherein each expression is produced at two different levels of emotional intensity (normal and strong) along with an additional neutral expression. This project focuses on more frequently observed emotions like happy, sad, angry, neutral.

4.2 Variance using Principal Components

On applying MFCC feature extraction, 180 features are extracted. In order to reduce the number of features, principal components are calculated by computing eigenvalues and eigenvectors from the covariance matrix. 55 principal components are required to achieve 95% of variance as shown in fig 4. In the graph plotted below x axis represents the 180 MFCC features extracted from the dataset and y axis represents the cumulative variance for each feature. As observed from the graph, 95% variance is achieved in first 55 features which are considered for further processing.

The number of principal components are calculated by combining computational efficiency and performance of the classifier. Two eigenvectors are chosen for illustration purpose as data is plotted using a two-dimensional scatter plot. The first two components alone contribute around 33% of information in the model as shown in Fig 5.

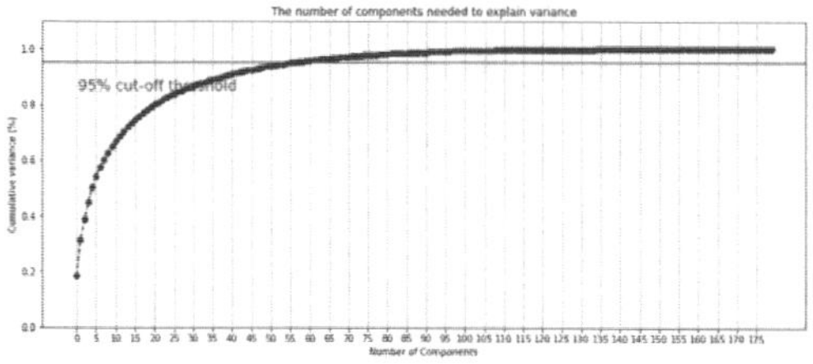

Figure 4. Plot indicating the number of components needed to illustrate variance

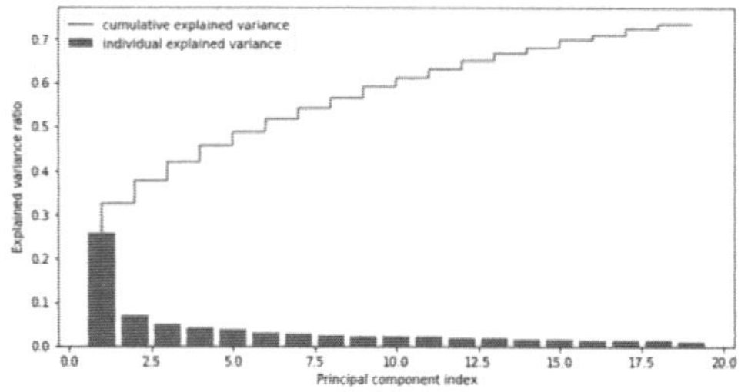

Figure 5. Plot of 'Explained variance ratio' vs 'Principal component index'

4.3 Scatterplot for gender

Fig 6 shows the scatter plot for gender of the first two principal components. The x and y axes represent the first two principal components respectively. The data is divided into two clusters, each for male and female represented by distinguishing colours in the figure below depending on the first PC. Few outliers are also observed.

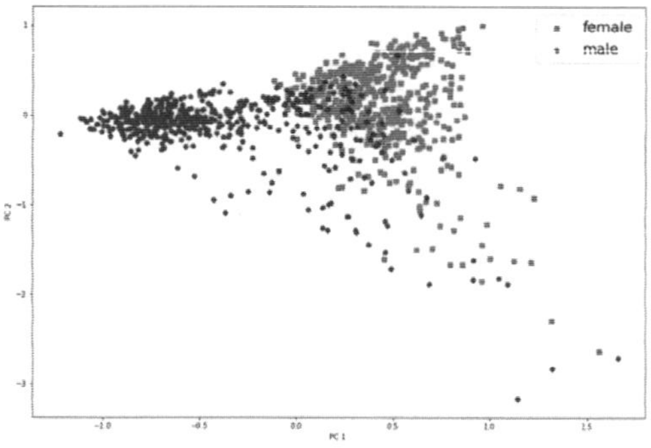

Figure 6. 2-dimensional PCA scatter plot for Gender

4.4 Scatterplot for emotion

Fig 7 shows the scatter plot for emotion of the first two principal components. The x and y axes represent the first two principal components respectively. The data is divided into four clusters, each for happy, sad, angry and neutral

 https://doi.org/10.1051/itmconf/20214003008

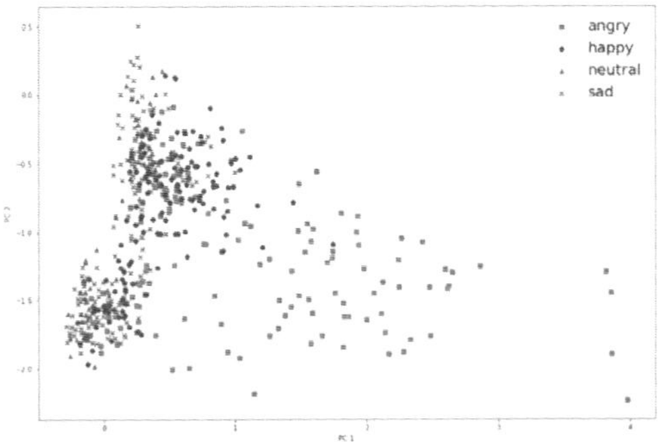

Figure 7. 2-dimensional PCA scatter plot for Emotion

Table 1. Evaluation metrics for gender recognition

Gender	Precision	Recall	F1-score	Support
Female	0.98	1.00	0.99	171
Male	1.00	0.98	0.99	183

Table 2. Evaluation metrics for emotion recognition

Emotion	Precision	Recall	F1-score	Support
Angry	0.80	0.84	0.82	51
Happy	0.66	0.66	0.66	44
Neutral	0.69	0.76	0.72	33
Sad	0.71	0.60	0.65	40

represented by distinguishing colours in the figure below. Overlapping features are observed for happy, sad and neutral emotions whereas the angry emotion is easily differentiable with few outliers.

4.5 Implementation and Evaluation Measures

The dataset is split into training and testing sets, 75% is used for training the system and 25% for testing. The system is evaluated with the help of various parameters such as Precision, Recall, F1-score and Support which is calculated for each gender and emotion individually. The results of which are displayed in Table I for gender recognition system and in Table II for emotion recognition system.

The accuracy score obtained by the gender recognition system is 98.88% where the train accuracy score is 100% and test accuracy score is 98.88%.

The accuracy score obtained by the emotion recognition system is 72.02% where the train accuracy score is 100% and test accuracy score is 72.02%.

4.6 Evaluation Results

The output of the system is analysed using a confusion matrix. In contrast to other machine learning classification metrics like "Accuracy" give less useful information, as Accuracy is simply the difference between correct predictions divided by the total number of predictions. Fig 8 represents the confusion matrix as obtained for the gender recognition system and Fig 9 represents the confusion matrix for emotion recognition system.

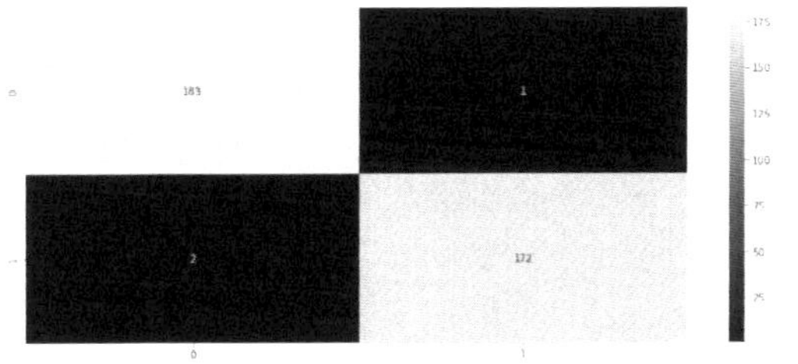

Figure 8. Confusion matrix for gender recognition system

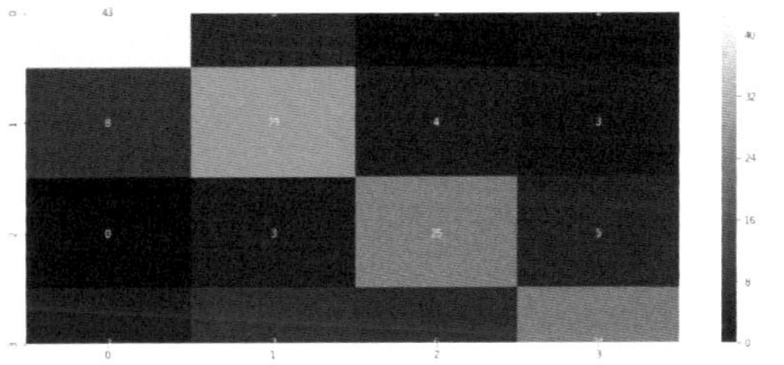

Figure 9. Confusion matrix for emotion recognition system

Conclusion

This proposed system uses 1440 voice samples in .wav format and produces 98.88% accuracy for gender recognition and 72.02% accuracy for emotion recognition. This improvement in accuracy is achieved due to better preprocessing of data with the help of VAD and better feature extraction by using MFCC and PCA.

Assorted evaluation parameters are considered while computing the final result. Inclusion of kernel PCA will produce more reliable emotion classification results. Additional speaker identification features along with multiple categories of emotions can be added, which may potentially improve the system's results and performance.

Acknowledgment

It is our privilege to express our sincerest regards to our supervisor Dr. Madhu M. Nashipudimath for the valuable inputs, able guidance, encouragement, whole-hearted cooperation throughout the duration of this work. We deeply express our sincere thanks to Head of the Department, Dr. Sharvari Govilkar and Principal, Dr. Sandeep M. Joshi, Pillai College of Engineering, New Panvel for support and boost.

References

[1] Sharma, Gyanendra & Mala, Shuchi, *2020 10th International Conference on Cloud Computing, Data Sci-*

ence Engineering (Confluence), "Framework for gender recognition using voice", 32-37, (IEEE, India, 2020).

[2] Koduru Anusha, Hima Bindu Valiveti, and Anil Kumar Budati, International Journal of Speech Technology **23**, 1 - 11 (2020).

[3] Mr. Sundar Ka, Sadagopan E.Nb, Chandran Mc, Aswin Raja S, "Emotion Recognition Using Support Vector Machine." (2020).

[4] Gumelar, Agustinus Bimo, et al, *2019 IEEE 7th International Conference on Serious Games and Applications for Health (SeGAH)*, "Human Voice Emotion Identification Using Prosodic and Spectral Feature Extraction Based on Deep Neural Networks" (IEEE, Japan, 2019).

[5] Jiang, Wei & Wang, Zheng & Jin, Jesse & Han, Xianfeng & Li, Chunguang."Speech Emotion Recognition with Heterogeneous Feature Unification of Deep Neural Network. Sensors" (2019)

[6] Aggarwal, Gaurav, and Rekha Vig, *2019 Amity International Conference on Artificial Intelligence (AICAI)*, "Acoustic Methodologies for Classifying Gender and Emotions using Machine Learning Algorithms.",(IEEE,United Arab Emirates, 2019), 672 - 677 (2019).

[7] Jain, Manas & Narayan, Shruthi & Balaji, Pratibha & Bhowmick, Abhijit & Muthu, Rajesh. "Speech Emotion Recognition using Support Vector Machine", 45-55 (2018).

[8] Poonam Rani, and Ms Geeta, International Journal of Electronics Engineering (ISSN: 0973-7383), **10** • Issue 2, 165-174 (2018).

[9] Hossain, Nazia & Jahan, Rifat & Tunka, Tanjila, International Journal of Software Engineering & Applications, **9**, 37 - 44 (2018).

[10] Kerkeni, Leila & Serrestou, Youssef & Mbarki, Mohamed & Raoof, Kosai & Mahjoub, Mohamed, 10th International Conference on Agents and Artificial Intelligence, 175 - 182 (2018).

[11] Alshamsi, Humaid & Këpuska, Veton & Alshamsi, Hazza & Meng, Hongying, *2018 9th IEEE Annual Ubiquitous Computing, Electronics Mobile Communication Conference (UEMCON)*, (IEEE, USA ,2018), "Automated Speech Emotion Recognition on Smart Phones", 44 - 50 .

[12] Wang, Zhong-Qiu & Tashev, Ivan. Learning utterance-level representations for speech emotion and age/gender recognition using deep neural networks, (2017).

[13] Sengupta, Saptarshi & Yasmin, Ghazaala & Ghosal, Dr.Arijit, International Conference on Computing, Communication and Networking Technologies (ICCCNT 2017) "Classification of Male and Female Speech Using Perceptual Features" (2017).

[14] Pahwa, Anjali & Aggarwal, Gaurav, International Journal of Image, Graphics and Signal Processing, **8**, 17 - 25 (2016)

[15] Xavier, Arputha rathina, International Journal of Computer Science, Engineering and Applications, **2**, 99 - 107 (2012).

[16] Paulraj, M. P., et al. "A speech recognition system for Malaysian English pronunciation using Neural Network." (2009).

[17] Rong, Jia & Li, Gang & Chen, Yi-Ping Phoebe, Information Processing & Management **45**, 315 - 328 (2009).

[18] Rosenberg, Aaron & Sambur, Marvin, IEEE Transactions on Acoustics, Speech, and Signal Processing **23**, 169 - 176 (1975).

https://doi.org/10.1051/itmconf/20214003009

Freehand to Digital Circuit Reconstruction Using HOG and SVM

Shristi Mittal[1, *], *Rhutuja* Satpute[1, **], *Shubhamm* Mohitte[1, ***], Dr. *Leena* Ragha[1, ****] and Ms. *Dhanashri* Bhosale[1, †]

[1]Department of Computer Engineering Ramrao Adik Institute of Technology (Affiliated to University of Mumbai) Dr. D. Y. Patil, Vidyanagar, Sector 7, Nerul, Navi Mumbai, Maharashtra

Abstract: Sketches are commonly used in the fields of engineering and architecture, especially for the early design phases. Engineers spend considerable time setting up initial designs using pencil and paper, and then redrawing them to any software. This problem can be solved by using the idea to scan the circuit sketch with android device which is drawn on the paper and translate it into standard layouts and run circuit simulations. The scanned image will be pre-processed and further segmented. The segmented image will be used to extract the features which are in turn given for classification. Recognizing sketches may seem so quick and intuitive to humans but it is really a big challenge for the machine. In this proposed work the aim is to achieve high precision trainable electronic circuit component recognizer for sketched circuits with fast response time and simple extensibility to new components.

1 Introduction

In real life, it is often very handy to draw an electronic circuit with various components on paper. However, paper is not a reliable media for storing information. On the other hand, if the authors want to try things out and test on paper if the sketched circuit is functional or not which is almost impossible. In sketch recognition, a difficult task is to maintain a good balance between the ability to draw and the significance of recognition. A sketch recognition system would save a great deal of time for engineers to redraw these into professional applications. With the help of this system, they can provide scanned image as the input which is hand drawn on paper and directly get the results. The project focuses on developing a system that will reduce human efforts to redraw the circuits on desktop and directly provide a scanned image as the input. Thus, by using this the authors can redraw to check whether the system is functioning properly or not and also to check out the results of the circuit drawn. Digital circuits area unit less liable to noise or degradation in quality than analog circuits. It is conjointly easier on digital signals to perform erro detection and correction using digitized circuits. Electronics sector is immensely classified because it produces equipment for industries like telecommunications, electronic elements, producing and manufacturing equipment with growth within the semi-conductor business has become price over four hundred billion dollars globally.[1] product created by this sector area unit wide utilized in a spread of client and industrial physics product, which incorporates PCBs (printed circuit boards) within those electronic product. PCB circuits can also cause errors like potential distance variations between the written lines, line thickness dissimilarities, false alarms owing to dirty atmosphere, etc. though the PCB business has enforced advances, these days constant reasonably errors continue to exist, and it's still developing examination determination. Hand drawn circuit recognition project may well be a bridge between paper drawn circuit and generated digital circuit, which might eventually scale back time used for making a digital circuit for PCB. Similarly, this projectmay well be helpful for education systems like for physics department for teaching students, process those fashioned circuits yet as any drawn circuit may well be digitized to see if the operating of the circuit is correct.

1.1 Objectives:

The main objectives of this paper are that the recognition of hand-drawn circuit schematics with the application of digital image process. The person utilizing this project will discover hand drawn circuit elements and varied kindsof circuit diagrams into the machine.

*e-mail: shristivmittal@gmail.com
**e-mail: satputerhutuja50@gmail.com
***e-mail: mohite.shubhu61@gmail.com
****e-mail: leena.ragha@rait.ac.in
†e-mail: dhanashri.bhosale@rait.ac.in

1.2 Motivation:

The fields of electronics have been relatively much slow in developing due to the in feasibility and inconvenience of its physical nature in the fast-paced digital world. Engineers have to manually feed them to machine and convert them from an image diagram to machine understandable format. This system aims to assist electrical engineers, students, examiners, professors in detailed detection and re- drawing circuits of the free hand drawn electrical circuit making it easier to mobilize and practice them.

The organization of the paper is as follows. Section 2 discusses the survey and analysis of the existing systems. Section 3 provides the proposed system. The results are discussed in section 4 followed by conclusion and references.

2 Literature Survey

2.1 Survey of Existing Systems:

The paper of B. Edwards, V. Chandran with concept of "Machine recognition of hand-written circuit diagrams" [1] has advantage of 92 percent accuracy for resistors and conductors. Small spots are often ignored in most systems of circuit recognition. However, in spite of the accuracy there is wrong classification of capacitors and voltage sources.

"Proposed a system of Hand-drawn logical circuit component recognition using Support Vector Machine and Fourier Descriptor" by Mayuri D. Patare and Madhuri S. Joshi has Support Vector Machine (SVM) [2] whichat the same time minimizes error in the empirical classification and maximizes the geometric margin. The shape signature is chosen which makes it more precise and effective using complex coordinates. The overall accuracy of circuit detection is reached by 83 per cent. But the device does have minimal detection of components.

Computer Vision Approach is extremely useful as it uses syntactic simulation in "Automatic contextual recognition of hand-drawn content" by Williams, Kyle, Filho, Milton Ribeiro, Renshaw and Megan [3]. It does classification using a com- bination of invariant moments pixel-distribution options that are scalar and straight-line vector representations in two-dimensional figure representations.

Srikanth Sridar's and Krishnan Subramanian's, "Circuit Recognition using netlist proceedings" [4] is a vital project because it focuses on a brand- new methodology victimization Artificial Neural Network (ANN) to form amachine that may directly browse from a hand drawn cir- cuit image any electrical symbols. It reduces the errorsby combining with back propagation so it modifies theweights of the assorted layers.

"Segmentation and Recognition of Electronic components in hand-drawn circuit diagrams" by Momina Moetesum, Syed Waqar Younus, Muhammad Ali Warsi and Im- ran Siddiqi [5] concluded that more components are detected and a combination of traditional pre-processing methods with modern novel methods are used to create a Netlist for simulation. The authors see that input symbols and data need not be provided.

""Pre-processing Techniques in Character Recognition" by Yasser Alginahi [6] gives us another way of recognizing freehand drawn circuits using Convolution Neural Net- works (CNN). A network is trained using huge amounts of data set of 28 thousand images. It gives an accuracy of
91.85 percent in recognizing handwritten digits but slow computing time makes it difficult.

2.2 Analysis of Existing Systems:

On surveying the assorted systems, the authors have understood the downsides and understood the areas they'd like to work on to create an efficient and successful system. They understand that a robust data-set with varying features was missing in most systems leading to less accuracy of versatile input. They must aim to gather and build a versatile and real data from experts and amateurs as their project would be utilized by the users. Thus, to form a decent size of training and testing dataset to figure on various classifiers. They see drawbacks in system because of ignorance of noise, small spots, shadows and ranging illumination, color and large size data where the photographs don't seem to be processed properly before giving as input causing poor feature extraction.[4] Pre-processing is an especially vital method to enhance the results of any classifier. The image must be clear, its partsand components precise for input. They realize that they want to settle on the proper set of features by creating a strong segmentation procedure.[2]

The low accuracy within the existing systems is additionally because of feature extractor.[1] They see that HOG (Histogram of Oriented gradients) has worked more effectively as compared to SURF (Speeded up robust features) [2]. In the systems they tend to see that ANN, CNN and SVM dominate among others pretty much as good classifiers. Neural networks like ANN and CNN perform higher on terribly massive datasets however they increase complexity. The system victimization ANN gets too advanced and isn't scalable. [3] CNN provides the simplest accuracy of around 91 [5] and SVM around 92 percent [1]. On additional survey, they've realized that the system operating on CNN needs an enormous dataset and still performs slower. Their goal is to form this application for users in real life to be used existing handheld devices and not on GPU powered computers that CNN wouldn't be able to fulfil.

3 Proposed System

The proposed system has the following steps.
- Collection of data
- Preprocessing of Data
- Segmentation
- Feature extraction
- Classification
- Detection and Redrawing

Each of these stages is explained in detail in the following sub sections.

3.1 Dataset

A decent quantity of data is needed to train Support Vector Machine. Most on-line data-sets don't seem to be accurate as they do not capture the varied features of hand-drawn circuits as real data does. They have collected the data from varied students and colleges from completely different departments in our university for flexibility and versatility in the data. The data is hand-drawn complete circuits. These pictures once captured square measure either blur, of low image quality, and additionally contain varied noise like paper lines, sunlight, and shadow. the photographs captured additionally could contain over one circuits and additionally text.

The training data-set consists of three files of 400 images for components for resistor, inductor, diode cropped from each circuit. These images are in jpeg format. The testing data-set has 100 images for the three components. The data-set also contains the components in all possible vertical and horizontaldirections and the images are also rotated in three directions 90,180,270 to get all possible angles.

The sketch of the circuit is a composition of component symbols and connection lines, every symbol requires a specified recognizer that should be trained with a data set. The interest here is restricted to "capacitor, inductor, diode, resistor, voltage source and ground components". They alsomade some ideal conditions and assumptions given below,
1. The connection lines are only horizontal and vertical direction wires.
2. Open lines represent a component symbol such as capacitorand ground.

These assumptions make it easier to segment symbols and connection. Recognition task consists of segmentation of component symbols and recognition of these symbols.

3.2 Pre-Processing

The proposed system takes the scanned image of hand-drawn circuits and supports the jpeg format. The images takenwith a digital camera should be of a minimum of two pixels. The images of hand-drawn circuits obtained not only havelots of noise and illumination variations but are also of varied sizes and colors. Also, the circuits are almost scribbled thus having lots of errors, and hence they need to be pre-processed to create a clean and relevant data-set to feed as input toSVM for better classification rate. As their focus in this project is the outline and components of the circuit, the background data is irrelevant and hence increases size and noise when fed in for feature extraction and classification. Thus, by applying image pre-processing techniques they have reduced the trainingtime drastically and increased the accuracy of the classifier to a huge extent. Hence, they have focused on pre-processing.

3.2.1 Resizing

The images that are captured using a camera are usually of different sizes due to the exposure and quality of the camera system. The pixel values are between 0 and 255 recorded in integers. Generally small weight values are used by most classifiers and neural networks because slowing down or disrupting the resizing process is usually caused by large weight or integer values. First, these images are resized to a (100,100) pixel size which seemed the best small size to extract features from using HOG without too much loss and feed as input to SVM.

3.2.2 Cropping

The image can then be cropped to avoid other text or circuits, as per the choice of the user. The user selects the start and endpoint using the cursor and selects the image he requiresand crops it.

3.2.3 Converting to Binary

Then, they've converted the image from RGB to grayscale and further the image is converted into binary form to reduce the data for easier processing. Since, they have different lighting conditions on our images on different areas, for such varying illuminations on our images they've used Adaptive thresholding method for better results.

3.2.4 Noise Removal

Thresholding results in a binary image, but it consists of lotsof noise that needs to be removed to not lead to confusionin ratio and cuts during feature extraction. Some of the foreground bright pixels from the edges and unimportant regions of bright pixels were removed by applying opening morphology. It preserves the circuit while removing the noise. They've also tried erosion but it was destructive and had the disadvantage that if affect all regions with bright pixels (including the circuit) indiscriminately. The results are shown, the effect of the opening is easily visible and they see that all pixels which can be covered with structuring element within the bright region will be preserved but those that cannot be covered will be eroded.

3.2.5 Dilation

Opening morphology thus left the authors with some broken circuit lines and thin joins that got eroded as noise. It becomes very important to join these lines to not confuse them with components and a ground to be confused as a diode. They've use dilation to increase the circuit lines area so it joins the broken parts of the circuit and retaining the circuit while the noise is gone.

3.2.6 Image Thinning

The dilated image is too wide and may cause joining of two lines such as between capacitor which would cause confusion and reduce the efficiency of the classifier. The dilated imageis then thinned, giving them a noiseless, connected circuit with thin lines

3.2.7 Skeletonization

Ideally, the final image should have the minimum and most important data for better feature extraction. By using skeletonization, we convert the image into a simple yet compact representation of the circuit while capturing its essential features. This process converts the circuit lines in single-pixel width with high density, so they get a bright circuit and also reduce the size and unnecessary data away. By doing this they've created better data and got the exact points where the circuit changes direction and also got the endpoints of the other components. Endpoints are extremely important as features during feature extraction for classification. Finally,with this step, they've created the right dataset to give as input for segmentation.

3.3 Segmentation

"Segmentation is the method of breaking apart the image into items that measure sufficiently small to be detected" [1]. Endpoints from pre-processing take us to the segment capacitor, voltage source and ground symbols as they contain open lines. The capacitor, voltage source, ground components vary in two properties – the length ratio of lines, the number of lines they have. Hence, they do not require a trained recognizer for these components.

They can be identified using the below unique properties:

1. The length ratio of the two lines in a capacitor is 1.
2. The length ratio of the two lines in a voltage source is 0.5.
3. Ground varies from the other components with the numberof lines.

The horizontal and vertical connection lines are detected with the help of the line segment detector algorithm [4]. These connection lines according to the assumption are only horizontal and vertical. Then, the detected connection lines are removed and components are segmented using adaptive thresholding which is already done in the pre-processing stage. They've applied the morphological closing operation to fit the image for the contour detection [5]. It produces separated foreground regions. The small regions with thresholding contours are ignored using a threshold T by region area. As these small regions are probably noise due to illumination differences or remaining of connection lines. The regions remaining can be potentially circuit components that can be recognized with the recognition procedure.

3.4 Feature Extraction

After pre-processing and segmentation on the image of the circuit, features of every element are extracted. This step is the heart of the system as this step has a larger impact on the detection rate. This step helps to classify the elements according to their features. Feature extraction is that the name was given to a bunch of procedures for the measurement of the relevant form data contained during apattern so the task of classifying the pattern is created simply by a proper procedure. Histogram Oriented Gradients [HOG] feature extraction techniques are generally suitable with the classifier Support Vector Machine (SVM). The authors have used the Histogram Oriented Gradients [HOG] algorithm to acquire the feature vector for every training sample.

i) [HOG] algorithm gives the region of image by creating a feature vector that utilizes two features the gradient magnitudeand pixel direction.

ii) HOG divides the cells into 2*2-pixel cells.

iii) For a vector with x and yMagnitude is calculated as

$$M = \sqrt{dx^2_{(x,y)} + dy^2_{(x,y)}} \quad (1)$$

The equation can also be expressed as

$$M = \sqrt{f_x^2 + f_y^2} \quad (2)$$

Direction is given by

$$D = \tan^{-1} \frac{dy_{(x,y)}}{dx_{(x,y)}} \quad (3)$$

The equation can also be expressed as

$$D = \tan^{-1} \frac{f_y}{f_x} \quad (4)$$

where x and y are components of the vector;
M = magnitude;
D = direction;
dx and dy = differentiation factors of x and y;
fx = function of x;
fy = function of y;

It is a rotation and scale invariant feature description algorithm. Histogram Oriented Gradients [HOG] feature vector received and the training sample label was taken for training the Support Vector Machine (SVM).

The features extracted by HOG for the various components are given below. The authors have studied that the features for resistor are continuous curved lines and for resistor differ as they are sharped hill continuous lines. Similarly, the magnitude and direction for capacitor is two lines having ratio 1.0 and for voltage source a ratio of about 0.5 units.

The features extracted for ground are three small lines and the diode can be classified having a closed triangular shape feature. As these images may have cuts, noise and the lines may be slant, they have trained the extractor by giving it a number of circuits for it to better understand and identify the lengths of the ratio.

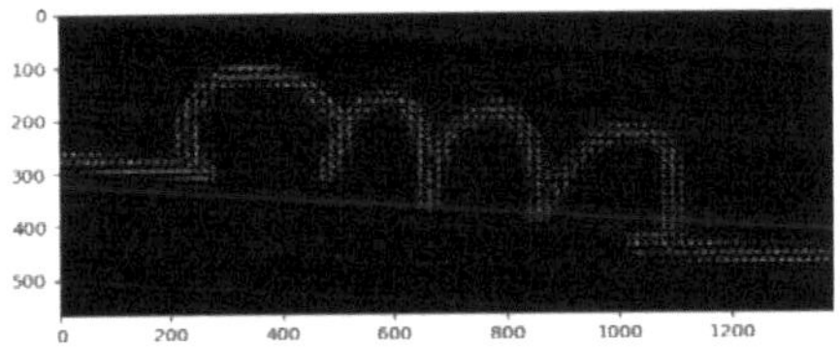

(a) HOG of Inductor

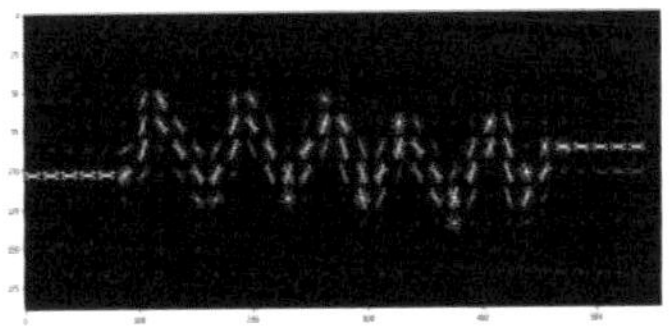

(b) HOG of Resistor

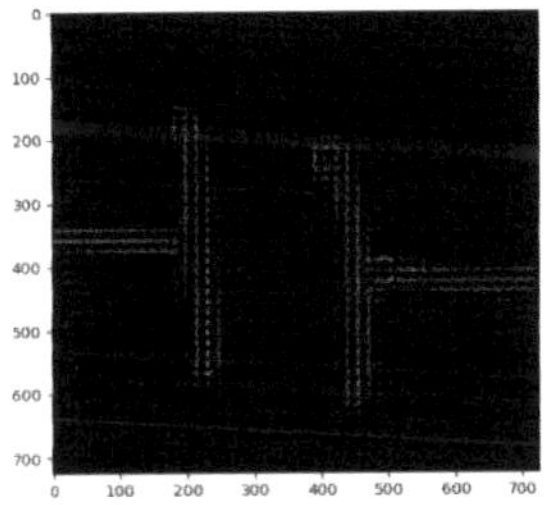

(c) HOG of Capacitor

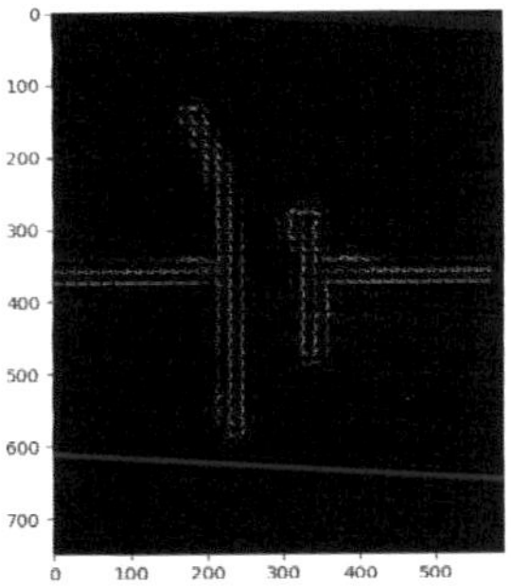

(d) HOG of Voltage

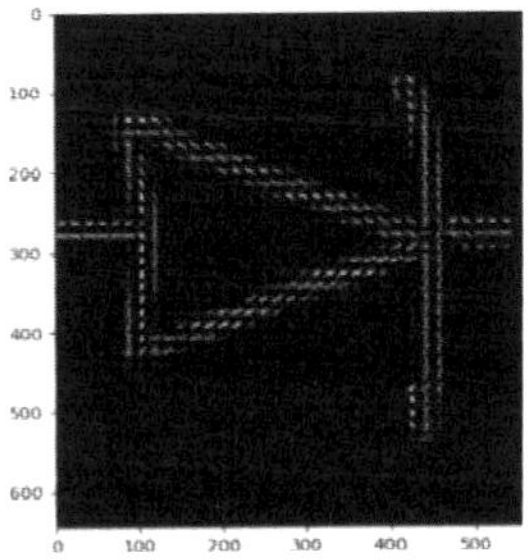

(e) HOG of Diode

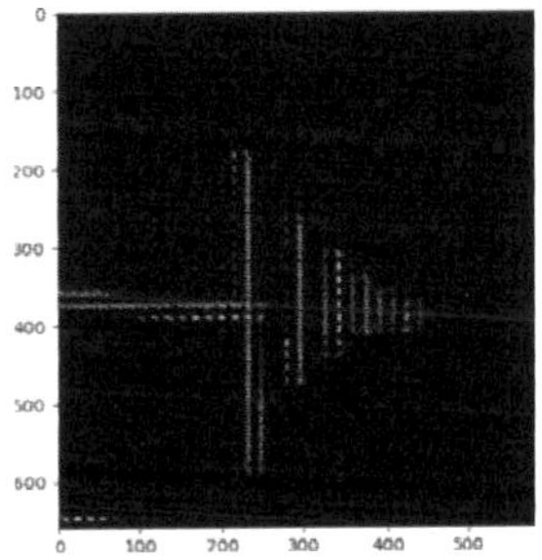

(f) HOG of Ground

3.5 Classification

Classification is the process of labelling of objects into one of a number of predefined categories. The authors have a supervised data. Thus, they can classify using supervised learning algorithms. As analyzed in the literature survey CNN also seem to be competitive with SVM. But they see the benefits of SVM and see it could solve their problem. So they try this classification model. Support Vector Machine (SVM) minimizes error in the empirical classification and maximizes the geometrical margin.

Originally, Vladimir N. Vapnik and Alexey Ya. Chervonenkis, in 1963, invented SVM algorithm. Learning algorithms which analyze data used for classification and analysis of regression are compatible supervised learning models (such as SVM-support vector machine) that relies on support-vector networks. A set of hyperplanes or single hyperplane inside a high- or infinite-dimensional space makes up as support vector machine. Processes like classification, regression and tasks like outlier's detection use these hyperplanes. Experimental test results only after three to four relevance feedback rounds have shown that SVMs attain significantly higher search accuracy as compared to traditional queries. For image segmentation systems and for systems using modified version SVM, this holds true. Also, it benefits from proper feature selection and extraction techniques.

 https://doi.org/10.1051/itmconf/20214003009

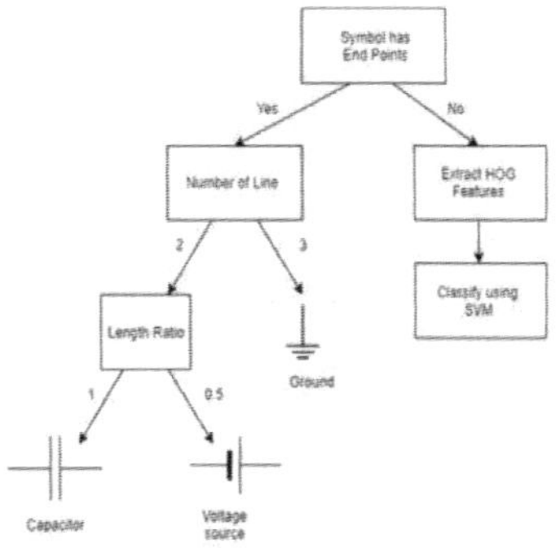

Fig. 1: Classification Flowchart

They've used linear Support Vector Machine (SVM), to get an output of three different classes - Resistor, Inductor and Diode. The algorithm they've used for SVM is as follows:

1. Import the training dataset.
2. After calculating histogram of features, scale the features.
3. Fit the SVM classifier to the training set.
4. Save the SVM.

The Fig 3 represents the flow of classification of our system. Each component is checked if it has end points.If yes, the number of lines is checked in the component and it is further classified as ground for 3 lines. For 2 lines it is classified as capacitor or voltage source on the basis of length ratio. If the component has no end points it is given the HOG to extract features and further to classify using SVM.

3.6 Detection and Redrawing

The detection goal is to interpret the circuit taken from the scanner. In the identification of the same sequence of operations the circuit is detected by comparing the extracted features with simulation of the trained classifier. Then, the components are identified and redrawn based on the Support Vector Machine (SVM) classifier.

4 Results and Discussion

The system working on parameters depending on features would provide the result indicating the separation of components and wired nodes from the circuits. All the processes from image acquisition to classification of circuit and redrawing a digital circuit will be in a same place. The resultant data will be the product of duly trained Support Vector Machine (SVM) which is trained under several similar data to provide precise accurate results more than 90%. Hence, a system is developed to detect the components, nodes and gates of the provided drawn circuit.

Some of their results:

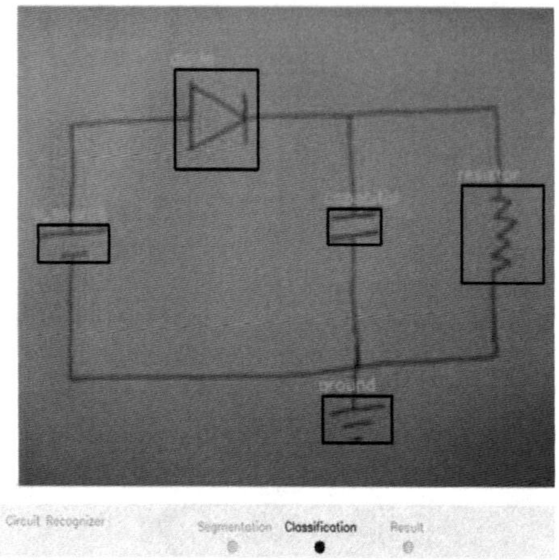

Fig. 2: Classification of components

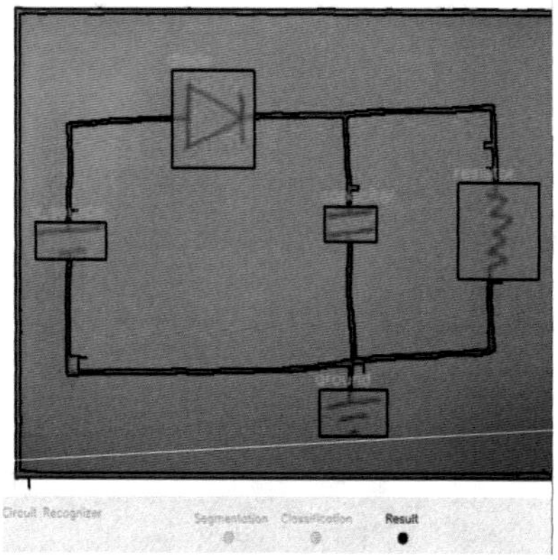

Fig. 3: Final Result of reconstructing lines of the circuit

4.1 Confusion between Components

They've observed that a few components are confused by their SVM trainer and results in wrong detection.

1. The resistor gets confused as inductor due to the unclear round curves of the inductor and the lines of the resistor and vice-versa.

2. Similarly, the resistor and ground are confused in some cases due to the three continuous lines and the

spaces not being accurately understood and hence recognizing them as the resistor's up and down lines.

3. The diode is a vertical line with a triangle, this triangle gets confused as the resistor's hill peak line.

Hence, these confusions get reduced and the results keep getting accurate as the author's SVM trainer gets more experience.

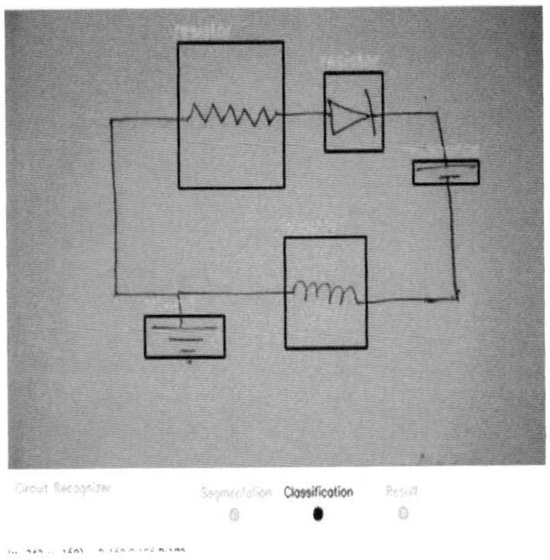

Fig. 4: Diode and Inductor detected as Resistor

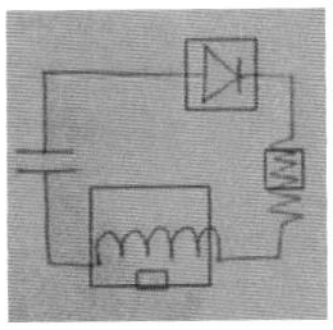

Fig. 5: Inductor detected as Resistor and Voltage Source

4.2 Performance Measure

Table 1: Performance measure of component detection

	Resistor	Inductor	Diode	Capacitor	Voltage Source	Ground	Not Detected
Resistor	90	5	0	0	0	5	10
Inductor	10	90	0	0	0	0	10
Diode	5	0	92	0	0	3	8
Capacitor	0	0	0	96	4	0	4
Voltage Source	0	0	0	5	95	0	5
Ground	5	0	0	0	0	95	5

The above **table 1** describes the percentage of performance of the recognized components as they have used training and testing ratio as 80 is to 20, 80-training data and 20-testing data which gave 95% accuracy for Resistors 90% for Inductors and 100% for Diode, 90% for capacitor, 95% for Voltage source and 95% for Ground as well.

5 Conclusion

The Automation of circuit component recognition and generating the whole circuit for further processing using simulators to test them is a challenging task. In this paper we attempted to recognize the various components with good accuracy. The future work includes identification of many more components and the generation of the whole circuit with confidence.

6 Reference Papers

1. B. Edwards and V. Chandran, "Machine Recognition of Hand-Drawn Circuit Diagrams", Research Concentration in Speech, Audio and Video Technology, School of Electrical and Electronic Systems Engineering Queensland Universityof Technology, Brisbane, Australia.

2. Mayuri D. Patare (Post graduate student) and Madhuri S. Joshi (Professor of CSE Department), "Hand-drawn Digital Logic Circuit Component Recognition using SVM", Jawaharlal Nehru Engineering college, Aurangabad, Maharashtra, India.

3. Williams, Kyle; Filho, Milton Ribeiro; and Renshaw, Megan, "Automatic contextual recognition of hand-drawn content", Technical Disclosure Commons, March, (2018).

4. Srikanth Sridar and Krishnan Subramanian, Circuit Recognition Using Netlist Proceedings of the 2013 IEEE Second International Conference on Image Information Processing (ICIIP-2013).

5. Momina Moetesum1, Syed Waqar Younus1, Muhammad Ali Warsi1 and Imran Sid-diqi1. Segmentation and Recognition of Electronic Components in Hand-Drawn Circuit Diagrams. Research Article: EAI.EU, 10.4108/eai.13-4-2018.154478.

6. Yasser Alginahi (2010), "Preprocessing Techniques in Character Recognition", Character Recognition, Minoru Mori (Ed.), ISBN: 978-953- 307-105-3, Talibah University,Kingdom of Saudi Arabia.

Offline handwritten signature verification using various Machine Learning Algorithms

Chinmay Lokare[1,*], *Rachana* Patil[1,**], *Saloni* Rane[1,***], *Deepakkumar* Kathirasen[1,****], and *Dr. Yogita* Mistry[1,†]

[1]Ramrao Adik Institute of Technology, Nerul, India

Abstract. In today's world it is necessary to protect one's authenticity in order to ensure the protection of personal information that only the authenticate credentials of a person can have access to. Nowadays there is an increase in number of malpractices like signature forgery to access the important information of a person. To encounter signature verification problem, there have been a number of advances in verifying the authenticity of signature using various techniques including Machine Learning and Deep Learning. This paper introduces a novel approach to verify the signatures using difference of gaussian filtering technique, gray level co-occurrence matrix feature extraction technique, principle component analysis and kernel principal component analysis associated with various machine learning algorithms. The publicly available Kaggle offline handwritten signature dataset is used for training. This article compares the accuracy of the dataset on various machine learning algorithms. After training datasets the lowest accuracy achieved is 56.66% for Naive Bayes algorithm. The highest accuracy achieved is 82% for K-Nearest Neighbour (KNN) and 81.66% for Random Forest using principle components and kernel principle components of the dataset.
Keywords: Difference of Gaussians(DoG), GLCM, Principle Component Analysis (PCA), KPCA, K- Nearest Neighbour(KNN)

1 Introduction

The process of biometric verification is utilized for verifying the identity of people dependent on their special characteristics. It has gotten a famous norm for admittance to high security systems. One of the execution of biometric approval is Signature verification, which expects to perceive whether or not a given signature is certifiable(genuine) or fashioned(forged).

Authentication should be possible in two distinct approaches: Online and Offline based on how the signature is obtained. In the online strategy, the signature is captured while writing in this manner providing the dynamic data including area, speed, acceleration, pen pressure, pen lift, pen down, edge and time. Within the offline methodology, the signature is scanned once the signature is written so resulting in the static image of the signature known as the scanned signature. It's tough to differentiate a signature within the offline mode than in online mode that provides a lot of estimations.

Handwritten signatures have diverse dimensions and shapes and the varieties in them are so large at a time that it's miles difficult to test the authentic people. Also, the signature of an individual differs from time to time. Small varieties are intrinsic and these can be endured by the verification system. Here we have examined the effective usage of Offline signature verification using various filtering techniques.

2 Literature Survey

The field of handwritten signature verification has gotten a lot of attention in recent decades, but it's still a work in progress. The goal of offline signature verification, which is characterised by using images of signatures that are static in nature, is to determine whether a signature was created by the claimed person or by an impostor. This section gives an overview of how the issue has been approached by various researchers over the years, as well as recent advances in the field. Batista et al. [1] proposed a hybrid generative–discriminative ensemble of classifiers for building a writer-dependent HSV framework that dynamically selects the classifiers. During the generating stage, the signatures are separated into a grid structure and numerous independent left-to-right directions. To function at various levels of perception, HMMs(Hidden Markov Models) are trained with varying numbers of states and codebook sizes. Each enrolled signature's HMM likelihoods are then computed and aggregated into a feature vector, which is then used to build a pool of two class classifiers using a specialised Random Subspace Method (discriminative stage).

One-Class SVM (OC-SVM)[2], which aims to alleviate the difficulties associated with large user populations. The proposed approach models only one class as a one-class classification problem (genuine signatures). This is a use-

*e-mail: chinmaylokare9@gmail.com
**e-mail: rachanakp22@gmail.com
***e-mail: salonirane29@gmail.com
****e-mail: deepakkathir15@gmail.com
†e-mail: yogita.mistry@rait.ac.in

ful feature because, in most cases, the device only has legitimate signatures for each writer to train the classifier with. Nonetheless, a significant obstacle remains the low number of genuine signatures.

Offline Handwritten Signature Recognition using Biometrics, which refers to identifying an entity based on physiological or behavioural characteristics, has the potential to reliably differentiate between an approved person and an imposter, according to Gulzar A. Khuwaja and Mohammad S. Laghari [3]. The device is trained with low-resolution scanned signature images and uses a neural network to recognise offline handwritten signatures.

Off-Line Signature Recognition Systems were suggested by V Bharadi [4]. Handwritten signatures are one of the most commonly used biometric traits for document and individual authentication. The performance metrics of typical systems, as well as their feature extraction mechanisms, are contrasted.

Bertolini et al. [5] suggested a writer-independent method for handwritten signature verification in the WI scenario. Dissimilarity representation and SVMs as classifiers are used in this method [5]. The writers' two major contributions are as follows: I develop a new collection of graphometric features centred on the curvature of the most important segments, which is simulated with Bezier curves. (ii) Using an ensemble of classifiers structure to boost anti-forgery resistance.

3 Proposed Methodology

The authors have imported the handwritten signature dataset from Kaggle which consisted of 700 images for the training purpose and 300 images for the testing purpose which are further divided into genuine and forged signature images in both the training dataset and the testing dataset. Further the authors have pre-processed the images where they have reduced the noise from the gausssian denoising filter and further took the difference of gaussian. Now extracted the features from these noise reduced signature images using Gray-Level Co-occurrence Matrix or GLCM and then reduce the dimensions of these extracted features using the Principle Component analysis (PCA) and Kernel Principle Component analysis (KPCA). Further the accuracies are predicted on the test dataset on the basis of training of the train dataset on 10 different machine learning algorithms.

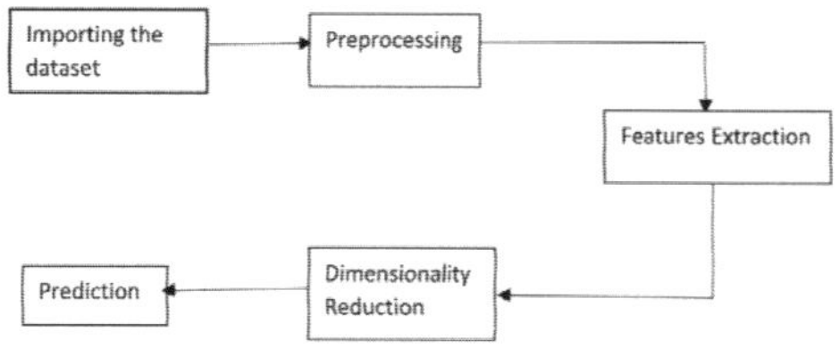

Figure 1. Proposed Methodology

3.1 Gaussian Denoising Filter

Gaussian blurring is a smoothing technique which makes use of a low pass filter whose weights are derived from a Gaussian function. It basically blurs the images and removes noise. The kernel represents a Gaussian hump(bell-shaped). The core concept is to use this 2D distribution as a "point-spread" function, which is accomplished using convolution. Because the image is stored as a collection of discrete pixels, first discrete approximation to the Gaussian function must be created before proceeding with the convolution. Gaussian smoothing is achieved using normal convolution methods after a sufficient kernel has been computed. A denoised image can be produced by performing Convolution of a smoothing kernel with required noisy images. Different denoising results can be obtained on the basis of the property of these kernels. Gaussian smoothing is achieved using normal convolution methods after a sufficient kernel has been computed. The Gaussian Distribution in 1D has formula:

$$G(x)=\frac{1}{\sqrt{2\pi}\sigma}e^{\frac{x^2}{2\sigma^2}}$$

where σ is the distribution's standard deviation. The authors have also concluded that the distribution has a mean of zero (that is, it's based on x=0). Figure below illustrates the distribution.

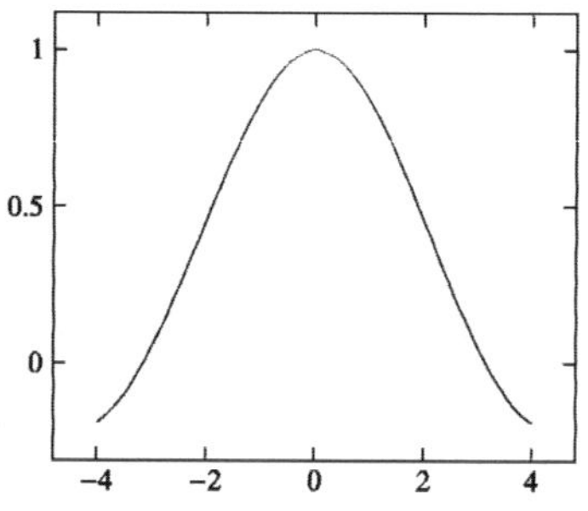

Figure 2. Gaussian distribution with 0 mean

The Gaussian Distribution in 2D has formula:

$$G(x, y) = \frac{1}{2\pi\sigma^2}e^{-\frac{x^2+y^2}{2\sigma^2}}$$

where,

σ = 1 and mean=(0,0)

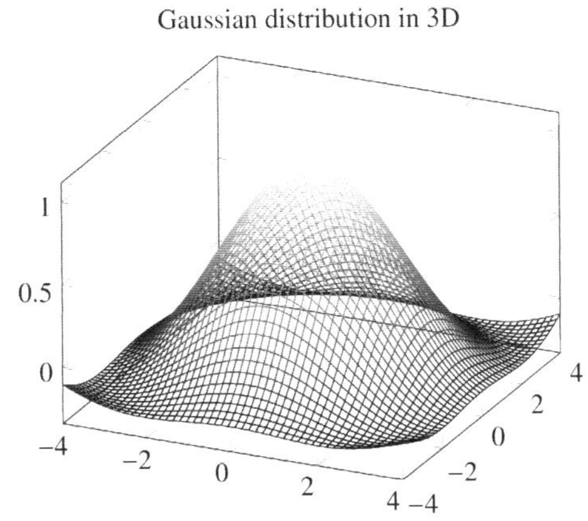

Gaussian smoothing is achieved using convolution, which involves employing this 2-D distribution as a point-spread function. Because the image is stored as a series of discrete pixels, first a discrete approximation to the Gaussian function must be built before proceeding with the convolution. The Gaussian distribution is non-zero everywhere in principle, requiring an infinitely large convolution kernel, but it is essentially zero beyond about three standard deviations from the mean in reality, allowing us to truncate the kernel at this stage.

3.2 Difference of Gaussians(DoG)

As stated earlier the Gaussian Blurring uses Gaussian kernels for image smoothing, further the authors have highlight certain high-frequency parts in an image. In Gaussian Blurring the standard deviation of the Gaussian affects the degree pf smoothing. More high frequency components will be suppressed with larger standard deviation that is there will be more blurring. Thus, if two Gaussian Kernels are considered having different standard deviation of the same image and when the difference of them is taken, an output is received with a particular high-frequency components based on the standard deviations used. The basic logic is to remove high-frequency components by blurring that represent noise, and by taking the difference some low-frequency components of the homogeneous areas of the image are removed. Here, the difference of Gaussian acts like a band-pass filter. Here the authors have taken two Gaussian kernels with a Standard Deviation. $\sigma 1 \geq \sigma 2$ Here the kernel with signa1 when is convoluted with the signature image, the high frequency components will be blurred more as compared to the other kernel. When the subtraction of these is done the information that lies between the frequency range which is not blurred or suppressed is gained.

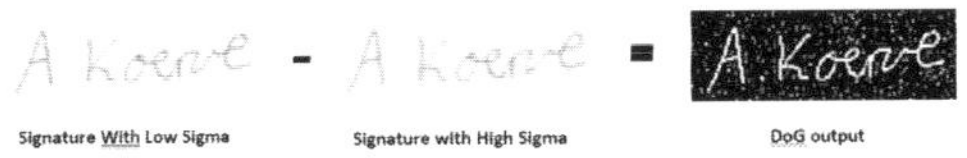

Figure 3. DoG Output

3.3 GLCM

Also referred to as co-occurrence distribution, a co-occurrence matrix is the distribution of co-occurring values at a given offset which is defined over an image or we can say it represents the distance and angular spatial relationship of specific size over an image. A gray-scale image is used to build the GLCM. It determines how frequently a pixel with the gray-level value x appears horizontally, vertically, or diagonally to adjacent pixels with the value y. Horizontal (0 degree), vertical (90 degree), diagonal with bottom left to top right (-45 degree), and top left to bottom right (-45 degree) are all possible directions in GLCM (-135). P0, P45, P90, and P135, respectively, are the directions. For example:
Lets take a point P(x,y) Where x and y are the gray level values (tone) in the image

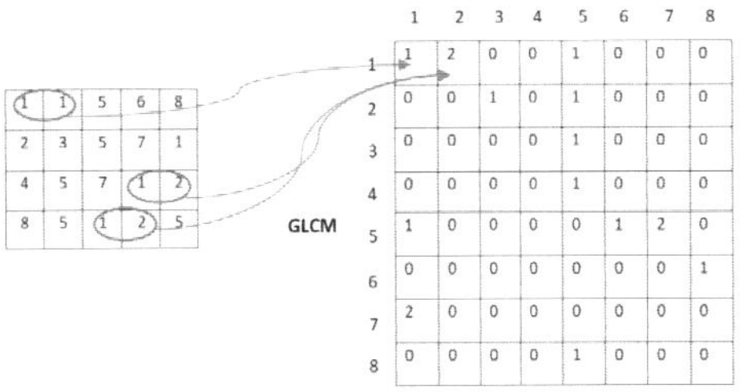

Figure 4. process used to create the GLCM

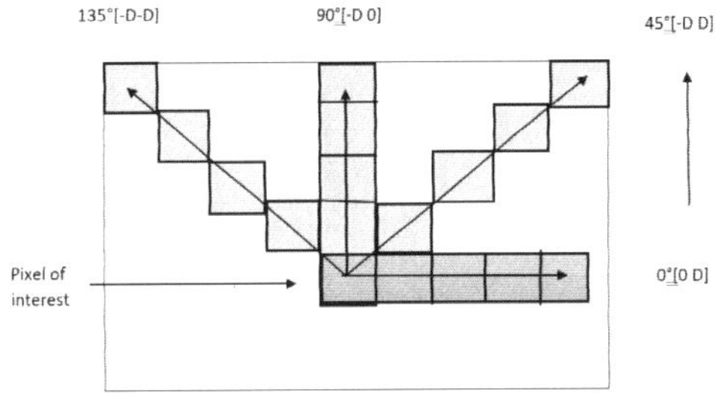

Figure 5. GLCM angles

Here you can see the element (1,1) in GLCM matrix has a value 1 as in the original matrix there is only one because there is only one scenario in the input image when two horizontally adjacent pixels have the values 1 and 1, the GLCM matrix element (1,1) has the value 1 because the original matrix only had one. Element (1,2) in the GLCM has the Because there are two, the value is two occurrences of two pixels that are horizontally adjacent with the values 1 and 2, whereas in the GLCM, Element

(1,3) has the value 0 since there are no horizontally contiguous pixels with the values 1 and 3.

GLCM features:
The authors have used 5 GLCM features namely Energy, Correlation, Homogeneity ,Contrast and Dissimilarity. And are used on 4 different offset distances and 3 angles combination. Finally total of 25 different features with these combinations were gained. Let's have a look at these features one by one.

- **Energy :-**

Energy(ene) =

$$\sum_x \sum_y g_{xy}^2$$

where,

$$\sum_x = \sum_{i=1}^{N_g}$$

$$\sum_y = \sum_{j=1}^{N_g}$$

Ng = Number of distinct gray levels in the image

$$g_{xy} = (x, y)^{th} \text{entry in GLCM}$$

$$g_{xy} = \sum_{y=1}^{N_g} g(x, y)$$

It is also known as Uniformity or Angular second momentum which measures the textural uniformity, that is pixel pair repetition. It is responsible for the detection of disorders in textures. The maximum value Energy reaches to is 1.

- **Correlation –**

It is the measure of gray tone linear dependencies present in the image.

Correlation (cor) = $\frac{\sum_x \sum_y (xy) g_{xy} - \mu_i \mu_j}{\sigma_i \sigma_j}$

where,

$$\sum_x = \sum_{i=1}^{N_g}$$

$$\sum_y = \sum_{j=1}^{N_g}$$

Ng = Number of distinct gray levels in the image

$$g_{xy} = (x, y)^{th} \text{entry in GLCM}$$

$$g_{xy} = \sum_{y=1}^{N_g} g(x, y)$$

μ_i, μ_j, σ_i andσ_j are the means and standard deviations of g_i and g_j

- **Homogeneity –**

Homogeneity (hom) = $\sum_x \sum_y \frac{1}{1+(x-y)^2} g_{xy}$

It's also known as Inverse Difference Moment, and it's a measurement of image homogeneity that assumes greater values for smaller grey tone variations between pair elements. In the presence of near diagonal elements in GLCM its sensitivity is high. When all the elements in the image are same it has a maximum value. GLCM Homogeneity and GLCM contrast are substantially but inversely associated in terms of equivalent distribution in the pixel pairs population. This means that homogeneity decreases if contrast increases while the energy is constant. where,

$$\sum_x = \sum_{i=1}^{N_g}$$

$$\sum_y = \sum_{j=1}^{N_g}$$

Ng = Number of distinct gray levels in the image

- **Contrast:**

Contrast measures the spatial frequency of an image
Contrast (con) =$\sum_x \sum_y (x - y)^2 g_{xy}$ where,

$$\sum_x = \sum_{i=1}^{N_g}$$

$$\sum_y = \sum_{j=1}^{N_g}$$

and is difference momentum in GLCM. It is a difference between the highest and the lowest values of a contiguous pixel sets. The amount of local variations in and image is measured by Contrast. Low spatial frequency and a low contrast image exhibit the GLCM concentration term around the principle diagonal.

- **Dissimilarity-**

It is basically a measure of distance between the pairs of pixels. dissimilarity= $\sum_x \sum_y |(x - y)| p(x, y)$

Below are 5 features of two signature images with distance 3 and angle 0.

	Energy	**Corr**	**Diss_sim**	**Homogen**	**Contrast**
0	0.5706	0.3507	34.0768	0.6689	8328.8668
1	0.5874	0.4373	28.0754	0.7051	6905.2941

3.4 PRINCIPLE COMPONENT ANALYSIS (PCA)

PCA is a dimensionality reduction technique which is used to reduce the dimensionality of larger dataset. It does so by transforming a large dataset of variables into smaller one that usually retains most of the information. The

 https://doi.org/10.1051/itmconf/20214003010

purpose of reducing the dimensions is to make a dataset more simple because it is usually much easier to visualize smaller datasets and explore them that the smaller ones which eventually helps for faster learning for the Machine Learning algorithms. The Principle Components are the new variables that are created by combining the basic variables in a linear fashion. The calculation of these combinations is done in such a manner that the newly generated variables, or Principle Components, are not correlated, and the majority of the information contained in the starting variables is squeezed or compressed into the first components. Steps to perform PCA over a dataset.

1. Standardization – Standardization of the range of the original continuous variables from 0 to 1 so that they all contribute equally to the analysis is the primary purpose of this step.
2. Covariance Matrix Computation – The purpose of this step is to be aware of how the input variables are varying from mean with respect to each other. This is done because in some cases the variables might contain redundant information as they are correlated on a higher degree.
3. Computation of Eigen Vectors and Eigen values – The authors have computed the Eigen Vectors and Eigen Values from the covariance matrix to determine the Principle components.
4. Feature Vector – A feature vector is a matrix with the eigenvectors of the components as columns.
5. Recast the Data – The data is reoriented to the ones represented by the Principle Components using the Feature Vector from the original axes. This is accomplished by multiplying the original dataset's transpose by the feature vector's transpose.
   We have got 25 components in the GLCM. We further reduce these 25 into 5 components.

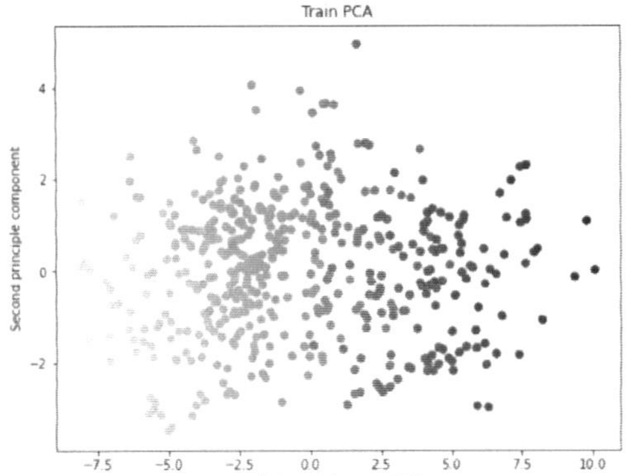

Figure 6. Train PCA

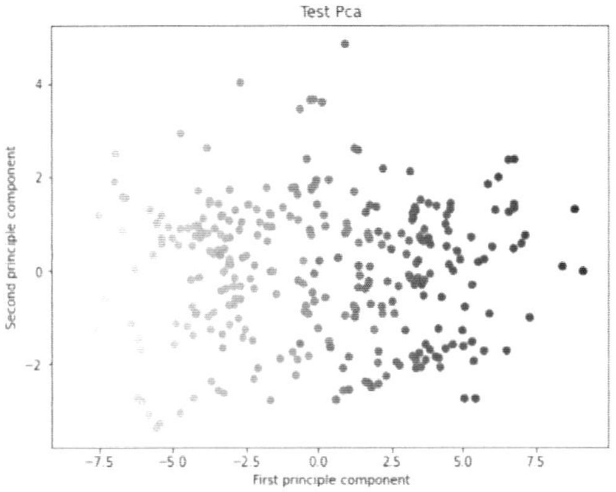

Figure 7. Test PCA

3.5 KERNEL PRINCIPLE COMPONENT ANALYSIS (KPCA)

Since the data has been retrieved from images there is also certain non-linearity in the structure of the extracted features. The PCA is used to reduce the dimensions of the linearity in the structure of the images. Further KPCA is used to reduce the dimensions of the non-linear structure of the data. The theory behind the KPCA is that by projecting data into a higher dimension space, many data that are not linearly separable in space can be made linearly separable. Simple arithmetic operations on the original data dimensions are used to create the extra dimensions. Here also the authors have reduced the 25 components into 5 components retrieved from the GLCM.

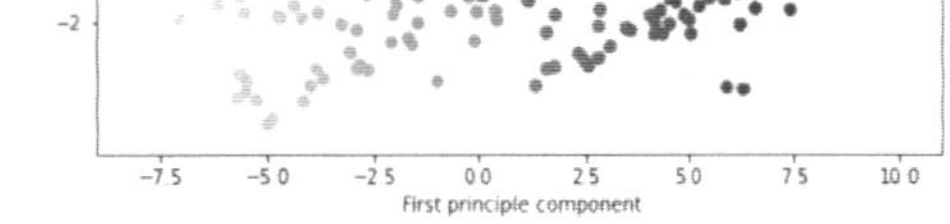

Figure 8. Train Kernel PCA

4 Result

Standardizing the GLCM data results are received in a scaled dataset. Further this scaled dataset is applied to the PCA, KPCA and received reduced dataset. Later these

Table 1. Result

	Scaled dataset	PCA Dataset	Kernel PCA Dataset
ANN	76.00	74.67	74.00
SVM	70.00	70.33	70.33
Logistic Regression	63.00	62.00	62.00
KNN	**82.00**	75.66	75.66
Decision Tree	62.66	74.33	75.33
Bagging Decision Tree (Ensemble Learning I)	75.33	75.33	62.66
Boosting Decision Tree	62.66	75.33	75.33
Naive Bayes	56.66	62.33	62.33
Random Forest	67.66	**81.66**	**81.66**
Voting classification	69.66	71.66	72.00

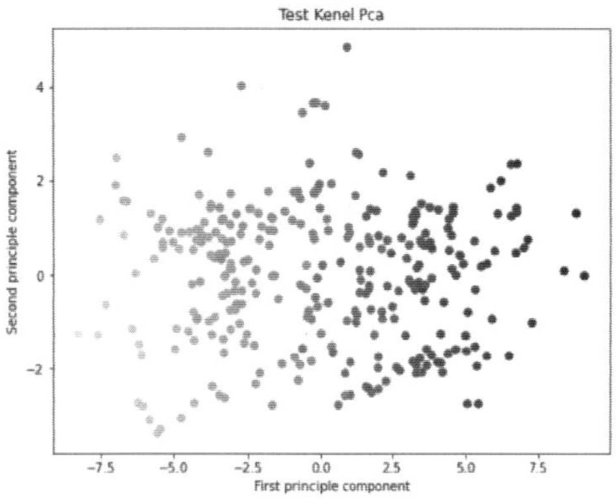

Figure 9. Test kernel PCA

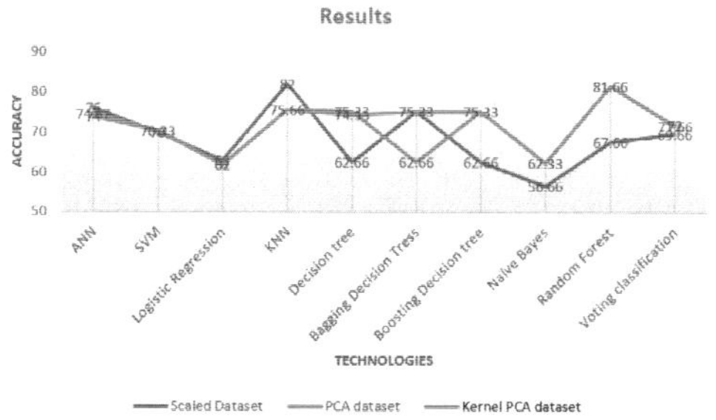

Figure 11. Technologies vs Accuracy %

three datasets are applied to 10 different Machine Learning algorithms and got the results which are displayed in Table 1. Amongst these, the scaled dataset performed well with the accuracy of 82% on the KNN and the PCA, KPCA datasets performed almost the same with the Random Forest.

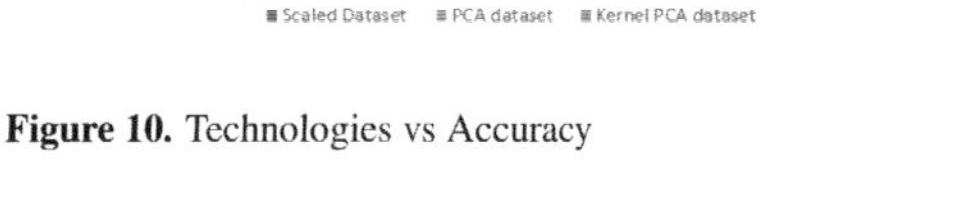

Figure 10. Technologies vs Accuracy

5 Conclusion

This paper proposed a novel approach to verify the signatures using difference of gaussian filtering technique, gray level co-occurrence matrix feature extraction technique, principle component analysis and kernel principal component analysis associated with various machine learning algorithms. Signature images are trained with image processing algorithms, machine learning and deep learning algorithms. The K-Nearest Neighbour algorithm is assessed under the following conditions:varying circumstances, increasing or decreasing the quantity of multiple signing sessions, signature training database with a variety of inks and expanding the number of possible combinations signatories. By comparing with the existing techniques, KNN gives better result with 82% accuracy. In this field, numerous methodologies are used, but accuracy has to be improved. The accuracy achieved thus far by existing systems is not particularly high, and much more study into offline signature verification is required. Future work will be focused on improving the accuracy and determining the ideal approach.

References

[1] L. Batista, E. Granger, and R. Sabourin, "Dynamic selection of generative– discriminative ensembles for offline signature verification," Pattern Recognition, vol. 45, no. 4, pp. 1326–1340, 2012.

[2] Y. Guerbai, Y. Chibani, and B. Hadjadji, "The effective use of the one-class svm classifier for handwritten signature verification based on writer-independent parameters," Pattern Recognition, vol. 48, no. 1, pp. 103–113, 2015.

[3] Gulzar A. Khuwaja and Mohammad S. Laghari Offline Handwritten Signature Recognition using Biometrics April 2011 at International Scientific Academy of Technology and Engineering

[4] V A Bharadi Off-Line Signature Recognition Systems Proceedings of the International Conference and Workshop on Emerging Trends in Technology,ICWET 2010

[5] D. Bertolini, L. S. Oliveira, E. Justino, and R. Sabourin, "Reducing forgeries in writer-independent off-line signature verification through ensemble of classifiers," Pattern Recognition, vol. 43, no. 1, pp. 387–396,2010.

[6] A.Bhanu Srothara , M. Hanmandlu , Offline Signature Verification using CNN,International Journal on Future Revolution in Computer Science amp; Communication Engineering Volume: 4 Issue: 9,September 2018.

[7] Sameera Khan, Avinash Dhole, A Review on Offline Signature Recognition and Verification Techniques Vol. 3, Issue 6, June 2014

Lane Detection Techniques using Image Processing

Vighnesh Devane, Ganesh Sahane, Hritish Khairmode, Gaurav Datkhile
Department of Electronics Engineering, Ramrao Adik Institute of Technology, Mumbai University, India.

Abstract

Lane detection is a developing technology that is implemented in vehicles to enable autonomous navigation. Most lane detection systems are designed for roads with proper structure relying on the existence of markings. The main shortcoming of these approaches is that they might give inaccurate results or not work at all in situations involving unclear markings or the absence of them. In this study one such approach for detecting lanes on an unmarked road is reviewed followed by an improved approach. Both the approaches are based on digital image processing techniques and purely work on vision or camera data. The main aim is to obtain a real time curve value to assist the driver/autonomous vehicle for taking required turns and not go off the road.

Index Terms

Lane detection, autonomous navigation, unmarked road, digital image processing, vision, real time

I. Introduction

Autonomous vehicles are the next big step in the transportation sector due to their efficiency and safety benefits. A self-driven car needs to understand its surroundings so it can navigate its way in streets with minimal human assistance. Driver and passenger safety systems are one of the rising topics in automotive development. Lane keep and Steering assist are some of the systems that allow vehicles to detect lanes and help in keeping the car in an ideal position with respect to the road and assist in turning.

There are several works presented on the well structured roads having proper lane markings, but fewer on unmarked roads. Two of such approaches are reviewed in this paper. These are based on Digital Image processing techniques.[5] It involves using various computer algorithms for manipulating images. This is performed to obtain an image that is according to desired requirements which can be with improved clarity, sharpness, details, reduced noise for efficient extraction and analysis.

The common steps in both approaches are Thresholding, Warping and Region of Interest(ROI). Then further processing includes Pixel Summation(Histogram), Gaussian Blur, Image dilation,Canny Edge Detection, and Sliding Window algorithm which are subject to specific methods.

II. Literature Review

Although our goal is detecting lane on unmarked roads, there are several vision-based techniques deployed for well structured roads i.e marked roads.

Mr. Mustafa Surti [1], used Hough transform for lane detection. Marked lane lines are detected by applying Canny edge detection on a custom Region Of Interest. They demonstrated the lane detection system for lane departure warning on an embedded development board like RaspberryPi.

Using Gaussian Based road model [2] lane could be detected using a multi stage algorithm. The processes involved in these stages are histogram generation, fitting and normalisation. Different ordered polynomial functions are used for lane fitting.

In the approach used in [3], curvature is detected using image processing. The captured image is converted into binary space and segmented into three sections. The curvature direction is determined on the basis of the section having the most number of non-zero pixels. Approach I follows similar logic, but for calculating curve values average values[8] of the histogram of the lower part of the binary image are used instead of performing image segmentation.

The approach [4] presents an efficient version of sliding window algorithm for lane curve fitting. One of the advantages of this improved version is that it allows sliding window to be applied even on uneven lane markings. In the second approach, this method is applied onto the road edges from the binary image. The coordinates of the road edges are then processed to obtain the curve value and the vehicle offset.

Few of the popular algorithms includes GOLD system developed by Broggi which uses an edge-based lane boundary detection algorithm [9]. Remapping of acquired image is done to represent in bird's eye view of the road. Extraction of quasi vertical bright lines that concatenated into specific larger segments is done using specific adaptive filtering.

The RALPH system [10], used to control the lateral position of the vehicle. Lane's curvature and lateral offsets are determined by a template to the averaged scan line intensity profile which is adaptively adjusted and aligned using a matching technique. The same university developed another system called AURORA which tracks the lane markers present on structured road using a color camera mounted on the side of a car pointed downwards toward the road [11].

A. *Benefits of Lane Detection*

- Lane Detection System can provide assistance and certain details to the drivers and as well as pedestrians.
- Such systems can allow drivers to drive safely and also help them to stay on their particular lane by providing some kind of indicators.

- According to research presented by WHO, [6] about 1.35 million human deaths were caused due to road traffic injuries. Autonomous vehicles could bring this number down by some extent.
- According to McKinsey's report in 2015, the mass adoption of autonomous vehicles could possibly reduce road traffic accidents by 90%.
- Since software will drive the car, [7] the modern vehicle can now be programmed to reduce emissions up to 60% according to Ohio University.

III. Lane detection techniques in detail

For detection of lane and getting curve values, among the two approaches, initial steps are essentially the same i.e Thresholding and Warping. All the processing will be done on the video frame captured by the camera which is supposed to be mounted on the vehicle.

A. Approach I

1) Thresholding: : Since the approach is for the unmarked roads, Colour thresholding is done in order to detect the colour of the road instead of detecting the lane markings. The thresholding is not applied directly on the image obtained from video feed. Images obtained from video feed are in BGR (Blue Green Red) format, and are converted into HSV(Hue Saturation Value) [12] since it is better for object or colour detection. HSV color space represents how colors behave under light. Mathematically HSV consists of three matrices for 'Hue', 'Saturation', and 'Value' with ranges lying between 0-179, 0-255 and 0-255 respectively. The main application for HSV is color based image segmentation which is used in both of the approaches for separating the road from its surroundings. Thresholding will convert the image into a binary image i.e the area where the color of the path is detected will be white and the rest of the part of that image will be kept black.

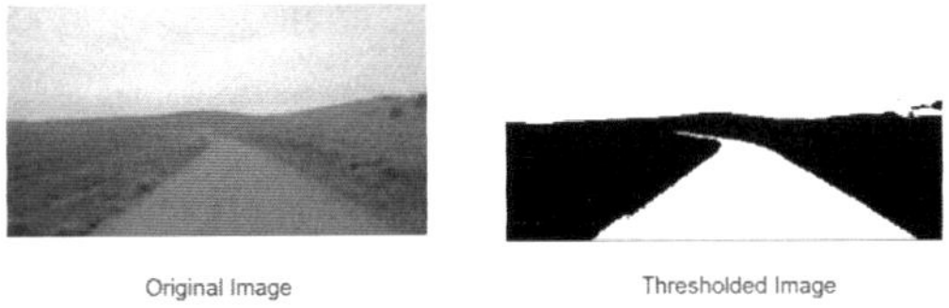

Fig. 1: Thresholding

2) Warping: : The image obtained via the camera module has an angled perspective. The lane edges cannot be precisely located using this image as is. In order to fix this image warping concept is used. Image warping is used to get a birds eye / top view of the path by changing the perspective of the image. In Perspective Transformation[14], the perspective of an image or a video frame is changed in order to get certain insights in a better way. For warping a set of points is provided over the image. The values of these points will determine the Region of Interest (ROI). With these two sets of points the perspective of the image or a video frame is transformed. It is done by passing these coordinates to OpenCV library functions. The resultant would be an image which would be cropped as per the ROI and having a bird's eye perspective. Warping is one of the most common preprocessing techniques used in lane detection systems as it provides an image upon which further calculations would be easier as compared to original image obtained through camera. The basic idea is to get a rectangle shape when the road is straight.

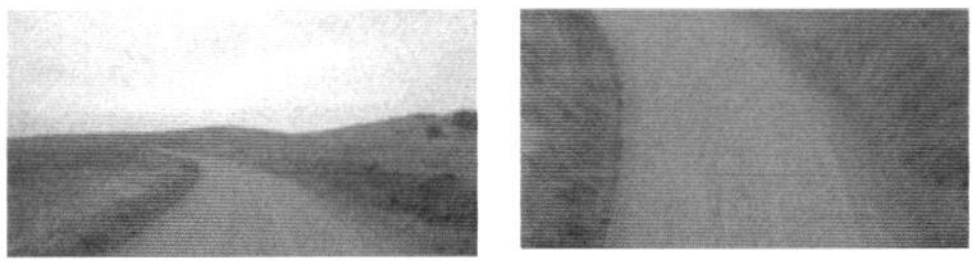

Fig. 2: Warping sample

3) Pixel Summation: : To find the curve of the path, summation of pixels[8] is done which involves the following steps :

- The Warped and Color Thresholded image is binary i.e it has only black and white pixels.
- The next step involves locating and mapping the lanes in the processed image obtained previously using a histogram. With respect to computer vision applications, a histogram is used for graphically representing the color tone distribution of an image.
- For this case, the image is thresholded in binary form so the histogram plots the black(zero) and white(non-zero) variations. The non- zero pixels represent the lane geometry, the position of the starting pixel from the bottom left of the image along the x-axis denotes the starting position of the lane.
- It is possible some of the pixels in the image are due to noise. To avoid them during calculation, a certain threshold value is set which would helps to determine if a particular column is a part of the path or it's just noise.
- The average value of histogram of the lower 1/4th part of the image will be considered as the base point i.e centre of the path. The average value of histogram of the complete image can be termed middle point. Thus subtracting the middle point with the base point will give us a curve value which can be calibrated for further use such as obtaining a steering angle/radius of curvature.

B. Approach II

In this approach, instead of using the entire road pixels only the edges of the path are considered. The initial steps of Capturing the image, Warping and Thresholding are the same as the previous approach. A few image pre-processing procedures are performed to get a clearer view of the path edges. The extracted edge geometry is then used to determine the curvature using sliding window algorithm.

1) Gaussian Blurring: : Due to uneven road conditions, the edges of the thresholded image are too sharp. This may

Fig. 3: Pixel Summation Output

lead to improper edge extraction. Thus the imperfections are smoothened out by applying a blur function to the image. Various methods are available like averaging, bilateral filtering, Gaussian blurring and median blurring[13]. For the test case in consideration, Gaussian blurring provided the best balance in reducing noise and smoothening out the sharp edge imperfections.

This algorithm scans the entire image pixels, followed by recalculating the pixel values. The intensity of this operation is adjusted by a kernel. The greater the kernel size greater is the blur effect but at the same time can lead to decreased details.

Gaussian blur mixes the values of the neighbouring pixels depending on the distance to the reference pixel. The pixels right next to the reference pixel get more of the colours from the reference pixel while the pixels farther away get less of the colours. This can be viewed graphically wherein the values closer to 0 on the x-axis have higher magnitude on the y-axis which gradually decreases as we go further away from the centre.

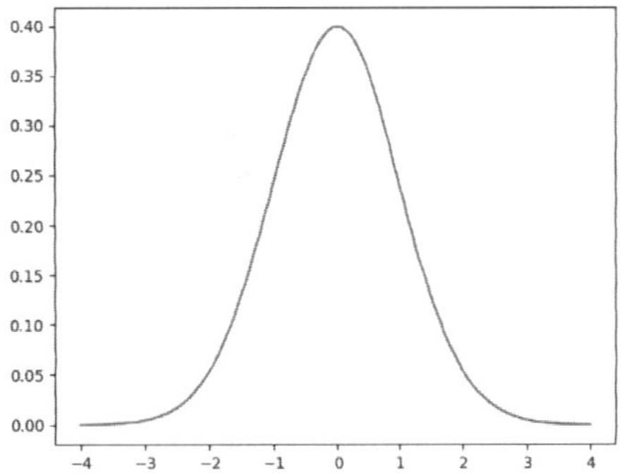

Fig. 4: Gaussian Curve

Thus the uneven edges get smoothened out as the pixels don't have a too sharp change in their values after blurring.

2) Canny Edge Detection: :

The next step is to extract the lane boundaries from the thresholded binary image. For straight lines, this can be achieved by Hough transform. However, the proposed test case needs to detect curved edges as well so Canny edge detection algorithm is used. The Canny edge detector uses a multi-stage algorithm[15] to detect edges in images by identifying large changes in gradients of the pixels. The steps involved are:

- The noise present in the image is reduced with a 5x5 Gaussian filter.
- The intensity gradient of the image is found by Sobel kernel which detects horizontal, vertical and diagonal edges.
- A scan of the complete image is performed followed by removing any unwanted pixels which may not represent the edge; this is also known as Non-maximum Suppression.
- The final stage i.e Hysteresis Thresholding decides which all edges are really edges and which are not. For this, two threshold values - min and max are used. Any edges having intensity gradient more than the maximum value are accepted as edges and those below minimum value are are discarded. The remaining which lie between these two values are classified based on their connectivity.

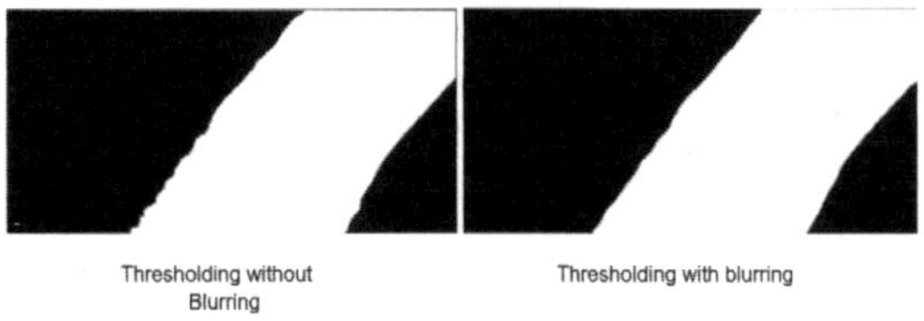

Fig. 5: Thresholding Comparison

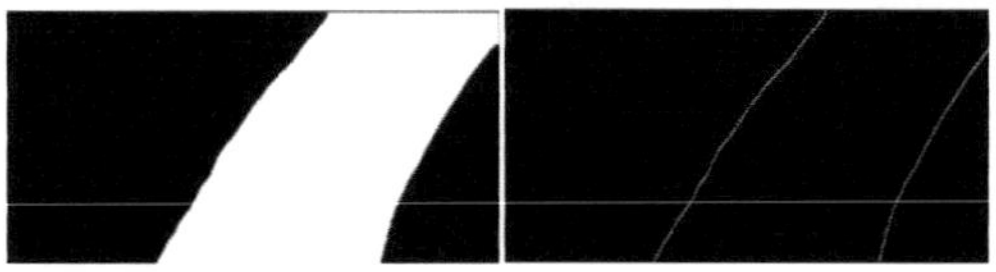

Fig. 6: Canny edge detection

3) Image dilation: : The result of the canny detection function is an image with just the edge geometry of the path with the remaining image features removed. This needs to be passed to a searching algorithm for detection of this geometry. But before doing so image dilation is performed which makes the edges more clear and filled with less breaks. Dilation expands the image pixels and adds pixels to object boundaries.

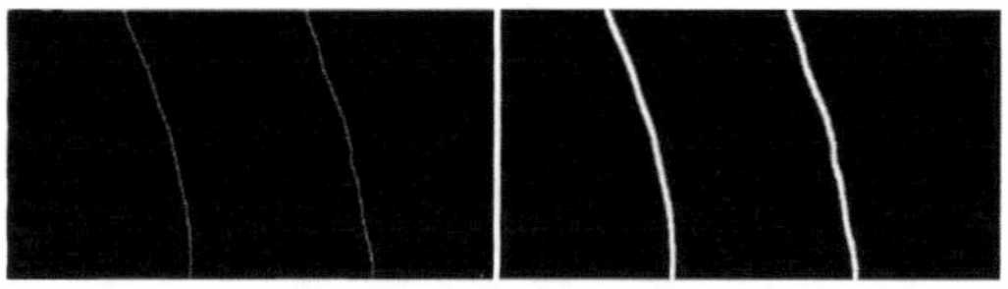

Fig. 7: Image Dilation

https://doi.org/10.1051/itmconf/20214003011

4) Sliding Window Algorithm: : The next step involves detecting the position where the edges are actually present in the Cannied image.[4]

- The image is split into horizontal slices, and a search window i.e a rectangular region of fixed width and height is slid across each slice finding areas of highest frequency.
- For this application, a window search takes a histogram of the lower half of the binary image, revealing the neighbourhood about where the lane lines begin.
- Once this is found for both left and right edges, a fitting function is applied to get a best fit curve on the line. Window searching can be computationally expensive. To save time for subsequent frames, a margin search can be performed . i.e it only searches in a tight margin around the previously established lane line.
- The values of positions obtained for both left and right edges are stored and then subtracted with each other to find the curvature value.
- Similarly the vehicle offset i.e the deviation of the vehicle from the ideal center is obtained by calculating the difference between the ideal center and the edge position.

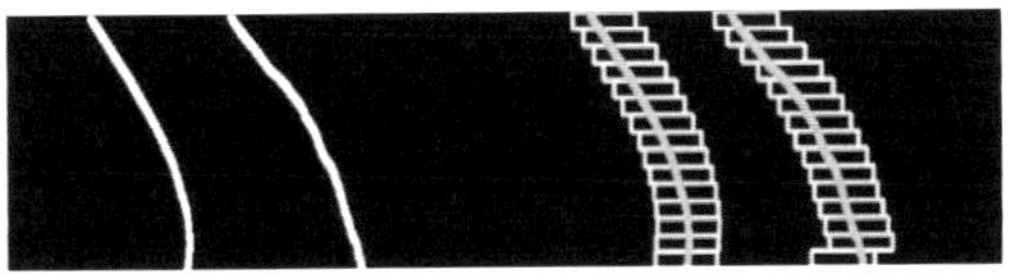

Fig. 8: Sliding Window Output

Fig. 9: Approach II Output

IV. Differences between outputs and their ideal use cases:

A. Approach I:

- As it provides the approximate turning values without using the edge positions, offset value can't be calculated.
- This can be used as a fallback method for traditional methods wherein the edges are too disoriented.

B. Approach II

- It can be applied for both situations - lane markings or no lane markings (using edges of road).
- It is more accurate than the 1st approach as using the edge geometry gives a precise value of the lane position.
- Other important measures such as vehicle offset can be measured which are not possible using approach 1
- The downside of this method is if the edges are too uneven or obscured by vegetation the curve values may get disturbed.

V. Conclusion

The overall purpose of this project is to enable lane detection in poor road conditions wherein the lane markings are not clearly visible or worn out. This project approaches for the implementation of lane detection with the help of image warping, thresholding and approaches such as pixel summation and sliding window algorithm for an intelligent autonomous vehicle. Finally, the advantages and shortcomings of the above methods along with their ideal use case scenarios have been discussed.

References

[1] Mr.Mustafa Surti,Prof. (Dr.) Bharati Chourasia, "Real time lane detection system using Python and OpenCV on Raspberry Pi ",*International Journal for Research in Engineering Application Management (IJREAM)*, Vol-05, issue-06, 2019.

[2] Boon-Chin Yeo, Way-Soong Lim Heng-Siong Lim, "Lane Detection In The Absence Of Lane Markings For Roadway Surveillance With Thermal Vision", *ICIC International*, Vol-12, Number-03, 2016.

[3] Anuja Ihare, "Image Processing to Detect Curvature of Road", *Article on medium.com*, 2020.

[4] Landon Bentley, Joe MacInnes, Rahul Bhadani Tamal Bose, "A Pseudo-Derivative Method for Sliding Window Path Mapping in Robotics-Based Image Processing", 2019.

[5] GurjyotKaur and Gagandeeop Singh, "A Review Of Lane Detection Techniques",*International Research Journal of Engineering and Technology (IRJET)* , Vol-02, Issue-03, 2015.

[6] World Health Organization, "Road traffic injuries", *Article on who.int*, 2020.

[7] Ohio University, "The Future of Driving", *Article on ohio.edu*, 2021.

[8] M. Hassan, "Self-Driving Car using Raspberry Pi", *Article on computer-vision.zone*, 2020.

[9] B. M, Broggi, "GOLD: A parallel real-time stereo Vision system for generic obstacle and lane detection", *IEEE Transactions on Image Processing*, 1998.

[10] Pomerleau D. and Jochem,"Rapidly Adapting Machine Vision for Automated Vehicle Steering", *IEEE*, 1996.

[11] M. Chen., T. Jochem D. T. Pomerleau, "AURORA: A Vision-Based Roadway Departure Warning System", *In Proceeding of IEEE Conference on Intelligent Robots and Systems*, 2004.

[12] Sertap Kamçı, Dogukan Aksu, Muhammed Ali Aydin,"Lane Detection For Prototype Autonomous Vehicle", *AIRCC Pulishing Corporation*, Vol 09, Number 13, 2019.

[13] Open Source Computer Vision,"Smoothing Images", *Article on docs.opencv.org*, 2021.

[14] Adrian Rosebrock,"4 Point OpenCV getPerspectiveTransform Example", *Article on pyimagesearch.com*, 2014.

[15] Aryan,"Implement Canny Ede Detector in Python using OpenCV", *Article on geeksforgeeks.org*, 2020.

https://doi.org/10.1051/itmconf/20214003012

Detection of Phishing Websites Using Ensemble Machine Learning Approach

Dharani M.[1], *Soumya* Badkul[1], *Kimaya* Gharat[1], *Amarsinh* Vidhate[1], and *Dhanashri* Bhosale[1]

[1]Computer Engineering Dept., Ramrao Adik Institute of Technology, Navi Mumbai ,India

Abstract. In this paper, we propose the use of Ensemble Machine Learning Methods such as Random Forest Algorithm and Extreme Gradient Boosting (XGBOOST) Algorithm for efficient and accurate phishing website detection based on its Uniform Resource Locator. Phishing is one of the most widely executed cybercrimes in the modern digital sphere where an attacker imitates an existing - and often trusted - person or entity in an attempt to capture a victim's login credentials, account information, and other sensitive data. Phishing websites are visually and semantically similar to real ones. The rise in online trading activities has resulted in a rise in the number of phishing scams. Cybersecurity jobs are the most difficult to fill, and the development of an automated system for phishing website detection is the need of the hour. Machine Learning is one of the most feasible methods to approach this situation, as it is capable of handling the dynamic nature of phishing techniques, in addition to providing an accurate method of classification.

1 Introduction

1.1 Overview

The internet has seen rapid growth in the last decade. With the internet connecting billions of people globally, it is critically important to acknowledge the fact that the safety and privacy of internet users are not optimal. The rate of cyber-crime is on the increase and leads to great financial losses each year.

Phishing attacks account for more than 80% of reported security incidents. According to Verizon's 2021 Data Breach Investigations Report (DBIR) [1], 3,841 phishing incidents were reported till May 2021, wherein data disclosure was confirmed for at least 50% of the cases. Number of breaches involving phishing shows an 11% increase in 2021 as compared to the previous year. 95% of these attacks were financially motivated causing a loss of thousands of dollars per minute.

Many programs and seminars are conducted across the world to educate users and spread awareness about phishing scams. A Uniform Resource Locator (URL) is a unique identifier that locates a resource on the internet. It carries various parts such as protocol, domain name, port, path, query, etc. There are certain aspects in the URL of a phishing website that may be used to distinguish it from legitimate ones. However, it is not always possible to identify a website's legitimacy just by looking at the URL from a human perspective. Fortunately, Machine Learning algorithms prove to be an accurate and efficient method in recognizing these features, and in predicting whether a given website is a phishing website or a safe one.

1.2 Background

1.2.1 Lexical Features of URL

Lexical Features are the textual properties of a word. The most significant lexical features of a URL are length, hostname, and path. These features can be used to judge the legitimacy of a website without actually visiting it.

1.2.2 Ensemble Methods

Ensemble machine learning methods train multiple classification models using the same learning algorithm which generally consists of a pool of Decision Trees. Since each constituent learning algorithm will have its individual output, the final result is obtained with the help of a combining mechanism such as:

1. Voting (majority wins)
2. Weighted voting (some classifier has more authority than the others)
3. Averaging the results

2 Related Work

A very extensive range of studies and research has been carried out on the subject of phishing detection. General classification approaches include page-based, content-based, domain-based, and URL-based methodologies. In addition to user education, email regulation and classification could aid in the reduction of phishing incidents. "Anti Phishing Simulator" software was developed at Firat University, Turkey to facilitate the detection of phishing and spam emails by examining the email content [2].

The use of Convolutional Neural Networks for classification has also yielded positive outcomes, according to this study [3].

Most of the work done on the subject made use of traditional machine learning algorithms like Naïve Bayes, Support Vector Machine and Decision Trees [4].

The Remove Replace Feature Selection Technique (RRFST) aims at reducing the number of features by excluding insignificant features. It randomly selects a feature to test its effect and replaces it with another if it has a negative effect [5].

This study [6] made use of the XGBOOST algorithm for phishing detection which evidently outperformed other popular machine learning methods.

3 Limitations of Existing System

In most cases, a large number of features have been used in training the classification model, leading to an increase in the search space dimension. According to Hughe's effect, also known as the Curse of Dimensionality, the efficiency of a classifier demonstrates a steady increase only up to a certain threshold dimensionality, after which a decline is observed. To overcome this problem, a feature selection technique must be used.

The use of traditional machine learning algorithms may produce fair results but are highly susceptible to underfitting and overfitting, and may not always produce optimum results. Ensemble machine learning algorithms may be used to address this problem.

Even though email is the most common method of approaching potential phishing victims, it is not the only one. There are other forms of communication, such as social media, advertisements, text messages, telephonic conversations, etc., which can also lead an unsuspecting internet user to a phishing website. Thus, the scope of detection must not be limited to email-based phishing attacks.

Phishing is not only financially motivated but could also have identity theft or cloning as its incentive. Hence, in addition to securing banking and e-commerce websites, it is equally important to focus on generic sites as well, to ensure the overall safety and security of an internet user.

4 Proposed Methodology

4.1 System Architecture

The input is collected from the user in the form of a Uniform Resource Locator string which is treated as raw data. It is then processed to extract the lexical features and other characteristics. The processed input data is further passed to the trained model which predicts whether the URL entered is legitimate or fraudulent.

The training dataset for the classifier is collected from the standard machine learning repository made available for use by the University of California, USA, which contains over 2000 records with 30 attributes. The information present in this dataset is collected from a global community called 'Phishtank' which collectively reports and verifies the presence of suspicious activities on a potential phishing website. After data pre-processing and feature selection, the dataset is used to train the classification models which are based on bagging and boosting ensemble machine learning algorithms.

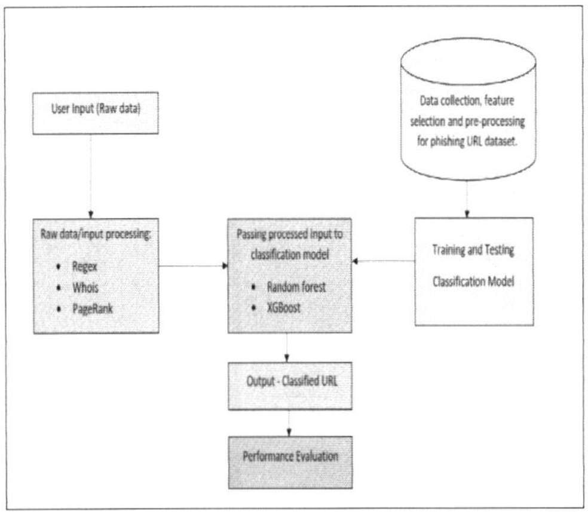

Figure 1. System Design

4.2 Raw input processing

The following pre-processing techniques and tools were used to transform the raw URL input text into predictable and analyzable features:

- Regex: A regular expression (Regex) is a sequence of characters that define a search pattern. It is used to determine the following features of the input URL: '@' symbol, '//' redirect symbol, Prefix, Suffix, Protocol, Subdomain etc.
- whois : whois is a protocol which can be used to calculate the domain age by retrieving the creation date.
- PageRank: The PageRank algorithm indicates the importance of a website on the internet based on its popularity determined by the number of visitors.
- Prefix or Suffix : Prefixes or suffixes are generally added to domain names, usually separated by a hyphen, giving the illusion of legitimacy.
- Long URL : Phishers might use long URL to hide malicious file locations within multiple sub-directories.
- Domain length : Maximum length of domain can be 63 characters. Phishers might use long domain names to confuse the users.

4.3 Feature selection

Feature selection is a pre-processing step in which the insignificant features present in the data are eliminated.

Chi-Square test, a filter method, is used to evaluate the relationship of a feature with the target variable. Chi-Square value is given by (eqn. 1):

$$\chi_c^2 = \sum_{i=1}^{n} \frac{(O_i - E_i)^2}{E_i} \quad (1)$$

where c is the degree of freedom, O is the observed value, and E is the expected value.

Clearly, Chi-square value is small if the observed value is close to the expected value. Therefore, features are more independent if Chi-Square value is small.

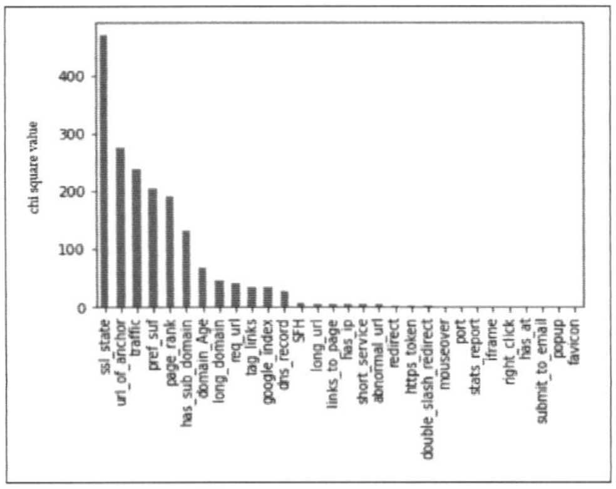

Figure 2. Chi-Square value against feature set

Fig.2. shows the plot of features against the chi-square value. The top 15 features having high Chi-square value were selected for training the classification model.

4.4 Classification model

The next step is to train a machine learning model for classifying a website as a legitimate or a phishing website. It is crucial to choose the right classification method which adheres to the application's needs.

There are 3 main types of ensemble methods: Bagging, Boosting, and Stacking. Bagging stands for Bootstrapping and Aggregating. It is designed to avoid overfitting by the random selection of data. Boosting is an ensemble approach that aims to convert weak learners into strong learners by reducing bias. Stacking uses a meta-classifier to combine the predictions of heterogeneous weak learners. This project proposes the use of two classification algorithms, Random Forest Classifier and XGBoost.

4.4.1 Random Forest Algorithm

Random forest is a popular bagging ensemble machine learning algorithm, which aims to reduce the variance by training the model on different parts of the same training set using multiple deep decision trees, whose individual results are then averaged to obtain the final classification. To decide branching of nodes on a decision tree during classification, the Gini Index is used, which is calculated as given below (eqn. 2):

$$Gini = 1 - \sum_{i=1}^{c} (P_i)^2 \quad (2)$$

where, Pi is the relative frequency of the observed class, and c is the number of classes.
Gini Index is calculated by subtracting the sum of the squared probabilities of each class from 1. Thus, the class with the least value of Gini Index is given the highest preference, as its probability of being wrong is less. The use of Gini Index is advantageous as it favours large partitions and has a simple implementation.

Random Forest Algorithm (Algo. 1):
Step 1 - Random samples are selected from the dataset.
Step 2 - Decision trees for individual samples are constructed. The algorithm will then obtain the prediction result from the formed decision trees.
Step 3 - Every predicted result will go through voting.
Step 4 - Result with the majority of votes will become the final prediction result.

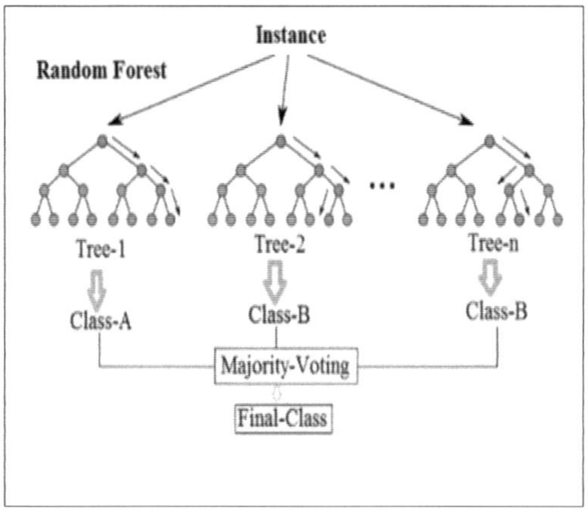

Figure 3. Random Forest [7]

4.4.2 XGBoost Algorithm:

XGBoost (Extreme Gradient Boosting) is an optimized distributed gradient boosting algorithm designed to be highly efficient, flexible, and portable. It uses parallelization, tree pruning, and weighted classifiers to produce superior results.

XGBoost Algorithm (Algo. 2):
Step 1 - Regression trees are the weak learners and each regression tree maps an input point to one of its leaf nodes.
Step 2 - An initial model F1, that will be associated with the residual (F1-y), is defined to predict the target variable y.
Step 3 - A new model T2 is trained with independent variables and residual errors from the previous step as the data to get the predictions.
Step 4 - F2, a combination of F1 and T2, is the boosted version of initial model F1. It is calculated with the additive predictions and residual errors along with some learning rate from output predictions obtained previously from the model.
Step 5 - Steps 3 and 4 are iterated for M number of times

until the required number of models are built.
Step 6 - The final prediction from boosting is the additive sum of all the previous predictions made by the models.

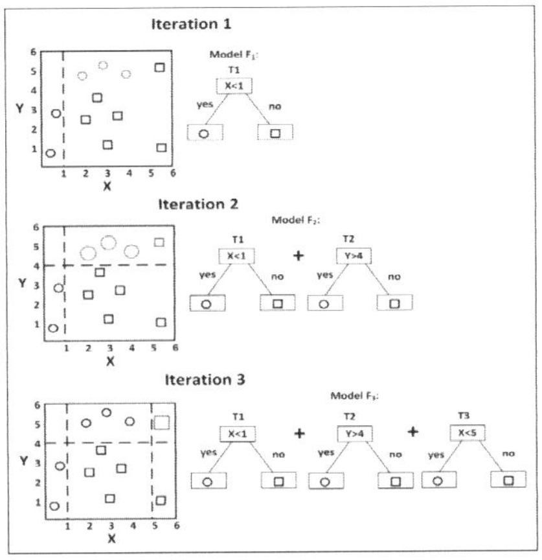

Figure 4. Gradient Boosting Visualization [8]

Bagging methodologies like Random Forest are known to reduce the variance and thus solve the overfitting problem, but may lead to high bias or underfit. Boosting algorithms rectify each learner's errors, thereby solving the problem of underfitting, but may lead to overfitting. Bagging and boosting when used in conjunction, will lead to a satisfactory bias-variance trade-off. XGBoost algorithm may be used to improve the performance of the Random Forest model, when applied for Phishing Detection as in this case, the predictive accuracy is far more important than model interpretability.

4.4.3 Performance measures

The predictive efficiency is indicated by evaluation metrics as discussed below. Accuracy is defined as the number of correct predictions over the total number of predictions made by the model (eqn. 3). Precision is the number of true predictions made by the model (eqn. 4). Recall is the number of positives returned by the model (eqn. 5). F-score is the mean of precision and recall (eqn. 6).

$$Accuracy = \frac{TP + TN}{TP + TN + FP + FN} \tag{3}$$

$$Precision = \frac{TP}{TP + FP} \tag{4}$$

$$Recall = \frac{TP}{TP + FN} \tag{5}$$

$$F - score = \frac{2 * Recall * Precision}{Recall + Precision} \tag{6}$$

Table 1. Confusion Matrix

	Positive	Negative
Positive	TP	FP
Negative	FN	TN

These metrics are used to create the Confusion Matrix illustrated in Table 1, which describes the model's performance on test data for which the true values are known.

5 Result and Discussion

Using the Random Forest Classifier, a prediction accuracy of 91% was obtained. Using XGBoost, a prediction accuracy of 93% was obtained.

The values of performance measures for both the algorithms are shown in Table 2.

Table 2. Summary of test results

Algorithm	Accuracy	Precision	Recall	F-score
Random Forest	0.911	0.883	0.914	0.898
XGBoost	0.937	0.938	0.949	0.928

The Confusion matrix values of Random Forest and XGBoost are shown in Table 3.

Table 3. Confusion Matrices values of Random Forest and XGBoost

n=461	Random Forest		XGBoost	
	Positive	Negative	Positive	Negative
Positive	239	24	224	19
Negative	17	181	10	188

For obtaining the final result, the website's legitimacy is predicted individually by both the models and then an average is drawn. This method produces three possible values - Zero (0) indicating that the website is safe, Half (0.5) indicating that the website has a 50% risk of being malicious, One(1) indicating that the website is phishy, and must be avoided.

6 Conclusion

As the number of internet users is on a steady rise, the number of unregulated websites today is more than ever. Phishing evolves with time, as illegitimate websites do not exist permanently, and undergo changes frequently.

Using the URL of a website for phishing detection with the aid of Ensemble Machine Learning Algorithms like Random Forest and XGBoost with a curated feature set is expected to produce highly accurate predictions with a satisfactory bias-variance trade-off in a timely and secure manner.

References

[1] Verizon, Data Breach Investigations Report (2021)

[2] M. Baykara, Z. Z. Gürelr, 6th International Symposium on Digital Forensic and Security, 1 (2018)

[3] A. Vazhayil, R. Vinayakumar, K. P. Soman, 9th International Conference on Computing, Communication and Networking Technologies, 1 (2018)

[4] R. Kiruthiga and D. Akila, Int J Recent Technol Eng, **8** , 111 (2019)

[5] H.S. Hota, A.K. Shrivas, Rahul Hota, Procedia Comput. Sci., **132**, 900 (2018)

[6] H. Musa, A.Y Gital, F.U Zambuk, A. Umar, A.Y. Umar, J.U Waziri, J Theor Appl Inf Technol , **97**, 1434 (2019)

[7] V. Jagannath, Random Forest Template for Spotfire, TIBCO (2020)

[8] Z. Zhang, G. Mayer, Y. Dauvilliers, G. Plazzi, F. Pizza, R. Fronczek, J. Santamaria, M. Partinen, S. Overeem, R. Peraita-Adrados, et al., Sci. Rep, **8**, 1 (2018)

Image plagiarism detection using GAN - (Generative Adversarial Network)

Kaustubh Gayadhankar[1], Rishi Patel[2], Hrithik Lodha[3], and Mr. Swapnil Shinde[4,]*

[1]Department of Information Technology, Ramrao Adik Institute Of Technology, Nerul, Navi Mumbai, India
[2]Department of Information Technology, Ramrao Adik Institute Of Technology, Nerul, Navi Mumbai, India
[3]Department of Information Technology, Ramrao Adik Institute Of Technology, Nerul, Navi Mumbai, India
[4]Department of Information Technology, Ramrao Adik Institute Of Technology, Nerul, Navi Mumbai, India

Abstract – In Today's date plagiarism is a very important aspect because content originality is the client's prior requirement. Many people on the internet use others' images and get publicity while the owner of the image or data won't get anything out of it. Many users copy the data or image features from the other users and modify it a little bit or create an artificial replica of it. With sufficient computational power and volume of data, the GAN models are capable enough to produce fake images that look very much similar to the real images. These kinds of images are generally not detected by modern plagiarism systems. GAN stands for generative adversarial network. It has two neural networks working inside. The first one is the generator which generates a random image and the second one is the discriminator which identifies whether the image being generated is a real or a fake image. In this paper, we have proposed a system that has been trained on both fake images (GAN Generated images) and real images and will help us in flagging whether the image is plagiarised or a real image.

Keywords – plagiarism, GAN models, artificial replica, Generator, Discriminator, real image

1. Introduction

There are a lot of images that are uploaded daily by content creators. Numerous people use the internet daily, they share a lot of images and gain plenty of information. Content creators play a very important role to develop unique content and people also like their work that's why people invest more time on the internet. Images used in an e-commerce website for fashion-related stuff or image of products, in such areas plagiarism needs to be avoided.

Content creators work hard daily to develop unique content. But few people on the internet, take the images which are created by content creators, and by making few modifications or creating an artificial replica of it, they use that image for their purpose which jeopardizes the aspect of content originality.

There is no system available to catch such plagiarism. It is very necessary to avoid such plagiarism. In this paper, the main focus is on image plagiarism. There are some methods available such as Fourier Based Image Plagiarism Detection Technique. In the Fourier-based Image Plagiarism Detection Technique, the source image is taken and converted into its discrete form, and also all the images in the database are converted into their discrete form and then, Fourier transform is applied to the source image as well as the database image, so at the output it's going to be a matrix. The source image is searched by the searching method into the database and then comparison is made between both images F1 components by computing distances.

But in this paper, the GAN model is used for detecting Image plagiarism, GAN stands for Generative Adversarial Networks. One part of GAN can generate new images and the other part is used for discrimination. One part creates new images and the other part is used only for detecting whether the generated image is real or fake. This Adversarial technique is shining in the industries and gained the attention of academia as well. GAN has effective generating new image samples technique. GAN has improved a lot of Computer Vision-related tasks and achieved remarkable success in that area.

https://doi.org/10.1051/itmconf/20214003013

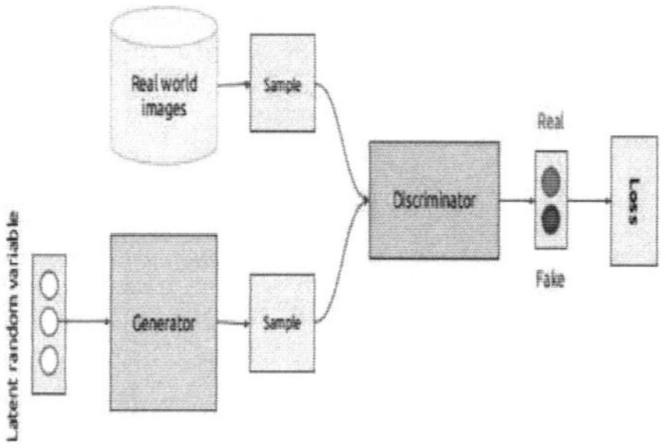

Fig. 1. GAN model architecture

In the proposed system, a CNN model (CNN stands for Convolutional Neural Network) will be used that will be trained on both the real image dataset and the fake image dataset that has been created by the GAN model. This will help in identifying whether the image is plagiarized or not. In the last part of the proposed system, a Flask-based Application is developed. In that application, the user will register and log in to use this system. Then the user will upload an image. The image will undergo feature extraction, and then the generated model results will tell whether the input image is plagiarized or not.

The remaining part of the paper is organized as follows: in section 2, a survey regarding the related work is reviewed. In section 3, a thorough explanation of the proposed methodology along with a system flow diagram and analysis is given. In section 4, results are displayed. In section 5, overall understanding and outcomes are summarized and concluded. In section 6, the future Scope of the system is mentioned. In the last section, references are mentioned.

2. Related Work

This section reviews various research done in the field of GAN models and image plagiarism. Many research people have applied various methods and algorithms in the process to make the system more accurate. For building the proposed system, various works have been studied and gathered information that could make the proposed system more efficient and accurate.

Liang Gonog et al. [1] The authors have introduced the background of the GAN, theoretic models, and extensional variants of GANs, and analyzed whether the variants can further optimize the original GAN or change the basic structures. The authors have also explained the typical applications of GANs. At last, the authors have also summarized the existing problems of GANs and future work of GANs.

Petr Hurtik et al. [2] The authors have proposed a method to search a plagiarised image in a database. The authors have used a tool called FTIP which is based on the F1 transform technique. The algorithm was very fast. It was able to achieve a one hundred percent of success rate for non-cropped images. The authors have claimed that they can achieve the same success rate on cropped images as well but at the expense of high computational time.

Yining Lang et al. [3] The key challenge that the authors have faced is plagiarized clothes are generally modified in a certain region on the real design to escape the supervision by traditional retrieval methods. For this, the authors have proposed a network named Plagiarized-Search-Net (PS-Net) which is based on regional representation, where the authors have utilized the landmarks to assist the learning of regional representations and compare fashion items region by region.

Yang Jie-Cao et al. [4] The authors have analyzed that the adversarial training concept is much more robust in both feature learning and representation. The authors have also mentioned that GAN also has some problems, such as non-convergence, model collapse, and uncontrollability due to a high degree of freedom. Various GAN models like stackGAN, InfoGAN, WGAN, LSGAN, and output are shown in this paper.

Tao Xu et al. [5] The authors have proposed an Attentional Generative Adversarial Network (AttnGAN) that allows attention-driven, multi-stage refinement for a fine-grained text-to-image generation. With the help of this neural net, the model was able to process fine-grained details at separate subregions of the image by paying attention to the relevant words in the natural language description. On top of that, a deep attentional multimodal similarity model is also proposed to calculate a fine-grained image-text matching loss for training the generator.

3. Proposed Work

3.1 Overview of the proposed approach

The dataset that was have used in this paper was created by Oxford University which has 8200 different flower images categorized into 102 different categories. Each category inside this dataset has somewhere between 40 to 258 different flower images. Each image in this dataset has its class label and along with its description. Each set of these flower images inside a particular category will have many overlapping features with some minute differences so that the model can cover up many variations of a specific flower to add variance to the data.

The overall approach of the proposed system is depicted in fig 2. In the proposed system, a flask interface was created for the user to communicate with the system. The user can first log in to the system if the user exists or he/she can sign up if not. After logging into the system, the user can upload the photo for a plagiarism check. After uploading the picture, the pre-processing and feature extraction on that image will be performed wherein all the local and global features of the image are extracted. And then a CNN algorithm is applied on top of it which is completely trained using the GAN or fake image dataset. This algorithm will then decide whether the uploaded image is a GAN generated i.e. artificially generated or not. The result will be displayed on the website itself. The images in the fake image dataset were of size 8x8 i.e. 64 images in total, this was done so that the user gets more choices to choose his/her image according to their convenience since GAN along with the replicas also creates some noise as a part of output while producing the image. So to make the replicas distinguishable from the noise the images were being generated in the format of 8x8 i.e. 64 images in total.

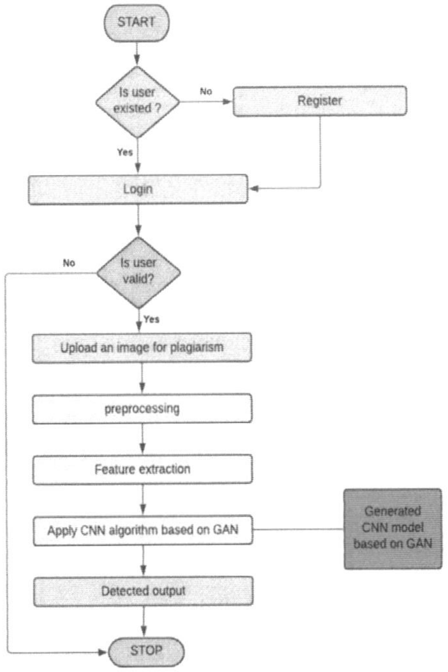

Fig. 2. Flow diagram of the system

3.2 Requirement Analysis

1. Software Requirements

- Operating System: Windows 10, 7, 8.
- Python
- Anaconda Launcher
- Spyder, Jupyter Notebook, Google Colab, Flask
- MySQL

2. Hardware Requirements

- Processor: Intel i5 or above
- RAM: 8 GB or more
- Hard disk: 250 GB or more
- GPU RAM: 4 GB or more

3.3 Design of the system

The architecture of the system is depicted in fig. 3. The architecture is divided into two parts. First is the GAN image generation part. The second is the plagiarism check part. The steps involved are mentioned below:

- In the GAN model, two networks are working inside it. The first is Generator, which produces random images from random values initialized during the first epoch.
- These random images are then sent to the discriminator to identify whether the image generated is fake or real.
- If the discriminator classifies the image as fake, the values inside the latent random variable are adjusted according to the loss calculated in the previous epoch. Based on these new values, the generator now creates a new random image.
- This entire process of creating an image, and then classifying it, is repeated again and again until and unless the generator can fool the discriminator into believing that the image generated is a real image.
- Using this GAN model, a fake image dataset was created which will serve as a dataset for the next phase (plagiarism part).
- In the second phase i.e. plagiarism phase, images from the fake image dataset and real image dataset will be taken one by one, and then pre-processing and feature extraction would be performed.
- A CNN algorithm will be trained and tested on top of it, which help us in classifying the images uploaded.
- On the user end, the user would be requested to upload an image for plagiarism check.
- Then the CNN algorithm which is generated using a fake image dataset classifies whether the image uploaded is a GAN generated or a real image.

In the proposed system, the approach of StackGAN mentioned in [12] was followed with some minor changes for building the GAN model. Fig 4. depicts the architecture of the model mentioned. At first, the text description is converted to text embedding using a text encoder network. This embedding is used for generators and discriminators in both stages as conditioning variables. Before this embedding is sent to the generator network, it's made to pass through a conditional augmentation network. The main for using this CA network is to convert the text embedding into a conditional latent variable. This conditional variable provides some extra information on what to judge and what to create. In this network, the embeddings are broken down and sampled from a vector of means and a vector of variances. A reparameterization trick is also applied to assist in the process of gradient descent (in finding gradients).

Architecture diagram :

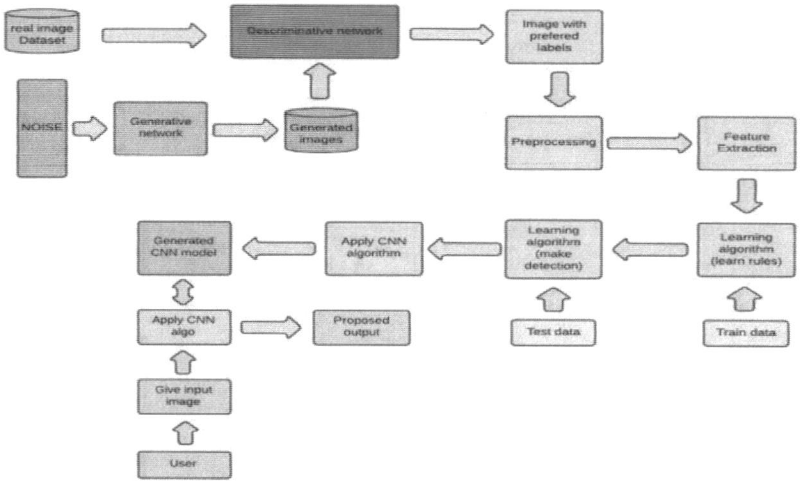

Fig. 3. The architecture of the system

A conditional variable from the previous step is sent to the generator model along with some random noise which is normally distributed with a mean of 0 and variance of 1. This is upsampled in the generator to create stage 1 images or low-resolution images of the format 64x64. These images along with the real images are compared and downsampled and then sent to the discriminator. In the discriminator, the output of the downsampling block is concatenated with the spatially compressed embeddings and sent to the classifier to classify whether the image is a real or fake image.

Low-resolution images created in stage one are sent to the stage two generator along with the text embedding. The second layer has its conditioning augmentation layer. This layer creates its own conditional variable by passing the text embedding through it. The generator in the second layer uses both i.e. stage one images and the conditional variable to encode and then decode the final high-resolution image. At last, stage two generated images, and the real images are compared and downsampled and then passed to the discriminator. The output of the downsampling block is concatenated with the spatially compressed embeddings and sent to the classifier to classify whether the image is a real or fake image.

The loss function that was mentioned in the original paper [6] was the minimax GAN loss function. There are two loss functions in this process. First is the Discriminator loss function which maximizes the average of the log of probability for real images $D(x)$ and the log of the inverse probability for fake images $G(z)$.

$$maximize \log D(x) + \log(1 - D(G(z))) \quad (1)$$

The second is the Generator loss function which minimizes the log of the inverse probability predicted by the discriminator for fake images $G(z)$.

$$minimize \log(1 - D(G(z))) \quad (2)$$

The generator loss had a problem, the loss function for the generator saturated very quickly i.e. the generator cannot learn as quickly as the discriminator and if the discriminator job was very easy, the discriminator would win the game within no time and the model cannot be trained efficiently because of it. So to remove this problem, there was a small change that was made in the generator loss function. Now the generator would maximize the log of the discriminator probabilities for generated images $G(z)$ instead of minimizing it. So now the Eq 2. would be reformulated as:

$$maximize \log(D(G(z))) \quad (3)$$

With the help of this, better gradient information could be supplied while updating weights, and the training phase would become more stable.

The loss function that was used in this work is Wasserstein Loss Function. It's a modified version of the GAN loss function in which the discriminator does not classify instances i.e. real or fake, rather it outputs a number. It tries to make the number bigger for real instances than for fake instances and there's no constraint on the range of this number. It also includes two loss functions. First is the discriminator loss function which maximizes the difference between its output on real instances $D(x)$ and its output on fake instances $G(z)$.

$$maximize (D(x) - D(G(z))) \quad (4)$$

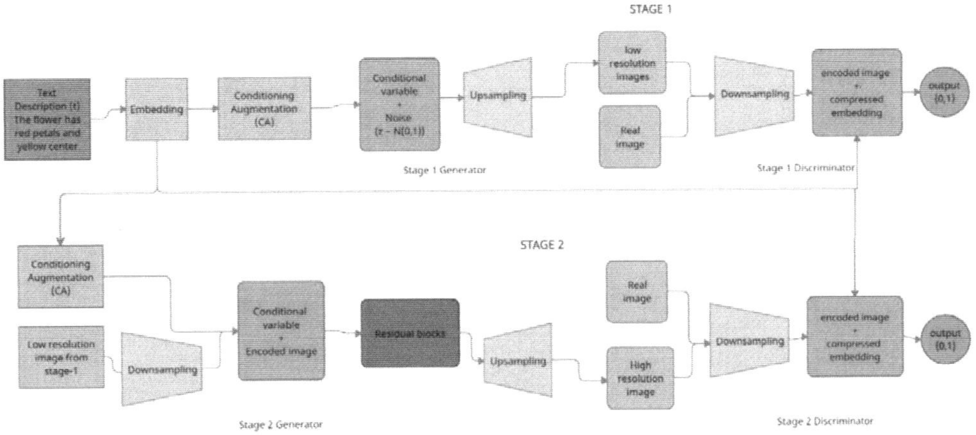

Fig. 4. Stacked GAN model architecture

For the generator loss function, it maximizes the discriminator's output for its fake instances $G(z)$.

$$maximize\ D(G(z)) \qquad (5)$$

This loss function is less vulnerable to getting stuck than the minimax-based GAN loss and it also avoids problems with vanishing gradients. One more significant advantage this has is rather than using cross-entropy as a metric it uses earth mover distance as a metric which is considered as a true metric for measuring the distance in a space of probability distribution. The generator is trying to maximize the value because it would only then help the generator to convert its fake image into a real image since the higher is the number, the higher is the probability of the fake image being classified as a real image.

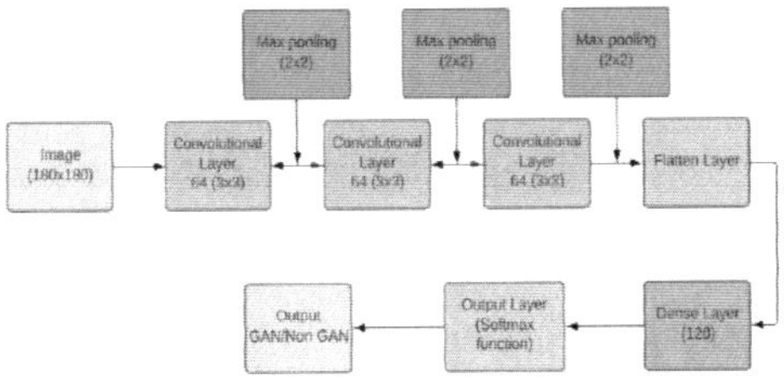

Fig. 5. CNN model used in the second part of the system i.e. plagiarism system

Fig 5. depicts the CNN model used in the plagiarism system. The images given from the first part of the system i.e. GAN part will serve as a training dataset for the second part of the system. The dataset for this part includes both GAN-generated images as well as the real images (grouped in 8x8 format). These images are then pre-processed. In the pre-processing step, the images are read from their respective file paths and then rescaled to the size of 180x180 format. After loading the images, the data is divided into two sets – images and their class labels i.e. GAN or Non-GAN. The images are then scaled down by a factor of 255 to reduce the complexity of the training process (which helps in easing out the calculations).

Then a model with three convolutional layers and three max-pooling layers, a dense layer as hidden layers, and ReLU as the activation function was created. For the output layer, a softmax activation function was used which will help us in classifying whether the image is a GAN image or a Non-GAN image. The training data is then fitted on top of this model. In the proposed system, "Adam" as the optimizer function was used. The loss function used is "sparse categorical cross-entropy". The validation split was chosen 0.2 and the model was trained for seven epochs.

3.4 Preprocessing

At first, all of the class ids are supposed to be loaded which are stored in a pickle file. Then, load the filenames, which are also stored in a pickle file. Then, the text embeddings that have been created using a text encoder network are supposed to be loaded which are also present in a pickle file. Next, the bounding boxes are created to extract objects from the crude images. This function will return a dictionary of file names and their corresponding bounding boxes. After this, a function is created to load the image and crop it around the bounding box according to the image size passed as an argument to that function. In this work, an image size of 64 has been used as the argument for this function. At last, all of the above methods are combined i.e. all the images, their class labels, and corresponding embeddings to get the dataset which will then be used for the training process.

3.5 Training phase

At first, the values of the hyperparameters required for this training process are initialized. The learning rate for generator and discriminator was initialized to 0.0002. Adam optimizer is defined for the training. The learning decay rate was set to 600. Then, two tensors with labels real and fake were created. These are required for the training of generator and discriminator. In this work, label smoothing was also used in this step. Next, the dataset from pre-processing step is used for training purposes. For the next step, a "for" loop was created which should execute for the number of times given in the number of epochs variable. Inside this loop, another "for" loop will execute based on the number of batches specified. Next, a mini-batch of data is sampled for the current iteration. In this iteration, a noise vector is initialized, a batch of images and batch of embeddings is selected and images are also normalized.

After this, the generator network is used to generate fake images. For this step, first, the Stage-I generator is used to generate low-resolution fake images, then the Stage-II generator network is used to generate high-resolution images. Then a compressor model is used to compress the embedding. Next, train the discriminator model on the fake, the real images. Following this, training of the adversarial model will be done, which is a combination of generator and discriminator models. And then, after every epoch, images are generated and saved and even the losses are computed and saved for evaluation purposes. At last, the model (weight values) is saved as well.

4. Results

4.1 Phase – I (Text to Image)

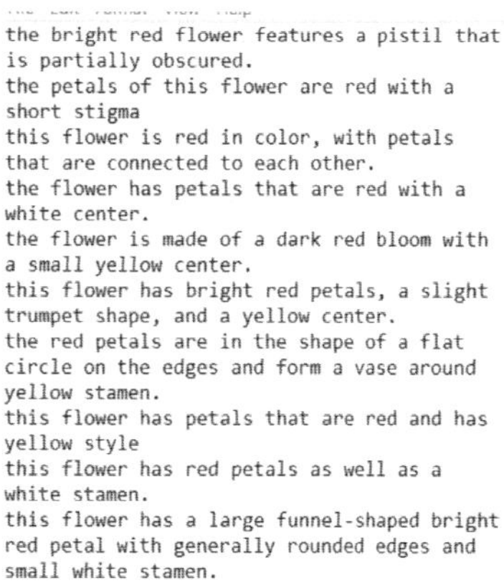
the bright red flower features a pistil that is partially obscured.
the petals of this flower are red with a short stigma
this flower is red in color, with petals that are connected to each other.
the flower has petals that are red with a white center.
the flower is made of a dark red bloom with a small yellow center.
this flower has bright red petals, a slight trumpet shape, and a yellow center.
the red petals are in the shape of a flat circle on the edges and form a vase around yellow stamen.
this flower has petals that are red and has yellow style
this flower has red petals as well as a white stamen.
this flower has a large funnel-shaped bright red petal with generally rounded edges and small white stamen.

Fig. 6. Text description

Fig. 7. Corresponding flower image i.e. for the above text description

Fig. 8. GAN generated image for the above text description

Fig 6. depicts text descriptions that will be given as input to the GAN model. Fig 7. depicts the corresponding image for the text description. Fig 8. Depicts 8x8 GAN generated image that will be produced by the GAN model as the output.

Some of these images are very similar to the original image i.e. we can see that some flowers have the same dark red petals and yellow center as the original one, and even the shape and the distribution of these petals around the center is very similar to the real one. Some flowers are not exactly similar to the original flower but have some features in common like the yellow center and the vase-like distribution of petals around the center. But there are some flowers in this plot that are not at all similar to the real flower, these are noises that the GAN model has generated. These noises can be reduced to a significant extent by increasing the number of epochs in the training phase.

4.2 Phase – II (Plagiarism check)

Fig. 9. 8x8 Real image

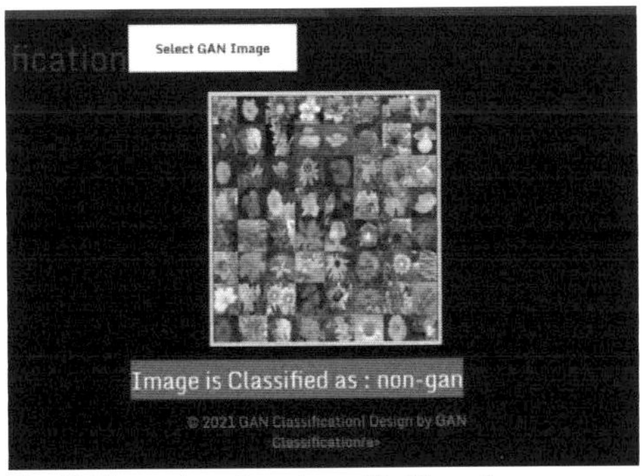

Fig. 10. The result on the web app (Real Image)

Fig 9. depicts 8x8 real image i.e 64 real images clustered together in 8x8 format. Fig 10. depicts the result of the web app for the real image. We can observe from the above result that the image has been classified as a non-gan or not artificially produced image.

Fig. 11. 8x8 GAN generated image

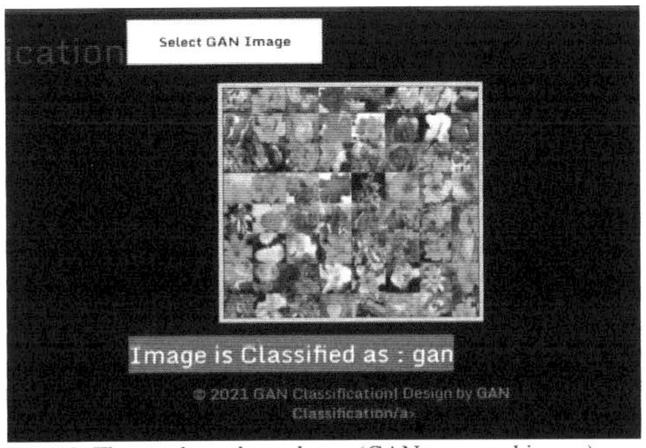

Fig. 12. The result on the web app (GAN generated image)

Fig 11. depicts 8x8 GAN generated image i.e 64 images clustered together in 8x8 format. Fig 12. depicts result on web app for the GAN generated image. We can observe from the above result that the image has been classified as GAN or artificially produced image.

The CNN (Convolutional neural network) algorithm was trained for a plagiarism check system with about 120 images from both the fake image dataset (created by the GAN model) and the real image dataset. The validation split was set to 0.2 i.e. 20 percent of these images were taken in the validation set. The validation accuracy at the end of the training was observed to be **97** percent. Accuracy here is defined as the number of correctly predicted points out of all the data points. The data points here belong to two classes i.e GAN and non-GAN.

5. Conclusion

In this paper, the major focus was on the image plagiarism problem, i.e. checking whether the image is artificially produced (GAN produced) or not. We proposed plagiarism detection system based on GAN using the CNN technique which was mainly used in the pre-processing step. In this project, we first apply the GAN to the dataset and this output will be sent as an input to the CNN. Then this model will predict whether the image is a GAN generated or not. This will help a lot in identifying plagiarized images since many modern plagiarism systems are not able to detect such kinds of artificially created images.

6. Future Scope

We have trained the GAN model on the Flowers dataset for this project but the same architecture can be extended to generate and detect any category of the image by training the model on that category of data. The proposed system can be used on any platform where there is a high chance of using plagiarized images i.e. artificially generated images, specifically on blog websites and in the fashion industry. We were able to achieve good results by training the GAN model on our machine but the accuracy of the system can also be further extended by a significant amount by training the model on machines that have very high computational power.

References

[1] L. Gonog and Y. Zhou, "A Review: Generative Adversarial Networks," 2019 14th IEEE Conference on Industrial Electronics and Applications (ICIEA), pp. 505-510, (2019)

[2] P. Hurtik and P. Hodakova, "FTIP: A tool for an image plagiarism detection," 2015 7th International Conference of Soft Computing and Pattern Recognition (SoCPaR), Fukuoka, Japan, , pp. 42-47, (2015)

[3] Y. Lang, Y. He, F. Yang, J. Dong, H. Xue, "Which is Plagiarism: Fashion Image Retrieval based on Regional Representation for Design Protection," Proceedings of the IEEE/CVF Conference on Computer Vision and Pattern Recognition (CVPR), pp. 2595-2604, (2020)

[4] Y. J. Cao, L. L. Jia, Y. X. Chen, N. Lin, C. Yang, B. Zhang, Z. Liu, X. X. Li, H. H. Dai, "Recent Advances of Generative Adversarial Networks in Computer Vision," in IEEE Access, vol. 7, pp. 14985-15006, (2019)

[5] T. Xu , P. Zhang , Q. Huang , H. Zhang , Z. Gan, X. Huang, X. He, "AttnGAN: Fine- Grained Text to Image Generation with Attentional Generative Adversarial Networks," 2018 IEEE/CVF Conference on Computer Vision and Pattern Recognition, pp. 1316-1324, (2018)

[6] I. J. Goodfellow, J. P. Abadie , M. Mirza, B. Xu, D. W. Farley, S. Ozair, A. Courville, Y. Bengio, "Generative Adversarial Nets," arXiv preprint arXiv:1406.2661, vol: 1, (2014)

[7] K. Ak, A. Kassim, J. Lim and J. Y. Tham, "Attribute Manipulation Generative Adversarial Networks for Fashion Images," 2019 IEEE/CVF International Conference on Computer Vision (ICCV), pp. 10540-10549, (2019)

[8] I. Amerini, L. Galteri, R. Caldelli and A. Del Bimbo, "Deepfake Video Detection through Optical Flow Based CNN," 2019 IEEE/CVF International Conference on Computer Vision Workshop (ICCVW), pp. 1205-1207, (2019)

[9] M. Chen, Y. Qin, L. Qi and Y. Sun, "Improving Fashion Landmark Detection by Dual Attention Feature Enhancement," 2019 IEEE/CVF International Conference on Computer Vision Workshop (ICCVW), pp. 3101-3104, (2019)

[10] N. Dalal and B. Triggs, "Histograms of oriented gradients for human detection," 2005 IEEE Computer Society Conference on Computer Vision and Pattern Recognition (CVPR'05), vol: 1, pp. 886-893, (2005)

[11] K. E. Ak, J. H. Lim, J. Y. Tham and A. A. Kassim, "Efficient Multi-attribute Similarity Learning Towards Attribute-Based Fashion Search," 2018 IEEE Winter Conference on Applications of Computer Vision (WACV), pp. 1671-1679, (2018)

[12] H. Zhang, T. Xu, H. Li, S. Zhang, X. Wang, X. Huang and D. Metaxas, "StackGAN: Text to Photo-realistic Image Synthesis with Stacked Generative Adversarial Networks," 2017 IEEE International Conference on Computer Vision (ICCV), 2017, pp. 5908-5916, (2017)

https://doi.org/10.1051/itmconf/20214003014

Deep Learning based Human Action Recognition

Ritik Pandey[1], *Yadnesh Chikhale*[2], *Ritik Verma*[3] and *Deepali Patil*[4]

[1,2,3,4]Ramrao Adik Institute of Technology, Information Technology Department, 40076 Nerul, India

Abstract. Human action recognition has become an important research area in the fields of computer vision, image processing, and human-machine or human-object interaction due to its large number of real time applications. Action recognition is the identification of different actions from video clips (an arrangement of 2D frames) where the action may be performed in the video. This is a general construction of image classification tasks to multiple frames and then collecting the predictions from each frame. Different approaches are proposed in literature to improve the accuracy in recognition. In this paper we proposed a deep learning based model for Recognition and the main focus is on the CNN model for image classification. The action videos are converted into frames and pre-processed before sending to our model for recognizing different actions accurately..

1 Introduction

The recognition of human activity is a primary issue in the field of computer vision. A human activity can be as simple as throwing a ball or brushing your teeth. At present, researchers have been working on this issue since it has received sensible attention. This work has its focus on recognizing an individual action. Deep Learning based human action recognition has been proposed in this work. The existing approaches are computationally more expensive so that makes the architecture more computationally efficient. Many different techniques can be used to recognize human action and research on this topic increased since the rise of popularity in artificial intelligence and machine learning. The basics of human action recognition are extracting features and prediction of action from an image or video.

CNN(Convolutional Neural Networks) is a particular deep learning method used for extracting and learning about features from Images. Currently there is no particular method for Video Classification. In this proposed system, the use of a CNN model to predict action which works by learning of features from individual images. It is difficult to perform human action recognition on a still image because of the limited resource of information, so using a collection of images from every video is proposed in our system where Video can be converted into a sequence of images. This work can be used in scenarios where a human activity is to be recognized by a computer.[3] Activity recognition is used in applications such as surveillance, anti-terrorists and anti-crime securities as well as life logging and assistance which reduces the cost of human resources. It is difficult to recognize human activity by a machine. This is an important challenge to the field of computer vision. The implementation of proper data pre-processing techniques has a high effect on the learning process of CNN model. These properties make the proposed method more suitable for action recognition in videos.

2 Related Work

Du Tran et al. have trained their data on large video datasets. They have used UCF101 Dataset for training the data from which they have achieved 52.8% accuracy with 10 classes [1].

Fernando Moya Rueda evaluated a novel CNN architecture for HAR (Human Action Recognition) using multichannel time-series acquired from body-worn sensors- IMUs. CNN-IMU [3]. This method is costly compared to normal non-sensors based recognition.

Most of them have emphasized their work mostly on human action recognition(HAR).In [2] they have to take a low-resolution life-logging camera data and predicted the action capture the camera, before that they have trained a large dataset Imagenet for the getting maximum percent of accuracy [20].

Y. Du, W. Wang, and L. Wang, et al presented a skeleton based action recognition for an end-to-end hierarchical recurrent neural network and they first split the human skeleton into five parts and then give them to five subnets. Only from the skeleton joins the alike human actions are not easily distinguished [4][5]. This method

* Corresponding Author: ritikpandey5@gmail.com,yadnesh1806@gmail.com ,vermaritk@gmail.com, deepali.patil@rait.ac.in

is computationally heavy and more complex.
Sometimes people get robbed or violence breaks down at a crowded place, at that moment it gets difficult to find the culprit or to keep an eye on the culprit. For keeping track of people at crowded places deep learning model are used for crowd management and for keeping track on suspicious activity[6].
According to the search, most of the research on action recognition is done on the state-of-the-arts [7] [14] and human action recognition (HAR) which is further used for the prediction of activity. The plotting of the graph is also done on the basis of the accuracy of the data [8] [9]. It explains the formation of the deep architecture with the help of Graph Estimation Procedure [10].
In depth based action recognition they have used 120 different classes for 3D based Human action recognition in theses they have evaluated the activity analysis [11]. It has datasets of 10s which were taken from youtube, in these there are various different action classes which have more interaction between humans with other objects. They have explained the statistics of dataset and performance for neural networks and they trained and tested the action classes dataset [13].
Gundong Guo et al. presented an overview of the state-of-the-art methods of still image-based action recognition and described and categorized on many high-level cues and low-level features for action based analysis for the still images. Still image-based human action recognition mainly focuses on recognizing a person's behaviour or action based on a single image [15]. In this work action recognition from still images is done and is not preferable for movement involving Human Action videos.
Various methods are used by researchers in the field of action recognition. These methods are of different types like Sensors based recognition [16], Machine Learning based action recognition and Deep Learning based action recognition.
Hong-Bo Zang & group has presented a review on the human action recognition method and also provided comprehensive overviews of this approach. They also included progress in hand-designed action features in RGB [17].
Fangyu Liu comes up with a 3DCNN-DQN-RNN method that combines all three methods to get the efficient semantic parsing of large-scale 3D point clouds. The main feature of their method, it provides the automatic process that plots the raw data to the classified results [18]. This method is a computationally expensive method and requires strong hardware to work.
Jun Liu, introduces a large set of data for RGB+D human action recognition. Their model helps in the long-term temporal correlation of the action for each part of the body and helps in providing them better action classification[19]. In this paper, skeleton data in the form of graphs is used which is a different form of data compared to video and images.

3 Methodology

To proceed with our proposed method of implementation, there are few basic steps. Fig 1 shows the flow of the steps to be carried out for the implementation of our proposed algorithm.

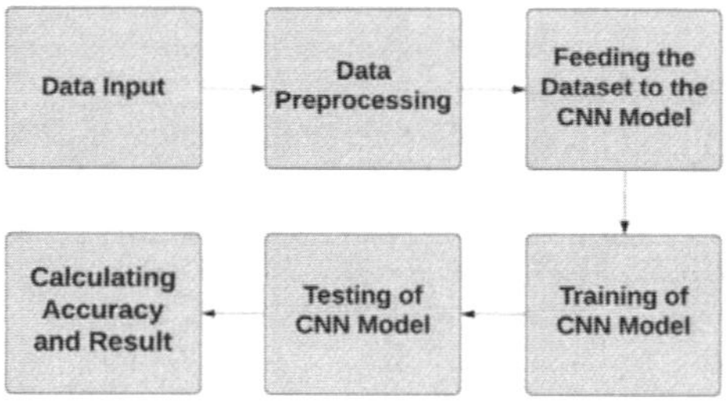

Fig.1. Flow Chart of Proposed System

A. Data Pre-processing: The first step is learning about the UCF101 dataset [1]. It is done by creating a data frame which has information about every video file in the dataset and which human activity class it belongs to. After that, the videos in the dataset are split into frames(images). The video which is used from the UCF101 dataset has 25 fps property. While converting into images, every 3rd frame for the 25 frames in 1 second is considered and saved. Later, the Image Data Generator by Keras rescales, shears, zooms and horizontally flips the frames so it'll be better for the learning process in the CNN model [12].

B. Training Model using Convolutional Neural Networks (CNN): Convolutional Neural Networks are used for learning and finding patterns in images. The CNN model is perfect for image classification. Each single image is a combination of pixels which are represented in the form of a matrix. The operations are applied on this matrix. A CNN has various layers namely, convolutional, pooling, fully connected, etc. In each of these layer's unique operations are applied in the process of learning. For image processing, our propose to use Tensor Flow and Keras which will be used for training the classifier. For other mathematical calculations and arrays, NumPy is used in Python. Pandas, matplotlib, and cv2 are used for handling data frames, graphs and dealing with images respectively.

The structure of our CNN Model is:

1. Two sets of convolutional and max pooling layers.
2. Flatten layer.
3. Two dense layers.

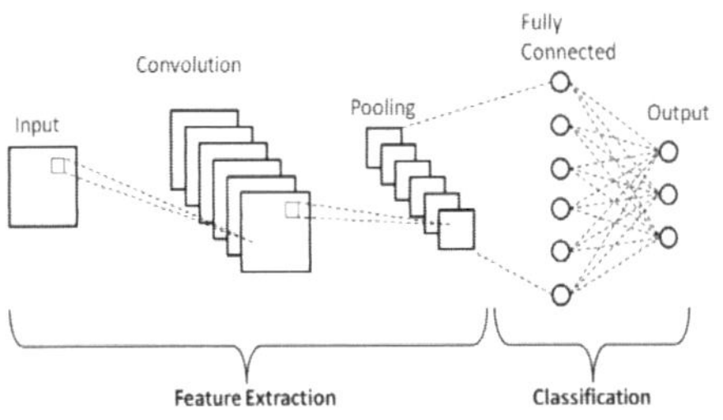

Fig.2. Sequence of layers in CNN Model

The image passes through multiple filters in two sets of convolutional 2D layers. The size of filters is kept constant and the number of filters is increased in the next convolutional 2D layer. In both these convolutional 2D layers, Relu or Rectified Linear Activation Function is used.To calculate the losses, Categorical_Crossentropy is used. Adam optimiser is used on the data to improve learning and increase accuracy.

C. Predicting Results: Accuracy and loss factors are calculated in the evaluation of the Train and Test dataset to see how efficient the model is. Training accuracy and loss graphs are made using matplotlib to know the structure of learning by epochs. The softmax probability output from the last dense layer is used to create a prediction probability matrix. Accuracy of every class is calculated separately. Random images of a few action classes are used from Google to test on the model. The model gives 80-88% accuracy currently.

4 Experimental Results

The paper gave information about the importance of Data Pre-processing. The dataset chosen for this is UCF101. This dataset has 13,000 videos which are divided into 101 categories or action classes. In total, the dataset is 27 hours. Each video has a length of 4 to 11 seconds. The variety of data available in the UCF101 dataset helps the model on testing data which are from outside of the dataset. The action classes in the dataset are Cricket Bowling, High Jump, Long Jump, Playing Piano, Yoyo, Playing Guitar, Ice Dancing, Punch, Drumming, Horse Race, Pull Ups, Volley Ball, Base Ball, Sumo Wrestling, Diving, etc. Noise data reduces the efficiency of the model and cleaning it can improve the accuracy. Using the techniques mentioned an accuracy of 80-88% on our CNN model is achieved. Different combinations of 6000 to 12000 images in datasets have been used in training and testing by CNN model. Various Video to Images conversion patterns are followed in pre-processing and the best learning and highest accuracy was found when the number of images of a video per second was 20-40% of the frame rate (eg. For a 30fps video, each frame that is a multiple of 3 is chosen to be saved as an image).

It was seen that as the number of epochs increased, the training accuracy increased and training loss decreased [10]. They both stop increasing and decreasing after a certain number of epochs.

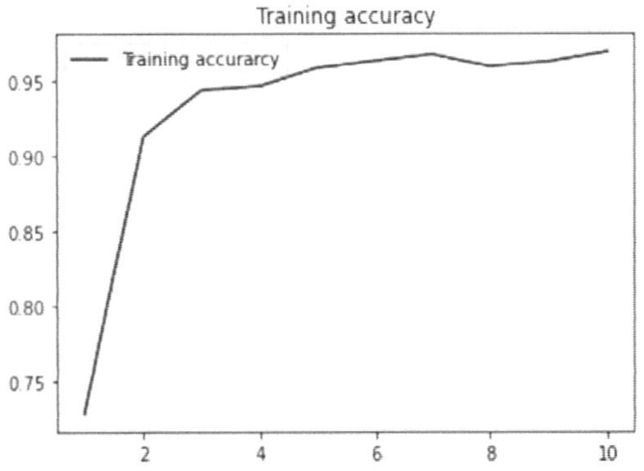

Fig.3. Training accuracy graph for 8000 images train set with 10 epochs

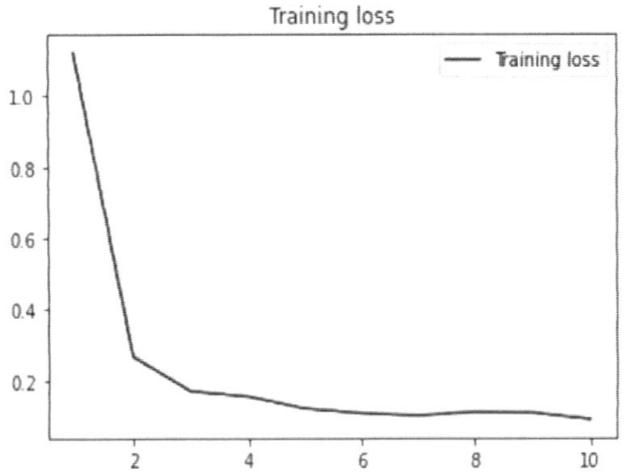

Fig.4. Training loss graph for 8000 images train set with 10 epochs

After the training of CNN model on various combinations of train datasets from 4000 to 9000 images, accuracies ranging from 80-88% were found on Evaluation.

Later, accuracies of all action classes were calculated and it was found that classes which had lower resolution and quality of video had lower accuracy. It was also found that Action Classes which has some kind of same and small similar activity in between, had an slight error in differentiating classes (eg. Athletic events like Pole Vault, Long Jump and High Jump have the same first half i.e. Running, in the similar background environment

```
model.fit(train_generator, epochs= 5)

Epoch 1/5
109/109 [==============================] - 831s 7s/step - loss: 2.3894 - accuracy: 0.5453
Epoch 2/5
109/109 [==============================] - 55s 501ms/step - loss: 0.3157 - accuracy: 0.8967
Epoch 3/5
109/109 [==============================] - 55s 501ms/step - loss: 0.2059 - accuracy: 0.9327
Epoch 4/5
109/109 [==============================] - 56s 510ms/step - loss: 0.1580 - accuracy: 0.9506
Epoch 5/5
109/109 [==============================] - 55s 506ms/step - loss: 0.1503 - accuracy: 0.9489
<tensorflow.python.keras.callbacks.History at 0x7fd26a8ed450>

[ ] model.evaluate(test_generator)

36/36 [==============================] - 237s 7s/step - loss: 0.3927 - accuracy: 0.8837
[0.3926926553249359, 0.8837004303093219]
```

Fig.5. Accuracy of 88.37% on Evaluation of test dataset

To be sure about the efficiency of the model, random images of the respective class from Google were given for prediction to our CNN model. These random images were predicted correctly as shown in Figure 6, 7 and 8.

```
img = cv2.imread(os.path.join("/MyDrive/",'pool.jpg'))
imgplot = plt.imshow(cv2.cvtColor(img, cv2.COLOR_BGR2RGB))
plt.show()
```

```
print(predictionz)

['Diving']
```

Fig.6. Class Prediction on a random Diving image from Google

```
img = cv2.imread(os.path.join("/MyDrive/",'17-1.jpg'))
imgplot = plt.imshow(cv2.cvtColor(img, cv2.COLOR_BGR2RGB))
plt.show()
```

```
print(predictionz)

['CricketBowling']
```

Fig.7. Class Prediction on a random CricketBowling image from Google

https://doi.org/10.1051/itmconf/20214003014

```
img = cv2.imread(os.path.join("/MyDrive/",'850.jpg'))
imgplot = plt.imshow(cv2.cvtColor(img, cv2.COLOR_BGR2RGB))
plt.show()
```

```
print(predictionz)

['SumoWrestling']
```

Fig.8. Class Prediction on a random Sumo Wrestling image from Google

5 Future Scope

In the proposed system, the videos are converted into frames and are fed to the CNN Model. CNN is a deep learning neural network which can learn and find patterns in images. It has been seen that the videos were treated as separate images. Further works can be on adopting an approach that treats the video data more like a video and not like an image, eg. A combination of CNN and RNN. In the future for more accurate prediction, mapping of the Human Body in the form of graphs can be done using computer vision algorithms. Skeleton data in the form of graphs can be used. There are also different sensors which are used to create data which can help in better accuracy. 3D CNN is a different field which is also growing at present times. Study and work in these interesting Video Classification fields will help to make accurate predictions.

6 Conclusion

In this paper, Human Action Recognition Algorithm by using frames from Human Action videos is proposed. First, the Data Preprocessing takes place where videos are separated into frames. Then various operations are done on the images to make it ready for the CNN model. Features are extracted from the video frames, which are then used to help determine the accuracy of the model, evaluation, and prediction of classes of videos. Here, the output for small images are combined as a final output of the class. The variety of data in action classes of the UCF101 dataset helps in better prediction of images outside of the dataset. Since the whole image which includes both human character and background in the frames are used in CNN model, essential information is extracted. Because of the huge variety of data in the used UCF101 dataset, the proposed model can be tested on random videos which are separate from UCF101.

References

1. Du Tran, Lubomir Bourdev, Rob Fergus,Lorenzo Torresani, Manohar Paluri "Learning Spatiotemporal Features with 3D Convolutional Networks", IEEE International Conference on Computer Vision (ICCV), 2015.

2. Tsai, Jen-Kai & Hsu, Chen-Chien & Huang, Shao-Kang. " Deep Learning-Based Real-Time Multiple-Person Action Recognition System", Sensors. 20. 4758. 10.3390/s20174758, Aug 2020.

3. Romaissa, Beddiar & Nini, Brahim & Sabokrou, Mohammad & Hadid, Abdenour, "Vision-based human activity recognition: a survey", Multimedia Tools and Applications. 79. 10.1007/s11042-020-09004-3, Aug 2020.

4. Y. Du, W. Wang, and L. Wang, "Hierarchical recurrent neural network for skeleton based action recognition", In IEEE Conference Paper on Computer Vision and Pattern Recognition (CVPR), 2015, pp. 1110– 1118.

5. S. Song, C. Lan, J. Xing, W. Zeng, "An End-to-End SpatioTemporal Attention Model for Human Action Recognition from Skeleton Data" in AAAI, pp. 4263–4270, 2017.

6. Riddhi Sonkar, Sadhana Rathod, Renuka Jadhav Deepali Patil et al. "Crowd Abnormal Behaviour Detection using Deep Learning", ITM Web of Conferences, 2020

7. T. S. Kim and A. Reiter, "Interpretable 3d human action analysis with temporal convolutional networks," in The IEEE Conference on Computer Vision and Pattern Recognition (CVPR), 10.1109/CVPRW.2017.207

8. M. Niepert, M. Ahmed, Konstantin Kutzkov et al. "Learning convolutional neural networks for graph," in International Conference on Machine Learning (ICML), 2016.

9. J. Bruna, W. Zaremba, Arthur Szlam, Yann LeCun et al. "Spectral Networks and Locally Connected Networks on Graphs," in International Conference on Learning Representations, 2014.

10. M. Henaff, J. Bruna, Yann LeCun et al. "Deep convolutional networks on graph-structured data," arXiv:1506.05163 [cs.LG], 2015

11. Jun Liu, Amir Shahroudy, Mauricio Perez, Gang Wang, Ling-Yu Duan, Alex C. Kot, "NTU RGB+D 120: A Large-Scale Benchmark for 3D Human Activity Understanding", IEEE Transactions on Pattern Analysis and Machine Intelligence (TPAMI), 2019. 10.1109/TPAMI.2019.2916873

12. F. Caba Heilbron, B. Ghanem, J. C. Niebles, et al. ."A large-scale video benchmark for human activity understanding". IEEE Conference on Computer Vision and Pattern Recognition (CVPR), 2015, pp. 961-970

13. W. Kay, J. Carreira, K. Simonyan, B. Zhang, C. Hillier, S. Vijayanarasimhan, F. Viola, T. Green, T. Back, P. Natsev, and others, "The Kinetics Human Action Video Dataset," arXiv:1705.06950 cs. CV, May 2017.

14. Yu Cong ,Yun Fu "Human Action Recognition and prediction: A survey", Computer Vision and Pattern Recognition, Cornell University, June 2018.

15. Gundong Guo, Alice Lai et al. "A survey on till image based human action recognition" west virginia university, may 2014.

16. Chen Chen, Roozbeh Jafari, Nasser Kehtarnavaz et al. "A survey of depth and initial sensor fusion for human action recognition" , 2017.

17. Hong Bo Zhang, Yi-Xiang Zhang, Bineng Zhong, Qing Lei, Lijie Yang, Ji-Xiang Du and Duan-Sheng Chen et al. "A Comprehensive Survey of Vision-Based Human Action Recognition Methods" feb 2019.

18. Fangyu Liu, Shuaipeng Li, Liqiang Zhang, Chenghu Zhou , Rongtian Ye , Yuebin Wang , Jiwen Lu et al. "3DCNN-DQN-RNN: A Deep Reinforcement Learning Framework for Semantic Parsing of Large-scale 3D Point Clouds" Tsinghua University.

19. Amir Shahroudy, Jun Liu, Tian-Tsong Ng and Gang Wang et al. "NTU RGB+D: A Large Scale Dataset for 3D Human Activity Analysis" Singapore Institute for Infocomm Research, april 2016.

20. Aksasse H., Aksasse B., Ouanan M.,"Deep Convolutional Neural Networks for Human Activity Classification",. In: Jain L., Peng SL., Alhadidi B., Pal S. (eds) Intelligent Computing Paradigm and Cutting-edge Technologies. ICICCT 2019. Learning and Analytics in Intelligent Systems, vol 9. Springer, Cham, 2020.

https://doi.org/10.1051/itmconf/20214003015

Foreign money transfer using blockchain

Nupur Giri[1], *Dheeraj Singh Jodha*[2], *Yash Goyal*[2], *Akshay Thite*[2], and *Abhay Tiwari*[2]

[1]HOD, Dept. of Computer Engineering, Vivekanand Education Society's Institute of Technology, Mumbai, India
[2]Student, Dept. of Computer Engineering, Vivekanand Education Society's Institute of Technology, Mumbai, India

Abstract. This paper focuses on using blockchain technology to enhance efficiency and cost reduction in foreign transactions through the banking system. Blockchain provides crucial features such as immutability of records and decentralization, which is then used to carry out foreign transactions. This process could be a major change in the transactions carried out by removing middle banks during transactions thereby speeding up the process. This paper aims at designing an Ethereum decentralized Banking Application for foreign money transfer using blockchain technology. Smart Contract helps to eliminate middle banks in the process by acting on rules specified in the contract. Every participant needs to abide by these rules which make it trustworthy and maintain the authenticity of the process.

1 Introduction

This paper aims at proposing an Ethereum based decentralized Banking Application for money transfer using blockchain technology. Some bank users do not prefer online baking as they perceive it as unsafe. Online banking is vulnerable to frauds such as phishing. In phishing, the user hands over their password unexpectedly to the fraudster. Online banking is insecure for computer illiterates.

Many frauds occur in traditional banking (Offline banking) too. These include credit card frauds where the fraudster uses the stolen credit card to make payments, signature forgery where the fraudster falsely replicates the signature of the other person to carry out transactions, etc.

So, if an online banking application is designed using blockchain technology, the transactions are stored permanently in the blockchain and every transaction can be traced back to when and where the transaction took place without revealing the identity of the user. Here, the past deposits and withdrawals cannot be changed and present deposits cannot be hacked as all the transactions are verified by all the nodes in the network. The current foreign trading system is centralized, which gives rise to many market risks, and the traditional process of sending money across borders is slow as it has many banks in the middle as authorities or intermediaries, hence adding up to the cost, and making this process inefficient for sending small sums of money.

This paper aims to build a decentralized system that would act as a platform for the fast, cost-efficient, and user-friendly transfer of money across the borders, without any unnecessary intermediary banks. Currently, at each step of traditional currency exchange, we have to spend a small amount of money but we still use these services because we don't have many alternatives to pick from. Today, the most significant financial market is the foreign exchange market, but the main entities comprise only large banks and companies, all of whom charge hefty fees for their services. With the right perspective, Blockchain technology can help reduce cost, enable faster transactions, and help to secure the complete infrastructure of any financial transaction. By undressing Blockchain technology into its components, we can utilize the ideas that work great, and abandon the ones that are still immature. We can leverage the relevant features of Blockchain technology, and make new systems on top of the existing financial services that are faster, cheaper, and more predictable. Together with existing technologies, we can build a system that is not as visionary as existing Blockchain solutions, but yet more realistic, scalable, adaptable, and most importantly, feasible and sustainable.

2 Literature survey

We leveraged a Distributed ledger to store and replicate the transactional data on each node in the network. The predefined ledger rules are responsible for removing conflicts and inaccuracies in the data stored at various nodes, as written in Blockchain White Paper [1].

There's already a study done on how Blockchain technology can be used in the banking sector [2]. In this paper, they used a trusted third party that issues coins to users, which customers can then use for transactions on the blockchain platform, which in turn lead to a reduction in money-laundering and financing of terrorism, as each transactional record is stored in a tamper-proof distributed ledger, with which the provenance of each fund can be easily verified. The system that this paper talked about, is cost-effective as the banks can use the encrypted applications in place of encoding and issuing the ATM cards. It is a highly secure system and reduces financial fraud. It is almost not possible to hack this system. Hence, we used a bank authority in our network which issues initial currency to the users which can be used for future transactions.

The white paper of Bitcoin [3] introduced an electronic system for the transfer of money that can be designed where there is no trusted third party but the system is secure. So, we referred to this paper and reduced the trust among the users and the banking authorities, as the transactions on the blockchain platform are based on cryptographic proof and not on trust, which completely removed any kind of trust between the parties. This also solved the problem of double-spending. And hash-based proof-of-work prevents any kind of tampering of transaction records without redoing the proof-of-work (PoW).

There's been a lot of studies that have been done on how Blockchain technology can be leveraged in the Foreign Exchange system. More specifically, the Bank of Canada (BOC) and Monetary Authority of Singapore (MAS) started their projects Jasper and Ubin [4] respectively, to see how Distributed Ledger Technology can be used to clear settlements of payments. We referred to this paper to study the scope of Blockchain in the Foreign Exchange system, and find the problems that it can solve with it. The various demerits associated with the current cross-border transfer technologies are differences in currency values with the sender and the receiver and the involvement of a correspondent bank to act as an intermediary. To minimize the risks associated with the settlements, the concept of "two-phase commit" is employed to help two or more processes coordinate to decide to commit or abort all the processes of the transaction.

Norwegian University of Science and Technology designed a private blockchain for international currency exchange [5], which pointed out that the current international currency exchange system is slow and expensive due to many factors. Every intermediary involved takes a certain amount of fees to complete the transaction. So, we referred to this paper to study how a blockchain-based system can be developed for foreign money transfers that have all the advantages of blockchain technology and how this system would be faster, secure, realistic, and scalable than the traditional system.

We found out that [6], regulatory and compliance issues are the biggest factors that contribute towards the resistance to the adoption of blockchain. A blockchain-based system can be designed to carry out bank transactions. This system can solve many problems associated with the traditional banking system. It is also much more secure as compared to the current banking system as the system would be decentralized, which forms an important feature of our system.

To measure the efficiency of our system, we investigated the link between transaction and computation costs in the blockchain environment [7]. We measured the computational cost (measured in units of gas) and used our insights to make changes in our smart contracts by writing code that is optimal in terms of time and space, which in turn reduced the computational costs.

Table 1. Denomination of ethers converted into US Dollars and Indian Rupees (Accurate as of 24 April 2021)

Unit	Wei Value	Wei	US Dollar value	INR value
Wei	1 Wei	1	\$2.26e-15	₹1.7e-11
Kwei	1e3 Wei	1e3	\$2.26e-12	₹0.17e-9
Mwei	1e6 Wei	1e6	\$2.266e-9	₹0.17e-6
Gwei	1e9 Wei	1e9	\$2.266e-6	₹0.17e-3
Microether	1e12 Wei	1e12	\$2.266e-3	₹0.169
Miliether	1e15 Wei	1e15	\$2.26642	₹169.82
Ether	1e18 Wei	1e18	\$2,266.42	₹1,69,82 3.98

The above table depicts the denominations of ethers converted into US Dollars and Indian Rupees, the data source is coinmarketcap.com and is accurate as of 24 April 2021 since the conversion rates of Ethereum to USD/INR is highly volatile. We referred to this table to calculate approximate results to measure the efficiency of our system.

3 Methodology

The steps involved in developing the local blockchain-based system involves the following:

3.1 Designing of the smart contract :

Designing the smart contract to carry out the successful transaction is the major part of the project. A smart contract contains a set of rules which are to be satisfied to carry out the transaction. Here, the smart contract checks whether the amount to be transferred is less than or equal to the user's account balance. If that condition is satisfied then the transaction takes place successfully. The smart contract for the transfer will be designed using the Solidity language. The smart contracts will consist of appropriate rules to carry out the transfer.

3.2 Designing of the user interface :

User Interface will consist of a home page that will have links to the Login and Registration Pages. The registration page is for registering the new user. The user has to log in and will enter the web page through which the transaction takes place. Here, the user can view the balance of their account, carry out transactions by entering the amount to be transmitted, and the receiver's address.

3.3 Linking of user interface and smart contract

The User Interface and Smart Contract need to be linked so that, when the transaction is initiated using the UI the conditions need to be checked and results are updated in the accounts. Smart Contract will be linked to the user interface using the Web3.js API.So, when the transaction is initiated, the smart contract will be executed and verification of the transaction will be carried out at every node in the network. If the transaction is successful then it would be reflected in the bank account balances of the sender and the receiver.

3.4 System design

3.4.1 Block diagram

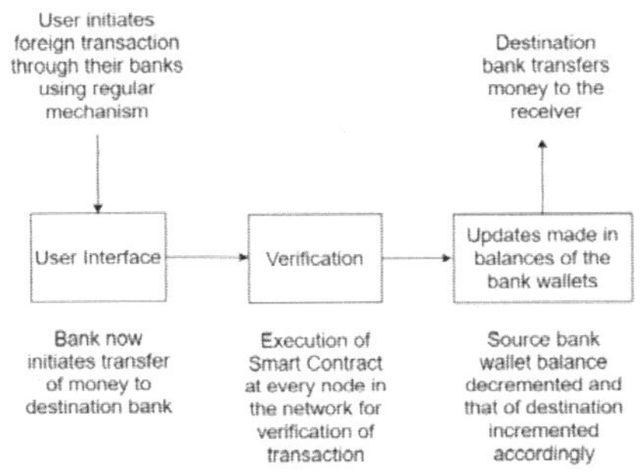

Figure 1: Block diagram

In the proposed Block Diagram in Fig. 1, the user initiates foreign transactions through their banks using the regular mechanism, then the bank initiates the transfer of money to the destination bank. This is followed by a verification process, where a smart contract is executed at every node in the network for verifying the transactions. The balances are updated accordingly in the bank accounts, and then finally the receiver receives the money from the destination bank.

3.4.2 Modular diagram

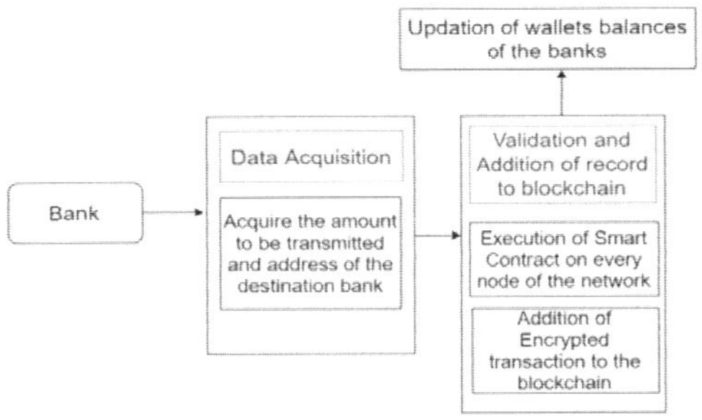

Figure 2: Modular diagram

In the Modular Diagram in Fig. 2, the user initiates the transaction, and the Data Acquisition module acquires the amount to be transmitted and the address of the receiver. This is then followed by validation and addition of records to the blockchain by executing smart contracts and adding encrypted transactions to the blockchain. Then finally, the bank balances are updated.

3.4.3 Flow chart

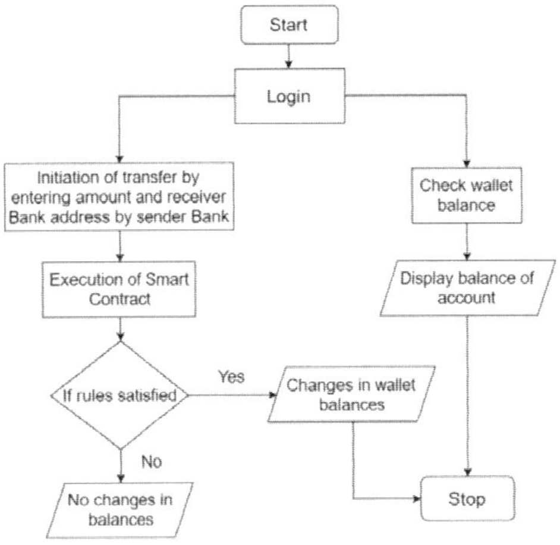

Figure 3: Flow chart

The proposed Flow Chart (Fig. 3) allows users to log in and check their balance, along with that they can also initiate a bank transfer to a particular address, which is followed by the organizing execution of smart contracts, and balances are updated only if smart contracts' rules are fulfilled. This makes sure that the system works with a set of strictly followed rules which prevents any fraud activities.

4. Implementation and results:

4.1 Ganache and Metamask:

Ganache is a personal blockchain for rapid Ethereum application development. To develop, deploy, and test dApps in a safe environment, Ganache can be used across this entire development cycle. To interact with the Ethereum blockchain we use a software cryptocurrency wallet called MetaMask. The following figures show the output after confirming the transaction on Metamask and executing it.

Figure 4. Live Blockchain on Ganache

Figure 5. Metamask Wallet Transaction Details

4.2 Relation between Gas Price and Mining Time:

In Fig. 6, the graph shows the relationship between the amount of gas price (in Gwei) VS the time it takes for a transaction to be mined. These numbers are accurate as of 24 April 2021 and might fluctuate based on many different factors daily. There's a lower limit to the amount of gas price assigned to a transaction, and any gas price below that value would result in a failed transaction as the miners won't mine that transaction due to its low gas price.

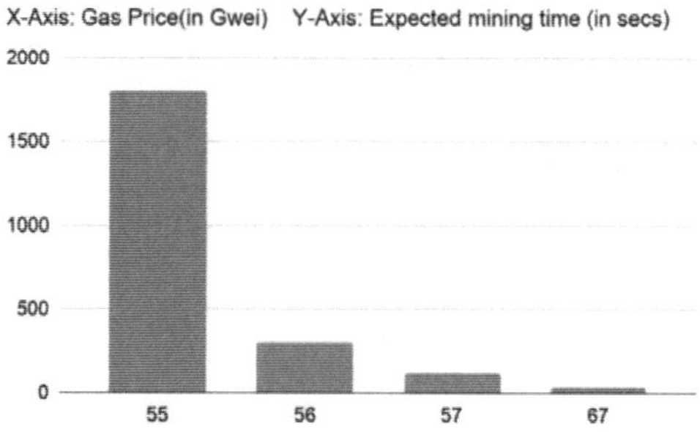

Fig 6. Gas Price v/s Mining Time

4.3 Cost analysis of various functions of the smart contract:

The smart contract of this project deals with the logic of sending the money from the sender's wallet to the receiver's wallet, along with updating the balances to both the party's wallets accurately. There are 3 main functions inside the smart contract, which are mentioned in Table 2.

We executed the transactions several times and noted the total ether expended during each of these function calls. We kept the Gas Price fixed at 20 Gwei, and executed the functions multiple times, and noted the number of gas units that were used by them, which further helped us to calculate the cost (in INR) associated with each function call, as shown in Table 3. Adding the cost of each function call, we found out that our system takes around ₹869 to transfer money from sender to receiver's wallet.

Table 2. List of smart contract functions

Function ID	Function Name	Purpose of function
F1	sendToForeignBank()	To transfer money
F2	addMoneyToWallet()	adds rupees to the wallet
F3	debitMoneyFromWallet()	debits money from wallet

*Ether to INR conversion rate is correct on the date accessed(24 April 2021). We fixed the gas price to be 20 Gwei.

Table 3. Average approximate costs across the smart contracts' functions

ID	Avg gas used	Avg gas limit	Total Cost (INR)*
F1	28397	42595	₹93.30
F2	207779	311668	₹679.85
F3	29373	44059	₹95.96

4.4 Comparison between transaction amount and fees for our system v/s Western Union :

Western Union is one of the popular choices for international money transfers. So, we did a comparative study between Western Union and our system in terms of how much transaction fees do they cost for a particular amount of money that is to be transferred internationally. We used the online service of Western Union to find out the transaction fees charged by them and compared it with that charged by our system, which we calculated based on the gas price and gas used by each of our transactions. However, this value doesn't give the exact amount, it's just an approximate amount taken by our system as we presented in Table 3.

From the graph in Fig. 7, it is concluded that, as the amount to be transferred, is increased, the transaction fees associated with it increase as well. Hence, making Western Union and other similar services, not an ideal choice when it comes to sending high amounts of money internationally. Our system, on the other hand, charges a fairly constant amount of transaction fees throughout.

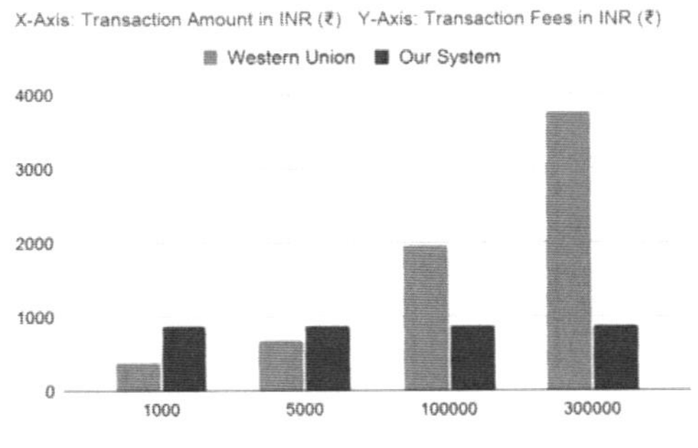

Figure 7. Transferred Amount v/s Transaction Fees

4.5 Risks Involved :

Few risks involved in the Foreign Exchange are as follows:

4.5.1 Counterparty Risk

It is the likelihood/probability that one of the parties involved in a transaction might be a defaulter on its contractual obligation. Investment, credit, and trading transactions are where this risk may exist. This system mitigates the counterparty risk by leveraging the smart contract protocol. This protocol is intended to automatically execute, document relevant events(legal), or control and perform actions according to the terms of the contract.

4.5.2 Replacement Risk

It occurs when counter-parties of a failed bank or a broker find they are at risk of not receiving their payment from the bank that has failed. We leverage the smart contracts' fallback function which sends back any ethers to the sender's account in case of a failed transaction due to insufficient funds.

4.5.3 Transaction risk:

It refers to the severe effect that foreign exchange rate changes may have on a transaction that is complete before the settlement takes place. The exchange rate is associated specifically with the time delay between entering into a trade or contract and then settling it. Forex trade occurs at any time between the 24 hour period which can cause currency's exchange rates to change before they are settled. This causes fluctuations in the forward rate of the forex. This system mitigates this risk by taking the timestamp of the transaction while it is initiated on the sender's side and fixes the exchange rate then, which is the considered exchange rate at the receiver's side.

5. Scope of blockchain:

5.1 Lacuna in Existing Systems :

The existing systems for Foreign Transfer have many drawbacks. Some of them are as follows :

- Current systems used for foreign exchange are the SWIFT technology, Transferwise.
- Slow transfer
- High transaction fees
- Intermediary banks involved
- Problems may arise due to the time zones.
- No fixed time duration for transfer
- Central organizing unit for transfer
- May not support the certain currency

These drawbacks can be eliminated to a great extent using Blockchain technology.

5.2 Benefits of Blockchain :

The proposed system using Blockchain has many advantages over the existing systems. Some of them are as follows :

- Blockchain-based system.
- Fast transfer
- Low transaction fee
- No intermediary banks
- No problems because of time zones.
- Fixed time duration for transfer
- Decentralized system
- Supports all currencies

6. Conclusion:

Thus, the proposed system using Blockchain will be better than the existing systems. The transfer will be less time-consuming. The transaction charges would be lesser than the transaction charges in the existing systems like Swift, TransferWise, etc. as there are no intermediary banks involved in the process. There would be no problems because of the time zones and the transfer time would be fixed. The proposed system will be flexible and can be made to support all the currencies. The deployed smart contract will give the functionality to transfer money from one account to another, making sure that the additions and deductions of money are correctly reflected in both the accounts, following the rules and the regulations of the Foreign Transfer. Due to the decentralized nature of the system, all the risks, as discussed above like Counterparty risk, Replacement risk, Transaction risk can be mitigated.

7. References:

[1] "*Blockchain White Paper*", National Archives and Records Administration (2019)

[2] "*Blockchain-Based Banking Application*" S. A. Patil, A. S. Mohire, A. U. Vibhute, R. S. Shaikh, J. A. Mali (2019)

[3] "*Bitcoin: A Peer-to-Peer Electronic Cash System*" S. Nakamoto, (2008)

[4] "*Enabling Cross-BorderHigh-Value Transfer Using Distributed Ledger Technologies*" The Bank of Canada (BOC) and the Monetary Authority of Singapore (MAS) (2019)

[5] "*Private Blockchain system for international currency exchange and hedging*" K. A. Andersen (2018)

[6] "*Cross Border Money Transfer using Blockchain-Enabled by Big Data*" R. Achanta, Infosys (2018)

[7] "*Investigating the link between transaction and computational costs in a blockchain environment*", A. Jabbar & S. Dani (2020)

[8] "*Global Perspective: Landscape of Blockchain usage in the continuous linked settlement in Banking*", A. Giri, H. Solanki (2020)

https://doi.org/10.1051/itmconf/20214003016

A Relative Analysis of Genetic Algorithm and Binary Particle Swarm Optimization for Finding the Optimal Cost of Wind Power Generation in Tirumala Area of India

Prasun Bhattacharjee[1,*], *Rabin K.* Jana[2], and *Somenath* Bhattacharya[1]

[1]Department of Mechanical Engineering, Jadavpur University
[2]Operations & Quantitative Methods Area, Indian Institute of Management Raipur

Abstract: Although India presently holds the global fourth-biggest instated Wind Power Generation (WPG) capability, it necessitates advancing more rapidly to satisfy the rising energy requirement of its evolving economy while restraining the consequential greenhouse gas emission. To accomplish the impressive target of setting up 140 GW WPG competence by 2030 as proposed by the Government of India, a greater number of financially viable wind farms are required to function all over the country without further ado. This paper focuses on finding the optimal cost for WPG in the Tirumala area of Andhra Pradesh. Genetic Algorithm (GA) and Binary Particle Swarm Optimization (BPSO) have been employed concurrently with four randomly chosen terrain conditions. The research outcomes demonstrate the superior capability of BPSO to attain the most optimal cost of energy.

1 Introduction

The ceaseless release of CO_2 with other greenhouse gases to the environment owing to diverse social activities is intensifying the mean air temperature and aberrant weather conditions which further initiate global climate change [1]. Renewable energy resources propose affluent alternates amid the mounting international apprehension for the restricted stock of fossil fuels and their hazardous penalties on the ecosystem [2]. Even during the Covid-19 pandemic in 2020, the usage of renewable energy grew by 3% while the requirement of every other fuel dropped universally [3].

Being the fourth-prominent power expending nation, India has an enormous capability to curb climate change. An ambitious goal of setting up the capability to generate 450 GW from renewable energy resources by 2030, including 140 GW wind capacity, has been already communicated by the Central Government of India. Global Wind Energy Council (GWEC) Report for 2021 testifies that 39 GW of wind power capacity could have been instated in India till February 2021. Being the second most populated nation on the Earth, it turns out to be enormously crucial for India to utilize more renewable resources to propel its developing economy in an eco-friendlier manner [4]. Globally, the cost of generating electricity from wind has shrunken sharply over the years [5]. That is why it becomes particularly essential for India to channelize added endeavors for meeting its incessantly mounting energy demand through wind power generation.

The National Institute of Wind Energy (NIWE) has already confirmed the wind power potential of 302 GW at 100 m for India [6]. As of 31 January 2021, India retains 10.3% of its 377260.67 MW total installed capacity from Wind Power Generation (WPG) units [7]. One of the most appealing features behind the prospect of wind power in India is that WPG the Levelized cost of electricity is almost 35% lower than most of the coal-fired power generation units and this is likely to further falloff by 7% by 2022 [8]. The progress of WPG in India over the past five financial years has been displayed in Fig. 1.

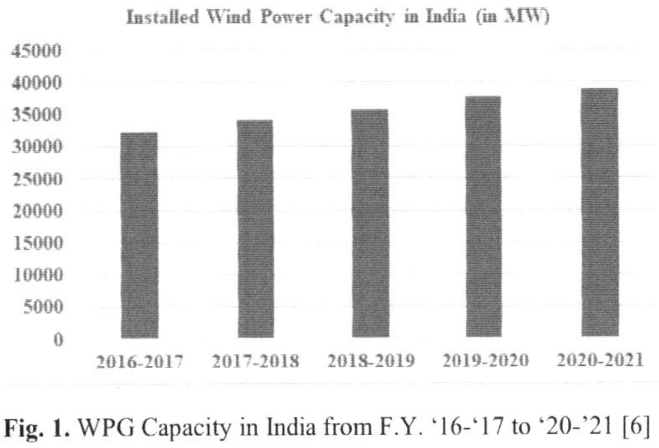

Fig. 1. WPG Capacity in India from F.Y. '16-'17 to '20-'21 [6]

* Corresponding author: prasunbhatta@gmail.com

To check the upsurge of the global mean temperature much lower than 2 degrees Celsius as proposed by the Paris agreement of 2016 which has been ratified by 195 signatories including India, greenhouse gas emissions are needed to be cut through the rapid green energy transition. India, being principally a coal-dependent country, can achieve its emission cut goals and lower its fossil fuel dependency by identifying more profitable opportunities for WPG [9].

This paper focuses on identifying the optimal cost of WPG for the Tirumala area of Andhra Pradesh. NIWE has identified the WPG capacity of Andhra Pradesh as 44228.60 MW WPG of which 4077.37 MW could have been commissioned up to 31 October 2019 [6,10]. To find the optimal cost of WPG, several critical facets like terrain condition, wind velocity, and various cost components related to Wind Turbine (WT) are to be evaluated. As the wind flow over a region is inherently arbitrary, the optimal placement of WTs in a wind farm requires extensive computational effort to trade-off between the power output and the Cost of Energy (CoE). Artificial Intelligence (AI) techniques like Genetic Algorithm (GA) and Binary Particle Swarm Optimization (BPSO) have been engaged simultaneously to minimize the optimal CoE per unit electricity. The optimization outcomes from both AI-based methods are compared to evaluate their relative efficiency.

2 Problem Formulation

2.1 Wind Power Calculation

According to the theory of aerodynamics, the kinetic energy absorbed by the WT is calculated as per Eq. (1).

$$P = \frac{1}{2}\rho A \vartheta^3 C_p cos\theta \quad (1)$$

where P is the kinetic energy accessible for extraction by WT, ρ is the air density, A is the wind wheel area, v is the speed of the wind flow, C_p denotes the power coefficient [as stated by the Betz's law, which is of the extreme value of 0.593] and θ is the error of yaw i.e., the angular deviation between the wind wheel revolution axis and the wind flow direction [11]. The wind energy obtainable for power generation for a solo WT positioned at (x_i, y_i) has been expressed in Eq. (2).

$$E = \int_\theta p(\theta) \int_v p_v^\theta (v; c_i, k_i|\theta)\eta(v) \quad (2)$$

where $p(\theta)$ denotes the possibility of airstream flow from directional angle θ, p_v^θ $(v;\ c_i,\ k_i|\theta)$ $\eta(v)$ is the Weibull distribution. The energy captured by individual WTs is added to estimate the yield of the wind farm [12].

2.2 Objective Function

Wind farms are required to stay economically viable through proficiently handling the CoE which is calculated as per Eq. (3) which has been recommended in the 22nd Genetic and Evolutionary Computation Conference held in 2015 [12].

$$CoE = \frac{\left(C_t N + C_s floor\left(\frac{N}{m}\right)\right)\left(\frac{2}{3}+\frac{1}{3}e^{-0.00174N^2}\right)+(C_{om}N)}{(1-(1+r)^{-y})/r} * \frac{1}{8760*P} + \frac{0.1}{N} \quad (3)$$

where C_t signifies the expense due to a WT and is considered as USD 750000. C_s characterizes the expenditure due to a sub-station and is considered as USD 8000000. N indicates the count of WTs of the wind farm and m indicates the number of WT per sub-station is considered as 30. C_{om} stands for the yearly operational and maintenance charge and is considered as USD 20000 per year. r denotes the fraction of interest and is considered as 3%. y indicates the expected operational life of the wind farm and is considered as 20 years [12].

2.3 Wind Flow Pattern and Terrain Condition

In the current research work, Tirumala (13°41′30.4″ N 79°21′34.4″ E) area of southern Indian state Andhra Pradesh has been selected for finding the optimal CoE for WPG [4]. The wind pattern from 2012 to 2017 as reported by Kumar et. al. has been displayed in Fig. 2.

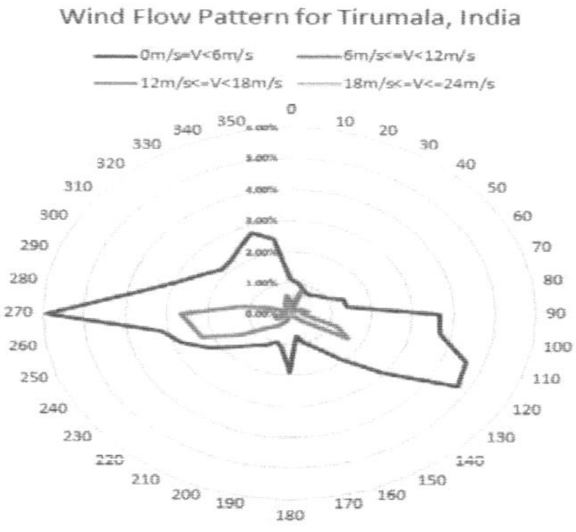

Fig. 2. Wind flow velocity distribution for Tirupati, India [4]

Four arbitrarily chosen terrain layouts have been considered for evaluation. Three layouts of them are with obstacles and the remaining one has no obstacles present in the wind flow path. The considered terrain conditions have been displayed in Figs. 3 – 6.

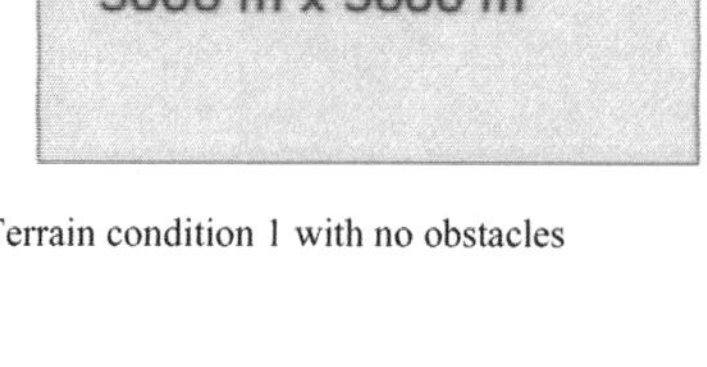

Fig. 3. Terrain condition 1 with no obstacles

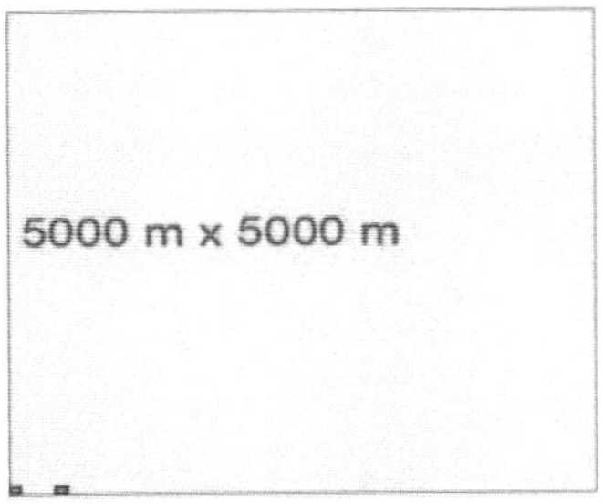

Fig. 4. Terrain condition 2 with two obstacles

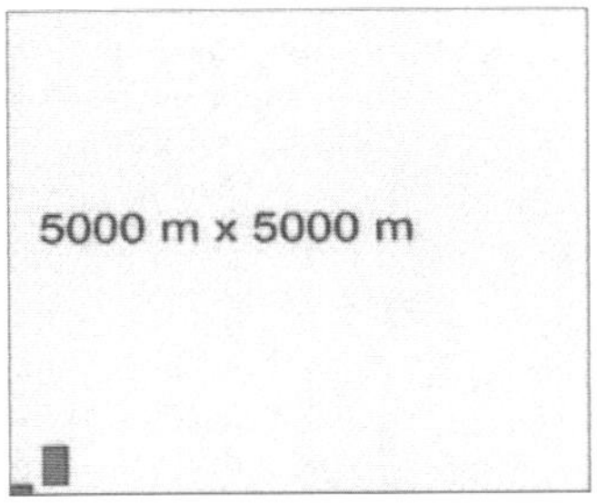

Fig. 5. Terrain condition 3 with two obstacles

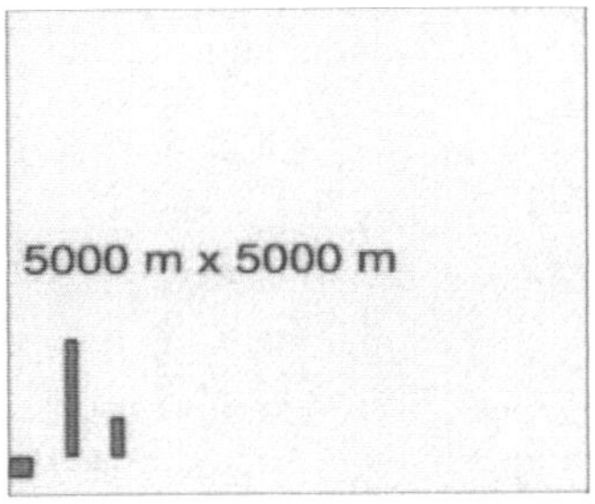

Fig. 6. Terrain condition 4 with three obstacles

3 Optimization Algorithms

In the present work. GA and BPSO have been engaged simultaneously to find the optimal CoE for the previously mentioned terrain conditions. GA is an AI-based search technique and has been extensively used in various engineering domains [13]. GA can be presented as follows.

1. Prepare the factors like population size, iteration count, chances for crossover, and mutation.
2. Set the population randomly.
3. Calculate the aptness of each chromosome.
4. Perform the arithmetic crossover process as follows:
   - 4.1 Choose a number arbitrarily within 0 and 1. If it is less than the chance of crossover, choose the parental chromosome for the crossover method.
   - 4.2 Instigate the crossover procedure.
   - 4.3 Examine the viability of the offspring.
   - 4.4 If the offspring is feasible, then integrate them into the recent population.
5. Complete the mutation method as follows:
   - 5.1 Choose a number arbitrarily within 0 and 1. If it is less than the chance of mutation, pick the chromosome for the mutation method.
   - 5.2 Institute the mutation procedure.
   - 5.3 Validate the fresh chromosome.
   - 5.4 If the created chromosome is feasible, integrate it into the recent population.
6. Test the aptness of the fresh entities created through crossover and mutation methods.
7. Pick out the most excellent result following the decision maker's predilection.

Particle Swarm Optimization (PSO) is another AI-enabled searching methodology that emulates the social actions of bees through communicating the information related to the universal and limited finest solutions [14]. The BPSO is a variant of PSO that considers every particle as a bit string. The site of a 'particle' can be amended by exchanging between 0 and 1 consistent with the particle velocity [15]. For the eth bit of fth particle, the velocity v_{ef} can be presented using Eq. (4)

$$v_{ef} = wv_{ef} + c_1r_{1f}(p_{ef} - x_{ef}) + c_2r_{2f}(g_f - x_{ef}) \quad (4)$$

where w indicates the inertia weight and can be determined using Eq. (5).

$$w = w_{max} - (w_{max} - w_{min})\frac{l}{l_{max}} \quad (5)$$

where w_{max} and w_{min} are the highest and lowest limits of inertia weight correspondingly. l is the existing count of recurrence and l_{max} is the highest count of reiterations. c_1 and c_2 are acceleration parameters. r_{1f} and r_{2f} are random variables ranging between 0 and 1. p_{ef} specifies the fth bit of the distinct most excellent position of the eth particle. g_f denotes the fth bit of the collective best position. The transfer function which is utilized to revise the bit value is formulated as per Eq. (6).

$$s(v_{ef}) = \frac{1}{1+e^{-v_{ef}}} \quad (6)$$

$$x_{ef} = \begin{cases} 1, & if\ rand() \leq s(v_{ef}) \\ 0, & otherwise \end{cases}$$

where *rand* is a function that arbitrarily chooses a number with uniform distribution [15].

4 Results and Discussion

For assessing the relative performance of GA and BPSO, a similar objective function of minimization of CoE has been considered. CoE has been measured in USD/kW. Both the optimization techniques have been iterated 50 times. Population size has been considered as 20 for GA and BPSO. 1.5 MW WT of radius 38.5 m has been employed for the current research. To reduce the wake loss effect, the gap between two adjacent WTs has been kept as 8 times the WT radius. The cut-in and cut-off velocities for the considered WT are 3.5 m/s and 20 m/s respectively. The research outcomes have been shown in Fig. 7.

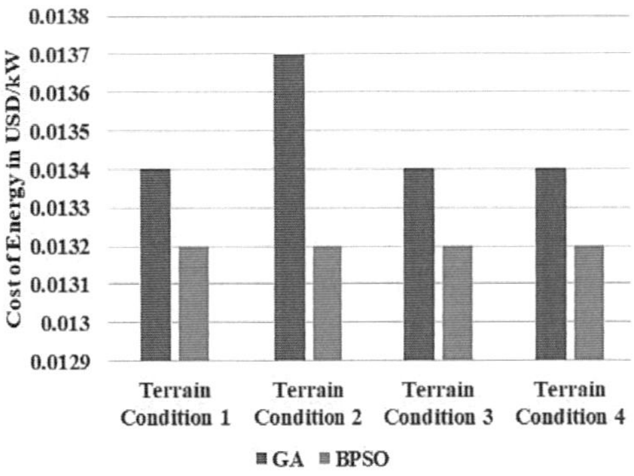

Fig. 7. Comparison of CoE for different terrain conditions

The plot displayed in Fig. 7 demonstrates that BPSO is more effective than GA for finding the optimal CoE for every terrain condition assumed with wind flow pattern for Tirumala. The BPSO algorithm is capable of finding CoE as low as USD 0.0132/kW consistently for every chosen terrain condition.

5 Conclusion

In the current paper, AI-based efficient techniques have been presented to find the optimal CoE for WPG in the Tirumala area of Andhra Pradesh state of India. GA and BPSO have been utilized concurrently to assess their relative effectiveness. The research outcomes prove the better suitability of BPSO in finding the optimal CoE for four randomly chosen terrain conditions. The present study will initiate novel opportunities for investigating the financial viability of WPG for different geographical locations worldwide.

The first author would like to express to the TEQIP department of Jadavpur University for providing a research grant for the current study.

References

1. B. Obama, Science **355 (6321)**, 126-129(2017)
2. P. K. Chaurasiya, V. Warudka, S. Ahmed, Energy Strategy Reviews **24**, 342-357(2019)
3. https://www.iea.org/articles/the-impact-of-the-covid-19-crisis-on-clean-energy-progress
4. M.B.H. Kumar, S. Balasubramaniyan, S. Padmanaban, J.B. Holm-Nielsen, Energies **12(11)**, 2158(2019)
5. R. Sitharthan, J.N. Swaminathan, T. Parthasarathy, *2018 National Power Engineering Conference* (IEEE, Madurai, 2018)
6. https://en.wikipedia.org/wiki/Wind_power_in_India
7. https://cea.nic.in/renewable-generation-report/?lang=en
8. https://gwec.net/india-wind-outlook-towards-2022-looking-beyond-headwinds/
9. https://www.downtoearth.org.in/blog/climate-change/the-world-is-warming-and-we-don-t-know-how-much-carbon-can-it-take-67251
10. https://nredcap.in/PDFs/Tenders/RE_Status_31_10_2019.pdf
11. Z. Wu, H. Wang, Energy Procedia **16**, 53 – 57(2012)
12. D. Wilson, S. Rodrigues, C. Segura, I. Loshchilov, F. Hutter, G. L. Buenfil, A. Kheiri, E. Keedwell, M. Ocampo-Pineda, E. Özcan, S. I. V. Peña, B. Goldman, S. B. Rionda, A. Hernández-Aguirre, K. Veeramachaneni, S. Cussat-Blanc, Renewable Energy **126**, 681-691(2018)
13. R. K. Jana, P. Bhattacharjee, International Journal of Design Engineering **7(2)**, 77(2017)
14. A. Duggirala, R.K. Jana, R. V. Shesu, P. Bhattacharjee, (2018). Sādhanā **43(1)**, (2018)
15. J. Liu, Y. Mei, X. Li, IEEE Transactions on Evolutionary Computation **20(5)**, 666–681(2016)

 https://doi.org/10.1051/itmconf/20214003017

Successive Image Generation from a Single Sentence

Amogh Parab[1], Ananya Malik[1], Arish Damania[1], Arnav Parekhji[1], and Pranit Bari[1]

[1] Dwarkadas J. Sanghvi College of Engineering, Mumbai, India

Abstract - Through various examples in history such as the early man's carving on caves, dependence on diagrammatic representations, the immense popularity of comic books we have seen that vision has a higher reach in communication than written words. In this paper, we analyse and propose a new task of transfer of information from text to image synthesis. Through this paper we aim to generate a story from a single sentence and convert our generated story into a sequence of images. We plan to use state of the art technology to implement this task. With the advent of Generative Adversarial Networks text to image synthesis have found a new awakening. We plan to take this task a step further, in order to automate the entire process. Our system generates a multi-lined story given a single sentence using a deep neural network. This story is then fed into our networks of multiple stage GANs inorder to produce a photorealistic image sequence.

Keyword - Deep Learning, Generative Adversarial Networks, Natural Language Processing, Computer Vision, LSTM, CNN, RNN, GRU, GPT-2.

1. INTRODUCTION

Tell me and I will forget, Teach me and I will remember, Show me and I will learn.

Imagining a visual story from a text description is the ultimate test of semantic, visual, and spatial world knowledge. Artificial Intelligence's arrival can be effectively used to create expert systems that will exhibit intelligent behaviour, provide solutions to complicated problems, and further help to develop stimulations equivalent to human intelligence within machines.With the advent of technologies within the NLP (Natural Language Processing) and CV (Computer Vision) communities, there exists a strong initiative to promote vision and language research. Until now this field has been heavily populated with research focused on the vision to language stream that has promoted visual perception to understanding diverse linguistic representations. With our work we intend to focus on the language to vision path which will allow vision understanding and offer new avenues of visual content creation. We intend to be able to formulate this tool to communicate and express stories, and correspondingly data better.

We propose generating a story and it's visualisation in the form of a sequence of images with a single textual input. Our system uses a transformer to generate a multi-lined story from the input. This story is then fed into a network of multiple two staged StackGANs to develop a sequence of images.

2. LITERATURE SURVEY

We have used Yitong Li et al [1] paper as our base paper, as it aims to solve a similar problem, that of story visualisation. The paper suggests that given a multi sentence paragraph, the need to generate an image storyline that must accurately represent the story, scene and context of the given input and focuses less on the continuity of the motion between images. In order to maintain the intertextuality between the images, the paper proposes the use of a Deep Context Encoder that identifies the context. It makes the use of a sequence of GANs which are modified two accommodate two discriminators at the image level and the story level. The main aim of the paper is to be able to generate images on a multi sentence paragraph input that comprises the entire story. The uniqueness of this paper lies in the initial fact that it proposes an entirely new task of Story Visualisation. The challenges faced by this system is to maintain consistency in the images as each sentence might not depict the entirety of the scene and to keep the layout of the scene similar, if not same throughout the entire sequence.

Zhang et al. [2] proposes the StackGAN to convert a single line of text to a singular image. This model uses a stacked architecture of GANs. They propose two techniques to implement the same - using a two stage architecture and multi-stage architecture. The two stage architecture is a conditional architecture using two GANs, the multi stage architecture advocates use for conditional and unconditional generative tasks by suggesting a tree like structure in the Stage II. The proposal of StackGANs [2] was a pathbreaking move as it allowed visuals of higher resolution to be generated with conditional GANs. The conditioning augment module allows conditional smoothening, that is it focuses on the details of the image as well. The two layer architecture is useful in our application of Story Visualisation, as it provides a two level discriminator check to ensure that the image doesn't lose the contextual consistency. Unlike the implementation in StackGAN v1 where both level 1 and level 2 discriminators are fed with the entire text embedding in order to ensure that the background text is not lost, our application demands a contextual support. Hence we are able to modify the stack based architecture proposed to ensure that the sentence is fed in discriminator 1 maintains the textual intricacy and the stage 2 discriminator can yield to increasing a higher resolution output.

Ouyang et al. [3] have posed a generation of realistic images from text description by cumulating two neural network structures of LSTM and GAN. Their approach surmounted the problem of creating high resolution images from text descriptions, which required multiple training of models. This new network thus helps to generate a sequence of images depicting evolution of image based on the sentence. This paper is distinctive from the other image generating ones because it blends two separate neural networks LSTM and Conditional GAN, thereby giving more detailed images. The problem solved by them was to create a sequence of images on a text description. Our model does the same thing, but we are to produce output of a sequence of images on the generated story. We are proposing to replace the LSTM in this paper with the GPT-2 transformer to help maintain more context along the generated sentences. The main challenge will be to not only check the generated image is acceptable but withal ascertain that the background story stays intact and does not divert from the actual scene. To do so we are using stacked GANs.

Clark et al. [4] represents an alternate approach to language modelling for neural text generation. It uses the concept of context to maintain consistency throughout the generated story. This context is maintained by the use of entities, whose representations are stored as vectors and updated regularly.This paper works on combining entity recognition with neural text generation. This is what sets this paper apart from other basic text generators using LSTMs or GRUs. It uses an additional vector for the entities which are then combined with the context of the previous sentence. The model provides better results than other models which are verified by Mention Generation and Pairwise Sentence Selection in addition to Human Evaluation.

Pawade et al. [5] covers automatic text generation which would be required to create more text from a given sentence. Their paper aims to generate a new story based on a series of inputted stories. It makes use of Recurrent Neural Networks - Long Short-Term Memory (RNN-LSTM) to generate a new sequence of text, that is, it forms sentences based on a provided input text. The generated stories received an accuracy rating of 63% by humans rating the stories on the basis of grammar, event linkage, interest level and uniqueness. However, the limitations of the system were that the stories generated were monotonous and repetitive.

3. PROPOSED ARCHITECTURE WITH MODULAR DESCRIPTION

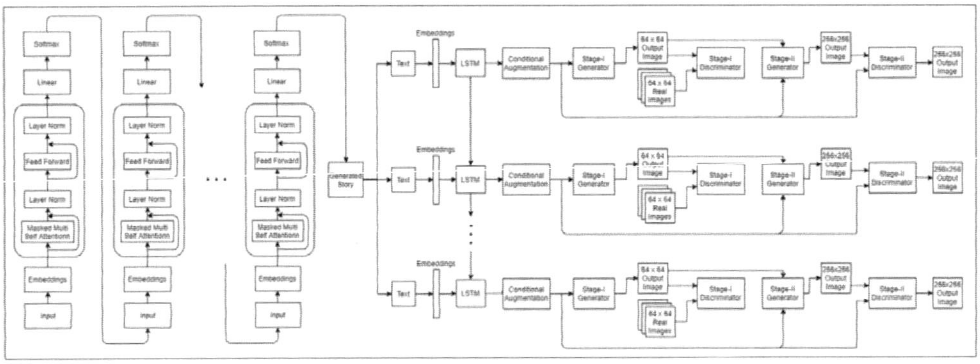

Fig. 3.1. Illustration of proposed model.

Fig 3.1 is the complete system architecture of the proposed model. It comprises GPT-2 decoder and StackGAN model which will be explained in detail in the following sections.

The user inputs a sentence, which will be used to generate a story. The sentence is mapped to vectors which are passed on to the GPT-2 decoder. The output of the decoder is passed through the convolution layer and then a softmax function is applied on it to predict the next word. This procedure continues till a story is generated. After the story is generated, it is passed on to the StackGANs model which will be used to generate images.

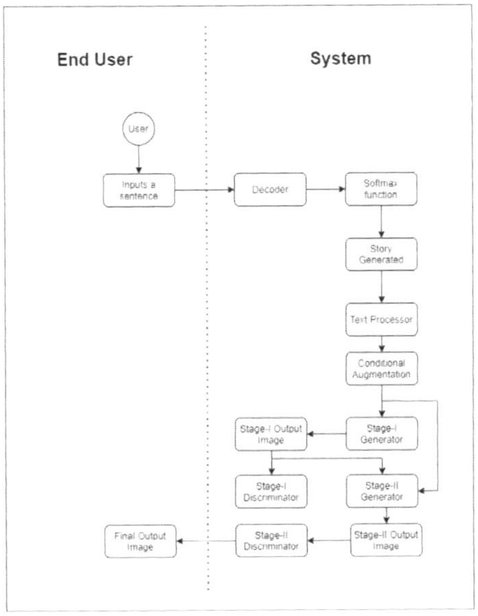

Fig. 3.2. Illustration of system flow.

The generated story is passed on through a text processor which keeps the story's background intact. A condition is augmented to the output of the processor and then is passed as an input to the StackGAN Stage-I generator. The generator generates an output which is verified with the Stage-I discriminator. These outputs are then passed on as input to the Stage-II generator along with the processed text. The output of the stage-II generator is validated with the stage-II discriminator and then the final image is returned. This process continues till the story ends. After the end of the story, sequences of images describing the story are presented to the user.

4. EXPERIMENTATION AND RESULT

Image Generation Module: Since we employ StackGAN to generate images from text, we experimented on the Bird Dataset by Caltech-UCSD 2010-11 dataset. Our method aims to use the two GAN architecture, like the StackGANS++. The first level, or the first GAN structure aims to model the scene, i.e. sketch the basic shape and colour, leading to the production of a low resolution image. In order to obtain a more detailed image in higher resolution, we now use the previously obtained image and sentence as input and generate the higher resolution image.

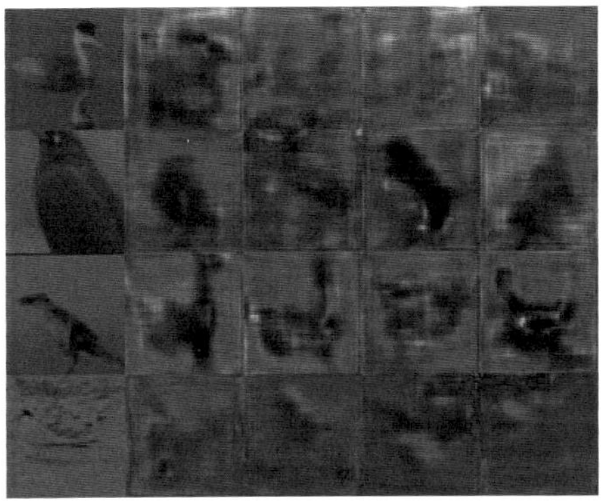

Fig. 4.1 Generated images using StackGAN.

Text Generation Module: We are using the pre trained GPT-2 transformer to perform the task of language generation. Transfer learning is used to fine tune the transformer architecture on our dataset to generate the new sentences. GPT2 is a decoder only transformer. It makes use of the concept "auto-regression". The GPT2 transformer uses masked self attention. We made use of a Flintstones Dataset comprising 25,184 video clips of the animated TV show The Flintstones.

```
Input text:
Fred is standing in a room. He is looking at something intently.

Generated story:

Fred is standing in a room. He is looking at something intently. In
the middle of the room a man sits down, and looks at the light that
shines on it. It is a single, blue-yellow. It is not the light of a
car, it is a light of his. He takes out his camera.
```

Fig. 4.2. Generated Story using GPT-2 transformer.

After working separately on the two modules, we have started integrating them. Experimenting on Flintstones dataset. we expect the following outputs.

```
Input text:
Dino is in the living room.

Generated Story:

Dino is in the living room. He lays on a rock chair in front of the
television. He looks up and into the other room. He jumps off of the
chair and lays on the floor.
```

Fig. 4.3. Expected Text Output on Flintstones Dataset.

 https://doi.org/10.1051/itmconf/20214003017

Fig. 4.4. Expected Image Output on Flintstones Dataset.

We will be using the PORORO-SV dataset which contains a video clip with a written description. A compilation of 40 video clips form a story, and in total there are around 16,000 video clips of 408 movie stories. Thus the entire dataset comprises 15,336 description-image pairs. We have preprocessed the data and is ready to use for training.

```
0-------------------------------------------------------
pororo is trapped on an ice pillar. the ice pillar is very high and in the middle of the air.
pororo is trapped on an ice pillar with pororo sled. pororo is scared and doesn't know what to do.
eddy is telling pororo not to move. eddy is leaving to call other friends.
eddy and friends hurriedly came back. pororo and eddy  are standing on a steep ice cliff.
eddy is asking rody to help. rody is stretching rody arms and legs to reach the ice pillar.

1-------------------------------------------------------
Pororo says be quiet to Crong. Crong holds a book out to Pororo.
Pororo asks what the book is while Pororo says that Pororo is sleepy.
Even if Pororo says Pororo is sleepy Pororo takes and opens the book to read it.
Pororo starts to read a book Crong sits beside Pororo.
Pororo skips the middle of the book so as to finish it quickly.

2-------------------------------------------------------
poby grabs a cookie and give a try.
harry eats a cookie and finds out that something is wrong.
pororo's face turns red. and crong drops cookies.
petty and loopy look at each other and wonder why.
pororo suggests another cookie to crong. and crong denies.

3-------------------------------------------------------
pororo found loopy easily. pororo is surprised to find loopy so easily. loopy drooped.
loopy is telling pororo and petty is okay. loopy dusting the earth off loopy clothes.
pororo is happy that poby found loopy. loopy is feeling sorry for loopy.
pororo is happy that poby found loopy. but loopy is feeling sorry for loopy.
pororo is entering into a large tree. harry and petty are watching pororo entering. harry and petty are on the top of the
staircase.
```

Fig. 4.5. Generated Story.

Fig. 4.6. Stage I Low Resolution Image

Fig. 4.7. Stage II High Resolution Image

Fig. 4.8. Ground Truth

5. CONCLUSION

We thus have been able to study story generation and visualisation as one concentrated task. The proposed model deals with the task jointly by generating the story from a single sentence and simultaneously generating the images for visualisation. Unlike other systems, this proposed solution allows the generation of the story and images under one system. Generation of text using GPT2 increases the controllability of the text and using StackGAN allows accurate depiction and photo-realistic images. The sequential generation model followed here allows context to be maintained. Using statistical and human evaluation factors we are able to determine that the proposed system is an improvement over current baseline models.

Today, story generation and visualisation are viewed as separate tasks, primarily because the maintenance of context across the streamlined procedure is tough. Thus, by developing more state-of-the-art context-maintenance and embedding systems we can greatly improve the efficacy of this task. By expanding the dataset beyond the realms of the Pororo cartoon we will be able to include stories that are applicable to current events. This will allow the applications of such a system to extend to media companies for the generation of stock images and in the publishing of children's books and comics.

REFERENCES

[1] Yitong Li, Zhe Gan, Yelong Shen, Jingjing Liu, Yu Cheng, Yuexin Wu, Lawrence Carin, David Carlson, Jianfeng Gao. StoryGAN: A Sequential Conditional GAN for Story Visualization. 2019

[2] Han Zhang, Tao Xu, Hongsheng Li, Shaoting Zhang, Xiaogang Wang, Xiaolei Huang, Dimitris Metaxas. StackGAN++: Realistic Image Synthesis with Stacked Generative Adversarial Networks. 2018

[3] Xu Ouyang, Xi Zhang, Di Ma, Gady Agam. Generating Image Sequence from Description with LSTM Conditional GAN. 2018

[4] Elizabeth Clark, Yangfeng Ji, Noah A. Smith. Neural Text Generation in Stories Using Entity Representations as Context. 2018

[5] Dipti Pawade, Avani Sakhapara, Mansi Jain, Neha Jain, Krushi Gada. Story Scrambler - Automatic Text Generation Using Word Level RNN-LSTM. 2018

[6] K.-M. Kim, M.-O. Heo, S.-H. Choi, and B.-T. Zhang. Deepstory: Video story qa by deep embedded memory networks. 2017.

[7] Ashish Vaswani, Noam Shazeer, Niki Parmar, Jakob Uszkoreit, Llion Jones, Aidan N. Gomez, Łukasz Kaiser, Illia Polosukhin. Attention Is All You Need. 2017

[8] Tanmay Gupta, Dustin Schwenk, Ali Farhadi, Derek Hoiem, Aniruddha Kembhavi. Imagine This! Scripts to Compositions to Videos. 2018

[9] Van-Khanh Tran, Le-Minh Nguyen and Satoshi Tojo. Neural-based Natural Language Generation in Dialogue using RNN Encoder-Decoder with Semantic Aggregation. 2017

[10] Xu Tao, Pengchuan Zhang, Qiuyuan Huang, Zhang Han, and Xiaodong He. AttnGAN: Fine-Grained Text to Image Generation with Attentional Generative Adversarial Networks. 2017

[11] Scott Reed, Zeynep Akata, Xinchen Yan, Lajanugen Logeswaran, and Honglak Lee. Generative Adversarial Text to Image Synthesis. In International Conference on International Conference on Machine Learning. 2016

[12] Zizhao Zhang, Yuanpu Xie, and Yang Lin. Photographic Text-to-Image Synthesis with a Hierarchically-nested Adversarial Network. 2018

[13] Fuwen Tan, Song Feng, and Vicente Ordonez. 2018. Text2Scene: Generating Abstract Scenes from Textual Descriptions. 2018

[14] Bowen Tan, Zichao Yang, Maruan Al-Shedivat, Eric P. Xing, Zhiting Hu. Progressive Generation of Long Text with Pretrained Language Models. 2021

[15] Zihang Dai, Zhilin Yang, Yiming Yang, Jaime Carbonell, Quoc Le, and Ruslan Salakhutdinov. Transformer-XL: Attentive language models beyond a fixed-length context. 2019

[16] Angela Fan, Mike Lewis, and Yann Dauphin. Hierarchical neural story generation. 2018

[17] Tingting Qiao, Jing Zhang, Duanqing Xu, Dacheng Tao. MirrorGAN: Learning Text-to-image Generation by Redescription. 2019

ITM Web of Conferences **40**, 03018 (2021) https://doi.org/10.1051/itmconf/20214003018
ICACC-2021

SpeakQL Natural Language to SQL

Dhairya Shah, *Aniruddha* Das, *Aniket* Shahane, *Dharmik* Parikh, and *Pranit* Bari

[1]Dwarkadas J Sanghvi College of Engineering

Abstract. Incorporating SQL questions from normal language is a long-standing open issue and has been drawing in extensive intrigue as of late. Natural Language Interface (NLI) is the confluence of Natural Language Processing (NLP) and Human-Computer Interaction, which allows interaction between humans and computers through the utilization of Natural Language. Here we are gonna deal with the problem of automatic generation of Struc- tured Query Language (SQL) queries. SQL is a database language for querying and manipulating relational databases. Despite the spectacular rise in the acceptance of relational databases, there is a fundamental limitaion to the ability to fetch data from those databases. One of the major reasons for this is the fact that the users of these relational databases need to comprehend convoluted structured query languages. In this body of work, we present an interface that allows users to interact with the databases using Natural Lanaguage as opposted to the conventional structure query languages.

1 Problem Definition

A vast amount of today's information is stored in relational databases. Writing and executing SQL queries is an integral part of relational database courses. Natural Language Processing (NLP) is one of the most active techniques used in Human- Computer Interaction. It is a branch of Artificial Intelligence (AI) that is used for information retrieval, machine translation and linguistic analysis. The main objective of NLP is to allow communication between humans and computers without memorizing commands and complex procedures. Posing questions to databases in the form of natural language is a very handy and straighforward technique of accessing data, especially for beginners who do not fully comprehend convoluted query languages such as SQL. This system focuses on the solution of the problems aris- ing in the analysis or generation of Natural language text or speech. The main purpose of Natural Language Pro- cessing is that the computer must be able to interpret a speech signal and perform the required action. For casual users who do not understand database query language like SQL, an easy method of data access is asking questions to databases in natural language and we will handle the com- plex queries for them. Hence it is not necessary for the user to have prior knowledge about the SQL queries. NLP is a technique that makes the computer understand the languages used by humans. While natural language may be easy for people to learn and use, it has been proved to be hard for a computer to master. Despite such challenges, natural language processing is regarded as a promising and important endeavour in the field of computer research. The goal is to inculcate the computer with the ability to understand and generate natural language so that the user can address his computer through text as though he was addressing another person. We all want to make better decisions by using data, which means we need to overcome major obstacles such as tedious dashboards or SQL scripts. One way to solve it would be to convert natural language to SQL like Google Translating is translating Japanese to English which is exactly what we aim to achieve.

2 Literature Survey

Initially [1] a seq2seq(encoder-decoder) approach along with reinforcement learning to reward the model based on the generated output was used. The input sentence is broken down into tokens which are used to generate the SQL query by the augmented pointer network. The input sequence is the concatenation of the column names, required for the se- lection column and the condition columns of the query, the question, required for the conditions of the query, and the limited vocabulary of the SQL language such as COUNT, SELECT etc. Three functions are utilized in order to produce the SQL query.

Each word is assigned a type as an entity from either the number, column or knowledge graph, by a type aware model [2] . It employs a sketch-based approach and views the task as a slot filling problem. By grouping dif- ferent slots in a reasonable way and capturing relationships between attributes. It uses an input encoder consisting of 2 bi-directional LSTMs to encode word and pair type. To encode word and type pairs of the question, they concatenate embeddings of words and their corresponding types and input them to BI-LSTM.

Natural Language is mapped to executable programs by focusing on Semantic Parsing [3]. This technique employs pointer networks, which creates the SQL query by encoding the question into continous vectors, using three channels. Gradually, the model understands when to generate an SQL keyword, a cell or a column name.In order to mitigate the ill-formed outcomes, the model integrates column cell relation as well. Pointer networks are originally introduced [4] which takes a sequence of elements as the input and out- puts a sequence of discrete tokens corresponding to posi- tions in the input sequence.

A different approach [6] where instead of generating SQL query it focuses on output cells containing data as answers to the given input natural language question. It leverages semantic parsing for question answering sys- tems over databases. It uses BERT to implement this ques- tion answering system. BERT is a new language represen- tation model, which stands for Bidirectional Encoder Rep- resentations from Transformers. BERT is a pre-trained deep model which uses a large amount of plain text to pre-train. Subsequently, we can fine-tune the pre-trained BERT representations with an supplementary output layer to produce state-of-the-art models for a wide range of tasks, such as text classification and question answering. First, it introduces the idea of finding the answer from a database without semantic parsing to solve the problem of hard label work for semantic parsing. Sec- ond, it uses a BERT-based model to implement the idea and achieve a baseline level experiment result.

3 Proposed Architecture with Modular Description

One of the most common approaches to converting natural language to SQL queries is to use a sequence-to- sequence style model, usually with reinforcement learning. But the most common issue that is faced in this kind of a model is the fact that order of the generated sql query matters. Acommon question is why does the order matter? - A very easy explanation can be given to explain the issue. Let's take a set of random numbers and we desire the indices of the numbers in such a way that they are sorted according to some given rule. The rule could be outputting them in ascending order, descending order or any arbitrarily fixed permutation. This yields exactly n! permutations, all of which are valid. If we generate our training set by randomly picking out any of those permutations, all the n! final output configurations will have equal probability for the same input vector X. Thus, this formation is much less statistically efficient. Thus it was found that restricting the order as much as possible for the outputs was always a better approach

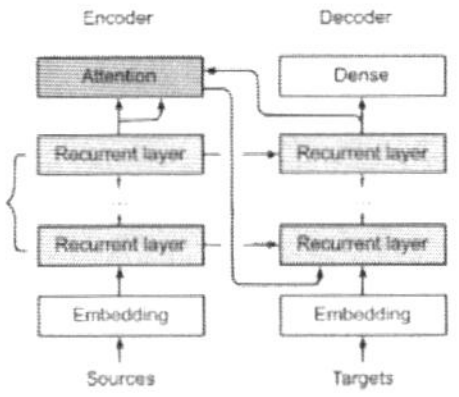

Figure 1. Sequence to Sequence model

To mitigate the issues of order in sequence-to- sequence-style models, usually reinforcement learning is used. The model trained using a policy gradient algo- rithm. But improvement due to the implementation of reinforcement learning is often limited. A sketch-based approach is used in this case to generate an SQL query from a sketch. The most challenging part is to generate the WHERE clause. The issue with sequence-to-sequence decoder is that the prediction of the next token depends on all previously generated tokens. The sketch-based ap- proach uses two devices: sequence-to-set and column at- tention. The former is designed to predict an unordered set of constraints instead of an ordered one, and the later is designed to capture the dependency relationship defined in the sketch when predicting. SQLNet only needs to fill in the slots in the sketch rather than to predict both the output grammar and the content. The sketch will highly align with the SQL grammar. The prediction of the value of one slot is only conditioned on the values of those slots that it depends on. Each dependency is a directed edge in the sketch. Thus our model can be viewed as a graphical model based on this dependency graph and the query syn- thesis problem as an interface problem on the graph. To synthesize more complex SQL queries, we simply employ a sketch that supports a richer syntax.

Sequence-to-set: Instead of generating a sequence of column names, we can simply predict which column names appear in this subset of interest. We compute the probability, Pwherecol(col|Q). Two LSTMs encode the column names and the questions do not share their weights. The dimension of column vectors and embed- dings of the column name and the natural language ques- tion have the same dimension of their hidden states of the LSTM. Column attention: Basically selecting which col- umn is the most relevant to predicting the real column. Eg. "number" is more relevant to predicting the column "no." but the token "player" is more relevant to predicting the column "player".

https://doi.org/10.1051/itmconf/20214003018

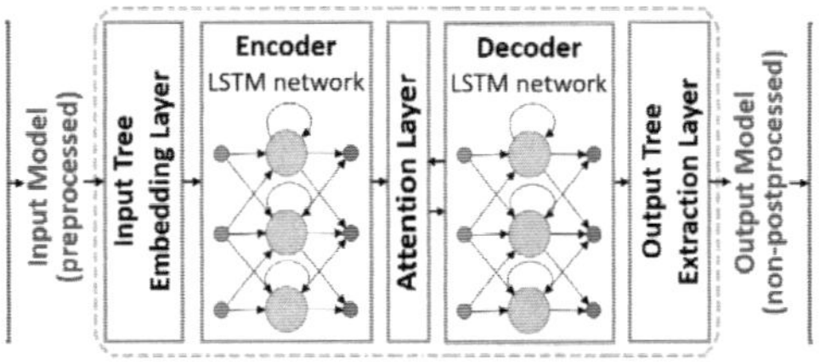

Figure 2. Encoder-Decoder with Column Attention Architecture

Selection of column slot: After the computations of the probabilities of columns being included in the set are done, SQLNet needs to decide which columns to include in the WHERE. In particular, a network is used to predict the total number K of columns to be included in the subset, and choose the top-K columns with the highest probability to form the column names in the WHERE clause. Selection of operator slot: For each column in the WHERE clause, predicting the value of its OP slot is a 3-way classification: the model needs to choose from =, <, >. Column attention is also used in selecting the operator for the slot. Selection of value slot: Here we need to use a sequence-to-sequence structure to generate the sub-string. The order of the tokens in the VALUE slot indeed matters. Selection of column slot: The selection of the column slot for the SELECT clause is the same as the WHERE clause except there's only one column.

The column names and natural language descriptions are treated as a sequence of tokens. Each token is represented as a one-hot vector and fed into a word embedding vector before feeding them into the bi-directional LSTM. The GloVe embedding is used in this case. Input encoding model details: Column names and natural language descriptions are treated as a sequence of tokens using the Stanford CoreNLP tokenizer to parse the sentence. Each token - one-hot vector and fed into a word embedding vector before being fed into the bi-directional LSTM. The "GloVe" word embedding is used. Weight sharing details: Components in SQLNet only share the word embedding. The model, also, has multiple LSTMs for predicting different slots in the sketch. Different LSTMs can be used for predicting different slots simultaneously, which will yield a better performance. Training the word embeddings during training yielded better results.

4 Expected Outcome

The output expected is the data obtained from the database for the given input. The output will be in the form of table similar to output generated from a SQL query. SQL query generated for the natural language will also be provided. Moreover, user's also have a quick access to recent queries entered by the user. This could as well be a step forward to making database usability accessible to people who have no knowledge how a database query works.

It can handle up to two levels of complexity (nested loops). Outcomes involving algebraic SQL functions like *avg, sum,* etc and most queries work smoothly via this approach. A primary issue with the approach is identifying correct tables to join in cases of semantic ambiguity. In addition to that, the use of well-defined constraints, prevents the system from generating queries involving select all operations using the asterisk (*) operator. While the results returned may still be the same, the query generated would be more elaborate relatively.

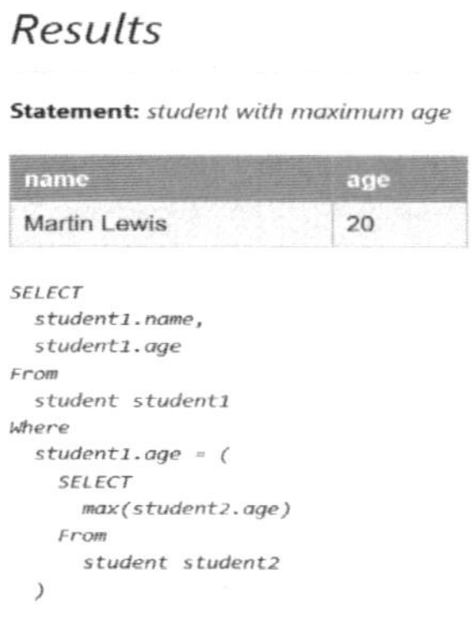

Figure 3. Generated Output (1)

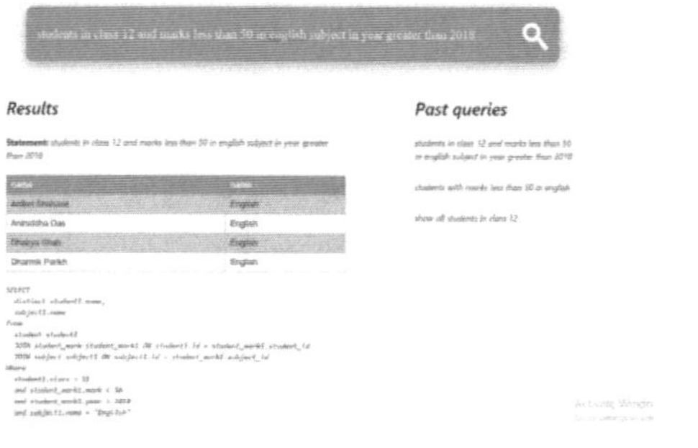

Figure 4. Generated Output (2)

References

[1]Victor Zhong, Caiming Xiong, Richard Socher, Seq2SQL: Generating Structured Queries from Natural Language using Reinforcement Learning (2017)

[2]Wonseok Hwang, Jinyeong Yim, Seunghyun Park, and Minjoon Seo: Exploration on WikiSQL with Table-Aware Word Contextualization. (2019)

[3]Tao Yu, Zifan Li, Zilin Zhang, Rui Zhang, Dragomir Radev, TypeSQL: Knowledge-based Type-Aware Neural Text-to-SQL Generation (2018)

[4]Yibo Sun, Duyu Tang, Nan Duan, Jianshu Ji, Guihong Cao, Xiaocheng Feng, Bing Qin, Ting Liu, Ming Zhou Semantic Parsing with Syntax- and Table-Aware SQL Generation (2018)

[5]Yibo Sun, Duyu Tang, Nan Duan, Jianshu Ji, Guihong Cao, Xiaocheng Feng, Bing Qin, Ting Liu, Ming Zhou

https://doi.org/10.1051/itmconf/20214003018

Semantic Parsing with Syntax- and Table-Aware SQL Generation (2018)

[6]Po-Sen Huang, Chenglong Wang, Rishabh Singh, Wen-tau Yih, Xiaodong He Natural Language to Struc- tured Query Generation via Meta-Learning (2018)

[7]Tong Guo, Huilin Gao Table2answer: Read the database and answer without SQL (2019)

[8]Ion Androutsopoulos, Graeme D Ritchie, and Peter Thanisch Natural language interfaces to databases–an introduction. (1995)

[9]Jonathan Berant, Andrew Chou, Roy Frostig, and Percy Liang Semantic parsing on freebase from question-answer pairs (2013)

[10]Ronghang Hu, Jacob Andreas, Marcus Rohrbach, Trevor Darrell, and Kate Saenko. Learning to reason: End-to-end module networks for visual question an- swering (2017)

[11]Christopher D. Manning, Mihai Surdeanu, John Bauer, Jenny Finkel, Steven J. Bethard, and David McClosky. The Stanford CoreNLP natural language processing toolkit. In Association for Computational Linguistics (ACL) System Demonstrations (2014)

Chatbot for Mental Well-being

Vivek More[1], *Viren* Koli[2], *Vignesh* Vijaykumar[3] and *Vimla* Jethani[4]

[1]Ramrao Adik Institute of Technology,Nerul,Navi Mumbai
[2]Ramrao Adik Institute of Technology,Nerul,Navi Mumbai
[3]Ramrao Adik Institute of Technology,Nerul,Navi Mumbai
[4]Ramrao Adik Institute of Technology,Nerul,Navi Mumbai

Abstract. With the world becoming more and more competitive every passing day, the number of people suffering from stress and other mental health issues is exponentially increasing. Even school children and senior citizens are becoming victims of stress and pressure in today's world. The mental health of an individual is as important as physical health is. But unfortunately, there is a lack of awareness and proper mental health facilities in today's society, which demands the individuals to fight against their odds without any support. Therefore, we propose a solution in the form of a chatbot that will act as a medium for the users to communicate with, let out their feelings and thus get relieved of the stress that is clogged up. A chatbot application will be available at all times at the user's expense and will also keep track of the user's mood over a span of time.

1 Introduction

Just like physical health, mental health also plays a vital role in determining the lifestyle of an individual. A healthy mind is crucial towards leading a healthy lifestyle. But unfortunately, due to lack of proper awareness and social stigma, mental health is not given the due importance it deserves. A major portion of the population succumbs to stress and pressure, due to the perpetual magnification of competition in day-to-day life. A large population, ranging from school kids to senior citizens, suffer from mental health disorders and a major portion of the same are either devoid of proper medical facilities and awareness or are reluctant to make use of the same. A poor mental health hampers the daily functioning of an individual and at times are accompanied by comorbidities like hypertension, heart disease, suicidal intentions, etc. So it is really important to maintain mental health as it is to maintain physical health in the modern era

In this project, we propose to build a therapy chatbot with the aim of providing a solution to the discussed dilemma. A therapy chatbot will provide an easily accessible medium that will provide a hassle-free service that is always at one's disposal. While a chatbot cannot replace a human therapist entirely, it can provide a platform for individuals to let out their emotions and get moral support whenever they feel distressed.

The motivation behind this project is that there are several situations wherein a person feels the need to talk to someone and share his/her feelings but cannot do so due to various reasons like lack of time, personal issues, non-availability of a listener, etc. Most of them are reluctant to go through therapy due to social or monetary obligations. We intend to make a chatbot which will be easy to chat with and available 24x7, so that the user can share his/her emotions while he/she feels stressed out or worried, which will help him/her improve their mood.

2 Related Work

2.1 Existing System and their Limitations

It is difficult for individuals suffering from panic attacks, anxiety attacks or undergoing any stress-related scenario, to find someone to have a conversation with in order to relieve the stress. It is practically unfeasible to find an ally to talk to every time an individual suffers from unpredictable panic attacks. Also, most individuals don't feel comfortable talking about their issues with a real person due to various reasons.

From the chatbot perspective, it would be apt to say that, presently, it is quite a herculean task to develop a chatbot that can mimic human conversation. Chatbot's performance almost always ends up deteriorating when it comes to long conversations. Most of the chatbots available in the market are restricted to a particular use-case and are mainly retrieval-based models.

2.2 Literature review

Miriam Romero, Cristina Casadevante y Helena Montoro [1] is a paper in which the fundamental aspects of designing a psychologist chatbot, along with the advantages and disadvantages of chatbots and the scientific guarantees that they need to fulfill. Eileen Bendiga,Benjamin Erbb Lea Schulze-Thuesinga, Harald Baumeistera,Karger publishers [2] is an IEEE paper in which Comparative analysis of most relevant psychological chatbots and their area of interest along with methodology used.

Takeshi Kamita ,Tatsuya Ito,Atsuko Matsumoto, Tsunetsugu Munakata and Tomoo Inou, Hindawi publishers [3] is an IEEE paper in which a different way to approach this with self-guided mental healthcare course and a system using a chatbot based on Digital-SAT method. Cameron Megav Bond O Neil [4] is a paper which includes chatbot for mental health counselling, its basic overview and future development possibilities for the same.

Kien hoa ly Annmaria Ly Gerhard Andersson [5] is an IEEE paper in which A study/analysis of a group of people who used SHIM chatbot which combines different techniques along with its limitations. Kyo-Joong Oh DongKunLee ByungSoo Ko Ho-Jin Choi [6] is a paper which focuses on emotion recognition, monitoring and conversation understanding on chat assistant.

2.3 Recurrent Neural Network (RNN)

A Recurrent Neural Network (RNN) is a type of Neural Network that recollects the past information that it has learnt and utilizes the same in its decision making process. Apart from recollecting the information that they have learnt during their training phase, RNNs also allow output of the previous step to be fed to the current step input. Therefore, unlike traditional neural networks wherein inputs and outputs are assumed to be independent, in RNN information from the previous inputs play a significant role in the decision making for the current step. Because of this ability of RNN to recollect and use the previous input, they are generally used when dealing with sequence or temporal data. Now, Vanilla RNN suffers from the 'Vanishing Gradient' problem due to which, the gradients (used to update the weights) shrink over time as they backpropagate and this affects the weight updates for long input sequences, often leading to a loss in the early context of the input. Therefore, variants of RNN like LSTM or GRU are preferred over RNN when dealing with long input sequences.

2.4 Long Short-Term Memory (LSTM)

LSTM (Long Short-Term Memory) network was proposed in 1997. It is a type of RNN that tries to overcome the drawbacks faced by Vanilla RNN by introducing the concept of gates and cell states. The information travels between the LSTM cells in the form of cell states. Each cell has three gates, viz. Forget gate, Input gate and Output gate The forget gate is used to decide what information from the previous cell state should be retained. Basically, the previous cell's hidden state and current input are passed through a sigmoid function that returns a vector with values in the range of 0 to 1. This is multiplied with the previous cell state that helps determine what information to forget and what to retain (where value closer to 0 means forget and value closer to 1 means retain). In the input gate, the previous hidden state and current input are passed through the sigmoid function and the output is multiplied with the output obtained by passing previous state and current input through the tanh function. This output determines which information is essential. This is then pointwise added with the updated previous cell state to get the current cell state. In the end, we have the output gate that helps determine what the hidden state of the current cell should be.

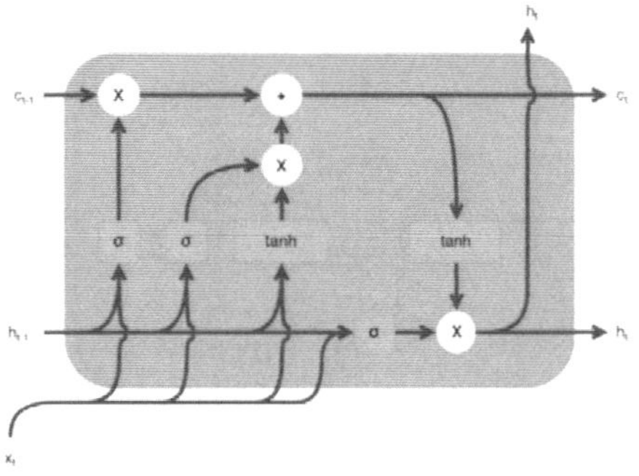

Source: Image from Wikipedia under CC BY-SA 4.0 License

Fig 1: LSTM Architecture

3 Proposed Work

The primary goal of the proposed project is to develop a generative chatbot that will communicate with the users with the aim of helping ease their state of mind and keep track of their mood over a span of time. The chatbot will act as a medium for users to let out their emotions and will be available 24/7 at the user's expense. The motive behind this project is to help people relieve their stress by providing a medium to talk to.

The proposed project will be implemented using the following two major modules:

SVM classifier: Will detect the mood parameter based on the emotional alignment of the input

Seq2Seq model: It makes use of two RNNs in the form of encoder and decoder to take in the tokenized input and generate an apt response for the same as output.

3.1 Dataset

For the mood classification model, we have used the Empathetic Dialogues dataset. The dataset is divided into 2 parts. The training file consists of a novel dataset of 12,424 conversations grounded in emotional situations. It contains 79,190 lines. Along with that, it contains 32 emotion labels mapped to each conversation.

The test file consists of 10,956 lines. This file is used in the evaluation of the model. Further while preprocessing, we have considered two main columns which are named as Emotion and Text for training and testing purposes.

For the seq2seq model along with the above dataset we have used the Counselchat dataset.

The CounselChat corpus has 10 unique columns which includes the question title, question text, answer text from multiple therapists along with the url for the same are available.

There are a total of 1659 lines of question-answers with the therapist.

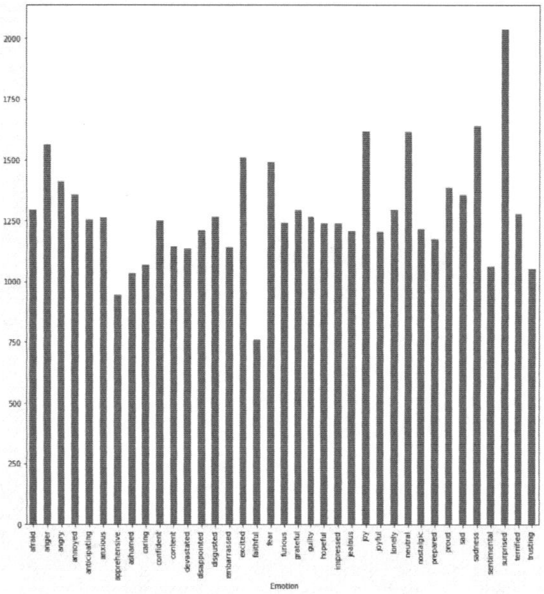

Fig 2: Empathetic Dialogues dataset

3.2 Data Preprocessing

First, both the training and the test file have been appended together, and then the sample function was used, to shuffle the sorted dataset randomly. Then, each of the 32 Emotions was mapped to a unique number. For preprocessing, regular expressions are used to remove the special symbols, punctuation, links, etc. After that, the strip function was used to remove the blank spaces from the processed text and remove the stop words and then Word level tokenization was used on each sample. Snowball Stemmer is used on the tokenized samples to further improve it. This preprocessed data is then passed to a vectorizer which converts the data into a sparse matrix. In our case, we used a Term Frequency - Inverse Document Frequency vectorizer. Along with vectorization, samples are converted into unigrams+bigrams. Then the vectorized data is passed on to a balancer which balances the data using the random oversampling technique. For this, we used a RandomOverSampler provided by the imblearn library. Then the balance data is then split with a ratio of 80% training and 20% testing data.

We have used Regular Expressions to clean the dataset, like removing punctuations, resolved clitics into their base terms, etc. Then, we have appended to the input and target sequence lists, sentences from the dataset with lengths less than or equal to 30. Next, we have appended <SOS> and <EOS> tags to the end and beginning of the target list entry, respectively. Then from these lists, we have created a vocabulary dictionary of words having a frequency greater than 5 in the input and target sequence lists. The vocabulary assigns a unique number to each word. Inverting this vocabulary, we get the inverse vocabulary that the model uses to generate the sentences. We have used 'GloVe (Global Vectors for Word representation) Embedding' to create the embedding layer. Word embedding in simple terms is a representation of text wherein words with similar meanings have similar representation, which means in the vector space, two words having similar vectors will be placed closely. The GloVe is an unsupervised learning algorithm in which the ratio of co-occurrence probability of two words is encoded as vector differences in the vector space.

3.3 Architecture

We have used Keras to develop the Seq2Seq model and linear SVM model of the Sklearn library. We have also incorporated the additive Attention mechanism proposed by Bahdanau et al., to improve the efficiency of the Seq2Seq model.

3.3.1 SVM Classifier

Support Vector Machine is a Supervised learning algorithm, which primarily helps with the classification problem. In layman terms, the basic functionality of SVM is to find out from the many possible decision boundaries in an N-dimensional space that classifies the input, the one that has the maximum distance from data points of both the classes (maximum margin). Such a decision boundary is referred to as the Hyperplane. So, in our project, the SVM algorithm will be generating a predetermined numerical output that corresponds to a particular mood based on the input provided by the user, thus determining the mood of the user's chat.

3.3.2 Seq2Seq Model

The Seq2Seq model is an encoder-decoder model developed to work with data with some sequential order, like sentences and time series, and convert the same into another sequence of information required by the user. It makes use of Recurrent Neural Networks internally for converting the input sequence to the required output sequence. But, Vanilla RNNs are not often used, because of the vanishing gradient problem they suffer from, which affects the weight updates of the same for long input sequences, often leading to a loss in the early context of the input. Hence, variants of RNN like LSTM or GRU are used when dealing with long input sequences.

The Seq2Seq model has an encoder-decoder architecture, where both the encoder and decoder comprise stacked RNN cells or its variants (like LSTM or GRU). The embedded version of the input is fed to the encoder, which then operates upon the same and outputs the context vector, a fixed size feature vector encoding the context of the input sequence. This is then fed as input to the decoder, wherein each decoder cell takes in the previous decoder cell's hidden states and output as input, works upon the same to generate the output, that is fed to the softmax layer. Then the probability distribution over the entire vocabulary is generated, and finally, using the argmax function, the index of the desired output word is determined.

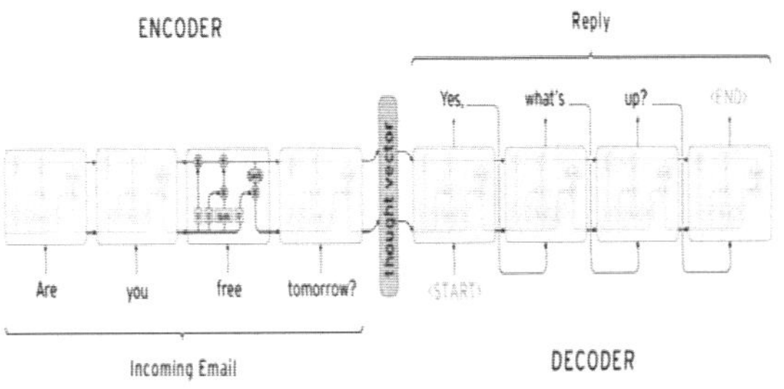

Source: https://www.haptik.ai/tech/taking-neural-conversation-model-production/
Fig 3: Seq2Seq Architecture

3.3.3 Attention Model

Now, the encoding of the entire input sequence into a fixed-length context vector leads to the loss of a significant amount of information when dealing with a long input sequence. To deal with this drawback, we make use of a mechanism known as the Attention mechanism. We make use of the Attention mechanism proposed by Bahdanau et al., which is an additive attention mechanism. Instead of creating the context vector from the output of the last LSTM cell of the encoder, for every decoder cell, a context vector is generated as the weighted sum of all hidden states of the encoder and the attention weights calculated for the decoder cell. This helps the decoder cell focus on the apt part of the input sequence rather than working on a fixed-sized encoded version of the input. For every decoder cell, the alignment score tells how well the jth position input and output at position "i" match, calculated based on the previous decoder's hidden state ($s_{(i-1)}$) and the hidden state h_j of the encoder.

$$e_{ij} = a(s_{i-1}, h_j) \qquad (1)$$

Then, using the softmax function, the attention weights are calculated from the alignment scores as

$$\alpha_{ij} = exp(e_{ij}) / \sum_{k=1}^{Tx} exp(e_{ik}) \qquad (2)$$

The context vector is calculated as the weighted sum of the hidden states of the encoder and the attention weights as

$$C_i = \sum_{j=1}^{Tx} a_{ij} h_j \qquad (3)$$

In this way, the decoder can focus on the local context of the input part it needs to pay attention to, rather than working with a fixed-length encoding of the entire input, wherein significant information might get lost for long sequences.

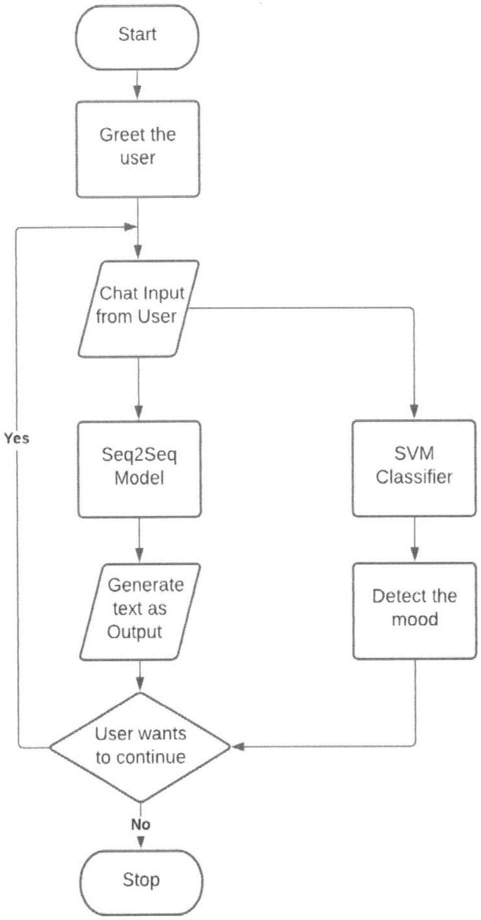

Fig 4: Flowchart of the model

4. Result

Presently, we have implemented the mood classifier module with the help of SVM classifier. Since we have implemented the mood classifier as a standalone module for our current implementation, the code first does the task of tokenization using TfidfVectorizer to obtain individual words from the provided input and produce a vector of weights where each weight is associated with a word in input. The classifier is then fed with this vector, based on which it outputs the mood of the input.

Table 1: Training details of Seq2Seq Model with Attention

Number of Epochs :	30
Accuracy	96.53
Validation Accuracy	90.69

https://doi.org/10.1051/itmconf/20214003019

The Seq2Seq model has been trained for 30 epochs and it gives an accuracy of 96.53 and validation accuracy of 90.69.

The overall working of the system is in the following manner:

1. The user first provides the system with the chat input.
2. The system tokenizes the input to get the n-gram array which is then further modified to get the vector of weights linked to each pair
3. The vector is fed as input to the mood classifier, which determines the mood of the input given by the user.
4. The user input is fed as input to the Seq2Seq model, which, based on its training with a dedicated dataset, generates an apt response for the given input. The same will then be provided to the user as the output of the system.

Table 2: Comparison of model with and without Attention

Seq2Seq Model without Attention (Number of Epochs: 30)	Seq2Seq Model with Attention (Number of Epochs: 30)
Bot: Initial Greeting! User: i am feeling sad Bot: i am so sorry User: i had a fight with my friend Bot: oh wow what did you meet User: will you be my friend Bot: i love her	Bot: Initial Greeting! User: i am feeling sad Bot: why User: I had a fight with my friend Bot: thats terrible User: will you be my friend Bot: yes

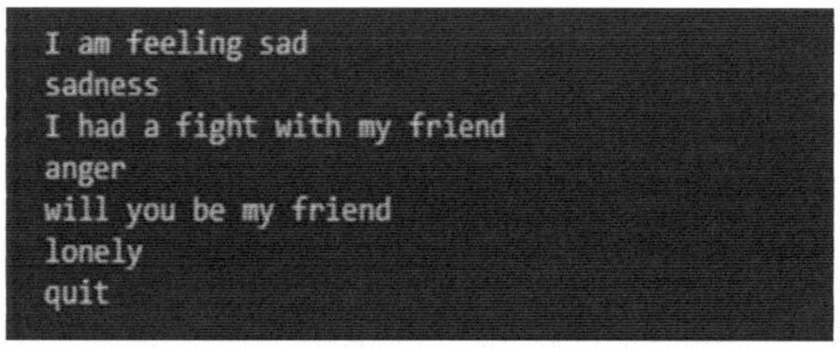

Fig 5. Output of SVM model

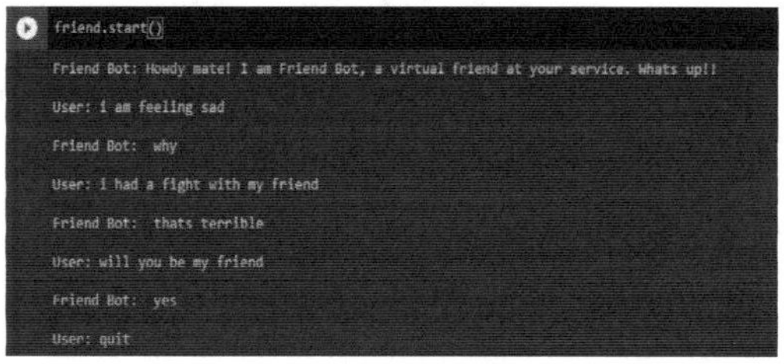

Fig 6.Seq2Seq Output

5 Conclusion

This project serves the primary goal of developing a virtual intelligent environment in the form of a chatbot for the users to communicate with and express their emotions, with the intention of helping the user relieve of the bottled stress and improve their mood to a certain extent. The chatbot will provide a simple interface for the users to put in the input chats, to which the chabot will do mood classification and generate an apt response using the Seq2Seq model that will then be displayed to the user with the aim of helping the user ease his/her mood. Another hallmark of the system is its 24/7 availability at the user's expense.

From the future work perspective, the system can generate a weekly report of how the user's mood varied throughout the week which will give the user an idea of how much he/she is improving over time. Furthermore, a more distinctive dataset with real therapy conversations will further improve the working of the system.

6 Reference

[1] HOW TO CREATE A PSYCHOLOGIST-CHATBOT Miriam Romero, Cristina Casadevante y Helena Montoro Universidad Autónoma de Madrid, 2020 https://www.researchgate.net/publication/338320380_COMO_CONSTRUIR_UN_PSICOLOGO-CHATBOT

[2] The Next Generation: Chatbots in Clinical Psychology and Psychotherapy to Foster Mental Health – A Scoping Review S. Eileen Bendiga,Benjamin Erbb Lea Schulze-Thuesinga, Harald Baumeistera, 20 August, 2019 https://www.karger.com/Article/Pdf/501812

[3] Towards a chatbot for digital counseling Cameron • Cameron • Megaw • Bond • Mulvenna • O'Neill July 2017

[4] A Fully Automated Conversational Agent for Promoting Mental Well-Being: A pilot RCT using mixed methods Kien Hoa Ly, Ann-Marie Ly, Gerhard Andersson,2017 https://www.researchgate.net/publication/320369339_A_Fully_Automated_Conversational_Agent_for_Promoting_Mental_Well-Being_A_pilot_RCT_using_mixed_methods

[5] A Chatbot for Psychiatric Counseling in Mental Healthcare Service Based on Emotional Dialogue Analysis and Sentence Generation Kyo-Joong Oh, DongKun Lee, ByungSoo Ko, Ho-Jin Choi, 2017

[6] Dataset collect and information about Counsel Chat from source-https://github.com/nbertagnolli/counsel-chat/tree/master/data

[7] Towards Empathetic Open-domain Conversation Models: a New Benchmark and Dataset Hannah Rashkin, Eric Michael Smith, Margaret Li, Y-Lan Boureau https://arxiv.org/abs/1811.00207

[8] NEURAL MACHINE TRANSLATION BY JOINTLY LEARNING TO ALIGN AND TRANSLATE Dzmitry Bahdanau Jacobs University Bremen, Germany KyungHyun Cho Yoshua Bengio∗Universite de Montr ´

[9] Jeffrey Pennington, Richard Socher, and Christopher D. Manning. 2014. GloVe: Global Vectors for Word Representation. [pdf] [bib]

Algorithmic trading for a buy-sell platform: study and comparison

Aishwarya Sakhare[1], *Nihar* Mhaskar[2], *Vicky* Mishra[3], *and Madhuri* Chavan[4]
[1]Department of Information Technology, RAIT, Nerul Email: aishwarya.sakhare56@gmail.com
[2]Department of Information Technology, RAIT, Nerul Email: niharmhaskar266@gmail.com
[3]Department of Information Technology, RAIT, Nerul Email: vickymishra1967@gmail.com
[4]Department of Information Technology, RAIT, Nerul Email: madhuri.chavan@rait.ac.in

Abstract. With the high-paced change in the world, the market trends are changing fast and so do the technological innovations. The market is constantly moving and being affected by a number of external factors, making it difficult for investors to make decision. To solve this problem, a technique called Algorithmic Trading is being implemented widely. It refers to use of computer algorithms to make trading decisions without human intervention. In this paper, the algorithm used to predict the stock prices is the LSTM model which is a methodology of neural networks.The existing models are less accurate as they do not take technical factors into consideration. LSTM model, having error of only 0.00036%, overcomes this limitation and also takes temporal factor into consideration.

Keywords: Algorithmic Trading, LSTM, Neural Network, Ridge Regression

1 Introduction

Algorithmic Trading, also known as automated trading, refers to use of computer algorithms to make trading decisions without human intervention. This computer algorithm uses a defined set of instructions to carry out a trade. These instructions are based on some mathematical model involving price of stock, timing, volume, etc. Not only does this make trading instant and accurate but also reduces the chances of mistake by ruling out human intervention. Algorithmic Trading is categorized in two types - High Frequency Trading (HFT) and Low Frequency Trading (LFT). High Frequency Trading is a type of algorithmic trading that is dependent on ultra-fast connections to monitor the market quotes and adjust orders accordingly. It is stated that HFT is done by firms located near data centres when large orders are exchanged. Low Frequency Trading is a slower form yet isfaster than trading floor specialists and other non- algorithmic techniques. LFT focuses on design theory to implement a strategy. The three factors to design an LFT are sentiment analysis, technical analysis and machine learning techniques. [1, 2]

Traditionally, traders calculate gain and loss based on market trends. Also, there are methods which depend on statistical considerations. The Spread or Pair trading method uses two financial measures that are co-related. Another method called Multi-Spread Trading is based on quantitative algorithms and does not depend on market trends. [3] One more methodology widely used is the Machine Learning model. It refers to a set of algorithms and rules that apply mathematical computations to learn from big sets of data and improve its efficiency. [4] This is based on various trend indicators such as Exponential Moving Average (EMA), Average Directional Movement Index (ADX), Relative Strength Index (RSI), Average True Range (ATR), Moving Average Convergence Divergence (MACD), Moving Average Convergence Divergence (MACD Signal). [5]

To predict the future stock prices with maximum accuracy, Long Short Term Memory (LSTM) model is implemented. This is a technique that uses Recurrent Neural Network (RNN).Artificial Neural Networks are information processing models that are inspired by the

https://doi.org/10.1051/itmconf/20214003020

human brain. The neuron collects signals from input channels named dendrites, processes information in its nucleus, and then generates an output in a long thin branch called axon.[6] A RNN is a type of ANN that is designed to take temporal dimension into consideration by having a memory (internal state) (feedback loop). RNN contains a temporal loop in which the hidden layer not only gives an output but it feeds itself as well. An extra dimension is added which is time. [7]

2 Methodology

The work presented has taken into consideration the technical indicators that affect the Stock Market. Examples of such technical indicators include Exponential Moving Average(EMA), Moving Average Convergence Divergence (MACD), and Slow Moving Average (SMA), etc.[8]. The work demonstrates the use of the LSTM model considering the technical indicator, SMA. As shown in Fig.1, the results obtained by this prediction model are used by the trading algorithm to decide whether to sell the stock or buys it, based on a threshold value.

Technical indicators are mathematical calculations carried out on history of stock price. They are generally used as visual aidsto help identify the direction the market is going to change in. These technical indicators are also fed to the model through secondary input branches. Since, this model allows to provide more than one input, the data of technical indicators can be provided as well. Hence, the model works better in terms of consideration of sudden changes in the market.

Recurrent neural networks are good at working with time series data. This also applies to LSTM due to their ability to remember the past. This is done by using recurrent nodes to store the old states. RNNs and LSTMs can work on current information as well as information the network had seen in previous steps. Also, implementation of a LSTM cell is handy as they're a default layer in TensorFlow's Keras.

This work has used a basic technical indicator called Slow Moving Average (SMA). To calculate the simple moving average of a stock, the average of the closing price of the stock of previous n days is taken. The processed data containing the calculated SMA goes through the data transformation as required for the model of LSTM. The data is scaled to fit within the values 0 to 1.

SMA is not a time-series piece of data, hence, it is not passed through the LSTM. Instead it is fed to the model at the last step,in before the final prediction is made. It is given as input it in to the penultimate 64-node dense layer.So, we will need a model with two inputs, one concatenation layer and an output. As we had divided the data into training and test dataset, similarly the technical indicator's data is split. The inputs in this model are the original training data set and the training dataset of technical indicator's data. The output of these two models is then given as concatenated input to the third model which gives the final output.

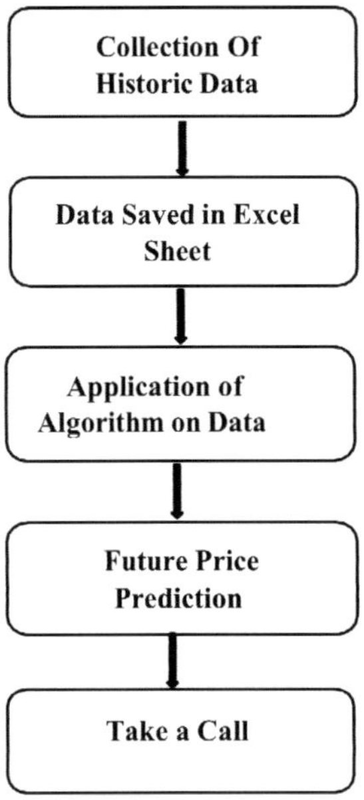

Fig. 1. Formulation and Working

These predictions are then used to perform asset allocation. This refers to process of dividing a given capital in a set of assets such that risk is reduced and profit is increased. For this, the CAPM formula is used. Finally, a trading algorithm named Bollinger bands is used to depict when the system will buy or sell the stock.

Bollinger bands are an indicator which uses slow moving average as the middle band and two other bands named Upper band and Lower band. Lower band is calculated as difference of mean and two times the standard deviation. Similarly, Upper band is calculated as mean summed with two times the standard deviation. [9]

This strategy is used to identify buying and selling position based on the position of stock price as compared to the bands. When the position of stock price is less than the lower band the algorithm will long the stock, while it will short the stock when the position of stock price is greater than the upper band. [10]

 https://doi.org/10.1051/itmconf/20214003020

3 Results

The methodology used in this work helps to predict the close price of different stocks on n^{th} day (where n is pre-decided value).
On evaluating the model, the evaluation results show an error of0.00036655570147559047 which will magnify on upscaling the predicted data.

The model is evaluated for various stocks. The graph shown inFig.2 depicts the result for close price of Apple.

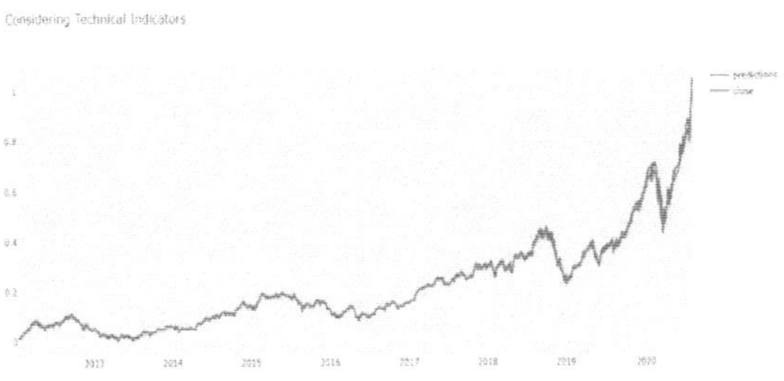

Fig. 2. Prediction using LSTM including technical indicators.

The prediction of close price of Apple stock and comparison with original price. This prediction model uses technical indicators and works well with training data and test data. It also manages to avoid the shift problem that occurs in existing prediction methods.

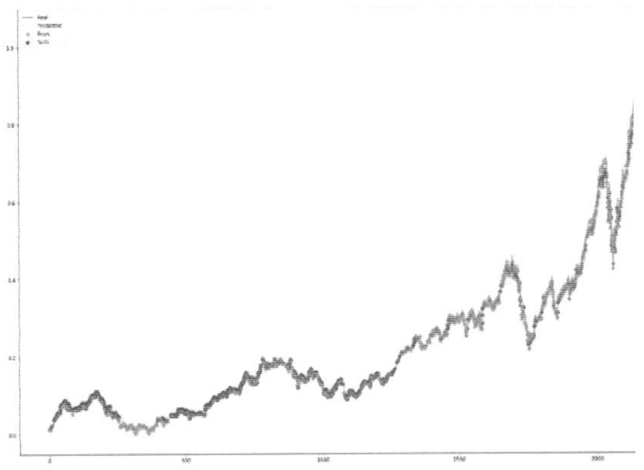

Fig. 3. Buying and Selling points

The green points, as shown in Fig.3 are plotted when the application buys the particular stock and red when it sells the stock. This is done using an algorithm for trading based on a threshold value.

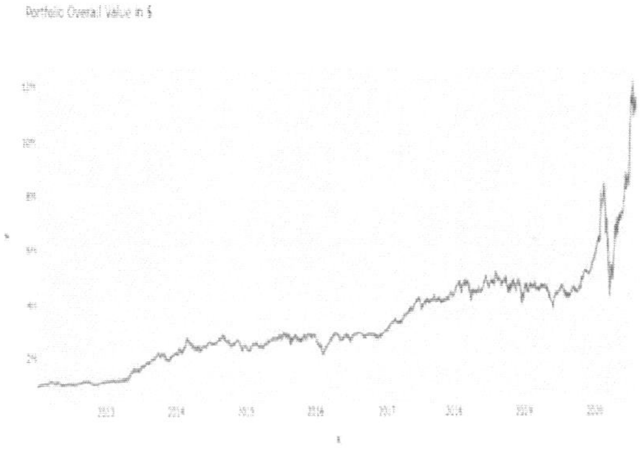

Fig. 4. Portfolio Overall value

The graph shown in Fig.4 depicts how asset allocation formula used performs in 8 years. It shows the net value ofportfolio at the end of time span as compared to the beginning. This basically helps users to understand how much return is earned on the given portfolio.

Table 1. Comparison of Algorithms

1.	Algorithm	Ridge Regression	LSTM Time Series Model	LSTM using Technical Indicators
2.	Performance Parameters	'Close' value of stock	'Close' value of stock	Slow Moving Average'Close' value of stock
3.	Score	(Accuracy) 0.7950028030821767	(Error) 0.000197624322026968	(Error) 0.00036655570147559047
4.	Advantages	This model reduces the problem of overfitting.	The prediction results obtained using the LSTM model are accurate to a large extent and are better than the earlier model.	This model proves better than previous model as it avoids the right shift as seen earlier.
5.	Disadvantages	The model becomes less sensitive to the variations of the independent variable.	The model faces the problem of being shifted to right by a constant amount.	The drawback of this model is it is not too much efficient in capturing the sudden changes. [11]

3.1 Comparison with Ridge Regression

In this work, the first model used for predicting was the Ridge Regression Model which reduced the problem of overfitting. In ridge regression, 'Least sum of squares' is applied to obtain the best fit line. Ridge regression works by applying a penalizing term (reducing the weights and biases) to overcome overfitting. It works by attempting at increasing the bias to improve variance (generalization capability). This works by changing the slope of the line, the model performance might be little poor on the training set but it will perform consistently well on both the training and testing datasets. Alpha is part of the penalty term in the equation of ridge regression. As Alpha increases, the slope of the regression line is reduced and becomes more horizontal. The model becomes less sensitive to the variations of the independent variable. [12]

On implementing this model, the score defined by the model is as follows-

Ridge Regression Score: 0.7950028030821767

3.2 Comparison with LSTM

This model uses cells, called as long short term memory cells. These are small models that are structured to store memory in large neural networks. Inside the LSTM cell, a recurrent node is
present that allows this function. The recurrent node has an edge with a weight equal to one. The edge is looping backon itself to ensure that at every feedforward iteration the cell retains the information from all the previous steps.

Feedforward Neural Networks (vanilla networks) map a fixed size input (such as image) to a constant size output (classes or probabilities). A drawback in Feedforward networks is that they do not have any time dependency ormemory effect. RNN can recall what happened in the previous time stamp so it works great with sequence of text. LSTM networks work much better compared to vanilla RNN since they overcome the vanishing gradient problem. [13, 14]

On evaluating the model, the result shows an error value of 0.000197624322026968 on normalized values. On upscaling the values, the error is magnified as well.

4 Conclusion

Share market is a mode of investment for buyers and sellers across the globe. A software which predicts the rise and fall of stock prices can help investors plan their investment wisely. Hence, to fulfil the above requirement, it is important to predict the future trend of a stock using the large pre-existing database. To do this successfully, various factors that determines the state of market need to be understood. To automate this process, algorithmic trading is implemented, which is basically an execution process based on a written algorithm.

As shown in Table 1, this work depicts a comparison of three methods, starting with the traditional Machine Learning Model that uses Ridge Regression. Further to increase the accuracy, time series model called Long Short Term Memory is used. This performs better as the technique is a part of Recurrent Neural Network that imitates the working of brain. Even though the error given by this method is very less, the model fails to capture real-life events. Hence, to make it efficient in terms of predicting future trends, the LSTM Model is fed more data using the technical indicators.

References

[1] F. Obeid Algorithmic Trading: High Frequency & Low Frequency Trading, FERM 6303: Financial Assets and Markets, (2018).

[2] E. Boehmer, K. Fong, J. Wu, International evidence on Algo. Trading, Journal of financial and quantitative analysis,(2015).

[3] M. Bogliardi, F. S. Canepa, G. Frisina , Multiple Spread Trading 60, An investment methodology neutral to financial market trends English Version, (2019).

[4] P. Rys, R. Slepaczuk, ML in algo. trading strategy opt. implementation and efficiency, Faculty of Economic Sciences, University of Warsaw, (2018).

[5] E. Sorhun, How Is a Machine Learning Algorithm Now-Casting Stock Returns? A Test for ASELSAN, Springer Science and Business Media LLC, (2019).

[6] N. Budhani, Dr. C. K. Jha, S. K. Budhani—Prediction of Stock Market Using Artificial Neural Network, International Conference of Soft Computing Techniques for Engineering and Technology (ICSCTET), (2014).

[7] M. Adya and F. Collopy, How effective are neural networks at fore-casting and prediction, A review and evaluation, J. Forecasting, vol. 17, (1998).

[8] M. R. Vargas, C. E. M. dos Anjos, G. L. G. Bichara and A. G. Evsukoff, "Deep Leaming for Stock Market Prediction Using Technical Indicators and Financial News Articles," International Joint Conference on Neural Networks (IJCNN), (2018).

[9] Lo, Andrew W., H. Mamaysky and J. Wang. "Foundations of Technical Analysis: Computational Algorithms, Statistical Inference, And Empirical Implementation," Journal of Finance, (2000).

[10] F. Allen, R. Karjalainen, using genetic algorithms to find technical trading rules, Journal of financial economics, vol. 51, issue 2, (1999).

[11] K. Chen, Y. Zhou and F. Dai —A LSTM-based method for stock returns prediction: A case study of China stock market, IEEE International Conference on Big Data (Big Data), (2015).

[12] H. Duzan, NSBM. Shariff, Ridge regression for solving the multicollinearity problem: review of methods and models, Journal of Applied Sciences, 15, (2015).

[13] T. Gao, Y. Chai and Y. Liu, "Applying long short- term memory neural networks for predicting stock closing price," 8th IEEE International conference on software engineering and service science(ICSESS), (2017).

[14] D. Wei, "Prediction of Stock Price Based on LSTM Neural Network," International Conference on Artificial Intelligence and Advanced Manufacturing (AIAM), (2019).

ITM Web of Conferences **40**, 03021 (2021) https://doi.org/10.1051/itmconf/20214003021
ICACC-2021

Deep Learning techniques for effective diagnosis of Alzheimer's disease using MRI images

Prajakta Tambe[1], *Rutuja Saigaonkar*[1], *Nidhi Devadiga*[1] *and Ms. Pallavi H. Chitte*[1]

[1] Department of Computer Engineering, University of Mumbai, Ramrao Adik Institute of Technology, India.

Abstract— The determination of Alzheimer's disease (AD) from neuroimaging data such as MRI has been immensely researched over the last few years. Recent advances in deep learning from a computer perspective have advanced in that research. However, the general limitations of such algorithms depend on the large number of training images, as well as the need to carefully perform the construction of deep networks. In past few days deep learning strategies have found great achievement in the analysis of medical imaging. But very little has been done in the use of deep learning strategies to turn up and differentiate Alzheimer's disease. We are building a deep convolutional network and demonstrating performance on the ADNI-Alzheimer's Disease Neuroimaging Initiative Dataset. We present a deep convolutional neural network to recognize Alzheimer and differentiate according the current phase of the disease.

I. INTRODUCTION

Senile dementia commonly known as Alzheimer's Disease, is a typical cause of dementia in the elderly. Although there isno current cure, research is underway with the development of new drugs. Alzheimer's disease has an impact on people in many ways [1]. Patients struggle with memory lapses, disorderliness, they struggle when they speak, read or write. In due course, they may not even be able to remember their relatives. They may fail to do daily tasks like brushing their teeth or even combing hair. Due to this, it causes that person to become troubled or angry or to lose their way to home. This disease can also cause death in adults. There are 3 main stages of Alzheimer's disease - moderate, mild and very mild. Diagnosis of Alzheimer remains inaccurate till the patient reaches moderate Alzheimer. However, advanced recognition as well as categorizing of Alzheimer is important for proper medical care and prevention of brain tissue impairment. More recently, deep learning strategies have led to develop high level performance in a wide range of fields also taking into consideration medical imaging analysis. We present a deep convolutional neural network for the diagnosis of Alzheimer using MRI brain data analysis. The biggest advantage in MRI scans is that they generate high- resolution localizations and, the image details are prominent for diagnostic purposes. The most obvious characteristic of Alzheimer's disease is neuron loss, followed by cerebral atrophy extending from the signature regions of the Alzheimer's disease (e.g., hippocampus and amygdala) to the entire cortical region, which MRI scan can identify.

II. BACKGROUND STUDY AND RELATED WORK

Making Alzheimer diagnostic classification and detection models is a very difficult task. Yet there is some remarkable research add this area. In [1] they proposed a method of multiclass classification between NC, MCI, and AD patients using multi-categorical data using deep learning and used both an ANN and a 1D CNN for this purpose. They demonstrated the effectiveness of multi-categorical data in classification as well as to compare ANNs with CNNs in this type of task. In [2] paper introduces a deep learning model for the determination of Alzheimer using Multi-Class Alzheimer's Disease and segmentation using Brain MRI Data. Design a deep convolutional network and illustrate performance in the OASIS- Open Access Series of Imaging Study database. The model is inspired by the Inception V4 network. Post preprocessing, the input is transferred to the stem layer. They rearrange the last layer of SoftMax for the diagnosis of Alzheimer's disease. The SoftMax layer consists of 4 distinct classification categories: moderate, non- demented, mild and very mild Alzheimer. The network utilizes an MRI scan as input and removes layer intelligence element representation from the first to the final exit layer. [3] This study consists of two main tests. They therefore validated an existing and highly effective method of acquiring Support Vector Machine (SVMs) as an initial trial. As those results were unsuccessful, a comprehensive learning process (Convolutional Neural Network) was proposed as a second test. [4] This study shows a group of deep-belief networks is made when the ultimate prediction is decided by the electoral system. 2 basic frameworks in deep learning and 4 different voting schemes are used and compared, due to which the development of strong structures within which discriminative elements are calculated in an unregulated manner. The resulting method was tested on a large dataset from the ADNI- Alzheimer's disease Neuroimaging Initiative.

 https://doi.org/10.1051/itmconf/20214003021

III. PROPOSED APPROACH

In this project we are using the dataset from ADNI -Alzheimer's Disease Neuroimaging Initiative database. The data obtained for this project has a directory structure with four different classes, that is, Moderate Demented, Very Mildly Demented, Very Mildly Demented, Non-Demented. On the input data preprocessing and data augmentation process is applied. In this process, we first crop the right and left part of the hippocampus present in the brain MRI images. The training dataset includes 8192 sample MRI.

images, whereas the validation dataset includes 1023 samples. The test dataset includes 1279 sample MRI images. To generalize the very specific MRI images, we apply augmentation operations in the form of rescaling, rotation, width shift, height shift, shearing, brightness changes, zooming and horizontal flipping.

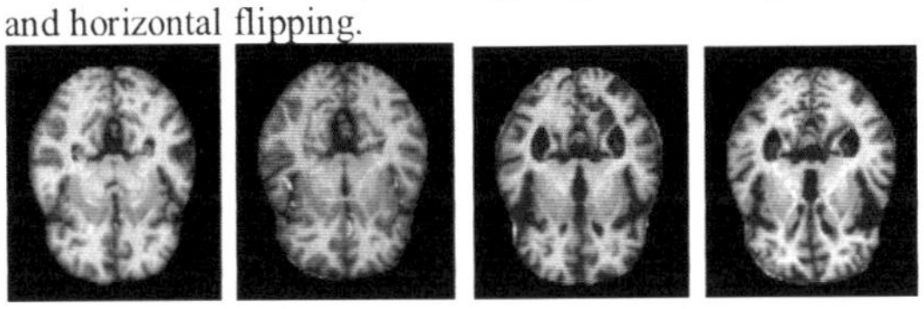

Fig 3.1: MRI Images (ADNI dataset)

For the training purpose using CNN algorithm a model is built in which every layer such as convolution layer, Relu layer, pooling layer from the model performs feature learning on the augmented data to obtain a feature map. It extracts the features using the feature detector and puts them in the feature map and preserves spatial relationship between pixels. The pooled feature map matrix is flattened into vectors and it is fed to the fully connected neural network.

Convolution Neural Network

Neural Networks are wont to classify various functions like image, sound, words. differing types of Neural Networks are used for a range of purposes, for instance to predict the sequence of words we use Recurrent Neural Networks accurately as LSTM, similarly for image classification using the Convolution Neural Network. In the next section, we are going to build a basic CNN blockchain. A standard Neural Network consists of 3 layers Input layer: The layer where we provide inserts in our model. the quantity of neurons during this layer equals the entire number of elements in our data (the number of pixels within the case of an image)

Hidden Layer: Inputs from the first layer is inserted into this layer. There may be many hidden layers counting on our data size and model. Hidden layers can have different numbers of neurons usually larger than the number of features. The output of every layer includes the matrix duplication of the previous layer by the readable weights of that layer then by inserting the readable bias followed by the activation function that produces the network out of line.

Output Layer: The output within the hidden layer was then incorporated into an entry function like a sigmoid or SoftMax that converts the output of every phase into a possible point for every phase. Data is then entered into the model and extracted from each layer is found this step is termed feed forward, and so calculates the error using error function, other common functions of cross entropy error, square loss error etc. After that, we return and detached the model with the output of the output. This step is termed back propagation which is employed to cut back losses.

CNN's biggest advantage compared to its predecessors is that it automatically acquires key features without any human guidance.

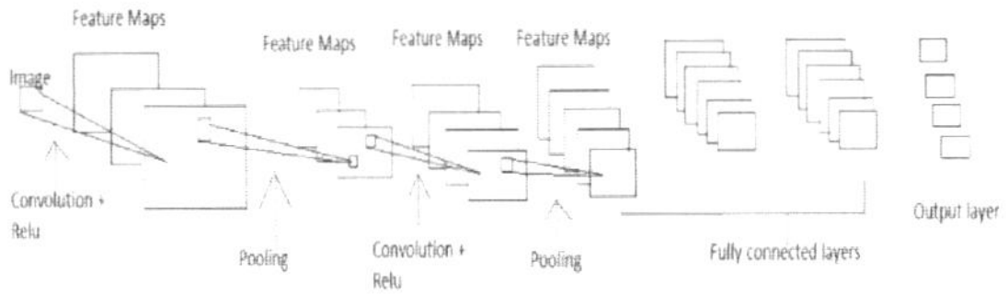

Fig 3.2: Convolutional Neural Network

Based on the extracted features it is used to classify Alzheimer Disease and its types which are: 1. No dementia 2. Very mild AD. 3. Mild AD. 4. Moderate AD

In this project, the algorithm has 3 steps. the initial step is preprocessing and then augmentation of data, the next step is extracting features from input images, and then the dementia classes are classified. The classification of dementia stages is done using a CNN based approach inspired by VGG19 for our fundamental model. The VGG19 was revised and one extra Conv layer was inserted within the model which proved efficient to know the maximum features from a small dataset. Within the algorithm, 2 modified VGG19 layers were working parallel with fourteen Conv layers, 5 max-pooling, 3 batch normalizations and 3 Gaussian noise. The rationale for the model is that the parallel work to bring out the vital features. Successive lateral layers enhanced the classification accuracy.

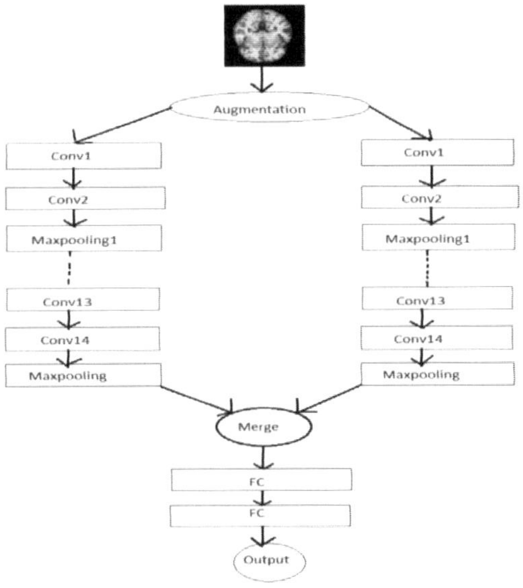

Fig 3.3: Augmentation

We implemented the dataset on VGG19 andDensenet121 and ResNet50.After evaluating the best accuracy which was obtained in modified VGG19, we deployed the model using TensorFlow, and used it in a flask-based server application for easy access for users.

https://doi.org/10.1051/itmconf/20214003021

IV. EXPERIMENTS AND RESULTS

Model One: For the first model, we built a complex CNN architecture. It consists of three simultaneous layers of Conv2D and Max Pooling followed by a Flatten layer, which is again followed by two Dense Layers.

Model: "sequential"

Layer (type)	Output Shape	Param #
conv2d (Conv2D)	(None, 222, 222, 32)	896
max_pooling2d (MaxPooling2D)	(None, 111, 111, 32)	0
conv2d_1 (Conv2D)	(None, 109, 109, 64)	18496
max_pooling2d_1 (MaxPooling2	(None, 54, 54, 64)	0
conv2d_2 (Conv2D)	(None, 52, 52, 128)	73856
max_pooling2d_2 (MaxPooling2	(None, 26, 26, 128)	0
conv2d_3 (Conv2D)	(None, 24, 24, 128)	147584
max_pooling2d_3 (MaxPooling2	(None, 12, 12, 128)	0
flatten (Flatten)	(None, 18432)	0
dense (Dense)	(None, 128)	2359424
dense_1 (Dense)	(None, 4)	516

Total params: 2,600,772
Trainable params: 2,600,772

Fig 4.1: CNN Architecture

The model was trained for 40 epochs at the end of which it had a loss of 0.0934, and a categorical accuracy of 0.9643 for the training dataset. The graph for it as shown below:

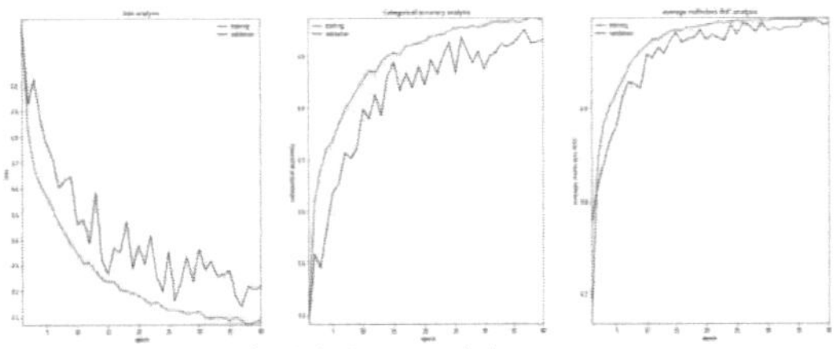

Fig 4.2: Loss and Accuracy analysis

Model Two: Next we used the Transfer Learning model. For this model, we used ResNet50 followed by several layers to modify. It had the below architecture:

resnet50(Functional)	(None, 7, 7, 2048)	2358771:
dropout (Dropout)	(None, 7, 7, 2048)	0
flatten_2 (Flatten)	(None, 100352)	0
batch_normalization (BatchNo	(None, 100352)	401408
dense_4 (Dense)	(None, 64)	6422592
batch_normalization_1 (Batch	(None, 64)	256
activation (Activation)	(None, 64)	0
dropout_1 (Dropout)	(None, 64)	0
dense_5 (Dense)	(None, 64)	4160
batch_normalization_2 (Batch	(None, 64)	256
activation_1 (Activation)	(None, 64)	0
dropout_2 (Dropout)	(None, 64)	0
dense_6 (Dense)	(None, 64)	4160
batch_normalization_3 (Batch	(None, 64)	256
activation_2 (Activation)	(None, 64)	0
dropout_3 (Dropout)	(None, 64)	0
dense_7 (Dense)	(None, 32)	2080
batch_normalization_4 (Batch	(None, 32)	128
activation_3 (Activation)	(None, 32)	0
dropout_4 (Dropout)	(None, 32)	0
dense_8 (Dense)	(None, 32)	1056
batch_normalization_5 (Batch	(None, 32)	128
activation_4 (Activation)	(None, 32)	0
dense_9 (Dense)	(None, 4)	132

Total params: 30,424,324
Trainable params: 6,635,396
Non-trainable params: 23,788,928

Fig 4.3: Resnet50

This model was also trained for 40 epochs and showed a training loss of 0.9700 and categorical accuracy of 0.5541 which was less than what we expected so we scrapped this model for its low performance.

Model Three: This was also a Transfer Learning based model which had VGG19 as base model. It had the followed architecture:

Model: "sequential_3"

Layer (type)	Output Shape	Param #
vgg19 (Functional)	(None, 7, 7, 512)	20024384
dropout_5 (Dropout)	(None, 7, 7, 512)	0
flatten_3 (Flatten)	(None, 25088)	0
batch_normalization_6 (Batch	(None, 25088)	100352
dense_10 (Dense)	(None, 2048)	51382272
batch_normalization_7 (Batch	(None, 2048)	8192
activation_5 (Activation)	(None, 2048)	0
dropout_6 (Dropout)	(None, 2048)	0
dense_11 (Dense)	(None, 1024)	2098176
batch_normalization_8 (Batch	(None, 1024)	4096
activation_6 (Activation)	(None, 1024)	0
dropout_7 (Dropout)	(None, 1024)	0
dense_12 (Dense)	(None, 4)	4100

Total params: 73,621,572
Trainable params: 53,540,868
Non-trainable params: 20,080,704

None

Fig4.4: VGG19

After training for 40 epochs, we encountered 0.2194 training loss and categorical accuracy of 0.9138 which was fairly more decent than Resnet50.

Model Four: In this, we used Densenet121 as our base model with the following architecture:

Model: "sequential_1"

Layer (type)	Output Shape	Param #
densenet121 (Functional)	(None, 7, 7, 1024)	7037504
dropout (Dropout)	(None, 7, 7, 1024)	0
flatten (Flatten)	(None, 50176)	0
batch_normalization (BatchNo	(None, 50176)	200704
dense (Dense)	(None, 64)	3211328
batch_normalization_1 (Batch	(None, 64)	256
activation (Activation)	(None, 64)	0
dropout_1 (Dropout)	(None, 64)	0
dense_1 (Dense)	(None, 64)	4160
batch_normalization_2 (Batch	(None, 64)	256
activation_1 (Activation)	(None, 64)	0
dropout_2 (Dropout)	(None, 64)	0
dense_2 (Dense)	(None, 64)	4160
batch_normalization_3 (Batch	(None, 64)	256
activation_2 (Activation)	(None, 64)	0
dropout_3 (Dropout)	(None, 64)	0
dense_3 (Dense)	(None, 32)	2080
batch_normalization_4 (Batch	(None, 32)	128
activation_3 (Activation)	(None, 32)	0
dropout_4 (Dropout)	(None, 32)	0
dense_4 (Dense)	(None, 32)	1056
batch_normalization_5 (Batch	(None, 32)	128
activation_4 (Activation)	(None, 32)	0
dense_5 (Dense)	(None, 4)	132

Fig 4.5: DenseNet

After training the model for 40 epochs, we recovered a loss of 0.7075 and accuracy of 0.8411 which is better than Resnet50 but still poorer than VGG19.

V. OUTPUT

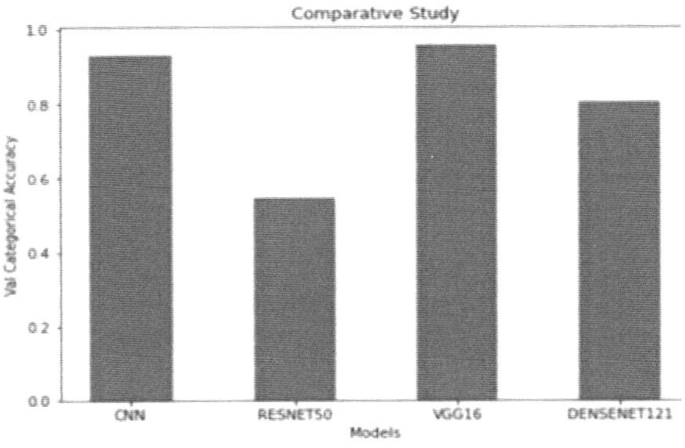

Fig 5.1: Validation Categorical Accuracy Analysis

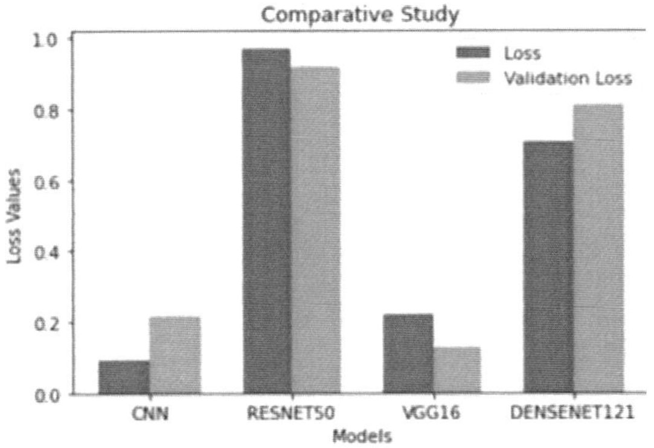

Fig 5.2: Loss Analysis

From the above graph it shows that VGG19 has the highest accuracy. Now we deploy the best model using TensorFlow so that we can use it in our website. We have created a User Application for patients, we have used Python's web framework called Flask which helps us in creating an end-to-end application for patients. We have deployed TensorFlow model here to predict the outcome for the MRI images with Alzheimer's.

VI. CONCLUSION

Alzheimer is a common neurodegenerative disease among older people. Therefore, its early detection is necessary for proper treatment and to prevent mis happenings. This project helps to automate Alzheimer detection using deep learning. The main goal of this project is to come up with an efficient way for people to take the necessary and suitable precautions from having Alzheimer's disease. Using this project, we can make its diagnosis accessible to everyone if brain MRI scan is available. Deep Learning is a rapidly evolving field, and its application can be especially be used in the healthcare field to maximize the benefit of the people.

REFERENCES

1. Cohen, David & Carpenter, Kristy & Jarrell, Juliet & Huang, Xudong. (2019). Deep learning-based classification of multi-categorical presenile dementia data. Current Neurobiology. 10. 141-147.

2. Yanqing, Jyoti & Zhang, Islam. (2017). a unique Deep Learning Based Multi-class Classification Method for Alzheimer's Disease Detection Using Brain MRI Data. 213-222.10.1007/978-3-319-70772-3_20.

3. N. D. Kodikara, R. N. Rajapakse and K. A. N. N. P. Gunawardena "Applying CNN for pre-detection of Alzheimer's disease from structural MRI data" (2017) 24th International Conference on Mechatronics and Machine Vision in Practice (M2VIP), 2017, pp. 1-7, doi: 10.1109/M2VIP.2017.8211486.

4. Ramírez J, Górriz JM , Munilla J, G, Ramírez J Ortiz A. Ensembles of Deep Learning Architectures for the first Diagnosis of the presenile dementia. Int J Neural Syst. 2016 Nov;26(7):1650025. doi: 10.1142/S0129065716500258. Epub 2016 Apr 4. PMID: 27478060.

5. Zeng An, Jia Longfei, Frizzell Tory, Pan Dan Huang Yin, Song Xiaowei, Early Detection of Alzheimer's Disease Using resonance Imaging: a unique Approach Combining Convolutional Neural Networks and Ensemble Learning, Frontiers in Neuroscience (2020), DOI=10.3389/fnins.2020.00259

6. I. X. Jiang, T. Nguyen, L. Chang, R. S. Turner and Almubarak, "Early Detection of Alzheimer's Disease Using Patient Neuropsychological and Cognitive Data and Machine Learning Techniques," 2019 IEEE International Conference on Big Data (Big Data),Los Angeles, CA, USA, 2019.

7. Jyoti & Zhang, Islam Yanqing. Brain MRI analysis for Alzheimer's disease diagnosis using an ensemble system of deep convolutional neural network. Brain Informatics. (2018). 5-10.1186/s40708-018-0080-3.

8. L., Gong, X., K., Mao, Chen, M., Li, J., Nandi, Yue, & Li, M. (2018). Auto-Detection of Alzheimer's Using Deep Convolutional Neural Networks. 2018 14th International Conference on Natural Computation, Fuzzy Systems and Knowledge Discovery (ICNC-FSKD).

9. Qin Z, Lan T, Ding Y Zhang C, Luo C, , Xiao Z. Brain MR Image Classification
for Alzheimer's Diagnosis supported Multifeature Fusion.Comput Math Methods Med. (2017); 2017:1952373. doi:10.1155/2017/1952373. Epub 2017 May 22

10. Kim Oh, Chung K., YC., K.W. et al. Classification and Visualization of Alzheimer's Disease using Volumetric Convolutional Neural Network and Transfer Learning. Sci Rep 9, 18150 (2019).

11. R N Rajapakse , N D Kodikara and K A N N P Gunawardena. Applying Convolutional Neural Networks for Pre-detection of Alzheimer's Disease from Structural MRI data. Conference Paper November 2017 DOI: 10.1109/M2VIP.2017.8211486.

12. Naimul Mefraz Khan ,Marcia Hon. Towards Alzheimer's Disease Classification through Transfer Learning. arXiv:1711.11117v1 [cs.CV] 29 Nov 2017.

13. Anjal D ,Vindhya G B, Medha Mansi, , Muskan Kedia Mahera Alam,. Prediction of Alzheimer's Disease using Machine Learning Technique. IRJET-International Research Journal of Engineering and Technology Volume: 07 e-ISSN: 2395-0056 Issue: 05 | May 2020.

14. G Stalin Babu, S N Tirumala Rao, R Rajeswara Rao. Exploring of Classification Methods for Early Detection of Alzheimer's Disease. International Journal of Engineering and Advanced Technology (IJEAT).8958, Volume-8 Issue-6, August 2019.

15. Chandni Naidu, Dhanush Kumar, N Maheshwari, M Sivagami, Gang
Li. Prediction of Alzheimer's Disease using Oasis Dataset. IRJET-International Journal of Recent Technology and Engineering Issue-6S3 April, 2019.

16. Suhad Al-Shoukry1,2, And Taha H. Rassem1 (Senior, IEEE), Nasrin M. Makbol3 1 Faculty of Computing, University Malaysia Pahang, College of Computing and Applied Sciences, 26300 Kuantan, Pahang, MALAYSIA Alzheimer's Diseases Detection by Using Deep Learning Algorithms: April 2020.

17. Mr. Amir Ebrahimighahnavieh, SuhuaiLuo Raymond Chiong The University of Newcastle, University Drive, Callaghan 2308, Australia. Deep learning to detect presenile dementia from neuroimaging: a scientific literature review. 2019.

18. Tong L, Venugopalan J, Wang MD, Hassanzadeh HR. Multimodal deep learning models for early detection of Alzheimers stage. doi 10.1038/s41598-020-74399-w Sci Rep. 2021Feb5;11(1):3254.

19. Lucia Billeci, Lorenzo Bachi, Alessandro Tonacci and Asia Badolato. Machine Learning for the Classification of Alzheimer's Disease and Its Prodromal Stage Using Brain Diffusion Tensor Imaging Data: a scientific Review September 2020.

20. Aunsia Khan and Muhammad Usman Dept. of Computing, Shaheed Zulfikar Ali Bhutto Institute of Science and Technology (SZABIST), Islamabad, Pakistan. Early Diagnosis of Alzheimer's Disease using Machine Learning Techniques A Review Paper. KDIR 2015 - 7th International Conference on Knowledge Discovery and data Retrieval.

21. Kwangsik Nho, Taeho Jo, and Andrew J. Saykin. Deep Learning in Alzheimer's Disease: Diagnostic Classification and Prognostic Prediction Using Neuroimaging Data. Front Aging Neurosci. 2019.

22. Alzheimer disease facts and figures. Available online 20 March 2018.

Deep Learning based Automatic Extraction of Student Performance from Gazette Assessment Data

Rohan Nikam[1,*], *Ritesh* Pardeshi[2,**], *Yash* Patel[3,***], and *Ekta* Sarda[4,****]

[1]Department of Computer Engineering, Ramrao Adik Institute of Technology, India
[2]Department of Computer Engineering, Ramrao Adik Institute of Technology, India
[3]Department of Computer Engineering, Ramrao Adik Institute of Technology, India
[4]Department of Computer Engineering, Ramrao Adik Institute of Technology, India

Abstract. Everyday millions of files are generated worldwide containing humongous amounts of data in an unstructured format. Most of us come across at least one new document every week, which tells the large volume of data associated with documents. All the data in these documents is in unstructured format which makes it difficult for further processing. The extraction of data from this documents still remains largely a manual effort resulting in higher processing time. A system that could extract the required fields from documents and store them in a structured format automatically will be of much significance. In this paper, we have described an approach for extracting the data from Exam Result Gazette document and then storing it in a CSV file. Mask RCNN model having a backbone of ResNeXt-101-32x8d and Feature Pyramid Network(FPN) has been hypertuned for detecting the required fields. Then PyTesseract Optical Character Recognition System has been used for extracting the data from detected fields. Our proposed system is trained on custom data set created by us and then evaluated on test data to extract the required fields. The overall accuracy of our system is 98.69%. The results indicate that the system could be used for efficiently extracting the required fields from given exam result gazette document.

1 Introduction

Documents are the primary tools to accumulate information in each and every domain [1]. Few of the domains which heavily rely on documents are Healthcare domain having Medical records, Medical test reports of patients, Banking domain consisting mainly of Credit card statements, Account statements, Tech domain consisting of Resumes, Education domain consisting of Registration forms, and Marksheets. At present, manual work is mostly involved in processing these documents. For extracting the data from these documents and storing it in database organisations employ a separate team of individuals. So this becomes a time consuming and expensive process [2]. All the documents present in every domain have valuable data locked in them. If this data is extracted and stored in a structured format then it could be utilized for fast and efficient searching. Various analytic algorithms could be applied to data for business processing to get deeper insights out of it. A system that can automatically extract all these data can drastically improve the efficiency of many business processes by avoiding error-prone manual operations. In this paper, we used a college Exam Result Gazette Document or we can use any documents to extract the features, for example, as shown in Figure 1. We used the Mask RCNN model [3] with ResNeXt-101-32x8d as the backbone [4] and the Feature Pyramid Network (FPN)[5]. To detect the relevant fields from Exam Result Gazette Document, we performed hyper-parameter tuning of the model. COCO annotation format [6] is used as input to the model. In order to train the model from accessible Exam Result Gazette Document, we created unique training data in COCO annotation format.

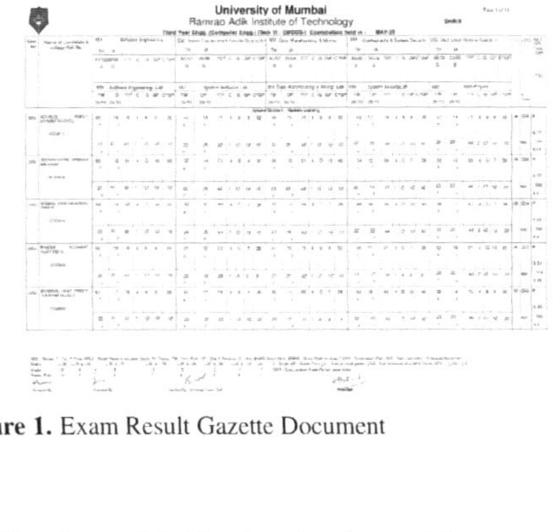

Figure 1. Exam Result Gazette Document

The detected fields given by the model are extracted and are then passed to Optical Character Recognition system (OCR) to extract the text from detected image and is then stored in a CSV file.

*e-mail: rohananilnikam1@gmail.com
**e-mail: riteshnleo@gmail.com
***e-mail: yashpatel7025@gmail.com
****e-mail: ekta.sarda@rait.ac.in

2 Literature Survey

B. Majumder [7], defined representation learning to identify target fields. In a document, not all the data present is of use. So hence a model was proposed which studies the relationship and relevancy between labels and data. The limitation of this model was that it was generalized to only specific fields like numbers in the document. **T. Denk [8]**, proposed a Combined Grid-based approach with BERT-based text encodings. BERTgrid, based on Chargrid, is a text-based grid that embeds vectors, making its structure and semantics accessible to the neural processing network. The inconvenience of that particular article was that it was more expensive computationally because the grid approach treated raw pixels.

V. Sunder [9], defined two techniques, The first is neural learning which uses a pre-trained deep learning model for reading and converts document images into structured form by adding a predefined database schema and the second is reusable logic programs for the extraction of entities from a document using the entities and primitive links identified through neural learning to synthesize. A template-free solution has been learned to detect pre-printed text and enter text/handwriting and to predict pair-wise relationships. The pre-printed text and input text lines were detected via a convolutional network method. Then functionality from the detection network is integrated to classify potential connections in a semantic way. **B. Davis [10]**, proposed combination method exceeds heuristic regulations and that visual characteristic is essential to achieving high accuracy.

S. Paliwal [11], for both table detection and structure recognitions, describe deep learning end-to-end models. The model takes advantage of the interdependence between the two tasks of table detection and table structure recognition, which are divided between table and column regions. This is followed by semantic rule-based row extraction from the table subregions identified. Content-based, noise-resistant, and generalized machine-learning approach for unseen document formats. This addressed the task of removing important fields from a variety of document formats that are potentially unseen. SYPHT – a machine learning solution for the extraction of field documents – was introduced in this paper **X. Holt [12]**. SYPHT combines OCR, heuristic filtering, and the supervised document ranking model in order to predict the field level so that image quality, skew, orientation, and content layout changes can be robust.

X. Zhao [13], used the grid text in Convolution Neural networks where text is embedded into the document as a feature containing both semantic meaning and spatial distribution. After studying this paper it is learned that semantic feature extraction needs to be improved along with taking into consideration the image level features. **A. Katti [14]**, a new type of text representation has been introduced which preserves a document's 2D layout. It proposed a novel paradigm for processing and understanding structured documents. Instead of serializing a document into a 1D text, the proposed method, named chargrid, preserves the spatial structure of the document by representing it as a sparse 2D grid of characters. The model predicts a segmentation mask with pixel-level labels and objects bounding boxes to group multiple instances of the same class.The chargrid paradigm is applied on an information extraction task from invoices and demonstrates that this method is superior to both, state-of-the-art NLP algorithms as well as computer vision algorithms.

G. Sehgal [15], developed a framework that recognizes handwritten and printed text, removes noisy effects, identifies documents and visual entities, such as tables, linens, and boxes. One limitation of this is that if there is background noise in the image, it does not function well when recognizing handwritten text. **R.Palm [16]** describes a system that used RNNs and LSTM to capture the context of data in the document. On the basis of the context of words in the document, the model tries to generalize onto unseen templates. The model presented in this paper does not take into consideration the spatial layout of the document and follows a left to right order to extract key, values pairs.

3 Proposed Work

After doing the literature survey we have observed that there are some limitations of existing systems. Some existing systems that we have studied only work well with numerical data and some systems are not able to extract the data accurately which is in tabular format.

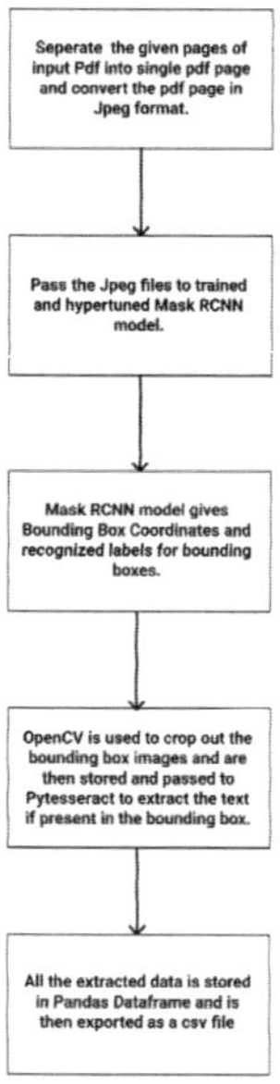

Figure 2. Flow Diagram

In the Figure 2, as we have mentioned that our system can work with textual as well as numerical data present in the document. As shown, the system first takes an input of Exam Result Gazette PDF, separates each page of the PDF, converts it into a JPEG format and then passes it to the Mask RCNN model having a backbone of ResNeXt-101-32x8d and Feature Pyramid Network(FPN). Hyper-tuned Mask RCNN model gives the bounding box coordinates and recognized labels for each corresponding bounding box.

OpenCV is used to crop out the bounding box images and then these images are passed onto rightly configured Optical Character Recognition System PyTesseract to extract the text if present in the bounding box. Finally, all the extracted data is stored in Pandas Dataframe and exported as a CSV file.

3.1 Techniques

For creating the data for training open source Dataturks Annotation tool has been used. Dataturks annotation tool gives annotation in COCO Format which stores annotation in JSON Format. In the document, there are various fields like Name, Roll Number, Total Marks, Average Pointer, etc. Bounding boxes are drawn along these fields.

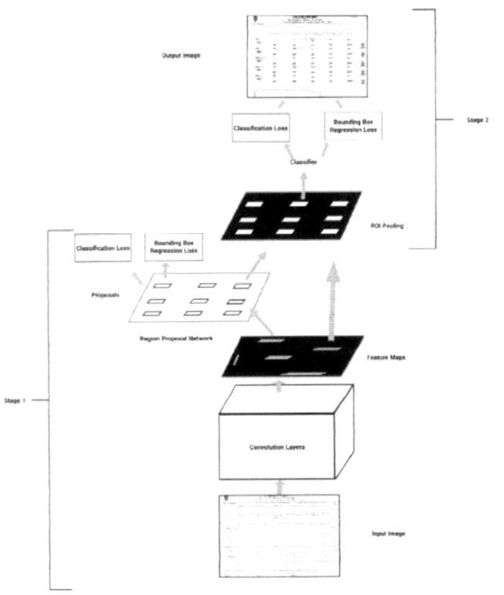

Figure 3. Working of Model

Mask RCNN model having a backbone of ResNeXt-101-32x8d and Feature Pyramid Network(FPN) has been used for detecting and recognizing various fields present in the document. Figure 3 shows the working of the model. Hyper-parameter Tuning has been done in Mask RCNN for increasing the accuracy of detecting and recognizing various fields present in the Exam Result Gazette document. Mask RCNN model works in two stages, the first stage is the region proposal stage, and the second stage is a network to predict most probable bounding boxes, classes of bounding boxes and mask for each bounding box. In the first stage, two networks are used backbone network and the region proposal network. Both these networks' backbone and region proposals run once per image to give a set of region proposals. The backbone network consists of a CNN to generate multiple Regions Of Interest using a lightweight binary classifier. For this, the model uses nine anchor boxes over the entire image. The classifier returns a score on the basis of whether an object is present in the proposed boxes or not. On the basis of threshold scores, some proposed boxes are selected and are warp into a fixed dimension. In the second stage, this fixed dimension of features is passed to a fully connected layer to make classification using the Softmax function. Warped features are also passed to a Mask Classifier. Mask Classifier is a CNN model which outputs a Mask for each Region Of Interest. In both, the stages of Mask RCNN Classification Loss and Bounding-Box Regression Loss have been used.

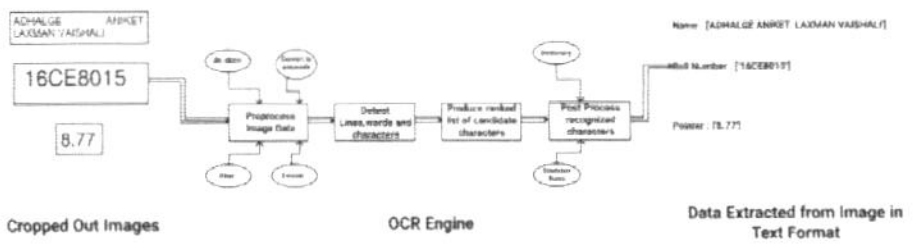

Figure 4. Working of OCR

After predicting the boxes from the document, the whole image, which is passed to the optical character recognition system (OCR), is cut into these boxes. The general working of OCR is shown in Figure 4. OCR scans each of the cropped images pixels by pixel to predict the text sent inside the cropped image. Then all the data recognized with OCR technique is exported to a CSV file.

3.2 Implementation Details

The Dataturks annotation tool was used to create the first custom data set of documents. Dataturks annotation tool after annotating the images gives the output in COCO format. For implementing the whole model, the Google Colaboratory platform has been used. Graphical Processing Unit(GPU) available on Google Colaboratory has been used For training the hyper-tuned model. PyTorch and Detectron2 Framework have been used.

Mask RCNN model having a backbone of ResNeXt-101-32x8d and Feature Pyramid Network(FPN) has been used for detecting and recognizing various fields present in the document. Model is trained using a custom training set created by Dataturks annotation tool. Various hyper-parameters in the model are adjusted in order to get high accuracy in detecting and predicting the fields of the documents. Once the fields are detected in the document, the coordinates of bounding boxes enclosing the detected fields and their respective labels are given as output by the model. These coordinates along with the image of gazette document are passed to OpenCV library for cropping out the bounding boxes from whole image. All the cropped images are stored according to their predicted labels in respective folders in order to pass to Optical Character Recognition System.

For extracting the text from the cropped images, the PyTesseract Optical Character Recognition Open Source library has been used. To correctly extract text from cropped images, various PyTesseract configurations were tested, and the one with the highest accuracy was chosen

https://doi.org/10.1051/itmconf/20214003022

as the final configuration. Different configurations have been used for extracting the numerical data like marks and textual data like name. Data of students with only unique roll numbers are stored. This data is stored in a Pandas DataFrame and is then exported as a CSV file.

4 Result and Analysis

The hyper-tuned Mask RCNN model was trained on Exam Result Gazette Document consisting of 25 pages, each page having 5 records. The trained model was evaluated on Exam Result Gazette Document consisting of 15 pages, each page having 5 records. Each record has 15 fields, such as Name, Roll Number, Pointer, Average Pointer, Total Marks, Subject1 Marks, Subject2 Marks, Subject3 Marks, Subject4 Marks, Subject5 Marks, Lab1 Marks, Lab2 Marks, Lab3 Marks, Lab4 Marks, Mini Project Marks.

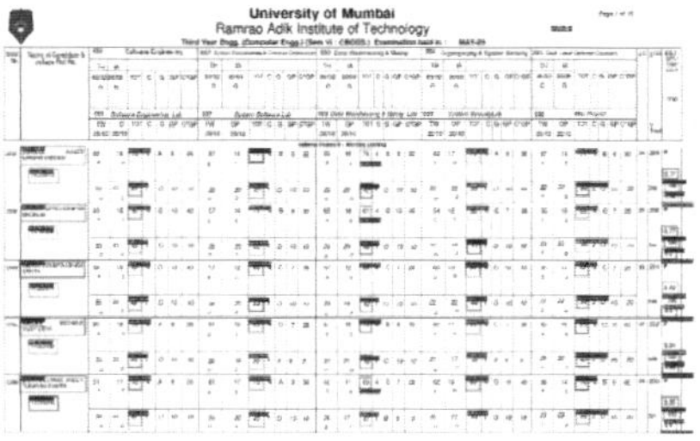

Figure 5. Detected Fields

Figure 5 shows the fields detected by the model for one single page of Exam Result Gazette PDF. Along with the detected fields and their respective labels the confidence percentage is also mentioned beside the boxes. While testing we have kept a threshold of 80% that means only the fields with confidence percentage greater than 80% are finally given as output.

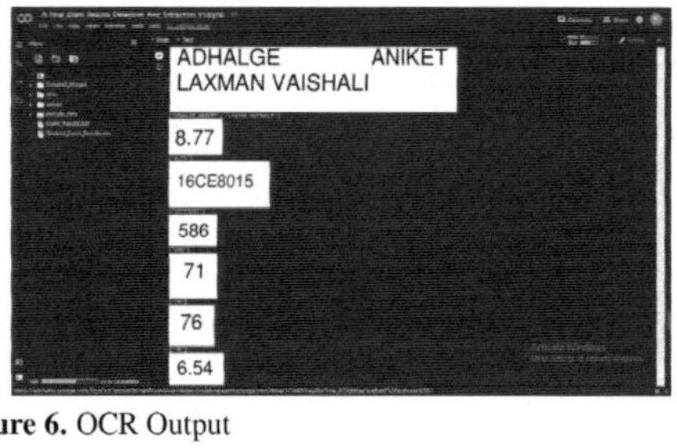

Figure 6. OCR Output

The detected fields from Exam Result Gazette Document are cropped with OpenCV library and are then passed to PyTesseract Optical Character Recognition System to extract the data as shown in Figure 6.

In this paper we have worked with Exam Result Gazette PDF consisting of 15 pages having total 71 student records. After giving the whole PDF to our system, after all processing, a CSV file is generated, which could then be downloaded and accessed locally as shown in Figure 7.

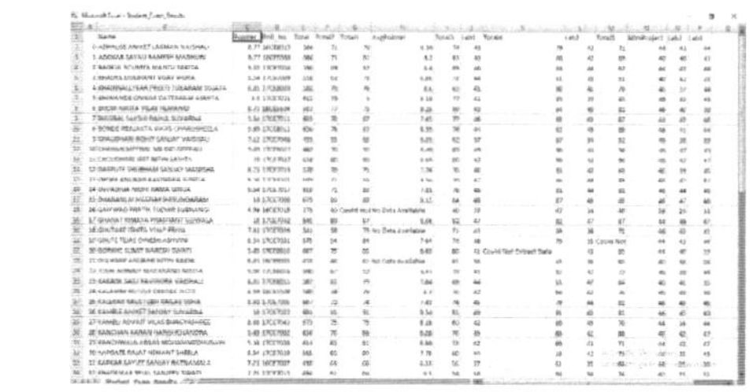

Figure 7. Output CSV File

The total fields that the model had to detect and recognize from the test set were around 1605 fields, out of which 1584 fields were detected and recognized by the model correctly. The metric that we used for evaluating our whole system consisting of detection and recognition part is Accuracy. Precision and Recall metrics have not been used because the system gives the complete extracted data as the output but if some of the digits in a detected field are wrong it will result in overall wrong prediction of the field. So this renders Precision and Recall as unsuitable evaluation metrics. The formula used to calculate overall accuracy after detection and recognition is given in Equation 1.

$$Accuracy = \frac{Total\ Correct\ Predictions}{Total\ Number\ of\ Predictions} \quad (1)$$

The system has 98.69% accuracy. Name, Pointer, Roll Number, Lab1 Marks, Lab2 Marks, Lab4 Marks fields have an accuracy of 100%.

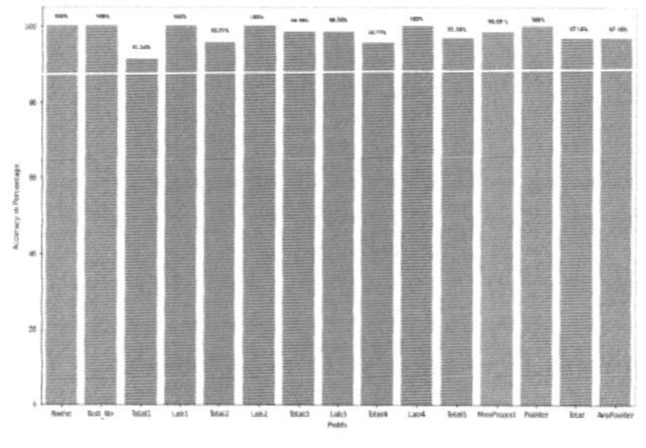

Figure 8. Accuracy Of Fields

Figure 8 shows the Accuracy of model across all the fields detected and recognized from the Exam Results Gazette PDF.

5 Conclusion And Future Work

Documents have valuable information locked in them, which could be utilized for fast and efficient searching and for business processing to get insights out of them, but one caveat is that it is largely a manual effort. A system that can extract all this data automatically can significantly improve the efficiency of many corporate workflows.

Here, we have studied recent research papers and carried out a comprehensive literature survey based on the topic of structured data extraction from documents. We have trained and hyper-tuned a model for automatically detecting the values of required fields from the documents. We worked with the Exam Result Gazette Document. We got an overall accuracy of 98.69% on unseen Exam Results Gazette document. In future work, this system can be expanded to detect and recognize fields across various business domains.

References

[1] A. J. Sellen, R. H. Harper, "The Myth of the Paperless Office", MIT Press. (2003)

[2] B. Klein, S. Agne, A. Dengel, "Results of a Study on Invoice Reading Systems in Germany", Springer. 451–462 (2004)

[3] K. He, G. Gkioxari, P. Dollár, R. Girshick, "Mask R-CNN", CVPR. (2018)

[4] S. Xie, R. Girshick, P. Dollár, Z. Tu, K. He, "Aggregated Residual Transformations for Deep Neural Networks", CVPR. (2017)

[5] T. Lin, P. Dollár, R. Girshick, K. He, B. Hariharan, S. Belongie, "Feature Pyramid Networks for Object Detection", CVPR. (2017)

[6] "Common Objects In Context", https://cocodataset.org, Accessed: 2020-12-10

[7] B. Majumder, N. Potti, S. Tata, J. Wendt, Q. Zhao, M. Najork, "Representation learning for information extraction from form-like documents", ACL. 6495–6504 (2020)

[8] T. Denk, C. Reisswig, "Bertgrid: Contextualized embedding for 2d document representation and understanding", NeurIPS. (2019)

[9] V. Sunder, A. Srinivasan, L. Vig, G. Shroff, R. Rahul, "One-shot information extraction from document images using neuro-deductive program synthesis", Neural-Symbolic Learning and Reasoning at IJCAI. (2019)

[10] B. Davis, B. Morse, S. Cohen, B. Price, C. Tensmeyer, "Deep visual template-free form parsing", ICDAR. (2019)

[11] S. Paliwal, V. D, R. Rahul, M. Sharma, L. Vig, "Tablenet: Deep learning model for end-to-end table detection and tabular data extraction from scanned document images", ICDAR. (2019)

[12] X. Holt, A. Chisholm, "Extracting structured data from invoices", ALTA. (2018)

[13] X. Zhao, E. Niu, Z. Wu, X. Wang, "Cutie: Learning to understand documents with convolutional universal text information extractor", CVPR. (2019)

[14] A. Katti, C. Reisswig, C. Guder, S. Brarda, S. Bickel, J. Höhne, J. Faddoul, "Chargrid: Towards understanding 2d documents", EMNLP. (2018).

[15] V. D, R. Rahul, G. Sehgal, Swati, A. Chowdhury, M. Sharma, L. Vig, G. Shroff, A. Srinivasan, "Deep reader: Information extraction from document images via relation extraction and natural language." Asian Conference on Computer Vision, Springer. (2018)

[16] R. Palm, O. Winther, F. Laws, "Cloudscan-a configuration-free invoice analysis system using recurrent neural networks", ICDAR. **Vol. 1.** IEEE. (2017)

Text Summarization using Extractive and Abstractive Methods

Saurabh Varade[1,*], *Ejaaz* Sayyed[1,**], *Vaibhavi* Nagtode[1,***], and *Dr. Shilpa* Shinde[1,****]

[1]Department of Computer Engineering, Ramrao Adik Institute of Technology, India

Abstract. Text Summarization is a process where a huge text file is converted into summarized version which will preserve the original meaning and context. The main aim of any text summarization is to provide a accurate and precise summary. One approach is to use a sentence ranking algorithm. This comes under extractive summarization. Here, a graph based ranking algorithm is used to rank the sentences in the text and then top k- scored sentences are included in the summary. The most widely used algorithm to decide the importance of any vertex in a graph based on the information retrieved from the graph is Graph Based Ranking Algorithm. TextRank is one of the most efficient ranking algorithms which is used for Web link analysis that is for measuring the importance of website pages. Another approach is abstractive summarization where a LSTM encoder decoder model is used along with attention mechanism which focuses on some important words from the input. Encoder encodes the input sequence and decoder along with attention mechanism gives the summary as the output.

1 Introduction

Internet contains vast amount of data. World Wide Web provides huge variety of data from which the user can find useful data according to their needs and purpose. So, it is highly beneficial to use an efficient text summarization in this hectic world so that only user required and useful data is provided to the user in a lesser time.

Manual summarization of documents and web-contents is too time-consuming this paved the need for automatic summarization [8, 10] of text documents and web-contents. Summary created with the help of main points which are the sentences form the original document with the help of a software is known as automatic summarization.

Text Summarization can be classified into different categories:

- Single document and multiple document
- Generic and query-based approach
- Extractive and Abstractive

Text summarization can be applied to single document as well as it can be extended to multiple documents. The summarized version of the original set of documents should be context based. LexRank like ranking algorithm can be used for multiple document summarizations. TextRank algorithm can be used for single document summarization.

Text summarization can be classified based on the purpose, input, and method that are opted. Automatic summarization has two types of approaches which are extractive and abstractive summarization. Selecting some word, phrases or sentences form the original text to form a summary is known as extractive summarization [11]. Abstractive summarization works by building a semantic similarity representation and using the techniques of natural language generation to generate a summary which is similar to manual summary.

A reader and an identifier are required to select between prime and unnecessary words/sentences in the text document to generate a summary of a large text document. Clustering similar document and presenting a summary is also provided by the document summarization. A good summary generator should be capable of reflecting the different fragments from the document while keeping the redundancy level low.

In today's world where there is an over-abundance of data and lack of manpower with less time to interpret the data, an automatic method for text summarization is necessary. A system is proposed named Data Summarization using Unsupervised Learning which will summarize the data from these documents or article and will provide a summary for that data of random length.

The main objective of the project is to implement a system which summarizes the data in a proper manner preserving its overall meaning and contents also without changing its context. Aim is also to reduce the time and provide a better result for the end user.

The organization of report is as follows. Section 2 will provide an introductory and brief review of related works along with some important facts. Section 3 will elaborate details of the system (Data Summarization Using Unsupervised Learning) and its implementation. Section 4

*e-mail: saurabhvarade123@gmail.com
**e-mail: ejaaz.sayyed.common@gmail.com
***e-mail: vaibhavinagtode1@gmail.com
****e-mail: shilpa.shinde@rait.ac.in

will present experimental results obtained and the analysis work. Section 5 will provide the conclusion for our work.

2 RELATED WORK

A survey of the research done for data summarization and the currently existing systems, give the following results.

A. **Automatic text summarization using semi-supervised learning [1]**

Idea mentioned in this paper includes development of summarization which is fully automated. This is the main idea and it uses unsupervised and semi-supervised learning. Linear classifiers which are simple have been used for implementation. Increase in performance can be clearly seen in experiments on Reuter's news-wire.

Half of the performance increase, which is allowed by a fully supervised system can be reached by unsupervised learning. This is realistic for related applications. The way how query based summaries are used is helpful.

B. **Review Paper on Extractive Text Summarization [2]**

It gives an over-all idea of Extractive Summarization with its features. A summarization machine is given which provides an idea about how the summarization process works. Query driven and generic are the two basic types update of the summary.

The extractive text summarization methods are preferred depending on TF-IDF (Term Frequency-Inverse Document Frequency), cluster based methods, the single value decomposition is preferred by the Latent Semantic Analysis (LSA) and the vector space model is preferred by the concept based summarization.

Advantages:

- Precise information can be helpful to understand the document more effectively and efficiently.
- Summarization can make searching easier.

Limitations:

- Extraction of crucial information or optimization of the whole document in order to minimize the time required to review the document.
- Generation of summary with minimum redundancy, maximum relevancy and referring elements of document in the summary.

C. **Extractive Text Summarization Using Graph Based Ranking Algorithm And Mean Shift Clustering [3]**

An introduction to Abstractive and Extractive summarization is given along with Extractive summarization types. A brief introduction to Graph based ranking algorithm is mentioned. Concepts used are Text Rank (Single document), Lex Rank (Multiple documents), Co-Rank, Mean-shift Clustering and ROUGE Evaluation.

Advantages:

- Mean Shift Algorithm is general, application-independent tool.
- Model-free, doesn't assume any prior shape on data clusters.

Limitations:

- Output depends on window size.
- Computationally expensive.

D. **Summarization Using Clustering and Classification: Spectral Clustering Combined with k-Means Using NFPH [4]**

In classical k-means, the cluster centroid's initialization method can be replaced by this system. This can solve some of the limitation which occur in k-means algorithm.

Advantages:

- From the output it is cleared that the accuracy was improved by 2

Limitations:

- The processing time has increased from 1 to 6 seconds . This is due to fact that an additional step is required for performing the NFPH based initialization of k-means.

E. **Enhancing unsupervised neural networks based text summarization with word embedding and ensemble learning [5]**

Bag of words method and sentence to vector representation are the two implementation models used. Each word is represented as vector in word2Vec. Similarly, in sentence2Vec, each sentence in the text document is represented as vector in an embedded low-dimensional space.

Two summarization frameworks are developed based on unsupervised deep learning. This is called as Auto-Encoder and Variation Auto-Encoder. Computer vision and bioinformatics has been successfully applied with extreme learning machine which has become the state of the art learning framework.

Promising results have been gained in English and Arabic languages by models applied in different languages. Graph model is the approach used for the summarization for the Arabic newspaper.

Advantages:

- Word2Vec approach (already published by Google) improves the results obtained by unsupervised deep learning models.
- Data fusion(voting) improves the quality of the output summary as the fusion is of the BOW and Sentence2Vec transformation which further re-ranks sentences.

Limitations:

- Needs big corpus with a large number of text documents in order to build powerful models with a more discriminative feature space.
- It is very hard to find the optimal parameters for the adopted neural network models.
- Learning process is time-consuming because the models are trained on a large number of documents.

F. **Unsupervised text summarization using sentence embeddings [6]**

There are many applications of Dense vector representation in NLP. There was text summarization in the past which used the clustering of sentences in a high-dimensional space. But, those systems used TF-IDF instead of sentence embeddings.

Latent semantic indexing to identify sentences that explain the latent concepts very nicely in the document were used by another class of vector based methods. Work of Skip thought vector space has the main goal of encoding sentences in vector using RNN with LSTM.

Unsupervised text summarization approach is proposed by clustering the sentence embeddings trained to embed paraphrases near each other. A summary is generated from the cluster of sentences by selecting a representative from each cluster.

The text whose embedding is the nearest according to Euclidian distance that is, to the centroid of the cluster, are chosen by the extractive method.

The embeddings are decoded into sentences by the decoder in abstractive method. This is done by using the Recurrent Neural Network (RNN) with Long Short Term Memory (LSTM). Skip thought embeddings have less precision than the system which uses Paragram embeddings.

Advantages:

- Systems using Paragram embeddings have a higher precision than those using Skip-thought embeddings.
- K-Means proved to be better than Mean shift clustering performance wise on the tested datasets.

Limitations:

- The decoder tended to generate a fair number of <UNK> tokens, which is possibly because the tested dataset is a scientific papers, which has a number of words that do not occur frequently enough in the dataset itself, leading to poor parameter estimates in the decoder.
- No clear trend between the different types of abstractive systems that is consistent across clustering methods.

G. **A survey on LSTM memristive neural network architectures and application [7]**

Long short-term memory (LSTM) consists of state memory and multilayer cell structure which is a special type of recurrent neural network (RNN). The paper has explored LSTM architecture and different variants of lstm architecture. Various applications such as prediction, pattern classification, analysis, etc are mentioned with lstm models.

3 METHODOLOGY

3.1 Proposed Work

3.1.1 Extractive Summarization

The proposed system uses vectorization of sentences, TextRank algorithm, cosine distance calculation.

The proposed system comprises of three main modules:

- First the text preprocessing, cleaning and similarity construct.
- Second the Graph based ranking (in this case) with appropriate ranking algorithm and techniques.
- Lastly, the selection of the k highly ranked sentences as the summary of the text input.

3.1.2 Abstractive Summarization

The proposed system uses a LSTM [9] encoder decoder model along with attention mechanism.

The Proposed system comprises of main modules:

- First the text preprocessing and tokenization
- Second comes an encoder model which encodes the input text and pass it to the decoder.
- Third, the decoder and the attention mechanism produce the output that is the summary of the input text by selecting the representation from the final states.

3.2 Techniques

3.2.1 Extractive Summarization

The preprocessing and cleaning step consist of removing the stop words. Stops words are the words which are commonly used in a language. Removal of stop words helps in the model to focus on ranking based on the important words or words with significant usage thus can be beneficial in reducing the variance. Then, the sentences are vectorized. The previous proposed systems use Bag of Words (BOW), which is extended in this model to Vector of sentences, so as to achieve the extractive nature of the model.

The similarity construction of the sentences is done with cosine distance calculation. Similarity between two vectors of an inner product space is measured by cosine similarity. The vectors are plotted on a two-dimensional plane and the cosine of the angle between the vector gives how close or similar these vectors could be. Similarity will be one if cosine of the angle is calculated between two same vectors.

3.2.2 Abstractive Summarization

The preprocessing and cleaning step consist of removing the stop words, URL's, html tags, digits, punctuations, contractions. Doing this helps in the model to focus on contextual information. This model is using Many-to-Many Seq2Seq model which contain an encoder and a decoder.

Here, long term dependencies are captured using LSTM (Long Short Term Memory) as an encoder and decoder components. This helps on overcoming the problem of vanishing gradient. Converting the input sequence into a vector is done by encoder and the output sequence is predicted by the decoder with the help of attention mechanism.

Attention mechanism helps the decoder by focusing more on some words in the input sequence.

3.3 Design of the System

Figure 1 shows the flow of extractive summarization and Figure 2 shows the flow of abstractive summarization.

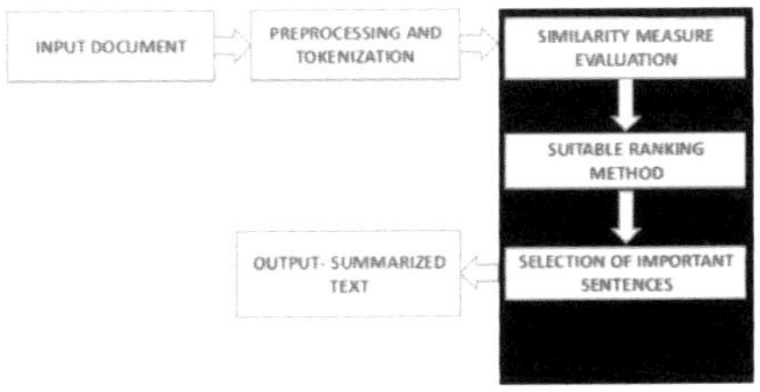

Figure 1. Flow of Extractive summary generation

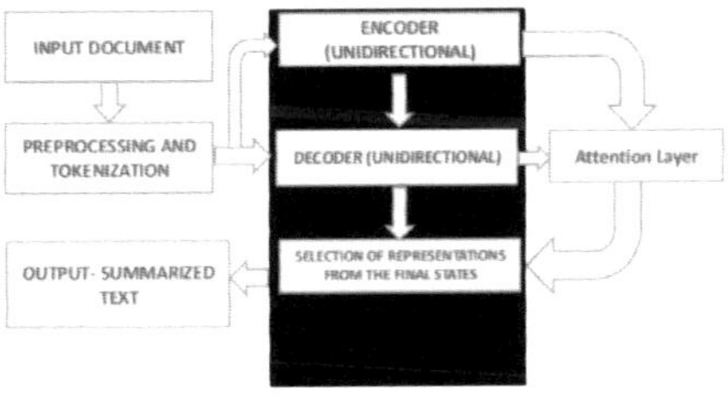

Figure 2. Flow of Abstractive summary generation

4 RESULT AND ANALYSIS

The techniques used in the extractive summarization, that is, the cosine distance and the TextRank algorithm(in NLP, also called text Rank algorithm) makes the system computationally efficient and can be used for many NLP based applications. The time complexity of the implemented extractive summarization is linear, that is it requires time of O(n). The system can randomly generate any number of summary sentences, depending on the size of the text given as input. The accuracy of the abstractive summarizer model tends to be around 60% for unseen data. For the extractive summarization, the accuracy varies as the sentences in the summary may or may not be much reliable.

The loss incurred for the validation or test data in the training phase is decreasing with a plateau like line. Soon after 18 epochs, the model starts to learn noise data in the training data, indicating possibility of overfitting. The model then breaks the training phase and returns best parameter values of the model to have been achieved. Figure 3 shows the graph depicting the loss of the model during the training phase for abstractive summarization. The X-Axis shows the number of epochs and the Y-Axis shows the loss function value during the training of the model. Figure 4 shows the graph depicting the accuracy of the model during the training phase for abstractive summarization. The X-Axis shows the number of epochs and the Y-Axis shows the Accuracy metric during the training of the model.

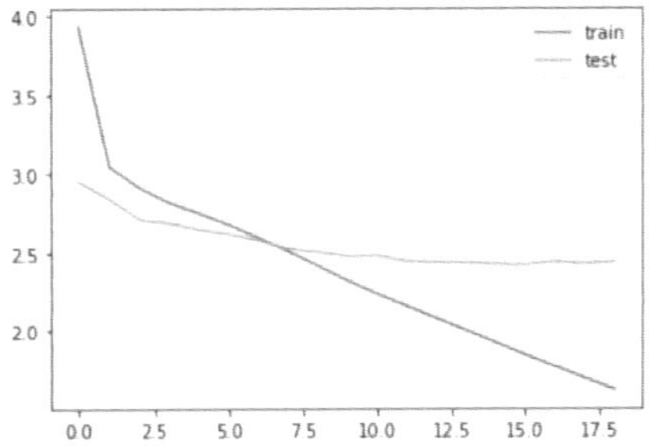

Figure 3. Loss graph for abstractive summarization

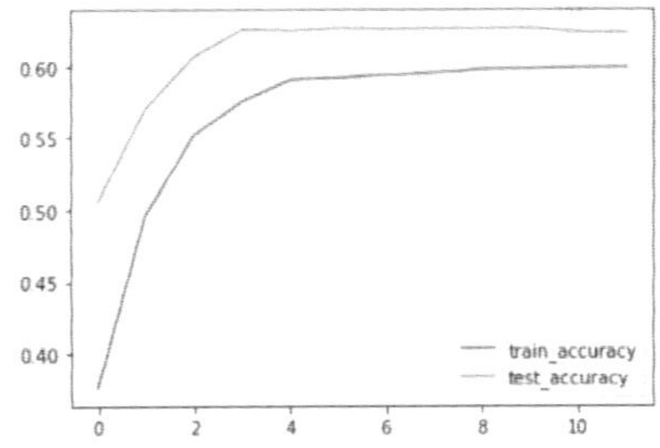

Figure 4. Accuracy graph for abstractive summarization

5 CONCLUSION AND FUTURE WORKS

Comparing the two-systems performance-wise, the result of extractive summarization will be more efficient and but less accurate than abstractive summarization, as the dependency of the abstractive summarizer will be more on its model and hence might take a significant amount of time to process a small amount of data. The accuracy of the abstractive summarizer model tends to be around 60% for unseen data. For the extractive summarization, the accuracy varies as the sentences in the summary may or may not be much reliable.

In future, implementation of the current system in different domains like android application, standalone software or integration with multi-tools applications can be done. A working system for different research papers that can handle all type of notions from different streams like medical, astronomy, etc. can also be implemented. Further, more fine tuning and improving the abstractive summarizer for long documents and improving the performance for small documents for better understanding and intuition of working of the model can be achieved.

6 ACKNOWLEDGMENT

The authors thank the many people who have done lots of help for us and provided useful guidance for our paper.

References

[1] A. Zamanifar, B. Minaei-Bidgoli, M. Sharifi, " Automatic text summarization using semi-supervised learning", ACIS International Conference on Software Engineering, Artificial Intelligence, Networking, and Parallel/Distributed Computing, 635 (2008)

[2] A. Sahoo, Dr.A. Kumar Nayak, " Review Paper on Extractive Text Summarization", IJERCSE, **5**, (2018)

[3] Ramesh, Reema, Rajan, Binu"Extractive Text Summarization Using Graph Based Ranking Algorithm And Mean Shift Clustering", ICRTCCNT, (2019)

[4] N. Sapkota, A. Alsadoon, P. W. C. Prasad, A. Elchouemi, A. K. Singh, " Data Summarization Using Clustering and Classification: Spectral Clustering Combined with k-Means Using NFPH", COMITCon, 146-151 (2019)

[5] N.Alami, M. Meknassi, N. En-nahnahi, " Enhancing unsupervised neural networks based text summarization with word embedding and ensemble learning", Expert Syst. Appl., **123**, 195-211 (2019)

[6] Padmakumar, Aishwarya, A. Saran, " Unsupervised Text Summarization Using Sentence Embeddings", (2016)

[7] Smagulova, Kamilya, James, Alex, " A survey on LSTM memristive neural network architectures and applications", The European Physical Journal Special Topics, **228**, 2313-2324 (2019)

[8] "Introduction to summarization in machine learning", https://towardsdatascience.com/a-quick-introduction-to-text-summarization-in-machine-learning-3d27ccf18a9f, Accessed: 2020-12-02

[9] "Understanding LSTMs", https://colah.github.io/posts/2015-08-Understanding-LSTMs/, Accessed: 2021-01-15

[10] "Automatic Summarization", https://en.m.wikipedia.org/wiki/Automatic_summarization, Accessed: 2020-10-06

[11] "Introduction to textrank in python", https://www.analyticsvidhya.com/blog/2018/11/introduction-text-summarization-textrank-python/, Accessed: 2020-09-06

https://doi.org/10.1051/itmconf/20214003024

Sound Classification Using Python

Swapnil Jadhav[1], *Sarvesh* Karpe[1,*], and *Siuli* Das[1]

[1]Dpartmnt of Instrumentation Engineering, Ramrao Adik Institute of Technology, Nerul, Navi Mumbai, India

Abstract. Sound assumes a significant part in human existence. It is one of the fundamental tangible data which we get or see from the climate and their components which have three principal credits viz. Sufficiency (Loudness of the sound), Frequency (The pitch of the sound), Timbre (Quality of the sound or the personality of the sound for example the Sound contrast between a piano and a violin). It is an event generated from the action. Humans are highly efficient to learn and recognize new and various types of sounds and sound events. There is a lot of research work going on Automatic sound classification and it is used in various real-world applications. The paper proposes an examination of an establishment disturbance classifier reliant upon a model affirmation approach using a neural organization. The signs submitted to the neural association are depicted through a lot of 12 MFCC (Mel Frequency Cepstral Coefficient) limits routinely present toward the front finish of an adaptable terminal. The introduction of the classifier, assessed as far as percent misclassification, show an exactness going between 73 % and 95 % relying upon the term of the choice window. Transmitting sound using a machine and expecting an output is considered a highly accurate deep learning task. This technology is used in our smartphones with mobile assistants such as Siri, Alexa, Google Assistant. In the case of the Google Speech recognition data set over 94 percent accuracy is obtained when trying to identify one of 20 words, silence or unknown. It is a very difficult task to recognize audio or sound events systematically and work on it for identification and give output. We are going to work on it using python programming language and some deep learning techniques. It's a basic model that we are trying to develop, taking the next step to the innovative model that can help society and also which represent the innovative ideas of Engineering Students.

1 Introduction

Use Automatic environmental sound classification is a quickly changing and developing space of exploration with various applications. While there is a ton of exploration in related sound fields like discourse, voice acknowledgment, and music, there has been recognizable that the work done is exceptionally less on the order of natural sounds. In like manner, noticing the new progressions in the field of picture grouping where convolutional neural organizations are utilized to arrange pictures with high precision and at scale, frames the topic of appropriateness of these methods in different spaces, like sound characterization. Spectrograms are a useful system for imaging the scope of frequencies of a sound and how they change during an outstandingly short period of time. We will use a tantamount technique known as Mel-Frequency Cepstral Coefficients (MFCC) [1]. The essential differentiation is that a spectrogram uses a straight isolated repeat scale (so every repeat repository is scattered an identical number of Hertz isolated), while a MFCC uses a semi logarithmic partitioned repeat scale, which is more similar to how the human hear-capable structure estimates sounds [2].

Sound order is the way toward tuning in to and breaking down sound accounts. It is otherwise called sound grouping, this cycle is at the core of an assortment of present-day AI innovation including remote helpers, programmed discourse acknowledgment, and text to discourse applications. You can likewise discover it in prescient support, shrewd home security frameworks, and media ordering and recovery. Sound portrayal projects like those referred to above start with clarified sound data. Machines require this data to sort out some way to hear and what to tune in for. Using this data, they encourage the ability to isolate between sounds to complete unequivocal endeavours. The remark communication consistently incorporates requesting sound reports subject to project-unequivocal prerequisites through the help of committed sound portrayal organizations. Natural Sound Classification: Just as the name construes, this is the gathering of sounds found inside different conditions. For example, seeing metropolitan sound models, for example, vehicle horns, roadwork, alarms, human voices, and so forth This is utilized in security frameworks to recognize seems as though breaking glass. It is likewise utilized for prescient upkeep by recognizing sound disparities in processing plant apparatus. It is even used to separate

* Corresponding author: sarveshssuk@gmail.com

creature calls for natural life perception and conservation [3].
Sound acknowledgment issue comprises of three distinct stages as pre-handling of signs, extraction of explicit highlights, and their grouping. Signal pre-handling isolates the info signal into various fragments which are utilized for separating related highlights. Highlight extraction decreases the size of information and addresses the mind-boggling information as highlight vectors. Intersection rate, pitch, and casing highlights utilized in discourse acknowledgment applications were arranged utilizing different classifiers like choice trees, irregular woods, and k closest neighbor. Environmental sound classification (ESC) partakes as a fundamental and central development of Smart sound recognition (SSR). The key focal point of ESC is to decisively perceive reality class of an obvious sound, similar to doorbell, horn and jackhammer. With the helpful employments of SSR in solid watch out for skewer structures, splendid contraption applications and clinical consideration, the ESC issue has taken a great deal of interest as of late. For modified talk affirmation Automatic Speech Recognition (ASR) and music data affirmation (MIR), it has been cultivated extraordinary enhancements with advances in Artificial Intelligence (AI). As a result of amazingly non-fixed characteristics of natural sounds, these signs can't be orchestrated as talk or music so to speak. Toward the day's end, the models included for ASR and MIR will be powerless when applying to ESC issues [4].
Programmed differencing of ecological sounds, similar to cow cutting, canine bark, and vehicle alarm, can be utilized in applications, for example, distant reconnaissance for wellbeing purposes and home computerization. A motivating application is the utilization of home observing gear which recognizes and isolates various sounds that are created in the home climate and it additionally cautions the client appropriately. Instances of such sounds are child crying, forced air system, and glass breaking. The acknowledgment of such homegrown sounds, whenever executed on a cell phone, can prompt new and significant applications. Natural sounds comprise of different non-human sounds (barring music) in ordinary everyday life. During the previous few years, numerous endeavors to perceive ecological sounds have been made. By and by, there is an expanding center around ordering ecological sounds utilizing profound learning methods. The upgrades in the field of picture characterization lately are driving specialists to begin utilizing pictures when arranging sounds [5].
Natural or establishment disturbance is a lot of sounds created in an environment and recorded using a mouthpiece. A common model is the signs recorded by a PDA or land line when an individual is tuning in. By strategies for a Voice Activity Detector (VAD) it is plausible to isolate between voice activity and second when there is simply establishment noise, and from following assessment it is possible to recognize the sort of commotion present during a conversation. Even more overall, the use of a sound affirmation structure can offer generous potential in a couple of use circumstances: fixed and versatile correspondence, talk affirmation, lawful speaker recognizing verification, acoustic sensors, and perception and security applications [6].
Various courses of action have been proposed over the two or three years to make talk dealing with estimations more vivacious to establishment uproar. Regardless of the way that their display within the sight of racket has been improved fundamentally, they are as yet far from offering a comparable show levels as are obtained in clean conditions. There is at this point a wide edge for advancement in the life to establishment fuss of most talk dealing with estimations [7].
We investigate the hypothesis and execution of profound learning in the Python programming language. Spotlights on uses of profound learning for sound and music, yet examines general calculations and standards pertinent to any issue [8].
Environmental sound classification (ESC) [9] is also called Sound Event Recognition (SER). SER can recognize and identify the context of the information available in the different audio streams and also sounds available in the environment [10]. There is some distortion present in the acoustic sources and microphones, also a noise or the sound have their representation in form of waves and amplitudes which makes them difficult to understand with their various frameworks. Also, the data present or provided to the system contains a huge amount of audio files and recordings and the mixing of some of these audio files makes the ESC system more difficult to identify and segregate accordingly. Transmitting sound using a machine and expecting an output is considered a highly accurate deep learning task. We are using this technology in our smartphones with mobile assistants such as Siri, Alexa, Google Assistant. In the case of the Google Speech recognition data set over 94 percent accuracy is obtained when trying to identify one of 20 words, silence or unknown. It is a very difficult task to recognize audio or sound events systematically and work on it for identification and give output. We are going to work on it using python programming language and some deep learning techniques.
It showed that the consideration system added to the sound order. The rest of the paper is as follows: Section 2: Related Work, Section 3: Problem Statement, Section 4: Methodology, Section 5: Simulation., Section6: Result and discussion, Section 7: Conclusion followed by Acknowledgement and References.

2 Related Work

Each sign in the data base is first partitioned into diagrams with a fragmentary front between consecutive edges. Every plan of Mel-Frequency Cepstral Coefficients (MFCCs) are removed by applying the discrete cosine change to the log-energy which leads to Melscaling channel bank [11]. These features used to segregate the supernatural condition of a sign. Vector Quantization (VQ) was applied for planning stage. The planning with stage is performed by a Feed-Forward neural organization arranged by a Resilient Back

Propagation computation. The final result is blessed to a reasoning that chooses the class of racket as demonstrated by the most significant worth in the neural organization.

3 Problem Statement

Our work demonstrates the application of Deep Learning techniques which will be useful for the classification of the sounds available in the surrounding environment. By focusing on identifying the particular animal or the human-made sounds which are used for secretive communication using sounds during undercover operations/ ambush attacks. There is some distortion present in the acoustic sources and microphones, also noise or the sound have their representation in form of waves and amplitudes which makes them difficult to understand with their various frameworks. Also, the data present or provided to the system contains a huge amount of audio files and recordings and the mixing of some of these audio files makes the ESC system more difficult to identify and segregate accordingly. Thus, use the python programming language and some deep learning techniques for the identification and classification of sound.

4 Methodology

It turns out maybe the best part to isolate from sound waveforms (and advanced signs generally speaking) has been around since the 1980's is still top tier: Mel Frequency Cepstral Coefficients (MFCCs), introduced by Davis and Mermelstein in 1980. Underneath we will go through a specific discussion of how MFCCs are produced and why they are useful in solid investigation [12]. This part is genuinely particular, so before we make a dive, what about a few key terms identifying with automated sign preparing and sound assessment. Here we separated two highlights mel-spectrogram and mfcc as demonstrated in the figures which contains sound sign wave structure and spectrograms of those sound signal waveforms [13]. Attempted to utilize both element for preparing and mfcc performed well indeed. Along these lines, here we will utilize mfcc for preparing. Be that as it may, prior to getting yield we need to learn and see approximately couple of terms like Sound waveform and Spectrogram which have talked about as follows:

4.1 Fourier transform

The Fourier Transform (equation 1) decays a component of time (signal) into constituent frequencies. Essentially a melodic agreement can be conveyed by the volumes and frequencies of its constituent notes, a Fourier Transform of a limit shows the plentifulness (proportion) of each repeat present in the principal limit (signal) gives the numerical articulation of Fourier Transform.

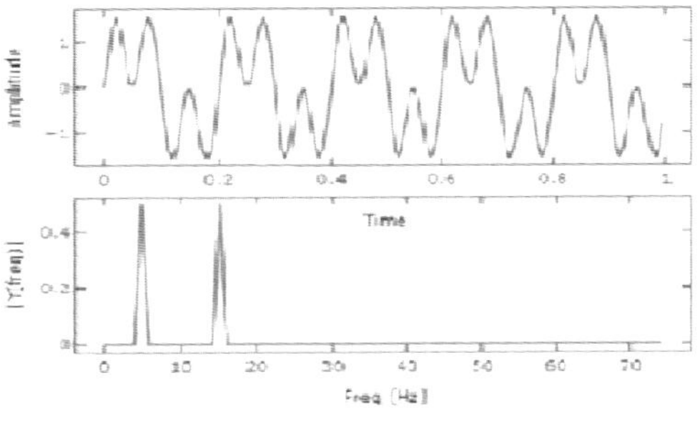

Fig. 1. Fourier Transform.

$$STFT\{x(t)\}(\tau,\omega) \equiv X(\tau,\omega) = \int_{-\infty}^{\infty} x(t)w(t-\tau)e^{-j\omega t}\,dt \qquad (1)$$

There are varieties of the Fourier Transform (Figure 1) including the Short-time Fourier change, which is done in the Librosa library and incorporates separating a sound sign into edges and a short time later taking the Fourier Transform of each edge. In strong planning all around, the Fourier is a rich and significant way to deal with break down a sound sign into its constituent frequencies. Figure 1 shows the qualities of Fourier change as far as plentifulness, recurrence and time.

4.2 What is sound waveform?

These sound models are regularly tended to as time course of action, where the y-rotate assessment is the abundancy of the waveform. The abundancy is ordinarily assessed as a component of the change of squeezing factor around the enhancer or authority contraption that at first got the sound. Except if there is metadata related with your sound examples, these time arrangement signs will frequently be your solitary information for fitting a model. The figure 2 is the example of dog barking sound waveform.

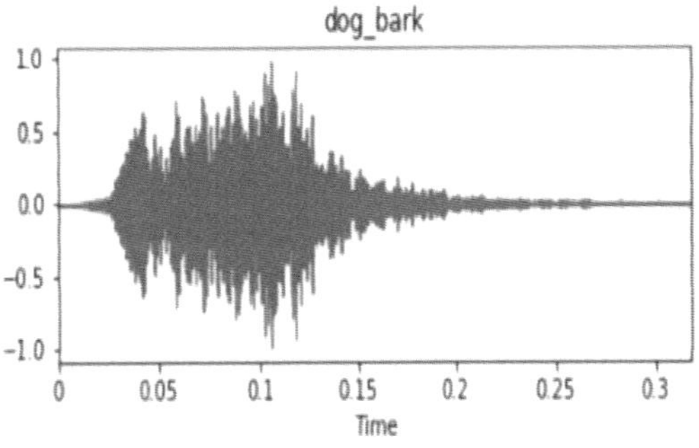

Fig. 2. Audio Signal Waveform.

https://doi.org/10.1051/itmconf/20214003024

4.3 What is Spectrogram?

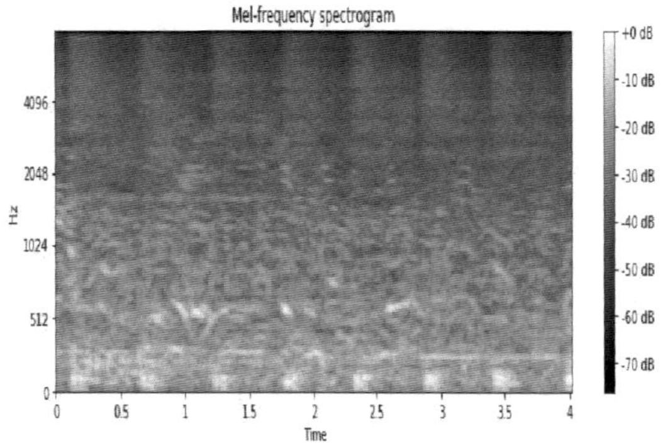

Fig. 3. Spectrogram

A spectrogram is a visual depiction of the scope of frequencies of a sign as it contrasts with time. A wonderful strategy to consider spectrograms is as a stacked point of view on periodograms across some stretch of time electronic sign. The figure 3 represents the Mel Frequency Spectrogram graphical representation.

4.4 What is Mel Frequency Cepstral Coefficients (MFCCs)?

MFCC's, as referred to above, stay a top tier contraption for isolating information from sound models. Notwithstanding libraries like Librosa giving us a python joke to enrol MFCC's for a sound model, the essential math is to some degree frustrated, so we'll go through it thoroughly and join some significant associations for extra learning. Steps for determining MFCC's for a given sound model:

1) Cut the sign into short housings (of time).

2) Figure the periodogram check of the power range for each edge.

3) Apply the mel channel bank to constrain spectra and entire the energy in each channel.

4) Take the discrete cosine change (DCT) of the log filter bank energies.

5 Simulation

5.1 Importing Libraries

This kernel is useful for any audio classification task.The following libraries are used in this kernel.

1.Tensorflow (for model making and training)

2.sklearn (for splitting the data into train, test, validation)

3.librosa (for loading and feature extraction of audio signals)

4.pandas (for reading csv file)

5.matplotlib (for plotting)

5.2 Loading and Pre-processing

Out of 50 classes, 10 classes are used. Data frame has column "esc10" which contains 10 classes. So, will be using these 10 classes only.

CSV FILE PATH = "esc50.csv" this command actually help to the system to read the csv file and its date to determine the audio files. While the DATA PATH command is the command that gives the path to the folder which contains audio files which are going to execute and determine and classify accordingly.

df = pd.read csv(CSV FILE PATH) this command is working as reading command and writing the output in the form of csv file format which gives us more information about our audio data set and their containing audio files to recognize their categories and segregate accordingly.

5.3 Visualization

Taking one sample from each of 10 classes for visualization. For this step the audio files were segregated and here the libraries like librosa and matplotlib comes in place where they perform the operations on these audio files as audio signals and it takes as an input, then extracting it and plotting them in the manner of Audio signal waveform and mel-spectrograms using MFCC techniques.

The figure 4 and 5 shows the different graphical representation in terms of Dog Audio Signal Waveform and Dog spectrogram respectively.

The figure 6 and 7 shows the different graphical representation in terms of Clock Tick Audio Signal Waveform and Clock Tick spectrogram respectively. Here we extracted two features mel-spectrogram and mfcc. Both features are used for training and mfcc. It performed very well. So, here we are going to use mfcc for training.

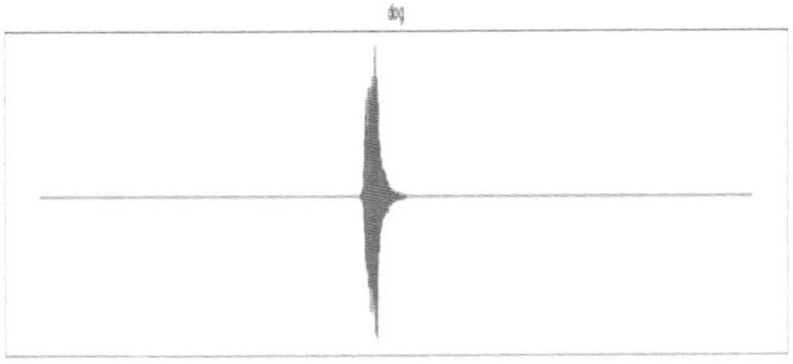

Fig. 4. Dog Waveform

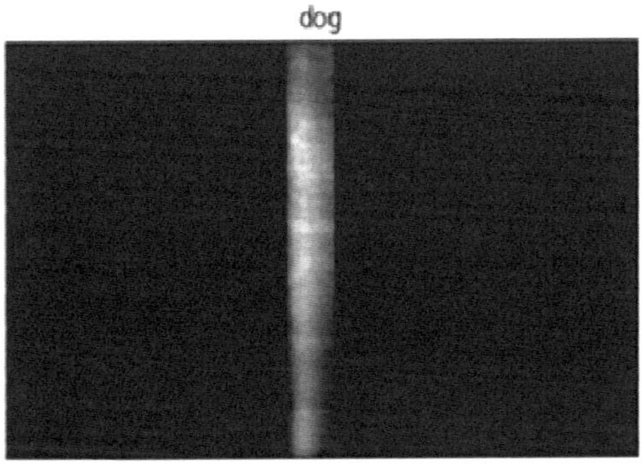

Fig. 5. Dog Spectrogram.

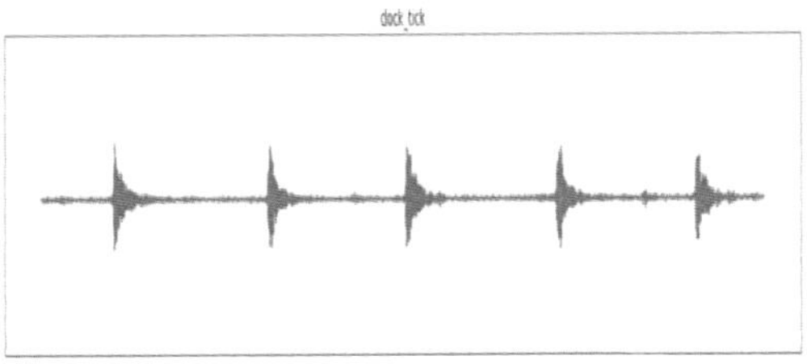

Fig. 6. Clock TIck Waveform.

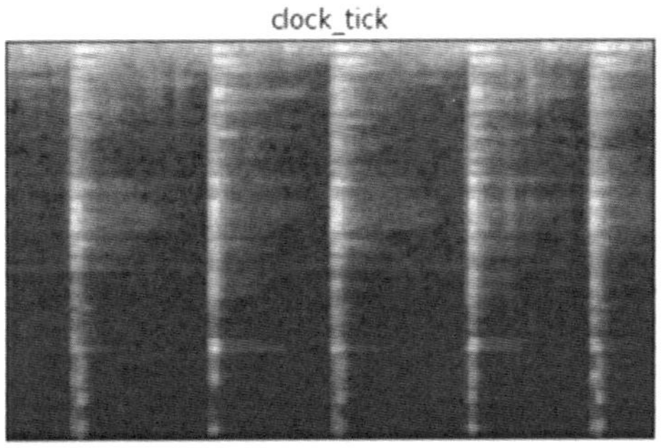

Fig. 7. Clock TIck Spectrogram.

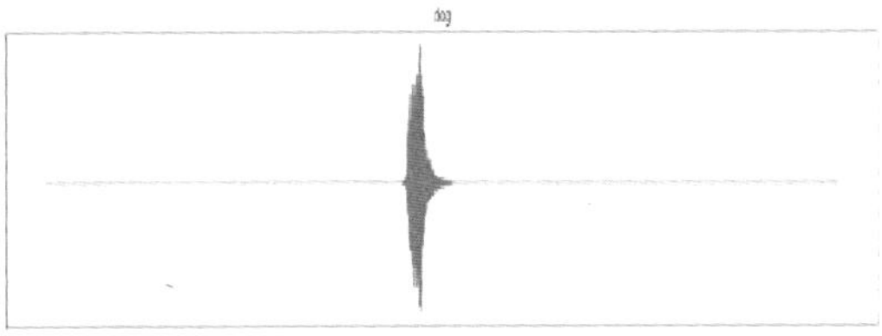

Fig. 8. Dog spectrogram without any distortion (audio clip with removal of external noise)

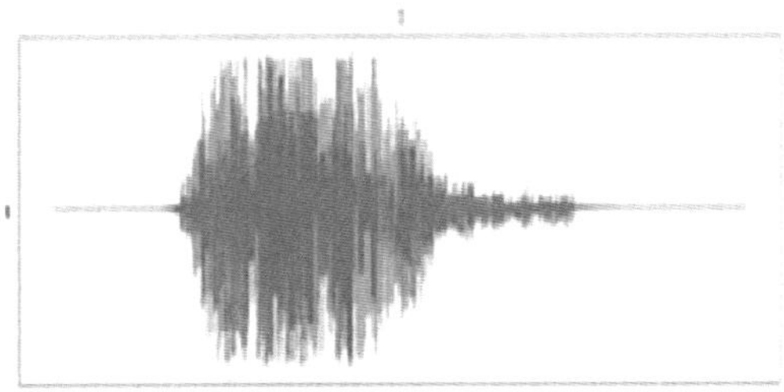

Fig. 9. Dog spectrogram with distortion (audio clip containing external noise)

6 Result and Discussion

In the starting of our work, the errors were mostly in the programming. After solving the errors there was still one disadvantage, that all the required audio files (2000 sound files) could not be uploaded in one go on the google colab platform. so, we have uploaded the most important audio files required for the project output, and we got the spectrogram and audio signal waveform

While testing, we have compared and verified our spectrogram output with the outputs in already published research paper available on the web. On comparing with the outputs from research paper, we are getting the similar spectrogram outputs in our project as shown in figure 8, but there is some noise in the output as there is differences in the audio files and frequency is also different for the audio files which is shown in figure 9.

So, the outputs will never be exact copy of each other's but will be similar.

7 Conclusion

This paper inspects particular sound features, the blends of different sound features, and the effect of sound plans on ESC. To explore the effect of isolating various features on solid affirmation, the part blend procedure for Mel-MFCC, LM-MFCC and T-M is proposed. LM-MFCC is by and large utilized for the sound part input. Consolidating the LM with MFCC highlights, it develops the broadness of sound data. The assessments show that LM-MFCC is even more astounding and uses less cycles to accomplish a steady impact. Its confirmation influence is genuinely higher than that of other part inputs. As the information highlight of the GRU, it accomplishes a 0.92 F1-score on Urban Sound 8K. It performs best among the entirety of the highlights. From the point of view of sound get-together R-MFCC and FR-MFCC, it is shown that the impact of sound arrangement change on strong confirmation is near nothing. Regardless, the progress and modify mix of sound groupings can likewise barely improve the precision of confirmation. The idea system is gotten along with the GRU relationship to patch up the loads. The key of the model in this paper is to make the LM thought weight and MFCC thought weight as close as could really be expected and weight the last suspicion results straightly. The assessment results show that our model shows power.

In coming about examination, we will proceed to study and zero in on the work in the two viewpoints. sound highlights are improved persistently to accomplish pleasant portrayal and be fitting to recollect extraction for huge learning a more appropriate classifier is proposed for sound depiction.

The authors are very much grateful to Ramrao Adik Institute of Technology for providing support with various amenities required to carry out this work properly. An excessive thanks to Dr. Siuli Das who helped to gain knowledge about the system, how to design it and solved many queries.

References

1. S. P. Dewi, A. L. Prasasti, B. Irawan, *2019 IEEE International Conference on Signals and Systems (ICSigSys)*, (2019).
2. https://mikesmales.medium.com/sound-classification-using-deep-learning8bc2aa1990b7
3. https://lionbridge.ai/articles/what-is-audio-classification
4. A. Khamparia, D. Gupta, N. G. Nguyen, A. Khanna, B. Pandey, P. Tiwari, IEEE Access **7**, (2019)
5. V. Boddapati, A. Petef, J Rasmusson, L. Lundberg, *Procedia Computer Science*, **112**, (2017).
6. F. Demir, D. A. Abdullah, A. Sengur, IEEE Access **8**, (2020)
7. M. E. Rahaman, S. M. S. Alam, H. S. Mondal, A. S. Muntaseer, R. Mandal, M. Raihan, *10th International Conference on Computing, Communication and Networking Technologies (ICCCNT)*, (2019).
8. https://www.kaggle.com/doofensmirtz/85-validation-accuracy-tensorflow
9. Zohaib Mushtaq, Shun-Feng Su, *Applied Acoustics*, **167**, (2020).
10. F. Beritelli, R. Grasso, *2nd International Conference on Signal Processing and Communication Systems*, (2008)
11. https://doi.org/10.1155/2019/5803184
12. S. Virkar, A. Kadam, N. Raut, S. Mallick, S. Tilekar, IJERT, **9**, (2020).
13. https://towardsdatascience.com/recognizing-speech-commands-using-recurrentneural-networks-with-attention-c2b2ba17c837

File Fragment Classification using Content Based Analysis

Anirudh Bhat[1,*], *Aryan* Likhite[1,**], *Swaraj* Chavan[1,***], and *Leena* Ragha[1,****]

[1]Ramrao Adik Institute Of Technology, 400706 Nerul, Navi Mumbai India

Abstract. One of the major components in Digital Forensics is the extraction of files from a criminal's hard drives. To achieve this, several techniques are used. One of these techniques is using file carvers. File carvers are used when the system metadata or the file table is damaged but the contents of the hard drive are still intact. File carvers work on the raw fragments in the hard disk and reconstruct files by classifying the fragments and then reassembling them to form the complete file. Hence the classification of file fragments has been an important problem in the field of digital forensics. The work on this problem has mainly relied on finding the specific byte sequences in the file header and footer. However, classification based on header and footer is not reliable as they may be modified or missing. In this project, the goal is to present a machine learning-based approach for content-based analysis to recognize the file types of file fragments. It does so by training a Feed-Forward Neural Network with a 2-byte sequence histogram feature vector which is calculated for each file. These files are obtained from a publicly available file corpus named Govdocs1. The results show that content-based analysis is more reliable than relying on the header and footer data of files.

1 Introduction

There are several ways to restore files. If the file system index is available, standard methods use it to restore the blocks of the deleted file. In computer forensics, file carving is an act of extracting data from a disc drive or other storage device without using the file system that created the file with originality. It is a tool that recovers files in an unallocated space without any file information and is used to retrieve information and performing a digital forensic investigation. It's commonly referred to as "carving," which is a term for the process of organizing raw data using format-specific features present in the format. As a forensics technique that recuperates records dependent on file structure and content and with no matching file system meta-data, file carving is regularly used to recover files from the unallocated space in a drive. Unallocated space alludes to the zone of the drive which no longer holds any document data as demonstrated by the record framework structures like the document table. On account of harmed or missing document framework structures, this may include the entire drive. In straightforward words, numerous file systems don't zero-out the information when they erase it. All things being equal, they essentially eliminate the information on where it is. File carving is the way toward reproducing documents by checking the crude bytes of the circle and reassembling them. This is typically done by inspecting the header (the initial not many bytes) and footer (the last couple of bytes) of a record.

*e-mail: anirudh.b.123@gmail.com
**e-mail: aryanlikhite@gmail.com
***e-mail: swarajchavan540@gmail.com
****e-mail: leena.ragha@rait.ac.in

2 Motivation

File carvers that use file metadata are inherently faulty. Data are fragmented on modern hard drives which imply that to identify the block, the header and footer data can not always be present. File carvers that use the principle of graphs for file fragment recognition are limited to dealing with only bi-fragments. If the header data is tampered with then the 'file' command available in Linux-based file systems fails to recognize the file.

As there are statistical variations between different file formats, the different file format fragment has to be distinguished by using the knowledge that the byte sequence and byte frequency are not the same in each file format. Hence using these features it is feasible to classify different file types using algorithms that would provide high accuracy.

This study can then facilitate a means to overcome the primary limitation of modern file carvers which is the inability to identify the fragments in the hard drive which are not the header or footer fragment and hence can help improve them.

3 Related Work

3.1 Survey of Existing Systems

Many file carvers use "magic numbers" to recover files. A popular tool "Scalpel", available for both Linux and Windows OS, uses such a technique.[5]. The Unix file command is perhaps the most widely used file identification command in history. The file command compares particular regions of a file (usually the file's header and footer) to a database and returns the first match. When dealing with full data, this command is fairly effective. When a file fragment is presented with a file fragment, it means that it is a "data" file. Some complex file carvers use the graph theory to achieve file carving capabilities.[4] In the process of the Format Validation phase, several techniques are studied as well.

Non-header and non-footer based approach was first used in [10]. They used historgrams of ASIC codes which was given to a clustering algorithm. The approach yield an accuracy of 27.7% for byte frequency analysis and 46% for byte frequency cross-relation

In [9] various classifiers were used such as SVM, k-NN, k-means, decision trees and neural networks. Classification speed was the main goal. Also, instead of using hand-engineered features, they used sampled file blocks. They also used the first block of every file to boost their accuracy as the first block contains the magic numbers

Fitzgerald[3] used NLP Techniques for the identification of file fragments. They selected 24 file fragments for their classification. The dataset they created was obtained from the publicly available dataset. They extracted several features which include 1-gram, Shannon Entropy, and Hamming weights. For classification, they went with the route of using a Support Vector Machine combined with a bag-of-words model which is used in NLP. Their approach yielded an accuracy of around 47%

F. Wang[6] used Sparse Coding for N-Gram Feature Extraction. For their dataset, they used the publicly available dataset of Govdocs1. They used the approach of creating a sparse dictionary for N-Grams of higher-order such as N = 4,8,16,32,64 etc. They then used these dictionaries to create N-Gram frequencies of the given file fragment. Support Vector Machine (SVM) was used for the classification and generation of the model. They decided to include 18 file formats in their studies and achieved an identification accuracy of around 61.31%.

Q. Chen et al.[2] took the approach of classification of file fragments by converting fragments to gray scale images and applying deep learning techniques. They took fragments of 512 bytes. These fragments of 512 bytes were then converted into 64x64 greyscale images. These greyscale images were then used to train a convolution neural network (CNN) and train a model. The generated model was trained and tested on 16 file formats from the public dataset of Govdocs1. The accuracy they achieved was around 71%.

In another research paper by F. Wang [7], they used the approach of classification using Convolution Neural Networks. They formalized the task as a multi-class classification problem. The goal was to locate specific byte patterns that belong to an identifiable file type, similar to magic numbers. To achieve this, they used convolution neural networks. The dataset they used was from Govdocs1. They performed tests on 20 different file formats. They performed experiments with several different fragment sizes such as 64 bytes, 126 bytes, 256 bytes, etc. Their results showed that as we increase the size of the fragments used for classification, the accuracy is increased. Also, if the fragment belongs to the start or the end of the file, the classification accuracy is higher as compared to a fragment taken from the middle of the file

Bhatt, M[1] used a different classification technique than most others. Their approach was a mixture of clustering and classification. They separated the file types into complex and simple file types. The simple file types were further differentiated into text files and image files. They extracted a total of 10 features. They used information gain which is a feature ranking metric. They then used Support Vector Machine for the classification of the file fragments. The accuracy they measured was around 67.78%.

3.2 Limitations of Existing Systems

File carvers that use file metadata are inherently flawed as files in modern hard drives are fragmented. This means that the header and footer data may not always be present to classify the block.

File carvers that use the graph theory for the identification of file fragments are limited to working with only bi-fragments[4]. This means that they are incapable of detecting more than 2 fragments. Also, the existing 'file' command available in Linux-based file systems, fails to identify the file if the header data is tampered with. This is because it uses header and footer content for the identification of the file.

Fitzgerald[3] which uses the NLP techniques manages to get an accuracy of around 47.5%, which is less than desirable for use in real-world data. In a real scenario, if a file carver was able to correctly identify less than half of the files, there would be a huge time loss for the forensic expert.

F. Wang[6] used Sparse Coding. In this approach, they created a sparse dictionary of N-Grams. While this method is suitable when the value of N is kept low, the time for calculating the feature vector increases exponentially as we increase the value of N. Their results show that higher values of N yield far better accuracy than lower values of N. So, to achieve acceptable accuracy's, we need to increase the value of N, but increasing the value of N makes the processes of feature calculation more time consuming

Q. Chen et al.[2] converted the file fragments into greyscale images and achieved an accuracy of around 70.4%. While the accuracy is high compared to most other studies, converting fragments into greyscale images and passing them through a CNN, is not feasible in real-time applications. Real-time applications need to be fast and should be able to work on a relatively large database.

Bhatt. M[1] used the Hierarchy-Based File Fragment Classification method. They classified simple file correctly but for complex file, often it failed to classify the specific type of complex file accurately. Moreover, the time required for the feature extraction for one file was very high, and this time can add up when working with a large sum of data in real-world situations.

Besides all this, none of the studies have provided a codebase to test their results or use them as a base for further research in this domain. The experiments are done on different datasets and parameters, so the results and not comparable.

https://doi.org/10.1051/itmconf/20214003025

4 Proposed Approach

4.1 Proposed Work

The first step was to gather a bunch of files for training and testing purposes from the Govdocs1 dataset. *Govdocs1* is a publicly available corpus of files. Govdocs1 contains approximately 900,000 files with 213 different file types. But the distribution of these file types is not uniform. Some file types appear more than 150,000 times while some file types have less than 10 files. Hence there is a huge disbalance between the different file types. For an accurate and fair test, it is ideal that every file type in the training and testing data-set should have an equal count of files. This is required so that the models do not become biased to one file type due to an excessive amount of training on particular file types. To overcome this problem, it was decided to select 13 extensions based on their frequency in the Govdocs1 data-set and their relative popularity.

The Govdocs1 corpus provides pre-packaged "threads". These threads contain 1000 files randomly sampled from the main corpora specifically meant to be used for research purposes. The process began by downloading multiple such threads and collecting all the files in a single location. Then, out of the 13 selected file formats, 300 random files were sampled for each of the 13 file types, 200 for training, and 100 for testing. This formed the main data-set for training and testing the models. The 13 file formats used were as follows

PDF	HTML	BMP	JPG
RTF	PNG	DOC	TXT
XLS	GIF	XML	PS
CSV			

For any machine learning algorithm, feature extraction is the most important step. In [1] a variety of features were tested and ranked using a feature ranking metric called information gain. The results showed that Unigram Count Distribution or Byte Histogram, was the most dominating feature with an information gain score of 3.807 while other features had scores in the low 1's and 0's. A byte histogram of a file is the distribution of occurrence of each byte in the file. Since there are 8 bits in a byte, there is 256 such byte that every file can contain.

Figure 1, visualizes the Byte histogram of several file types by converting the histogram into 16x16 grey-scale images. From the image, one can observe that the distribution of bytes in different file formats varies vastly. This visualization shows us that using a byte histogram as a feature vector is possible and can yield good results.

Although using single bytes as a feature vector has proven to be useful[6], a lot of information is lost when considering only a single byte. There are certain sequences of bytes that occur more in a specific file type. For example, in JPG files, before the byte sequence of "FF" the byte sequence of "00" is stuffed as it is used as an escape character. This makes the frequency of byte sequence "00" much higher as compared to other file types. Another example that can be given is that of HTML files. HTML

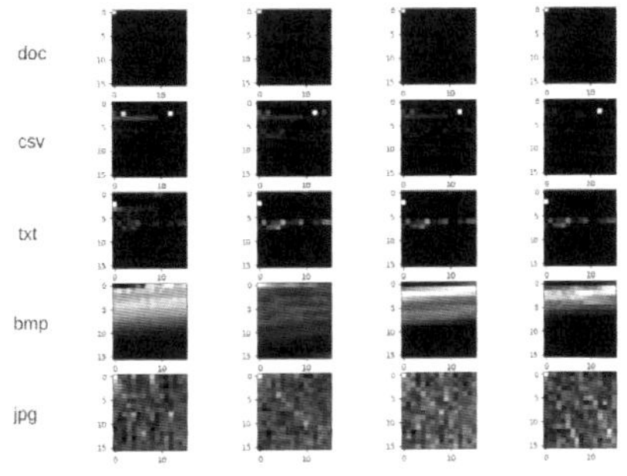

Figure 1. Byte Histogram of several file types converted to Greyscale Images

files contain a large number of "</" character sequences, so subsequently, the byte sequence of "</" is higher in HTML files than most other files. Hence instead of considering single bytes, it is better to consider a sequence of 2 bytes to gain more information and possibly increase the accuracy. To test this, we ran multiple tests with SVM as well as MLP with uni-gram and bi-gram as data input. The results are discussed in Section VI. Based on our results, instead of using a single byte as a feature vector, 2-byte sequence histograms are used as the feature vector. As there is a 256 possible byte, considering a sequence of 2 byte makes the possible feature vector of size 256*256 that is 65536.

4.2 Proposed Methodology

As discussed in the above section, a 65536 sized array is used as the feature vector. A data set of 13 file types with 200 files for training and 100 files for testing for each of the 13 file types is created. This brings us to a total of 1300 files. For each of these 1300 files, a 2-byte sequence histogram is generated. The actual process of generation of this histogram is explained in detail in Section 5.1. After generation of the histogram, they are saved in a CSV file separated by training and testing histograms respectively. This data is then tested on a Feed-Forward Neural network as well as a Convolution Neural network to test which neural network performs better.

Feed-Forward Neural Networks is a kind of Artificial Neural Networks. They are made of the basic building block of computational units called neurons. A feed-forward network is generated by using multiple neurons in multiple layers. They are arranged in a way that the output of each neuron of one layer is fed as input to every neuron in the next layer. Each of these neurons also has an activation function, which ensures a degree of non-linearity in the model. Each layer can have a different number of neurons and there can be different number of layers depending on the capacity of the learning needed. However, the number of neurons in the first input layer and the last input layer are fixed. The number of neurons in the first input layer must match the number of features and the number

https://doi.org/10.1051/itmconf/20214003025

of output neurons must match the number of possible outcomes. Furthermore, the choice of activation function can make a big difference. The network was tested with several hidden layers and in the end, it was decided to settle with 4 hidden layers based on performance speed and accuracy. As for the activation function, after testing with multiple ones, it was found that "RELU" appeared to be the best, and the least time-consuming choice.

For testing our data with a Convolution Neural Network, the data was first reshaped into a 256x256 matrix. This data was then passed through 2 layers of a convolutional network followed by a feed-forward neural network.

After a model is generated using the training data, the model is then saved and used to power a command-line utility. A command-line utility will provide ease of access to the end-users and will act as a replacement for the standard "file" command used by Digital Forensic Experts.

5 System Design

The design of the system is mainly divided into 4 major components, namely Data Collection, Data Processing, Training and Testing, and the Command Line Tool. Each of these phases is explained in detail below

5.1 Data Collection

The first step in data collection is downloading the threads from the Govdocs1 dataset. These threads are a collection of 1000 files. 9 such threads were downloaded and then finally collected in a single folder. The next thing to decide is how many file formats to consider. Based on the number of files present for each file format, 13 of the most commonly occurring file formats are picked. These 13 file formats should have an equal number of files for each format. Hence randomly, 300 files for these 13 formats are sampled from the set. From these 300 files, 200 files are sampled for the training set, and the remaining 100 files are reserved for the testing set. With this, an equal number of files for each file format is present so that the training data is not dominated by only 1 single file.

5.2 Data Processing

In the data processing phase, these files are needed to be converted into their respective 2 pair byte histogram. To achieve this, first, start with a file and load it into memory. Now a random starting point is selected from the file such that the starting point is at least 4096 bytes behind the end of the file. After picking the starting point, then read the next 4096 bytes. This represents the fragment. 4096 bytes are selected because that is the most common fragment size in modern hard drives. For every sampled fragment, the raw hex data is analyzed and the 2 pair Byte Histogram is generated. This forms the feature vector and is saved in a file separated by training and testing data. This process is repeated for every such file.

5.3 Training and Testing

The train.csv file is first loaded into the memory. Then a Feed-Forward Neural network represented as FNN in the diagram is created based on the architecture discussed above. All the data from the train.csv file is used to train the created Feed Forward Neural Network. After successful training, load the test.csv file into memory which has the testing data. This data is then fed into the trained FNN, which classifies the file. This classification is then compared with the real file format and hence the accuracy is measured. The generated model is saved for powering the command-line tool.

5.4 Command Line Tool

The command-line tool is used as the final product which users can use to classify the fragments. When the command-line tool is used, first, the saved model is loaded into memory. The user specifies a file for which they wish to know the extension. This file is then read and then a 2 pair byte histogram is generated for the file. This generated Byte Histogram is fed into the FNN which classifies the file into one of the 13 selected formats. This classification result is then shown to the user

6 Results

Figure 3 shows the results of the tests of unigram vs bigram. In total, 4 test runs were conducted with different datasets to make the results more authentic. The two classifiers used were SVM and MLP and the features were chosen were Unigram and Bi-Gram. From the bar graph, it can be seen that MLP with Bi-Gram features was the most successful. This proves that the theory for going with the Bi-Gram feature as it retains more information was correct because it provided us with more accuracy as compared to going with Unigram features.

Architecture	Accuracy
65536x1024x512x13	90.32
65536x1024x256x13	89.92
65536x512x256x13	89.43
65536x256x128x13	89.00
65536x512x13	88.7

Table 1. List of architecture and their accuracy

Table 1 represents the result of our tests conducted with different architectures with MLP. It can be observed that the accuracy is more the less similar even after using different architectures. This shows that the architecture of the Neural Network doesn't affect our results significantly. The best performing architecture was the one with 2 hidden layers of 1024 and 512 neurons respectively

The tests were also conducted on a Convolution Neural Network. The result of Convolution Neural Networks was found be around 75% - 80%. This is because histograms do not posses any spatial data. CNN's are better used for

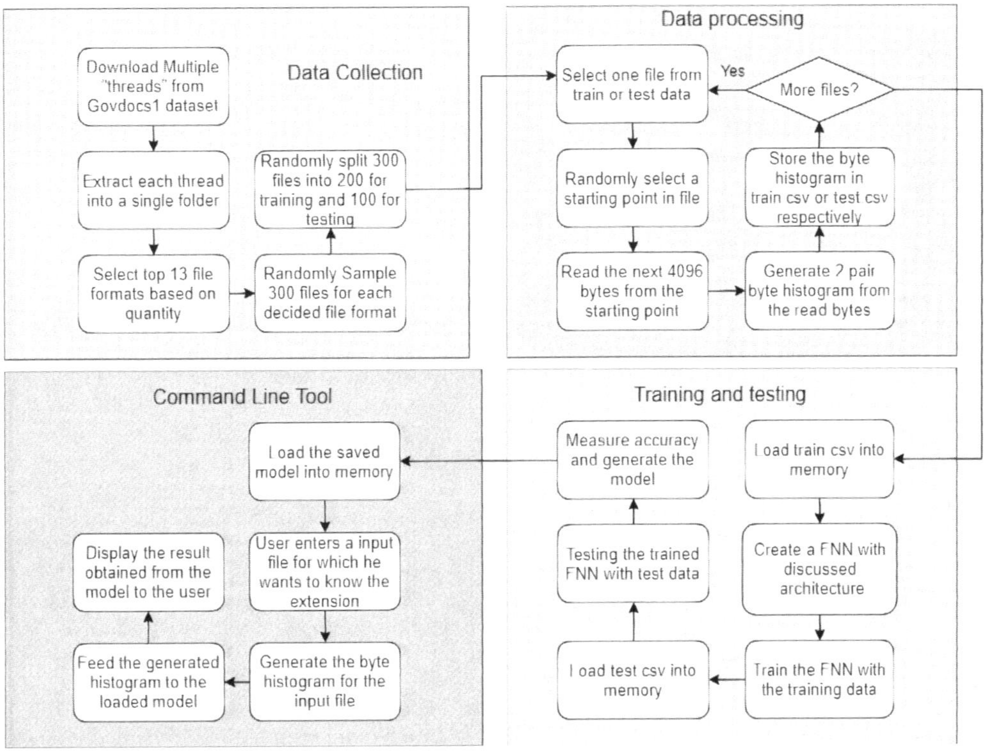

Figure 2. System Design which shows the workflow of the entire system

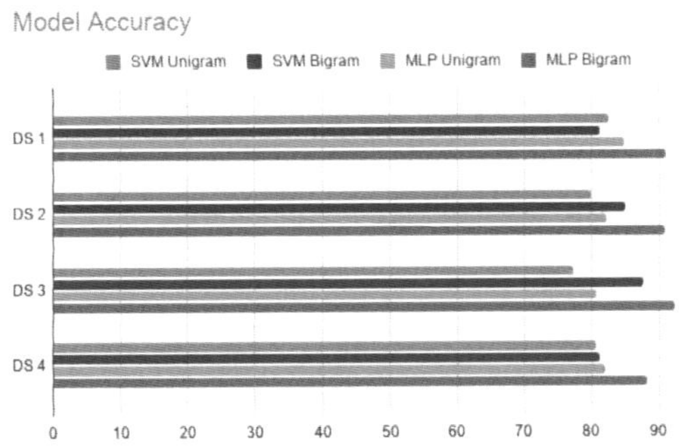

Figure 3. Accuracy of models used on different data sets

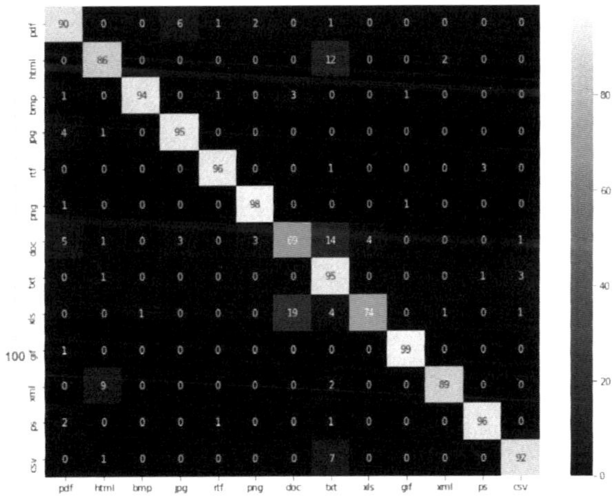

Figure 4. Accuracy of models used on different data sets

image datasets. Hence in comparision with CNN, MLP performs much better.

Figure 4 shows the confusion matrix of our results for the MLP Classifier. It is evident that for the majority of the file formats, the MLP Classifier works well and classifies the files correctly. We can see that for HTML files, some of the HTML files were classified as TXT files. The predictions are made on a 4096-byte fragment. Due to the small nature of the fragment, it is quite possible that the particular fragment selected from the HTML files contained only pure text part and hence it classified them as TXT files. We also observe that many of the XML files were classified as HTML files. These two file formats have a lot in common, such as both of them use "tags" and contain a lot of text. The similarity between these file formats is the reason that XML files are getting classified as HTML files.

https://doi.org/10.1051/itmconf/20214003025

Upon examination of PDF files, it was observed that a large portion of PDF files contained images. The images in PDF files are stored as a separate object that contains the raw binary data of the image. Similar to the HTML and TXT confusion, due to the small size of the fragment, it is quite possible that the fragment extracted from the PDF files contained only the pure image raw bytes and hence some PDFs are classified as jpg and png

The worst performing file types were doc and xls. Both these file types can be considered "complex" file types. A file type can be considered complex when it combines elements from different file formats like text and images. Looking at the results of XLS files, we can see that a lot of XLS files were classified as doc files. The results may seem confusing at first as at a surface level, we do not find any similarities between these two file formats. After opening these two file formats in a hex editor and taking a look at their hex data, it was observed that the byte sequence between these two file formats is very similar. In both the file formats, doc as well as xls, there is a big portion where the bytes "FF" or "00" are repeated for long stretches. Naturally, the bigram frequency of the bytes "FF" and "00" are high in both these file formats as compared to any other file format. Due to such similar nature of raw binary data between these two file formats, many of xls files are classified as doc files. The worst performing file format was DOC. The majority of doc files were classified as txt files. This behavior is expected as the majority of doc files are filled with long stretches of texts, so the fragment extracted from these doc files during the training and testing period might only contain text. Some doc files are classified as xls files. The justification for this is the same as to why xls files are classified as doc files. Some doc files are also classified as png and jpg files. Doc files often contain images as well along with the text. Hence the reason for some doc files getting classified as jpg or png is that the particular fragment tested might only contain raw image bytes.

Overall, the confusion matrix shows us that the current classification method works well with simple file formats which only contain one specific type of data and works a little worse for complex file formats which usually contain multiple types of data.

7 Conclusion and Future Work

After several trials, it was found that using single bytes can cause a lot of information to be lost, making the accuracy a bit low. Using 2 bytes helps to retain much more information hence improving the accuracy. It can also be concluded that Feed Forward Neural network works much better than SVM for this particular scenario

The main objective of this paper was to utilize all the research work that has been done in the field of file fragment classification and create a tool that is available for everyone to use on their local machine. In these experiments, only the top 13 of the most frequently used file formats were considered, due to hardware and resource constraints. In an ideal world, the training should be done on a large dataset with many varieties of file formats. This is where the open-source part of the project comes to use. Anyone with more resources or a server with better hardware can use this codebase to train more datasets, and possibly find more features to extract and improve on the original codebase.

References

[1] Bhatt, Manish, Avdesh Mishra, MdWasi Ul Kavbir, S. E. Blake-Gatto, Rishav Rajendra, Md Tamjidul Hoqueand Irfan Ahmed. "Hierarchy-Based File Fragment Classification." *Machine Learning and Knowledge Extraction* 2, no. 3 (2020): 216-232.

[2] Chen, Qn, Qing Liaoe L. Jiang, Jun Fang, Siuming Yiu, Guikai Xi, Rong Li et al. "File fragment classification using grayscale image conversion and deep learning n digital forensics." In *2018 IEEE Security and Privacy Workshops (SPW)*, pp. 140-147. IEEE, 2018.

[3] Fitzgerald, Simran, George Mathews, Colin Morris, and Oles Zhulvyn. "Using NLP techniques for file fragment classification." *Digital Investigation* 9 (2012): S44-S49.

[4] Pal, Anavndabrata,v and Nasir Memon. "The evolution of file carving." *IEEE sigvnal processing magazine* 26, no. 2 (2009): 59-71.

[5] Poisel, Rainer, Simon Tjoa, and Paul Tavolato. "Advancevd file carving approaches for multivvmedia files." J. *Wirel. Mob. Networks Ubiquitous Comput. Dependable Appl.* 2, no. 4 (2011): 42-58.

[6] Wang, Felix, Tu-Thach Quach, Jvason Wheeler, James B. Aimone, andv Conrad D. vJames. "Sparse coding for n-gram featurve vextravction and training for file fragmenvt cvlassifvicativon." v*IEEE Transactions ovn Information Forensics and Security 13*, no. 10 (2018): 2553-2562.

[7] Wang, Yanchao, Zhongqian Su, and Dayi Song. "File fragment type identification wvith convolutional neural networks." In Proceedings of the 2018 International Conference on Machine Learning Technologies, pp. 41-47. 2018.

[8] Lee, Seokjun, and Taeshik Shon. "Improved deleted file recovery technique for Ext2/3 filesystem." The Journal of Supercomputing 70, no. 1 (2014): 20-30.

[9] Ahmed, Irfan, Kyung-Suk Lhee, Hyun-Jung Shin, and Man-Pyo Hong. "Fast content-based file type identification." In IFIP International Conference on Digital Forensics, pp. 65-75. Springer, Berlin, Heidelberg, 2011.

[10] McDaniel, Mason, and Mohammad Hossain Heydari. "Content based file type detection algorithms." In 36th Annual Hawaii International Conference on System Sciences, 2003. Proceedings of the, pp. 10-pp. IEEE, 2003.

https://doi.org/10.1051/itmconf/20214003026

Various Approaches of Machine Translation for Marathi to English Language

Nilesh Shirsath[1,3,*], *Aniruddha* Velankar[2,**], and *Ranjeet* Patil[3,***] *Dr.Shilpa* Shinde[3,****]

[1]Department of Computer Engineering, Ramrao Adik Institute of Technology, India

Abstract. Machine Translation (MT) is a generic term for computerised systems that generate translations from one natural language to another, with or without human intervention. Text may be used to examine knowledge, and turning that information into pictures helps people to communicate and acquire information.There seems to be a lot of work conducted on translating English to Hindi, Tamil, Bangla and other languages. The important parts of translation are to provide translated sentences with correct words and proper grammar. There has been a comprehensive review of 10 primary publications used in research. Two separate approaches are proposed, one uses rule based approach and other uses neural-machine translation approach to translate basic Marathi phrases to English. While designed primarily for Marathi-English language pairs, the design can be applied to other language pairs with a similar structure.

1 Introduction

Machine Translation (MT) is a common name for computerized systems which are responsible for generating, with or without human assistance, translations from one natural language into another.It is part of Natural Language Processing (NLP) where translation from the source language to the target language is conducted, preserving the same meaning of the phrase To help them make text and speech into another language, humans can use Machine Translation Systems. The program can run without any human intervention. The conventional approach is achieved for the MT to translate large quantities of knowledge involving terms that could not be interpreted. The MT performance level can differ considerably , MT programs need "training to improve the quality of the outcome in the relevant domain and language pair.

As MT can be classified into different categories

- Rule-based Systems: uses a combination of language and grammar rules
- Statistical Systems: learn to translate by analyzing large amounts of data
- Neural Machine Translations (NMT): learn to translate through one large neural network (multiple processing devices modeled on brain)

Different MT providers like Google translator, Yandex Translator etc. provide some translation tools with ethics to customer like:

- Confidentiality – There is no confidentiality in the content translated by online MT platforms such as Google and Microsoft translators. Translated data is stored by the owners of the platform and may later be reused.
- Notifying the Client about MT Use - Whether a translation company should notify customers about the use of MT for their projects is a point of debate in the industry. Many are in pursuit of informing the customer of the use of MT and others may not disclose the use of MT. If you have questions about MT use, be sure to ask your provider.

1.1 Objective

The main Objective of the project is to implement a tool which translates Marathi sentences to English without changing the meaning of the sentence using the Rule Based and Neural Based System

1.2 Motivation

As the steady progress in the field of technology, the Internet's growth has also increased at a tremendous rate. With globalization, the official language of the globe has been English. In Marathi literature, there are approximately 71 million Marathi speaking individuals and various works. However, Marathi language is comprehensible for a very small group of people so a system is proposed, which takes the input Marathi sentence and translates it into English which is an understandable language.

2 RELATED WORK

A survey of the research done for data summarization and the currently existing systems, give the following results.

A. **Transmuter: An Approach to Rule-based English to Marathi Machine Translation.[1]**

*e-mail: Nileshshirsath2389@gmail.com
**e-mail: aniruddhav25@gmail.com
***e-mail: patilranjeet3699@gmail.com
****e-mail: shilpa.shinde@rait.ac.in

The basic method used to implement this framework is Rule Based Machine Translation.The model is built on a generalised approach based on the categories/domains to which a term belongs.The proper spelling of words The target language was created based on the parse tree's basic traversals.The architecture is partly applied in the context of a Machine Translation method.

A generalized approach based on the categories/domains to which a word belongs is the model.In order to achieve better quality translations, the number of rules formed is high for the target language generation.The consistency of the translation of this method depends on The size of the knowledge base of grammar. If the size exceeds this threshold, due to contradictory laws, the consistency can decrease.

B. **Script Translation System For Devanagari To English. [2]**

The proposed scheme will translate more than one Devanagari word to English using the rule-based approach. This is accomplished by understanding the different parts of speech in Marathi phrases. In a bilingual dictionary, tokenizing and describing each word's English meaning and obtaining a clear translation.

The machine translation based on the rule includes generating a number of rules and handling their exceptions as well.Compared to dictionary-based methods that include word-to-word translations, rule-based machine translation provides better translation quality.Given the number of laws to be used in the scheme, flawless translations for each and every sentence will not be done.

C. **Part of Speech Tagger for Marathi Language.[3]**

The rule-based element of the speech tagger, which uses a set of handwritten rules to apply words to all potential tags.The system uses a morphological analyzer to identify the root word and compares it to the corpus to assign appropriate tags.Where an expression has more than one suffix, grammar rules are used to reduce ambiguity.Dictionaries are required in order to assign appropriate tags to each expression.

The basic standard of useful instructions aids in avoiding ambiguity.Because of the lack of corpus for statistical analysis, POS tagging is difficult for the Marathi language. When there is a broad range of meaningful rules to prevent disambiguation, the rules-based POS tagger will achieve greater accuracy.Additional meaningful rules need to be provided to improve the performance of the system

D. **Inflection Rules for English to Marathi Translations[4]**

When it comes to getting the right translation, inflection is crucial.Inflection is the process of changing the meaning of a word by adding the proper suffix to it according to the structure of the sentence.The implementation of the Inflection for English to Marathi Translation is presented in this paper.

The inflection of nouns, pronouns, verbs, and adjectives in a sentence is determined by the other words and their qualities. The rules for inflecting the above Parts-of-Speech are presented in this document.

E. **Marathi to English Neural Machine Translation with Near perfect corpus and transformers [5]**

To translate Marathi sentences into English, the device uses the Neural Machine Translation (NMT) method.It focuses on transformer based architecture.All of the transformers' findings were equivalent to Google Cloud API-V2. In MAE and RMSE, it also outperformed Google API. This study suggests that this is the case.

From the results and examples it is observed that the proposed transformer-based model was able to outperform Google Translation with limited but almost correct parallel corpus.The system produced satisfactory results by making use of word piece tokenizer but in order to improve the performance sentence piece tokenizer must be implemented and the results need to be compared.

F. **Challenges in Rule based machine translation from Marathi to English [6]**

In the field of machine translation, translation divergence is a difficult problem to solve. For proper understanding and identification of divergence issues in machine translation, a thorough investigation is needed. It's a time-consuming process to use Rule Based Machine Translation. The development of a large number of laws necessitates a great deal of human effort.To accomplish these translations, a set of rules must be created.

They explained the different forms of divergence patterns in the Marathi and English language pair in this article. In addition, these divergence trends must be identified and classified.This method's consistency is determined by the scale of the grammatical information base. The scale and depth of the information base grows as more exceptional cases are treated. When the scale of the information base grows, so does the precision, up to a certain point.If the size exceeds a certain threshold, the accuracy can suffer as a result of contradictory laws. As a result, a scheme relying on this method must strike a balance between precision and the number of exceptions it can accommodate.

G. **A survey on LSTM memristive neural network architectures and application [7]**

The first generation of machine translation methods used dictionary-based methods to do word-to-word translations. Its flaws prompted the development of the second generation, which used rule-based and transfer-based techniques.

It has been discovered that rule-based machine translation necessitates the development of a large number of rules as well as the treatment of their exceptions. The system is feasible up to a point, but this approach would have higher translation accuracy.The emphasis of this paper is on Marathi to English translation based on rules.

The Marathi to English translation system will aid in the automation of the process of translating documents and scripts, as well as the reduction of manual translation work.In some sentence translations, there may be some disambiguation. The translation rules, on the other hand, will be framed in such a way that generic sentences or sentences from other domains will be translated.

H. **A Baseline Neural Machine Translation System for Indian Languages [8]**

The research provides a Neural Machine Translation system for Indian languages that is both simple and effective.It establishes a firm baseline for future research by demonstrating the viability of numerous language pairs.

Even if there aren't enough resources, this method, which uses cutting-edge machine learning techniques, produces competitive outcomes for many language pairs.They investigate the multilingual teaching scenario for the Indian language, employing a variety of tried-and-true strategies to get competitive results.These tasks entail calculating embeddings for later classification or sentiment analysis tasks.Indian languages will benefit from the ability to transfer learning from well-tested embeddings, such as English.

I. **On the Properties of Neural Machine Translation: Encoder-Decoder Approaches [9]**

An encoder and a decoder are frequently used in neural machine translation models.From a variable-length input text, the encoder extracts a fixed-length representation, from which the decoder creates an accurate translation.

This paper analyses the properties of neural machine translation using two models: RNN Encoder–Decoder and a newly designed gated recursive convolutional neural network in this research.They show that neural machine translation performs rather well on short phrases with few unknown terms, but that its performance rapidly degrades as the sentence length grows.It also demonstrates that the suggested gated recursive convolutional net-work automatically learns the grammatical structure of a phrase

J. **Neural Machine Translation using Recurrent Neural Network [10]**

The constructed system integrates Recurrent Neural Network models to achieve the maximum accuracies in 10 epochs.It was discovered that using many models at the same time resulted in a better model with improved overall accuracy for the system.

According to the findings of the studies, bidirectional Recurrent Neural Networks with encoding algorithms have a higher accuracy than the other Recurrent Neural Network models. Single Recurrent Neural Network models, such as the basic RNN, only RNN with encoding, only bidirectional RNN, and RNN with encoder and decoder, have lower accuracy than RNN models with encoder and decoder.

K. **BLEU: a Method for Automatic Evaluation of Machine Translation [11]**

According to the report, BLEU will speed up the MT RD cycle by allowing researchers to quickly zero in on useful modelling concepts.A new statistical investigation of BLEU's association with human judgement for translation into English from four distinct languages (Arabic, Chinese, French, and Spanish) representing three different language families supports this viewpoint.

The strength of BLEU is that it has a strong correlation with human assessments since it averages out individual sentence judgement errors over a test corpus rather than attempting to discover the exact human judgement for each sentence: quantity leads to quality.Finally, because MT can be thought of as the generation of natural language from a textual environment, the BLEU might be used to assess MT tasks.

L. **Sequence to Sequence Learning with Neural Networks [12]**

On a large-scale MT challenge, it was demonstrated that a big deep LSTM with a limited vocabulary and essentially no assumptions about problem structure can outperform a traditional SMT-based system with an unlimited vocabulary. Given the success of the simple LSTM technique on MT, it should work well on a variety of other sequence learning problems as long as there is adequate training data.

It was determined that finding a problem encoding with the highest number of short-term dependencies is critical since it makes the learning problem considerably easier.They were unable to train a standard RNN on the non-reversed translation problem using this method, but they believed that when the source sentences were reversed, a standard RNN should be easily trainable.

2.1 Limitations of Existing System

- As Marathi language is less researched and progressed on the grammar and translation front the existing systems fail to incorporate all the rules necessary for translation.
- In the Machine Learning approach only a few translation models are implemented which are not providing satisfactory and correct translation in the testing phases due to the lack of the data.

3 METHODOLOGY

3.1 Proposed Work

3.1.1 Rule Based Machine Translation:-

A system is proposed using rule-based machine translation which follows a sequential procedure of taking input Marathi sentences followed by the machine translation operations such as Tokenization, Tagging etc. Which on the successful implementation tries to predict the most accurate translation of the given sentence.

3.1.2 Neural Machine Translation:-

An Encoder Decoder model will used, which helps to analyzes corpus of data and by understanding pattern between them and for generate the translations.

3.2 Techniques

3.2.1 Rule Based Machine Translation:-

Rule-based translation mostly depends on different built-in linguistic rules and millions of bilingual dictionaries for each pair. Translations are done on vast and revolutionary linguistic rules. Automatic translation systems are focused on linguistic knowledge about the source and target languages, essentially derived from dictionaries and grammars (unilingual, bilingualor multilingual) covering the key semantic, morphological and syntactic regularities of each language.

An RBMT method produces output sentences (in some target language) for input sentences(in some source language) based on morphological, syntactic, and semantic study of both the source language and the target involved in a specific task of translation.

3.2.2 Neural Machine Translation:-

In NMT model, a single system can be trained directly on source and target text. Unlike other system NMT works cohesively to maximize its performance and it also used vector representation for words and internal state. The NMT uses a bidirectional recurrent neural network,also called an encoder, to process a source sentence into vectors for a second recurrent neural network, called the decoder, to predict words in the target language. This process, while differing from phrase-based models in method, prove to be comparable in speed and accuracy

3.3 Design of the System

3.3.1 Rule Based Machine Translation:-

1. **Source Text**: Takes Marathi sentences as an input text.
2. **Tokenization**: Sentence is broken down into token words.

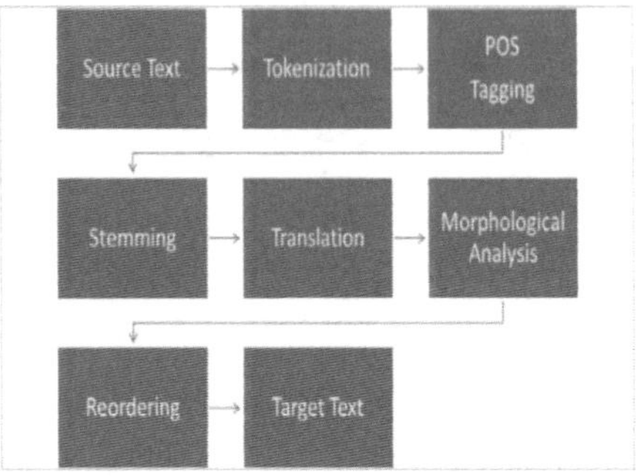

Figure 1. System Design of Rule Based Machine Translation

3. **POS tagging:** Part of speech of each token is determined.
4. **Stemming:** From the token root word is stemmed out for translation.
5. **Translation:** Translation of Marathi words is found using a bilingual dictionary
6. **Morphological Generation:** Grammatically correct words are generated according to the suffixes.
7. **Sentence Reordering**: Morphologically Generated words are reordered according to the written grammar rules.
8. **Target Text**: Translated English Sentence is obtained.

3.3.2 Neural Machine Translation:-

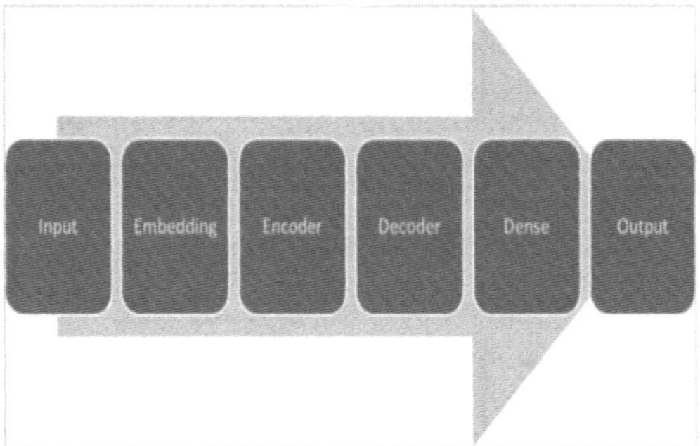

Figure 2. System Design of Neural Machine Translation

1. **Input**:Source sentence which is to be translated into an integer encoded form.
2. **Embedding:**In this layer we will map each integer encoded word to an vector which will act as input for next layer.There are various word embedding techniques which map(embed) a word into a fixed length vector.

3. **Encoder:**The encoder reads the input sequence and summarises the data in what are known as internal state vectors (in case of LSTM these are called as the hidden state and cell state vectors). The encoder outputs are discarded, leaving only the internal states.
4. **Decoder:**The initial states of the Decoder LSTM are set to the final states of the Encoder LSTM. The decoder begins generating the output sequence using these initial states.
5. **Dense:**It's used in this network explicitly to adjust the LSTM layer's dimensions to the one we like. In this case, the network's output is expected to be a probability distribution over all of the vocabulary terms.
6. **Output:**Predicted translation of input sentence into one hot encoded form.

4 RESULT AND ANALYSIS

After analyzing the papers and implementing RBMT approach we were able to extract a few linguistic rules related to Marathi language which are helpful for translation of sentences.After studying many Marathi sentences we realized the derived words are formed by adding suffixes in most of the cases.We collected some of the common suffixes and separated on the basis of POS tags for the simplification purpose.System is able to translate the sentences having the structures and rules which were analyzed but the system struggles to translate the sentences which occur having the different and complex structure.

On the other hand NMT achieves decent translation whose accuracy can be judged by BLEU score.It does not requires dedicated POS tagging or stemming since the model learns it while training.To achieve great quality of translation using NMT large corpus of cleaned data is required and also the large computational power is necessary to train the model.NMT is able to use algorithms to learn linguistic rules on its own from statistical models.The biggest benefit to NMT is its speed and quality.NMT makes faster translations than the Rule Based method and has the ability to create higher quality output.

BLEU-1	0.865143
BLEU-2	0.800385
BLEU-3	0.758851
BLEU-4	0.660873

Accuracy of model on training data:

BLEU-1	0.623246
BLEU-2	0.507332
BLEU-3	0.455952
BLEU-4	0.341381

Accuracy of model on testing data:

5 CONCLUSION AND FUTURE WORKS

The RBMT approach shows how exactly translation works and give almost perfect translation of given sentence if rule for it is written correctly and development of a such large number of laws required great human effort.As the rule increases the consistency of result will decrease because of treatment of some exceptions but it produces the result correctly and it's also depends on scale of grammatical information base.The system is feasible up to a point, but this approach would have higher translation accuracy.

About NMT approach its not necessary to have deep linguistic knowledge, but we required good amount of cleaned data so the model can train properly and develop its own rule in order to give proper translation.The BLEU-4 score of about 0.7286, provides an upper bound on what we might expect the efficiency from this model. A BLEU-4 score of about 0.3510 was achieved, providing a baseline skill to improve upon with further changes and improvements to the model.

The accuracy of Rule Based Translation may be improved by developing better rules, which may be achieved through more language research and assistance from linguists or language experts. A huge corpus of high-quality labelled data for the Marathi language can help enhance machine translation utilizing a neural-based methodology.

6 ACKNOWLEDGMENT

The authors thank the many people who have done lots of help for us and provided useful guidance for our paper.

References

[1] G V Garje,G K Kharate,Harshad Kulkarni,"Transmuter: An Approach to Rule-based English to Marathi Machine Translation",(2014).

[2] Jayshri A.Todase and Sushama Shelke ,"Script Translation System For Devnagari To English",(2018).

[3] Sharvari Govilkar, Bakal J W and Shubhangi Rathod,"Part of Speech Tagger for Marathi Language",(2015)

[4] Inflection Rules for English to Marathi Translation , IJCSMC, Vol. 2, Issue. 4, April 2013, pg.7 – 18

[5] Swapnil Ashok Jadhav,"Marathi To English Neural Machine Translation With Near Perfect Corpus And Transformers",(2020).

[6] Namrata G Kharate and Dr. Varsha H. Patil,"Challenges in rule based machine translation from marathi to english",(2019).

[7] G V Garje,Adesh Gupta,Aishwarya Desai,Nikhil Mehta,Apurva Ravetkar,"Marathi to English Machine Translation for Simple Sentences",(2014).

[8] A Baseline Neural Machine Translation System for Indian Languages,Jerin Philip, Vinay P. Namboodiri, C.V. Jawahar",(2019).

[9] Cho, Kyunghyun van Merriënboer, Bart Bahdanau, Dzmitry Bengio, Y.. (2014). On the Properties of Neural Machine Translation: Encoder-Decoder Approaches. 10.3115/v1/W14-4012

[10] Neural Machine Translation using Recurrent Neural Network,International Journal of Engineer-

https://doi.org/10.1051/itmconf/20214003026

ing and Advanced Technology (IJEAT)ISSN: 2249 –8958,Volume-9 Issue-4, April 2020

[11] BLEU: a Method for Automatic Evaluation of Machine Translation , Proceedings of the 40th Annual Meeting of the Association for Computational Linguistics,(2002).

[12] Sequence to Sequence Learning with Neural Networks,Ilya Sutskever, Oriol Vinyals, Quoc V. Le",(2014).

 https://doi.org/10.1051/itmconf/20214003027

Deep-Facial Feature-Based Person Re-identification for Authentication in Surveillance Applications

Pranjal Borse[1,*], *Aishwarya Gaikwad*[2,**], *Nachiket Dabhade*[3,***], *Harsha Saxena*[4,****]
[1]Department of Computer Engineering, Ramrao Adik Institute Technology, India
[2]Department of Computer Engineering, Ramrao Adik Institute Technology, India
[3]Department of Computer Engineering, Ramrao Adik Institute Technology, India
[4]Department of Computer Engineering, Ramrao Adik Institute Technology, India

Abstract. Nowadays, a large network of cameras is predominantly used in public places which provide enormous video data. These data are monitored manually and may be utilized only when the need arises to ascertain the facts. Automating the system can improve the quality of surveillance and be useful for high-level surveillance tasks like person identification, suspicious activity detection or undesirable event prediction for timely alerts. In this paper, we proposed a model that can Re-identify a person from a single camera tracking environment. This system will automatically extract face features of the person and generate the Unique Id for each person when it enters for the first time in the monitored area. Its face features are stored in the database which will help to Re-identify the person whenever the same person appears again. The challenges faced by the system are occlusion, pose, light conditions, and face orientation. The proposed system highlights, effect of different deep neural networks for Person Re-identification and compares based on the accuracy, GPU usage, Speed, Number of faces detected by overcoming the challenges like illumination and occlusion. The advantage of the system is it doesn't require the database of people in advance for recognition and it will be helpful for criminal identification for crime control and prevention.

Keywords: Person Re-identification, Face identification, CNN, HOG, MTCNN

1 INTRODUCION

Deep Facial Feature-Based re-identification is the technique of recognizing the person based on his Id which was assigned based on his previous appearance. Our human brain can recognize the unknown person which was seen previously. Similarly, we can develop a system which can recognize a person with his unique id based on his previous appearance in camera automatically. Humans can recognize strangers by face, height, body structure, skin color and hair color. Out of all people's faces, the most reliable feature for recognition. Earlier re-identification was not possible due to Real-Time challenges like variation in viewpoints, illumination conditions, Occlusion, Crowded environment, and different camera views. Few of the challenges are resolved due to immense growth in technology like capturing video in low illumination.

Person Re-identification came into the scene due to the incidents where technology couldn't help. The following three main incidents that lead to research on this topic are: First one is the London's Subway bomb blast on July 7, 2005. After this incident investigators had to parse the entire city's CCTV footage and yet there was no outcome. Second one was the Boston Marathon bombing on April 15th, 2013 and the Third one at Bangalore, a brutal attack in November of 2013 on an ATM. With the increase of crime rate, it is important to catch criminals and send them behind bars. Our security officers are struggling to catch all of them, but finding a few criminals in a huge crowd is not feasible. Security and crime control concerns are the motivating factors for the deployment of intelligent video surveillance systems.

The Proposed system has been intended to overcome the drawbacks of the past surveillance systems and to enhance security, adaptability, and efficiency. The main aim of this research is to automatically detect and Reidentify the person in public places. The model is designed for matching face structure with either existing one or set as new data. Initially there will be no data and then the system stores data as the new face is targeted. The system will have to work to detect human movements and recognize face efficiently, even in low lighting, all-natural environments. The system will reduce the cost of making the database and create a Real-Time database in public places automatically. It also helps to detect suspicious people and notify the Admin/Police as soon as the person is detected.

The paper's organizing is done as follows: Section 2 discusses the currently present techniques, analysis of previous research associated with our proposed methodology. The associated research is defined as a base for our approach. Section 3 presents models of computation and theoretical tools we adopted to Reidentify persons. Experimental results and also the comparative analysis of varied algorithms is shown in Section 4. Section 5 has Conclusions of the outcomes obtained by using our work.

*e-mail: pranjalborse158@gmail.com
**e-mail: aishwaryagaikwad21@gmail.com
***e-mail: nachiketsdabhade2@gmail.com
****e-mail: harsha.saxena@rait.ac.in

2 LITERATURE SURVEY

Many existing methods of Person Re-identification construct robust features that are, at the same time distinctive, and as robust as possible to describe a person's appearance in a variety of settings, some extensive survey and research s have been developed since then.we briefly introduce some research in the context of the proposed model done in person re-ID history.

To our knowlage the first work done in the feid of Multi-camera tracking with explicit "Re-identification" was done by W. Zajdel, Z. Zivkovic[1] in 2005,The method is designed as a unique, hidden, label - for-person dynamic Bayesian network defined to encode probabilistic relationships between symbols and signs(colors and spacetime) signals from tracklets. The user's person ID, which is determined by the reverse side of the signs, distributions are calculated using a different Bayesian deletion algorithm.

Many surveys[2][3] have been done where the features of a person are extracted either by using hand-crafted methods or the Deep neural networks methods in both image-based and video-based Re-Identification.

The handcrafted features technique[4] like Generalized Maximum Multi Clique optimization ,Local Maximal Occurrence Representation(LOMO) , Iterative Hankel Total Least Squares(IHTLS) provide less accuracy than the deep neural network proposed by D. Yi[5].This is the first work to apply deep learning in Re-id based on image classification.They use a Siamese neural network to determine whether a pair of login images belong to the same user identity. The reason for choosing the Siamese cat model is that the number of training samples for each of the personalities is limited (usually there are two). With the exception of some parameter variations, the main differences are that it adds that extra cost to network operation when a thinner partition is used at the same time. Experimental data that do not match each other, and the two methods do not directly compare. Even if the performance is not stable, even with small amounts of data, deep learning methods have become a very popular option for re-ID. Bai, Xiang[6] proposed part-based feature representation to learn pedestrian feature.The pedestrian is described as a succession of body components from head to foot using Long Short-Term Memory (LSTM) in an end-to-end manner.They achieve a 90.84%mAP on Market-1501 which is a image based dataset.The image-based dataset didnt provide better accuracy when it is used in Real-Time environment.

Video-based Re-ID is suitable for the analysis of video in real world as it increase the accuracy of the system.Kasturi R, Ekambaram R.[7] focuses on the recent work done in the Re-identification research area using images and videos.They also perform Re-identification on videos and discuss the challenges faced by the system such as Camera calibration,non-overlapping views. E. Ristani and C. Tomasi[8] learnt features using CNN for Multi-Target Multi-Camera Tracking and Re-identification on Market-1501 and DukeMTMC-ReID dataset and achieved the training accuracy of 63.27% and 74.81% respectively.

The previous papers used the body features for Re-identification and acquired less accuracy. P. Li, M. L. Prieto[9] used face features for identification in real world by using DNN architectures and patch matching techniques. They achieve the Testing accuracy of 78.5% using Siameses network and overcome the challenge of low resolution. V. Mathew, T. Toby[10] also work on the face feature extracted from CNN model for Re-identification in Real time videos. They used Viola Jones algorithm and MTCNN model to detect face and train these models on ChokePoint dataset. The drawback of this paper is they tested this model on very few videos.

We have also done research on Generic Object Detection[11]and face detection techniques[12][13][14] using Deep learning which help to identify multiple person faces in a video-based data. Based on the research we concluded to compare the performance of three well known methods Histogram of Oriented gradients(HOG),Convolution Neural Network (CNN) and MTCNN with FaceNet for face identification and SVM model for recognition using faces embeddings

3 PROPOSED METHODOLOGY

The proposed methodology focuses on developing an efficient system to Reidentify the person in public places as shown in Figure 1. System takes video as the input and extracts frames to detect multiple people in the environment. By applying Various Deep learning Architecture like CNN, HOG and MTCNN, systems extract features from faces. Then, it compares facial features with known faces in the database. Initially, At the start of the system the database is empty. The face that appears first will be provided with Unique Id 0. After this the next face will be compared with the faces present in the database. If the person was recognized, which means it appears gain in the monitored area system will display its old id. If the person is not matched with the existing one, the system will store the new face and assign a new Unique Id to it. In the database file face features are stored along with its Unique ID as shown in Figure 1.

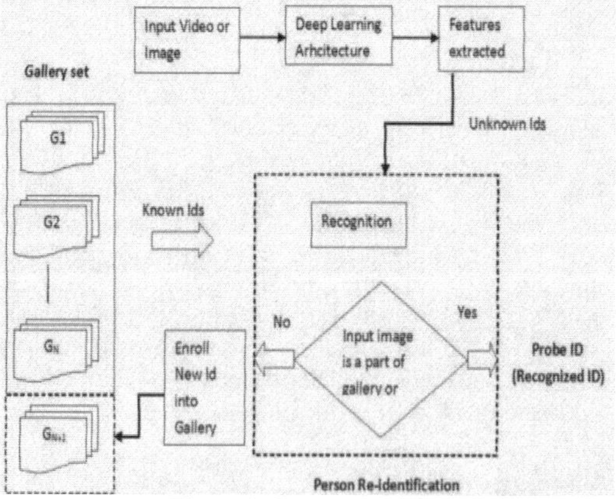

Figure 1: System Design

We have compared the working of three deep learning methods based on various parameters such as performance measure, accuracy, processing time, speed, GPU usage, side face detection and illumination conditions in Real-Time environment, which are as follows:

Method-1 Convolution Neural Network :

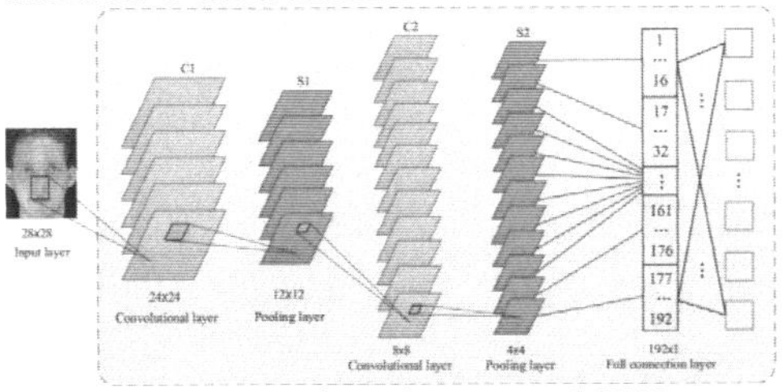

Figure 2 : Layers of CNN

CNNs work on multilayer perceptrons. Multilayer perceptrons are nothing but fully connected networks, in which each neuron in one layer is connected to any or all neurons within the next layer. CNN works in different types of layers. During a regular Neural Network there are various varieties of layers as shown in figure 2.

The model's structure is similar to that of the classic LeNet-5 model, however some parameters, such as the data file, network dimension, and full affiliation layer, are completely different. Two convolutional layers (C1 and C2), as well as two pooling layers, make up the developed CNN (S1 and S2). As seen in Figure 2, these layers are structured alternatively in the C1-S1-C2-S2 style. Within the input layer, there is only one feature map that is used to feed the normalised facial picture into the CNN model. C1 is the first convolutional layer, which has six feature maps and convolutes each somatic cell with a 5 to 5-size randomly produced convolution kernel. S1 is that the output of the previous layer was supported by the initial pooling layer, whose output is six calculated feature maps. Every component in the feature map is linked to the mean convolution kernel of the associated feature map in the C1 layer, ensuring that the weather's receptive fields do not overlap. The C2 and S2 units, respectively, are the second convolutional layer and the pooling layer, with twelve feature maps and similar calculation procedures as their predecessors. In addition, between the S2 layer and the output layer is a fully connected single-layer perceptron. The final word output might potentially be a 40-dimensional vector for face recognition of N people, as shown in Figure 2, when the sigmoid process is used for multi-label classification.

Method-2 Histogram of Oriented Gradient:

Histogram of Oriented Gradients (HOG) may be a feature definition that wants to process a picture, especially to detect an object. The fundamental premise of the Histogram of Oriented Gradients will then enter the way we calculate histograms and the way these feature vectors, found within the HOG dictionary, are employed by the editor as SVM to see the article involved as shown in figure 3.

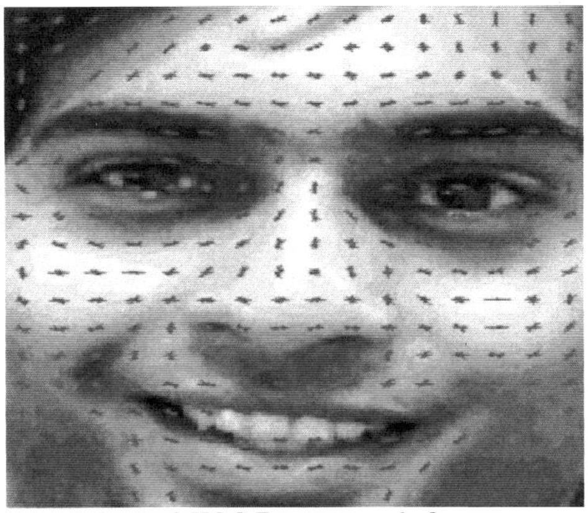

Figure.3 HOG Feature sample face

The HOG calculates image gradients per block.The Block is a pixel grid in which gradients are constituted from the value and route of alternate in the intensities of the pixel inside the block. As shown above, figure the sample image is fed to a feature descriptor which extracts useful information. In above example x and y are respectively the horizontal and vertical components of the change in the pixel intensity.The descriptors are calculated over blocks of pixels with 8 x 8 dimensions. These descriptor values are quantized into nine containers for each pixel over 8 × 8 blocks, with each bin representing a directional gradient attitude in that bin, which is the sum of the magnitudes of all pixels with the same perspective. The histogram is then normalised over a 16 × 16 block length, which implies four 8 × 8 blocks are normalised together.Similarly, the histogram is then normalized over a 16 x 16 block length, this means that four 8 × 8 blocks are normalised together to reduce the severity of mild problems. Because of the change in light, the system's accuracy suffers. For many faces, the SVM model is trained to use some HOG vectors.

Method-3 MTCNN with FaceNet:

Firstly we'd like to make sure whether the mtcnn library is installed correctly. After installation check the version. We used this library to form a face detector and extract the faces from the input images . This can be further integrated with Facenet face detector models. Initially we load a picture as a Numpy array. We create a numpy array using the PIL library and therefore the open() function. Further conversion of image to RGB is performed if the image is black and white or the alpha channel. We then create a MTCNN face detector class and use it to detect the faces from the loaded image. The output is the list of bounding boxes, each box states a lower-left-corner of bounding. It also defines the width and height of the box.We take absolutely the value of the coordinates if the library returns a negative pixel index. We will use these coordinates to extract the face. PIL library is employed to resize small frames to the desired size because the model expects the square input faces with the form 160 x 160. The MTCNN architecture we've got used contains four convolution layers and two fully-connected layers with parameters Θ for feature learning. W= [Wa, Ws] from layer 6 to layer 7 are learned separately for every task respectively. Three stages of mtcnn are shown in figure 4.

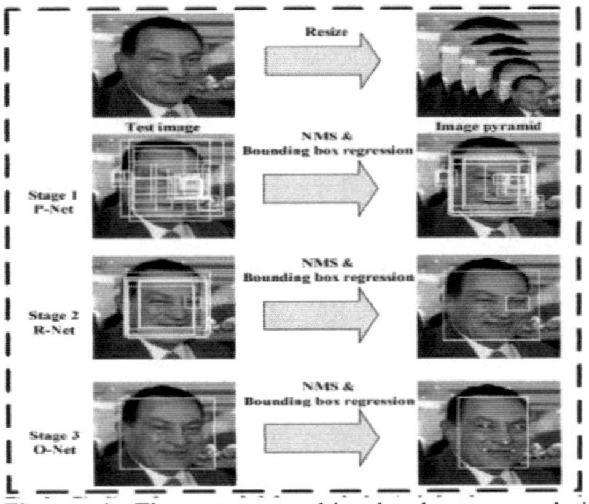

Figure 4 Three-stage multi-task deep convolution networks

We also produce a face embedding, which is a vector that represents the extracted features from a face and can be compared to other vectors of other faces. We create a classifier model that predicts identity using facial embedding as input. The embedding for a particular input frame will be generated by the Facenet model. Next we normalise the vector of every face embedding. It's important to normalise because we use them to check using distance metric. Vector normalisation means scaling the values until the magnitude is of unit length(1) of that vector. We set 'kernel' to 'linear' in a trial to suit linear SVM to schooling statistics. This can be performed using the SVC magnificence in scikit-analysis. We use face embedding as enter to form prediction with the fit version.

4 IMPLEMENTATION RESULTS AND ANALYSIS

The Table 1 highlights, effect of different deep neural networks for Person Re-identification and compares based on the accuracy, GPU usage, Speed, Number of faces detected by overcoming the challenges like illumination and occlusion. The comparative analysis shows that the MTCNN with facenet provides better accuracy in Real world and also handles issues like face orientation/angle variation and illumination as shown in Figure 5 and 6 respectively.

The testing results of the proposed models are shown below, When the first person appears, there is no data to match so system gives him a pickle file with ID '0' as shown in figure 7. When the second person appears, there is no data to match so system gives him a pickle file with ID '1' as shown in figure 8. When these person reappear in the system they will be Re-identified and assigned with the ID stored in the database. As shown in Figure 9, in input frame two faces are found and when the matching algorithm works this Reidentify people and matches up with ID '0' and ID '3' respectively.

Table 1 Comparison of different Algorithms

parameters	Method 1 (CNN)	Method 2 (HOG)	Method 3 (FaceNet)
1.Face Detection Algorithm	Pretrained models of Dlib or OpenCV	Pretrained models of CNN	Pretrained models of MTCNN
2.Face Recognition Algorithm	Face recognition module based on CNN	Face recognition module based on CNN	Linear Support Vector Machine (SVM)
3.Side Face Detection	Not Detected	Not Detected	Detected
4.Testing Accuracy	83-85%	80-82%	85-87%
5.GPU Usage	No	No	Yes
6.No of face Detected at a time	4-5	3-4	4-5
7.Result	Accurate result	Not accurate as CNN	More accurate result
8.No OF Frames Generated	Less or equal to 30 Frames/ Sec	Less or equal to 25 Frames/Sec	Less or equal to 10 Frames/ sec
9.Speed	Slower than method3	Faster than method1 and method 2	Slower than method3 but faster than method 1
10.Occlusion	It can handle Occlusions.	It can not handle Occlusions.	It can handle Occlusions

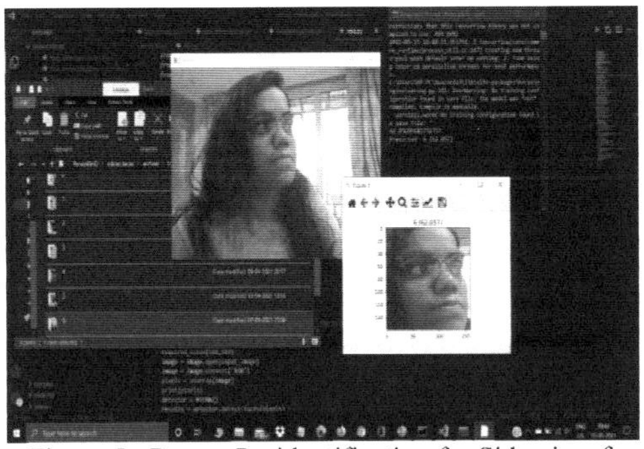

Figure 5 : Person Re-identification for Side view face

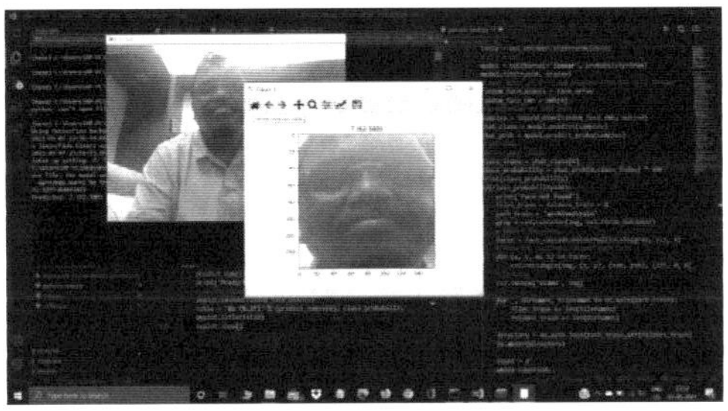

Figure 6 : Person Re-identification under different lighting condition

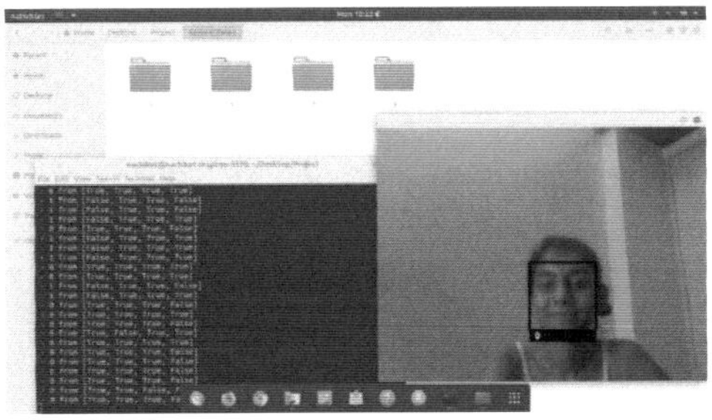

Figure 7 : Results of Unique ID '0' generation for First person enter in the system

Figure 8: Results of Unique ID generation for next new person

Figure 9 : Results of Person Re-identification

5 CONCLUSION

Person re-identification has gained a lot of attention in recent years. The Proposed model gives us accuracy around 87% and the HOG model gives accuracy 82% whereas the previous model gave accuracy under 85%. The system presented an approach to the challenges faced by the system are occlusion, pose, light condition, face orientation. Currently we are taking input as a single shot image from a single camera, in future we can extend this work by taking input across multiple cameras. The advantage of the system is it doesn't require the database of people in advance for recognition and it will be helpful for criminal identification for crime control and prevention. To increase performance of the system in future, we can use Hybrid models and effective feature learning techniques such as face matrices for face identification.

REFERENCES

[1] W. Zajdel, Z. Zivkovic, and B. Krose, "Keeping track of humans:Have i seen this person before?" in Proceedings of the 2005 IEEE International Conference on Robotics and Automation. IEEE, 2005,pp. 2081–2086.

[2] S. karanam, M. Gou, Z. Wu, A. Rates-Borras, O. Camps and R. J. Radke, "A Systematic Evaluation and Benchmark for Person Re-Identification: Features, Metrics, and Datasets," in IEEE Transactions on Pattern Analysis and Machine Intelligence,
vol. 41, no. 3, pp. 523-536, 1 March 2019, doi: 10.1109/TPAMI.2018.2807450.

[3] Liang Zheng and Yi Yang and Alexander G. Hauptmann, Person Re-identification:Past, Present,

Future,CoRR,abs/1610.02984,2016,http://arxiv.org/abs/1610.02984
[4] Liu, Wenqian and Camps, Octavia and Sznaier, Mario. (2017). Multi-camera Multi-Object Tracking.
[5] D. Yi, Z. Lei, S. Liao and S. Z. Li, "Deep Metric Learning for Person Re-identification," 2014 22nd International Conference on Pattern Recognition, 2014, pp. 34-39, doi: 10.1109/ICPR.2014.16.
[6] Bai, Xiang & Yang, Mingkun & Huang, Tengteng & Dou, Zhiyong & Yu, Rui & Xu, Yongchao. (2017). Deep-Person: Learning Discriminative Deep Features for Person Re-Identification. Pattern Recognition. 98. 10.1016/j.patcog.2019.107036.
[7] Kasturi R., Ekambaram R. (2014) Person Reidentification and Recognition in Video. In: Bayro-Corrochano E., Hancock E. (eds) Progress in Pattern Recognition, Image Analysis, Computer Vision, and Applications. CIARP 2014. Lecture Notes in Computer Science, vol 8827. Springer, Cham. https://doi.org/10.1007/978-3-319-12568-8_35
[8] E. Ristani and C. Tomasi, "Features for Multi-target Multi-camera Tracking and Re-identification," 2018 IEEE/CVF Conference on Computer Vision and Pattern Recognition, 2018, pp. 6036-6046, doi: 10.1109/CVPR.2018.00632.
[9] P. Li, M. L. Prieto, P. J. Flynn and D. Mery, "Learning face similarity for re-identification from real surveillance video: A deep metric solution," 2017 IEEE International Joint Conference on Biometrics (IJCB), 2017, pp. 243-252, doi: 10.1109/BTAS.2017.8272704.
[10] V. Mathew, T. Toby, A. Chacko and A. Udhayakumar, "Person re-identification through face detection from videos using Deep Learning," 2019 IEEE International Conference on Advanced Networks and Telecommunications Systems (ANTS), 2019, pp. 1-5, doi: 10.1109/ANTS47819.2019.9117938.
[11] Liu, Li & Ouyang, Wanli & Wang, Xiaogang & Fieguth, Paul & Chen, Jie & Liu, Xinwang & Pietikäinen, Matti. (2018). Deep Learning for Generic Object Detection: A Survey.
[12] Murat Taskiran, Nihan Kahraman, Cigdem Eroglu Erdem,
Face recognition: Past, present and future (a review),Digital Signal Processing,Volume 106,2020,102809,ISSN 1051-2004,
[13] Florian Schorff, Dmitry Kalenichenko, James Phinbin,." FaceNet: A Unified Embeddings for Face Recognition and Clustering",2015,IEEE
[14] Kaipeng Zhang, Zhangpeng Zhang , Zhifeng Li , Joint Face Dectection and Alignment using Multi-task Cascade Convolution Networks, 2016,IEEE

 https://doi.org/10.1051/itmconf/20214003028

Virtual Dressing using Augmented Reality

Omkar Jadhav[1,*], *Anish Patil*[1], *Jeffy Sam*[1], *and Kiruthika M*[2,*]

[1]BE Students, *Department of Computer Engineering*, Fr Conceicao Rodrigues Institute of Technology, Vashi Navi Mumbai, India
[2]Associate Professor, *Department of Computer Engineering*, Fr Conceicao Rodrigues Institute of Technology, Vashi Navi Mumbai, India

Abstract- With the rise in Virtual Reality technology now it is possible to bring the same experience of offline shopping in the online platform. Virtual Dressing using Augmented Reality is a platform where users can try clothes virtually. As users cannot physically try the dress in online mode this application will help the user to make a better judgement on how the dress will look on him. The proposed system will consist of an application where users can create their 3D model with just their pictures and can apply the garment on their model, which can be viewed in AR space to make a judgement of how the model looks on a person. ***Keywords: Online platform, Virtual Reality, 3D model, AR space***

1 Introduction

Due to the pandemic situation, there is a rise in online shopping platforms. People engage more in online shopping. But in the clothing section, users hesitate to buy as they will not get the same experience as with offline shopping because they can't try the clothes in online mode. This is a problem for both, the customers and the sellers. So, there is a need for a virtual trial room for online platforms.

With the rise in Virtual Reality and Augmented Reality technology now it is possible to bring this functionality to the online platform. A system could be devised where users can create their 3D model with just their pictures and can apply the garment on their model, which can be viewed in AR space to make a judgement of how the model looks on a person. Furthermore, we can suggest the appropriate measurement which fits the person.

3D virtual garment model design has gained large popularity in the computer-aided design (CAD) community. It is very important in the development of high-quality computer-generated video games since the quality of garment models on characters can play an important role in the quality of the game. With recent advancements in clothing simulation, it is possible to give dynamic behaviors to virtual garments. The garment model is designed to fit a virtual human body as a separate mesh on top of it.

2 Related Works

M.R. Minar et al. [1] propose an approach to try clothes on a 2d image of a person. Here a common processing pipeline for this setting has two stages: first, the try-on cloth is warped around the target human (GMM), and then the warped cloth in blend according to the pose of human (TMM). The cloth warping stage obtains the complete body silhouette area from the target human image. Then other body parts are masked leaving just the target clothing area. Further human parts are retained in the blending stage, the warped cloth mask is applied to the target cloth area. The first iteration of this model was proposed by Bochao Wang et al. [2] as CP-VTRON further was refined and called CP-VTRON+.

N. Wang et al. [3] proposes a method to transfer clothing texture on human avatars called Pix2Surf. Pix2Surf is capable of transferring the texture of clothing images to 3D human models in form of 3D garments. These 3D garments are worn on top of the 3D human model. This model helps to transfer the texture from the front and back garment images to the garment mesh. Pix2Surf maps the garment texture selected from a large number of clothing items available on online websites onto the human model in real-time. This model is useful to generate a virtual try-on.

S. Saito et al. [4] propose an effective representation that aligns the pixels of 2D images corresponding to the 3D object. This method is called PIFu(Pixel-Aligned Implicit Function). It maps the dress onto the 3D object. PIFU is an end-to-end deep learning method used to form a digital clothed avatar. PIFu infers the underlying 3D surface and texture from a single image or can opt for multiple images. Highly detailed shapes, such as clothing and variations can be digitalized uniformly. T. Alldieck et al. [5] propose an effective model to reconstruct people's clothing from a single RGB image. Using it, PIFu is capable of generating high-resolution surfaces such as the back of the person from the single front view. This method achieves a state-of-the-art performance and outperforms the prior work of clothed

* Corresponding author: omkarjadhav1114@gmail.com, kiruthikafcrit123@gmail.com

human digitization from a single image. Texture reconstruction focuses on predicting texture at any point on the surface. It can make use of multiple views of the sample model and incorporates them into predicting an accurate human model. Further, the model was improved in defining texture and generating the back of the image. It is called PIFuHD [6].

3 System Architecture

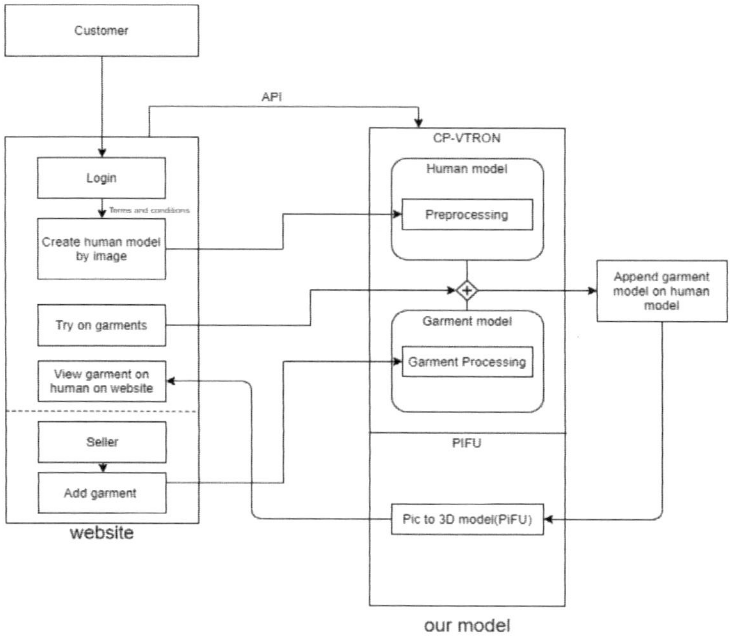

Fig. 1. System Architecture

The main objective is to design a service for e-commerce fashion websites. The Figure 1 shows the overall System architecture. The fashion websites will provide the customers with the try-on option on their website. The fashion website will send user images along with the garment image to the back-end which then processes the user and garment image to append garment on the user image and output the 2D image and 3D user model to the fashion website.

4 Methodology

.In the prescribed method, the garment image is applied on a user image then pipeline the user image to generate a 3D model. So, it consists of two separate parts: Garment on 2D image and transforming 2d image into 3d model. The first part i.e garment on 2D image is based on the CP-VTRON+ model. Before applying CP-VTRON+[1] the pre-processing should be done using CIHP_PGN[7] to segment the input user image. Further, the user pose is determined using Openpose[8] so the heatmap can be drawn to identify different body parts of the user image. The CP-VTRON+ has two stages: The clothing warping stage and the Blending stage.

The cloth warping stage is used to separate the target clothing area from other body parts and generate the pose structure to blend the cloth according to the user pose. The blending stage retains the original body parts and fits the wrap cloth onto the target clothing area. The model is based on CNN geometric matching [9]. The second part is based on the PIFu model. In PIFu(Pixel-Aligned Implicit Function) the task is to generate the complete 3d geometry structure using a single user image. Here the output of the CP-VTRON+ model is fed to PIFu as an input. Given an input image, it reconstructs the surface geometry and predicts the textured surface of a clothed human. Further, the texture is preserved by storing the RGB values of the image and transforming them into RGB vectors which are applied using the ray-tracing algorithm.

5 Scope

Virtual dressing using Augmented Reality is a platform that can be integrated into different e-Commerce and fashion websites. The basic motive is to give the people a tool using which they will be able to see the real-time simulation of different clothes without having to wear them physically. It's a Web-based application integrated into another website or can work as a standalone application to create a human 3D model. This system automatically transforms 2D pictures into a 3D model.

This model is further used by users to try clothes on different websites. This model can be viewed in AR space to take a better judgment on the looks of the clothes.

6 Implementation

6.1 Dataset

The images used for training the model are from VITON [10] and renderpeople website [11]. The Figure 2 shows dataset consists of cloth images. Figure 3 shows the target human images.

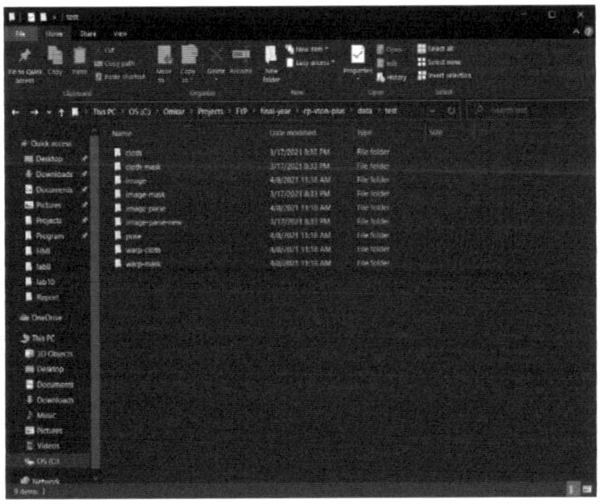

Fig. 2. Dataset by VITON

Fig. 3. Target human images

6.2 Data-Preprocessing

The cloth images are preprocessed by CIHP-PGN [7] network to get a cloth mask. The background of target human images is removed using removebg API [12]. The target human image along with the human image mask shown in Figure 4 is passed onto Openpose [8] to generate a pose heatmap shown in Figure 5.

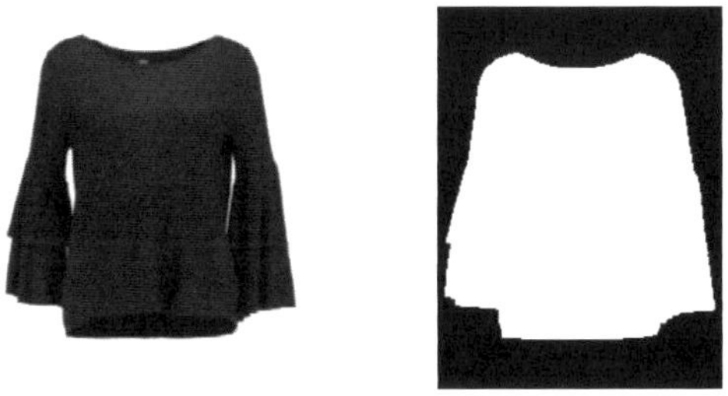

Fig. 4. CIHP-PGN input output

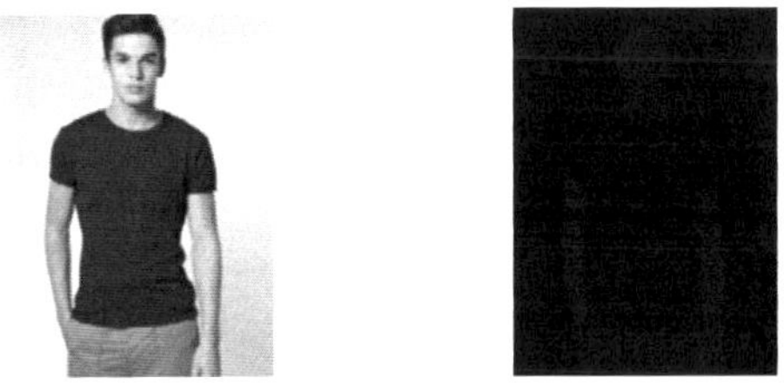

Fig. 5. Openpose input and outputs

6.3 Model Architecture

The model creates a pipeline of the process of creating a 3D model from a 2D image. The Figure 6 shows the flow of execution. The dataset provided by VITON [10] is processed by CIHP-PGN to generate the image mask and Openpose to generate a human pose map. Removebg is used to remove background from cloth and target human image. The cloth image along with cloth mask and a human image along with pose image is passed to CP-VTRON+. The pose grid is used to generate a warp cloth image in the cloth warping stage. The wrap cloth is appended on the target human image in the blending stage. Further, the output image from CP-VTRON+ is scaled to 512x512 pixel resolution to pass into the PIFu model. The PIFU model is used to generate the 3d model of the human image.

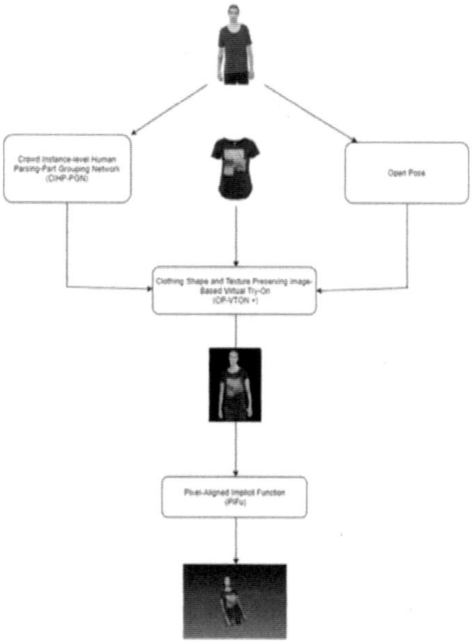

Fig. 6. Flow of execution

7 Results

7.1 Login and Register Page

The user registers using valid email id and password shown in Figure 7. The login interface where the user puts valid email id and password to login into application shown in Figure 8.

Fig. 7. Register Page

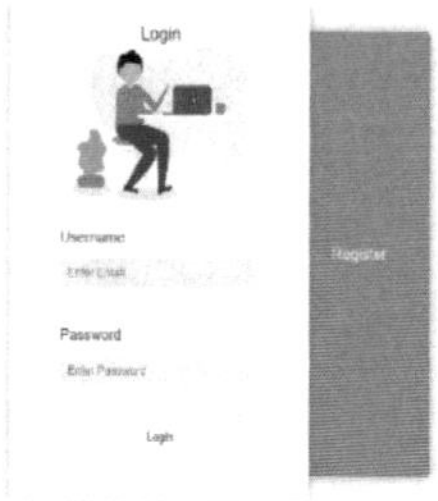

Fig. 8. Login Page

7.2 Home Page

The main dashboard of the application. Here the user can upload the image as well as view the original uploaded image. The default image is provided so user can try-on a default image shown in Figure 9. The showcase gallery of garment where the user can select the garment to be worn on the image shown in Figure 10.

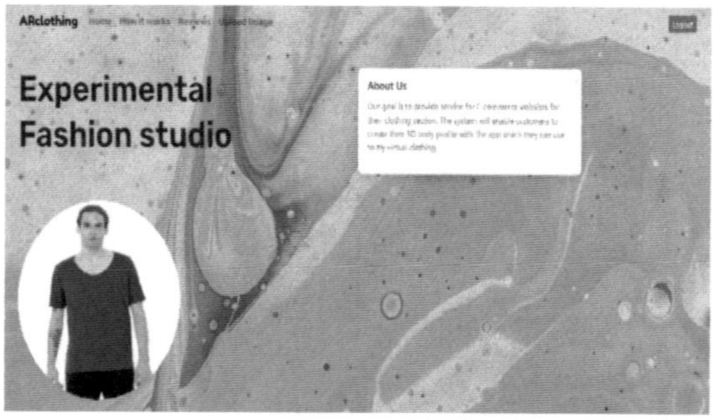

Fig. 9. Main Dashboard

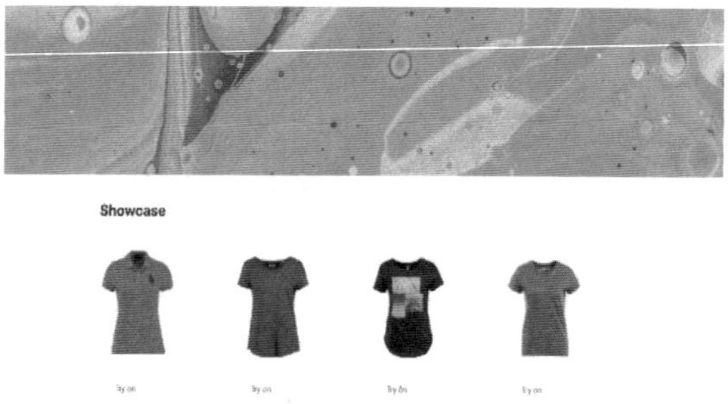

Fig. 10. Clothes Gallery

7.3 Result Page

The output of the model shown in Figure 11 is viewed as a 2d image as well as a 3d model. The 3d model is rendered using Threejs [13] library so the user can freely rotate the image in the 3d view.

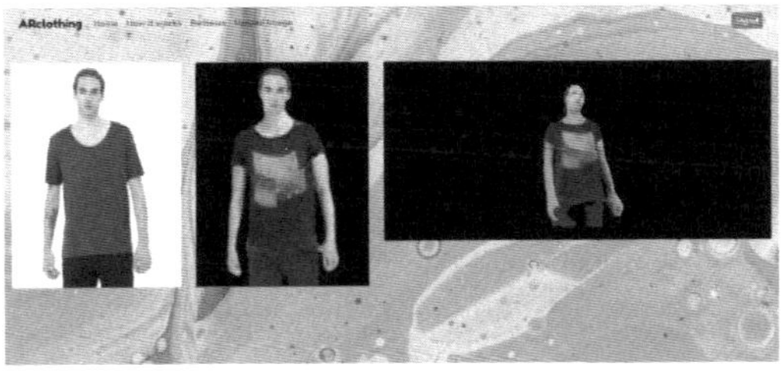

Fig. 11. 2D,3D result

8 Conclusion

The primary motive of the paper is to solve the user problems while shopping on the online fashion platform. As the user cannot try clothes in online mode so they do not get the same experience as offline mode. Hence this will bridge the gap by giving the user an option to try the cloth on their image and the user can get the feeling of wearing them without actually trying the cloth. Further, the user can take a look in 3D space at how the cloth looks on the image. For now, this product is only addressing the issue of the shirt but further, it can be extended to support all kinds of clothing accessories.

References

1. Minar, Matiur Rahman and Thai Thanh Tuan and Ahn, Heejune and Rosin, Paul and Lai, Yu-Kun, "CP-VTON+: Clothing Shape and Texture Preserving Image-Based Virtual Try-On," 2020
2. Bochao Wang, Hongwei Zhang, Xiaodan Liang, Yimin Chen, Liang Lin, and Meng Yang, "Toward characteristic-preserving image-based virtual try-on network", 2018
3. Aymen Mir, Thiemo Alldieck, Gerard Pons-Moll "Learning to Transfer Texture from Clothing Images to 3D Humans", March 2020
4. S. Saito, Z. Huang, R. Natsume, S. Morishima, A. Kanazawa, and H. Li, "Pifu: Pixel-aligned implicit function for high-resolution clothed human digitization,"arXiv preprint arXiv:1905.05172, 2019.
5. T. Alldieck, M. Magnor, B. L. Bhatnagar, C. Theobalt, and G. Pons-Moll, "Learning to reconstruct people in clothing from a single RGB camera", 2019
6. Saito, Shunsuke and Simon, Tomas and Saragih, Jason and Joo, Hanbyul, "PIFuHD: Multi-Level Pixel-Aligned Implicit Function for High-Resolution 3D Human Digitization", 2020
7. Ke Gong, Xiaodan Liang, Yicheng Li, Yimin Chen, Ming Yang, and Liang Lin, "Instance-level Human Parsing via Part Grouping Network,"arXiv preprint arXiv:1905.05172, 2018.
8. Z. Cao ,G. Hidalgo Martinez, T. Simon, S. Wei and Y. A. Sheikh, "OpenPose: Realtime Multi-Person 2D Pose Estimation using Part Affinity Fields," 2019
9. Ignacio Rocco, Relja Arandjelovic, and Josef Sivic, "Convolutional neural network architecture for geometric matching", 2017
10. Han, Xintong and Wu, Zuxuan and Wu, Zhe and Yu, Ruichi and Davis, Larry S, "VITON: An Image-based Virtual Try-on Network",2018
11. Renderpeople, 2018. https://renderpeople.com/3d-people.
12. Remove Background API, https://www.remove.bg
13. ReactThreeObjectLoader, https://www.npmjs.com/package/three-react-obj-loader

Early Detection of Depression Indication from Social Media Analysis

Maharukh Syed[1*], *Meera Narvekar*[2]

[1]D.J .Sanghvi College of Engineering,mahrukhsyed11@gmail.com, Mumbai.

[2]Department of Computer Engineering, D.J.Sanghvi College of Engineering,meera.narvekar@djsce.ac.in, Mumbai.

Abstract: Depression that stems through social media has been steadily growing since the past few years but with the current inclination towards social media reliance it is highly imperative to detect the early signs. Continuous observation of a user's social media interests and activities may highlight suspicious and negative thoughts. This observation can help in understanding their future course of action and also indicate any suicidal thoughts and behaviors. By using the machine learning models, early indications of depression detection can be addressed. This work studies different word embedding techniques for early detection of depression from social media posts. Further, this work develops a model using various NLP processes in order to address the issue of early detection. The recommendations can be useful as a Decision Support System for counselors, psychologist and also can be of good use by the cyber-crime cell department for criminal investigations.

Keywords: — Social Media Analysis, Natural Language Processing, Embedding System, Tag Cloud and CNN.

1. INTRODUCTION

Social media is a very powerful tool for communicating ones thought process. The dependence on these platforms have grown to a point where an individual feels the inherent need to be sharing every thought and feeling they have with their social followers for a sense of validation. However, when an individual witnesses their peers doing better, it also causes a sense of envy leading to self-depreciating thoughts and low self-esteem. Studies have also shown that excess use of social media leads to stress and negative mental health. It also promotes negative experiences like depression & anxiety, inadequacy about your life or appearance, self-isolation, cyber bullying, fear of missing out, etc. While depression and other mental illnesses may lead to social withdrawal and isolation, it was found that social media platforms are indeed increasingly used by affected individuals to connect with others, share experiences, and support each other. Using the machine learning method, this report focuses on ways to classify early signs of depression in textual data in its premature stage. A complete analysis of the various algorithms is hence studied in detail and by describing the previous and current work with respect to text classification; this research can formulate the practical base of this work.

For text classification tasks in the analyzed papers, various Natural Language Processing's like GloVe, Word2vec, FastText [1] have been used to address the issue.

Elaborate results in word and content based tasks which are related to mental health behavior understanding, for example are achieved using the Convolution Neural Network [8] an efficient technique to model words and sentences.

Fig.1. A depressed user pretends to be happy.

The rest of the paper is organized as follows. Literature review is explained in section 2. Proposed system is explained in section 3. Experimental results are presented in section 4. Concluding remarks in section 5 and future scope are given in section 6.

2. LITERATURE REVIEW

Marcel Trotzek, Sven Koitka, and Christoph, M. Friedrich [1] predict early signs of depression using machine learning algorithms in particular the Convolution Neural Network based on various word embedding techniques. Further ERDE (Early Risk

Detection Error) as a metric is taken for detection of depression. Context free word embedding models such as Word2vec suggested by (Peters et al., 2018) and GloVe, Fast Text proposed by (Trotzek et al, 2019) are the popular Natural Language Processing techniques been used to address the issue.

Ranjana Jadhav, Shaurya Jadhav, Hitesh Sachdev, Vinay Chellwani [2] proposed a Decision Tree classifier for Mood Disorder Questionnaires. This research using Machine Learning Approach basically determines the most significant feature in the data set and makes that as a deciding factor for detection of Bipolar Mental Health Disorder.

Jane H. K. Seah, et al [3] have carried out the research based on the data collected from Reddit for detection of suicidal tendencies of users on social platforms. This is performed using LDA and data mining approach of topic modeling is used to discover different context in which suicidal thoughts occur.

Akkapon Wongkoblap et al [4] have developed a prediction model to classify users with poor mental health using machine learning algorithm along with data from social media platforms.

Hao Wang et al [5] in their "Sentiment Expression via Emoticons on Social Media" have highlighted that emoticons play a vital role in understanding sentiments. They have analyzed the widely used emoticons along with performing feature extraction using word2Vec conversion and have classified sentiments with and without emoticons using Bayes Classification model along with Bag-of-Words.

Walter Gerych et al [6] explain that depression diagnosis is difficult as the percentage of afflicted users in most populations is small compared with those unaffected, leading to severe class imbalance. They have proposed a multi-stage machine learning pipeline. This is done using Auto-encoders and SVM Algorithm.

Sharath Chandra Guntuku, David B Yaden, Margaret L Kern, Lyle H Ungar, Joannes C Eichstaedt [7] in their findings have shown that social network data can be used to predict mental illness using various machine learning modules like Logistic Regression and Support Vector Machine (SVM).The prediction can be based on multiple scenarios where it can be self declared on twitter it can also be based on responses of online surveys, based on annotated posts, forum membership etc.

Y. Zhang et al [8] in their A Sensitivity Analysis of Convolution Neural Networks for Sentence Classification has helped understand CNN model in detail for sentence classification.This also helps aim is to distinguish between important and comparatively inconsequential design decisions for sentence classification. Classifiying negative and positive comments from the data set can be done using the CNN model.

Matthew Peters et al [9] in their research show how contextual word representation have provided significant improvements in the state of art for a wide range of NLP tasks. They have shown an emperical study of how Neural Architecture influences accuracy and qualitative properties of the representations that are learnt.

2.1 Gaps Identified

Having a left-to-right flow, the addressed models are unidirectional in their pre-training. This restricts their ability to understand the actual meaning of pre-trained representations. Models like OpenAI GPT (Trotzek et al, 2019.) [1], used left-to-right architecture and ELMo (bidirectional but shallow) concatenates forward and backward language models do not fulfill the precise needs as depression detection is an emotionally challenged condition which cannot be detected directly. Hence there is a need for Contextual language modeling to be taken into consideration. Contextual representations can further be unidirectional or bidirectional. A language model can be trained forward or backward, but it continues to be unidirectional as the prediction of future characters or words is supported by past seen data.

Analyzing the various papers it was seen that various word embedding techniques were used in order to classify text for further processing. A deep study of different models was carried out. A lot of research papers have been gone through for understanding depression detection but very few of them are applicable and appropriate to the particular topic.

3. PROPOSED SYSTEM

Fig.4.Steps involved in proposed system.

The proposed system aims at detecting depression indications using deep learning techniques applied on word2Vec embedding methods [1]. The steps include preprocessing of the data which means removing unwanted and utilizing only the information that is necessary. Data cleaning is performed in this step. Further the paper performs visualization of data where it creates word clouds and graphical representation of data set is seen to understand data distribution. The next step includes dividing the training and testing dataset. A 60% - 40% ratio is taken for training and testing the data set. The last step includes applying the algorithm as shown in module II and lastly performing testing of the model.

3.1 Module I: Word2Vec

Pre-trained word descriptions, as seen in essay, can be context-free (i.e., word2vec, GloVe, fast Text) [1], which means that a single word description is formulated for each text in the vocabulary, or can also be contextual (i.e., ELMo and OpenAI GPT), [9] on which the word descriptions depend on the context where that word occurs, meaning that the same word in different formats can have different meanings. Let's understand the modules briefly. Trained to reconstruct linguistic descriptions of word, Word2Vec is a shallow, two-layer neural network. Typically taking in an oversized corpus of text, it creates a vector space of huge dimensions, and each unique word within the corpus is assigned a corresponding vector in the space. Words that share common contexts within the corpus are located in close presence of each other, within the space owing to the word vectors being positioned the certain way.

Being distinctly computationally-efficient predictive model for learning word embedding's from raw text; Word2Vec comes in two flavors viz. the Continuous Bag of Word (CBOW) & the Skip-Gram model. These models are similar in their algorithmic sense.

3.2 Module II: Convolution Neural Network

3.2.1 System Design using CNN

This is a deep learning approach mostly used for analyzing data with grid-like structure other than image classification filtering [8]. It is a feed forward neural network which contains input data, multiple hidden layers and an output layer. The main aim is to learn a specific property from the given data and find out whether the user is depressed or not.

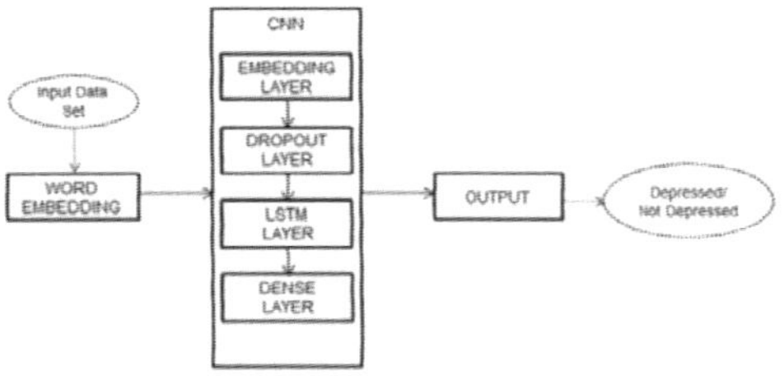

Fig.2. System Design using CNN.

Convolution neural networks have the following hidden layers:

I. Embedding Layer

The embedding layer will learn embedding for all the words from the dataset. It is basically a low dimensional space which in this case tries to take the large dataset and convert it into an output of 2D vector form, having 1 embedding for each word. Such kind of embedding makes it easier to perform machine learning on large social media data sets. It is also noted that such embeddings are flexible in nature and can be reused across the models. Let's consider a simple example to understand the use of Embedding Layer.

Suppose the input data contains various adjectives like joy,lonely,ugly,tensed,suicidal,horrible,gay,awful,beautiful,low,excited. From the above it can be categorized or clubbed the emotions into happy, sad or depressed. In this way a broader perspective for the emotion can be persuaded.

a. Dropout Layer

In deep learning neural network algorithms over fitting a training set is quiet common. For the same reason when you have a large network few nodes are dropped before every layer in order to avoid statistical noise. Prevention of over fitting is done here in order to remove unwanted data at a level of 50%. It is very important to remove the extra nodes from the neural network so as to get accurate predictions.

II. LSTM Layer

The Long Short Term Memory Layer has 2 layers known to detect complex layers. It is a type of recurrent network where it is capable of learning from its previous layer and using that information for sequence prediction. The LSTM layer supports active prediction. It also contains data from the previous layers hence drop out for recurrent data required here is with a level of 20%. When the CNN model is followed by the LSTM model it helps in defining two sub models namely the CNN model for feature extraction and LSTM model to predict the different features between the different layers of the model.

III. Dense Layer

This layer is always applied at the end in a CNN model which helps to connect all the layers to one another giving precise output and a fully connected layer. The neurons in each of the layers receive input from the previous layer in such format that all the layers are

https://doi.org/10.1051/itmconf/20214003029

tightly connected to each other. This helps in having a strong dense network.

4. EXPERIMENT AND RESULT

The steps involved in implementation of CNN based approaches for detecting depression is shown below. Here the algorithm is applied on sentiment 140 1.5 Lac record of dataset. The characteristics include the context id, timestamp, username and comments. The implementation and test environment used is Microsoft Visual Studio IDLE 3.6(64-bit). System configuration includes Intel(R) core(TM) i7,8 GB RAM and 64 bit OS. Algorithm flow is shown as below:

1. Users tweet is taken; next pre-processing is performed to do data cleaning by replacing all NA values to zero.
2. A matrix with user, tweet and the timestamp is created.
3. Algorithms are applied to separate positive negative and neutral comments and to make prediction of whether the user is in depression or not.
4. Here cross validation splits the initial dataset into training and testing with 60-40 criteria. Next evaluation metrics are used for performance evaluation of algorithms.

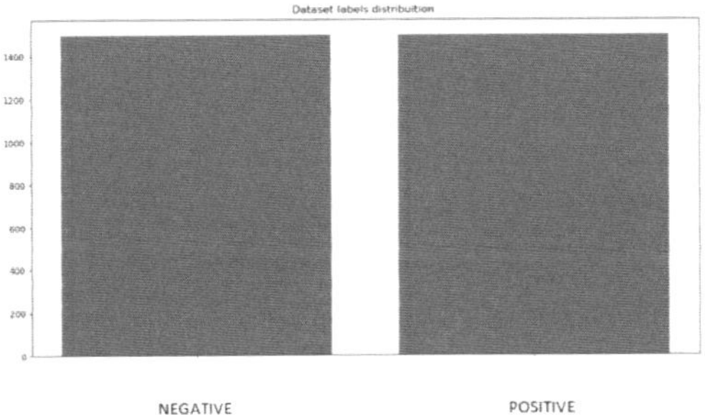

Fig.5. Figures shows execution of above mentioned algorithms to show positive and negative comments in the data set

I. Tag Cloud

Tags are single words which are of great importance and connection to each other generally used to detect metadata. It is a graphical representation of words which are of great prominence and appear more frequently in the dataset.

Fig .6.Word Cloud Generation for positive and negative words

Frequently occurring words appear in larger size compared to the words which occur rarely.

II. Accuracy and loss

It is the averaged squared difference between actual and estimated value. It is basically used to capture how accurate result the model has given. Lower value considered as best fit.

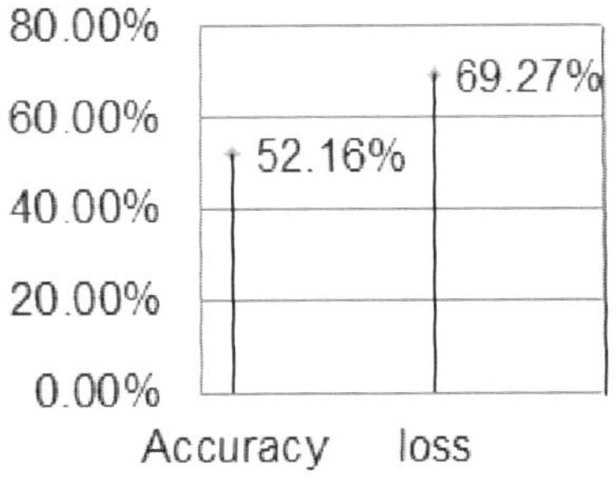

Fig.5.Accuracy and Loss using CNN.

5. CONCLUSION

In this research it is discussed that deep learning approach for early detection of depression from social media analysis can be done using deep learning algorithms. It has been observed that an accuracy of 52.1 % with loss of 69% is seen using Word2Vec along with CNN Model. Better results from future findings aims at using BERT as a word processing model along with Deep Learning models to find depression in a specific user over the period of time.

6. FUTURE WORK

This paper aims at doing a comparative study based on machine learning algorithms for prediction of depression using context-free model. In the future contextual bidirectional model can be considered for prediction in order to enhance the performance and which can provide better decision support systems to users.

7. REFERENCES

[1] Marcel Trotzek, Sven Koitka, and Christoph, M. Friedrich, :"Utilizing Neural Networks and Linguistic Metadata for Early Detection of Depression Indications in Text Sequences", DOI 10.1109/TKDE.2018.2885515, IEEE Transactions on Knowledge and Data Engineering ,2019.

[2] Ranjana Jadhav,Shaurya Jadhav,Hitesh Sachdev,Vinay Chellwani, :"Mental disorder detection.Bipolar disorder scrutinization using machine learning" IEEE,978-1-5386-5933-5,2019.

[3] Jane H. K. Seah,et Al, :"Data Mining Approach to the Detection of Suicide in Social Media"IEEE-International Conference on Big Data,2018.

[4] Akkapon Wongkoblap, Miguel A. Vadillo, Vasa Curcin, :"Detecting and Treating Mental Illness on Social Networks", IEEE-ICHI, DOI 10.1109/ICHI.2017.24,2017.

[5] Hao Wang, Jorge A. Castanon et Al, :"Sentiment Expression via Emoticons on Social Media" IEEE-International Conference on Big Data, 978-1-4799-9926-2/15/$31.00,2015.

[6]Walter Gerych, Emmanuel Agu, Elke Rundensteiner,"Classifying Depression in Imbalanced Datasets using an Autoencoder-Based Anomaly Detection Approach"IEEE-ICSC DOI 10.1109/ICSC.2019.00028,2019.

[7] Sharath Chandra Guntuku,David B Yaden,Margaret L Kern,Lyle H Ungar,Joannes C Eichstaedt,:"Detecting depression and mental illness on social media : an intergrative review", IEEE-ICHI, DOI 10.1109/ICHI.2017.24, 2017.

[8] Y. Zhang and B. Wallace, "A sensitivity analysis of (and practitioners' guide to) convolutional neural networks for sentence classification", IJCNLP,pages 253–263,2017.

[9] Matthew Peters, Mark Neumann, Luke Zettlemoyer, and Wen-tau Yih,:"Dissecting contextual word embeddings: Architecture and representation", pages 1499–1509,CEMNLP,2018.

Comparative Analysis of Deep Learning Techniques to detect Online Public Shaming

Mehdi Surani [1*], *Ramchandra Mangrulkar*[2]

[1]D.J.Sanghvi College Of Engineering, Mumbai, imsurani15@gmail.com
[2]Department of Computer Engineering, D.J.Sanghvi College Of Engineering, Mumbai, ramchandra.mangrulkar@djsce.ac.in

Abstract. Over the past years the exponential growth of social media usage has given the power to every individual to share their opinions freely. This has led to numerous threats allowing users to exploit their freedom of speech, thus spreading hateful comments, using abusive language, carrying out personal attacks, and sometimes even to the extent of cyberbullying. However, determining abusive content is not a difficult task and many social media platforms have solutions available already but at the same time, many are searching for more efficient ways and solutions to overcome this issue. Traditional models explore machine learning models to identify negative content posted on social media. Shaming categories are explored, and content is put in place according to the label. Such categorization is easy to detect as the contextual language used is direct. However, the use of irony to mock or convey contempt is also a part of public shaming and must be considered while categorizing the shaming labels. In this research paper, various shaming types, namely toxic, severe toxic, obscene, threat, insult, identity hate, and sarcasm are predicted using deep learning approaches like CNN and LSTM. These models have been studied along with traditional models to determine which model gives the most accurate results.

1 INTRODUCTION

The act of instant emotional expression through social media is becoming a common phenomenon in today's world. Platforms like Facebook, Twitter, Instagram, etc. allow one to freely communicate and express their thoughts and opinions. This can simply go from informing people about some accidental incident over the highway to divert traffic to commenting ill about some famous politician's speech. Social media can therefore range from being informative to offensive.

Maintaining decency and avoiding informal content is the current area of concern for all social media platforms. Such unhealthy acts have indirect effects on an individual's health. Lifelong traumatic effects like anxiety, depression, mental illness, and other psychological effects are seen to some extent. This causes users to isolate themself from such social platforms and can inculcate suicidal thoughts. Therefore, there is a need for establishing classification techniques and blocking mechanisms.

Many social platforms are working on this current issue where a user can be reported/ restricted or blocked if it is observed by many users around them. Such acts of shaming individuals on social media platforms need to be controlled sufficiently.

There is a need for automation in this area, which can help companies classify the different shaming categories and take actions accordingly. Many machine learning techniques have been deployed in the past years to work on this topic. However, this paper would like to implement deep learning techniques to find better results for the detection of Online Public Shaming. This research paper includes sarcasm as a shaming category where an ironic or satirical remark tempered by humour is mainly used to say the opposite of what's true to make someone look or feel foolish. Such comments are difficult to detect. Therefore, it is necessary to include this as a part of other shaming categories, namely, toxic, severe toxic, threat, obscene, insult, and identity hate.

The rest of the paper is organized as follows. The literature review is present in section 2. The Methodology is given in section 3. Experimentation and Results are present in section 4. Detailed discussion is given in section 5. Concluding remarks and future scope is given in section 6.

2 LITERATURE REVIEW

Most prior work in the area of online public shaming prediction has been spread across several overlapping fields. This can cause some confusion as different works may tackle specific aspects of shaming language; some may define the term differently or apply it to specific online domains only (Twitter, Online forums, Facebook, YouTube, etc.) to further complicate the comparison between approaches, nearly all previous work uses different evaluation sets. One of the contributions of this paper is to provide a public dataset to better move the field ahead. A public dataset that includes not only the existing shaming categories but also aspects where some users would post sarcastic comments in the same voice as the people that were producing toxic or identity hate content. Although online public shaming has been carried over a long time, their detection using Machine Learning has taken off only in recent years; in-depth research has been carried out on their detection in recent years. Hence, the authors have conducted an extensive survey to study the current findings and identify the gaps.

In "Online Public Shaming on Twitter: Detection, Analysis, and Mitigation" Rajesh Basak et al. Created a

* Corresponding author: imsurani15@gmail.com

web application that is based on the classification and categorization of shaming tweets. Here classification is performed using Support Vector Machine (SVM) and desired results are achieved. This [1] paper has the potential solution for countering the menace of online public shaming on Twitter by categorizing shaming comments. They have performed detection on a small dataset. Currently, the data available over the internet is in millions and hence we need to tackle large datasets. SVM, however, doesn't perform well when a huge dataset is taken, there is a need for a between model to work with a large amount of data.

In "A deeper look into sarcastic tweets using deep convolutional neural networks" S. Poria et al. For sarcasm detection, the system proposed by Cambria et al. [2] uses deep convolution neural networks. They have a deep understanding and study of emotions as well as sentiments for the detection of sarcasm. This is done by using predefined models for the extraction of features. Both sentiment and emotion feature for sarcasm detection and discussed the use of predefined models for feature extraction. This paper [2] identifies sarcastic text using the CNN model. But the performance of the model on large corpus and other domain-dependent corpora has not yet been explored. Other shaming categories have not been incorporated for sentiment classification.

In "An Effective Approach for Detection of Sarcasm in Tweets" Sreelakshmi K et al. proposed a system for sarcasm detection using Support Vector Machine (SVM) and Decision Trees for modelling the proposed system. Various features of the text like lexical, pragmatic, context incongruity, topic, and sentiment were taken into consideration. Sreelakshmi K et al. [3] have worked with SVM Models which again limits the usage of large datasets. Large corpora must be explored along with decision trees in order to find better results.

In "Classification of Abusive Comments in Social Media using Deep Learning", Mukul Anand et al [4] propose deep learning methods like CNN (Convolution Neural Network) and LSTM (Long Short Term) categorizing various shaming types like toxic, severe toxic, obscene, threat, insult, and identity hate. Online abuse is only identified by the user's report there is a need for automation in the detection of abusive comments in social networks which are carried out using deep learning techniques in this research work. Deep learning techniques were used for detecting online public shaming however there is a need to explore more shaming types and automation

In "Abusive Language Detection in Online User Content" Chikashi Nobata et al. [5] have proposed a system to identify hate speech on online user comments from two domains that outperform a state-of-the-art deep learning approach. They achieved this using standard NLP features along with different syntactic and embedding features. Naïve NLP techniques are implemented; in [5] there is a need for enhanced and new machine learning models to detect precise and accurate shaming categories.

In "Tackling Toxic Online Communication with Recurrent Capsule Networks" Soham et al. [6] have implemented Recurrent Neural Network and Capsule network as its backbone and captures contextual information for toxic online communication. Traditional methods have been overcome using neural networks. There is a need for multi-label toxicity categorization for different types of shaming.

In "Imbalanced Toxic Comments Classification using Data Augmentation and Deep Learning" Mai Ibrahim et al. [7] present various data augmentation techniques to overcome the data imbalance issue in the Wikipedia dataset. An ensemble of three deep learning models namely CNN, LSTM, and GRU (Gated Recurrent Units) were used to detect the type of toxicity present in the toxic comments dataset. Using multiple deep learning models can help us in understanding which technique will prove the best for online shame detecting. Findings from this can be of good contribution to this research work.

In "Audio and Video Toxic Comments Detection and Classification" Sangita et al. [8] model aim to apply the text-based Convolution Neural Network (CNN) with word embedding, using fastText word embedding technique improving the detection of different types of toxicity to improve the social media experience. Deep learning techniques were used for detecting online public shaming however there is a need to explore more shaming types and automation along with other models like RNN, LSTM, to detect which model gives better results.

In "Online Public Shaming Approach using Deep Learning Techniques" Mehdi et al. [9] have performed data visualization using text-based Convolution Neural Network (CNN) as the proposed model. They aim to categorize the dataset based on the different multi-labels, top words, confusion matrix, ROC curves for each label type, and classification report for the same. A deep understanding of the dataset is carried out in [8] using Convolution Neural Network.

In "A deep understanding of the dataset is carried out in [10] using Convolution Neural Network." Manav Kohli et al. used a wide range of Recurrent Neural Network models for the task of classifying abusive comments by users of Wikipedia. Findings include non-neutral baseline models based on TFIDF sentences. Gated Recurrent Unit (GRU) and LSTM results were compared which showed similar outcomes.

3 METHODOLOGY

This paper aims at detecting online public shaming using deep learning algorithms. Deep Learning is a subset of Machine Learning where data preprocessing and feature selection can be minimized without impacting the performance of algorithms. Deep learning models can extract relevant features on their own. The research is implemented in the following way. Machine or deep learning algorithms cannot process plain text as inputs. The system does not understand the human language. It becomes difficult or rather impossible for the algorithms to process such raw input data. To solve this issue there are Word embedding techniques available. Word embedding techniques provide a solution for this issue by transforming the plain text into a vector format which can be used by the machine learning model. Such transformations can also be used for finding semantic relationships between the associated words. The proposed model is using GloVE (global vectors for word representation) as a word Embedding Technique. The GloVE works by aggregating the global word-word co-occurrence matrix from a corpus. GloVE stresses that the frequency of co-occurrences is vital information and should not be "wasted" as additional training examples. Acquiring less training time GloVE outperforms Bag-of-Verbs and TFIDF. The position and semantic aspect of the word is also looked into.

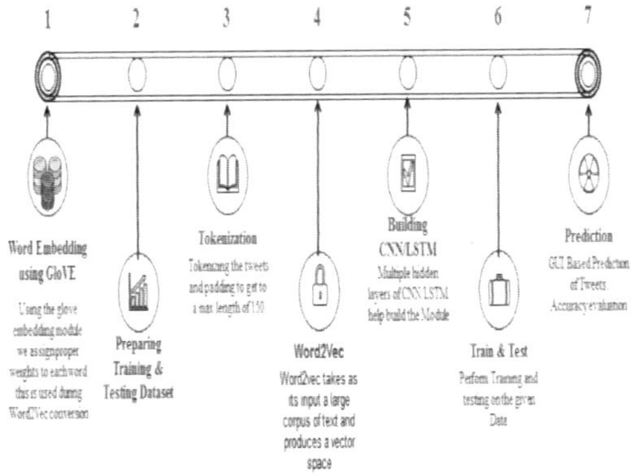

Fig. 1. Overview of Methodology

Word embedding is done using the GloVe (Global Vectors for Word Representation) embedding module. Here weights are assigned to each word this will then be used during word2Vec conversion. Training and testing data are separated and then prepared for tokenization. Tokenizing is the process of taking the tweets and padding them accordingly to maintain uniformity in the data set created. As it is known that the input set is very large and therefore there is a need to create vectors from the large text corpus hence forming a vector space. Once the vector space is formed, connecting the model is created using the following 2 algorithms: CNN and LSTM. Training and testing are performed on the above algorithms and accuracy evaluation along with a GUI Based prediction of Tweets is created. Further detailed understanding of the Deep Learning modules with their hidden layers is shown below

3.1. CNN (Convolution Neural Network)

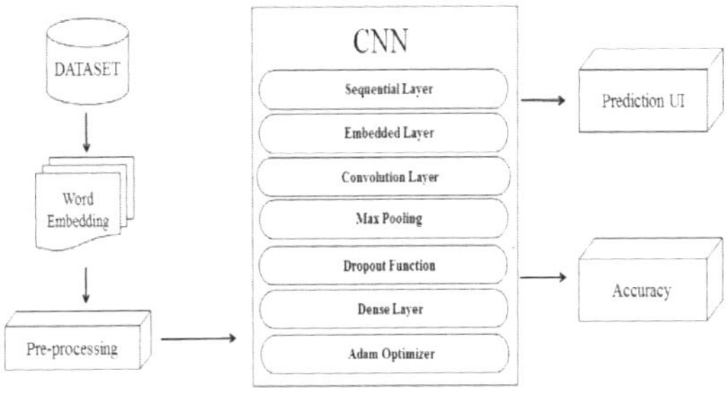

Fig.2. Convolution Neural Network and its Hidden Layers

3.2 LSTM (Long Short Term Memory)

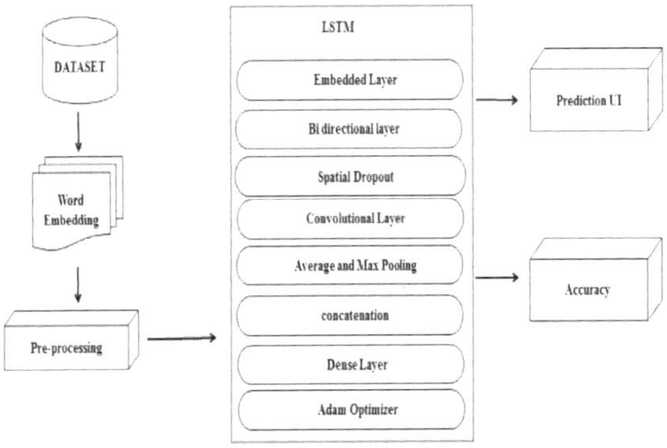

Fig.3. Long Short Term Memory (LSTM) and its Hidden Layers

With the recent advancements that are happening in data science, it is found that for nearly all those sequence prediction problems, LSTMs are observed as the most effective solution. LSTMs have a foothold over conventional feed-forward neural networks compared to RNN in some ways. This is often due to their property of selectively remembering patterns for long period (memory). LSTM's are nothing but improved RNNs as the data flows through a mechanism referred to as cell states and in this way, LSTMs can selectively remember or forget things. The data at a specific cell state has three different dependencies. These dependencies are often generalized as the previous cell state (i.e. the data that was present within the memory after the previous time step) The previous hidden state (i.e. this is often an equivalent as the output of the previous cell) The input at the present step (i.e. the new information that's being fed in at that moment). Every information from the previous state can be carried forward to the next state and whether that information is relevant to be kept or forget during training can be adjusted using the gates. For this very purpose, a Bi-Directional Layer is added, connecting two hidden layers generating a single similar output layer. Through this form of deep learning, a model which has information from the past and the future simultaneously can be achieved giving a more precise outcome.

https://doi.org/10.1051/itmconf/20214003030

4 DETAILED DISCUSSIONS

Online public shaming which targets to humiliate people over the internet publically is a form of individual harassment which is legally not acceptable. The serious consequences include ruining reputations, careers, and psychological damages. Social media being a vast platform makes it difficult to track each nuisance created by the audience. Naive Machine learning methods have helped achieve limiting these behaviors by blocking, restricting accounts or comments. The research work presented by Basak et al [1] and Shreelaxmi et al. [3] detected online shaming using the SVM model restricting the experiment to a small dataset. The research and implementation presented in this paper use advance deep learning techniques like CNN and LSTM which help explore large datasets, provide better results and accuracy, combine multiple datasets and predict multi-class labels which serve the objectives of this research. The Kaggle data set is combined along with the sarcasm dataset which helps us to expand the dataset and therefore achieve our first objective of using multiple datasets. This data set is then applied to the CNN and LSTM models to predict multiple shaming labels using a Graphical User Interface. This is trained and tested for more than 162070 entries. The model is trained in such a way that any new entry or tweet added to the user interface will predict the shaming categories hence, achieving the next objective of the research work. The shaming categories include toxic, severe toxic, threat, obscene, insult, identity hate, and sarcasm. Future implementations aim at identifying the severity of the shaming category and blocking/restricting the comment using deep learning models. Another evaluation matrix includes the Accuracy calculation. Logistic Regression used in [8] is a machine learning technique that is compared to the deep learning model of CNN and LSTM and a comparative analysis is carried out. Section 5 will discuss the results and experiment details. The main motive of the research aims at implementing the latest technologies available to achieve better results and find out how they outperform traditional methods of predicting online shaming.

5 EXPERIMENT AND RESULTS

The research paper shows the implementation of CNN and LSTM based approaches for detecting online public shaming. The experiment was carried out to achieve better accuracy results and predict the shaming labels. After the model is trained, Graphical User Interface was created which takes in tweets or comments as input and predicts the multi-label shaming categories. The resultant graph shows the sentiments namely toxic, severe toxic, obscene, threat, insult, identity hate, and sarcasm. This is carried out for CNN and LSTM models separately showing results in Tables 2 and 3.

The GUI-based model has an input screen where the tweet is added and on submission, it returns a chart that represents the multi-label shaming categories each representing the intensity of that sentiment. Let us understand this with an example. The tweet "*Trump Administration send 300 million NOTHING to Puerto Rico victims* "is a pure example of sarcasm along with mere toxicity against Donald Trump. The model implemented in this paper predicts the severity of the shaming type. A multi-label shaming category graph is created.

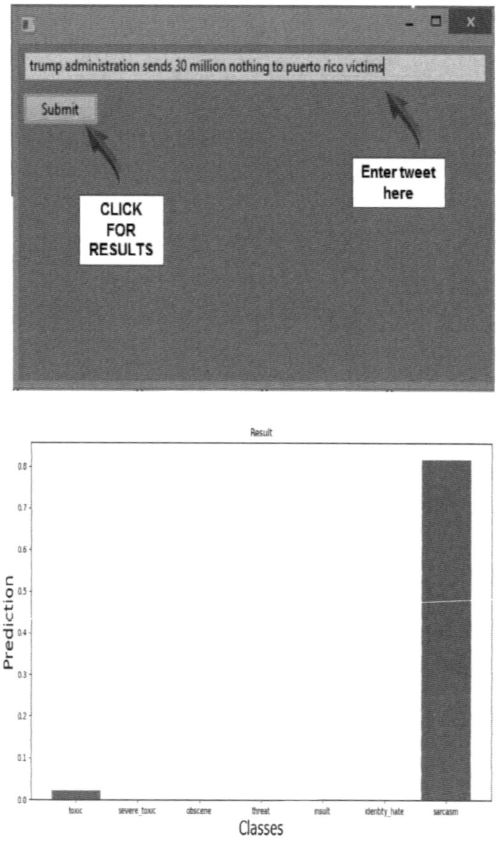

Fig.4.GUI Based shaming category prediction using LSTM

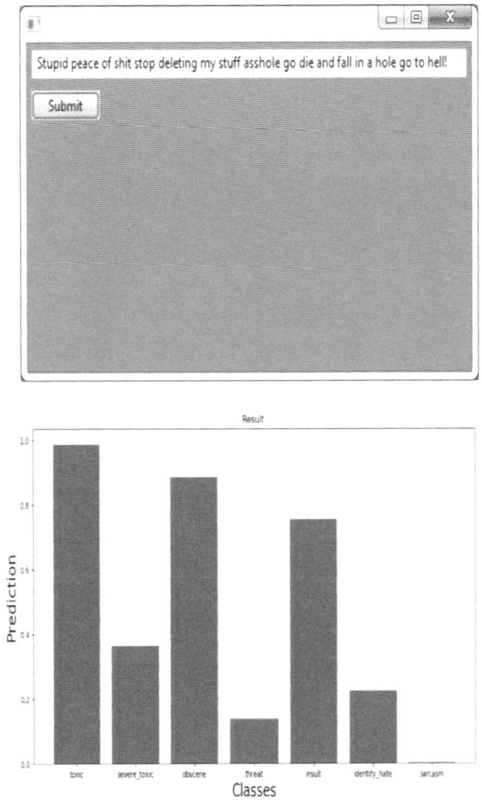

Fig.5.GUI Based shaming category prediction using CNN.

Sr.No.	Tweets from Dataset
1	Trump Administration send 300 million NOTHING to Puerto Rico victims
2	Are you threatening me for disputing neutrality? I know in your country it's quite common to bully your way through a discussion and push outcomes you want. But this is not Russia.
3	There must be some chemical imbalance in your brains. Are you ok? The gibberish you guys are writing concerning makes me feel sorry for you. Don't ever go to law school. You would do the legal profession badly
4	The government is so good, it ensures the citizens happily die of corona!
5	You are gay or antisemmitian?
6	F*CK YOUR FILTHY M*TH*R IN THE A*S, DRY!
7	I was merely apologising for my interference ;)

Table 1. Tweets selected to test CNN model.

Sr. No.	**Toxic**	**Severe toxic**	**Obscene**	**Threat**	**Insult**	**Identity Hate**	**Sarcasm**
1	0.021	0	0	0	0.009	0	0.819
2	0.305	0.001	0.001	0	0.002	0.0007	0
3	0.504	0	0.064	0.006	0.182	0.028	0.004
4	0.112	0	0.008	0.003	0.013	0.003	0.009
5	0.794	0.038	0.394	0.043	0.450	0.133	0.009
6	0.990	0.453	0.966	0.079	0.804	0.187	0
7	0.002	0	0	0	0	0	0.144

Table 2. Shaming type detection using CNN model.

Sr.No.	Tweets from Dataset
1	The JEWZ are Market Manipulators Hate them freaking Jews !! To bad Hitler did not nish them o !!!
2	300 missiles are cool! Love to see um launched into TelAviv! Kill all the gays there!
3	mom wants to know if you'll be free if she visits 14 months from now
4	Stupid peace of shit stop deleting my stuff asshole go die and fall in a hole go to hell!
5	You are gay or antisemitismstops?
6	F*CK YOUR FILTHY M*TH*R IN THE A*S, DRY!
7	Stupid peace of shit stop deleting my stuff asshole go die and fall in a hole go to hell!

Table 3. Tweets selected to test LSTM model.

Sr. No.	**Toxic**	**Severe toxic**	**Obscene**	**Threat**	**Insult**	**Identity Hate**	**Sarcasm**
1	0.767	0.004	0.769	0.004	0.2	0.694	0.01
2	0.642	0	0.018	0.056	0.053	0.56	0.005
3	0	0	0	0	0	0	0.951
4	0.986	0.361	0.885	0.137	0.756	0.2218	0
5	0.900	0.391	0.098	0.050	0.645	0.818	0
6	0.991	0.505	0.940	0.006	0.864	0.051	0
7	0.997	0.258	0.920	0.212	0.932	0.0344	0.006

Table 4. Shaming type detection using LSTM model.

https://doi.org/10.1051/itmconf/20214003030

Here the algorithm is applied on the Kaggle dataset which is 67MB. Combining this dataset with that of sarcasm we get a total of 7 shaming categories. This approach gives an aggregate accuracy of 98% in predicting the shaming type for the CNN and LSTM model. The authors aim to detect online shaming carried over Twitter and at the same time, increase the overall accuracy of classification through their approach. Figure 6 shows the accuracy achieved using CNN Model. Here an accuracy of 98.42% is achieved whereas figure 7 shows accuracy of 98.57% achieved by LSTM Model. A collective resultant of Machine Learning model and Deep Learning model is shown in figure 8. It clearly shows that CNN and LSTM outperform Logistic Regression and provide better results.

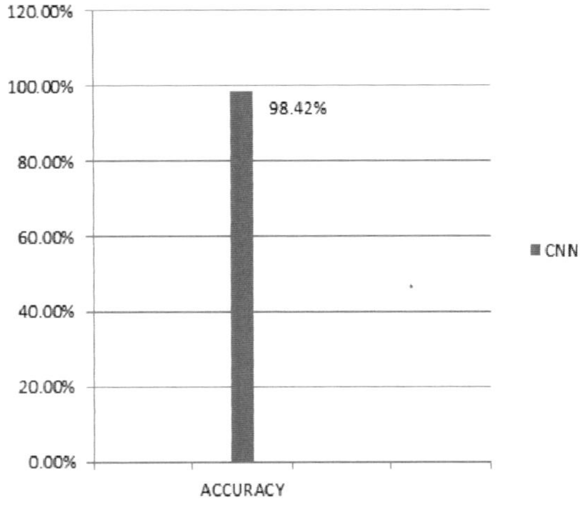

Fig6. Accuracy prediction using CNN.

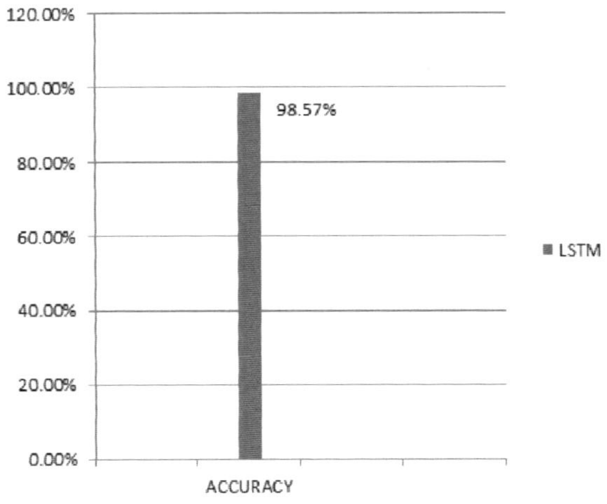

Fig.7. Accuracy prediction using LSTM.

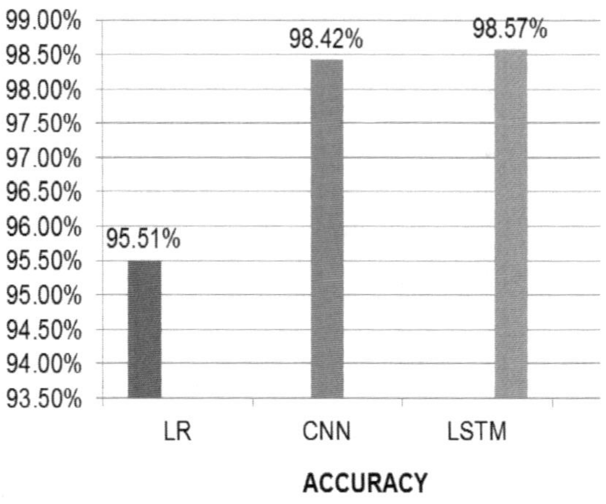

Fig.8. Accuracy comparison LR, CNN, LSTM.

The above is a comparative analysis of the 3 Models namely Logistic Regression [8], Convolution Neural Network, and Long Short Term Memory. The findings clearly show that Deep Learning models perform better than other machine learning models with respect to accuracy estimation

6 IMPACT AND CASE STUDIES

Being a victim of online public shaming can have multiple ill effects on a person's mental, physical, and emotional health. People are taking their own lives and affecting the lives of their fellow family members and friends. The ugly side to this digital world is that it gives easy access to people to pass on their opinion without giving them a second thought. Social media has given the platform for people to troll the latest memes circulating and comments from unknown users in an enormous volume on the victim's image take no time. Let's have a look at few case studies which had immediate actions taken for their action.

1. Case Study 1: Justine Sacco, 2013.

In the year 2013, Justine Sacco, a Public Relations person in an American Internet Company tweeted the below: *"Going to Africa. Hope I don't get AIDS. Just kidding. I'm white!"*. With a mere 170 followers, Justine faced criticism at an enormous rate and became one of the most discussed topics over the internet. Before she even landed, she lost her job.

2. Case Study 2: Aamir Khan, 2015.

In 2015, India's most famous Actor Aamir Khan commented on the rising intolerance in India and spoke about his wife Kiran Rao suggesting to move out of the country. The response from the general public over

Twitter and fellow actors seems to be very grinding and the consequences faced were that he was removed as Brand Ambassador of SnapDeal.

3. **Walter Palmer, Cecil the lion's killer dentist, endures the latest onslaught from social media mob, 2016.**

In the endless feedback loop of social media's shaming machine, Walter Palmer suddenly found himself with nothing to shoot and no place to hide. As soon as word got out that the Minnesota dentist had shot and killed an African lion. He received a lot of threats and hate over Twitter. Social media shaming goes so viral that even other users with the name Walter Palmer received more than 15 threat full tweets.

4. **Case Study 3: Melania Trump, 2016.**

In 2016 a Twitter user pointed out Melania Trump, spouse of the US President for plagiarism in one of her campaign speeches. There were media coverage and tweets regarding her speech being plagiarized from Michelle Obama. There was huge criticism and negative media coverage encountered immediately.

7 CONCLUSIONS AND FUTURE WORK

In this research implementation, it is observed that the deep learning approach for online public shaming has been carried out using CNN and LSTM. A comparative analysis with [1] it is observed that the LSTM model works better in predicting online public shaming. The graphical user interface also helps predict the shaming type, especially for sarcastic tweets that are difficult to predict. It is also seen that a large amount of data that could not be solved using SVM and other machine learning modules have shown better results using Deep Learning. This research shows that Deep Learning-based models outperform machine learning-based model. In the future complex deep learning models and hybrid algorithms can be studied and implemented and various Transformers can be used for the prediction of online public shaming.

8 REFERENCES

[1]Rajesh Basak, Shamik Sural, Niloy Ganguly, and Soumya K. Ghosh, "Online Public Shaming on Twitter: Detection, Analysis, and Mitigation.", IEEE Transaction -2329-924X,2019.

[2]S. Poria, E. Cambria, D. Hazarika, and P. Vij, "A deeper look into sarcastic tweets using deep convolutional neural networks," Creative Commons Attribution 4.0,arXiv preprint DOI:1610.08815, 2016.

[3]Sreelakshmi K Rafeeque P C, "An Effective Approach for Detection of Sarcasm in Tweets", International CET Conference on Control, Communication, and Computing (IC4)DOI-978-1-5386-4966-4,2018.

[4]Mukul Anand, Dr.R.Eswari, " Classification of Abusive Comments in Social Media using Deep Learning", IEEE DOI-978-1-5386-7808-4,2019.

[5]Chikashi Nobata, Joel Tetreault, Achint Thomas, Yashar Mehdad, Yi Chang, "Abusive Language Detection in Online User Content", IW3C2, DOI-10.1145/2872427.2883062,2019.

[6]Soham Deshmukh, Rahul Rade, "Tackling Toxic Online Communication with Recurrent Capsule Networks", IEEE-DOI: 10.1109/INFOCOMTECH.2018.8722433 2018.

[7]Mai Ibrahim, Marwan Torki and Nagwa El-Makky, "Imbalanced Toxic Comments Classification using Data Augmentation and Deep Learning", 2018.

[8]Sangita Holkar, S. D. Sawarkar, Shubhangi Vaikole, "Audio and Video Toxic Comments Detection and Classification", International Journal of Engineering Research & Technology (IJERT), Vol. 9 Issue 12, IISSN: 2278-0181,2020.

[9]Mehdi Surani, Ramchandra Mangrulkar," Online Public Shaming Approach using Deep Learning Techniques", Journal of the University of Shanghai for Science and Technology ISSN: 1007-6735,2021.

[10]Manav Kohli, Emily Kuehler and John Palowitch. "Paying attention to toxic comments." Stanford University,2018.

ITM Web of Conferences **40**, 03031 (2021) https://doi.org/10.1051/itmconf/20214003031
ICACC-2021

Pose Estimation and Correcting Exercise Posture

Rahul Ravikant Kanase [a]
rahulkanase2600@gmail.com

Akash Narayan Kumavat [a]
kumavatakash001@gmail.com

Rohit Datta Sinalkar [a]
rohit4sinalkar@gmail.com

Sakshi Somani [a]
sssomanisakshi@gmail.com

[a] *Department of Electronics Engineering, Ramrao Adik Institute of Technology, Nerul, Navi* Mumbai,400 706,India

Abstract: **Our posture shows an impact on health both mentally and physically. Various methods have been proposed in order to detect different postures of a human being. Posture analysis also plays an essential role in the field of medicine such as finding out sleeping posture of a patient. Image processing based and sensor based approach are the leading posture analysis approaches. Sensor based approach is used by numerous models to focus on posture detection in which the person needs to wear some particular devices or sensors which is helpful in cases such as fall detection. Image processing based approach helps to analyze postures such as standing and sitting postures. Fitness exercises are exceptionally beneficial to individual health, but, they can also be ineffectual and quite possibly harmful if performed incorrectly. When someone does not use the proper posture, exercise mistakes occur. This proposed application utilizes pose estimation and detects the user's exercise posture and provides detailed, customized recommendations on how the user can improve their posture. A pose estimator called OpenPose is used in this application. OpenPose is a pre trained model composed of a multi-stage CNN to detect a user's posture. This application then evaluates the vector geometry of the pose through an exercise to provide helpful feedback. Pose estimation is a method in which spatial locations of key body joints is calculated using image or video of the person. This computer vision technique detects human posture in images or videos and shows the keypoints such as elbow or knee in the output image.**

Keywords: CNN, OpenPose, pose estimation, posture detection.

I. INTRODUCTION

Various exercises such as dead lifts, squats and shoulder presses are favourable to human body fitness but it can also be very harmful if the exercises are performed improperly. Injuries to the muscles or ligaments can be caused due to heavy weights involved in these exercises. Due to the lack of training or knowledge many people do not follow the correct posture to be maintained while performing these exercises regularly. This may lead to muscle fatigue and muscle strain. In this course project, using the latest techniques in pose estimation we help people in performing exercises with correct posture by developing a project that detects the users pose while exercising and provides feedback and suggests improvements if necessary. The goal for this project is to prevent injuries and improve the form of human workout with just a computer and a camera. The initial step of the project is to use human pose estimation which is a highly applicable domain of computer vision. A trained model determines a person's joints as a list of skeletal key points from the given data which could be an RGB image or a depth map. Pose estimation plays an important role in solving problems related to human detection and activity recognition. It also helps in solving complex problems involving movement detection. OpenPose is used which utilizes neural network for the inference of this project. The latter part of this project involves detecting the quality of human pose for a given exercise. It is approached using heuristic based and machine learning models. The full application consists of two main components which can take a video of an exercise and provide feedback to the user.

https://doi.org/10.1051/itmconf/20214003031

II. RELATED WORK

Various methods have been studied for pose estimation that evaluate human poses using sensors, videos and machine learning approach. A few of them are mentioned below. A. Toshev and C. Szegedy [7] used neural network for the first time to improve pose detection using regression on CNN for finding the location of body joints. A stacked hourglass neural network architecture was introduced by A. Newell, K. Yang, and J. Deng [4] which works on bottom up and top-down approach for finding pose predictions. J. Shotton, A. Fitzgibbon and others [6] use single depth maps for prediction of 3D positions of joints using object recognition. F. Bogo, A. Kanazawa and others [2] used single RGB images to predict 3D pose and 3D mesh shape. Research work on detecting multiple human pose in a single frame has also been focused and worked on. G. Papandreou,
T. Zhu and others [5] used a two-stage process for detecting multiple poses, first step includes identifying people and the second includes detecting of their key points. For analysis of physical movements, P. Zell, B. Wandt, and B. Rosenhahn [9] used a method in which body is represented as mass-spring system to find the forces and torques that travel through the joints of the body.

Based on sequential prediction framework, S. Wei, V. Ramakrishna and others [7] propose a different architecture which uses multiple convolutional networks to clarify joint estimates over sequential passes and design a cascaded CNN network to represent texture and spatial information with convolution layers, sequentially incorporating global context to refine part confidence maps from previous iterations. During training period, due to the problem of vanishing gradients intermediate supervisions are added at the end of each stage. For reducing the information loss of neighbouring joints, the size of the receptive field is increased. To solve the problem caused by increasing the depth of network, intermediate supervision is used. This method also works against part occlusions. Each layer of a multi stage CNN becomes independent after training. These methods have high performances as compared to single CNN methods. For detection and regression, X. Chu, H. Li and others [13] use CNN's generalization. This model estimated human poses through combined detection and regression results by using shared CNN as the common input of detection and regression sub network. To identify multiple people in a real time frame, Z. Cao, T. Simon and others [3] used part affinity fields that extracted features with the first 10 layers of VGG-19 [9] without the need to detect individual person. It consists of three branch CNN architecture which predicts joint location, limb direction and orientation with part affinity field keeping the initial features. This method has improved accuracy of regression because it combines and enhances the output of three branches. For this application a simpler approach is used which includes the analysis of the angles of distances between joint key points to provide feedback to the user without a full physical simulation.

III. SYSTEM DESIGN

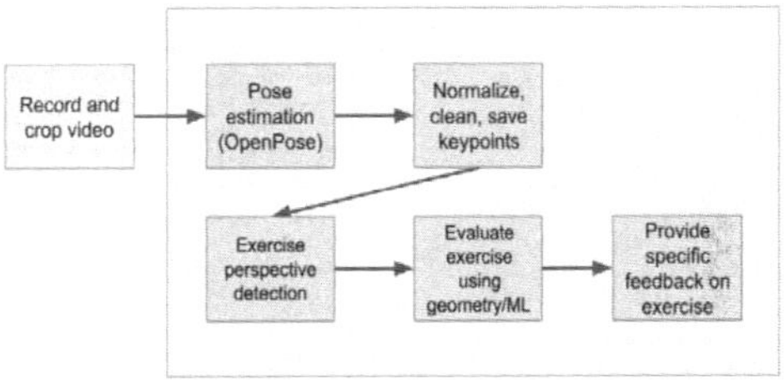

Fig 1. Block Diagram of Pose Estimation and Correction.

For the first step of the project, the users need to record a video performing a particular exercise with a certain point of view (front side, back side, etc.) which allows the exercise to be seen properly. There are no restrictions on the distance from the camera or the type of camera, the only thing that the user needs to take care is that their posture is visible properly. Further, the user should take care that the recorded video only contains the frames of exercise. The editing part of the video can be done using any software. OpenPose supports all common video formats that are available in modern computers and smartphones. Facial recognition includes facial key point detection which contains its own identifier and coordinates related to the image which is usually provided in pixels. Classification and localization are the important tasks in the training of key point detection in computer vision models. The same method is applied to the human posture for pose detection. It works in a much similar way which includes classification of body points, typically joints, determining the location within the video or image. Pose estimation methods are so highly developed that we can calculate how many pixels is the movement of the hand and if the height of the human body is known, it can be converted into distance. Velocity can be calculated with help of the frame rate of the video. Considering these measurements, we can say if the 5 body is in ascending or descending position. It also helps in classifying the activity or state of the human posture and analyzing what a person is doing and how they are doing it. OpenPose provides the output of key points in a frame-by-frame text in JSON format. Python uses pandas data frames to read the JSON files. Calculations are performed on key points provided with the help of Python. For providing feedback to the users, we generate video with the help of Open CV, the raw coordinates from OpenPose and our analyzed points.

IV. SYSTEM IMPLEMENTATION

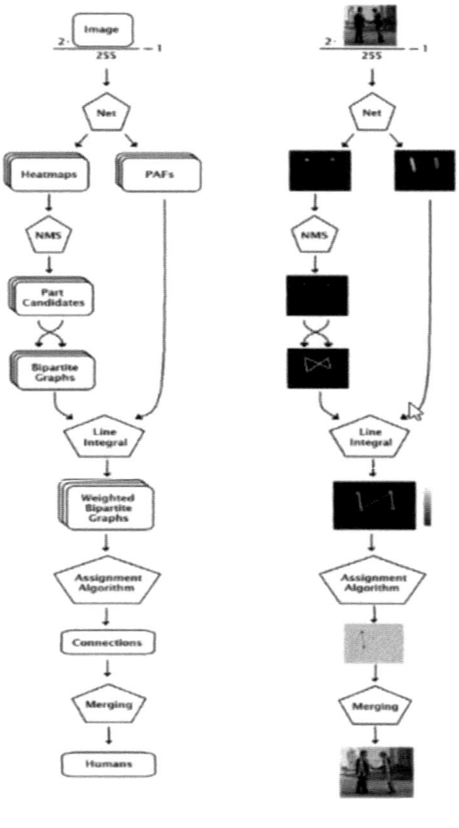

Fig. 2 Flowchart of Pose Estimation and Correction.

A. Pose Estimation

CNN is used for labeling RGB images in pose estimation. OpenPose software is used for posture detection which provides output in part affinity fields which shows the position and orientation of limbs. It has a high accuracy and efficiency as compared to other softwares. OpenPose is scalable and is easy for installation. It provides prediction of 18 key points including nose, neck, shoulders, wrists, elbows, hips, knees, and ankles.

B. Key point Normalization

Key point normalization includes list of predictions of key point for every video and part objects that store confidence of each key point. For representing full skeletal prediction of a human pose, joint key point is created to build a pose object of every frame in the video. A pose sequence object is created that combines each video frame for the entire video. The system is generalized considering different body length measurements, camera quality, distance from the camera and other factors. To compensate with these differences, the pose is detected based on torso length in pixels. The average of distance from neck key point to left and right hip key point is called as the torso length. It remains constant throughout the frames in the video.

C. Perspective Detection

A particular exercise has to be recorded from a particular camera perspective. For example, for front raise, the video has to be recorded from the side of the body and could be performed with left or right arm. With the help of the most visible key points, the system determines which arm is used.

D. Geometry Evaluation

In geometric evaluation, body vectors are calculated with the help of key points.

E. Example: Bicep Curl

In bicep curl, the angle between upper arm and torso vector is calculated which denotes the movement of the upper arm. If the upper arm does not move significantly, then it should be parallel to torso for the entire frames of the video. A large change in the angle between these vectors will indicate that there is a movement in the upper arm. The dumbbell should be brought up completely so that the bicep is fully contracted beyond 90° between upper arm and forearm. A minimum threshold is set for the angle between upper arm and forearm. It will point out for incorrect form due to heavy weight used etc. At the start, the angle between the upper arm and forearm is 180° which decreases when we the bicep is contracted upwards and again increases when the dumbbell is brought downwards. The minimum angle will be found at which the weight is lifted for the complete video frames. From the available data, a feedback will be suggested on the posture of the user regarding the rotation of the arm or the weight of the dumbbell. Making use of shoulder to help lifting the weight upwards denotes incorrect form of exercise which results in excess rotation of shoulder. (Refer Fig.3 and 4)

F. Example: Front Raise

In front raise, the user should lift the weights slightly above the shoulders and avoid using torso movement for lifting the weights. To identify the form the posture during this exercise, two things are identified: Maximum angle between the arm and torso and the horizontal range of motion of the back. To detect the swinging in the upper body between the frames of exercise, the change in vector difference is calculated for monitoring back motion. (Refer Fig.5 and 6)

G. Example: Shoulder Shrug

In shoulder shrug, the motion of shoulders should be in the full range and the elbows should not be bent. Also using arms for lifting indicates incorrect form. It is calculated by the angle between upper arm and forearm, torso length indicates the range of motion of the shoulders. Low shoulder movement indicates incomplete motion of shoulders and the small angle denote bending of elbows. (Refer Fig.7 and 8)

H. Example: Shoulder Press

In shoulder press, the weights should not be lifted far forward or backward. Moving the torso too much will indicate heavy lifting or improper lifting. The torso should remain constant and vertical throughout the exercise. The shoulder press movement is observed using the motion of the back which is calculated by the neck and hip key points. It also depends on the motion of arm calculated by elbow and neck key points and the maximum angle between forearm and upper arm vector. The feedback is provided to keep the back straight if large motion is observed for the back. During lifting of dumbbells, rolling of shoulders is observed if the elbow locates behind the neck indicating incorrect form of the exercise. If the angle between forearm and upper arm is small, a feedback is given suggesting to improve the range of motion while lifting the weight. (Refer Fig.9 and 10)

I. Machine Learning Evaluation

This approach includes detection of pose in a data driven format. As the recorded video can be of any length, it creates an error in vector length for each example. To solve this error, dynamic time warping (DTW) with the nearest neighbour classifier is used. DTW is used to measure nonlinear similarity between two time series. Euclidean distance fails because sequences are phase shifted in time. In DTW, we dynamically calculate key points to a given point in the next sequence that correlates with the first. A distance matrix is calculated where the matrix element represents the Euclidean distance between two key point sequence. The most ideal match of points is found and the distance between them is determined. The limitation of DTW is that its not strong enough against noise. To avoid this error in DTW measures, a key point sequence is passed two times through a five-median filter. For detecting correct or incorrect form based on the calculated DTW distances, a nearest neighbour classifier is developed.

V. RESULTS

Videos for different exercises are uploaded on this system. The results for four exercises are shown below:

A. Bicep Curl

1. Input:

Fig.3 Input for Bicep curl

2. Output feedback for bicep curl:

```
Bicep curl
Output on command prompt:

Exercise arm detected as: left.
Upper arm and torso angle range: 17.36865202142973
Upper arm and forearm minimum angle: 44.669291603102366
Exercise performed correctly!
Exercise performed correctly! Weight was lifted fully up,
and upper arm did not move significantly.
```

Fig.4 Output for bicep curl

https://doi.org/10.1051/itmconf/20214003031

B. Front Raise

1. Input:

Fig.5 Input for Front Raise

2. Output feedback for front raise:

```
Front Raise
Output on command prompt:

Exercise arm detected as: right.
Horizontal range of motion for back: 0.2816274951491309
Max angle between torso and arm when lifting: 99.16085343152268
Exercise performed correctly!
Exercise performed correctly! Weight was lifted fully up,
and no significant back movement was detected.
```

Fig.6 Output for Front Raise

C. Shoulder Shrug

1. Input:

Fig.7 Input for Shoulder Shrug

2. Output feedback for shoulder shrug:

```
Shoulder Shrug
Output on command prompt:

Range of motion for shoulders: 0.25897065694603627
Max upper arm and forearm angle: 5.201748222054894
Exercise performed correctly!
Exercise performed correctly! Shoulders went through range of motion,
and arms remained straight.
```

Fig.8 Output for Shoulder Shrug

D. Shoulder Press

1. Input:

Fig.9 Input for Shoulder Press

2. Output feedback for shoulder press:

```
Shoulder Press
Output on command prompt:

Exercise arm detected as: right.
Range of motion for back: 0.2658138515490096
Minimum distance between elbow and neck: -0.265854705762O2264
Max upper arm and forearm angle: 175.1610366573757
Exercise could be improved:
You are rolling your shoulders when you lift the weights. Try to
steady your shoulders and keep them parallel. You are not lifting
the weight all the way up. Extend your arms through the full range
of motion. Lower the weight if necessary.
```

Fig.10 Output for Shoulder Press

VI. CONCLUSION

In this project, an application is presented which provides feedback on human posture while performing exercises using pose detection, visual geometry and machine learning. The output of pose estimation is used to calculate human body key points from the video provided. Machine learning algorithm is used for deciding posture correctness and geometric algorithms for providing feedback on exercise improvement. Four exercises have been considered which can be extended to many other exercises for future work. For getting a feedback at any place, a mobile application can be developed that allows the user to record a video from the smartphone an upload on the application. A web page can be developed to display the output provided by the project. Feedback improvement can be increased suggesting targeted action and specific suggestions regarding the body part used and the weight of the equipments. The exercise performed with correct form can be represented using graphical simulation to highlight the mistake of the user and how to improve it.

VII. REFERENCES

1) D. Berndt and J. Clifford. Using dynamic time warping to find patterns in time series. In AAAIWS, 1994.
2) F. Bogo, A. Kanazawa, C. Lassner, P. Gehler, J. Romero, and M. Black. Keep it smpl: Automatic estimation of 3d human pose and shape from a single image. In ECCV, 2016.
3) Z. Cao, T. Simon, S. Wei, and Y. Sheikh. Realtime multiperson 2d pose estimation using part affinity fields. In CVPR, 2017.
4) A. Newell, K. Yang, and J. Deng. Stacked hourglass networks for human pose estimation. In ECCV, 2016
5) G. Papandreou, T. Zhu, N. Kanazawa, A. Toshev, J. Tompson, C. Bregler, and K. Murphy. Towards accurate multi-person pose estimation in the wild. In CVPR, 2017.
6) J. Shotton, A. Fitzgibbon, M. Cook, T. Sharp, M. Finocchio, R. Moore, A. Kipman, and A. Blake. Real-time human pose recognition in parts from single depth images. In CVPR, 2011
7) S. Wei, V. Ramakrishna, T. Kanade and Y. Sheikh, "Convolutional Pose Machines," in 2016 IEEE Conference on Computer Vision and Pattern Recognition (CVPR), Las Vegas, NV, USA, 2016 pp. 4724-
8) W. Rawat and Z. Wang. (2017). Deep Convolutional Neural Networks for Image Classification: A Comprehensive Review. Neural Computation. 29. 1-98. 10.1162/NECOa00990
9)] K. Simonyan and A. Zisserman, "Very Deep Convolutional Networks for Large-scale Image Recognition," Published as a conference paper at ICLR, Cornel University Library, 2015.
10) J. Tompson, A. Jain et al. "Joint Training of a Convolutional Network and a Graphical Model for Human Pose Estimation." Eprint Arxiv(2014):1799-1807.
11) X. Chen, and A. Yuille. "Articulated Pose Estimation by a Graphical Model with Image Dependent Pairwise Relations." Eprint Arxiv(2014):1736-1744.
12) T. L. Munea, Y. Z. Jembre, H. T. Weldegebriel, L. Chen, C. Huang and C. Yang, "The Progress of Human Pose Estimation: A Survey and Taxonomy of Models Applied in 2D Human Pose Estimation," in IEEE Access, vol. 8, pp. 133330-133348, 2020, doi: 10.1109/ACCESS.2020.3010248.
13) X. Chu, W. Ouyang, H. Li and X. Wang, "Structured Feature Learning for Pose Estimation," 2016 IEEE Conference on Computer Vision and Pattern Recognition (CVPR), 2016, pp. 4715-4723, doi: 10.1109/CVPR.2016.510.
14)] W. Yang, W. Ouyang, H. Li and X. Wang, "End-to-End Learning of Deformable Mixture of Parts and Deep Convolutional Neural Networks for Human Pose Estimation," 2016 IEEE Conference on Computer Vision and Pattern Recognition (CVPR), 2016, pp. 3073-3082, doi: 10.1109/CVPR.2016.335.
15) Steven Chen & Richard Yang (2018). Pose Trainer: Correcting Exercise Posture using Pose Estimation.

Implementation of Deep Learning Based Sentiment Classification and Product Aspect Analysis.

Shweta Dhabekar[1], *Dr. M. D. Patil*[2]

[1]M.E Student, Ramrao Adik Institute of Technology, D. Y. Patil Deemed to be University, Navi Mumbai, India.
[2]Professor, Ramrao Adik Institute of Technology, D. Y. Patil Deemed to be University, Navi Mumbai, India.

Abstract. With the increase in E-Commerce businesses in the last decade,the sentiment analysis of product reviews has gained a lot of attention in linguistic research. In literature, the survey depicts the majority of the research done emphasizes on mere polarity identification of the reviews. The proposed system emphasized on classifying the sentiment polarity and the product aspect identification from the reviews. Proposed work experimented with traditional machine learning techniques as well as deep neural networks such as Convolutional Neural Network (CNN), Recurrent Neural Network (RNN) and Long Short Term Memory(LSTM) Networks. The proposed system gives a better understanding of these algorithms by comparing the outcomes. The Deep Learning approach in the proposed work successfully provides a mechanism which identifies the review polarity and intensity of the reviews and also analyses the short form words used by people in the reviews. The experimental results in this work, applied on amazon product dataset, shows that the LSTM model works the best for sentiment analysis and intensity of reviews with 93% accuracy. This research work also predicts polarity for short-form word reviews which is the common trend these days while writing the reviews.

1 Introduction

In recent years, there has been an increase in research activities focused on analyzing sentiment in textual resources. The number of articles published on sentiment analysis has been growing in recent years. Sentiment analysis or opinion mining is one of the subtopics of this research, in which it computationally investigates people's views, assessments, attitudes and emotions regarding entities, individuals, situations, events, texts.

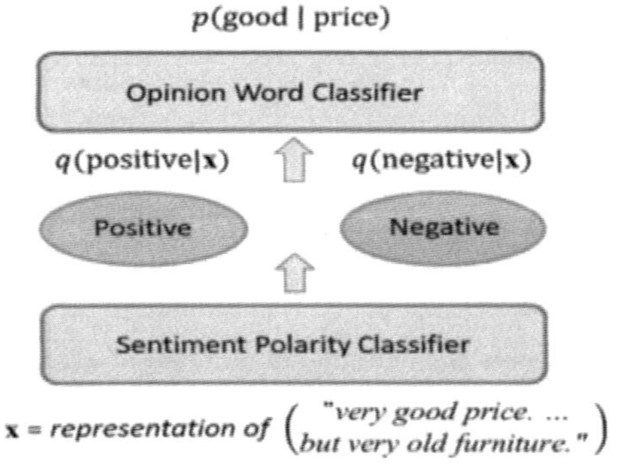

Fig. 1. Research carried out on sentiment analysis.

This approach has a wide range of applications. E-Commerce Businesses, for example, are constantly interested in consumer thoughts or opinions about their products and services. Before using a service or purchasing a product, most of the consumers are keen to know what other people think about it. Lastly, research scholars [2] utilize this data to conduct semantic analysis of trends in market and buyers sentiment, which might improve forecasting product quality. Despite this, monitoring online opinion sites, as well as extracting the information hidden in them, is still a difficult undertaking due to the proliferation of different sites. In extensive forum postings and blogs, each site often has a large amount of opinionated content that is not always easy to read. The ordinary reader may face trouble locating appropriate websites and precisely extracting the hidden details and viewpoints included therein [5]. Furthermore, teaching a computer to understand sarcasm is a difficult and time-consuming endeavour, considering that

* Corresponding author: author@e-mail.org

computers cannot yet think like humans. The goal of this work is to identify positive and negative opinions of various products and to develop a model which is a supervised model, having the capability to polarize enormous volumes of data. Customers' reviews and ratings, obtained from Amazon's Consumer Reviews, are used as a dataset for the proposed system. In proposed research, it retrieves the features from the input dataset and uses those features to build various supervised models. Traditional machine learning mechanisms and deep learning techniques like RNN, CNN and LSTM are included in these models. The accuracy for various such models were analyzed and a thorough evaluation of the obtained sentiments regarding the reviews was done. These models make use of 70% of the data for training and 30% of data serves as testing data.

Yi-Fan and Maria Soledad Elli gathered sentiments by analyzing reviews and evaluated the results to develop a business model in their paper[3]. Authors said that this model gained them a high level of precision. The major classifiers were Support Vector Machine (SVM) and Multinomial Naive Bayesian (MNB). Callen Rain[6] advocated expanding current Natural Language Processing (NLP) research. To determine whether a review was good or not, the Naive Bayes classifier was used by authors. In the field of sentiment analysis, deep-learning neural networks are popular too. For the sentiment labeling challenge, some researchers deployed a convolutional neural network with the purpose of avoiding unnecessary task-specific feature engineering. In contrast to this, [8] recommended employing recursive neural networks to gain a better grip over sentiment prediction.

2 Related Work

There have been several research papers published so far on product reviews, sentiment analysis and opinion mining. Wei Zhao et. el. [1] proposed a deep learning approach for product review sentiment classification, which uses existing ratings as input values. The work includes two steps: (1) to make the model understand the representation which evaluates the sentiment values for input sentences through ratings; (2) deploy the classification layer above the embedding layer and make the use of tagged sentences for supervised learning. Author proposes two types of low level network structures for determining review text, one is Convolutional Feature Extractors and the other is Long Short-Term Memory network. In [2], the author has given a deep survey of reviews and concluded that the internet is the most preferred method of learning, acquiring ideas and obtaining various product ratings. Millions of evaluations on a commodity, service, individual or location are posted on the internet daily. It's tough to read and evaluate such assessments one by one because of their sheer volume and quantity. As shown in Figure 2, there are three major types of sentiment classification based on access levels. If the classification is Document level, it's assumed that every single document is an opinion. If the classification is Sentiment Level, it checks if every sentence has some opinion expressed in the sentence. If the classification is Aspect level, it searches for the various aspects about which the sentiments are expressed in the opinion. Natural language processing (NLP), text analysis and determining the sentiment polarity of posted reviews are all part of the Sentiment analysis process. Its core objective is to recognize the sentiment behind the given review.

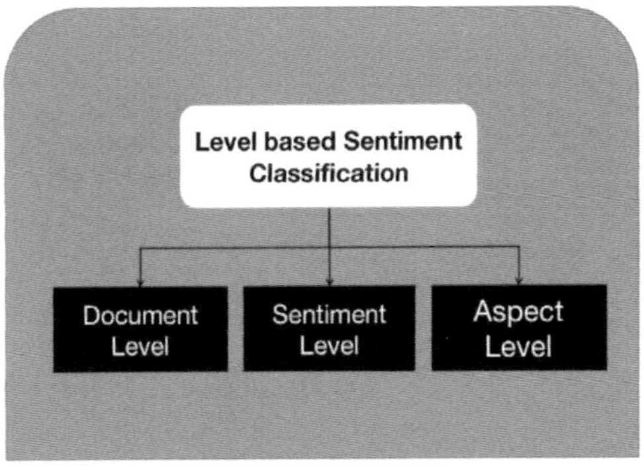

Fig. 2. Types of sentiment analysis.

https://doi.org/10.1051/itmconf/20214003032

Many algorithms exist to deal with NLP challenges. The categorization of reviews is presented in the proposed system. In comparison to Naive Bayes' and extreme entropy approaches, this research proves that SVM provides higher accuracy. On Yelp's rating dataset, Xu Yun[8] et al used current supervised learning techniques like supporting vector machine or perceptron algorithm to determine review's rating.

3 Proposed System

3.1 Proposed Work

Product review analysis is basically more concerned about analyzing the text to determine the polarity of the reviews. This is the most tedious task as the reviews are user-written and there is no specific pattern or rules to be mapped. As the data is highly temporal in nature, the use of artificial intelligence and right choice of AI model is presented in this research. Though traditional machine learning techniques like SVM behaved better in various NLP systems over the years, they have some limitations, which deep learning models easily overcome to a greater extent. Deep learning models have the advantage of requiring no hand-crafted characteristics to be carefully optimized. It takes word embeddings as input instead of features, which contain context information and the intermediate layers of the neural network learn the features during the training phase. Deep learning makes it possible to learn representations automatically. This eliminates the requirement for manual feature or word information construction. The proposed system comprises 3 major steps:

1. The system emphasizes on the polarity detection of the reviews.
2. The system emphasizes on also integrating the short form words used nowadays in reviews, which is not included in conventional data dictionaries.
3. The proposed system also emphasizes the polarity aspect of the product.

3.2 Proposed System Design

The proposed system first uses Meta Dictionary Mapping for short form words in reviews. Nowadays, there is a trend of people writing textual reviews by using short form words e.g. "gud" for "good" or "nyc" for "nice". So the standard dictionary does not have any mappings for such words and hence any review stating "the product is gud" won't be considered to be positive. Hence the proposed system maintains the key value pair of the short form words and actual words, thereby making the system capable of mapping the short form words as well for sentiment prediction.

The proposed system's overall flow is as follows:

- The proposed system first loads the dataset and cleans the dataset by removing unnecessary columns from the dataset.
- The cleaned dataset is then processed by VADER sentiment analysis after text pre-processing, for labeling the review's polarity (0,1).
- The Meta Dictionary is maintained for storing the short form words and their corresponding actual words, so that before the model is being trained, words are mapped to actual words.
- After labelling data for its polarity, the data is provided to the proposed LSTM model with some preliminary neural network configurations.
- The model training is then started. If the training output is not optimal, the configurations are varied and the model is retrained till the optimal results are obtained.
- Once the optimal results are obtained, the system then makes use of Part of Speech (POS) tagging on the input reviews. These input reviews are provided as test reviews for testing the model and product review aspects are extracted by segregating nouns from the tagged review tokens.

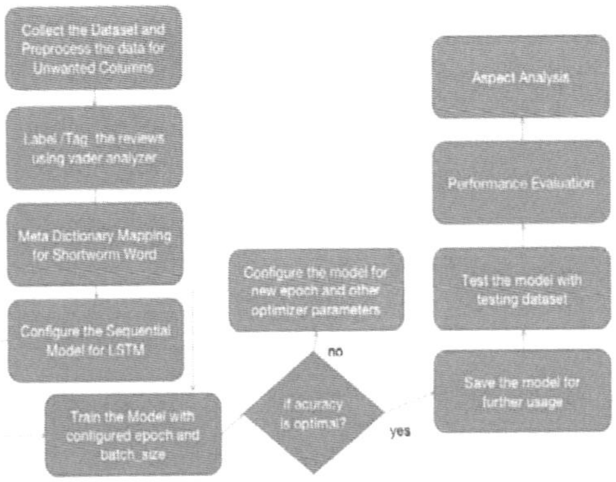

Fig. 3. Proposed system flow chart

3.2.1 Text Pre-Processing

Text Pre-Processing: To easily extract the features from the raw product reviews, this research first transforms our data into normalized unigrams. This research performs text pre-processing by removing all capitalization, tokenize the text by continuous sequences of alphabetic characters (remove whitespaces, symbols, punctuation and numbers). For each token, it filters out english stop words from the NLTK stopwords corpus. From this, it gives the summary statistics of the listed data such as total number of reviews, total reviews for different ratings values as seen in Figure 4.

3.2.2 Data Preparation

In order to avoid unwanted data, only some part of the entire data is used.

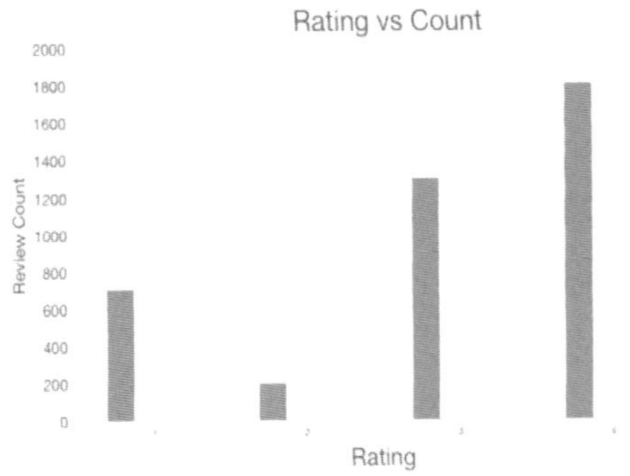

Fig. 4. Product rating v/s Count of Reviews

For example, reviews having a positive sentiment with ratings of 4 & 5 and reviews having a negative sentiment with ratings of 1 & 2 are considered, whereas reviews with neutral sentiments having a rating of 3 are dropped. The proposed systems' primary purpose is to analyze the reviews for its polarity whether it is positive or negative. There are two preliminary phases to this process. To convert a text into the corresponding numerical representation, it first locates a word embedding. Second, to identify whether proposed work makes use of machine learning methods or deep learning architectures to fit the numerical representations of text. Frequency-based embedding, such as the Bag of Words (BoW) model, is a typical technique for word embedding. The BoW model extracts a vocabulary list from a corpus and represents each text using various word counting approaches. This process depicts how well BoW works with supervised learning algorithms. The very first step in text pre-processing is to convert the raw reviews into cleaned reviews. This procedure entails eliminating non-character words such as numerals and symbols, removing stop words, converting all the review sentences to lower case and stemming words to obtain the root words. In Step 2, the CountVectorizer from the sklearn package is used to combine tokenization with occurrence counting in a single review. Tokenization in python is the process of splitting the entire text into smaller chunks or words. In Step 3, the proposed system converts the text in the review to the corresponding numerical representations (feature vectors) and fits these feature vectors to the supervised learning algorithm in Step 4. Tf - idf transform is used to scale down the impact of frequently appearing words. Tf stands for 'term frequency' and counts how many times a word has appeared in a single review. Df stands for 'document frequency and counts how many times a particular word has been used from a list of reviews.

 https://doi.org/10.1051/itmconf/20214003032

Table 1. Classification Report of Random Forest Classifier.

Polarity	Precision	Recall	F1-Score
0(negative)	0.87	0.82	0.84
1(positive)	0.94	0.96	0.95
Average	0.92	0.92	0.92

3.2.3 Applying Machine Learning Algorithm

As per the previous research done, various algorithms in machine learning are used for categorical prediction of review sentiments. Among various ML algorithms, Random Forest Classifier (RFC) has given better performance. So, to check the performance of RFC on the dataset, RFC with base_estimator as 'Decision Tree' and n_estimators = '100' was set to fit the model to achieve the prediction for review classification.

3.2.4 Applying Deep Learning LSTM

The proposed system makes use of the traditional Neural Network Mechanism for Creating the LSTM Network, but with the use of additional data pre-processing steps for cleaning the input dataset of reviews extracted from the site. There are time-stamps in LSTM, but the block of "MEMORY" in the LSTM cell is an additional feature. In general, LSTM cells consist of these building blocks 1. Input Gate "I", Candidate layer "C`", Forget Gate "f", Output Gate "O", Hidden state "H" and a Memory state "C". The diagram below shows the LSTM cell at time t. LSTM cell takes the old memory state Ct-1 and performs unit wise multiplication particularly with forget gate (f)

$$C_t = C_t\text{-}1 * f_t \qquad (1)$$

If it's found that the forget gate is equal to 0, then the old stored value in the memory state is straight away forgotten.

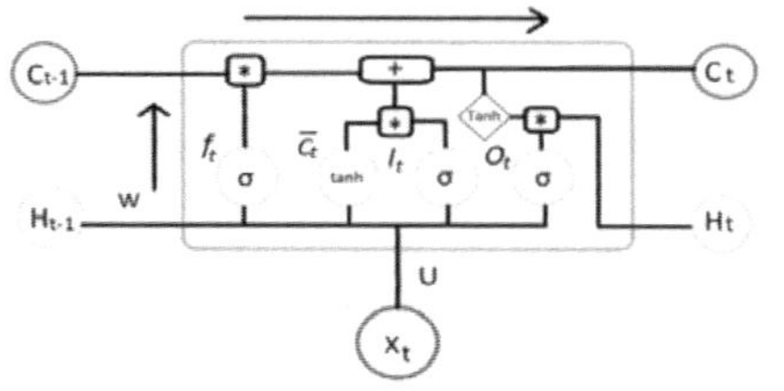

Fig. 5. LSTM Cell Overview

If it's found that the forget gate is equal to 1 then the old value in the memory state is straight away passed to the cell. So, with the present memory state C_t it calculates the new cell state in coordination with the C layer and the input state.

$$C_t = C_t + (I_t * C`_t) \qquad (2)$$

Where, C_t = present memory at time-stamp t which gets passed to next time-stamp. Fig 6. shows a flow diagram for C_t. The proposed system architecture has the standard Neural Network Blocks for generating the model for analyzing the sentiment from the reviews.

1) Long Short Term Memory networks are a special kind of Recurrent Neural Networks (RNN), capable of learning long-term dependencies.
2) Prepare X_train and X_test to 2D tensor and Train a simple LSTM
3) Load word embedding model which was trained in the above step.
4) Construct an embedding layer using an embedding matrix as weights and Train the LSTM model with embedding.
5) Compile and fit the model using the log loss function and ADAM optimizer.

The LSTM gate activation functions are often logistic functions. There are recurrent connections going in and coming out of the LSTM gates.

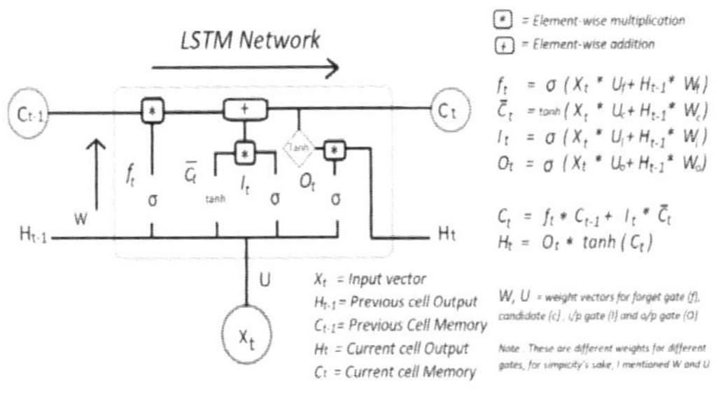

Fig. 6. LSTM Cell flow diagram

Table 2. Model History Report.

Model: "Sequential"		
Layer (type)	**Output Shape**	**Param #**
embedding (Embedding)	(None, 779, 128)	256800
lstm (LSTM)	(None, 196)	256800
dense (Dense)	(None, 2)	394
Total params: 511,194 Trainable params: 511,194 Non-trainable params: 0		
None (3355, 779) (3355, 2) (839, 779) (839, 2) Train on 2684 samples, validate on 671 samples		

Connection weights, which the model learns in the training phase, decides the operation of gates. Sequential model is created by using keras.models, the first layer which is an embedding layer is added to the sequential model with maximum features of data,128 embedding dimensions and input data shape. The LSTM layer is added to the model with 196 units given to the layer for dimensionality of the output space. Dense layer is added with an activation function which is "softmax" which means that the neural network has n neurons and the activation function will be "softmax". Then the model is compiled by model.compile() which Configures the model for training. The model is compiled with loss, optimizer and accuracy.

4 Results and Discussion

4.1 Experimental Setup and Dataset

The information includes Amazon reviews on mobile phones. The product feature names and the associated emotions are marked in the reviews. Clearly, each of these statements mention the entity being reviewed as well as the level of opinion held about it. After conducting a variety of experiments, this research concluded that the third party website reviews are not sufficient for the research purpose. Therefore, more than 5000 reviews were labelled manually. On a scale of -5 to 5, aspect ratings are labelled. This system has additionally pre-processed publicly available data, so as to fit the indicated element into corresponding buckets. By indicating value "1', it's seen that they are referring to portability with a positive attitude. This necessitated some word and sentiment translation manually. The dataset is a subset of a larger dataset of Amazon Product Reviews accessible on the Kaggle repository, as well as a synthetic version with certain changes (removal of some unnecessary columns from the original dataset). The proposed work presently will be predicting the intensity analysis of the review. The present model is able to determine the polarity of whether the obtained review is positive or negative. Also the proposed implementation consists of polarity intensity identification also, which will provide a more detailed prediction. e.g. "the product is good" - Positive, but "the product is extremely good" - Very positive. The Proposed work does not only determine the polarity of the reviews, but also determines the aspects of the reviews. Which means that if the review indicates some aspect of the product and the reason for its polarity, the proposed system will extract that information too e.g. "the product battery life is awesome", so the system will also determine that the review is positive and the aspect being "battery" of the product. The proposed system is also capable of identifying the polarity of reviews in which short form words are used for expressing the sentiments. If the person writes the review "the product is gud", the word "gud" is not a standard dictionary word, so the CNN model is deployed to predict the actual words from the trained dataset, which comprises of actual words and possible short form words that can be used in now a days. As seen in Fig 7, the input review uses the word "gud" which is a short form, but the final analysis shows the actual word "good" instead of the short form word.

```
Display is gud but battey is weak
Original 1
good
 (0, 1903) 1.0
[2]
Original 1
 weak
  (0, 4848) 1.0
[0]
[['display', ' good ', 'positive'], ['battey', ' weak ', 'Negative']]
```

Fig. 7. Product Aspect Analysis

The meta dictionary is maintained for mapping short-form words with actual words which are replaced while the pre-processing phase is computed.

4.2 Performance Evaluation

The proposed system is evaluated based on various parameters such as accuracy, precision recall and F1 Score. The Confusion Matrix is the base for all these values as shown below. All these performance metrics are dependent on true positive, true negative, false positive and false negative values:

a. True Positive TP: The expected value was positive and the predicted value was also Positive.

b. True Negative TN: The expected value was negative and the predicted value was also Positive.

c. False Positive FP: The expected value was negative and the predicted value was positive which is False prediction.

d. False Negative FN: The expected value was positive and the predicted value was negative which is False prediction

Precision: Precision talks about how precise your model is out of those predicted positives, how many of them are actual positives, calculated as Precision = TP/(TP+FP).

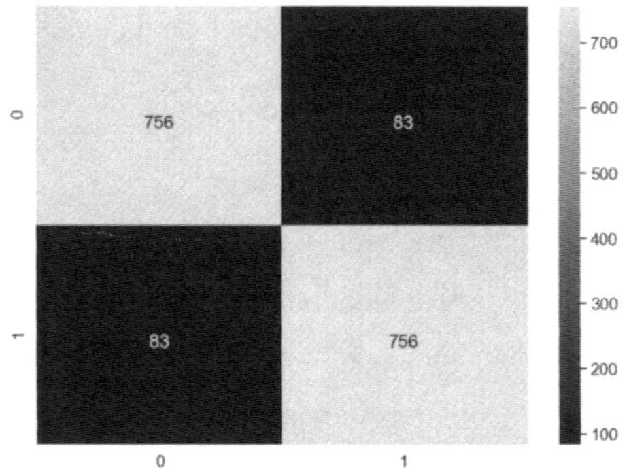

Fig. 8. Proposed Model Confusion Matrix

Table 3. Classification Report of proposed system.

Polarity	Precision	Recall	F1-Score
0(negative)	0.91	0.90	0.905
1(positive)	0.95	0.96	0.935
Average	0.93	0.93	0.92

Recall: Recall actually calculates how many of the Actual Positives our model captures through labeling it as Positive (True Positive),calculated as Recall = TP/(TP+FN).

F1 Score: F1 Score might be a better measure to use if its needed to seek a balance between Precision and Recall AND there is an uneven class distribution (large number of Actual Negatives),calculated as F1 Score = 2 x ((Precision * Recall)/(Precision + Recall)).

5 Conclusion

The obtained results show that the efficiency of the proposed LSTM model is better than the traditional machine learning models. Also, the proposed system does not restrict its scope over polarity or sentiment prediction, it digs into the polarity intensity such as whether the review is just positive or very positive and same for negative reviews. The proposed system results also show that the different aspects of the product about which positive is being said or negative is being said are also identified. The efficiency of the model is based on performance evaluation; it is seen that it outperforms Machine Learning model performance parameters. Hence it is observed that the proposed model gives more accurate and elaborative details about the reviews which is helpful in terms of analyzing the aspects of the products whose polarity is also identified. So because of this, consumers and service providers will get better clarification on products' market value and help them make important business decisions.

References

1. Wei Zhao, Ziyu Guan, Long Chen, Xiaofei He, Deng Cai, Beidou Wang, Quan Wang,Weakly-Supervised Deep Embedding for Product Review Sentiment Analysis, IEEE Transactions on Knowledge and Data Engineering, Vol. 30, No. 1, January 2018.
2. T. K. Shivaprasad,Jyothi Shetty,"Sentiment analysis of product reviews: A review", 2017 International Conference on Inventive Communication and Computational Technologies (ICICCT),doi 10.1109/ICICCT.2017.7975207
3. Y. Fang, Hai Tan, Multi-Strategy Sentiment Analysis of Consumer Reviews Based on Semantic Fuzziness, INSPEC Accession Number: 17745421, DOI: 10.1109/ACCESS.2018.2820025, IEEE, Page(s): 20625 - 20631.
4. W. Songpan, The Analysis and Prediction of Customer Review Rating Using Opinion Mining, 2017 IEEE 15th International Conference on Software Engineering Research, Management and Applications (SERA), NSPEC Accession Number: 17010009, 10.1109/SERA.2017.7965709, IEEE.
5. R. Abinaya, P. Aishwaryaa, S. Baavana, N.D. Thamarai Selvi, Automatic Sentiment Analysis of User Reviews, 2016 IEEE Technological Innovations in ICT for Agriculture and Rural Development (TIAR), NSPEC Accession Number: 16559951, 10.1109/TIAR.2016.7801231, IEEE
6. M. Kavousi, Sepehr Saadatmand, Estimating the Rating of the Reviews Based on the Text, First International Conference on Data Analytics & Learning 2018.
7. S. Zirpe; Bela Joglekar, Polarity Shift Detection Approaches in Sentiment Analysis: A survey, 2017 International Conference on Inventive Systems and Control (ICISC), INSPEC Accession Number: 17259793,DOI: 10.1109/ICISC.2017.8068737,IEEE
8. M. Kulkarni, Mayuri Lingayat, Effective Product Ranking Method based on Opinion Mining, June 2015International Journal of Computer Applications 120(18):33-37, DOI:10.5120/21331-4306
9. J. Fontana Rava; Gabriella Pasi; Marco Viviani, Feature Analysis for Fake Review Detection through Supervised Classification, 2017 IEEE International Conference on Data Science and Advanced Analytics (DSAA),INSPEC Accession No.17521817, 10.1109/DSAA.2017.51, IEEE
10. M. Chauhan, Divakar Yadav, Sentimental Analysis of Product Based Reviews Using Machine Learning Approaches, Journal of Network Communications and Emerging Technologies (JNCET)Volume 5, Special Issue 2, December(2015).
11. R. Collobert, J. Weston, L. Bottou, M. Karlen, K. Kavukcuoglu, and P. Kuksa. Natural language processing (almost) from scratch. Journal of Machine Learning Research, 12(Aug):2493–2537, 2011.
12. K. Dave, S. Lawrence, and D. M. Pennock. Mining the peanut gallery: Opinion extraction and semantic classification of product reviews. In Proceedings of the 12th international conference on World Wide Web, pages 519–528. ACM, 2003.
13. M. S. Elli and Y.-F. Wang. Amazon reviews, business analytics with sentiment analysis.
14. S. Hota and S. Pathak. Knn classifier-based approach for multi-class sentiment analysis of twitter data. In International Journal of Engineering Technology, pages 1372–1375. SPC, 2018.
15. B. Liu and L. Zhang. A Survey of Opinion Mining and Sentiment Analysis, pages 415–463. Springer US, Boston, MA,2012.
16. C. Rain. Sentiment analysis in amazon reviews using probabilistic machine learning. Swarthmore College, 2013.
17. R. Socher, A. Perelygin, J. Wu, J. Chuang, C. D. Manning, A. Ng, and C. Potts. Recursive deep models for semantic compositionality over a sentiment treebank, In Proceedings of the

2013 conference on empirical methods in natural language processing, pages 1631–1642, 2013.

18. Bo Pang and Lillian Lee, Opinion Mining and Sentiment Analysis, Foundations and Trends in Information Retrieval, 2008.
19. Theo M.V. Janssen, Frege, contextuality and compositionality, Computer Science, University of Amsterdam, 2002.
20. Xin Wang, Yuanchao Liu, Chengjie Sun, Baoxun Wang and Xiaolong Wang,Predicting Polarities of Tweets by Composing Word Embeddings with Long Short-Term Memory, Harbin Institute of Technology, Harbin, China, 2015.
21. Ziqian Zeng, Wenxuan Zhou, Xin Liu, Yangqiu Song, A Variational Approach to Weakly Supervised Document-Level Multi-Aspect Sentiment Classification, Computer Science, Computation and Language, 10 April 2019, arXiv:1904.05055.

https://doi.org/10.1051/itmconf/20214003033

System for Appropriation & Management of Vehicular Spare-Parts

Shreyank Karjigi[1], Rishi Rao[2], Viraj Kotian[3] and Ashish Jadhav[4*]

[1]Ramrao Adik Institute of Technology, Nerul, Navi Mumbai
[2]Ramrao Adik Institute of Technology, Nerul, Navi Mumbai
[3]Ramrao Adik Institute of Technology, Nerul, Navi Mumbai
[4]Ramrao Adik Institute of Technology, Nerul, Navi Mumbai

Abstract. This paper proposes a system that simplifies the procurement and sale of vehicular spare parts and is particular focused on automotive spare parts. Car manufacturers have immense pressure to provide spare parts for all vehicles. For a car in production spares are available until the term decided by the manufacturer. For a car that is off production spare parts are available through sources such as local market and third-party dealers. But, sometimes there can be a scenario where a customer wants to buy a part from the manufacturer but the part may be very expensive for the customer. The main objective of the work is to bring the manufacturer, supplier and buyer on a common platform and enable customers to purchase spare parts in an efficient and easy manner providing multiple options through a cloud based platform.

1 Introduction

Car sales have tremendously increased over the last few years, with some families owning more than one vehicle. Car manufacturers have immense pressure to provide spare parts for all these vehicles. For a car in production spares are available until the term decided by the manufacturer. And, for a car that is off production spare parts are available through sources such as local market and third-party dealers. But, sometimes there can be a scenario where a customer wants to buy a part from the manufacturer but the part may be very expensive for the customer. In this case the customer has no option but to buy the part from local market. This also applies for the person owning an off-production vehicle. Customers can face problem such as overpriced parts. Hence, there is a need for a system that provides customers with all the required information associated while buying parts from the local market. Customers should be able to get information about the parts such as recommended dealers, customer's reviews of the part, manufacturing information, pricing details and other related information regarding that part. This can help the customer while making the right choice.

1.1 A look at various scenarios

1.1.1 Buying spare parts for a car still in production

This covers all the vehicle that are currently in production and have spare parts available with the manufacturer. There can be a scenario that customer wants to buy a part from the original manufacturer but isn't satisfied with price of the part and decides to buy parts from local market. They can face a problem where multiple dealers have the same parts for different prices. They might get confused from whom to buy the parts.

1.1.2 Buying spare parts for a car off- production

This covers all the vehicle that are discontinued from production car that have exceeded 5 years after going off production and doesn't have spare parts available with manufacturer. Here also customer is forced to buy from the local market.

1.1.3 Buying spare parts for vintage cars

These are the car owners who deal with vintage car parts and want to sell parts for other vintage car owners. Finding spares for vintage car is hideous task as most of them aren't easily available in local markets.

1.2 Problems associated with the local market

As a mechanical geek there are many owners who would like to purchase the spare parts from local dealers and like to replace themselves using self-help videos available. This is especially true for non-critical body related spare parts. Some of the issues associated with purchasing car spare parts from local dealers in India are listed here.

* Corresponding author: Jadhav.ashish@gmail.com

1) Lots of dealers for the same part
2) Pricing issues with the dealers
3) No assurance for the quality of parts
4) No proper history of the parts
5) No proper warranty of the parts

To overcome these issues we have proposed a cloud based system in this paper. The major components of the system and its implementation aspects are discussed in section 3. We look at some of the work published in related areas in section 2.

2 Related Work

Fang Li, [1] in year 2015 have proposed that in China and the rest of the world, the full-vehicle industries are rather booming but there is also a need for additional spare parts management system alongside it. In this paper, they've concluded that there is a need for automobile spare parts management system to balance out the weak after sales service market of the full-vehicle industries. It is also beneficial for profit earning and cost saving. It will help ensure easy searching of products and timely delivery of products. With the help of information and network technology, application can be built for spare parts management. In this spare parts management model, they've used an IDEF method for system design.

Gong PingPing and Liu Feng, [2] in year 2009 have proposed that the current spare parts management system is under constant financial pressure, insufficient maintenance and lacks in a good design. This problem would be solved by running a spare parts management system with the help of computer/browser instead of using much manpower. In this paper, they've said that instead of "Client-Server" architecture, "Browser-Server" architecture can be followed. By making use of the internet and dealing with spare parts by the involvement of server, client, warehouse without much need for manpower, spare parts management system will be cost effective and can save time.

Nizam Baluch et.al. [3] give an operational perspective of sales part management for asset intensive industries and automotive manufacturing falls in this category.

Raghad Hemeimat et.al. [4] give an interesting analysis of the usage of statistical analysis for forecasting the demand for spare parts.

3 Proposed Methodology

We propose a cloud based system having a simple and user friendly user interface for purchasers, suppliers and administrator which can be accessed from any device. There is a backend database which stores the product information and is useful for processing of purchase, management of the items and providing various functional aspects to the users of the system.

The proposed system makes use of Flask framework which is a small web framework written in Python. It does not require any special tools or libraries and can be set up quite easily.

The products added are categorized by brands and category along with their proper attributes such as product origin, price, colour, etc. which the proposed system later use for building its recommendation module.

Recommendation module uses collaborative filtering, which finds similarity between two products of same category using the correlation obtained from the products rating and the numbers of rating users have given for a particular product.

The payment gateway used needs to be reliable, robust and trustworthy. Therefore, the system uses Stripe API. Using stripe, payment can be accepted online and users can track payments made.

3.1 System Design

The spare part appropriation and management system has been designed to be accessible from desktop as well as mobile devices, this is achieved by building a user interface that adapts to screen size being used.

Functionality provided by the system is as follows.

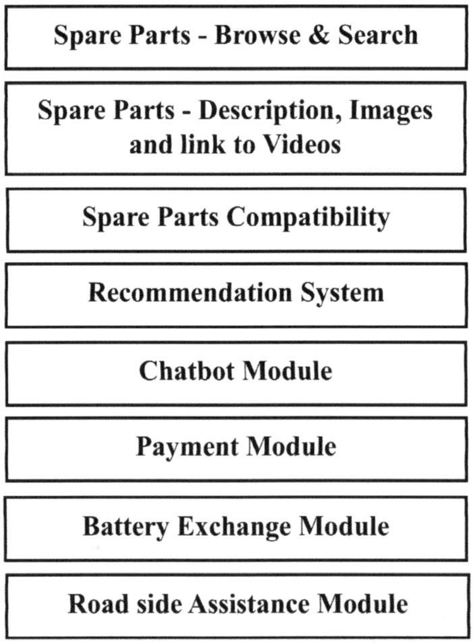

https://doi.org/10.1051/itmconf/20214003033

The interface scales according to number of products being added in the system, this is achieved without cluttering the screen. For a better browsing experience, the system needs to recommend products to the users based on what they are purchasing. This is achieved using recommendation system that provides product recommendation based on category they are browsing.

Overall system should be reliable for best user experience, this is achieved by using proper database technology in backend to process user orders. The database should have high scalability as well as proper security features to prevent against tampering of data.

In the designed application we have four databases namely

1) User Information
2) Product Information
3) Shipping Cart
4) Payment Information

User information database is used for storing user information such as their name, address, password etc. This is used during authentication.

Next, we have product information database, here all the product information such as product name, brand, category, pictures etc. are stored.

The shipping cart database is used to store all shipping related information such as customer shipping address, mode of payment, total amount and mode of shipping.

Payment information database is used to store payment information such as amount paid/due and transaction details.

Customer begins by directly browsing products on webpage, if they are interested in product they go for registration. Once registered they get confirmation on their email address. They can then login to the website using the registered credentials and then browse through the catalogue, if they like something they will add it to their cart and proceed for checkout. In checkout they have option to opt for cash on delivery (COD) or online payment such as net banking etc.

Once the payment is done, an invoice is sent to their registered email address.

Customers can also choose for installation help (if available) for their purchase.

In addition to this we have also implemented a battery exchange system. It provides an interface for the user to provide information and request for exchange of battery.

There is a module included to provide road-side assistance to users. This module uses the GPS location of the user to provide the facility of road-side assistance, where the service can be requested using the mobile during travel.

3.2 Recommendation System

Isinkaye et. al. [5] and Aggarwal [6] have given a very good introduction to the working and the types of generic recommendation systems.

The recommendation systems employ an algorithm to seek a recommendation based on the content searched, browsed or rated by a user based on the similarities. Many multi-millionaire companies also use recommendation systems in one or the other way.

A recommendation System majorly uses two techniques of filtering described in the following sub-sections.

3.2.1 Content Based Filtering Technique

Content-Based filtering uses the similarities between the products based on their attributes. For example, recommending a product based on its colour, type, sizes, etc.

3.2.2 Collaborative Filtering Technique

Collaborative filtering uses collaboration among different data sources, agents to give accurate results. It works on the fundamental principle of similarity.

The proposed system uses collaborative filtering to find the similarity between the products using correlation of the number of rating given by the users and ratings given.

Collaborative filtering works around the user interactions that the items have. Patterns can be found, which the data about the items, or users can't find easily, and, this can be done with these interactions. The main reasons why we used collaborative filtering for our development are as follows.

1) The features about the items or users to be known are not required by this type of filtering

 https://doi.org/10.1051/itmconf/20214003033

2) For a set of items of different types, this is best suited type of filtering

3) For our spare parts inventory where items of various categories can be added, this was most appropriate

4) In a set of similar items (e.g., oils) such as that of our spare parts inventory, though, known features like brands and for what vehicle (e.g., cars or bikes) this oil is used can be useful, and might benefit from content-based or hybrid approaches

5) Recommenders can be helped by this type of filtering to not overspecialize in a user's profile and completely different items can be recommended than what they have seen before

3.3 Chatbot System

E. Adamopoulou and L. Moussiades [7], [8] have given a very good discussion on the chatbot technologies.

A chatbot is a platform which is used as a texting agent, depicting a human, in a software and is used for chat conversations. It is an automatic conversing platform which gives a response every time the user texts something on it.

In this system, chatbot will be used by the users who wishes to obtain information and use our services as for example, roadside assistance, battery exchange program and installation services.

It was implemented using Twilio and Dialog flow. Twilio is a cloud communication platform-as-a-service company. By using its web API services, Twilio helps developers in making and receiving calls, sending and receiving text messages and performing other communication functions.

Dialog flow is a natural language platform which is used to create and design conversational user interface for a chatbot.

In this proposed system, we will be using WhatsApp as a chatbot platform. Benefits of WhatsApp as a platform are as follows.

1) It helps as assistance for customers on their preferred daily use platform and supports customers by responding to questions they ask in no time

2) Trust and loyalty are built with the customers

3) As per customer usage and their experience, the script of the WhatsApp chatbot can be changed and personalized

4) The end-to-end encryption in WhatsApp helps to make customer communication more comfortable and secure

3.4 Comparison of our System with Traditional E-Commerce Sites

A user survey of the prototype was conducted to see the satisfaction level for purchase of spare parts. From the feedback received the following table has been prepared.

SN	What Other Do?	What We Do?
1	Only General parts or accessories of car are available	Our system is purely based on spare parts. All types of spares such as general and complex are available
2	No proper testing of parts are done	Proper testing of product by team will be done
3	Authenticity of part are questionable	Certificate of authenticity will be provided
4	Verification of the dealers are not done	Strict verification of dealer will be done by checking their background
5	Battery Exchange not provided	Battery Exchange module provided
6	Road side assistance usually not provided by spare dealers in India	Road side assistance module included
7	Whatsapp based chatbot usually not provided	Whatsapp basedChatbot module provided

4 Conclusions and Future Work

We conclude that developing a cloud based system for spare parts online selling management works more efficiently than a conventional application for the same.
Various new components such as recommendation system, chatbot, roadside assistance and battery

https://doi.org/10.1051/itmconf/20214003033

exchange will be very useful for the customers and will help in increasing the marketability.

The proposed system makes use of Flask framework for implementation. All the products are categorized by brands and categories for easy product selection by users. Recommendation module is helping in recommending products to the customers.

The system is also made compatible for usage in various screen sizes. Stripe is being used as a secure payment gateway. The system is deployed using Scalingo.

To help automotive enthusiasts to be able to perform repairs using spares, there is a need to have more self-help videos. In fact for each part, multiple videos need to be created.

Incentives can be provided to local mechanics to purchase spare-parts. This would need to have a quick delivery mechanism in place.

References

1. Fang Li, Study on the spare part logistic management model facing the automotive industry, 3rd International Conference on Management, Education, Information and Control (MEICI 2015)
2. Gong Pingping, L. Feng, The Optimization and Application of Process Model for Spare Parts Management Information System, International Conference on Information Management, Innovation Management and Industrial Engineering, 2009
3. Nazim Baluch, Che Sobry Abdullah and Shahimi Mohtar, Evaluating effective spare-parts inventory management for equipment reliability in manufacturing industries, European Journal of Business and Management, Vol.5, No.6, 2013
4. Raghad Hemeimat, Lina Al-Qatawneh, Mazen Arafeh and Shadi Masoud, Forecasting spare parts demand using statistical analysis, American Journal of Operations Research, 2016
5. F. O. Isinkaye, Y. O. Folajimi and B. A. Ojokoh, Recommendation Systems: Principles, methods and evaluation, Egyptian Informatics Journal, Vol.16, No. 3, November, 2015
6. C. C. Aggarwal, Recommender Systems, Springer Publishing company, 2016
7. E. Adamopoulou and L. Moussiades, Chatbots: History, technology, and applications, Machine Learning with Applications Vol. 2, Elsevier, 2020
8. E. Adamopoulou and L. Moussiades, An overview of chatbot technology, IFIP International conference on artificial intelligence applications and innovations, AIAI 2020

https://doi.org/10.1051/itmconf/20214003034

Next Words Prediction Using Recurrent NeuralNetworks

Sourabh Ambulgekar[1,*], *Sanket* Malewadikar[2,**], *Raju* Garande[3,***], and *Dr. Bharti* Joshi[4,****]

[1]Ramarao Adik Institute of Technology Mumbai,Maharashtra
[2]Ramarao Adik Institute of Technology Mumbai,Maharashtra
[3]Ramarao Adik Institute of Technology Mumbai,Maharashtra
[4]Ramarao Adik Institute of Technology Mumbai,Maharashtra

Abstract. Next Word Prediction is additionally called Language Modeling. It is the undertaking of predicting what word comes straightaway. It is one of the major assignments of NLP and has numerous applications. Attempting to make model utilizing nietzsche default text record which will foresee clients sentence after the clients composed 40 letters, the model will comprehend 40 letters and anticipate impending top 10 words utilizing RNN neural organization which will be executed utilizing Tensorflow. Our Aim of creating this model to predict 10 or more then 10 word as fast as possible utilizing minimum time. As RNN is Long short time memory it will understand past text and predict the words which may be helpful for the user to frame sentences and this technique uses letter to letter prediction means it predict a letter after letter to create a word

1 Introduction

Natural Language Processing (NLP) is a significant part of artificial Intelligence, which incorporates AI, which contributes to finding productive approaches to speak with people and gain from the associations with them. One such commitment is to give portable clients anticipated "next words," as they type along within applications, with an end goal to assist message conveyance by having the client select a proposed word as opposed to composing it. As LSTM is Long short time memory it will understand the past text and predict the words which may be helpful for the user to frame sentences and this technique uses a letter to letter prediction means it predicts a character to create a word. As writing an essay and framing a big paragraph are time-consuming it will help end-users to frame important parts of the paragraph and help users to focus on the topic instead of wasting time on what to type next. We expect to create or mimic auto-complete features using LSTM. Most of the software uses different methods like NLP and normal neural networks to do this task we will be experimenting with this problem using LSTM by using the Default Nietzsche text file also known as our training data to train a model.

Next Word Prediction is also called Language Modeling that is the task of predicting what word comes next. It is one of the fundamental tasks of NLP and has many applications.

*e-mail: ambulgekars@gmail.com
**e-mail: malewadikarsanket088@gmail.com
***e-mail: garande.raju.16ce1029@gmail.com
****e-mail: bharti.joshi@rait.ac.in

2 Literature Review

This section describes the methodology adopted for the literature review. This paper represents an exploration of the contributions that have already been made in the academic field.

[1] multi-window convolution(MRNN) algorithm is implemented, also they have created residual-connected minimal gated unit(MGU) which is short version of LSTM in this cnn try to skip few layers while training result in less training time and they have good accuracy by far using multiple layers of neural networks can cause latency for predicting n numbers of words .

[2] This paper used RNN algorithm and also they have used GRU another form of RNN for code completion problem as RNN help to predict next code syntax for users. Authors claim that their method is more accurate compare to existing methods. They have separated next word prediction in two components: within-vocabulary words and identifier prediction. They have used LSTM neural language model to predict within vocabulary words. A pointer network model is proposed to identifier prediction.

[3] Authors worked on Bangla Language. They have proposed a novel method for word prediction and word completion. They have proposed N-gram based language model which predicts set of words. They have achieved satisfactory results.

[4]Authors have used LSTM for next word prediction for Assamese language. They have stored transcripted language according to International Phonetic Association (IPA) chart and fed to their model . They created a model for physically challenged people. This model uses Unigram, Bigram, Trigram, based Approach for next

word predic- tion and was found out average to predict the word but accuracy was around 30-40 percentage .

[5] In this paper they created a auto-next-keyword for Bengali language which was challenging and It was found out that it is hard to get good accuracy by using RNN algorithm as due to it's vanishing gradients and heavy recurrent NN take more time to train and test.

[6] They useded predicting next character high-lighter(PNCH) for Indian language it was more of text correction and less about next word prediction but was quite good to understand. The Method called hit and miss but accuracy is less and the model was not efficient for this kind of problem statement.

[7]This was first approach to tackle this kind of problem, the paper discusses about LM and perplexity algorithm which is base of natural language processing, This help us to make a 3D input data for our model.

[8] 1-degree feature patterns algorithm is useded to Solve the problem of Vanishing but provide less accuracy as basic sentence was useded in training and testing data like 'A man cries', 'A baby cries', 'A dog barks', and 'A cat mews'.

[9] As GPT is quite huge and costly model for this kind of task as word prediction is a simple project, GPT will only act like wasting useful resources for simple task.

Most of the paper is trying to create a model to predict next text, few of the paper was helpful, paper to predict next code using SVM and RNN, As this algorithm are helpful but new algorithm like LSTM may predict good results for this problem statement below are some limitations of the existing system.

- This process of predicting the next Word is quite complex because we have to predict words which user think so it is predicting the thoughts of the user so the accuracy is quite low compare to other ML project
- algorithms like SVM, Decision Tree, etc are not providing good results and take more time to predict the results as the task is quite complex.
- To be able to make useful predictions, a text predictor needs as much knowledge about language as possible so we have to keep training the neural network continuously on new languages and new data.

3 Proposed methodolgy

3.1 Data pre-processing

This proposed work in "Fig. 1" is an illustration to create a flexible model that can help users to detect next word while understanding user vocable in a fast and effective manner so user need to provide 40 letters then it passes this letter to LSTM NN and predicts N number of letters

- As shown in the above diagram, like in "Fig. 1" providing input data up to 40 letters later this sentence will pass through LSTM Neural Network
- Letter LSTM understand and learn every letter, letter by letter and create a score for the next letter.

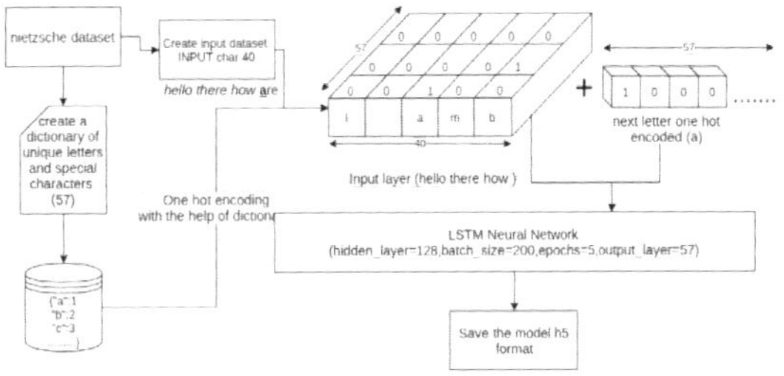

Figure 1. Our Proposed Work flow

```
Model: "sequential_2"
_________________________________________________________________
Layer (type)                 Output Shape              Param #
=================================================================
lstm_2 (LSTM)                (None, 128)               95232
_________________________________________________________________
dense_2 (Dense)              (None, 57)                7353
_________________________________________________________________
activation_2 (Activation)    (None, 57)                0
=================================================================
Total params: 102,585
Trainable params: 102,585
Non-trainable params: 0
_________________________________________________________________
```

Figure 2. Neural network Architecture

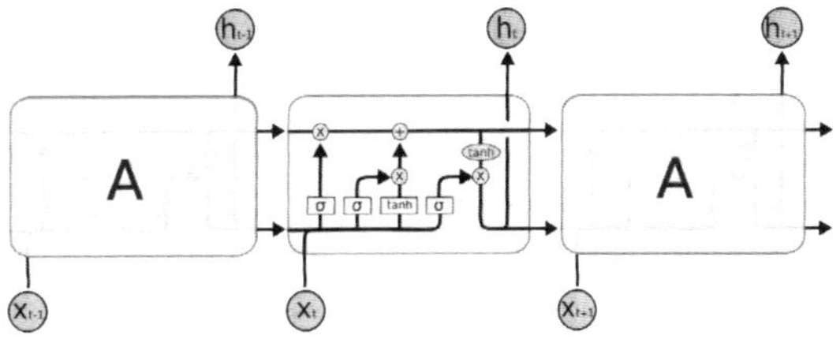

Figure 3. Recurrent Neural network

- This score then again will pass through the same LSTM and later it will predict a word letter by letter.
- Below is our neural network architecture plus our implementation methodology using the Tensorflow library.
  - Letter to bits, As computers, don't understand words so converting words to bits or array of bits using NumPy software.
  - Now creating a 3D array of all words it's like one-hot encoding for all letter and unique characters (200285, 40, 57) this was our training data
  - Later passing this X features to our model with input Neural node 40 and hidden node 128 then this will have an output layer with node equal to the input node.

3.2 LSTM Networks

Step 1: first import our helpful model Numpy pandas and other modules later importing Nietzsche default txt which is our dataset.

Step 2: The way to LSTMs is the cell express, the even line going through the highest point of the outline.

https://doi.org/10.1051/itmconf/20214003034

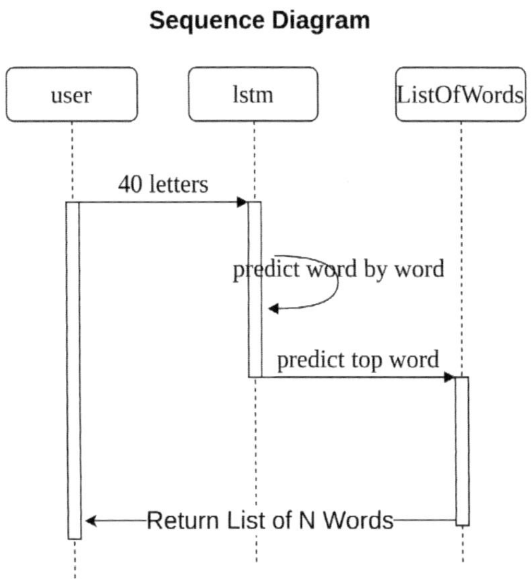

Figure 4. sequence diagram of application

The cell state is similar to a transport line. It runs straight down the whole chain, with just some minor direct communications. It's simple for data to simply stream alongside it unaltered.

The LSTM can eliminate or add data to the phone state, painstakingly managed by structures called entryways.

Gates are an approach to alternatively let data through. They are made out of a sigmoid neural net layer and a pointwise duplication activity.

The sigmoid layer yields numbers somewhere in the range of nothing and one, depicting the amount of every segment ought to be let through. A worth of zero signifies "let nothing through," while a worth of one signifies "let everything through!"

A LSTM has three of these gate, to secure and control the cell state.

4 Data Description and Data cleaning

The dataset contains 25,107 words from ebook author Franz Kafka. The Datasets for text data are easy to find and we can consider Project Gutenberg which is a volunteer effort to digitize and archive cultural works, to "encourage the creation and distribution of eBooks". From here we can get many stories, documentations, and text data which are necessary for our problem statement.

5 Experimental design

- User : Users job is to submit 40 input words for The LSTM Neural Network like in the "Fig. 4"
- LSTM: LSTM is train on default next word prediction text file it will calculate it's weight and predict a letter by letter for a top word.

1. "It is hard enough hi i am sanket whats y"
2. "which does not hurt us makes us stronger."
3. "i am not sad that you lied to me, i am upset that from now on I cannot have trust on you."
4. "those who were seen vibing were thought to be insane by those who could not hear the tune."
5. "is though enough to remember my opinions, without also remembering saurab my reasons for them!"
6. "not a lack of effection, but a lack of friendship that raju makes unhappy marriages."

Figure 5. Input test cases

- List Of word : This will holdall the words and if user want N number of top works it will count N word and return list of words to user.

5.1 Experimental result

- Above outline like in "Fig. 7", that the n-grams approach is inferior contrasted with the LSTM approach as LSTMs have the memory to review the setting from further, harking back to the substance corpus. While starting another endeavor, you ought to consider one of the current pre-arranged designs by looking on the web for open-source executions. Thusly, you will not have to start without any planning and don't need to worry about the arrangement cycle or hyperparameters.
- Trying to create model using Nietzsche default text file which will predict users sentence after the users typed 40 letters, the model will understand 40 letters and predict upcoming letter/words using LSTM neural network which will be implemented using Tensorflow.
- This product has more scope on social media for syntax analysis and semantic analysis in natural language processing in Artificial intelligence.
- We try to tune some input layers for different input layers we can see that whatever the input layer size may be the output prediction has accuracy of 54% to 55%
- for 10 input node the training accuracy is around 56% but the testing accuracy is around 54%
- for 20 input node the training accuracy is around 56% but the testing accuracy is around 55%
- for 30 input node the training accuracy is around 56% but the testing accuracy is around 55.5% same as 20
- for 40 input node the training accuracy is around 56.3% but the testing accuracy is around 54.9%

From the above example " Fig 5 " inputting 6 test cases and passing 5 in our params, the Neural network did a very great job in predicting, result as you can see in the second case in " Fig 6 " the model found out the string which comes after "str" (i.e stronger, strength)

From "Fig 7 " With the accuracy of around 56% our model did a very good job in real test case scenarios predicting these input test cases.

 https://doi.org/10.1051/itmconf/20214003034

1. it is hard enough hi i am sanket whats y
['a ', 'upiai ', 'e ', 'temtcid ', 'ic ']
2. it is which does not hurt us makes us str
['ength ', 'onger, ', 'iver ', 'ange ', 'ungarity ']
3. i am not sad that you lied to me, i am u
['pon ', 'nder ', 's ', 'tility ', 'ltimate ']
4. those who were seen vibing were tho
['ught ', 'se ', 're ', 'igh ', 'm ']
5. is though enough to remember my opinion
[' and ', ', ', 's ', '. ']
6. not a lack of effection , but a lack of
['the ', 'a ', 'man ', 'something ', 'his ']

Figure 6. N(5) number of outputs

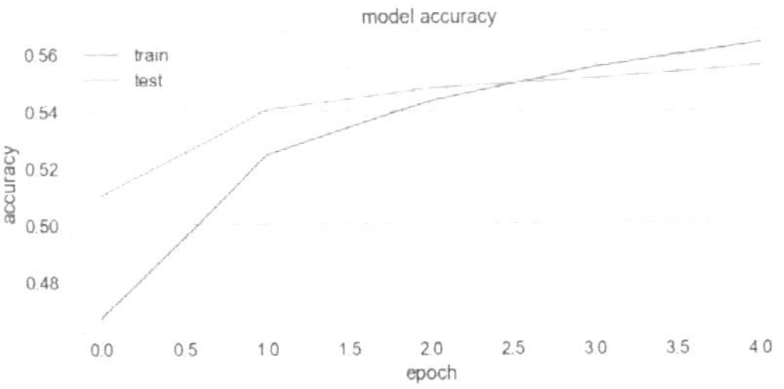

Figure 7. Accuracy on training and testing data

Future Work and Conclusion

- Understanding the paragraph using machine learning algorithms like RNN can help soon to understand and frame paragraphs and stories on their own.
- Creating lyrics and songs can be a major field in which this algorithm can help the end-users to predict the next phrase in songs considering the model is train on a music lyrics data set.
- As more data we can train the model which will reevaluate the weights to understand the core features of paragraphs/sentences to predict good results.
- Paraphrasing means formulating someone else's ideas in your own words. To paraphrase a source, you have to rewrite a passage without changing the meaning of the original text, so our algorithms can predict more number words considering a single sentence and help users to frame n number of sentences.

Standard RNNs and other language models become less exact when the hole between the specific circumstance and the word to be anticipated increments. Here's when LSTM comes being used to handle the drawn-out reliance issue since it has memory cells to recall the past setting. You can study LSTM Neural Net.

Our task in this project is to train and try an algorithm that best fit this task and mostly we are looking forward to implementing an LSTM to get good accuracy as this task is quite complex because we have to predict the user's future text which he will be thinking

At present we manage to understand the problem statement as this problem is unique, we created a 3d vector layer of input and a 2d vector layer for output and feed through to the LSTM layer having 128 hidden layers and manage to get accuracy to around 56% during 5 epochs

This paper presents how the system is predicting and correcting the next/target words using some mechanisms and using TensorFlow closed-loop system, the scalability of a trained system can be increased and using the perplexity concept the the system will decide that the sentence is having more misspelled and the performance of the system can be increased.

References

[1] J. Yang, H. Wang and K. Guo, "Natural Language Word Prediction Model Based on Multi-Window Convolution and Residual Network," in IEEE Access, vol. 8, pp. 188036-188043, 2020, doi: 10.1109/ACCESS.2020.3031200.

[2] K. Terada and Y. Watanobe, "Code Completion for Programming Education based on Recurrent Neural Network," 2019 IEEE 11th International Workshop on Computational Intelligence and Applications (IWCIA), Hiroshima, Japan, 2019, pp. 109-114, doi: 10.1109/IWCIA47330.2019.8955090.

[3] Habib, Md AL-Mamun, Abdullah Rahman, Md Siddiquee, Shah Ahmed, Farruk. (2018). An Exploratory Approach to Find a Novel Metric Based Optimum Language Model for Automatic Bangla Word Prediction. International Journal of Intelligent Systems and Applications. 2. 47-54. 10.5815/ijisa.2018.02.05.

[4] Partha Pratim Barman, Abhijit Boruah,A RNN based Approach for next word prediction in Assamese Phonetic Transcription,Procedia Computer Science,Volume 143,2018,Pages 117-123,ISSN 1877-0509,https://doi.org/10.1016/j.procs.2018.10.359.

[5] S. M. Sarwar and Abdullah-Al-Mamun, "Next word prediction for phonetic typing by grouping language models," 2016 2nd International Conference on Information Management (ICIM), London, 2016, pp. 73-76, doi: 10.1109/INFOMAN.2016.7477536.

[6] M. K. Sharma, S. Sarcar, P. K. Saha and D. Samanta, "Visual clue: An approach to predict and highlight next character," 2012 4th International Conference on Intelligent Human Computer Interaction (IHCI), Kharagpur, 2012, pp. 1-7, doi: 10.1109/IHCI.2012.6481820.

[7] Minghui Wang, Wenquan Liu and Yixion Zhong, "Simple recurrent network for Chinese word prediction," Proceedings of 1993 International Conference on Neural Networks (IJCNN-93-Nagoya, Japan), Nagoya, Japan, 1993, pp. 263-266 vol.1, doi: 10.1109/IJCNN.1993.713907.

[8] Y. Ajioka and Y. Anzai, "Prediction of next alphabets and words of four sentences by adaptive junction," IJCNN-91-Seattle International Joint Conference on Neural Networks, Seattle, WA, USA, 1991, pp. 897 vol.2-, doi: 10.1109/IJCNN.1991.155477.

[9] Joel Stremmel, Arjun Singh. (2020). Pretraining Federated Text Models for Next Word Prediction using GPT2

https://doi.org/10.1051/itmconf/20214003035

Balanced : An Application To Improve Mental Health

Mrunal Bhalerao[1,*], *Trisha* Lewis[1], *Raj* Salvi[1] , and *Rakhi* Kalantri[2]

[1] UG students, Computer Engineering Department, Fr. C. Rodrigues Institute of Technology, Navi Mumbai 400703, MH
[2]Assistant Professor, Computer Engineering Department, Fr. C. Rodrigues Institute of Technology, Navi Mumbai 400703, MH

Abstract. Mental health reflects a person's emotional, psychological, and social well-being. A report by World Health Organization suggests that India is the most depressed country in the world. Hence, the idea is to develop an end-to-end solution for identifying and treating mental-health issues. This paper aims to develop an application to keep track of user's behaviour, both online and offline, to understand and identify the possible mental health issues using various analytical and psychological methodologies. The target users of this application would be the individuals just above the age of 18 which fall under the category of working class. Post-identification, the user is recommended a suitable change in his routine and behavior which would lead to a better lifestyle.Mental illnesses detection is a challenging undertaking because a misdiagnosis can lead to catastrophic consequences.Hence, our android application will take the necessary steps to accurately identify and treat the mental health issue.

1 INTRODUCTION

In [1] **Mental Illness** is expounded as a mental, emotional, or behavioural disorder. The severity of the impact varies from no impairment to severe impairment, which significantly hampers or interferes with one or more important living activities. Thus, proper assessment of mental wellbeing is highly essential to understand and suggest therapies to individuals who are facing mental illness. While study suggests that 1 in every 4 adults is affected by mental health problems. However, recognising mental health concerns is crucial, as misinterpretation can result in life-threatening consequences. As a result, proper attention must be made to identify mental health disorders. Additionally, only 10 to 12 percent of those who are affected seek care. Hence, the absolute necessity of today is to create awareness about mental illness and normalize the concept of seeking help to overcome this dark phase [2].

Therefore, in this paper, we intend to develop an end-to-end solution for identifying the possibility of a mental illness, as well as, provide our user with a platform to deal with his/her condition and live a better life every day. We have developed an android application and a website to keep track of user's digital, physical and psychological activities, both online and offline. These activities are then perfectly analyzed by the Machine Learning model to predict the possibility of mental disorder. Post-identification, the user is recommended a suitable change his routine and behavior which would lead to a better lifestyle. Depending on the identification, if required, the user may also be connected to a psychologist, a yoga institute, gym etc., which may help him improve his mental health condition. In addition to this, weekly and monthly evaluations will be conducted for the users where they will be presented with a brief description of their habits and mental-health condition in an understandable visual format. Further lifestyle changes that they can undertake to improve upon their condition will be suggested.

2 LITERATURE REVIEW

In [3] U. S. Reddy et. al. has applied Logistical Regression, Random Forest and Decision Tree algorithms to determine which of these is the most accurate. A number of parameters in the dataset were compared to see how they related to one another.

In [4] M. P. Dooshima et. al. has applied Decision Trees and Naïve Bayes algorithm. Demographic, psychological, biological and environmental factors are used for prediction. To validate the obtained parameters, different health professionals were approached.. However, the Data is only for 30 patients.

In [5] M. Srividya et. al. made use of a questionnaire to obtain values for different attributes which could help in prediction of mental illness. This paper intended to analyse various algorithms and determine the accurate one. She used the labels from the collected data to compute a MOS. She then applied Decision Tree, Naïve Bayes and SVM algorithms to find the more accurate. In Conclusion, Support Vector Machine and K- Nearest Neighbors algorithms have similar accuracy results and show highest accuracy.

In [5] M. Srividya et. al. used Neural Networks to determine the psychological conditions of humans such as depression, stress, anxiety etc.

* Corresponding author: mrunal.bhalerao25@gmail.com

2.1 Data Mining algorithms for Privacy-Preserving:

2.1.1 Randomization

During Vertival Shuffling of Randomization records the attribute's semantic does not changes. Instead, the vertical position of the record is altered, concealing the true identity. To disguise the behaviour of individual records, noise can be added to the data. The aggregate behaviour of the data distribution is later rebuilt by removing the noise from the data[6].

2.1.2 k-anonymity

Data anonymization is used when it is necessary to make changes to the original data in order to remove any information that is directly traceable to any of the individuals. Here, the granularity of data representation is decreased by using suppression and generalization techniques [7].

2.1.3 l-diversity

The l-diversity model overcomes the foibles of the k-anonymity model. Thus, preserving identities at the level of k-individuals differs from protecting the corresponding values, particularly when sensitive values within a group are homogeneous. The anonymization scheme encourages intra-group variation in sensitive features [10].

For Chatbot, in [9], **"A Mental Health Chatbot for Regulating Emotions (SERMO) - Concept and Usability Test"**, mentions a mobile application integrated with a chatbot which developed using the Syn.Bot framework. It contains OSCOVA and Synthetic Intelligence Markup Language interpreter. It employs cognitive behaviour therapy (CBT) techniques to assist mentally ill individuals in regulating their emotions and dealing with their thoughts and feelings. Daily, it inquires of the user about events in their life and their emotions. And, using NLP and a lexicon-based technique, the app detects a user's core emotion from natural language input. A suitable measurement, such as activities or mindfulness exercises, is also advised.

In [11], we understand that mental health problems can be found at different ages. While, each category shows different signs, some show similar signs too. According to the paper, Mental illness can be categorized into 7 types: Personality illness, Compulsion illness, Eating illness, Emotional illness, selfish control illness, Nervousness Illness and Psychological illness.

In [8], **"Data Science in Public Mental Health: A New Analytic Framework"**, advanced methodologies were used to conduct a thorough investigation into public mental health issues using data science and to find remedies based on the findings of the data science publications. A complete comparison of those with conventional data analysis techniques is also presented. The study proposes a new framework for mental health professionals to tackle problems and challenges they face using data science. In order to get a good concept of the problem to be solved, the research also looks into predicting the time and resources required in any process.

In [12] The research **"Depression detection using emotion artificial intelligence,"** looks at depression as a significant cause of mental illness as well as an increased risk of dying young. Furthermore, depression is identified as a primary cause of suicide ideation, resulting in severe impairment in daily life, according to the study. It delves into emotion artificial intelligence as a subject of continuing research in emotion recognition, with a focus on text mining. They have applied NLP on Twitter feeds for demonstrating emotion analysis to focus mainly on depression. To detect depression tendencies, tweets are categorised as neutral or negative based on a chosen word list. Support vector machines and Naive-Bayes classifiers were employed in the class prediction process. The findings are determined using the F1-score, accuracy, and confusion matrix.

3 PROPOSED SYSTEM

Based on the literature, we propose a system whose major purpose is to create a website and an Android app, where the users will have to login through their google accounts and fill in a registration form. The form will allow us to have background information of the user. Once the user logs in, data including digital usage data, data containing mood and activities performed throughout the day, data from conversations with chatbot, Psychological health data and physical health data will be collected by the application and stored in the firebase. The data gathered will be pre-processed and analyzed. Based on this data, results about potential or possibility of mental illness will be available to the user. We will label and encode the data for better prediction. We plan on using Random Forest algorithm with Clustering for predicting a pattern within the data.

Once, a prediction is made the user will be presented with a proper treatment by connecting him/her with a psychologist, a yoga institute, a gym, etc. that may help ameliorate his mental health conditions. The user will also be suggested some to change his routine and suggest some behavioral changes which might help him improve his condition. In addition to this, we will provide in-app features including a Wellness Center, Habit Tracker, Scheduler, Music Player, connect with peers and Connect with professionals. Finally, weekly evaluations will be done and real time reports of the user's digital, physical and psychological behavior will be developed in layman a visual format.

 https://doi.org/10.1051/itmconf/20214003035

In this way, the solution will be an end-to-end solution for predicting the possibility of mental illness and also helping to lower the impact of mental illness.

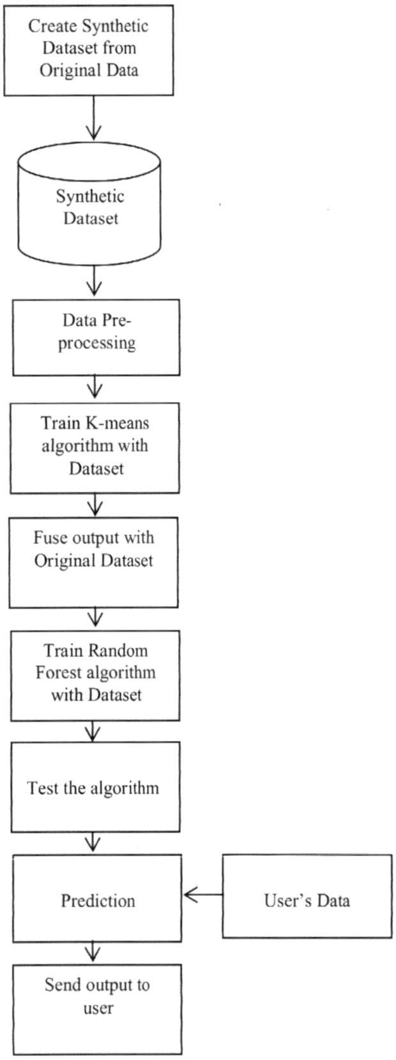

Fig. 1. System Block diagram

4 IMPLEMENTATION AND RESULT

4.1 Data Collection

The user will have to use the android/web application to login through their google accounts and fill in a registration form. The form will allow us to have background information of the user. Once the user logs in, data including:

4.1.1 Digital Information Data

Location, websites visited, battery levels, media playback time, applications used and duration of usage, data usage, network information, gyroscope information, accelerometer information, screen on time will be gathered and saved in the firebase.

4.1.2 Physical Information Data:

Number of steps walked, duration and schedule of sleep, the user will be prompted to mark a response for activities done throughout the day.

4.1.3 Psychological Information Data:

Responses to 4 psychometric quizzes suggested by a psychiatrist will be gathered, user will be prompted to mark a response for "how are you feeling?" multiple times a day, as shown in [9] conversations with the chatbot will be monitored and a basic emotion will be determined for every conversation.

4.2 Data Flow/Storage:

Data from the Web Extension will be stored on the Server (Heroku) and then on the Firestore. Data from the android application will be stored in the Firebase. In case of any discrepancy in data, it will be sent to a local server, pre-processed, and then sent to the database.

This data collected is used for prediction of possibility of mental illness. The visualization of collected data is shown to the user in a graphical format. This allows the user to verify and track changes in his behavior, which in turn satisfies are aim of evidence-based prediction. Figure 2 illustrates how real times graphs will be plotted. Figure 3 and Figure 4 are examples of plotted graphs.

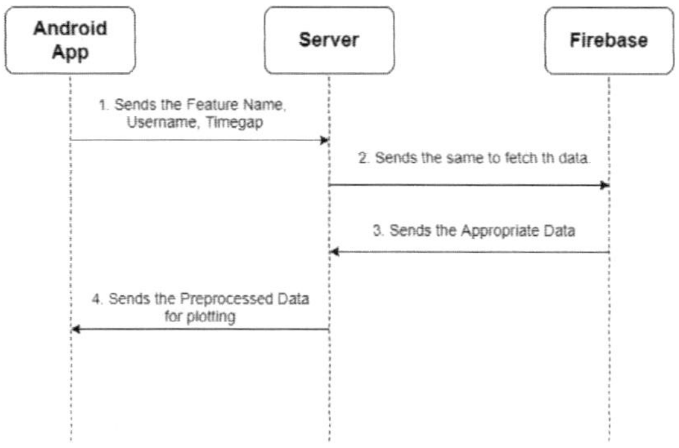

Fig. 2. Sequence Diagram

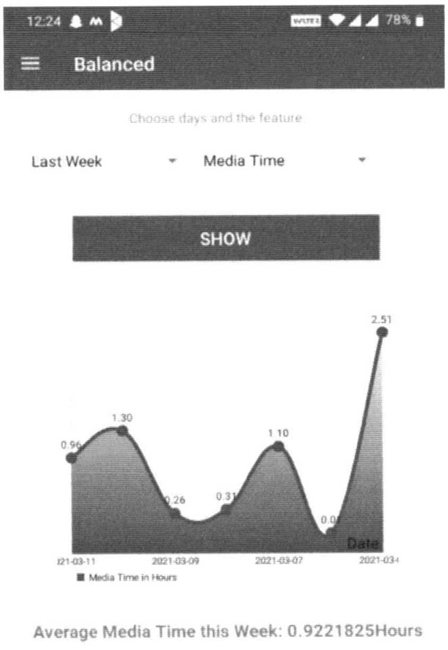

Fig. 3. Media Time

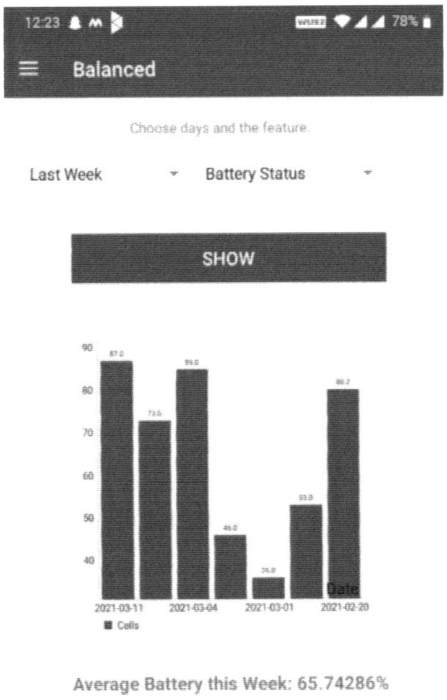

Fig. 4. Battery Status

5 DETECTION

We will forecast the user's mental ailment during the detecting phase. Some of the characteristics we are using are age, gender, family history of mental illness, results from 4 quizzes, their basic mood which is derived from the "how are you feeling?" and conversations with the chatbot, phone usage, applications used, etc. For the detection of possibility of mental illness, a synthetic dataset was formed by using the original dataset and some threshold values. The threshold values were carefully found out under guidance of a psychiatrist. Permutation and combination techniques were used to form the synthetic dataset from the original dataset.

Further, K-means clustering technique was employed to cluster the dataset into two classes: Possibility of Mental Sickness and No signs of Mental sickness. K-means clustering algorithm groups/clusters data having similar features by using the technique of Euclidian Distance. It also ranks the importance of each feature. In this scenario, quiz 3, quiz 4 and apps used are given highest priority as shown in the Table 1.

FEATURE	IMPORTANCE VALUE
Age	0.00198102
Emotional Quotient Test	0.00179368
Depression Test	0.00219795
Locus of control Test	0.50759175
Optimism Test	0.47584557
Screen on Time	0.00284181
Social Media	0.00250919
Youtube	0.0027524
Others	0.00248662

Table. 1. Ranking of Features

Now, the output of the clustering was fused with the original dataset. Thus, converting the un-supervised problem to a supervised task. This data was then fed into the Random Forest algorithm (RFT) for supervised learning. RFT is a combinations of multiple decision trees. Each individual tree determines a class prediction and the class with the most votes becomes the model's final prediction of whether the user is suffering from any Mental Illness or not.

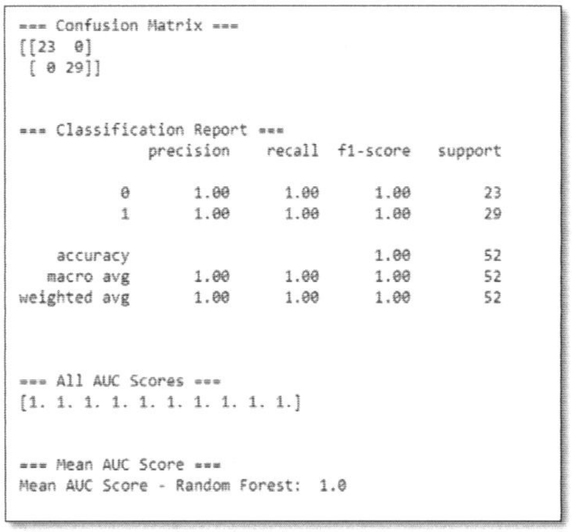

```
=== Confusion Matrix ===
[[23  0]
 [ 0 29]]

=== Classification Report ===
              precision    recall  f1-score   support

           0       1.00      1.00      1.00        23
           1       1.00      1.00      1.00        29

    accuracy                           1.00        52
   macro avg       1.00      1.00      1.00        52
weighted avg       1.00      1.00      1.00        52


=== All AUC Scores ===
[1. 1. 1. 1. 1. 1. 1. 1. 1. 1.]


=== Mean AUC Score ===
Mean AUC Score - Random Forest:  1.0
```

Fig. 5. Classification Report

Now, that the ML model is trained to predict the proper output, it was ready for deployment. We deployed our model on Heroku by linking the GitHub repository to the Heroku account. Heroku is nothing but a Platform as a Service. In other words, it allows to transfer the application from a local machine to the cloud. This, allows the deployment of a machine learning model in real-time.

5.1 Aid for a Better Mental Health

With a vision of providing guidance to improve our user's mental health condition, we have provided the following in-app features:

5.1.1 Digital Information Data

The Wellness Center mainly consists of the –

Virtual Gym: It consists of a carefully curated list of workouts which when performed boosts endorphin levels. They are classified on the basis of time.

Meditation Center: It consists of resources which can help the user deal with anxiety, depression, high stress levels, insomnia and anger. It will help to develop a overall better lifestyle with a positive attitude.

Learning Center: It has resources which able the user to utilize his/her time by learning something new or pursuing a hobby. This will allow the user to develop a habit of utilizing free time instead of wasting it in unproductive/disturbing activities which lead to mental sickness.

Peer Connect: This feature will allow the user to connect with other people who are willing to start a conversation. Thus, it will allow the user to socialize, which plays a vital role in boosting positive mental health.

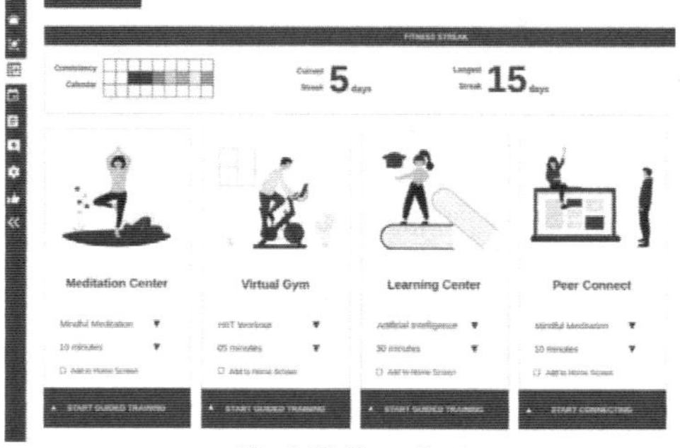

Fig. 6. Wellness Center

Fig. 7. Peer Connect

5.1.2 Habit tracker

It allows the user to keep a track of new habits he/she wants to inculcate. The user can select the target days for which he wants to follow the particular habit. Doing so ensures that the users are able to follow a healthy pattern in life.

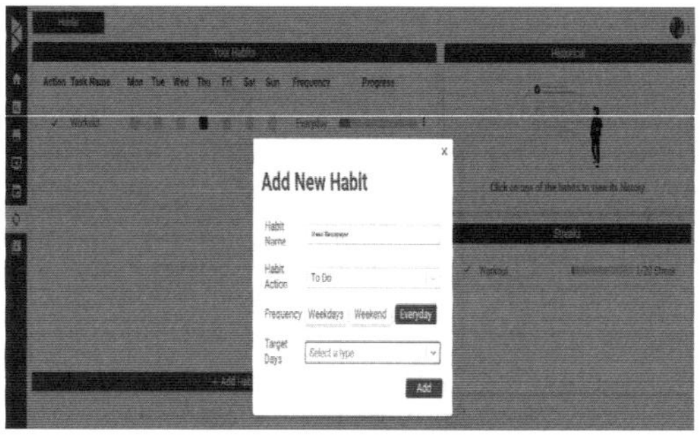

Fig. 8. Habit Tracker

5.1.3 Scheduler

The Scheduler is a Kanban board with a prioritization facility. Thus, the user can schedule and prioritize tasks he/she is supposed to perform. This will help our user to efficiently manage his tasks and be more organized. Which in turn will have positive mental health effects.

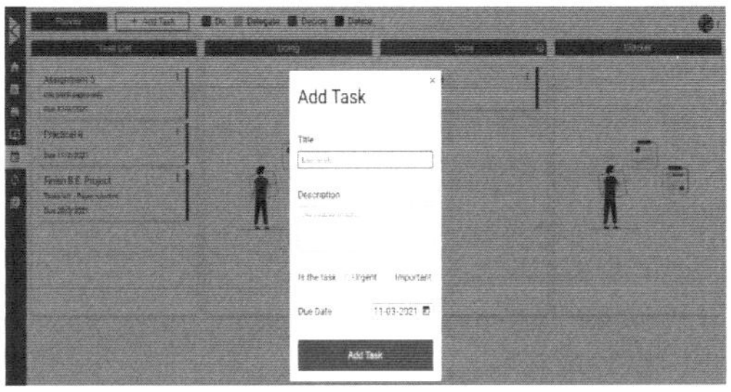

Fig. 9. Scheduler -1

Fig. 10. Scheduler -2

5.1.4 Seek Help

In this section, the user can talk to and connect with professionals like psychiatrists, psychologist, counsellor, gym instructor, yoga instructor, anger management programs, etc. and thus get help from experts to deal with mental sickness.

5.1.5 Chat-bot

A chatbot is created using Dialogflow, to provide our user with an AI friend. Here, the user can talk to "Balancia," the virtual assistant. The virtual assistant, will give the user suggestions to deal with his/her situation.

6 CONCLUSION AND SCOPE

Mental health directly affects the way we think, feel and act. It also has a tremendous effect on our physical health. Unfortunately, recognising mental health concerns is a challenging undertaking, and mistake can lead to significant consequences. More so, many taboos have developed around mental sickness and thus people prefer avoiding the issue rather than consulting professionals. Hence, through this paper, we would like to help our community by developing a user-friendly android application which would allow them to easily diagnose their mental health problems and will also help and support them to improve their mental health, right from the user's home. It will also expand the reach and utilisation of e-mental health resources, particularly in rural/remote locations where there is a lack of awareness and availability of experienced specialists. This software will provide evidence-based services to meet current demands while also enhancing client trust in e-mental health care. By utilizing advances in technology, we aim to detect unhealthy digital behaviors by track changes and help users improve their digital lifestyle by providing them smart and personalized health recommendations.

The scope of the paper includes:

- Detecting any existing mental health issues.
- Detecting unhealthy digital behaviour
- Providing real-time in-app graphical visualization of user's habits.
- Helping users improve their digital and physical lifestyle by providing them with smart and personalized recommendations and in-app activities.
- Through the application, connecting the user with professionals.
- Keeping track of user's mental health condition

In conclusion, the aim is to develop a web application and android application to provide an end-to-end solution for the diagnosis of possibility of mental health illness and treating it.

7 References

[1]Understanding Mental Illness: https://www.nimh.nih.gov/health/statistics/mental-illness.shtml

[2]How India Sees Mental Health: https://www.weforum.org/agenda/2018/04/5-charts-that-reveal-how-india-sees-mental-health/

[3] U. S. Reddy, A. V. Thota and A. Dharun, "Machine Learning Techniques for Stress Prediction in Working Employees," *2018 IEEE International Conference on Computational Intelligence and Computing Research (ICCIC)*, Madurai, India, 2018, pp. 1-4.

[4] M. P. Dooshima, E. N. Chidozie, B. J. Ademola, O. O. Sekoni, I. P. Adebayo, A Predictive Model for the Risk of Mental Illness in Nigeria Using Data Mining, *International Journal of Immunology*. Vol. 6, No. 1, 2018, pp. 5-16.

[5] M. Srividya, M. Subramaniam and B. Natarajan, "Behavioral Modeling for Mental Health using Machine Learning Algorithms" "Journal of Medical Systems" Vol. 42(5):88 May 2018.

[6] A. Kaur, "A hybrid approach of privacy preserving data mining using suppression and perturbation techniques," *2017 International Conference on Innovative Mechanisms for Industry Applications (ICIMIA)*, Bangalore, 2017, pp. 306-311.

[7] A. Kiran and D. Vasumathi, "A Comprehensive Survey on Privacy Preservation Algorithms in Data Mining," 2017 *IEEE International Conference on Computational Intelligence and Computing Research (ICCIC)*, Coimbatore, 2017, pp. 1-7.

[8] C. Silva, M. Saraee and M. Saraee, "Data Science in Public Mental Health: A New Analytic Framework,"

2019 IEEE Symposium on Computers and Communications (ISCC), Barcelona, Spain, 2019, pp. 1123-1128, doi: 10.1109/ISCC47284.2019.8969723.

[9] K. Denecke, S. Vaaheesan and A. Arulnathan, "A Mental Health Chatbot for Regulating Emotions (SERMO) - Concept and Usability Test," in IEEE Transactions on Emerging Topics in Computing, doi: 10.1109/TETC.2020.2974478.

[10] A. Evfimievski. 2002. Randomization in privacy preserving data mining. SIGKDD Explor. Newsl. 4, 2 (December 2002), 43–48.

[11] Pyae, Aung. (2018). Study of Mental Health Data Science (Systematic Literature Review).

[12] M. Deshpande and V. Rao, "Depression detection using emotion artificial intelligence," 2017 International Conference on Intelligent Sustainable Systems (ICISS), Palladam, India, 2017, pp. 858-862, doi: 10.1109/ISS1.2017.8389299.

Study on Design and Implementation of Distributed Multiple Camera Surveillance and Tracking System

Sureshkumar Jha [1*], *Rohan Sawant*[1], and *Parth Shinde*[1] *Rakhi Kalantri* [2] *Shagufta Rajguru*[2]

[1]*BE Students, Department of Computer Engineering, Fr Conceicao Rodrigues Institute of Technology, Vashi Navi Mumbai, India.*
[2]*Assistant Professor, Department of Computer Engineering, Fr Conceicao Rodrigues Institute of Technology, Vashi Navi Mumbai, India.*

Abstract. This work proposes a system to successfully track, identify and tag target objects or individuals in a real time environment. Consider a sequence of cameras installed in an environment or a facility. Through these cameras user can track those intruders who are aware of how the surveillance system operates and are actively trying to avoid getting seen by the surveillance system. This system tracks the movement of any person's movement and record and maintain that data throughout any facility in which such a solution is deployed. Movement logging of any individual person can also be done if it does not infringe his/her privacy. Design and analysis a distributed algorithm for the optimization of the recognition and mapping of the given subject/subjects on a User Interface. Finally, the performance and robustness of the distributed system is further analyzed through continuously training the algorithm or maybe real time demo.
Keywords— Movement Logging, Recognition, Distributed Systems, Real-Time Introduction.

1. Introduction

This Paper guides a stepwise walkthrough of an intelligent system that detects and captures motion information of moving targets for accurate object tracking in real-time. The classified object is being tracked for high-level analysis. This paper provides a detailed description on how to detect humans, recognize their faces, inference the unrecognized and recognized people and through that detect their movement. The human detection in a smart surveillance system aims at making distinctions among moving objects (i.e., specifically humans in this case) in a video sequence. Human detection may be a difficult task from a computer vision perspective because it is influenced by a good range of possible appearance thanks to changing articulated pose, clothing, lighting and background. Combined assumptions were made on the environment to be monitored, the cameras, and thus the targets. The environment of the facility is assumed to be divided into multiple layers (i.e., floors) and assume that each floor is a corridor and is often completely observed by a sequence of cameras. The matter of perimeter surveillance could also be a special case of this framework. The cameras are assumed to be subject to physical constraints, e.g., limited field of view and to be equipped with a low-level routine to detect intruders/targets that fall within the field of vision of a camera. Humans are assumed to be randomly appearing in the video sequences, within the sense that they have no idea of the cameras' configuration at any instant, and prediction of their trajectory is not possible. The gate camera must have proper vision of the person and possibly his face in an entry video sequence.

1.1. Problem Description and Formulation

This project deals with the implementation of a real-time dynamic closed-circuit television, which tracks both the known and unknown targets and classifies them accordingly. This classification and detection of such targets is visualized using a 2d Map or blueprint of the whole area. This helps in tracking of the targets within the facility.

The whole system relies upon the synchronization of the cameras to trace a specific target. This results in easy capture of any "red flag" activity of any target. 'RED FLAG' activities can constitute entry of the person through other means, rather than the designated entry, this may be immediately raised if the given target is an unknown entity. This flag is also raised when any valued object goes missing. The trajectory of the camera is additionally considered during the analysis and processing of the varied video streams coming from different cameras.

The facial data of any unknown person entering the power is merely relevant till a selected period of time, then its erased, only frequent visitors are logged into the database with their approval. This is done to make sure the privacy of the target is additionally not compromised

The visualization contains a map of the corridor or floor, or by making use of green or red bounding boxes, through which the inference of the algorithm is shown. This visualization concept also can expand vastly, depending on the sort of study or level of analysis the user requires.

1.2. Related Work

Some examples of the already existing software systems for the two are following: -

1.2.1. Categorization of Target Tracking Systems

First of all, target tracking systems can be classified into on-line real-time and off-line batch systems. Real time tracking system can be classified into different types. For the differentiation of the classes; Firstly, the video stream is provided to the software. Once the object is detected in the frame, the object is later classified on the basis of shape-based method, motion-based method and texture-based method, after running these modules it can determined whether the object in the frame is an object or a human.

1.2.2. Design and Implementation of Human Tracking using CCTV Cam

Closed Circuit Television (CCTV) technology made a big impact on how crimes were solved. Suppose a wanted person is in the vicinity, the authorities only have to give a dataset of the person's face. Once the person's face is seen in an ip camera node, it passes the data to the neighboring node thus tracking the person.

1.2.3. Intrusion Detection Using Image Processing Techniques

Through this monitoring system, user can easily detect the intrusion happening in a particular area. The software will be given a mapped data of the area and the people authorized in the area. If a person recognized by the software enters the area it would be safe, but if unauthorized personnel enter the area the system raises the intruder alert and stores the unknown face in the dataset and tracks the person in the overall premises.

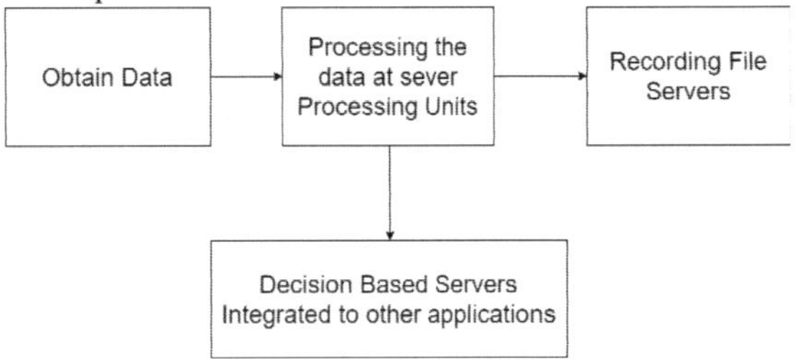

fig 1: Data collection and comparison for intrusion detection

1.3. Contribution

The above listed data are the various modules which are building blocks of the product that are going to be developed. The close circuit television (CCTV) technology is referred while developing the module to track the person's motion in the vicinity. Then the intruder alert system is referred to, which detects an unknown face in the vicinity. In the system, user is tracking a person in a particular vicinity and storing the data of the person in the server. This data is passed from one node to another node and tracks the person more efficiently, it also considers a person entering the area where the system is active. The person and the motion of the person is tracked by the yolov4 framing the bounding boxes around the body of the person. Then the identity of the person is determined using the deep sort and the mtcnn which stores the different characteristics of a person's face and creates bounding boxes around the face. Through this, it is determined whether the person is an intruder or regular by processing the face and streaming it simultaneously, this is achieved with the help of the multi-threading system where each work is considered as threads and different types of threads are executed simultaneously thus, reducing the system upgrades and speed of processing, and streaming the video. An unknown person will be red flagged and will be tracked and the face will be stored on the server for tracking the person through the vicinity.

2. System Overview

The system is mainly divided into a Subordinate, Superordinate Systems, and an external input (mobile app). The Superordinate system interacts with the actual raw data stream coming into the surveillance software, while the subordinate system generally deals with the internal flow and interaction between the various components of the superordinate system.

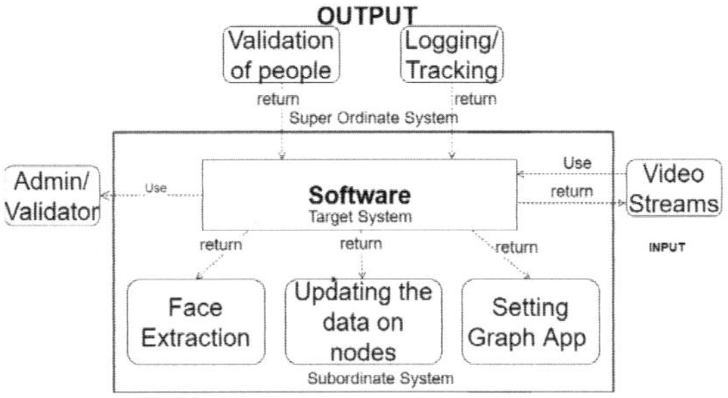

fig 2: basic architecture of the system where data is collected stored and processed.

The Superordinate System is a central object which receives the input video streams and the graph generated through the mobile app, it is responsible for preprocessing this data and passing it through various processing blocks such as feature extractor (can be both face and person, or either face only or person only), Feature Comparer, Tracker and Visualizer.it also handles the message/data passing between the Gate camera its children's nodes. Additionally, it performs regular checks if the graph needs setting or resetting whenever a new camera is added. It also checks whether a new face is added to the data pool and effectively pipelines all the processes with the respect to the newly updated data pool. It also sends alerts for validation to the User/Admin (i.e., Security Personnel of the facility). The Subordinate System is an object which handles the flow and interaction between all the different processes. It is responsible for maintaining proper flow of the system between its sub-modules. It thus helps the Superordinate maintain the proper pipeline for processing data and inferencing it to the output display

This system consists of the following processes:

1. Feature extractor
2. Feature Comparer
3. Visualizer

These also use sub modules such as Tracker and Parrallelizers

2.1. Flow of the System

To understand the Practical application of the project it is necessary to understand the flow of the system. The IP camera has its own Motion Detector and Face Detector.

This is used by the other successive processes to perform proper face recognition and human tracking.

 https://doi.org/10.1051/itmconf/20214003036

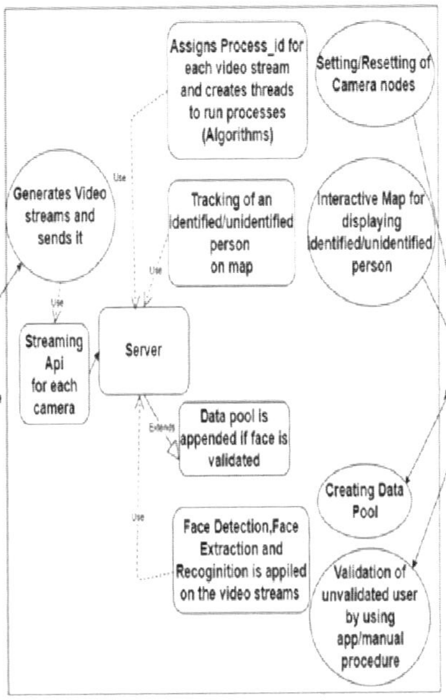

fig 3: interaction between user and admin

In an environment as specified in the Introduction, various video streams are given to System (Server). The Superordinate object initiates a camera object for each video stream, prioritizing the gate video stream object. First, since it is where the original data collection needs to start. It then catches each n (parameter for making batches) frames from the data stream and resizes it in proper format and batches it so it can be easily.

Be sent to the subordinate system for processing. The subordinate system enables the Tracker, Tracker Manager and the Parrallelizers and pipelines them accordingly before processing the frame. *

*Note-
The Tracker Object is generated exclusively for the gate camera and is only updated at the gate after every frame, this object is shared with its neighboring nodes using the Tracker Manager. The Tracker object traverses across the Graph which is created beforehand in the Graph Creation process.

The IP camera object stream the frames which goes through an object detector, it collects face and body data on the entry/point and updates the data pool and its indexes. If this frame is not from an entry point goes through the face/person detector (feature extractor) and then searches for the face/person in the data pool* using the feature comparer, once the face/person identified it can be tracked through the whole facility (Note: -The user can be enabling tracking of only the unidentified persons as well only identified targets or both, though only unidentified setting is recommended). The resultant data from both the Tracker and the Comparer are now fed to the Visualizer which in turn shows the output. The unidentified face/person is tracked throughout the facility for a specific time period. If the person is not validated it alerts the admin/security personnel about the person using his unique id. The unidentified person can also be validated by the admin in this time frame, if this is done, the face/person will be appended to the global data pool and given the identified status immediately.

Different algorithms are applied and explored for each process.

*Note-
The data pool mentioned here is a temporary one, it updates after each frame

2.2. Hardware and Software Requirements

Each process uses different resources namely CPU and GPU of the computer and therefore the software is heavily dependent upon the hardware system its running on. A pc or server with x86-64 (64-bit) compatible processor:3 GHz or a

better processor with 6 cores is recommended. Graphical processing unit (GPU): GTX 1080 GPU or a better is recommended at least 6 GB of VRAM is recommended. Compute Capability 3.5 or better should be supported by the GPU. At least 8 GB of RAM is required. Digital camera(s): Camera resolution may vary depending on the actual application. The recommended resolution is about 2 Megapixel.

CUDA 10.1 toolkit or newer is required. cuDNN 7.5 library is required.

2.3. Graph Creation

To create a Graph of cameras latitude, longitude, altitude of the camera is required. Cameras can be considered as the nodes in 3D space and by using distance vector algorithm the nearest neighbor graph is created. Purpose of this NN graph is to minimize the global search.

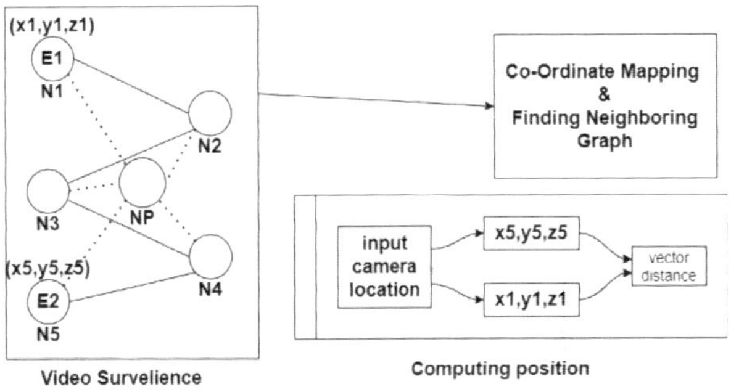

fig 4: camera interacting with each other passing the features from one node to another.

Whenever a camera node detects a person, it will extract the features and face of that person. Every node will have a local database in order to re-identify the person.

For re-identification purpose, Features and the faces of the person tracked will be stored and sent to neighboring nodes to look for that person with similar features and compare the face.

2.4. Feature Extractor

Feature extraction is particularly important in face recognition. Modules like FaceNet and TorchReid for face extraction and person feature extraction.

2.4.1. FaceNet

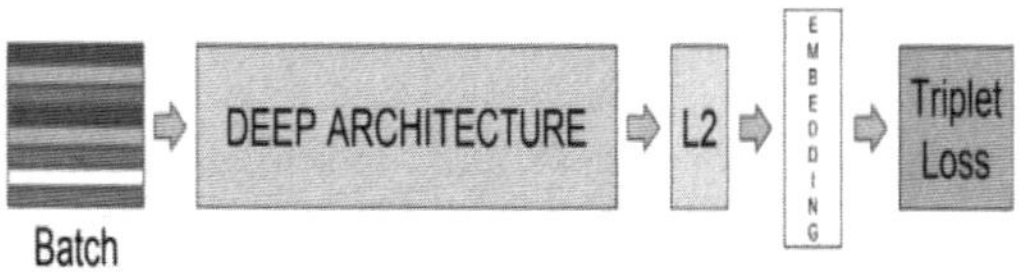

Fig5: FaceNet Architecture

Faces are stored in the form of 128 feature embeddings. Face recognition systems for face identification and verification using the VGGFace2 deep learning model are used to get 128 feature embeddings. The VGGFace refers to a series of models developed for face recognition and demonstrated on benchmark computer vision datasets. Face Identification involves calculating a face embedding for a new given face and comparing the embedding to the embedding for the single example of the face known to the system.

A one-to-many mapping for a given face against a database of known faces is called face Identification. For every batch of frames, faces are extracted and the face comparer is used to compare extracted faces with the already known faces from the global database. This data pool is populated by an entry flagged camera node and this global database is used by comparer to fetch the faces.

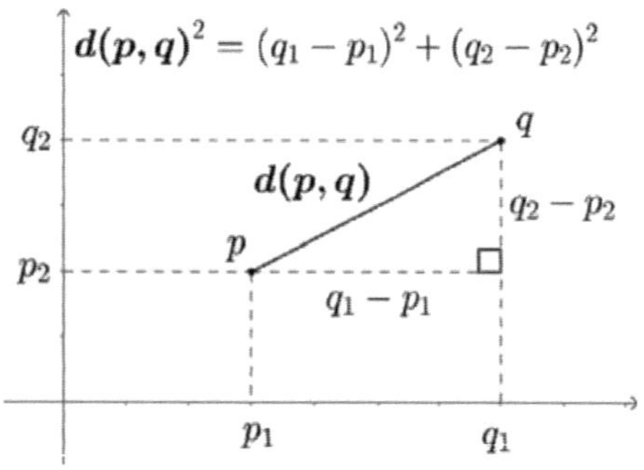

fig 6: Euclidean distance Formula

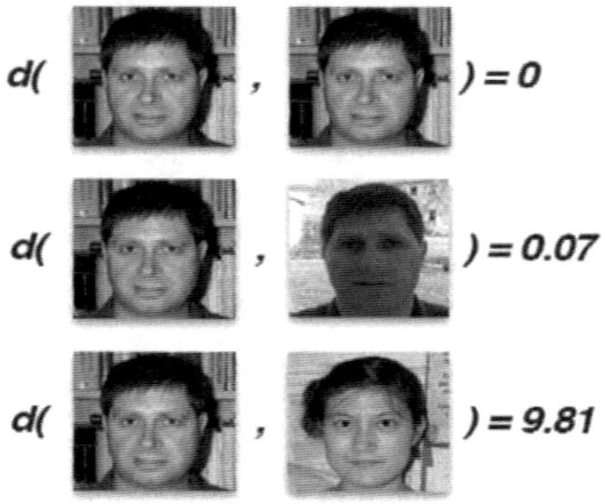

fig 7: Comparing the Euclidean distance determine same face

2.4.2. TorchReid

Torch Reid is a library for person identification written in PyTorch. provides a unified interface for all the ReID operations. The Pipelining of this framework is also quite simple and easy to use. It also provides easy access to various complicated tasks of person reidentification. In the context of a feature extractor for the system the TorchReid can be used to extract the features of a person from singular or multiple frames.

2.5. Face Comparer

A face embedding is a vector that represents the features extracted from the face. This can then be compared with the vectors generated for other faces. For example, another vector that is close (by some measure) may be the same person, whereas another vector that is far (by some measure) may be a different person. Typical measures such as Euclidean distance and Cosine distance are calculated between two embeddings and faces are said to match or verify if the distance is below a predefined threshold, often tuned for a specific dataset or application.

Here MTCNN is used to extract faces in the video stream. MTCNN (Multi-task Cascaded Convolutional Neural Networks) is an algorithm consisting of 3 stages, which detects the bounding boxes of faces in an image along with their 5 Point Face Landmarks (link to the paper). Each stage gradually improves the detection results by passing its inputs through a CNN, which returns candidate bounding boxes with their scores, followed by non max suppression.

Benchmark:

The specifications for the software to work are Intel i7-3621QM , TensorFlow 1.4.1.

The Pictures containing a single face:

Image Size	Total pixels	Process Time	FPS
460 x 259	119,140	0.118 seconds	8.5
561 x 561	314,721	0.227 seconds	4.5
667 x 1000	667,000	0.456 seconds	2.2
1920 x 1200	2,304,000	1.093 seconds	0.9
4799 x 3599	17,271,601	8.798 seconds	0.1

Fig 8: Benchmark of MTCNN

In this above context we use these modules to perform two mechanisms. Any of these methods may be used as a face comparison mechanism.

2.5.1. Iterative Method

In this method the face extracted from a batch of frames is compared to a global database of known faces one by one. Every vector of faces stored in the database is compared with the face to be identified. One-to-many comparisons are made by finding the Euclidean distance less than a particular threshold. If the Euclidean distance is less than a particular threshold then the face is identified as a known face and a green flag is given to that tracker object. Iterative method has to sequentially scan each entry in the database and find the Euclidean distance and compare it with the threshold. This is brute force technique which runs in linear time complexity of O(n) where n is the number of faces in the database. As the number of faces in the database increases the performance will reduce.

2.5.2. Approximate nearest neighbors

Nearest neighbor search of embeddings is the fastest and optimal way for searching through the database. Vectorized cosine distance is notably faster than the iterative method but still may be too slow. This is where approximate nearest neighbors' shines: returning results quickly by using annoy indexes. It reduces the time complexity to O (log n) where n is the number of faces in the database. If your use case requires creating and building indexes often, then search time is the key performance. On the other hand, if your use case requires rebuilding indexes often, then adding embeddings and index building times are important. The following table demonstrates the times I spent for those ann packages on 1M vectors. Performance of packages.

	Adding Embeddings	Index Building	Search	Total
Spotify Annoy	3.106772	7.398274	0.000715	10.505761
Facebook Faiss	0.225953	0.000000	0.244324	0.470277
Nmslib	0.805913	263.954751	0.003436	266.764100

fig 9: Benchmarks for annoy

We have used a better version called hnswlib which dynamically updates the given indexes for any features that are extracted.

2.6. Tracker:

A Tracker is used for the following reasons: -It is faster than Detection, it can help when detection fails, it preserves identity:

The tracker object consists of a person tracker and Person Reid module encapsulated together in a single class. We have used different trackers, some of them are: -

2.6.1. OpenCV inbuilt Tracker and Person Reid

In this module we use cv2.inbuilt tracker api for tracking a person/face in a singular video stream, and torch reid person re identification engine for re-identifying the person across the surveillance system. These two are bundled together in a Tracker class. The tracker we used in our system was: -

2.6.1.1. GOTURN tracker: -

This is a CNN based offline tracker. First, the tracker is trained using thousands of videos for general person/face tracking. Now, this tracker can be used to track most of the objects without any problem even if those objects were not part of the training set. The goturn algorithm is best for detecting already trained objects such as persons, cars etc. It works fast if it is trained to recognize a single object.

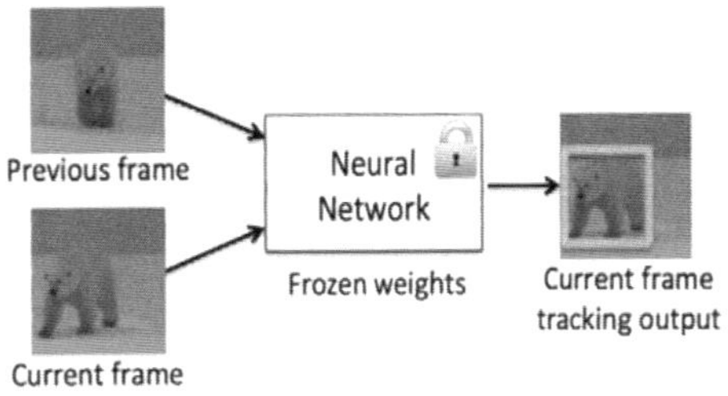

Fig10: Working of GOTURN

2.6.2. Tracker using deep sort algorithm through yolov4 and Person Reid

We use deep sort class and algorithm with the yolov4 for person tracking, this is bundled with torch reid person re identification. This object tracker coupled with person re identification engine of TorchReid can easily track human targets in multi camera multi target environment (limited only to humans).The TorchReid engine is uses the earlier extracted embeddings to recognize the person in the frame.

2.6.2.1. DeepSort

This is a great application for tracking, it contains a script to generate features for re-identification, which are appropriate to compare the visual appearance of the binding boxes for pedestrians using cosine simulations. It can also be used in standalone format without TorchReid.

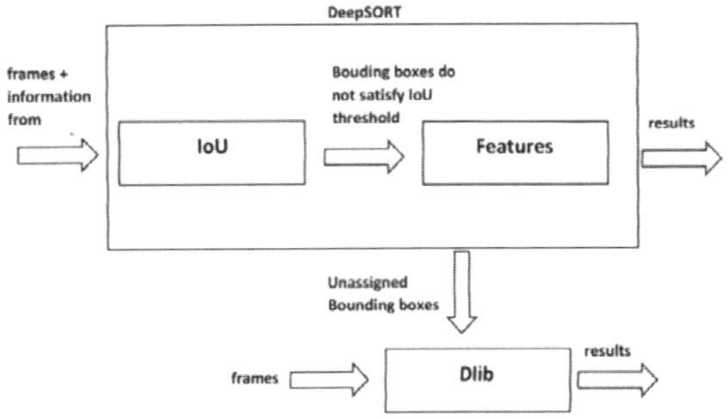

Fig11: DeepSort Architecture

2.6.2.2. YOLOv4

YOLO, as stated, stands for You Only Look Once, it is an object detection system in real-time that recognizes various objects in a single enclosure. Moreover, it identifies objects more rapidly and more precisely than other recognition systems. YOLOv4 outruns the existing methods significantly in both terms "detection performance" and "superior. The person tracker uses YOLO-v4 to perform person detection, which is used extensively for tracking.

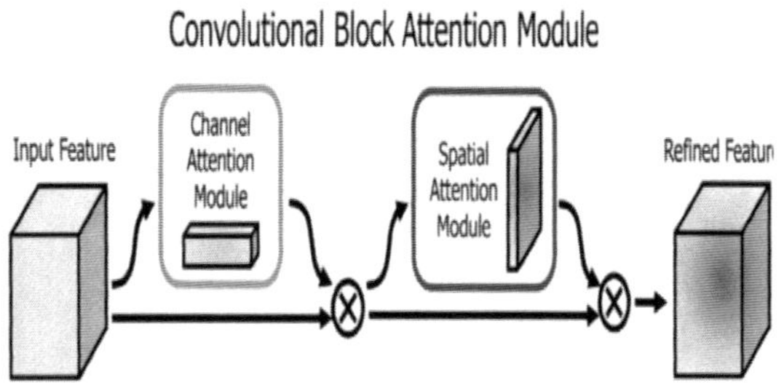

Fig12: How yolov4 framework processes images from video

There is a pre-trained YOLOv4 acquisition model that can accommodate up to 80 classes. To perform the tracking of an object using YOLOv4, we first convert the weights into a compatible TensorFlow model that will be stored in the test locations folder. After that all we need to do is run the script_ject_tracker.py script to use our object tracker with YOLOv4, Deep Sort and TensorFlow.

Fig13: creating bounding boxes around the humans in the frame

DeepSort implementation using Yolv4 is using the following parts: -

The Yolov4 gives us the Object detections and the original bounding boxes.

Kalman Filter: -It gives the bounding boxes of each person while tracking them using 8 parameters such as aspect ratio, height of the image etc. It helps us factor in the noise in detection and uses prior state in predicting a good fit for bounding boxes.

To verify our detections with the Kalman filter bounding boxes we have an association algorithm which can be quantified using a distance metric. The distance metric used here is the Hungarian algorithm. Deep learning is used by getting the appearance features of the person in a particular frame, since in a real-world situation these above methods are still not enough as they face problems such as occlusions and different viewpoints etc. This algorithm is used for multi camera multi target environments using TorchReid which helps in person reidentification in different videos.

2.7. Parrallelizers

These are the objects which help to optimize the performance of the system. There are various methods for such processes.

2.7.1. Multi-Threading

In this type of parallelism, there is a single thread for each of the camera video streams, it does all the jobs that are pipelined in the design flow of the system. This type of parallelism includes concurrently running threads, which have context switching and sharing of data enabled between them.

This type of parallelism ensures fast I/O work and less spawning time and resources for each thread. Since CPU resources are comparatively less, the overall process gets slowed down if multiple updates occur in the video processing part of the software.

2.7.2. Multi-Processing

In this type of parallelism, there is a single process for each of the camera video streams, it does all the jobs that are pipelined in the design flow of the system. This type of parallelism includes parallelly running processes on CPU/GPU cores, which have sharing of data enabled between them. This type of parallelism ensures less processing time for video processing, and since each process has comparatively. large amount CPU resources allocated; this helps in completing the video processing mechanisms such matrix multiplication inferences faster. Since CPU resources are comparatively more, this increases the spawn time required for each process and the I/O mechanism is also slowed down due to overall increased resource usage.

2.7.3. Multithreading coupled with Multiprocessing

In this Type of parallelism each video stream is assigned a thread for I/O operations (i.e., it opens and closes the stream) and each video processing jobs such as trackers, comparer, etc. are assigned individual processes for completing their jobs. Sharing is enabled among the internal processes for each stream and for among threads for different streams. This helps in smooth execution of both I/O as well as video processing operations. Message passing overhead increases this type of parallelism. Since a large number of execution units are created this may require more powerful hardware. These Parrallelizers help in scaling of the system for large facilities.

https://doi.org/10.1051/itmconf/20214003036

3. Main Results

We have implemented our system on a webapp and visualized its inference on the ip camera stream itself. The tracker tracks a person across different cameras and consistently persists the id of the person as well as the tag of recognized or unrecognized individual in the proximity of the multi-ca4mera system. Some results of the different combinations of the modules used are specified in the table below. The fps and the accuracy of the whole system is considered as an evaluation metric. We have used the following hardware specs for benchmarking purposes.

Graphics Card/GPU: NVIDIA GeForce GTX 1060

Processor: Ryzen 1700x

Motherboard: GIGABYTE AX 370

We have considered 4 video streams together currently; this can be easily extended to 6 with some negligible loss in fps and with a higher CPU and GPU can be extended to 10 video streams (the fps is expected to increase slightly)

Feature Extractor	**Compar-er**	**Tracke-r**	**Pers-on Per Fra-me**	**Parallel-izer**	**FP-S**	**Effici-ency**
Facenet/ MTCNN	Iterative	None	1-2	Multi thread	10-15	Good
Facenet/ MTCNN	Iterative	None	3-5	Multi thread	8-10	Medium
Facenet/ MTCNN	Iterative	None	3-5	Multi process	4-6	Bad
Facenet/ MTCNN	Annoy	None	3-5	Multi process	8-12	Medium
Facenet /MTCNN	Annoy	Deep Sort	3-5	Multi process	6-10	Medium
Facenet/ MTCNN	Hnswlib/ Frame Skipper	Deep Sort	3-5	Multi thread	10-20	Medium
TorchReid	Hnswlib/ Frame Skipper	Deep Sort	3-5	Multi thread	10-20	Good
TorchReid /Facenet MTCNN	Hnswlib/ Frame Skipper	Deep Sort	3-5	Multi thread	10-20	Best

Fig14: Benchmark of our system

Note*- The TorchReid module is relatively newly being added for experimentation in the system, more testing is required to give accurate analysis.

Fig15: Initial Setup screen

We have tested the results in public as well as private/Home environments, with a single camera on the top left corner of fig19 as the entry camera and all the others set as normal cameras. We have taken two positive targets, and two negative targets, which appear in the camera streams in regular intervals. They showed consistent results in the private environments and fluctuated in public spaces. We noticed uniform results according to the table in Fig17, though generally depending upon the consistent lighting of the given environment.

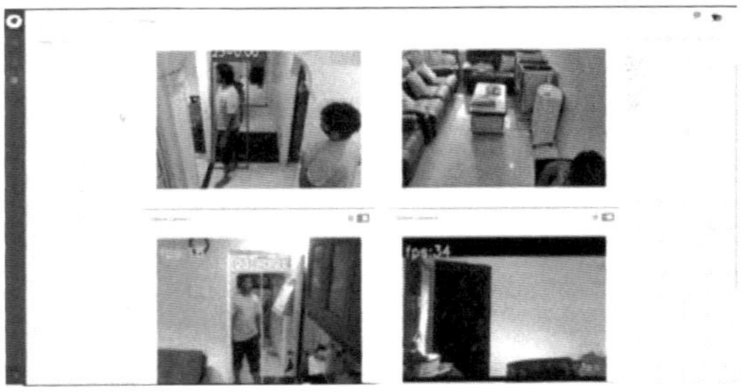

Fig16: Initialization of the system

Fig17: Green box and persistence of the person on multiple cameras

Fig18: Red Box and green box detection of negative as well as positive targets

4. Future Prospects

Since the results are experimental in nature, we continue to explore the combinations between different modules in our subsystem. The resultant table can be updated in the future. Our visualizing software could also be changed in the future (i.e., 2d map format), this would also have a significant impact on the fps, as it would generally increase in the absence of any extra creation of bounding boxes. In addition, we could easily deploy such a system on any cloud platform for a cheap subscription fee, so this can also be easily used in any practical situation in the future. Many applications can be derived using our proposed system such as attendance manager using computer vision, accurate footfall measurements in a distributed system environment, detection of crime scenes in multiple video streams, analysis of such crime scenes etc.

5. Conclusion

Detecting human beings accurately in a surveillance video is one of the major topics of vision research due to its wide range of applications. It is challenging to process the image obtained from a surveillance video as it has low resolution. A review of the available systems is also provided.

A proposed system and its design, flow and submodules are discussed, and various ways to increase the efficiency of the system are given to improve the human detection and tracking process in surveillance videos. These include adopting a new multi-view approach based on concentrating on developing an optimized system by experimenting on different combinations of the modules discussed.

6. Acknowledgement

Special thanks to Professor Ms. Rakhi Kalantari, Department of Computer Engineering, Fr. C. Rodrigues Institute of Technology, Vashi for her support and guidance.

7. References

This paper is citated from the below references:

[1] Design and Implementation of a Human Tracking CCTV System using IP-Cameras: https://www.researchgate.net/publication/328980042_Design_and_Implementation_of_a_Human_Tracking_CCTV_System_using_IP-Cameras

[2] Distributed multi-camera synchronization for smart-intruder detection: https: //ieeexplore.ieee.org/abstract/document/6314665.

[3] Paul, M., Haque, S.M.E. & Chakraborty, S. Human detection in surveillance videos and its applications - a review. EURASIP J. Adv. Signal Process. 2013, 176 (2013). https://doi.org/10.1186/1687-6180- 2013-176

[4] J. Celine and S. A. A, "Face Recognition in CCTV Systems," 2019 International Conference on Smart Systems and Inventive Technology (ICSSIT), Tirunelveli, India, 2019, pp. 111-116, doi: 10.1109/ICSSIT46314.2019.8987961.

[5] Todi, Manish. "Real-Time Machine Learning." Medium, Analytics Vidhya, 18 Oct. 2019, medium.com/analytics-vidhya/real-time-machine-learning-7aa55dafb2b.

[6] Wojke, N., Bewley, A., & Paulus, D. (2017). Simple Online and Realtime Tracking with a Deep Association Metric. In 2017 IEEE International Conference on Image Processing (ICIP) (pp. 3645–3649).

[7] Wojke, N., & Bewley, A. (2018). Deep Cosine Metric Learning for Person Re-identification. In 2018 IEEE Winter Conference on Applications of Computer Vision (WACV) (pp. 748–756).

[8] Facial recognition https: //github.com/BrandonJoffe/home_surveillance/tree/d5fd430dbc149b4 8a1424f44fc748c79ef971511

[9] Yolo motion detection: https://github.com/hunglc007/tensorflow-yolov4-tflite

[10] Xu Zhang, Xiangyang Hao, Songlin Liu, Junqiang Wang, Jiwei Xu, & Jun Hu (2019). Multi-target tracking of surveillance video with differential YOLO and DeepSort. In Eleventh International Conference on Digital Image Processing (ICDIP 2019) (pp. 701 – 710). SPIE.

[11] Zhou, K., & Xiang, T. (2019). Torchreid: A Library for Deep Learning Person Re-Identification in Pytorch. arXiv preprint arXiv:1910.10093.

[12] Zhou, K., Yang, Y., Cavallaro, A., & Xiang, T. (2019). Omni-Scale Feature Learning for Person Re-Identification. In ICCV.

[13] Zhou, K., Yang, Y., Cavallaro, A., & Xiang, T. (2019). Learning Generalisable Omni-Scale Representations for Person Re-Identification. arXiv preprint arXiv:1910.06827.

[14] The AIGuy. (2020) . yolov4-deepsort. https://github.com/theAIGuysCode/yolov4-deepsort.

[15] arunmandal53. (2020). facematch. https://github.com/arunmandal53/facematch.

[16] Mandal, Arun, et al. "MTCNN Face Detection and Matching Using Facenet Tensorflow." Pytorials.com, 16 Feb. 2018, www.pytorials.com/face-detection-matching-using-facenet/.

[17] Drake, Victoria. "Multithreaded Python: Slithering Through an I/O Bottleneck?" FreeCodeCamp.org, FreeCodeCamp.org, 15 Mar. 2021, www.freecodecamp.org/news/multithreaded-python/.

B. Amos, B. Ludwiczuk, M. Satyanarayanan,"Openface: A general-purpose face recognition library with mobile applications,"CMU-CS-16-118, CMU School of Computer Science, Tech. Rep., 2016

8. AUTHORS

Sureshkumar Jha –B.E. computers, Fr. C. Rodrigues Institute of Technology.

Rohan Sawant– B.E. computers, Fr. C. Rodrigues Institute of Technology.

Parth Shinde– B.E. computers, Fr. C. Rodrigues Institute of Technology

A Framework for Secure Storage and Sharing of Electronic Health Records using Blockchain Technology

Anurag Gharat[1,*], *Pratik* Aher[2,**], *Punit* Chaudhari[3,***], and *Bhavana* Alte[4,****]

[1]Ramrao Adik Institute Of Technology, Nerul, Navi Mumbai, India

Abstract. Public Healthcare issues have been the main focus of any governing body for the last decade. Each day a large amount of Healthcare data is generated by individual and Medical Organizations. This information is considered highly sensitive and private, which needed to be securely stored and protected against unauthorized access. In a traditional system, all the data is stored in a centralized system which comes with lots of drawbacks. This where blockchain technology comes in because of its Peer-To-Peer nature and Security. Several studies highlight the importance of blockchain and how it can be implemented in Electronic Medical Record(EMR) systems which face problems regarding privacy,security,decentralization and confidentiality.In this paper, We explain how blockchain technology may be utilised to improve EMR systems and how it could be a solution to these problems. We offer a framework for implementing blockchain technology in the healthcare industry for EMR.The aim of our proposed framework is first to implement blockchain technology for EHR and secondly to provide secure storage of electronic records by defining granular access rules for the users of the proposed framework. Moreover,this framework provides the EHR system with the benefits of a blockchain-based solution that is scalable, secure, and integrated.

1 Introduction

The worldwide Electronic Wellbeing Record (EHR) market is developing drastically and expected to reach ""39.7 billion by 2022[8]. To shield the security and protection of EHR, access control is a fundamental instrument for overseeing EHR information. Electronic clinical records (EMRs) are computerized records of patient's clinical history, medicines, actual assessments, medical problems, and so forth This data is viewed as profoundly delicate and private, which should be safely put away and secured against unapproved access. The medical care area confronted a pattern shift towards moving the current paper and archive based wellbeing record the board framework to advanced EHR frameworks that were intended to oversee electronic clinical records (EMR)[9]. These frameworks were utilized to store clinical notes and the research facility brings about its various parts. They were proposed to upgrade the well-being part of the patients by forestalling mistakes and expanding data access. The objective of EHR frameworks was to take care of the issues looked at by the paper-based medical care records and to give a productive framework that would change the condition of the medical care area. Even though utilization of EHR frameworks in the medical clinics or medical services was to improve the nature of medical services, these frameworks dealt with specific issues and didn't meet the assumptions related to them. These frameworks were hackable, information from such unified frameworks was effortlessly spilt and altered. Additionally, while this happened the proprietor had no admittance to it likewise had no clue about where his information is shared and who is utilizing it. Such frameworks couldn't give the proprietor the force and control to deal with his records. Utilizing a Blockchain controlled HIE can beat every one of the disadvantages of conventional HIE frameworks. While electronic health information exchange cannot replace provider-patient communication, it can greatly improve the completeness of patient's records, (which can have a big effect on care), as past history, current medications and other information is jointly reviewed during visits[11]. Also, appropriate and timely sharing of medical records will improve the decision-making process of Healthcare workers and also result in 1. Proper Medication 2. Avoid Medication Errors 3. Improved Security 4.Quality of Treatment. Such a system will benefit the patients by providing them with complete control of their data. He can add his records, update them and can grant access to anyone he wants to. He will have no limitation on the kind of data that has to be submitted. He can upload any format of data, whether an image, scanned prescription or a text-based brief history. Research and Analysis is another sector that can make use of such a platform effectively. They can request access for data from the owner through classified ways and as the data is always maintained by users themselves, the Analyst will always get true data that is always up to date. As all these transactions related to data will be logged directly on Blockchain,

*e-mail: anuraggharat55@gmail.com
**e-mail: ahe.pratik.17ce1014@gmail.com
***e-mail: punit.chau20@gmail.com
****e-mail: bhavana.alte@rait.ac.in

they will serve as hard and solid proof in future operations. A D-app (Decentralized App) based interface is implemented which utilizes Ethereum based blockchain. The addition of a new user to the Blockchain takes place through met mask ID. The user and doctor authentication along with other non-sensitive data are stored on a MongoDB cloud database. The D-app interacts with the Blockchain by adding a new Patient record, adding a new Healthcare worker in the access list and revoking access from the healthcare worker when the user wants. For the scope of our project, the Blockchain is tested on a local test net and the patient's medical records are stored on secure off-chain storage. In our case, we use InterPlanetary File System which is a peer-to-peer network for storing and sharing data in a distributed file system. Ganache is used to get Ethereum accounts pre-funded with ether for testing purposes. Metamask is a Chrome Plugin used to access the D-app. Ganache, Metamask and Ethereum are integrated using Truffle Framework. Proof of Work (POW) is a consensus algorithm used and SHA-256 hashing is used for creating transaction hashes. The Blockchain logic is written in Solidity programming language and tested on Remix IDE.

The remainder of this paper is arranged as follows. Section 2 will highlight our motivation while working on this project. We present related work and blockchain-based EHR studies in section 3. Section 4 is subdivided into our proposed work and the methodology we have used. In section 5, we present our system design. Results and the final output is described in section 6. In section 7, we have mentioned the conclusion and future work. Finally, we added the references in the last section.

2 MOTIVATION

Improve Transferability: Many patients have to change their doctors frequently due to conflicting schedules and high demand for doctors. Thus if all patients previous EMR are available his transfer from one doctor to another becomes easier.

Help Genuine Patients: Many impersonators use EMR records of unsuspecting patients for insurance frauds. This leads to genuine people not benefiting from insurance policies. Thus it will ensure that genuine people get benefits.

Improve Security: Blockchain renders records untameable due to hashing and thus it increases the security of records.

Ownership: The Healthcare provider can only access a patient's medical data with his consent. Hence the patient can be assured that his sensitive information cannot be accessed by any unknown entity.

Improve Healthcare: There have been so many cases where a patient receives an improper treatment, because of the old stale data that resides on a database which no one cares to update. In our approach, the healthcare worker will get a complete updated copy of data along with medical history.

3 RELATED WORK

The study of literature contains different principles and techniques from research papers that are used in system development. The study undertaken for the implementation of the intended framework involves an overview of different current applications and research papers. Such papers are being reviewed and analysed as part of the project's study survey.

3.1 A. Existing Systems

Zhang et al.[1] have given an analysis about which metrics and parameters should be considered while developing Blockchain Healthcare based Decentralized apps. They evaluated these metrics based on domain and technical perspectives.

Kumar et al.[2] mentioned the key requirements and challenges faced by Healthcare systems having Blockchains. They have also listed the applications in which Blockchain could be used.

Faisal Jamil et al.[3] proposed a way to store data on blocks for monitoring purpose. Data is directly stored in blocks in this approach.

Shahnaz et al.[4] mentions usage of IPFS in blockchain for maintaining records. Their proposed system works on role based architecture where a administrator assigns roles to users.

Vora et al.[5] have proposed system with combination of privacy preserving schemes. Here encryption of patients private data stored on Blockchain.

S. Ali et al.[6] proposed work is on permissioned blockchain . The paper highlights various areas in healthcare that can be benefitted with the use of Blockchain. The proposed methodology used DHT (Distributed Hash Tables) for access stored records on Blockchain.

Ehab Zaghoul et al.[7] has listed all the use cases of Blockchain in Healthcare. The paper presents ideas how using an offchain storage can be cost efficient. The paper compares all approaches in designing blockchain on cost basis.

4 PROPOSED APPROACH

4.1 Proposed Work

The Electronic Medical Records Management System will be built on public Blockchain platform Ethereum. It will be built with Truffle web framework which compiles the smart contracts and injects them into the Web App. All the users would necessarily have a Metamask account, which can be created using a installable chrome plugin. The Metamask plugin will allow us to connect to local Ethereum network and provide us account address to sign transactions. All the transaction fees will be paid using Metamask account. As our system is developed on a local test network it uses ganache for users account prefunded with 100 ethers. A New user can sign up on the blockchain using his Metamask Account. The D-APP is built on

MERN Stack. The entire front-end is developed using React and Backend is managed by Node-Express server. All the profile details which are non-sensitive will be stored on a MongoDB database. Once the successfully logs inside the D-APP he can upload his/her medical records in form of images. During upload process he will be needed to sign the transactions using his private key. The user documents are uploaded on an Off-chain storage securely. IPFS (Inter Planetary File System), a distributed file system, is used as an Off-chain storage. Once the records are uploaded on IPFS we store the hash of address of the stored documents on blocks of blockchain. So every time a new block is added on the blockchain, the IPFS hash of the record will be stored on the blockchain rather than the data. Hence, using this technique we reduce the space needed on blocks, which ultimately reduces the cost of every transaction.

4.2 Proposed Methodology

The EMR management system consists of three main parts:

1. IPFS (Inter Planetary File System): IPFS is a distributed system for storing and accessing files, websites, applications, and data. IPFS is a file sharing system that can be leveraged to more efficiently store and share large files. It relies on cryptographic hashes that can easily be stored on a blockchain. In IPFS, instead of addressing the content by the server it is stored on, it is address by the content itself from the nearby peer. This is necessary, if sensitive or personal data needs to be shared.

2. Smart Contracts: Smart contracts are actually similar to business contracts, all things considered. From Blockchain viewpoint, They add rationale to the Blockchain. Smart contracts are self-executing contracts containing the terms and states of an understanding among peers. The terms and states of the understanding are composed into code. The Smart contracts executes on the Ethereum blockchain's decentralized stage. The arrangements work with the trading of cash, offers, property, or any resource. In Layman terms, Smart contract is basically a program that runs on the Ethereum blockchain. It's an assortment of code (its capacities) and information (its express) that dwells at a particular location on the Ethereum blockchain. In our proposed procedure, Smart contract are utilized to transfer records, award admittance to medical care specialist to see your records just as repudiate access from medical care laborer..

3. Proof of Work (PoW): Proof of Work is a kind of consensus algorithm. As the name says for approving/validating an exchange a node ought to openly demonstrate that it did a specific measure of work. In Blockchain, this calculation is utilized to affirm exchanges and produce new squares to the chain. With POW, excavators go up against one another to finish exchanges on the arrange and get remunerated. Ethereum, as Bitcoin, at present uses an agreement convention called Proof-of-work (POW). This permits the hubs of the Ethereum organization to concede to the condition of all data recorded on the Ethereum blockchain, and forestalls specific sorts of financial attacks.

5 SYSTEM DESIGN

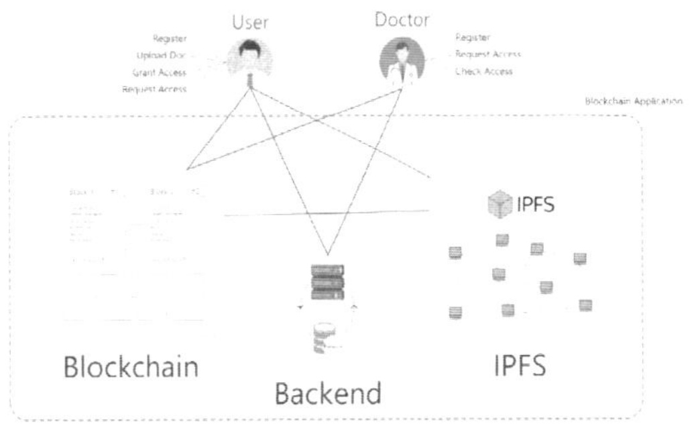

Figure 1. System Design

As shown in figure 1, doctors and patients act as entities or nodes in this blockchain network. Both the entities have their own login and signup pages. All their authentication details are processed on Node-Express server and then sent to MongoDB Cloud Database. Figure 2, shows the flow diagram for user and Figure 3 shows the flow diagram for doctors.

• The Patient User will need a Ethereum account using Metamask plugin to be a part of the Blockchain.

• Once done he can visit the signup page and enter his personal details and credentials. After passing all the authentication checks he must login into his profile.

• Once logged in he can upload his records in image format to the blockchain. For uploading he must pay some ethers in order to complete the transaction.

• There are several subpages created in user profile such as:

Dashboard: The dashboard will display all the records the user has uploaded till now.

Profile page: This page shows all his personal details mentioned during signup process.

Doctors Page: This page displays a list of all doctors who are a part of Blockchain. Here he can grant access to anyone after checking his profile.

Requests Page: This page has all the requests sent by doctors to access his medical records. He can either delete these requests or accept them.

Access List Page: This page mentions the list of all the Doctors that have access to view his records. He can revoke their access from here.

• When the user uploads his record, the record is sent to IPFS network which in turn returns us hash of the uploaded asset. We store this hash on the blockchain along

with other necessary details.
- The doctor needs to go through the same procedure of signup and then signin, once he enters in his profile he will be able to see the records if the user has granted him the permission. Else the dashboard will be empty.
- He can go to users page and find the user whose records he wants to check. He can request a grant access from the user, his request will go through the server and will be sent to the specific users dashboard.
- If he has access he can view the records from his dashboard.

Figure 2. Flow Diagram for user

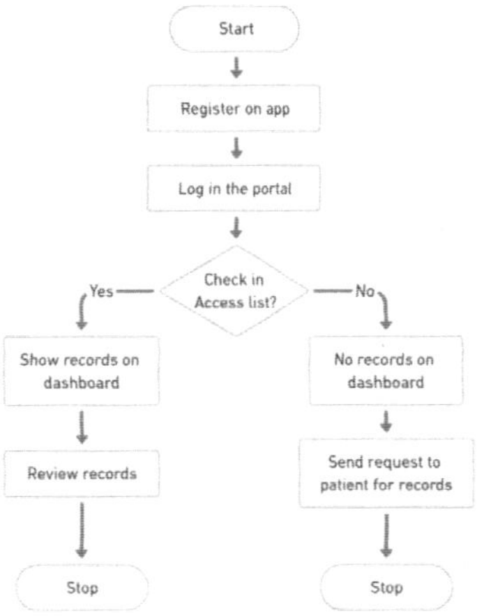

Figure 3. Flow Diagram for doctor

6 RESULTS

Figure 4 shows user dashboard where he can upload his reports as well as visit various sub-pages

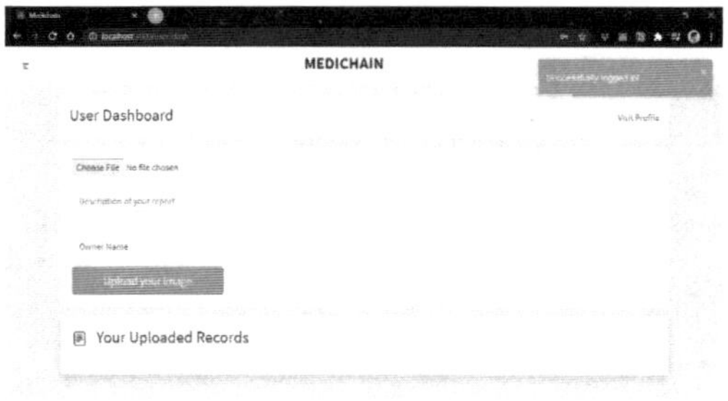

Figure 4. User Dashboard

User is uploading his record to IPFS and Blockchain is verifying the user by requesting his digital signature figure 5. The user pays some ethers to mine the block and save his record.

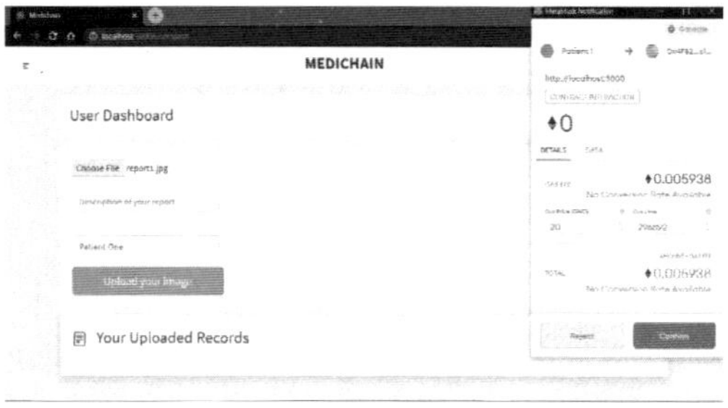

Figure 5. Transaction

Figure 6 shows how user provides Access to Doctor by checking his profile. While providing access he must sign the transaction using his private key for authenticity. Once done the doctors address will be added to user's access list.

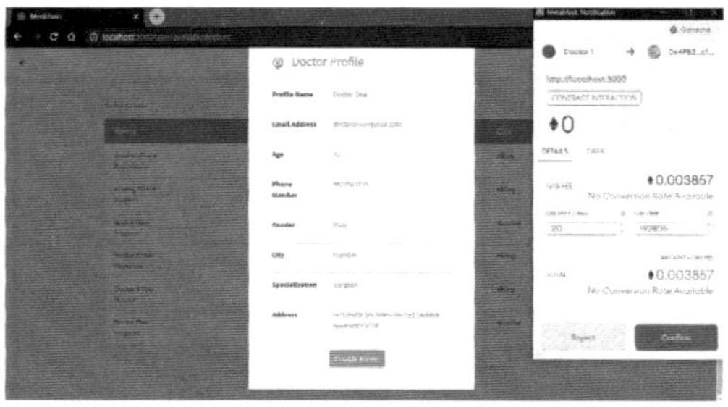

Figure 6. Providing Access to Doctor

On Figure 7 the user will have record of all the access provided by him along with timestamp. He can revoke access from any Doctor from here.

Figure 7. Access list

Figure 8 shows doctors dashboard where all the available records will be present. If he has no access, no records will be displayed.

Figure 8. Doctors dashboard

The Doctor from his profile can visit user's profile and send him a request to access his records figure 9. These requests will be displayed on User's Dashboard which user can either accept or reject.

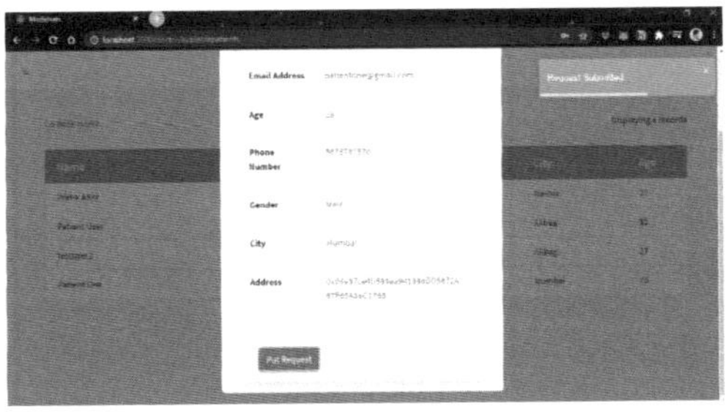

Figure 9. Patient's List

6.1 Result Analysis

Monetary Benefits: The cost of mining each block on a Blockchain is directly proportional to the content or variable data added on the block. Using the off-chain storage functionality to store images we were able to reduce the size of data added on blocks drastically, hence reducing the cost to mine each block.

Decentralization: The proposed Decentralized system has the power to overcome all the drawbacks that the traditional Centralized system possesses. Using Decentralization we are removing a central governing authority and distributing the trust among the peers of the network. Due to the secure nature of Blockchain, the chances of hacking or tampering is almost negligible.

Data Assets Sharing: Previously proposed systems displayed the sharing of only text-based data among the Blockchain peers. Whereas this implemented system allows to share and store asset-based records and data among peers. More types of data assets can be added in further versions.

User Authentication and other functionalities: All the studied Blockchain systems were more inclined towards using Blockchain and did not focus on other important aspects like - How will user Login to Portal?, How will the System flow work? How can he find suitable doctors? , How will the doctors request data? All these important aspects were studied and a suitable backend was added which would do all such tasks with ease. The Backend works as a supporting system to the Blockchain.

Record Ownership: One of the most important drawbacks of the Traditional System is handling the ownership of records. The proposed system provides the ownership of data in the hands of the user. The user himself is responsible to handle his data and maintain it instead of any other third party. Hence, reducing the risk of Illegal data leaks and sharing.

7 Conclusion and Future Work

Our current EMR management systems fails to maintain the secrecy and confidentiality of sensitive reports. With Blockchain based EMRs patients can play a more active role in the healthcare, and managing their medical and health data. This systems give individuals the complete control and hold on his/her personal data. As he no longer has to trust his Healthcare providers for keeping his data and there will be reduce in risk of sharing of data to untrusted parties. The proposed system succeeds in providing the power in the hands of the owner itself. Hence he can use, share and maintain his/her records as per his needs. It provides a far more superior ability to share and manage personal healthcare records than any traditional system. This system is scalable, hence adding new entities to the system is possible. Currently the system is deployed on localhost network using Ganache and Metamask. This is because our scope is limited to making a prototype of the real world system as deploying on the main Ethereum network requires real monetary transactions. But the system can be deployed on main network in future.

In future works we plan to add even more functionalities in the smart contracts like an Access Timer, Control on which documents to share, Edit/Update control to Doctors. Also the current system is limited to two entities namely Patients and Doctors. We plan to add more

 https://doi.org/10.1051/itmconf/20214003037

entities like Pharmacist, Insurance Providers and Medical Researchers. The release of Hyperledger Fabric v.2.0 has opened a new scope for us to try similar approach using a Private Blockchain.

References

[1] Zhang, P., Walker, M. A., White, J., Schmidt, D. C., Lenz, Metrics for assessing blockchain-based healthcare decentralized apps, (2017)

[2] Kumar, T.; Ramani, V.; Ahmad, I.; Braeken, A.; Harjula, E.; Ylianttila, M., Blockchain Utilization in Healthcare: Key Requirements and Challenges, (2018)

[3] Faisal Jamil, Shabir Ahmad , Naeem Iqbal and Do-Hyeun Kim, Towards a Remote Monitoring of Patient Vital Signs Based on IoTBased Blockchain Integrity Management Platforms in Smart Hospitals, (2020)

[4] Shahnaz, A., Qamar, U., Khalid, A.,Using Blockchain for Electronic Health Records., (2019)

[5] Vora, J., Nayyar, A., Tanwar, S., Tyagi, S., Kumar, N., Obaidat,S., Rodrigues, J.A blockchain-based framework for securing electronic health records., (2018)

[6] S. Ali, G. Wang, B. White and R. L. Cottrell, A Blockchain-Based Decentralized Data Storage and Access Framework for PingER, (2018)

[7] E. Zaghloul, T. Li, M. Mutka and J. Ren, d-MABE: Distributed Multilevel Attribute-Based EMR Management and Applications, (2020)

[8] TechSci Research, Global Electronic Health Records (EHR) Market (2020 to 2025), (2020)

[9] Peter Garrett, Joshua Seidman, EMR vs EHR – What is the Difference?, (2011)

[10] Idrees, S.M.; Nowostawski, M.; Jameel, R.; Mourya, A.K, Security Aspects of Blockchain Technology Intended for Industrial Applications, (2021)

[11] Professor Wilkins, Electronic health information exchange, (2019)

Cyber Bullying Detection on Social Media using Machine Learning

Aditya Desai[1], *Shashank* Kalaskar[2], *Omkar* Kumbhar[3], and *Rashmi* Dhumal[4]

[1,2,3]Student, Department of Computer Engineering, Ramrao Adik Institute of Technology, Nerul
[4]Assistant Professor, Department of Computer Engineering, Ramrao Adik Institute of Technology, Nerul

Abstract. Usage of internet and social media backgrounds tends in the use of sending, receiving and posting of negative, harmful, false or mean content about another individual which thus means Cyberbullying. Bullying over social media also works the same as threatening, calumny, and chastising the individual. Cyberbullying has led to a severe increase in mental health problems, especially among the young generation. It has resulted in lower self-esteem, increased suicidal ideation. Unless some measure against cyberbullying is taken, self-esteem and mental health issues will affect an entire generation of young adults. Many of the traditional machine learning models have been implemented in the past for the automatic detection of cyberbullying on social media. But these models have not considered all the necessary features that can be used to identify or classify a statement or post as bullying. In this paper, we proposed a model based on various features that should be considered while detecting cyberbullying and implement a few features with the help of a bidirectional deep learning model called BERT.

Keywords: Cyberbullying, Social Media, BERT, NLP, Semi-supervised learning , Twitter API.

1 Introduction

Millions of young people spend their time on social networking, and the sharing of information is online. Social networks have the ability to communicate and to share information with anyone, at any time, and in the number of people at the same time. There are over 3 billion social media users around the world. According to the National Crime Security Council (NCPC), cyberbullying is available online where mobile phones, video game apps, or any other way to send or send text, photos, or videos deliberately injure or embarrass another person. Cyberbullying can happen at any time all day, week and you can reach anyone anywhere via the internet. Text, photos, or videos of cyberbullying may be posted in an undisclosed manner. It can be difficult, and sometimes impossible, to track down the source of this post. It was also impossible to get rid of these messages later. Several social media platforms such as Twitter, Instagram, Facebook, YouTube, Snapchat, Skype, and Wikipedia are the most common bullying sites on the internet. Some of the social networking sites, such as Facebook, and the provision of guidance on the prevention of bullying. It has a special section that explains how to report cyber-bullying and to prevent any blocking of the user. On Instagram, when someone shares photos and videos made by the user to be uncomfortable, so the user can monitor or block them. Users can also report a violation of our Community and make Recommendations to the app.

As the social lifestyle exceeds the physical barrier of human interaction and contains unregulated contact with strangers, it is necessary to analyze and study the context of cyberbullying. Cyberbullying makes the victim feel that he is being attacked everywhere as the internet is just a click away. It can have mental, physical, and emotional effects on the victim. Cyberbullying mainly takes place in the form of text or images on social media. If bullying text can be distinguished from non-bullying text, then a system can act accordingly. An efficient cyberbullying detection system can be useful for social media websites and other messaging applications to counter such attacks and reduce the number of cyberbullying cases. The objective of the cyberbullying detection system is to identify the cyberbullying text and also take its meaning into consideration. One first analyzes the various aspects of a particular text and then applies the previous information or visuals to find the context of the text. There is a need to create a personalized system that can access such a text effectively and efficiently.

2 Literature Survey

M. Di Capua, et al. [1] proposes an unsupervised approach to develop a cyberbullying model based on an amalgam of features, based on traditional textual features as well as some "social features". The features were separated into 4 categories as Syntactic features, Semantic features, Sentiment features, and Social features. The author used a Growing Hierarchical Self Organizing Map (GHSOM) network, with a grid of 50 x 50 neurons and 20 features as the input layer. M. Di Capua, et al have applied the clustering algorithm k-means to classify the input dataset along with GHSOM on the Formspring dataset. The results of this hybrid unsupervised methodology surpassed the previous

* Corresponding author: adityadesai1703@gmail.com

results. The author then tested the youtube dataset with 3 different Machine Learning Models: a Naive Bayes Classifier, Decision Tree Classifier(C4.5), and a Support Vector Machine(SVM) with a Linear Kernel. It was observed that clustering results for the hate posts turned out to have a lower precision in the youtube dataset when compared to the FormSpring tests, as textual analysis and syntactical features perform differently on both sides. When this hybrid approach was applied to the Twitter dataset, it resulted in a weak recall and F1 Score. The model proposed by the authors can be improved and used in building constructive applications to mitigate cyberbullying issues.

J. Yadav, et al.[2] proposes a new approach to cyberbullying detection in social media platforms by using the BERT model with a single linear neural network layer on top as a classifier. The model is trained and evaluated on the Formspring forum and Wikipedia dataset. The proposed model gave a performance accuracy of 98% for the Form spring dataset and of 96% for the Wikipedia dataset which is relatively high from the previously used models. The proposed model gave better results for the Wikipedia dataset due to its large size g without the need for oversampling whereas the Form spring dataset needed oversampling.

R. R. Dalvi, et al.[3] suggests a method to detect and prevent Internet exploitation on Twitter using Supervised classification Machine Learning algorithms. In this research, the live Twitter API is used to collect tweets and form datasets. The proposed model tests both Support Vector Machine and Naive Bayes on the collected datasets. To extract the feature, they have used the TFIDF vectorizer. The results show that the accuracy of the cyberbullying model based on the Support Vector Machine is almost 71.25% that is better than the Naive Bayes which was almost 52.75%.

Trana R.E., et al. [4] goal was to design a machine learning model to minimize special events involving text extracted from image memes. The author has compiled a database containing approximately 19,000 text views published on YouTube. This study discusses the effectiveness of the three machine learning machines, the Uninformed Bayes, the Support Vector Machine, and the convolutional neural network used on the YouTube database, and compares the results with the existing Form databases. The authors further investigated algorithms for Internet cyberbullying in sub-categories within the YouTube database. Naive Bayes surpassed SVM and CNN in the following four categories: race, ethnicity, politics, and generalism. SVM has passed well with the inexperienced Naïve Bayes and CNN in the same gender group, and all three algorithms have shown equal performance with central body group accuracy. The results of this study provided data that can be used to distinguish between incidents of abuse and non-violence. Future work could focus on the creation of a two-part segregation scheme used to test the text extracted from images to see if the YouTube database provides a better context for aggression-related clusters.

N. Tsapatsoulis, et al. [5] a detailed review of cyberbullying on Twitter is presented. The importance of identifying different abusers on Twitter is given. In the paper, various practical steps required for the development of an effective and efficient application for cyberbullying detection are described thoroughly. The trends involved in the categorization and labeling of data platforms, machine learning models and feature types, and case studies that made use of such tools are explained. This paper will serve as an initial step for the project in Cyberbullying Detection using Machine learning.

G. A. León-Paredes et al.[6] have explained the development of a cyberbullying detection model using Natural Language Processing (NLP) and Machine Learning (ML). A Spanish cyberbullying Prevention System (SPC) was developed by applying machine learning techniques Naïve Bayes, Support Vector Machine, and Logistic Regression. The dataset used for this research was extracted from Twitter. The maximum accuracy of 93% was achieved with the help of three techniques used. The cases of cyberbullying detected with the help of this system presented an accuracy of 80% to 91% on average. Stemming and lemmatization techniques in NLP can be implemented to further increase the accuracy of the system. Such a model can also be implemented for detection in English and local languages if possible.

P. K. Roy, et al. [7] detail about creating a request for the discovery of hate speech on Twitter with the help of a deep convolutional neural network. Machine learning algorithms such as Logistic Regression (LR), Random Forest (RF), Naïve Bayes (NB), Support Vector Machine (SVM), Decision Tree (DT), Gradient Boosting (GB), and K-nearby Neighbors (KNN) has been used to identify tweets related to hate speech on Twitter and features have been removed using the tf-idf process. The best ML model was SVM but it managed to predict 53% hate speech tweets in a 3: 1 dataset to test the train. The reason behind the low prediction scale was the unequal data. The model is based on the prediction of hate speech tweets. Advanced forms of learning based on the Convolutional Neural Network (CNN), Long-Term Memory (LSTM), and their Contextual LSTM (CLSTM) combinations have the same effects as a separate distributed database. 10-fold cross-validation was used along with the proposed DCNN model and obtained a very good recall rate. It was 0.88 for hate speech and 0.99 for non-hate speech. Test results confirmed that the k-fold cross-validation process is a better decision with unequal data. In the future, the current database can be expanded to achieve better accuracy.

S. M. Kargutkar, et al. [8] had proposed a system to give a double characterization for cyberbullying. The system uses Convolutional Neural Network (CNN) and Keras for content examination as the relevant strategies at that time provided a guideless view with less precision. This research involved data from Twitter and YouTube. CNN accuracy was 87%. In-depth learning-based models have found their way to identify digital harassment episodes, they can overcome the imprisonment of traditional models, and improve adoption.

Jamil H. et al. [9] have described the implementation of a new social network model and its query language called GreenShip. They showed that with the support devices, GreenShip users can be more effective in the fight against online bullying, and loss of traditional, online, and social networks. The reputation of a management model that has been introduced to restrict access to harmful information, which focuses on the denial of the criminal code, the means for the dissemination of information to the users associated with the target. GreenShip has a reputation as a model that provides safe, "green" friends, due to the recognition of the different types of friendships on Facebook. The damage is as a result of bad friends and that was very limited, and the more complex, that there are many forms of friendship, and the communication lines are put aside for the sake of the benefits of privacy and control.

Rasel, Risul Islam, et al. [10] focuses on the removal of the comments made on social networks, and the analysis of the question as to whether these observations provide an offensive meaning. The reactions can be divided into three categories: offensive, hate speech, and neither of the two. The proposed model classifies the notes on the species), with an accuracy of more than 93%. Latent Semantic Analysis (LSA) has been used as a feature selection method to reduce the size of the input data. In addition to standard feature extraction methods such as tokenization, N-gram, TF-IDF was applied to detect the important notes. We made three different machine learning models, Random Forest, Logistic Regression, and Support Vector Machines (SVMs) to perform the calculation, analysis, forecasting, and a teasing comment.

3 Proposed Methodology

In this paper, a method to detect cyberbullying on social media is proposed that is not just based on the sentimental analysis but also considers the syntactic, semantic, and sarcastic nature of the sentence before classifying it as hate speech. To achieve our goal we start with the traditional sentiment analysis where we perform contextual mining of text to identify and extract the subjective information in the source material to understand the opinion, emotion, or attitude towards the topic. Later we introduce a group of "social" features that can highly affect and guide the cyberbullying detection process. We have divided all the features we have extracted into five categories:

- Sentimental Features
- Sarcastic Features
- Syntactic Features
- Semantic Features
- Social Features

All these features have been categorized based on the literature survey of the existing systems and each feature uniquely identifies the text. Choosing informative, descriptive, and independent features is a crucial stage for the effectiveness of the algorithms in pattern recognition and classification problems.

In sentimental features, we try to evaluate the sentiment(positive, negative) of a given text document. The research shows that human analysts tend to agree around 80-85% of the time and that is the baseline we have tried to consider while training our sentiment scoring system.

In sarcastic features, we try to consider the context incongruity. Incongruity occurs when a nonverbal behavior contradicts a person's word. A text may contain half of the objects in a congruent context which can be considered as expected context, whereas for the other half, objects were embedded in incongruent contexts. This can be a major factor in cyberbullying detection because the hidden nature of the sarcastic comment won't be detected in sentiment analysis because of the context incongruity. We also consider pragmatic features like emojis, mentions, etc. while detecting the sarcastic nature of the source material.

While considering the syntactic features we have identified in the lists of insults, we also monitor and take into consideration the number of such bad words or insults present in a single sentence and accordingly map a density to it. We have also validated the badness of the entire sentence based on certain parameters like density range. The emphasis of uppercase characters while making hate statements is also taken into consideration while generating syntactic features because it can be referred to as an act of shouting or attacking over social media platforms. Similarly, the use of special characters or patterns formed by them is also brought into consideration while deriving syntactic features.

Semantic Features can be used to determine the lexical relation which exists between two words in a language. The meaning of the word can be represented by Semantic features. Here we have tried to identify the trigram and the bigrams that occur while referencing something in the text format. Here usually the negation of the sentence is considered along with the mapping of different pronouns that can be implicitly or explicitly used to refer to another individual while harassing someone over social media.

Social features refer to the social behavior of the victim or the bully itself. The post itself won't be sufficient to detect the nature of the text. We have considered patterns in the behaviors of the bullies and identified a few features. We have considered the direct tagging of the victim while using hate speech. We also try to gain information regarding the context of the post based on the previous interactions between the bully and the victim. Profiling of the author can be done to discover its past interactions and involvement in similar malicious activities over social media platforms.

We proposed a cyberbullying detection model based on transformers. Similar to RNN, transformers can also be used to solve a wide variety of NLP(Natural Language Processing) problems like translation and text summarization as they can take sequential data as input. A recent improvement on the natural language task introduced the BERT. The BERT is a recent paper published by researchers at Google AI Language. BERT

 https://doi.org/10.1051/itmconf/20214003038

stands for Bidirectional Encoder from Transformers. It is a bidirectional model that is pre-trained on unlabeled texts from both left and right directions to understand the meaning of both contexts. BERT is a powerful model for NLP tasks because of the use of semi-supervised learning. We can use this model to create a state-of-the-art machine learning model for a specific task by applying an additional task-specific layer on top of the BERT architecture. BERT is a Bidirectional model which means it aims to understand the meaning of the word from both the left and the right context to derive a better meaning during the training phase.

We saw a **bat**.
This **bat** was given to me by my father.

Here in the first sentence if we focus on the context of the underlined word "bat" from the left till the word, it refers to the nocturnal animal. Whereas if we focus on the context of the word "bat" in the second sentence from right till the word it refers to the bat from the game of cricket. Thus a machine can face problems predicting the actual meaning of the word without considering both the context. This problem is solved by BERT as it is a bidirectional model.

BERT model requires its input in a preprocessed form as per the rules made by its developer. These rules have helped the model to achieve better performances. All inputs are embedded as a combination of the other 3 embeddings and given as an input to the model:

- Position embedding: BERT reads and uses existing embedding to express word order in a sentence.
- Segment Embedding: BERT can also take more than one sentence as input functions. It uses this embedding to understand the difference between two different sentences.
- Token Embeddings: This is the embedded text token from Word Piece token vocabulary.

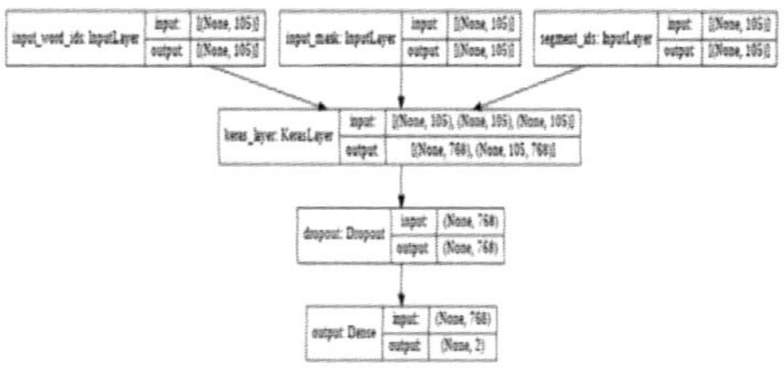

Fig.1. BERT model based on sentiment analysis

The Fig.1 depicts the BERT model we developed for sentimental analysis. The three separate embeddings are summarized together to give input to the internal layers of the model. The Fig.2 demonstrates the flow of the sentimental analysis process. To begin with, the final CLS token will output a matrix of hidden size. Furthermore it will be passed to a classifier layer. In conclusion the classifier layer will determine the sentiment of the input text.

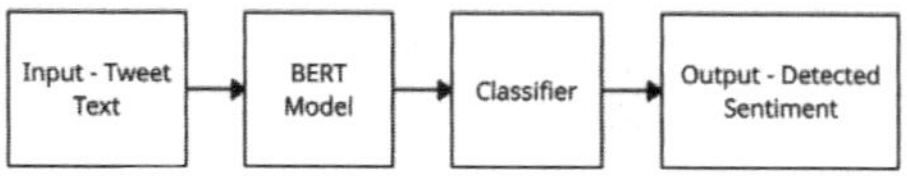

Fig.2. BERT model flow chart based on sentiment analysis

4 Result Analysis

The Fig.3 represents the input processing and prediction result that we performed during our testing. We used a tweet from Twitter with the trace of bullying and applied it to our model. Fig 4 shows the classification report based on our testing data. Here labels 0 and 1 represent Bullying and Non-Bullying respectively. The Fig.5 represents the confusion matrix based on the result of our testing data. Table 1 represents the accuracy of the SVM and Naive Bayes that is 71.25% and 52.70% respectively, when applied on the same dataset from [3]. Table 2 represents the accuracy of the BERT model on the same dataset. The result shows better accuracy when using the BERT model for sentiment analysis on the Twitter dataset. Our proposed model gave a better accuracy of 91.90% when applied to the Twitter dataset for the sentimental analysis which can be considered as a greater result when compared to the traditional machine learning models used on similar datasets.

Input preprocessing

```
[ ] tweet = ['And why do you care what I say???, lol looks like you have a crush you creepy Fag.']
    inputs = bert_encode(string_list=list(tweet),
                         tokenizer=tokenizerSaved,
                         max_seq_length=105)
```

Prediction

```
[ ] prediction = model.predict(inputs)
    print(prediction)
    print('Tweet is', 'Non-Bullying' if encoder.classes_[np.argmax(prediction)]==1 else 'Bullying')

    [[0.93678564 0.06321441]]
    Tweet is Bullying
```

Fig.3. Implementation Result

	precision	recall	f1-score	support
0	0.65	0.57	0.61	84
1	0.74	0.80	0.77	129
accuracy			0.71	213
macro avg	0.69	0.68	0.69	213
weighted avg	0.70	0.71	0.71	213

Fig.4. Classification report

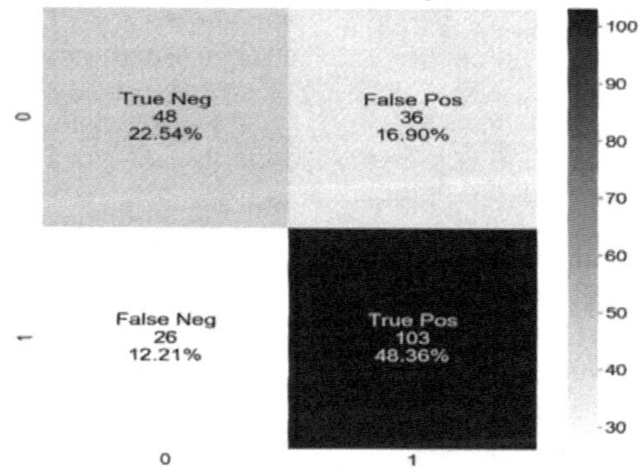

Fig. 5. Confusion Matrix

Table 1. Accuracy of SVM and Naive Bayes from [3]

Classifier	Accuracy in percentage
Naïve Bayes Classifier	52.70
Support Vector Machine	71.25

Table 2. Accuracy of BERT Model

Classifier	Accuracy in percentage
Pre-Trained BERT (testing)	70.89
Pre-Trained BERT (training)	91.90

5 Conclusion

We proposed a semi-supervised approach in detecting cyberbullying based on the five features that can be used to define a cyberbullying post or message using the BERT model. While considering just one of the features which was sentimental features the BERT model achieved 91.90% accuracy when trained over dual cycles which outperformed the traditional machine learning models. The BERT model can achieve more accurate results if provided with a large dataset. We can try to achieve even better results in the cyberbullying detection process if we consider all the features that we have proposed in this research paper. Based on all the features an application can be created to detect the bullying traces and thus help in detecting and reporting such posts. A combination of other models on top of the BERT model can also be used in the future to create a state-of-the-art model for the specific NLP tasks in detecting cyberbullying.

References

1. M. Di Capua, E. Di Nardo and A. Petrosino, *Unsupervised cyberbullying detection in social networks, ICPR*, pp. 432-437, doi: 10.1109/ICPR.2016.7899672. (2016)
2. J. Yadav, D. Kumar and D. Chauhan, *Cyberbullying Detection using Pre-Trained BERT Model*, ICESC, pp. 1096-1100, doi: 10.1109/ICESC48915.2020.9155700. (2020)
3. R. R. Dalvi, S. Baliram Chavan and A. Halbe, *Detecting A Twitter Cyberbullying Using Machine Learning*, ICICCS, pp. 297-301, doi: 10.1109/ICICCS48265.2020.9120893. (2020)
4. Trana R.E., Gomez C.E., Adler R.F. (2021) *Fighting Cyberbullying: An Analysis of Algorithms Used to Detect Harassing Text Found on YouTube*. In: Ahram T. (eds) Advances in Artificial Intelligence, Software and Systems Engineering. AHFE 2020. Advances in Intelligent Systems and Computing, vol **1213**. Springer, Cham. https://doi.org/10.1007/978-3-030-51328-3_2. (2020)
5. N. Tsapatsoulis and V. Anastasopoulou, *Cyberbullies in Twitter: A focused review*, SMAP, pp. 1-6, doi: 10.1109/SMAP.2019.8864918. (2019)
6. G. A. León-Paredes et al., *Presumptive Detection of Cyberbullying on Twitter through Natural Language Processing and Machine Learning in the Spanish Language*, CHILECON pp. 1-7, doi: 10.1109/CHILECON47746.2019.8987684. (2019)
7. P. K. Roy, A. K. Tripathy, T. K. Das and X. -Z. Gao, *A Framework for Hate Speech Detection Using Deep Convolutional Neural Network*, in IEEE Access, vol. **8**, pp. 204951-204962,, doi: 10.1109/ACCESS.2020.3037073. (2020)
8. S. M. Kargutkar and V. Chitre, *A Study of Cyberbullying Detection Using Machine Learning Techniques*, ICCMC, pp. 734-739, doi:10.1109/ICCMC48092.2020.ICCMC-000137. (2020)
9. Jamil, H. and R. Breckenridge. *Greenship: a social networking system for combating cyber-bullying and defending personal reputation.*, ACM : n. pag. (2018)
10. Rasel, Risul Islam & Sultana, Nasrin & Akhter, Sharna & Meesad, Phayung, *Detection of Cyber-Aggressive Comments on Social Media Networks: A Machine Learning and Text mining approach.* 37-41. 10.1145/3278293.3278303. (2018)

Mother's Lifestyle Feature Relevance for NICU and Preterm Birth Prediction

Himani Deshpande[1,*], *Leena Ragha*[2 **]

[1,2] Ramrao Adik Institute of Technology, Nerul, Navi Mumbai, India

Abstract. Maternal health plays an important role in defining the health of mother, child and childbirth experience. With the change in lifestyle over the decades, there have been many health challenges faced by woman, which makes it important for women to understand the impact of their lifestyle and physical health features on their wellbeing. In this study, we have realised the importance of mother's features with respect to preterm childbirth prediction and prediction for neonatal intensive care unit(NICU) facility requirement for newborn. Experiments are performed on MSF dataset which consists of records of 1000 women, 21 physical features and 78 lifestyle features are taken into consideration. Random forest based hybrid model using F-score and Mutual information is used to evaluate each features for their capability of True positive(TP) and False Negative(FN) predictions. For preterm birth prediction, out of all the features hypertension, diabetes, PCOS and consumption of outside food during teenage are found to be the most relevant features. While for NICU prediction diabetes, low amniotic fluid during pregnancy, exposure to air and noise pollution during teenage and consumption of alcohol after marriage are found to be relevant.

1 Introduction

Women are considered to be the key reason behind a happy family and a progressive nation. They are appreciated for carrying out multiple roles with great efficiency, one such responsibility is childbirth. Women are said to be the creators, and have been awarded with immense strength and ability to reproduce. She needs to be strong physically as well mentally to carry the baby in her womb and to enable a healthy pregnancy and childbirth experience. Her health has always been considered as one of the main goal of countries worldwide, as a healthy mother gives birth to a healthy child.

Woman's health is getting effected with rapid urbanisation as there are many challenges that women are facing which includes addition responsibilities and increased societal expectations[1]. Woman is supporting her family emotionally, physically and financially. She is breaking old stereotypes and boundaries and have proved her worth across all the domains. With all these comes a new set of challenges which results in change in her daily routine, stress levels, social environment and physical health status. A woman's health is closely associated with the way she lives through different phases of her life, which is called as her lifestyle. There have been changes in terms of her dietary pattern, sleeping patterns, travelling routine, job profiles, ignorance towards her health, delaying of pregnancy, dependence upon medicines, consumption of unhealthy substances like caffeine, alcohol, nicotine, etc. All these things are slowly effecting her health [2-5], according to a survey carried out by the Associated Chamber of Commerce and Industry (ASSOC-HAM), around 68% of women in the age group 21-52 years, who are working are afflicted with lifestyle related health issues like anxiety, diabetes, depression, reproductive issues, hypertension and obesity[6]. Woman's fertility and childbearing experience is influenced by her lifestyle[7,8] across her reproductive age which starts from the onset of menses during adolescent years. It is believed that infant's health is effected by the way a mother lives through different phases of her life specially during the time of pregnancy. Woman are advised to be careful and follow healthy habits during pregnancy. Thus making it important for her to understand her lifestyle and physical features which could have an effect on her health, child's health and overall childbirth experience. There are many things which together sum up her daily routine, she needs to realise till what extent these factors influence her reproductive health. For a healthier pregnancy along with preconception health care it is advised to understand the interconnection between various influencing factors [9]. Health complications can be minimised scientifically by machine learning computational models by finding the impact of features on woman and child health.

Machine learning has emerged as a scientific tool for understanding data and identifying the association of features with outcomes. Feature selection methods have

* Corresponding author: himani.deshpande@thadomal.org ** leena.ragha@rait.ac.in

gained popularity as machine learning models work accurately when informative data is provided as input while redundant and irrelevant data is filtered out. Feature selection helps to understand the importance of features while finding solutions. Feature selection methods are broadly divided into three categories namely filter method, wrapper method and embedded method. Woman's lifestyle and physical features could be evaluated using these methods, thus helping her to know where she needs to be careful in order to avoid future health complications.

The need of the hour is to create awareness among the female population of the society regarding their reproductive health as well as overall wellbeing and to educate them about the factors that affect their general and emotional wellbeing, and specifically the reproductive health. The aim is to prevent pregnancy complications to ensure a happy and meaningful life to mother and child.

2 Literature Review

Features plays an important role in medical domain, making feature selection a popular practice with medical data. For feature selection there is no significant work in maternal domain thus we have looked into research work across other domains. Yi- Wei et. al.[10], selected best few features using F-score, selected features were given as input to Random forest algorithm to further filter out relevant features. Paul et.al.[11] proposed GARF (Genetic Algorithm based on Random forest) for ranking features available in positiron emission tomography clinical data and images. Features were evaluated initially using spearman's correlation following by Genetic algorithm. Population selected by genetic algorithm is further weighted using Random forest misclassification rate, AUC and sparsity constraints. Peng et. al.[12], used minimum redundancy maximal relevance criterion for evaluating features, followed by backward and forward wrapper selection using multiple classification algorithms. Lee et.al.[13], first used genetic algorithm with dynamic parameter (GADP) settings for microarray data analysis. GADP differs from the traditional method in terms of setting the mutation and crossover rates. Huang et.al.[14] used a hybrid model, where two feature sets are generated using F-score and Information gain separately, these sets were then combined and given as input to SVM based wrapper method. P. Jaganathan et.al.[15], used modified F-score for more than two classes. The threshold is set to be mean of Improved F-score calculated for each feature. Huijuan et.al.[16], proposed MIMAGA-Selection algorithm for gene expression data. Harun et. al.[17], worked on text data using two-stage process, Information gain is used to evaluate text, most important features found are given as input to PCA and genetic algorithm for evaluation. Zheng et. al.[18], selected best features from the Diabetes Mellitus dataset using hybrid approach.

3 Methodology

This work focus upon understanding importance of mothers lifestyle and physical features in predicting the childbirth outcome being preterm or full-term and predicting if there will be a need to provide neonatal intensive care unit (NICU) for the newborn. In this section we will be looking into the methodology used to achieve this aim.

3.1 Dataset

Through the literature survey we realises that there are no standard datasets available and hence we created our own dataset named Mother's Significant features(MSF) [19] dataset which we used in this study. MSF dataset consists of 1000 records of mothers, this dataset comprise of detailed information on mother's lifestyle during her teenage, after marriage and during her pregnancy. Women residing in Mumbai metropolitan region of India, were interviewed by medical personnel just after childbirth. A total of 21 physical features and 78 lifestyle features were taken into consideration. Lifestyle features includes information about their home and work routine, economical status and stress related information. Out of the available records, 172 childbirth were born premature and 828 were full term deliveries. As per MSF dataset records, 243 newborn were availed NICU facility while 757 didn't avail this facility.

3.2 Pre-processing

MSF dataset consists of a mix of continuous and categorial features. Decision tree is biased towards continious features and features with more categories [20], for the same reason all the continuous features were converted into categorial with similar number of categories. These categories were decided after looking into literature and consulting gynaecologists and paediatricians.

3.3 VIBRF

The algorithm Variation and Information based random forest (VIBRF) is proposed by us and explained in detail in [21] is used for lifestyle relevant feature weighing in the proposed study. VIBRF is a hybrid feature selection model which combines filter method and embedded method to evaluate each feature. F-score is used to find the variation across the feature while entropy and mutual information estimates the information provided by the feature. Each feature is ranked looking into both the factors simultaneously. Random forest is formed using hybrid DF_trees, out of all these trees, most accurate were considered. Root nodes of these DF_trees were considered to be most relevant feature of that tree. Available features were found to be relevant if they appear to be root node of these trees, each features is assigned weights looking into the number of DF_Trees it is assigned as root node for.

3.4 Feature Importance Weighing

With this work we have assigned the weights to mother's features. Features are associated with their capability to predict occurrence of birth complications. Mother's physical health at the time of pregnancy is analysed using physical feature set of MSF dataset. Features defining mother's way of living, her family and environment conditions are covered under lifestyle sub-dataset of MSF dataset. Lifestyle features used for the study are spread over threes phases of her reproductive age that are firstly during her teenage, secondly after marriage and thirdly during pregnancy.

In the proposed model we have created random forests using 1000 DF_Trees, nodes of the tree is selected using the equation (1), where x and y are taken as 40% and 60% respectively, giving more weitage to mutual information achieved by the features while predicting outcome. FG_N is the Final Gian for Nth feature , F-$Score_N$ is the F-score for Nth feature and MI_N is entropy based mutual information. Equation (2), defines the entropy for Nth features used in equation (1).

$$FG_N = (x\%)\ F\text{-}Score_N + (y\%) MI_N \quad (1)$$

$$Entropy(N) = -p(yes)\ log2\ p(yes) - p(no)\ log2\ p(no) \quad (2)$$

In this study features are weighted on their capability of predicting true positive and false negative. VIBRF model is implemented twice, firstly random forest (VIBRF_TP) is formed using DF_Tree, best tress are selected having highest True Positive (TP) prediction and secondly VIBRF_FN consisting of best DF_Trees which are ones having highest False negative(FN) values. All features are given weights for being relevant considering root nodes of VIBRF_TP random forest and for being the misleading features looking into root nodes of VIBRF_FN. In the next section we will be discussing the weight given to each feature.

4 Results

Experiments are performed on mother's physical features and lifestyle features to find feature relevance for predicting preterm birth and NICU requirement for newborn. Features weightage using VIBRF_TP predictions were considered to be the relevance weight of features, while features weight using VIBRF_FN random forest is misclassification weight of feature. Fig. 1, shows the relevance and misclassification weight of physical features while predicting preterm childbirth.

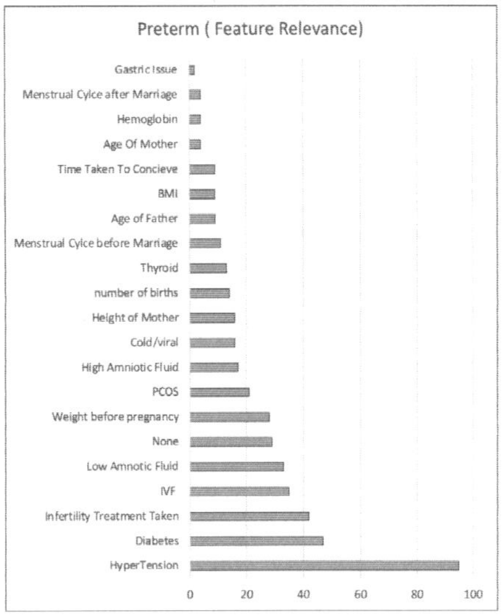

(a)

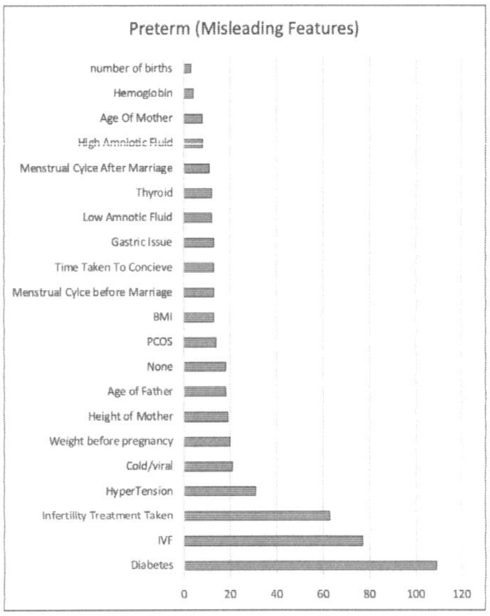

(b)

Fig. 1: Physical feature for Preterm birth prediction

Fig. 2(a),(b),(d),(e), shows the weights assigned to lifestyle features in terms of preterm childbirth predictions and NICU requirement prediction. Out of the 78 lifestyle features, few were never selected to be the root nodes in best DF_Trees for the same reason figure shows weight of features selected as root nodes. Remaining features were considered to be having no

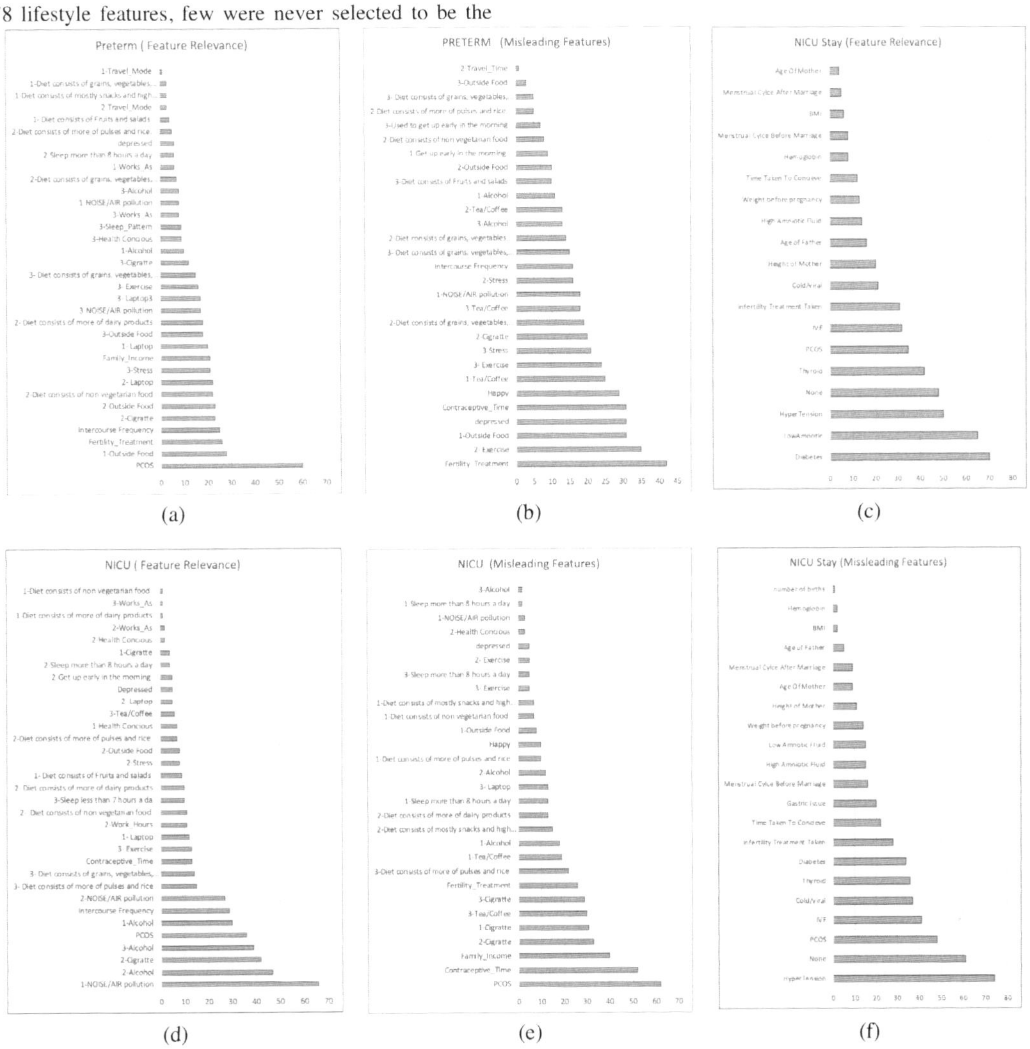

Fig. 2: Weight assigned to features (a),(b) lifestyle feature relevance and misclassification for preterm birth prediction, (d),(e) lifestyle feature relevance and misclassification for NICU prediction, (c), (f) physical feature relevance and misclassification for preterm birth prediction.

impact on prediction. Fig, 2(c),(f) shows the feature weight for relevance and misclassification of physical features for NICU requirement prediction. As the lifestyle features were spread across three phases of mother's life, in fig. (2), index 1 is used to depict teenage information, index 2 is used for after marriage information and 3 is used for information during pregnancy.

Looking at the weights assigned to physical features it has been observed that medical conditions like diabetes, hypertension, infertility treatment, amniotic fluid are more relevant while predicting NICU requirement and preterm childbirth. Among the lifestyle features PCOS, consumption of alcohol, cigarette and outside food, noise and air pollution during teenage , usage of laptop after marriage, stress level during pregnancy are some of the relevant features observed after experiments. Features which are identified as misleading, create bias towards the majority class in imbalance dataset are IVF treatment, diabetes, usage of contraceptive, family income.

5 Conclusion

Woman reproductive health is getting effected with the change in her lifestyle across different phases of her life. A female needs to be educated about the impact these changes may have on her health. With this study we have used hybrid random forest model to analyse mother's physical and lifestyle features and the importance of these features in association with NICU requirement of child just after childbirth and possibility of preterm child birth complications. It has been observed that health conditions like diabetes and hypertension have an impact on pregnancy outcome, while her lifestyle habits like consumption of alcohol, unhealthy food also effects the pregnancy outcomes namely preterm birth and NICU requirement. Woman understanding the features that can impact her maternal health and child's health, can bring changes in her lifestyle thus avoiding such complications.

Future Work

We suggest researchers to explore more pregnancy complications like low birth weight, caesarian delivery, induction of pain etc. in relation with woman physical and lifestyle features, using different machine learning techniques towards better solutions.

References

1. Series, R.D., Health Risk Factors in Rapidly Changing Economies.
2. Ratner, R.E., Christophi, C.A., Metzger, B.E., Dabelea, D., Bennett, P.H., Pi-Sunyer, X., Fowler, S., Kahn, S.E. and Diabetes Prevention Program Research Group, 2008. Prevention of diabetes in women with a history of gestational diabetes: effects of metformin and lifestyle interventions. The Journal of Clinical Endocrinology & Metabolism, 93(12), pp.4774-4779.
3. Goossens, J., Beeckman, D., Van Hecke, A., Delbaere, I. and Verhaeghe, S., 2018. Preconception lifestyle changes in women with planned pregnancies. Midwifery, 56, pp.112-120.
4. Anand, A. and Shakti, D., 2015, September. Prediction of diabetes based on personal lifestyle indicators. In 2015 1st International Conference on Next Generation Computing Technologies (NGCT) (pp. 673-676). IEEE.
5. Patil, M., Lobo, V.B., Puranik, P., Pawaskar, A., Pai, A. and Mishra, R., 2018, July. A proposed model for lifestyle disease prediction using support vector machine. In 2018 9th International Conference on Computing, Communication and Networking Technologies (ICCCNT) (pp. 1-6). IEEE.
6. Sharma, M. and Majumdar, P.K., 2009. Occupational lifestyle diseases: An emerging issue. *Indian journal of occupational and environmental medicine*, *13*(3), p.109.
7. Sharma, R., Biedenharn, K.R., Fedor, J.M. and Agarwal, A., 2013. Lifestyle factors and reproductive health: taking control of your fertility. Reproductive biology and endocrinology, 11(1), pp.1-15.
8. Coutinho, E.D.C., Silva, C.B.D., Chaves, C.M.B., Nelas, P.A.B., Parreira, V.B.C., Amaral, M.O. and Duarte, J.C., 2014. Pregnancy and childbirth: What changes in the lifestyle of women who become mothers?. *Revista da Escola de Enfermagem da USP*, *48*(SPE2), pp.17-24.
9. Hemsing, N., Greaves, L. and Poole, N., 2017. Preconception health care interventions: a scoping review. Sexual & reproductive healthcare, 14, pp.24-32
10. Yi-Wei Chen and Chih-Jen Lin. Combining SVMs with Various Feature Selection Strategies. Studies in Fuzziness and Soft Computing book series (STUDFUZZ, volume 207), 2006
11. Desbordes Paul RuanSuModzelewski Romain V auclin Sebastien V era Pierre Gardin Isabelle. Feature selection for outcome prediction in oesophageal cancer using genetic algorithm and random forest classifier. Computerized Medical Imaging and Graphics, Dec 2016
12. Hanchuan Peng, Fuhui Long, and Chris Ding. Feature Selection Based on Mutual Information: Criteria of Max-Dependency, Max-Relevance, and Min-Redundancy. IEEE Transaction on pattern analysis and machine intelligence Vol. 27, No. 8, August 2005
13. Chien-Pang Lee, Yungho Leu. A novel hybrid feature selection method for microarray data analysis. Elsevier Journal Applied Soft Computing 11 (2011) 208–213
14. Hui-Huang Hsu, Cheng-Wei Hsieh , Ming-Da Lu. Hybrid feature selection by combining filters and wrappers. Elsevier Journal Expert Systems with Applications 38 (2011) 8144–8150
15. P. Jaganathan, N. Rajkumar, and R. Kuppuchamy. A Comparative Study of Improved F-Score with Support Vector Machine and RBF Network for Breast Cancer Classification. International Journal of Machine Learning and Computing, Vol. 2, No. 6, December 2012
16. Huijuan Lu , Junying Chen , Ke Yan , QunJin , Yu Xue , Zhigang Gao , A Hybrid Feature Selection Algorithm for Gene Expression Data Classification, Neurocomputing (2017)
17. Harun Uguz, A two-stage feature selection method for text categorization by using information gain, principal component analysis and genetic algorithm, Knowledge-Based Systems. Volume 24, Issue 7,October 2011, Pages 1024-1032
18. Huilin Zheng, Hyun Woo Park, Dingkun Li, Kwang Ho Park, Keun Ho Ryu. A Hybrid Feature Selection Approach for Applying to Patients with Diabetes Mellitus: KNHANES 2013-2015. 2018 5th NAFOSTED Conference on Information and Computer Science (NICS)
19. Himani Deshpande, Leena Ragha, April 22, 2021, "Mother's Significant Feature (MSF) Dataset", IEEE Dataport, doi: https://dx.doi.org/ 10.21227/kq5k-b784.
20. Fang, L., Jiang, H. and Cui, S., 2017, July. An improved decision tree algorithm based on mutual information. In 2017 13th International Conference on Natural Computation, Fuzzy Systems and Knowledge Discovery (ICNC-FSKD) (pp. 1615-1620). IEEE.
21. Deshpande H., Ragha L., (*in press*). A Hybrid Random Forest based Feature selection model using Mutual Information and F-score for Preterm birth classification., International Journal of Medical Engineering and Informatics.

https://doi.org/10.1051/itmconf/20214003040

An intelligent Crop Price Prediction using suitable Machine Learning Algorithm

Ishita Ghutake[1], *Ritesh Verma*[1], *Rohit Chaudhari*[1], *and Vidhate Amarsinh*[1]

[1]Department of Computer Engineering, Ramrao Adik Institute of Technology, Nerul, Navi Mumbai, India

Abstract. Planning of crops for the next season has been a tedious task for the farmers as it is a difficult prediction about metrics of prices that their crop will fetch in a particular season which will be typically based on dynamic weather conditions. This leads to inaccurate prediction of crops' prices by farmers, and they happen to wrongly select the crops or in haste they happen to sell their crops early without storing and thus earning less than what the same crop would have fetched them in the future. This problem could be addressed by an ML model which will predict the prices of crops in advance showing the proper analysis of the crop and presenting their future scenario so that farmers can select the right crops to strategize crop production which involves crop selection, time of sowing deciding crop pattern and storage of harvested crops providing enough insights for predicting the appropriate price in the markets.
Keywords — Prediction, Random Forest Regression, Decision Tree Regression.

1. Introduction

India is known as an Agrarian country as about 55% of India's population depends on agriculture or related activities for their livelihood. Agriculture has a significant contribution to the primary sector of India's economy. At present farmers are facing massive loss, its reasons are infertile soil, climate change, oversupply in the market and there are a lot of other uncertainties involved. Prediction of expected price in the future is very much needed to manage and sell the products at the right time to maximize revenue and minimize loss. This prediction is based on data obtained from government sources and it uses Machine Learning techniques. This website can also potentially benefit other agriculture-based industries to strategize sourcing of raw material at a reasonable price. The research is based on machine learning which will provide farmers and people in allied profession with a prediction which they can access anytime and will present them with predicted price trends in various crops 12 months(monthly basis) in advance. Thus a ML model with good prediction accuracy can prove to be a great asset for our feeding farmers.

2. Motivation and Scope

Our economy is largely based on agriculture and allied professions (about 55% working professionals are employed in this) but still we are not very advance when it comes to analysis based on conditions and produce and smart techniques to have higher yield. The genesis of the research is the idea of crop planning which is very essential more than ever before to have a sustainable climate resilient food system. Crop production is highly subjective specifically the climate conditions. As in this we are considering rainfall as one of the main factors, on the basis of previous data of rainfall and the respective month's prices it would help us produce the accurate price in upcoming months on the basis of rainfall and correlated price. It's the high time we use technology we harness the potential at algorithms and help to mass increase their profit by selecting the most profitable crop and plan, sowing, harvesting and storage of the crops taking in consideration the predicted price monthly, so that they can select the crops in rotation and plan the storage of harvest so when the respective crops fetches the highest it can be sold.
This research will help farmers to strategize crop production which involves crop selection, time of sowing deciding crop pattern and storage of harvested crops. This also aims at helping people in agriculture allied professions in strategizing their raw material procurement and it is aimed at being very intuitive to the users.

3. Literature Survey

Several approaches have been used in order to improve financial output of agricultural produce. Some noteworthy systems were analyzed and considered in our system development. [1] Data to be used in this system is collected of concerned crops from local markets, online surveys using this collected data as dataset Machine Learning Models are trained. Price prediction is done using algorithms like Artificial Neural Networks, Partial least square & Autoregressive Integrated Moving Average. According to the results obtained by using above mentioned algorithms with the recent data for short and long duration prediction Partial least square and Artificial Neural Networks give better solutions when compared. [2] This paper was aimed at helping farmers take decisions based on ranking of suitability of a crop to concerned area. Prediction and ranking is done using supervised machine learning techniques such as K nearest neighbour regression algorithm and decision tree learning, [3] This paper provides a brief analysis of crop yield predictions using the Multiple Linear Regression process of the selected region.

It focuses on the agricultural analysis of organic and non-organic farming, timely crop cultivation, profit and loss of data and analysis of local business land in a particular area. It focuses on the organic, inorganic and real estate data sets where agricultural forecasts will be available. [4] This paper has shown how data mining techniques can be used in predicting crop yields according to the input parameters. Crop production is influenced by many agro-climatic input parameters. A System built for predicting the crop yield from the given input of climatic parameters indicate a trend of each crop being predominantly influenced by a particular climatic parameter. [5] This paper discusses various applications for data mining in solving various agricultural problems. It integrates the work of different authors in one place, so it is useful for researchers to get details of the current state of data mining techniques and applications depending on the agricultural field.

It also provides research on a variety of data mining techniques used in agriculture including Artificial Neural Networks, K-nearest neighbor, Decision Tree, Bayesian network, Fuzzy set, Support Vector Machine and K – means.

4. Proposed System Methodology

Its going to be website which could be accessed by farmers so that they can base on their financial conditions, need and feasibility and other metrics choose their desired crop. There will be multiple crops widely grown in the country. The dash will show the best doing crop along with the worst and the percentage by which they are soaring or trailing. Predictions will be till 12 oncoming months.
We are creating the Crop Price Prediction Website for crop forecasting were we take data from government of 20+ crops and represent the data in a structured manner representing the increase and decreasing the prices of crops per month and further showing the crops details like its type, location and export factors for the ease of the farmers to plan and manage their finances and sown/harvesting accordingly and also show data in the form of pie chart and graphs. It has user friendly interface and decision tree regression is used for prediction we do the In-Depth statistical analysis of previous data to create the refined platform for interaction which could be accessed by farmers so that they can on the basis of their prediction for 12 oncoming months.
Firstly from data.gov.in updated dataset will be taken comprising of rainfall and wholesale prices respectively month wise for every crop. After required pre-processing model will be trained and then aptly judged. If found suitable, front end and backend will be designed and the ML model will be deployed at the backend. Requisite updating will be timely done on the dataset model will be redeployed.
Here we are doing supervised learning because we have multiple inputs, an output and we are deriving a correlation between them.
The two options suitable for this are linear regression and decision tree regression because both can predict a range of values(continuous) based on multiple inputs and the one we choose is decision tree regression because here by observation in the given dataset there is no linear relation between the inputs and output. The algorithm will take inputs: The input parameters (months, year and monthly rainfall) Jupiter notebook for training ML model and Visual Studio to develop frontend and backend.
Data ingestion is to be done with the data collected from various sources. The injected data is to be prepared according to the requirement of the system. The Machine Learning model is to be designed and trained using the prepared data. Evaluation of the model is to be done using standard metrics. If the results are not as per the requirements, retrain the model. When the desired results are achieved deploy the system.

Details of Hardware/Software Requirement:

1.Jupyter notebook for training ML model and Visual Studio to develop frontend and back- end.
2.Python:
•Flask: It's a web framework used for backend development for the website and linking to html pages using predefined functions.
•Pandas: To be used to read the dataset and split it into independent, dependent variable and training and testing set
•NumPy: To shape the data as an array
•Scikit learn: To use regression algorithms
•Matplotlib: To plot the decision tree model for visual analysis
3.HTML to define the content of web pages
4.CSS to specify the layout of web pages
5.Java Script for scripting and programming the behaviour of web pages
6.csv files to store dataset.
7.Chart.js for flexible graphical presentation.

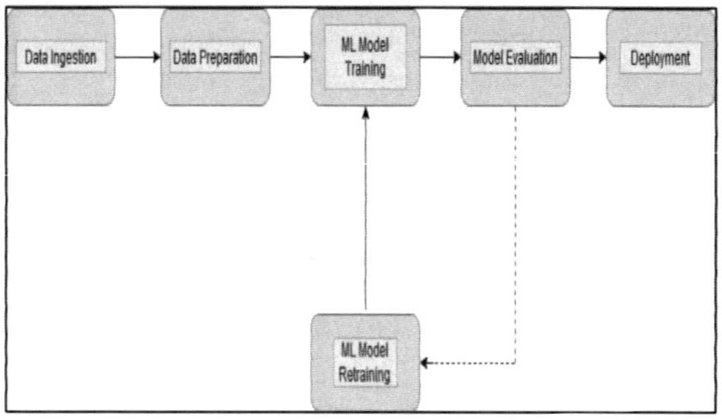

Figure 1: System Design

5. Data Pre-processing

It is a method in which the given raw dataset that is the dataset with unwanted attributes, having missing values, is converted into a processed dataset that is the dataset which is obtained after working on the raw dataset.

Algorithm Selection:

Decision Tree Regression: The dataset will be divided into multiple leaflets which are result of multiple decisions of yes and no and then the new data will be calculated based on what leaflet they land and then calculate the average of that leaflet.

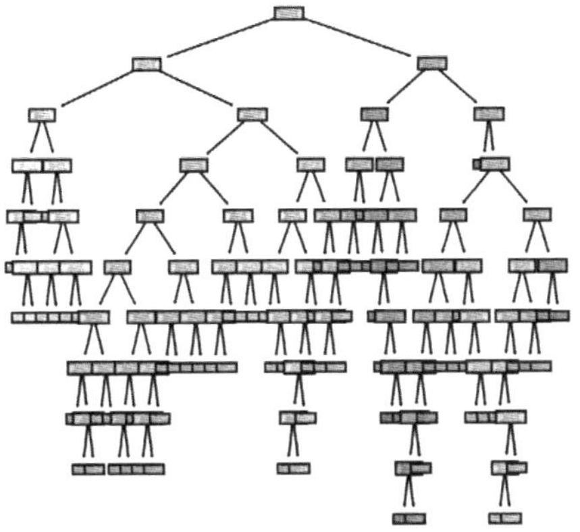

Figure 2: Implementation of Decision Tree Regression

Random Forest Regression:

It works on the basis of ensemble learning, which says that if we combine multiple algorithms or the same algorithm multiple times then we can create a superior algorithm. Random Forest makes use of multiple decision trees to give the output. As we have a huge dataset. Random forest will first extract a small chunk of that data feed it to one decision tree regression model and chain that model this process will be repeated multiple times. We can control by specifying the value of how many decision tree regressions we want. Now we got is a huge dataset on which multiple decision trees are trained. Now the testing data is provided to each of the decision tree, and they will give the output according to themselves and all the collection of output will be then averaged.

Training and Testing:

The dataset is split into two sets of training dataset and testing dataset. 80% for training the dataset & 20% for testing the dataset. The training dataset is a part of data taken for fitting model while the test dataset is also a part of data for the final testing of model on the training dataset. Using the train dataset, we will train our machine learning model. Our machine learning model will attempt to understand and learn on its own and then by using the test dataset we will test our model.

Evaluation:
On evaluating Regression Algorithms such as the Multiple linear, Random Forest and the Decision Tree and testing it on test dataset, made performance analysis of all the models with performance metric and then chosen Decision Tree algorithm.

About Website:

The home page contains various tabs with the names of all the respective crops so that it can be intuitively used by everyone. It's been kept very simplistic so as to make it more user friendly. After clicking on the tabs it will be redirected to the respective page of that crop which contains various details about it.

The second page has multiple sections which describe:

Percentage change in price over next 12 months

Two graphs indicating the price trends of comparing two consecutive years which are given in the below section.

Figure 3: Homepage

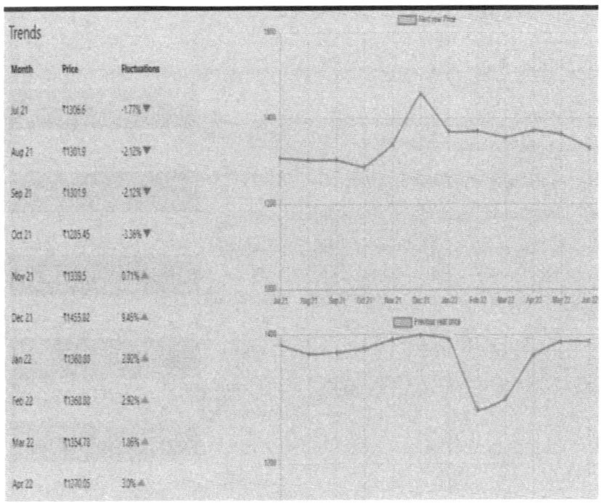

Figure 4: Redirected page

6. Result Analysis

Dataset taken from data.gov.in is analyzed and pre-processed. Two main regression techniques are used to build machine learning prediction model. To show the detailed analysis we consider wheat commodity.

- Current Price: - 1574.1/quintal (As of March 2020)
- Brief forecast (last 6 Months): -
- Min crop price:1476.9/quintal (19 Oct)
- Max crop price: 1672.65/quintal (19 Dec)
- Forecast Trends: We consider the prediction of a single month here to check the accuracy of our model.
- Price of wheat: 1495.8/quintal (Govt data of 19 Sept)
- Price of wheat predicted:1421.45/quintal (Predicted value)
- Change= -4.97% So, for the wheat crop price (19 Sept) we had approximately 95% Accuracy.

And overall, our model possess 92% Accuracy, varying slightly month to month.

7. Conclusion and Future Research

Detailed analysis has been conducted based on real time dataset using two different machine learning techniques. The research aims at a farmer friendly interactive website, predicting the price and forecast through web application and it is runs on efficient machine learning algorithms and technologies having an overall responsive interface to the users. The training datasets so obtained provides enough insights for predicting the appropriate price in the markets. Successfully predicting the price of the crops with 92% accuracy. The future objective is also to interface regional language interface, so that it should become easy to understand and communicate mutually between the software and farmers. The future enhancement also would be to add more features as we are considering rainfall as the factor we can further include temperature, soil fertility and regional use depending upon which the production of the crop varies from area and area so for that a better prediction model can be created and also further enhancement can done by focusing the accuracy of the model and increasing it to 97% or so.

8. Acknowledgement

Thanks to resources on govt site data.govt.in, Institute Management for providing the resources and helping us in all the possible ways. We also thank readers of this paper for showing interest in this topic and contributing towards the enhancement of this topic as well.

9. References

1. Yung-Hsing Peng, Chin-Shun Hs and Po-Chuang Huang,"Developing Crop Price Forecasting Service Using Open Data from Taiwan Markets,"IEEE 2017.
2. Md.Tahmid Shakoor, Karishma Rahman, Sumaiya Nasrin Rayta, Amitabha Chakrabarty,"Agricultural Production Output Prediction Using Supervised Machine Learning Techniques," IEEE 2015.
3. Vinciya, P., and Dr A. Valarmathi., "Agriculture Analysis for Next Generation High Tech Farming in Data Mining" (IJARCSSE) (2016).
4. S.Veenadhari, Dr. Bharat Misra, Dr. CD Singh. "Machine learning approach for forecasting crop yield based on climatic parameters," International Conference on Computer Communication and Informatics (ICCCI -2014), Jan. 03 –05, 2014, Coimbatore, India.
5. M.C.S.Geetha , "A Survey on Data Mining Techniques in Agriculture, " Journal of Innovative Research in Computer and Communication Engineering, Feb 2015

ITM Web of Conferences **40**, 03041 (2021) https://doi.org/10.1051/itmconf/20214003041
ICACC-2021

Algorithmic Trading Bot

Medha Mathur, Satyam Mhadalekar, Sahil Mhatre, Vanita Mane
Department of Computer Engineering, Ramrao Adik Insitute of Technology, Nerul, Navi Mumbai

***Abstract*—Algorithmic trading uses algorithms that follow a trend and defined set of instructions to perform a trade. The trade can generate revenue at an inhuman and enhanced speed and frequency. The characterized sets of trading guidelines that are passed on to the program are reliant upon timing, value, amount, or any mathematical model. Aside from profitable openings for the trader, algo-trading renders the market more liquid and trading more precise by precluding the effect of human feelings on trading. Our project aims to further this revolution in the markets of tomorrow by providing an effective and efficient solution to overcome the drawbacks faced due to manual trading by building an Algorithmic Trading Bot which will automatically trade user strategies alongside its own algorithms for day-to-day trading based on different market conditions and user approach ,and throughout the course of the day invest and trade with continuous modifications to ensure the best trade turnover for the day while reducing the transaction cost, hence enabling huge profits for concerned users be it Organizations or individuals.**

***Index Terms*— Algorithmic Trading, Finance, Random Forest Regressor, Intraday, Moving average, Multiple Data's, Gold Cross, Donchian.**

I. INTRODUCTION

Algorithmic trading is a technique for executing orders utilizing mechanized pre-modified trading guidelines representing factors like time, cost, and volume. This kind of trading endeavors to use the speed and computational assets of PCs comparative with human brokers.Just one of every five-day investor is productive. Algorithmic trading improves these chances through better technique configuration, testing, and execution

The USP of a trade bot is that it simplifies the work of traders and helps the trader to make quick money with the minimum efforts.Algo trading is now a 'prerequisite' for surviving in tomorrow's financial markets. Industry reports suggest global algorithmic trading market size is expected to grow from $11.1 bn in 2019 to $18.8 bn by 2024.[2] So the future of Algorithmic trading is yet to come. A lack of a "Simple yet Efficient Bot" for use by "Common-Man" has driven the need for this project.

Few Advantages of Algorithmic Trading –

1.Quick, Fast and Reduced Cost Trading

2.Enhanced Precision and Diversity in Trading

3.Backtesting enabling traders to assess and tweak a trading idea.

The global algorithmic trading market is expected to grow significantly between 2018 and 2026.

Our project aims to further this revolution in the markets of tomorrow by providing an effective and efficient solution to overcome the drawbacks faced due to manual trading like: Trades are executed at the best possible prices.

• Trade request situation is instant and precise (there is a high possibility of execution at the ideal levels).

• Trades are coordinated effectively and immediately to keep away from huge value changes.

• Reduced exchange costs.

• Simultaneous automated checks with different market scenarios.

• Reduced hazard of manual mistakes when trading.

• Algo-trading can be back tested utilizing historical and live data to check whether it is suitable for trading.

• Reduced the chance of errors by human traders as a result of emotional and psychological factors.

II. RELATED WORK

This section describes a literature survey of the various methods for algorithmic Trading with Machine Learning which are already proposed and implemented. It describes the survey of the existing system and software used for algorithmic trading with Machine Learning. The existing algorithmic trading with Machine Learning methods includes Only Random Forest , Random Forests and Probit regression , Genetic Algorithms like Deep MLP Neural Network , Support vector Machine Regression (SVR) and Random forests and Gradient boosted decision trees (using XGBoost) gives the summary of limitations of existing systems and software

Existing Softwares -

A few software's currently in use are [1]:

• Zerodha Streak: One of the most efficient trading platforms with Algorithmic Trading in India. The biggest benefit of Streak is that it lets the users perform algo trade without coding. The algos can be created even without the technical knowledge of programming.

• Omnesys Nest: It is one of the best algo trading platforms, provided by Thomson Reuters. It has all the excellent features of a state-of-the-art trading platform, including low latency rates and high levels of performance.

• Algonomics: It is a trading platform offered by NSEIT and is one of the best algo trading platforms. The differentiating feature of the platform is its ultra-low latency levels which are beneficial for high volume trades by the investment banks, fund managers and individual algo traders.

A. Using only Random Forest Algorithm [3]

Seasonality impacts and exact normalities in financial information have been very much archived in the monetary financial matters writing for more than seventy years. This methodology proposes a specialist framework that utilizations novel AI strategies to foresee the value return over these occasional occasions, and afterward utilizes these expectations to foster a beneficial exchanging technique.

In this methodology the creators present a mechanized exchanging framework dependent on execution weighted groups of irregular backwoods that improves the benefit and soundness of exchanging irregularity occasions.

An investigation of different relapse procedures is proceeded just as an investigation of the benefits of different strategies for master weighting. The outcomes show that recency-weighted troupes of arbitrary timberlands to create prevalent outcomes as far as both productivity and expectation exactness contrasted and other outfit strategies.

Figure 1 shows the diagrammatic representation of the system that was implemented.

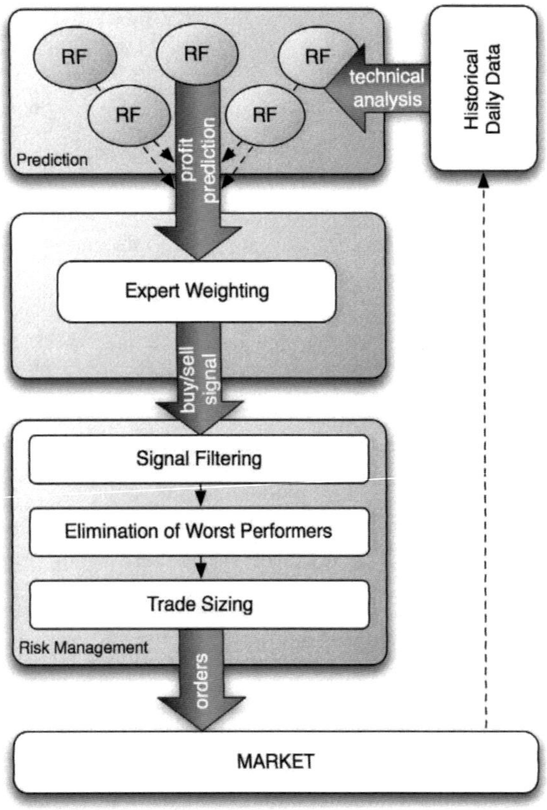

Figure 1 : Diagrammatic representation of the layered workings a fully automated expert trading system

B. Using Random Forest and Probit Regression [4]

In Forex there are numerous money sets and many exchanging individuals and each pair is not the same as the other, and every individual thinks in his own specific manner. Tracking down the best exchanging methodology is actually a mind boggling distraction. Their methodology was to present an expectation and choice model that produces beneficial intraweek venture procedure. The proposed methodology permits improving exchanging results intraweek high-recurrence exchanging. Such outcomes are promising for research on sequential mix of numerous calculations to Forex portfolio the executives.

It is presumed that algorithmic exchanging dependent on blend of arrangement and Probit relapse can be powerful in improving the forecast exactness. This blend assists with recognizing the fun occasions to purchase or to sell money sets.

Figure 2 shows the performance evaluation result of Random Forest plotted into a graph of real and predicted values.

Figure 3 shows performance evaluation of Probit Regression used in this paper.

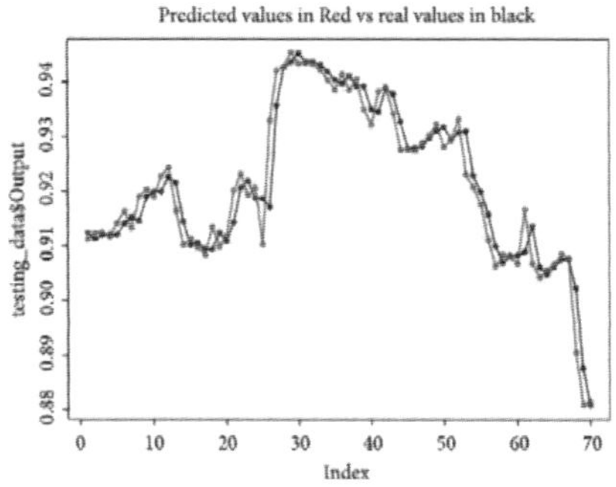

Figure 2 : Predicted values versus real values (predicted values in red, real values in black); for Random Forest regression using: 500 tree and 8 variables tried for each split.

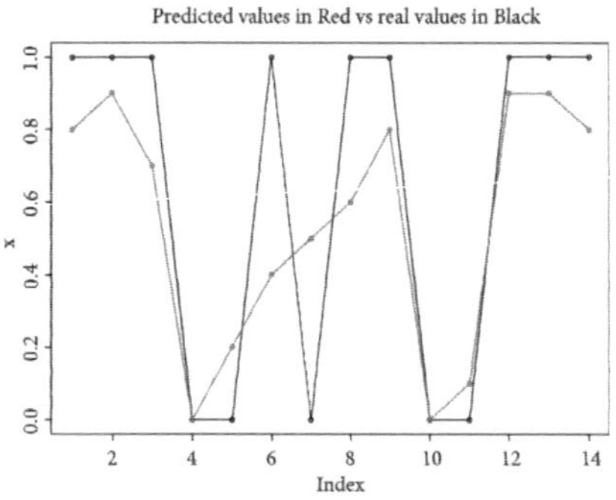

Figure 3 : Predicted values versus real values (predicted values in red, real values in black); for Probit regression.

C. Using Genetic Algorithms like Deep MLP Neural Network [7]

In this examination, we propose a stock exchanging framework dependent on advanced specialized investigation boundaries for making purchase sell focuses utilizing hereditary calculations. The model is created using Apache Spark huge information stage. Each Dow stock is prepared independently utilizing day by day close costs between 1996-2016 and tried between 2007-2016. The outcomes demonstrate that improving the specialized pointer boundaries upgrades the stock exchanging execution as well as gives a model that may be utilized as a choice to Buy and Hold

https://doi.org/10.1051/itmconf/20214003041

and other standard specialized examination models.

At that point, we utilized those streamlined component esteems as purchase sell trigger focuses for our profound neural organization informational index. We utilized Dow 30 stocks to approve our model. The outcomes show that such an exchanging framework produces practically identical or better outcomes when contrasted and Buy and Hold and other exchanging frameworks for a wide scope of stocks in any event, for generally longer periods.

Figure 4 shows the implemented system for the Genetic Algorithm as per the research paper.

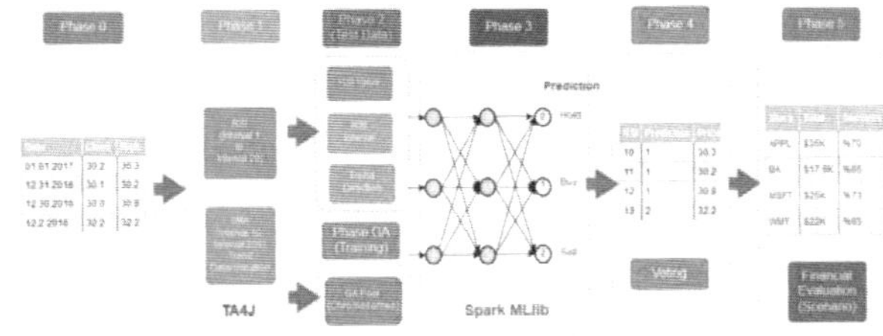

Figure 4 : Proposed Method (Genetic Algorithm and MLP)

D. Using only Support vector Machine Regression (SVR) [6]

This examination shows that utilizing a fixed training set on every day costs, it is feasible to acquire more modest forecast blunders in the test set than in the preparation set when utilizing a direct piece. Specifically, SVR acquired second rate prescient outcomes comparative with an arbitrary walk model for practically all stocks concentrated in regularly updated costs, utilizing fixed preparing, paying little heed to the embraced portion work. Steady model refreshing was additionally advantageous in the authorized value recurrence, and SVR models with direct and spiral bits accomplished preferable outcomes over the arbitrary walk model when this procedure was utilized. To accentuate the strength of the forecasts as time goes on, we prepared a 2-years up-to-the-minutes costs period for the chose Brazilian stocks, affirming better outcomes with a continually refreshed model.

The investigations introduced in this examination propose that occasionally refreshing the SVR model decreases the mean square mistake contrasted with utilizing an inflexible model without intermittent refreshing. A significant commitment of this investigation is an examination of value forecast consequences of the introduced SVR models with those of the arbitrary walk model, as per which markets are eccentric in the long haul. In this regard, the outcomes introduced here show that some SVR models, with occasional or fixed updates, may accomplish better compared to irregular prescient execution, particularly with the utilization of the direct piece. Another outcome which prompts further examination is the sign of a solid connection between SVR value forecast and unpredictability, thinking about a moving preparing window.

The outcomes consequently don't straightforwardly discredit the EMH. Given that the focal point of the investigation isn't the recognizable proof of buying or deals methodologies that take into consideration phenomenal additions, the examination doesn't resolve issues, for example, exchange expenses or portfolio hazard levels. Figure 5 shows how the SVR algorithm performed when evaluation is done on real stock in the research paper.

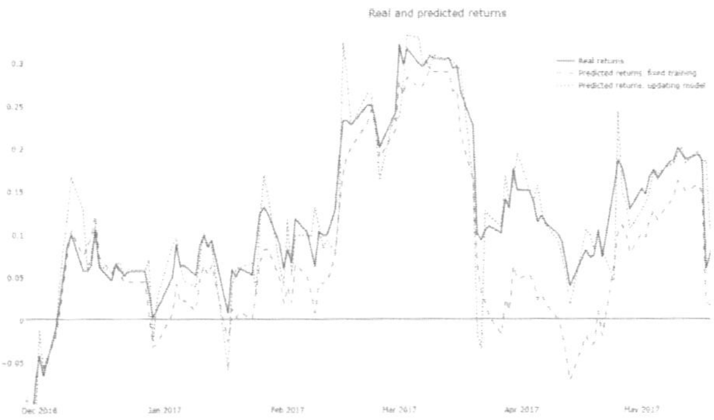

Figure 5 : Real returns and those predicted by the SVR model for the American stock BAC daily prices.

E. Using Random forests and Gradient boosted decision trees (using XGBoost) [5]

Creators think that utilization of AI procedures in stock value anticipating should be an all-around point of view and requests meticulously itemized execution. The proposed approach is a change in outlook in this class of issues by reformulating a customary estimating model as a characterization issue. Additionally, information disclosure from the examination ought to make new wildernesses or applications, for example, an exchanging methodology dependent on the qualities of the characterization exactness, researching the conduct of specific classes of stocks. In any case, outfit learning techniques have stayed unexploited in this field. In this paper, we have utilized Random Forests and XGBoost classifiers to fabricate our prescient model and our model has created amazing outcomes. The model is discovered to be powerful in anticipating the heading of stock development. The vigor of our model has been assessed by ascertaining different boundaries like exactness, accuracy, review, explicitness, and F-score.

Also, a significant piece of the curiosity of the current work lies in the cautious choice of specialized markers and their utilization as highlights. As the kind of the difficult that we're attempting to tackle is essentially that of monetary investigation, we enjoyed the benefit of adaptability of the use of different various highlights, each with its own understanding. Our model can be utilized for contriving new techniques for exchanging or to perform stock portfolio of the executives, changing stocks as per patterns expectation.

The proposed model is without a doubt a novel method to limit the danger of interest in financial exchange by anticipating the profits of a stock more precisely than existing calculations applied up until this point. Figure 6 shows comparison between different algorithms performance for trading. Figure 7 depicts how the used algorithm has performed on a real stock.

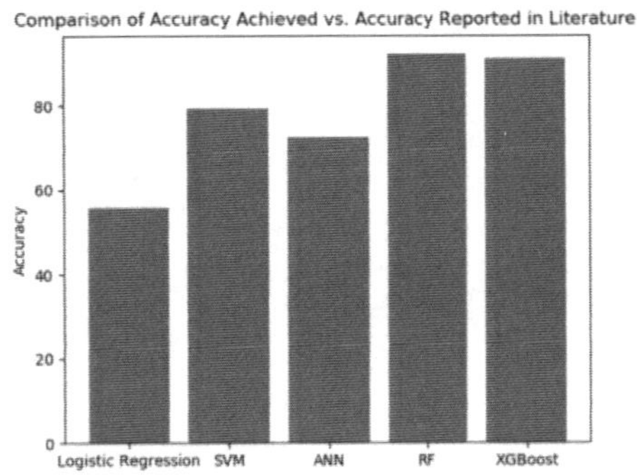

Figure 6 : Comparison of the accuracy achieved in this work to the accuracies achieved in available literature

Figure 7 : Trading Strategy suggested by the model on AAPL data

III. DATASET

Alpaca API and Yahoo Finance is used to fetch past data and put it into a dataset. The dataset comprises Date , Open Price , High Price , Low Price , Close Price and Volume traded for that particular Stock day wise.

A. Database Splitting

The dataset is split in 60:40 ratio. Four variables i.e., X_train, X_test (for inputs) and Y_train, Y_test (for outputs) are created.

B. Annotation Description

The dataset consists of various columns as mentioned above. The columns that we require for our Random Forest Regressor and prediction is only Date and Close Price for the particular stock. The Close Prices will help us get a trend or a Moving Average for our Intraday trading of that particular stock. This will be integrated with Financial strategies to boost performance with greater accuracy owing to predictive power of Random Forest Regressor.

IV. PROPOSED METHODOLOGY

The Architectural diagram of our proposed solution. We have two types of roles i.e. Trader and Bot. The Trader has access to trade orders, viewing market statistics, setting up a day trade strategy via the bot and manage their account. The Bot will be validating and placing trades as per market and user statistics, will be sending notifications, and have access to user wallet to execute trade orders. A few special features have been listed on top in the diagram.

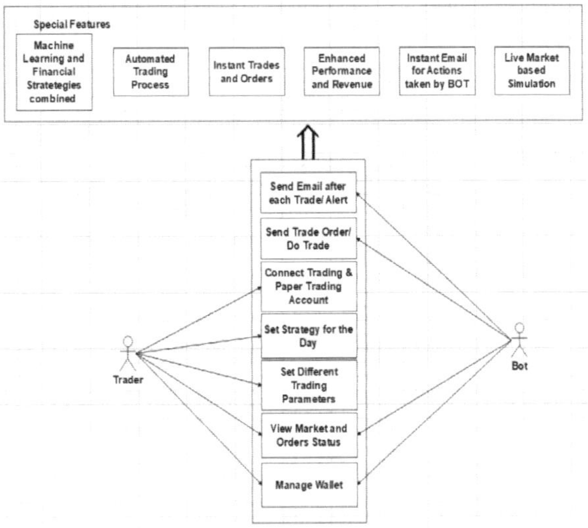

Figure 8 : Architectural Diagram for Algorithmic Trading Bot

A. Data Pre-processing

Data pre-preprocessing is applied on the dataset to get Intraday movements to pass into Random Forest Regressor.

a. We drop all other columns except Date and Close price.
b. To determine the actual trading signal, we assume that we traded on a prior day's close price, this is done by lagging the data by 1 day. We create a lag for 41 days.
c. We then clean the dataframe by dropping any NULL values.
d. Dataset is split as [0:33] data into X (inputs) and the rest into Y (outputs)

B. Splitting dataset into Test and Train dataset

Dataset spit into Training and Testing in the ratio 60:40. Four variables i.e., X_train, X_test (for inputs) and Y_train, Y_test (for outputs) is created.

C. Training the Random Forest Regression model on the training set

We import the RandomForestRegressor class and assign it to the variable regressor. We then use the .fit() function to

 https://doi.org/10.1051/itmconf/20214003041

fit the X_train and Y_train values to the regressor by reshaping it accordingly. Feature importance is calculated with regressor.feature_importances_ to help describe the importance of chosen features and to improve the model.

D. Predicting the Results

We predict the results of the test set with the model trained on the training set values using the regressor.predict function and assign it to 'Y_predicted'.

E. Visualizing the Random Forest Regression Results

A graph of Share Price (Both Y_test & Y_predicted) vs Date is plotted. The Actual values are plotted with "Red" color and the Predicted values with "Blue" color.

F. Integration of Financial Strategy Bot with Random Forest Model

Python Bot is coded which connects with a Paper Trading account via API. The strategy parameters are entered by the user , and once the Bot starts trading it will continue to do so until either Stop Loss is reached, Market is closed or User sends a Stop signal to Bot.

The Bot constantly checks Market conditions and current Positions in the market to decide its action. The Random Forest model is integrated as a joblib file with the bot and the Bot is made to take its decision on the basis of prediction from the model as well as the financial strategy.

V. Evaluation

Random Forest Regressor Model for Trading Analysis – Evaluation Metrics –

1. Explained Variance Score - Explained variance regression score function.

Best possible score is 1.0, lower values are worse.

$$explained\_variance(y, \hat{y}) = 1 - \frac{Var\{y - \hat{y}\}}{Var\{y\}}$$

2. R^2 Score - computes the coefficient of determination.

$$R^2(y, \hat{y}) = 1 - \frac{\sum_{i=1}^{n}(y_i - \hat{y}_i)^2}{\sum_{i=1}^{n}(y_i - \bar{y})^2}$$

3. Mean squared logarithmic error - computes a risk metric corresponding to the expected value of the squared logarithmic (quadratic) error or loss.

$$\text{MSLE}(y, \hat{y}) = \frac{1}{n_{\text{samples}}} \sum_{i=0}^{n_{\text{samples}}-1} (\log_e(1 + y_i) - \log_e(1 + \hat{y}_i))^2.$$

4. Random Forest Regressor Score - Return the mean accuracy on the given test data and labels.

regressor.score(X_test, y_test)

Mean accuracy of self.predict(X).y

VI. Result

1. Evaluation based on Metrics –

The Table shows the performance of our model against the evaluation parameters discussed earlier.

2. Random Forest Regressor Model:

Random Forest Regressor Model for Trading Analysis –

(Red: Actual Stock Price Movement, Blue: Bot predicted Stock Price Movement)

X axis: 'TSLA' Stock share prices of test dataset and predicted prices

Y axis: Dates of trade

The Fig 16 shows a plotted graph for Share Price vs Date depicting performance of the Random Forest Model.

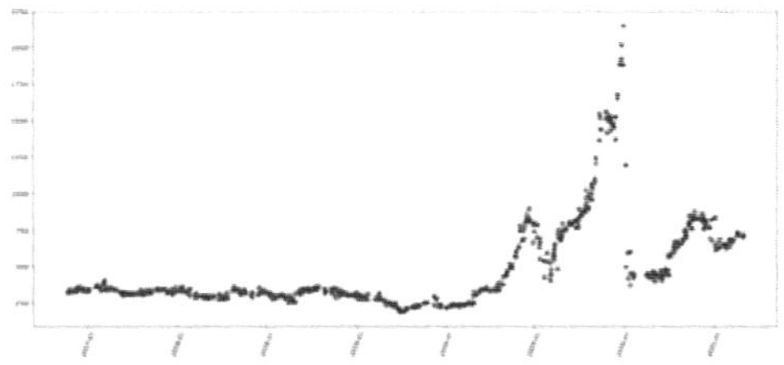

Figure 16: Share Price vs Date Graph

3. Backtesting **Moving Average Crossover** strategy –

Table 3 shows the Back Testing results against parameters of Strike Rate and Profit Earned for 1-year and 10-year duration.

Table 3: Moving Average Evaluation

DURATION	STRIKE RATE	PROFIT EARNED
1 year	77.78%	$ 820.8
10 years	53.85	$ 1993.43

The Fig 17 shows the plotted graph of Moving Average Strategy for 1-year and 10-year duration depicting the behaviour of bot against actual trade movement.

1 Year Chart

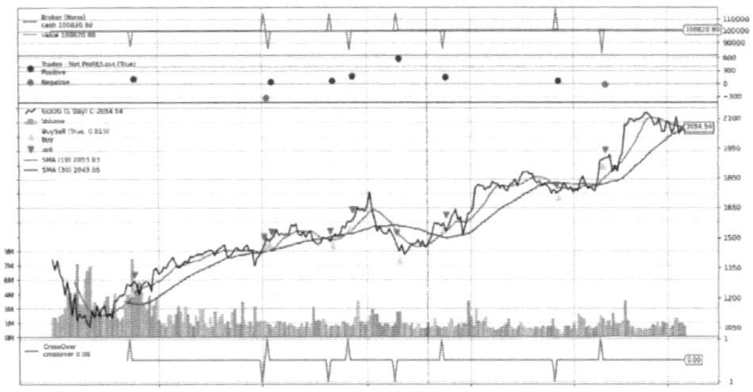

10 Years Chart

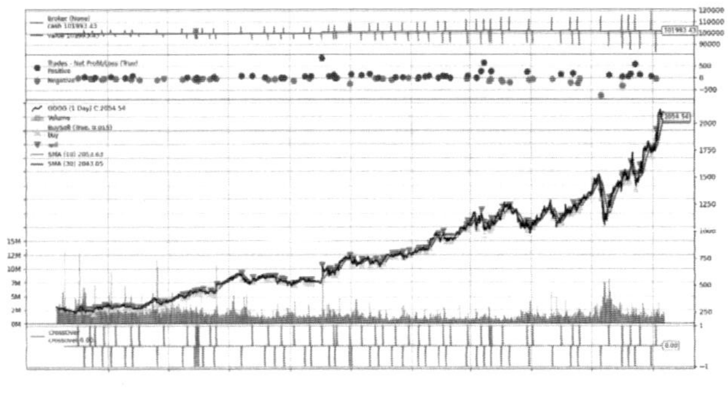

Figure 17: Moving Average Back testing

4. Back testing **Donchian** strategy –

Table 4 shows the Back Testing results against parameters of Strike Rate and Profit Earned for 1-year and 10-year duration.

Table 4: Donchian Evaluation

DURATION	STRIKE RATE	PROFIT EARNED
1 year	26.3157%	$ -2053.89
10 years	31.0924%	$ -7859.13

The Fig 11 shows the plotted graph of Donchian Strategy for 1-year and 10-year duration depicting the behaviour of bot against actual trade movement.

1 Year Chart

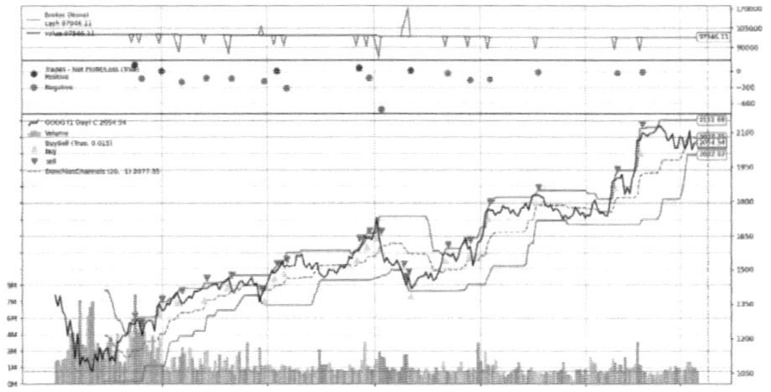

10 Years Chart

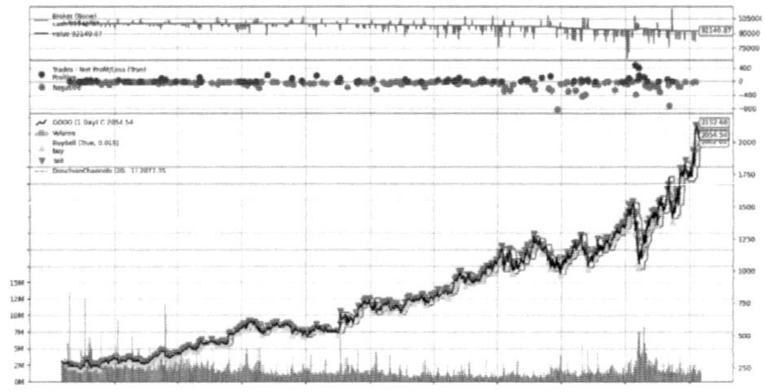

Figure 18: Donchian Back testing

5. Back testing **Multiple Data's** strategy –

Table 5 shows the Back Testing results against parameters of Strike Rate and Profit Earned for 1-year and 10-year duration.

Table 5: Multiple Data's Evaluation

DURATION	STRIKE RATE	PROFIT EARNED
1 Year	50%	$ 3874.1
10 Years	41.81%	$ 14802.73

The Fig 12 shows the plotted graph of Multi Data's Strategy for 1-year and 10-year duration depicting the behaviour of bot against actual trade movement.

1 Year Chart

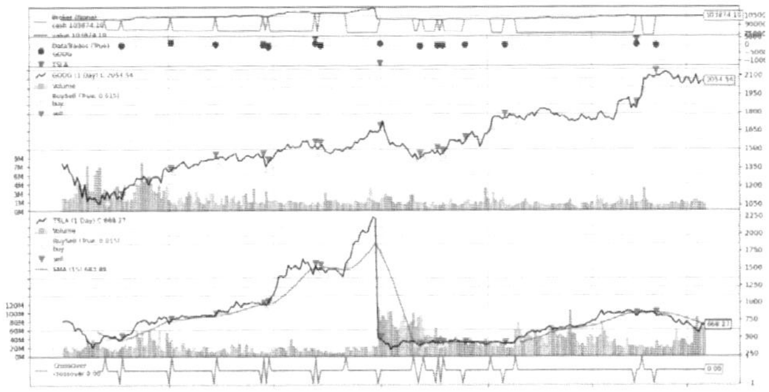

10 Years Chart

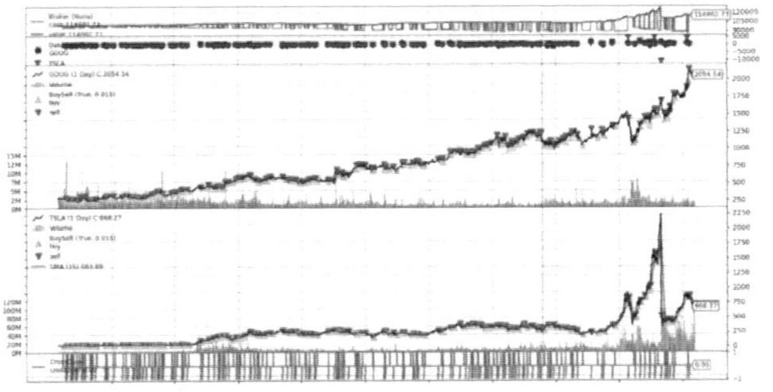

Figure 19: Multiple Data's Back testing

6. Back testing **Gold Cross** strategy –

Table 6 shows the Back Testing results against parameters of Strike Rate and Profit Earned for 1-year and 10-year duration.

Table 6: Gold Cross Evaluation

DURATION	STRIKE RATE	PROFIT EARNED
1 Year	NA	NA
10 Years	50%	$ 214.15

The Fig 13 shows the plotted graph of Gold Cross Strategy for 10-year duration depicting the behaviour of bot against actual trade movement.

10 Years Chart

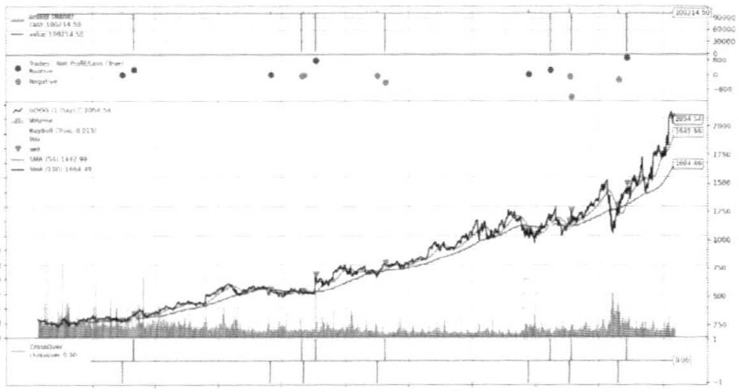

Figure 20: Gold Cross Back testing

VII. CONCLUSION

Algorithmic trading Bot not only provides Security, Cost, and Speed but is also a revolutionary technology for the future financial markets and economy. Algorithmic Trading Bot makes it easier for both new traders as well as established ones in getting profitable outcomes with minimized effort, time and loss.The integration of Financial Knowledge with Machine Learning is a demand of future Trading and enhances both Performance and Revenue.

VIII. ACKNOWLEDGMENTS

We would like to express our gratitude to our College, Ramrao Adik Institute of technology our Mentor, Dr. Vanita Mane Ma'am, and our Project Coordinator, Dr. Sangita Choudhary who have provided us with the opportunity to work on this project and given us support with guidance to make this project a success. We would also like to thank our teammates for their contribution and continued support and zeal towards this project. This project wouldn't be a success without their efforts.

REFERENCES

[1] *https://www.adigitalblogger.com/algo-trading/algo-trading-platforms/*

[2] *https://blog.quantinsti.com/growth-future-algorithmic-trading/*

[3] Ash Booth, Enrico Gerding, and Frank Mcgroarty. 2014. Automated trading with performance weighted random forests and seasonality. Expert Syst. Appl. 41, 8 (June, 2014), 3651–3661. DOI: https://doi.org/10.1016/j.eswa.2013.12.009

[4] Younes Chihab, Zineb Bousbaa, Marouane Chihab, Omar Bencharef, Soumia Ziti, and Miin-Shen Yang. 2019. Algo-Trading Strategy for Intraweek Foreign Exchange Speculation Based on Random Forest and Probit Regression. Appl. Comp. Intell. Soft Comput. 2019 (2019). DOI: https://doi.org/10.1155/2019/8342461

[5] Suryoday Basak, Saibal Kar, Snehanshu Saha, Luckyson Khaidem, Sudeepa Roy Dey, Predicting the direction of stock market prices using tree-based classifiers, The North American Journal of Economics and Finance, Volume 47, 2019, Pages 552-567, ISSN 1062-9408, https://doi.org/10.1016/j.najef.2018.06.013.

[6] Bruno Miranda Henrique, Vinicius Amorim Sobreiro, Herbert Kimura, Stock price prediction using support vector regression on daily and up to the minute prices, The Journal of Finance and Data Science, Volume 4, Issue 3, 2018, Pages 183-201, ISSN 2405-9188, https://doi.org/10.1016/j.jfds.2018.04.003.

[7] Omer Berat Sezer, Murat Ozbayoglu, Erdogan Dogdu, A Deep Neural-Network Based Stock Trading System Based on Evolutionary Optimized Technical Analysis Parameters, Procedia Computer Science, Volume 114, 2017, Pages 473-480, ISSN 1877-0509, https://doi.org/10.1016/j.procs.2017.09.031.

[8] Maragoudakis, M and Serpanos, D. (2010), towards stock market data mining using enriched random forests from textual resources and technical indicators. AIAI 2010, IFIP AICT 339, pp. 278-286. https://link.springer.com/chapter/10.1007/978-3-642-16239-8_37

[9] https://www.investopedia.com/terms/c/conditional_value_at_risk.asp

[10] https://www.investopedia.com/articles/trading/05/scalping.asp

[11] Manoj Thakur and Deepak Kumar. 2018. A hybrid financial trading support system using multi-category classifiers and random forest. Appl. Soft Comput. 67, C (June 2018), 337–349. DOI: https://doi.org/10.1016/j.asoc.2018.03.006

[12] M. Nabipour, P. Nayyeri, H. Jabani, S. S. and A. Mosavi, "Predicting Stock Market Trends Using Machine Learning and Deep Learning Algorithms Via Continuous and Binary Data; a Comparative Analysis," in IEEE Access, vol. 8, pp. 150199-150212, 2020, doi: 10.1109/ACCESS.2020.3015966.

[13] M. H. L. B. Abdullah and V. Ganapathy, "Neural network ensemble for financial trend prediction," 2000 TENCON Proceedings. Intelligent Systems and Technologies for the New Millennium (Cat. No.00CH37119), 2000, pp. 157-161 vol.3, doi: 10.1109/TENCON.2000.892242.

[14] R. D. Edwards, J. Magee, W. H. C. Bassetti, "Technical Analysis of Stock Trends, Chapter 1, The Technical Approach to Trading and Investing," 9 Edition, CRC Press, Boca Raton, 2007, pp. 3-7.

[15] M. A. H. Dempster and C. M. Jones, "A Real-Time Adaptive Trading System Using Generic Programming," Quantitative Finance, Vo. 1, No. 4, 2001, pp. 397-413. doi:10.1088/1469-7688/1/4/301

[16] L. Breiman, "Random Forests," Machine Learning Journal, Vol. 45, No. 1, 2001, p. 532.

[17] Wen Long, Zhichen Lu, Lingxiao Cui, Deep learning-based feature engineering for stock price movement prediction, Knowledge-Based Systems, Volume 164, 2019, Pages 163-173, ISSN 0950-7051,

https://doi.org/10.1016/j.knosys.2018.10.034.

[18] Viliam Vajda, Could a Trader Using Only "Old" Technical Indicator be Successful at the Forex Market?, Procedia Economics and Finance, Volume 15, 2014, Pages 318-325, ISSN 2212-5671, https://doi.org/10.1016/S2212-5671(14)00515-2.

[19] Patra, S., 2015. Techniques for time series prediction. International Journal of Research Science and Management, 2, pp.6-13. Google Scholar

[20] Wang, J., Wu, X. and Zhang, C., 2005. Support vector machines based on K-means clustering for real-time business intelligence systems. International Journal of Business Intelligence and Data Mining, 1(1), pp.54-64. Google Scholar

[21] W. Lv and R. Zhang, "A regression model on effective exchange rate of RMB based on Random Forest," 2011 International Conference on E-Business and E-Government (ICEE), 2011, pp. 1-3, doi: 10.1109/ICEBEG.2011.5882520.

[22] Cain Evans, Konstantinos Pappas, Fatos Xhafa, Utilizing artificial neural networks and genetic algorithms to build an algo-trading model for intra-day foreign exchange speculation, Mathematical and Computer Modelling, Volume 58, Issues 5–6, 2013,

Pages 1249-1266, ISSN 0895-7177, https://doi.org/10.1016/j.mcm.2013.02.002.

[23] A. Hirabayashi, C. Arahna, and H. Iba, "Optimization of the trading rules in forex using genetic algorithms," 2009, http://www.dollar.biz.uiovwa.edu.

[24] K.-J. Kim, "Financial time series forecasting using support vector machines," Neurocomputing, vol. 55, no. 1-2, pp. 307–319, 2003. https://doi.org/10.1016/S0925-2312(03)00372-2

[25] B. M. Henrique, V. A. Sobreiro, and H. Kimura, "Stock price prediction using support vector regression on daily and up to the minute prices," The Journal of Finance and Data Science,

vol. 4, no. 3, pp. 183–201, 2018.

https://doi.org/10.1016/j.jfds.2018.04.003

[26] S. Basak, S. Kar, S. Saha, L. Khaidem, and S. R. Dey, "Predicting the direction of stock market prices using tree-based classifiers," The North American Journal of Economics and Finance, vol. 47, pp. 552–567, 2019. https://doi.org/10.1016/j.najef.2018.06.013

[27] M. Kumar and T. M., "Forecasting stock index movement: a comparison of support vector machines and random forest," SSRN Electronic Journal, 2006. https://papers.ssrn.com/sol3/papers.cfm?abstract_id=876544

[28] A. Peltonen Tuomas, "Are emerging market currency crises predictable?-A test," 2006. https://link.springer.com/article/10.2307/3867664

[29] C. Brownlees, G. Gallo Financial econometric analysis at ultra-high frequency: data handling concerns Comput Stat Data Anal, 51 (4) (2006), pp. 2232-2245 https://doi.org/10.1016/j.csda.2006.09.030

[30] M.A. Goldstein, P. Kumar, F.C. Graves Computerized and high-frequency trading Financ Rev, 49 (2) (2014), pp. 177-202 https://doi.org/10.1111/fire.12031

Covid-19 emergency services and disease prediction system

Shreshtha Mankala[1], Nirzara Patil[1], Simran Rathore[1],and Mrs. Jyoti Joshi[2]

[1] Department of Information Technology, Ramrao Adik Institute of Technology, Nerul, India

[2] Mentor from Department of Information Technology, Ramrao Adik Institute of Technology, Nerul, India

Abstract- A medical emergency can be referred to as a medical or behavioral condition, which occurs suddenly and has severe symptoms, including severe pain, such that if a person delays medical attention it can cause: (1) loss of life;(2) serious impairment to the person's body; or (3) serious damage. Admitting a patient to a healthcare is a complex process which should be managed efficiently, which otherwise may cause serious consequences and patient dissatisfaction. The registration aspect of a patient admission is tedious and cumbersome, which is not at all suitable during a medical emergency. There is a need of a system through which user could fill the form for getting admitted to the hospital beforehand in order prevent delay in treatment. After the registration, the goal is to create a web application for hospital staff to manage the patients' data. The web application also analyses the types of patients in particular hospital and represent the data in the form of charts. The implementation of this system is carried out with the help of machine learning algorithms which also analyze Covid data and represent it continent wise, predict future cases in India, and conduct Covid detection by chest scan of a patient.

1. Introduction

A medical emergency can be referred to as a medical or behavioral condition, which occurs suddenly and has severe symptoms, including severe pain, such that if a person delays medical attention it can cause: (1) loss of life; (2) serious impairment to the person's body; or (3) serious and permanent damage.[1] Admitting a patient to a healthcare is a complex process which should be managed efficiently, which otherwise may cause serious consequences and patient dissatisfaction. Also, during an emergency, it is difficult to locate nearby hospitals and get the directions to it. This system consists of a mobile application through which the patients can register themselves and fill a particular hospital's admission form beforehand and can also get directions to the selected hospital. There can be a situation where after reaching the hospital, the patient comes to know that there is no availability of beds which will lead to delay in the treatment of the patient. Through this system, this confusion will also be avoided as the number of beds available, and the number of beds occupied for each hospital will be updated to a web application consisting of admin module. So, patient would not get the option to fill the form for a hospital which has no beds available. Covid-19 (Coronavirus) has affected the daily life and is alleviating country economy as much as a decade worth. This pandemic has affected many people, who are getting sick or get killed , many families are being destroyed. Many people have also lost their livelihood and jobs. Covid-19 has alleviated businesses, disrupted the world trade and movement across the globe.[2] Identification of the disease at an early stage can help in controlling the spread of the virus as well as for saving a life. This system uses machine learning algorithms such as K-means clustering, Polynomial regression, Inception -v3 for analyzing Covid-19 data, represent it continent wise, and conduct Covid detection through chest scan.

https://doi.org/10.1051/itmconf/20214003042

2. Motivation

Though there are computerized systems available for booking appointments in hospitals and healthcare and for maintaining records, there is no such existing system available where the patient can view the nearby hospitals and get admitted to any of them by filling the form previously. Through this system, not only can the patient book a bed in advance but can also get the directions to the hospital. Also, during this Covid-19 situation, it is important to manage related data in an efficient manner. Visualization of Covid cases, records in the web application part of the project will help in the efficient management of the data. Also, this application consists of a feature where through images of chest scan of a patient, it can be detected whether the patient has Covid which can help the doctors with the diagnostic and treatment procedures.

3. Literature Survey

History of Covid-19:
The Covid-19 pandemic has caused many Cancellations like clinical conferences , journey restrictions, social distancing, and different unprecedented preventive steps or procedures.

1. What are the Covid causing virus ? Coronaviruses are a huge own circle of relatives of zoonotic viruses. These viruses can motive illnesses like not unusual place bloodless to even extreme breathing illnesses like SARS and MERS. Zoonotic manner those viruses can transmit themselves from animals to human beings. There are numerous coronaviruses recognized to be circulating in one of a kind animal population which have now no longer but inflamed human beings.[3] Covid-19 is the maximum current one to contaminate human beings. Common symptoms and symptoms of Covid-19 contamination are similar to cold and consist of signs such as dry cough, fever, shortness of breath, and respiration problems. In extreme instances, it causes pneumonia infection, extreme acute breathing syndrome, kidney failure, or even death.

➢From where the Coronavirus came into existence? According to the medical experts, SARS-CoV-2 has originated in bat species. That's additionally how the coronaviruses at the back of Middle East breathing syndrome (MERS) and extreme acute breathing syndrome (SARS) were given commenced. SARS-CoV-2 commenced infecting human beings at one of Wuhan's open-air "moist markets."

➢Coronavirus Evolution: Scientists first found a human infecting coronavirus in 1965. It precipitated a not unusual place bloodless. Later that decade, researchers discovered a collection of comparable human and animal viruses. They had been named 'coronaviruses' because of their crown-like appearance. It was found that there are at least seven types of coronaviruses that can infect human beings. The one which reasons SARS emerged in southern China in 2002 and fast unfold to twenty-eight different countries. More than 8,000 humans had been inflamed via way of means of July 2003, and 774 died. A small outbreak in 2004 worried simplest 4 extra instances. This coronavirus reasons fever, headache, and breathing issues including cough and shortness of breath. MERS commenced in Saudi Arabia in 2012. Nearly 2500 instances had been of the individuals who travelled to Middle East. This coronavirus is much less contagious than SARS however extra dangerous, killing 858 humans. It has the equal breathing effects ,however it can also cause kidney problems. [4]

Following papers have been referred for the project:

3.1. Regression Analysis of Covid-19 using Machine Learning Algorithms :

The trend of Covid cases in India is analysed. The paper makes use of regression analysis as the dataset being analysed is continuous dataset. It aims at predicting the future Covid cases so as to take preventive measures. The analysis was performed using Support Machine vector and Polynomial regression both. SVM and Polynomial regression both are supervised machine learning algorithms. The paper also presents charts to visualize death rate, active cases, and total confirmed cases. The analysis performed raises some questions such as "Why is the mortality rate of India lowest as compared to other countries?".[5]

3.2. A Comprehensive Analysis of Covid-19 Outbreak situation in India :

In this study, trends and patterns were presented based on data from different state units of India and Ministry of Health and Family Welfare. The data was till 8th of April 2020. This study tried to answer six questions related to current coronavirus spread in comprehensive manner. The research questions were regarding the change in the country post lockdown, prediction of current cases, mobility changes in various regions of India, community outbreak spread due to conduction of religious and cultural events, prominent clusters formed with respect to Covid-19 outbreak in India, etc. The findings were mostly related to impact of mass events and social distancing on affected areas, short term predictions for infected cases ,network analysis and classification on infected patients.[6]

3.3. Deep learning-based detection and analysis of Covid-19 on chest x-ray images :

The paper explains how image screening of Chest scans and CT-scans can help detect Covid patients. An algorithm was developed using three different models which were

 https://doi.org/10.1051/itmconf/20214003042

Inception, Xception and ResNeXt models. These are Deep learning models which are best known for image recognition. The dataset was collected from Kaggle and had total of 6432 images. 5467 images were used for training and 965 for testing. The images were cleaned, resized and augmented for better results. Augmentation includes rotation, zoom size and some other factors. This paper has presented steps in which the detection can be done. First the image is preprocessed, then it is applied to a pretrained model and the output obtained is flattened to (n-1) and then passed through different layers depending on the model used. The accuracy of Xception model was found to be better than other two. [7]

4. Proposed System

The proposed system consists of following modules:

1. Mobile Application for patient to search for bed and get admitted in the nearest hospital in case of emergency.[8]
2. A web application for hospital staff to manage the patients' data generated from the mobile application. The web application also analyses the types of patients in particular hospital and represent the data in the form of charts.[9]
3. A Machine learning algorithm run on the backend of the web app which analyses the Covid data around the world and represents it continent wise. It also retrieves and analyses the current Covid cases in India from Jan 2020 to March 2021.
4. Another Machine Learning Web app is used to predict the cases of India in the future.
5. And lastly a web-based Machine Learning app is used for Covid detection from a patient's Chest scan.

The mobile app is made using android studio and firebase for taking user details and symptoms. The user can send emergency message to contact and make emergency call to helpline number. The web application is made using Vue.js and Vuetify. The user data is passed to the web app through firebase. The web app consists of different graphs representing total available beds and total occupied beds. The graph is displayed continent wise for total cases, recoveries, and deaths from a real time API.

The Covid patients data is clustered using k-means algorithm based on the number of days they require to discharge and age , which is used to determine the average number of days required to get discharged with respect to age. The number of clusters 'k' which is a hyper parameter is defined as 4 and the centroids were chosen accordingly. The data points have been clustered on the basis of squared distance between them.

The confirmed cases in India per day are put under polynomial regression model to determine the confirmed cases for near future days. The data is collected from Kaggle dataset. The data is preprocessed by reshaping and fit into polynomial feature of degree 5. The data is then fit into a linear regressor model and the accuracy of the model is found out to be 99.765%. The model is presented as a flask app which uses Html CSS and JavaScript and the result is rendered on the web app hosted on Heroku.

$$y = b_0 + b_1x_1 + b_2x_1^2 + b_2x_1^3 + \ldots\ldots b_nx_1^n \quad (1)$$

The deep learning algorithm Inception-v3 is used to train a model on ImageNet using Keras module.[10,11] The images were changed from BGR to RGB for better modelling. Normalization was formed so as to avoid overfitting. Then the data was split into training and testing data for Covid and Non-Covid chest scans .Label binarizer was used to convert the results into form of 0's and 1's.Finally the Inception-v3 model is applied on the prepared data, model is compiled using Adam optimizer. Image data generator is used to change the rotations and make all the images even. The model is built and now the model is trained by fitting with a batch size of 32 and epochs of 500.The model is then saved in a .h5 format and then used to make predictions.

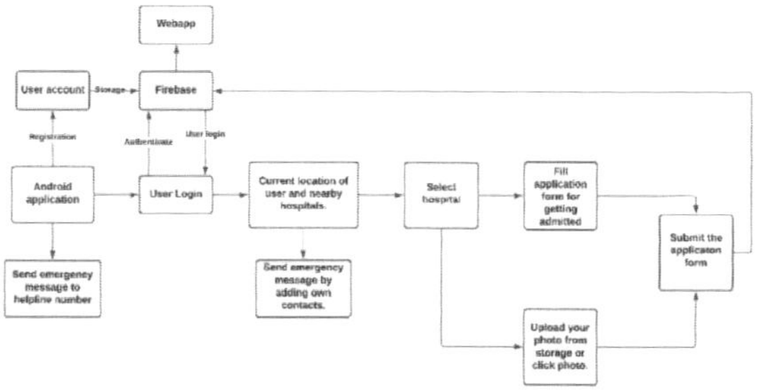

Fig 1. Block diagram of mobile app

The mobile app is connected to the Firebase. Firebase provides various services including registering a new account via email and password using Firebase authentication service or login into an existing account using the same. It also provides a real time database for storing retrieval f user data. The main page of app displays a map with current location of user and markers for hospitals nearby along with an emergency button. Emergency button is used to call the nearest helpline number and send SMS to user's personal contacts which can be set by user. Select hospital provides two services, uploading a photo from phone or clicking a picture in app as well as a patient form which is submitted to the selected hospital and uploaded to the firebase database. Firebase Database is further connected to the web app.

 https://doi.org/10.1051/itmconf/20214003042

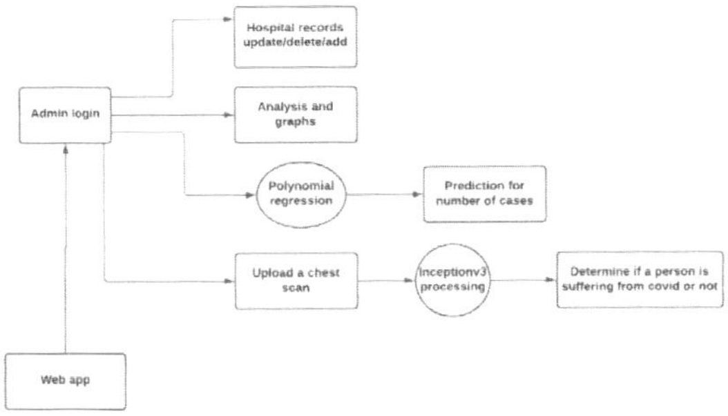

Fig 2. Block diagram of web app

The web app is connected to the Firebase like the mobile app. Firebase provides login into an existing staff account via email and password using Firebase authentication service. It also provides a real-time database for storing retrieval of user data. The admin or the hospital staff can handle multiple functionalities such as CRUD operations, Analysis and graphs, Covid cases prediction and Covid Detection system. Data generated by user from the mobile app can be or created, retrieved, updated, deleted using the web app. Analysis is done in the same data and represented in the form of graphs and charts. Covid Prediction system is Machine learning system which uses Polynomial regression to predict the Covid cases in the future. Finally Covid detection system is a Machine learning system which analyses the Chest can of patient and detects the existence of Covid-19 virus and generates a report.

5. Results

We have created an android app which can be used by the patient to get admitted to a nearby hospital by filling an application form. The user can also inform his emergency contacts and make an emergency call using the app in case of emergency. The app also shows the fastest route to the selected hospital using google map api. The number of available beds can be seen for any hospital.

The records of patients can be accessed using the web app by the admin. The web app shows many graphs relating to Covid updates such as everyday cases, deaths, ICU patients using live data from api. Through web app admin can change the status of patient from admitted to discharge or just to admit the patient. The number of available beds can also be updated.

In the web app we can see the clusters of Covid patients (as shown in fig. 10) to determine the average number of days required to discharge in association to age.[12] The web app can also predict number of confirmed cases of India for any day in the near future. The model has an accuracy of 99.65%. The web app can also analyze a chest scan to detect if a person is suffering through Covid-19 using Inceptionv3 and deep learning and the results will be displayed.

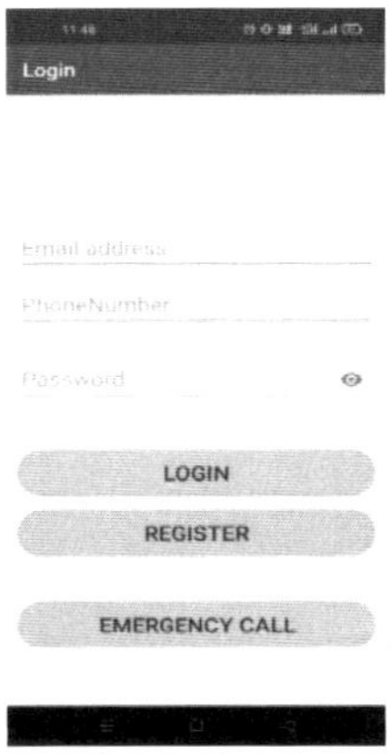

Fig 3. Login page

This is the first page that is displayed to the user on the application. The user can login or else the user can click on register button if he is first time user.

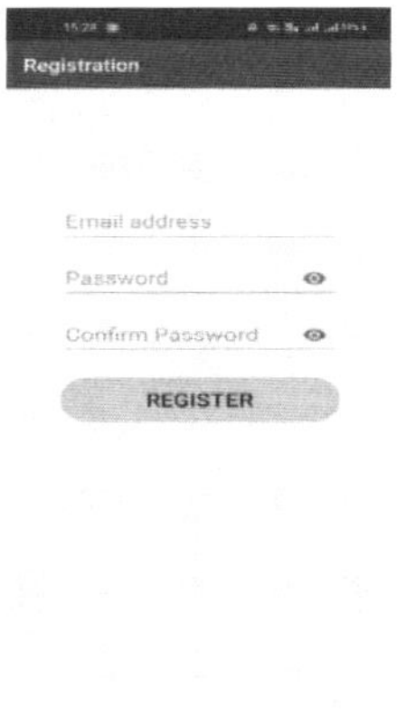

Fig 4. Registration page

The user can register using a valid email id and password.

Fig 5. Maps

https://doi.org/10.1051/itmconf/20214003042

As soon as the user logins the google maps are displayed indicating the user's current location and the nearby hospitals. The user can select one of the hospitals.

Fig 6. Add emergency contacts from phonebook.

The emergency contacts can also be added.

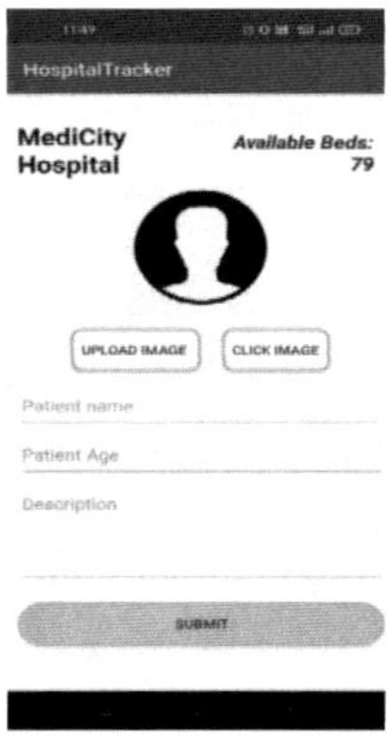

Fig 7. Application form

The user needs to fill the form and upload his image or take it using the camera hardware. The user then submits the application form and is redirected to the maps.

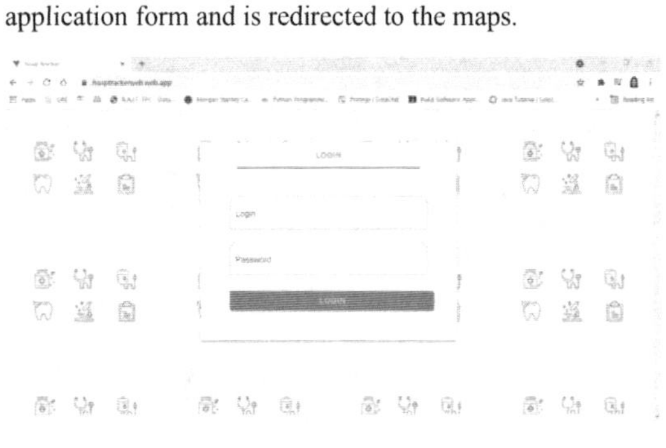

Fig 8.Admin login

Web app login for admin, uses email id and password as credentials.

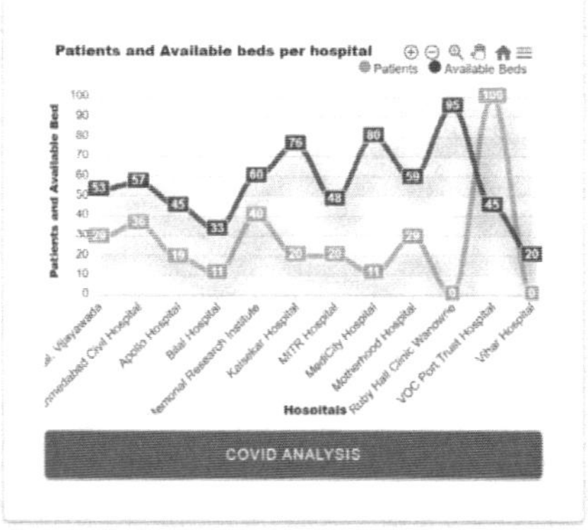

Fig 9. Graph for hospital occupancy

The graph shows the available and occupied number of beds hospital-wise.

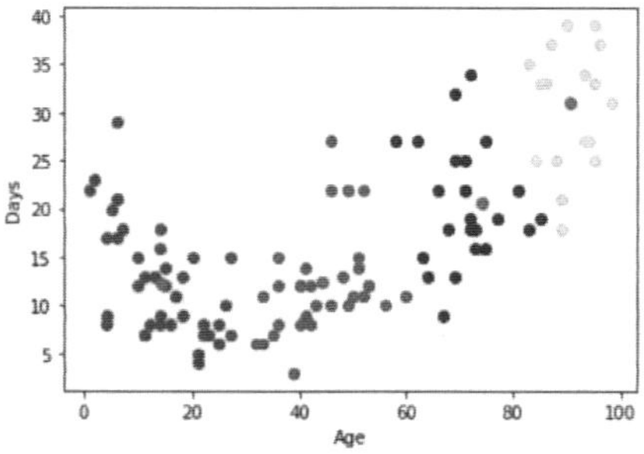

Fig10. Clustering graph

This graph shows clustering of patient data that depicts the average number of days required to discharge from the hospital with respect to age. X-axis depicts the patients age and Y-axis depicts the number of days required by different patients to fully recover. The bright red dot represents the average number of days to recover. This visualization can help the hospital management to plan accordingly and monitor the patients recovery. It also gives insight if the number of beds should be increased if the recovery time has increased, making the bed availability scarce.

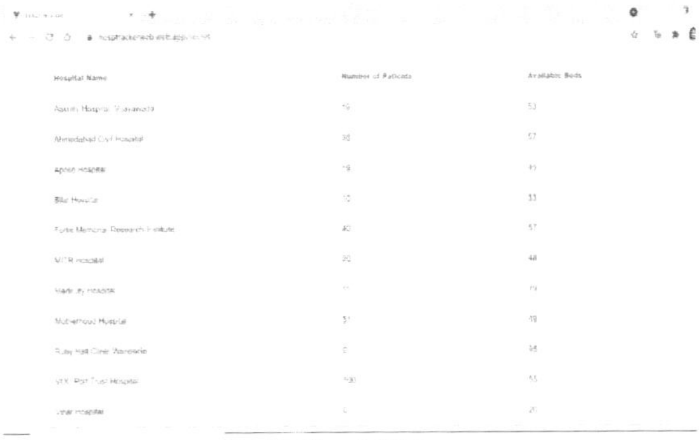

Fig 11. Hospital list

This is the list of names of hospital that have registered in the app and can be seen by users for getting admitted.

Fig 12. Records of patients

The records of patients of a particular hospital which also shows a pi chart for different disease count and male female pi chart. The records can be added, deleted, and updated.

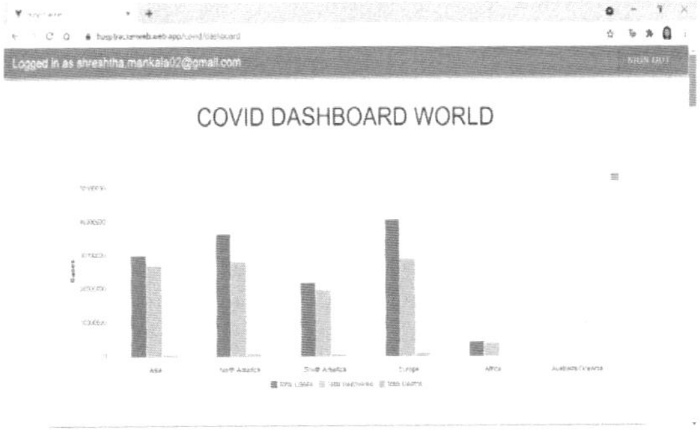

Fig 13. Covid dashboard

The total cases, recovered, and deaths for different continents can be seen and other graphs such as daily cases, everyday deaths, recovered, ICU patients for India can be seen.

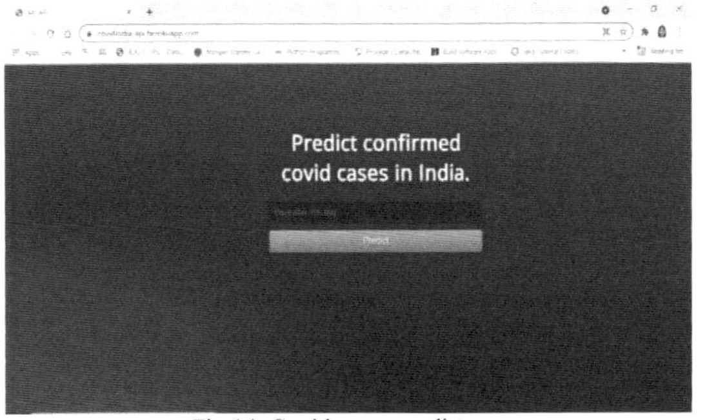

Fig 14. Covid cases predictor

Type the number of days and get an estimation of total confirmed cases of India after that many days.

Fig 15.Chest scan analyzer

The chest scan is to be uploaded by selecting an image from the system so as to make a decision if a patient is suffering from Covid or not.

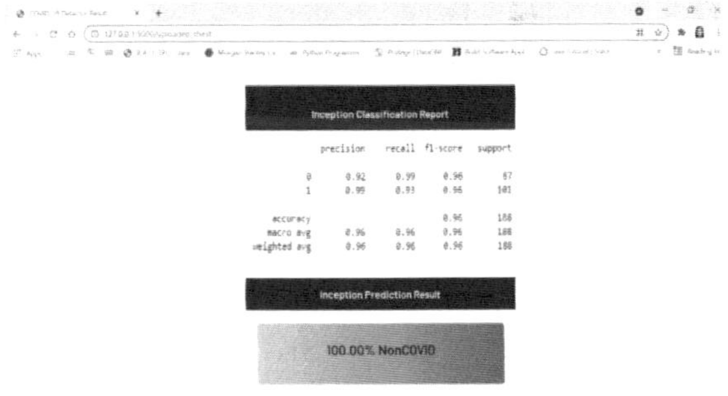

Fig 16. Chest scan result

The result or report of the scan is shown to decide if the patient is suffering from Covid or not. The accuracy of model is 96% as shown in Fig16. The Inception-v3 model uses 500 epochs and 32 batch size for getting best accuracy.

 https://doi.org/10.1051/itmconf/20214003042

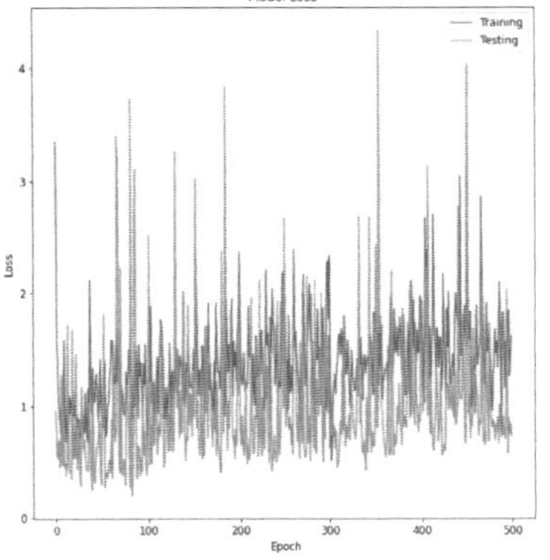

Fig 17. Model Loss

Fig 17 depicts the loss of Inception-v3 model during the training and testing phase. The loss in the testing data has declined significantly towards the 500th epoch as compared to the training data. The maximum loss observed is approximately 4% during the training phase and less than 3% during the testing phase, achieving the best results.

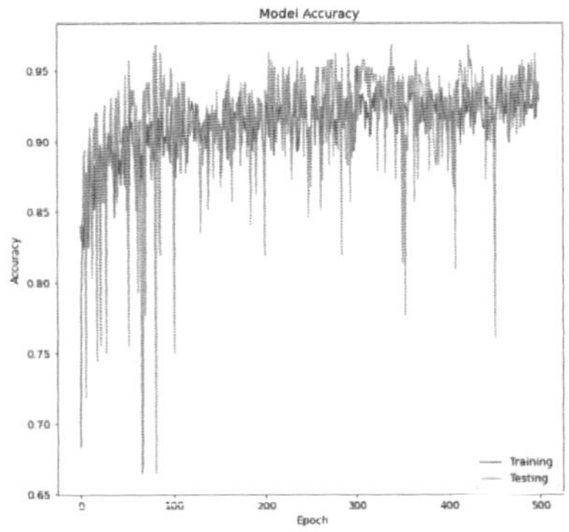

Fig 18. Model Accuracy

Fig 18 depicts the accuracy of Inception-v3 model during the training and testing phase. The model achieved the accuracy of 96% and got stable results while close to 500 epochs. X-axis shows number of epochs and Y-axis represents the accuracy metrics.

6. Conclusion

Covid-19 pandemic is a growing exponentially daily. With the ever-increasing number of cases, bulk testing of cases swiftly may be required. We have successfully categorized covid-19 scans, and it encourages the possible implementation of such techniques in the near future to automate diagnosis tasks. It is advised to consult medical professionals for any practical use case of this project. We do not intend to develop a perfect detection mechanism but only research about possible economically feasible ways to combat this disease. Such methods may be pursued for further research to prove their real case implementation.[7]

7. Future Scope

Currently the user has to provide login credentials in order to use the app or if the account does not exist the user has to register and then login. We can create a functionality where user can login using his/her google account. Secondly, we can also display the hospitals based on the emergency. For example, if the patient has a heart attack, then all the heart care hospitals with their location will be suggested.

8. References

1. https://www.columbiaindiahospitals.com/health-articles/medical-emergency-vs-urgent-care-medical-conditions
2. Mokter Hossain, The effect of the Covid-19 on sharing economy activities, (2020)
3. Dr.Ramesha.T ,Effect of COVID-19 and Historical Background – An Overview **7** , Issue 4 (2020)
4. https://www.webmd.com/lung/coronavirus-history
5. Ekta Gambhir, Ritika Jain, Alankrit Gupta, Uma Tomer, Regression Analysis of COVID-19 using Machine Learning Algorithms, **CFP20V90-ART; ISBN: 978-1-7281-5461-9** (2020)
6. Rajan Gupta, Saibal K. Pal and Gaurav Pandey, A Comprehensive Analysis of COVID-19 Outbreak situation in India (2020)
7. Rachna Jain, Meenu Gupta, Soham Taneja, Jude Hemanth D, Deep learning-based detection and analysis of COVID-19 on chest X-ray images(2020)
8. https://localpress.co.in/mumbai/doctor-launches-hero-app-help-people-locate-hospital-beds-mumbai/
9. Jürgen Stausberg, Priv-Doz Dr med, Dietrich Koch, Josef Ingenerf, Dr Rer Nat, and Michael Betzler, Prof Dr Med, Comparing Paper-based with Electronic Patient Records: Lessons Learned during a Study on Diagnosis and Procedure Codes, **10(5): 470–477** (2003)
10. Jia Deng; Wei Dong; Richard Socher; Li-Jia Li; Kai Li; Li Fei-Fei, ImageNet: A large-scale hierarchical image database (2009)
11. Alex Krizhevsky, Ilya Subserve, Geoffrey E. Hinton, ImageNet Classification with Deep Convolutional Neural Networks (2012).
12. Wenjing Ye Weiwei Lu, Yanping Tang, Identification of COVID-19 Clinical Phenotypes by Principal Component Analysis-Based Cluster Analysis. (2020)

LEAF DISEASE DETECTION OF CUCURBITS USING CNN

Isha Agrawal[1] ,Prada Hegde[2] Pooja Shetty[3], and Priyanka Shingane[4]

[1]Department of Information Technology, Ramrao Adik Institute of Technology, Nerul, Navi Mumbai, India
[2]Department of Information Technology, Ramrao Adik Institute of Technology, Nerul, Navi Mumbai, India
[3]Department of Information Technology, Ramrao Adik Institute of Technology, Nerul, Navi Mumbai, India
[4]Department of Information Technology, Ramrao Adik Institute of Technology, Nerul, Navi Mumbai, India

Abstract. Identification of plant disease is tough in agribusiness arena. If it is inaccurate, there occurs a tremendous damage in the production and economical price. Leaf Disease detection requires huge amount of work, knowledge, processing time in plant disease. The most used and edible vegetable all over the world is from cucurbitaceous family. The crops under this family have great economic value in the food industry and its production is done in large scale. This family consists of 965 species. If any of these plants catch disease then there would be a tremendous loss in the production of this field yields. Thus, treating them at early stage is best way to prevent such losses. Hence, Deep Learning Algorithm like CNN can be used to detect the diseases of the plants. The leaves of the plants would be used as primary material for identification of the disease, as they are much more visible on the leaves.

Keywords -- Leaf Disease Detection, Cucurbitaceae family, CNN, Deep Learning.

I. INTRODUCTION

Agriculture is an essential factor in the financial development of the nation. Certain factors affect the condition of the crops. Due to varying weather constraints these plants are highly prone to infections. Some of the plant species are huge family , thus if the diseases are undetected then it may lead to huge loss. Thus detecting and classifying each and every plant is mandatory , even a single infected plant can expand the disease. In rural areas, farmers lack certain knowledge about the diseases and better treatments for them. Hence we have developed a Deep Learning based solution to solve all these problems.

The proposed system named "Leaf disease detection of cucurbits using CNN" is going to be very helpful for farmers and agriculture practitioners. Using this system one can easily detect diseases occurred to the plants from the Cucurbitaceae family. This system will be created using an algorithm which will benefit the farmhand to discover the disease in the initial stage and can give the required treatment. There are many different systems that detect diseases for tomatoes, cotton, rice, maize, soya bean but there is no such system that detects diseases for Cucurbit family.

Isha Agrawal: ishaagrawal2000@gmail.com
Prada Hegde: pahegde99@gmail.com
Pooja Shetty: p.shetty1406@gmail.com
Priyanka Shingane: priyankashingane@gmail.com

As agriculture is the most important aspect, using this system will help to reduce the loss in agricultural yields of cucurbit family due to diseased plants. This system can also give the required treatment.

Most of the plant leaves are infected due to bacteria, fungi and atmospheric pollutants. For using this system one must use an image of a leaf, of a diseased plant. After uploading the image the disease will be detected and appropriate treatments will be suggested.

This project will help farmers to find out about the disease at its primary stage. This project will contain disease detection for some of the species of the family. The system detects the most common diseases The disease will classified based on texture , color and shape.

Fig 1. Pictures of some cucurbitaceae family fruits.

II. LITERATURE SURVEY

Different methods and classifiers are used by various researches to classify the diseases occurred. This paper is very precise for artificial intelligence and gives remedy for determining and categorizing discrete plant leaf diseases present, and uses convolutional neural network for classification grounds. It has four classifiers such as– logistic regression, KNN, SVM and CNN.[2]

K-means clustering is used for segmentation of images, GLCM is also used for texture feature extraction, comprising the mean for color feature extraction and KNN, Neural Network, and lastly SVM is for the final analysis of disease.[1]

Paper by Raghottam Kulkarni and Dr. A.V Sutagundar titled Plant Leaf Disease Management System (IEEE 2017). The main aim of the system was to analyze the plant species from its leaf and classify whether the leaf is in good physical condition or not and then if the leaf is detrimental, it warns about the disease the plant has and if it is in a good physical condition then it shows it is normal. Some of the main methods for detection of leaf diseases are: K-Means cluster, GLCM, and Neural Network Classifiers. Such methods successfully detect the healthy and unhealthy plant leaves. Hence, the project beneficial for identifying the plant disease.[3]

Disease detection was done on cotton plants using GLCM and SVM. To classify the diseases like Magnesium deficiency and Bacterial blight are differentiated by diverse class SVM classifier. Using SVM the features are pulled from the segmented images by considering the inputs given.[4]

The proposed methodology of image processing is achieved using MATLAB. Image processing is carried out on leaf images and features are pulled out from the diseased using k-means clustering algorithm. Variety of characteristics such as contrast, correlation, energy, homogeneity, mean, standard deviation and variance are taken out for tomato and cotton diseases.[5]

K-Medoid clustering and Random Forest classification algorithms are used to determine the diseases. The workflow of the project is: It stores an image dataset of unhealthy and healthy stalks. The whole database is separated into training and testing sets. Using this data the entire model is trained to detect the diseases.[6]

Based on RGB factors, the filters are put onto three channels. The LVQ are stored combining the resultant feature vector of convolution to train the network. The outputs confirm that the proposed methodology is very effective in recognizing the four varied types of tomato leaf diseases.[7]

The given algorithm distinguishes disease in plants and classify them in a very accurate manner as compared to the existing techniques.[8]

The system provides an accurate result generated by using various cluster sizes, enhanced the experiment with image segmentation. The system gives viable evaluation and prediction of the disease causing agent with necessary safety measures.[9]

Summary of Leaf-based plant disease detection systems: It is a systematic study to find and classify the leaf disease. This paper was written by Ravindra Jogekar , Dr. Nandita Tiwari.[10]

III. PROPOSED METHODOLOGY

The system proposed has the following phases:

- Image Collection
- Image pre-processing
- Segmentation/Feature Extraction
- Selection of Classifier
- Analyzing Results
- Predicting the Disease
- Treatment Suggestion

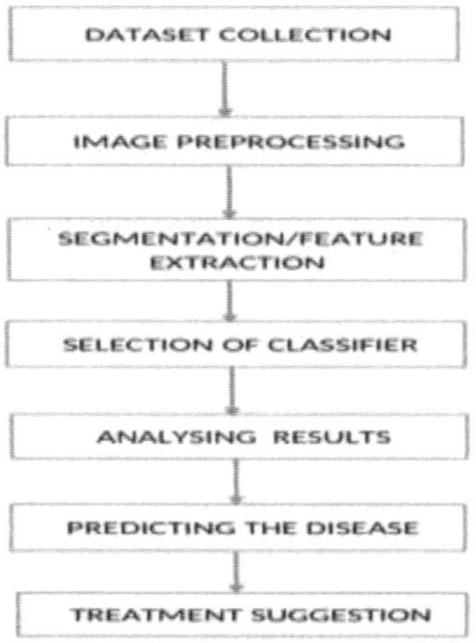

Fig 2. Workflow of the System

1. Image Collection:

In Image Collection, data-sets consist of both healthy and diseased leaves which has disorders like powdery mildew, downy mildew. Anthracnose, angular leaf spot, bacterial leaf spot. The images consist of cucumber, squash, pumpkin and watermelon. The pictures of the leaves were gathered from different online sites and were classified into healthy and diseased leaves.

2. Image Preprocessing:

In this, the images are resized into the resolution of 256x256 .The pre-processing is used to upgrade the image quality and to obtain the required features of images.

Formula for RGB to Grayscale:

Gray scale to RGB, have equivalent red, green and blue values.

$$B=G=R \quad (1)$$

For every image, pixels consisting red, green and blue of (R,G,B)

$$R' = G' = B' = (R+G+B) / 3 \quad (2)$$

3. Segmentation:

Segmentation of the image is carried out to separate the background from the leaves.

4. Selection of Classifier:

Here, the classification of the type of disease occurring on the leaves is done. There are many Deep Learning Algorithms which can be applied on the database. We have selected CNN as our classifier.

CNN (Convolutional Neural Network)

It is a complicated model that is used to categorize the diseases and is complex, thus it needs a good computational power. It is also a commonly used neural network which is referred to certain image classification problems. This system comprises of "Seven Layers" :

1. Input Layer (32 filters, Activation = 'relu')
2. Max pooling (strides = 2x2)
3. Convolutional Network Layer (64 filters, Activation ='relu')
4. Max pooling (strides = 2x2)
5. Flattening Layer
6. Fully connected Dense Layer (units = 120, Activation = 'relu')
7. Fully connected Dense Layer (units = 18, Activation = 'softmax')

The Database of the leaves from Cucurbitaceae family is created. It mainly consist of healthy and diseased leaves (i.e Angular leaf spot, Anthracnose, Bacterial leaf spot, Downey Mildew, Powdery Mildew) of Pumpkin , Cucumber, Watermelon and Squash. The images of these leaves are converted from RGB to Gray Scale for further process, as it helps in getting better accuracy. After the step of image pre-processing, using CNN a Deep Learning Algorithm training of the model is done.

The libraries used are open cv, keras, tensorflow, os, matplotlib.
The dataset was distributed in training set and validation set. Later the training data was generated by sending certain parameters like target_size, color_mode.

The the model was trained using Convolutional Neural network. There is a input layer with 32 filters and a convolutional network (2D) with 64 filters. Two max pooling layers of strides 2x2 are used, followed by a flattening layer. There are two fully connected dense layers with units 120 and 18.

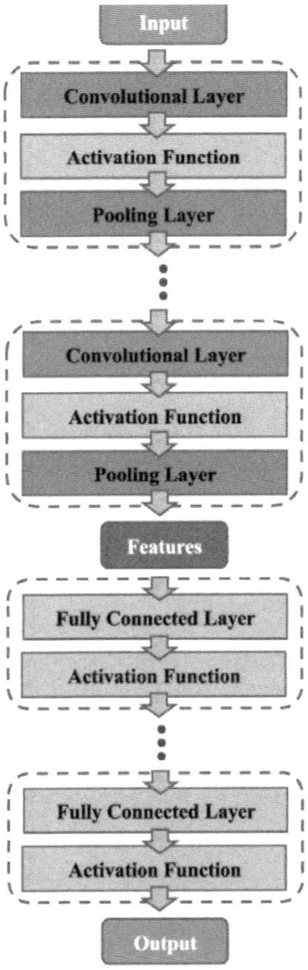

Fig 3. System Architecture

Max pooling in general, is an operation which calculates the maximum, or the largest value in each spot of feature map.

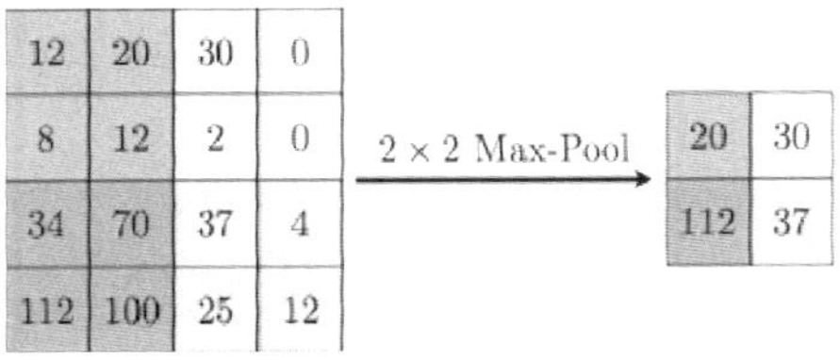

Fig 4. Example of Max Pooling

https://doi.org/10.1051/itmconf/20214003043

IV. EXPERIMENTAL RESULTS

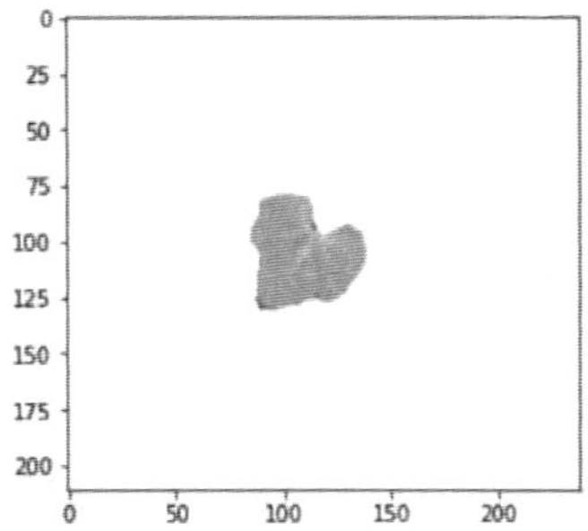

Fig 5. Image of leaf without background

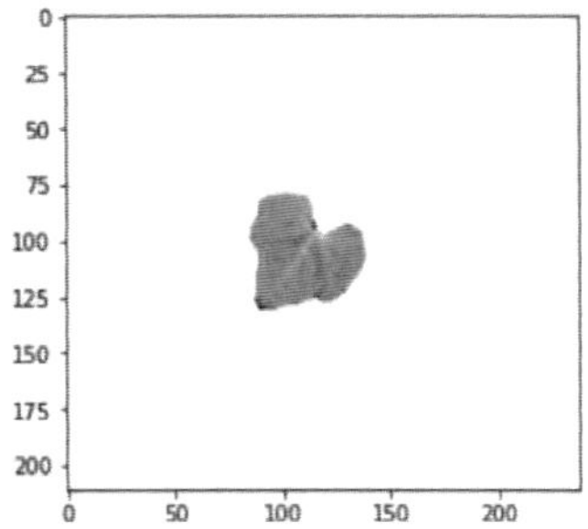

Fig 6. Conversion to grey scale

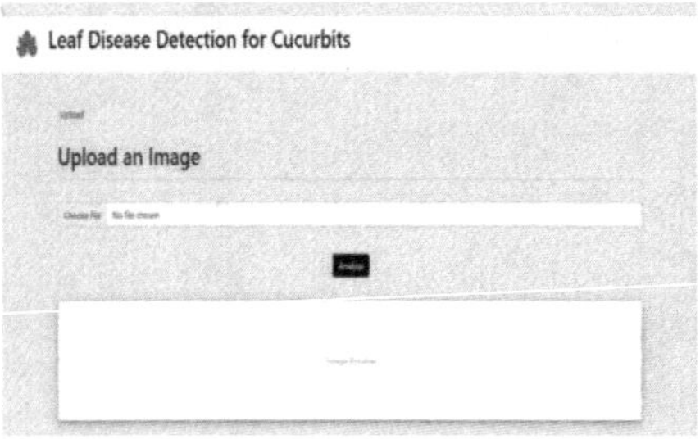

Fig 7. UI of Leaf Disease Detection of Cucurbits .

Fig8. Selecting an image for analysis (Diseased)

Fig 9. Displaying the result after analysis (Diseased)

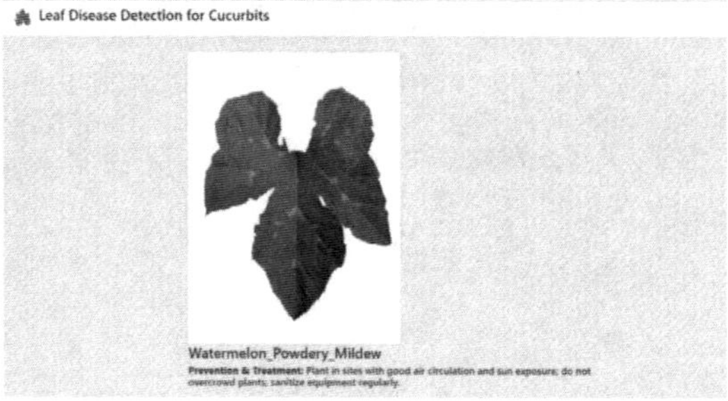

Fig 10. Treatment suggestion for the disease detected.

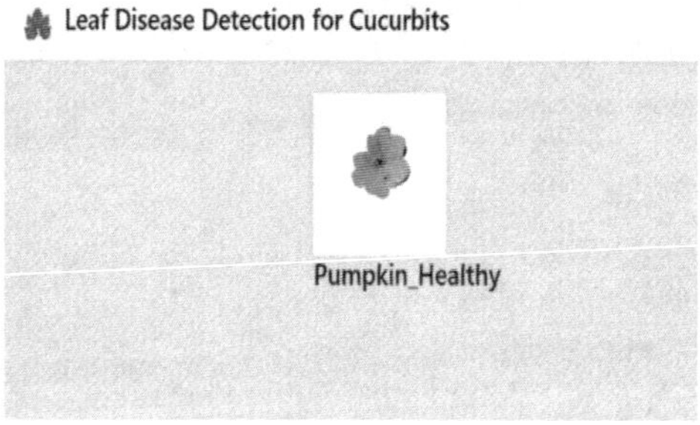

Fig12. Displaying the result after analysis (Healthy)

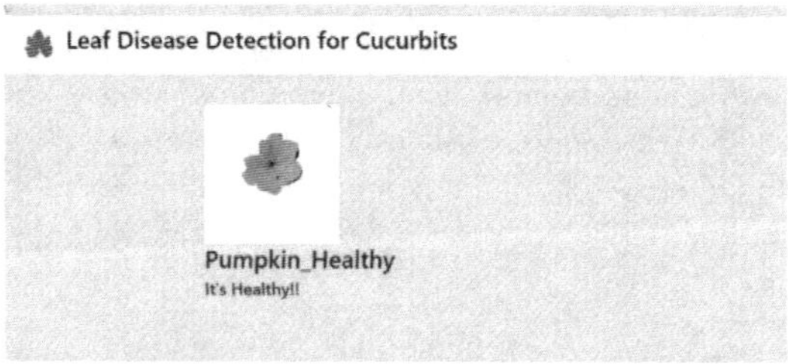

Fig 13. Suggestion for a healthy leaf

CODE SNIPPETS :

```
import tensorflow as tf
from tensorflow.keras import datasets, layers, models
model = tf.keras.models.Sequential()

model.add(layers.Conv2D(32,(3,3),input_shape = (256,256,1),activation= 'relu'))
model.add(layers.MaxPooling2D(pool_size=(2,2)))
model.add(layers.Dropout(0.05))

model.add(layers.Conv2D(64,(3,3),activation= 'relu'))
model.add(layers.MaxPooling2D(pool_size=(2,2)))
model.add(layers.Dropout(0.2))

model.add(layers.Flatten())

model.add(layers.Dense(120,activation= 'relu'))
model.add(layers.Dropout(0.3))

model.add(layers.Dense(18,activation= 'softmax'))
model.compile(loss = tf.keras.losses.CategoricalCrossentropy(),
              optimizer = 'adam',
              metrics = ['accuracy'])

history = model.fit(train_generator, epochs= 20,batch_size = 10, validation_data = test_generator)
```

Fig14. Layers applied in the model.

```
from tensorflow.keras.preprocessing import image
import numpy as np
import matplotlib.pyplot as plt
import cv2
def prediction(test):
    #plt.imshow(test)
    test7 = cv2.resize(test,(256,256))
    test8 = cv2.cvtColor(test7, cv2.COLOR_BGR2GRAY)
    test_img1=image.img_to_array(test8)
    test_img1=np.expand_dims(test_img1,axis=0)
    result=model.predict(test_img1)
    a=result.argmax()
    #print('a:',a)
    classes=train_generator.class_indices
    #print(classes)
    #print(len(classes))
    category=[]
    for i in classes:
            category.append(i)
    for i in range(len(classes)):
            if(i==a):
                output=category[i]
    return(output)
```

Fig15. Prediction Funtion

Name	Leaf Disease Detection using CNN	Other Systems
Remedies	In our system , treatment will be suggested according to the disease of the leaves at early stage.	In some systems, remedies are not suggested.
Detection	Disease detection is done on some of the species of cucurbitaceae family.	There is no disease detection system for Cucurbits
Accuracy	Moderate percent of accuracy is obtained	Less or moderate percent of accuracy is obtained.

Table 1 Comparison of our system with other existing system.

V. CONCLUSION

In the agriculture industry, there is great risk of the diseases that attacks the plants. Cucurbitaceae being the largest family and the most used vegetable , it is important to treat the plants in this family. This sums up the major image pre-processing and identification of leaf diseases using CNN Algorithm in the backend and UI is designed using Flask. The Training accuracy achieved is 72% and testing accuracy achieved is 12% .This approach can significantly helps in accurate detection of leaf disease.

By analyzing the amount of disease present in the leaf, one can use sufficient amount of fertilizers to control and treat the disease. Also the user can understand the affected area of leaf and solve the problem with ease.

VI. FUTURE SCOPE

The accuracy can be improved by trying different algorithms or applying some unsupervised learning technique such as k-means clustering. The quantity of dataset needs to be increased in order to achieve accurate predictions. This system can be improved by converting the proposed web app to android app in order to increase its usability . Additional species can also be added to cover the entire family.

REFERENCES

[1] Neha Kurale and Madhav Vaidya, "*Classification of Leaf Disease using Texture using feature and Neural Network* Classifiers", ICIRCA, 2018.

[2] Pushkar Sharma, Pankaj Hans , Shubash Chand Gupta, "*Classification of Leaf Disease using Machine Learning and Image Processin*", IEEE 2020

[3] Raghottam Kulkarni and Dr. A.V Sutagundar, "*Plant Leaf Disease Management System*", ICCMC, 2017.

[4] Namrata Bhimte and V.R Thool "*Disease Detection of cotton leaf spot using Image Processing and SVM Classifier*", ICICCS, 2018.

[5] Ch. Usha Kumari, S. Jeevan Prasad, G. Mounika, "*Leaf Disease Detection : Feature Extraction with K-means Clustering and Classification using ANN*", ICCMC, 2019.

[6] Indumati R, Saagari N, Thejuswini V, Swarnareka R "*Leaf Disease Detection and Fertilizer Suggestion*",ICSCAN, 2019.

[7] Melike Sardogan, Adem Tuncer , Yunus Ozen, "*Plant Leaf Disease Detection and Classification Based on CNN with LVQ Algorithm*", ICCSE (UBMK), 2018.

[8] Sandeep Kumar, KMVV Prasad , A. Srilekha, T.Suman , B. Pranav Rao , J Naga Vamshi Krishna, "*Leaf Disease Detection and Classification based on Machine Learning*", ICSTCEE, 2020

[9]Nikita Goel, Dhruv Jain , Adwitiya Sinha, "*Prediction model for Automated Leaf Disease Detection & Analysis*", IACC, 2018.

[10] Ravindra Jogekar, Dr. Nandita Tiwari, "*Summary of Leaf-based plant disease detection systems: A compilation of systematic study findings to classify the leaf disease classification schemes*", WorldS4, 2020.

 https://doi.org/10.1051/itmconf/20214003044

Automated Non-invasive Skin Cancer Detection using Dermoscopic Images

Shruti Kale[1], *Reema Kharat*[2], *Sagarika Kalyankar*[3], *Sangita Chaudhari*[4], *Apurva Shinde*[5]

[1]Dept. of Computer Engineering, Ramrao Adik Institute of Technology, Nerul, Navi Mumbai
[2]Dept. of Computer Engineering, Ramrao Adik Institute of Technology, Nerul, Navi Mumbai
[3]Dept. of Computer Engineering, Ramrao Adik Institute of Technology, Nerul, Navi Mumbai
[4]Dept. of Computer Engineering, Ramrao Adik Institute of Technology, Nerul, Navi Mumbai
[5]Dept. of Computer Engineering, Ramrao Adik Institute of Technology, Nerul, Navi Mumbai

Abstract. Skin Cancer is resulting from the growth of the harmful tumour of the melanocytes the rates are rising to another level. The medical business is advancing with the innovation of recent technologies; newer tending technology and treatment procedures are being developed. The early detection of skin cancer can help the chance of increase in its growth in other parts of body. In recent years, medical practitioners tend to use non invasive Computer aided system to detect the skin cancers in early phase of its spreading instead of relying on traditional skin biopsy methods. Convolution neural network model is proposed and used for early detection of the cancer, and it type. The proposed model could classify the dermoscopic images into correct type with accuracy 91.2%.

Keywords- tumour, melanocytes, medical practitioners, non-invasive, biopsy, dermoscopic.

1 Introduction

Skin Cancer rates are rising to another level. In 2018, 287,723 new incidents were reported according to intervals the planet and so the vary of deaths happened because of skin cancer is 60,712. at intervals the U.S.A., almost 96,480 new malignant skin cancers area unit aiming to be identified. The harsh UV radiation from the Sunlight affects melanoma. Malignant melanomas invariably tend to unfold into surroundings but if treated at its early-stage dangers can be avoided. Therefore, early identification and treatment of skin cancer is particularly important and crucial task. Generally, in automated system it can be done in two parts: (1) Pre-processing of input and effective feature extraction and (2) Accurate classification of the image.

Skin cancer is the irregular growth of skin cells and its most frequently grows on skin exposed to the sun. But this form of cancer can also happen on areas of your skin not normally unprotected to sunlight. Skin cancer is broadly categorised into Basal Cell Carcinoma (BCC), Squamous Cell Carcinoma (SCC) and Melanoma Skin Cancer (MSC). BCC is because of skin openness particularly found in the sun uncovered regions like the face, head, neck, arms, and legs. SCC is the disease of keratinocyte cells found on the external surface of the skin. It seems like a red firm knocks layered patches. Melanoma emerges from melanocytes, specific pigmented cells that are discovered. Risk can be reduced by preventive or avoiding revelation to ultraviolet (UV) radiation. Inspecting skin for doubtful changes can help to detect skin cancer at its earliest along with some computer aided diagnosis system. Early detection of skin cancer is an important step toward its fast growth in other parts of the body. Early detection of cancers and its type can be detected by dermatologists with the help of computer aided system. Thus, the objective of this paper is to construct a Computer Aided Diagnosis system that assists medical professionals detect skin cancer without the need of any expensive tools.

2 Related Works

About 1 million cases of non-cancerous and 288,000 of cancerous cases have been detected in 2018. The impact

* Corresponding author: reema97.kharat@gmail.com

of biological theory on continuation and value more stress the previously exhaust aid structure and lift the query of economic property. Carcinoma and particularly metric linear unit early spotting is difficult for each dermatologists. Dermatoscope is reserved into version the excellence of care, however in objective test dermatologist come through a restricted diagnostic sensitivity of 400 metric linear unit detection Thanks to the quality of graphic ideas rooted in an exceedingly dermoscopic image. General doctors appear to profit from use of a dermoscopic course, whereas several 51 accurately diagnosed lesions necessitate more enhancements. Likewise, specificity of diagnosing by dermatologists necessitate an extra improvement, as mirrored by a range of 28:1 to 9:1 variety of operations that require to be removed to spot one skin cancer and a 3:1 magnitude relation for general carcinoma.

Deep learning (DL) classifiers area unit an auspicious applicant for discovery of different types of skin cancers. With al, laboratory studies reportable a clinical sensitivity, a difference which could be credited to the standard of the dataset input, thus version knowledge as new. Lately, a primary potential scientific empiric education reportable on a 2-step method, count a second layer of sonification (visual information was noises) to a decilitre classifier to enhance accuracy of detection. This twin decilitre utilised a sophisticated dermoscope, a comparatively dear device, and a way extremely keen about medico expertise translation it fewer appropriate for extensive medical care doctor's use. Accordingly, the influence of copy superiority on correctness of diagnosing has been suspected. It had been determined to check a cheap device secret by its builder as a skin scientific instrument with polarized lightweight (SMP). Picture's quality paternal by SMP preclude in most cases a particular clinical diagnosing thanks to haziness and lack of fine high level dermoscopic patterns and diagnostic structures. Table 2.1 shows summary of existing approaches.

Table 2.1: Survey of Existing Systems

Authors	Image Acquisition	Image Pre-processing	Image Segmentation	Feature extracted	Classification
Ichim and Popescu [3]	ISIC and PH2	Elimination of noise, Dull Razor hair removal, contrast and normalisation	-	Texture features, shape and colour features	Neural network, Back propogation, AlexNet and ResNet
Demyanov et al. [4]	ISIC	-	-	K Means clustering and sparse coding with visual features from Dense SIFT and SURF	SVM with RBF Kernel and CNN
Majumder and Ullah[5]	PH2 Dataset	Dull-Razor algo- to remove dark hair from images. Median filter is also applied	Otsu's thresholding method	Asymmetry score across x and y-axis, Area/perimeter ratio, Compactness index, Average diameter of lesion	ANN with backpropagation algorithm
Jain et al.[6]	PH2 Dataset	Image resizing and contrast adjustment. Applied averaging filter to RGB input image.	Otsu's thresholding method is applied	Asymmetry score across x and y-axis, ABCD features, variegation and lesion diameter variation	Backpropagation Neural Network (BNN)
Eltayef et al.[7]	200 8-bit RGB color images with a resolution of 768*560	Directional Gabor filters are implemented to extract the hairs	Fuzzy c-means and MRF is used	Asymmetry, border, Color variegation and lesion diameter	Not specified
Isasi et al.[8]	40 RGB OF 500x500 pixels image catalogue by certified dermatologistes	Pattern recognition - globular	Globular, Recticular and blue-vein	ABCD feature alongwith features extracted from pattern recognition algo.	Classification is done by ABCD rule
She et al.[9]	Not specified	High pass filtering and gradients application	Snake-based detection technique is used to determine lesion boundry	Asymmetry score across x and y-axis, Area/perimeter ratio, Compactness index, Average diameter of lesion.	Principal component analysis is used for classification

3. Proposed System

Various attempts made by researchers to detect skin cancer in its early stage using image processing techniques, machine learning and deep learning techniques. As the problem of existing system was that the output was not accurate. Even if it's cancerous cell it was detected as non-cancerous or vice-vera. System Flow Diagram predicting malignant lesions is depicted in figure 3.1. It consists of four steps: (1) image pre-processing; (2) image segmentation; (3) feature extraction; (4) classification. The input image can have noise and other artefacts such as hairs, air bubbles, varied colour, and textures of the skin. To get rid of these factors, pre-processing need to be applied on input image before passing it to segmentation or feature extraction phase so that eventually classifier can classify these input lesions in correct classifications as malignant or benign and the type of cancer.

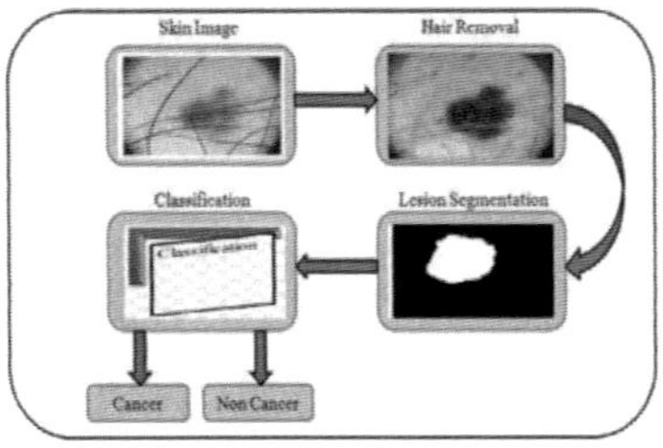

Figure 3.1: Skin Cancer Detection Process

In our proposed system, we are utilising modified Convolutional Neural Network (CNN). CNN consist of sequence of several convolutional layer with filters, pooling layers, fully connected layer. Convolutional layer is the initial layer which extract feature from an image. Pooling layers minimize the size of the activation map. Fully connected layer operates on a flattened input where each input is connected to all neurons. Figure 3.2 shows architecture of CNN.

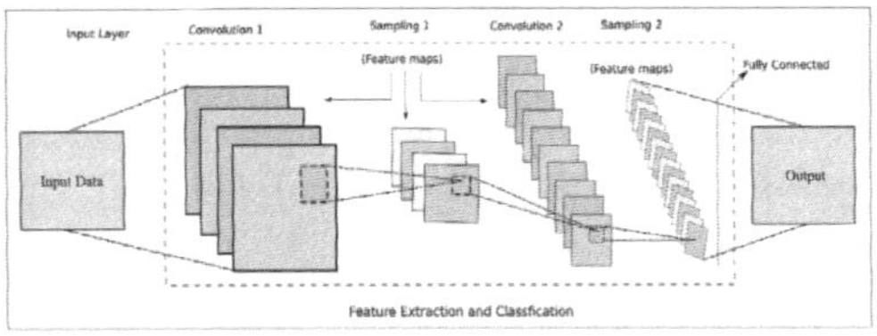

Figure 3.2: CNN Architecture

CNNs are convolution neural networks that are appeared to be terribly strong in areas like image recognition and classification. CNNs area unit a supervised learning technique and area unit thus instruct victimisation knowledge tagged with the various categories. primarily, CNNs grasp the link in the middle of the input items and the category tags and contains 2 elements: the invisible layers during which the options area unit drawn out and, at the top of the process, the totally connected layers that area unit used for the classification task. not like regular neural networks, the hidden layers of a CNN have a particular design. In regular neural networks, every layer is made by a collection of vegetative cells and one vegetative cell of a layer is connected to every neuron of the previous layer. The design of hidden layers in a very CNN is slightly completely different.

This restriction to native connections and extra merge layers sums up native vegetative cell product into one worth leads to Translation in variant options. This leads to an easier coaching procedure and a lower model quality. Figure 3.3 shows the steps followed in the proposed system. Data: Data is extracted for analysis towards skin cancer. In our project we have used over 8000 images. To test our system.

Data Visualization: It is a graphical presentation of knowledge of data. By applying perceptible segments like charts, graphs, maps to provide an approachable way to see and acknowledge direction, outliers, and patterns in data.

Image Pre-Processing: Is a procedure of making the rare information and creation it appropriate for a machine learning model. A filter is used to enhance the objects such as hair, marks in bright background.

CNN: Over here the system predicts if the image is cancerous or not, and if its cancerous then it predicts which type of cancer it is.

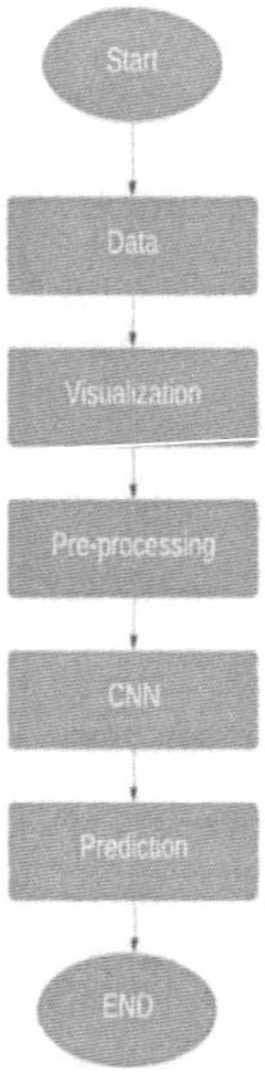

Figure 3.3: Flowchart for Proposed System

Performance Metrics for classifiers: A classifier set apart every object to a category. This piece of work is mostly not excellent, and entity is also assigned to the incorrect category to gauge the classification quality, the category—the category allocated by the classifier is likened with the class. this permits the substances to be alienated into the subsequent four subsections:

1. True positive (TP): Positive Class.

* Corresponding author: reema97.kharat@gmail.com

2. True negative (TN): Negative Class.
3. False positive (FP): Incorrect Positive Class.
4. False Negative (FN): Incorrect Negative Class.

4. Results and Discussion

The diagnostic system proposed in this work has been implemented in Python. To testing the system, a dataset of approximately 10000 dermoscopic images from PH^2 dataset are considered including mixed lesion images for Benign keratosis, melanocytic nevi, dermatofibroma, melanoma, vascular lesions, basal cell carcinoma, actinic keratoses [10].
The system could correctly classify the images as 'Benign' or 'Malignant' as well as the type of skin cancer as BCC, SCC or MSC. Figure 4.1 shows the sample statistics for different dermoscopic images. Figure 4.2 shows sample images for various types of skin cancer. Figure 4.3 and 4.4 shows localization and characteristics of lesion for some sample input images.

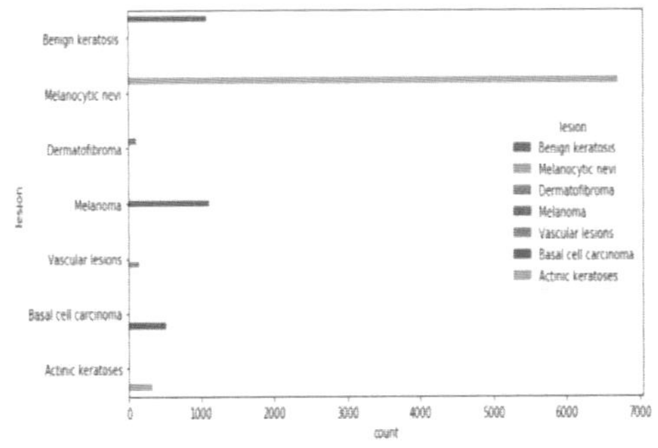

Figure 4.1 : Satistics of dermoscipic images used in experimentation

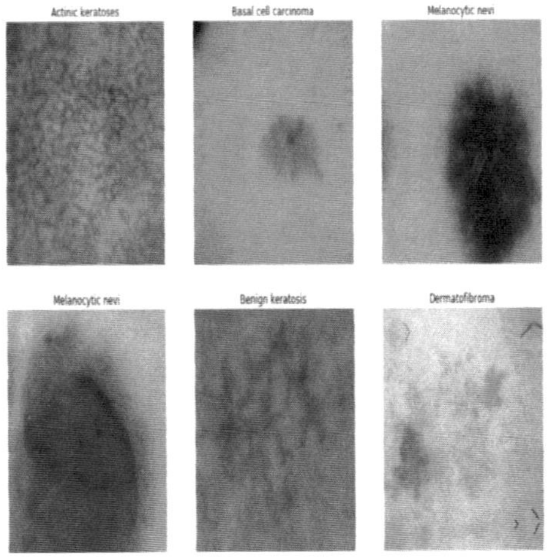

Figure 4.2 : Skin cancer dermoscopic images

	lesion_id	image_id	dx	dx_type	age	sex	localization
0	HAM_0000118	ISIC_0027419	bkl	histo	80.0	male	scalp
1	HAM_0000118	ISIC_0025030	bkl	histo	80.0	male	scalp
2	HAM_0002730	ISIC_0026769	bkl	histo	80.0	male	scalp
3	HAM_0002730	ISIC_0025661	bkl	histo	80.0	male	scalp
4	HAM_0001466	ISIC_0031633	bkl	histo	75.0	male	ear

Figure 4.3: Localization of lesions

	lesion_id	image_id	dx	dx_type	age	sex	localization	lesion
0	HAM_0000118	ISIC_0027419	bkl	histo	80.0	male	scalp	Benign keratosis
1	HAM_0000118	ISIC_0025030	bkl	histo	80.0	male	scalp	Benign keratosis
2	HAM_0002730	ISIC_0026769	bkl	histo	80.0	male	scalp	Benign keratosis
3	HAM_0002730	ISIC_0025661	bkl	histo	80.0	male	scalp	Benign keratosis
4	HAM_0001466	ISIC_0031633	bkl	histo	75.0	male	ear	Benign keratosis

Figure 4.4: Characteristics of lesions

Figure 4.5 show the user interface of the system and the predicted output as the type of skin cancer. There is 91.2 accuracy achieved at epoch as 19 for the CNN model. Figure 4.6 shows the confusion matrix for 200 sample which are used in testing the model.

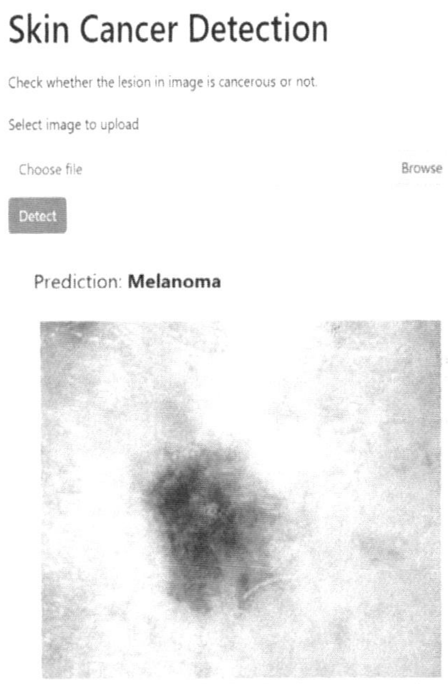

Figure 4.5: The predicted skin cancer type

https://doi.org/10.1051/itmconf/20214003044

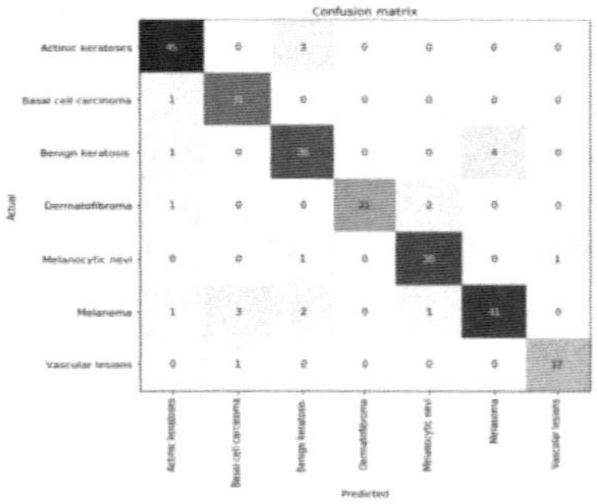

Figure 4.5: Accuracy assessment of the predicted result

5. Conclusion

In recent years, tremendous growth has been observed in different type of skin cancer in Asian as well as other countries. Non-invasive approaches are gaining popularity for early detection of skin cancers. A well-trained computer aided diagnosis system can correctly detect presence of skin cancer and the type of cancer. In the proposed system some pre-processing is applied on input images to enhance the quality of images and to remove hairs from those images. CNN model is utilised which could be able to correctly detect the skin cancer and classify it as one of the three types of cancer with an accuracy of 91.2%. However, overall accuracy can be affected in case of dermoscopic image of dark-skinned patients. Also, the large size dermoscopic image dataset would be used further to improve the classification results.

References

[1]. World Cancer Research Fund on Skin Cancer Statistics. https://www.wcrf.org/dietandcancer/cancer-trends/skin-cancer-statistics.

[2]. World Health Organization on Climate Change and Human Health. https://www.who.int/globalchange/climate/summary/en/index7.html

[3]. Ichim, Loretta, and Dan Popescu. "Melanoma Detection Using an Objective System Based on Multiple Connected Neural Networks." IEEE Access 8 (2020): 179189-179202.

[4]. Demyanov, Sergey, et al. "Classification of dermoscopy patterns using deep convolutional neural networks." 2016 IEEE 13th International Symposium on Biomedical Imaging (ISBI). IEEE, 2016.

[5]. Majumder, Sharmin, and Muhammad Ahsan Ullah. "Feature extraction from dermoscopy images for an effective diagnosis of melanoma skin cancer." 2018 10th International Conference on Electrical and Computer Engineering (ICECE). IEEE, 2018.

[6]. Eltayef, Khalid, Yongmin Li, and Xiaohui Liu. "Detection of melanoma skin cancer in dermoscopy images." Journal of Physics: Conference Series. Vol. 787. No. 1. IOP Publishing, 2017.

[7]. Jain, Shivangi, and Nitin Pise. "Computer aided melanoma skin cancer detection using image processing." Procedia Computer Science 48 (2015): 735-740.

[8]. Isasi, A. Gola, B. García Zapirain, and A. Méndez Zorrilla. "Melanomas non-invasive diagnosis application based on the ABCD rule and pattern recognition image processing algorithms." Computers in Biology and Medicine 41.9 (2011): 742-755.

[9]. She, Zhishun, Y. Liu, and A. Damatoa. "Combination of features from skin pattern and ABCD analysis for lesion classification." Skin Research and Technology 13.1 (2007): 25-33.

[10]. PH^2 Database. https://www.fc.up.pt/addi/ph2\%20database.html

An Improved Approach for Deep Web Data Extraction

Shilpa Deshmukh
Department of Computer Application
SIES college of Management Studies
Navi Mumabi, India
shilpad@sies.edu.in

P.P. Karde
Department of Information
Technology Government Polytechnic,
Amravati,,India.
p_karde@rediffmail.com

V.R.Thakare
Department of Computer Science,
SGB Amravati University,
Amravati, India
vilthakare@yahoo.co.in

Abstract
The World Wide Web is a valuable wellspring of data which contains information in a wide range of organizations. The different organizations of pages go about as a boundary for performing robotized handling. Numerous business associations require information from the World Wide Web for doing insightful undertakings like business knowledge, item insight, serious knowledge, dynamic, assessment mining, notion investigation, and so on Numerous scientists face trouble in tracking down the most fitting diary for their exploration article distribution. Manual extraction is arduous which has directed the requirement for the computerized extraction measure. In this paper, approach called ADWDE is proposed. This drew closer is essentially founded on heuristic methods. The reason for this exploration is to plan an Automated Web Data Extraction System (AWDES) which can recognize the objective of information extraction with less measure of human intercession utilizing semantic marking and furthermore to perform extraction at a satisfactory degree of precision. In AWDES, there consistently exists a compromise between the degree of human intercession and precision. The objective of this examination is to diminish the degree of human intercession and simultaneously give exact extraction results independent of the business space to which the site page has a place.

Keywords: www, deep web, crawler, deep web data extraction, heuristic approach
.

1. Introduction

The growth of World Wide Web (WWW), which is a huge repository of information, is unimaginable. Www opens the doors to the huge amount of data for the world. The search engines help us to retrieve the needed information from such a huge repository. It provides us with a list of links to the websites which might contain the requested information. Nevertheless large data remains invisible to the user because huge amount of information on the web is available today only through search interfaces. E.g. if the user wants to find the information about the flights she has to go to that airways web site and fill the details in a search form. It will display the flights and the available seats as a result, which are dynamically generated pages containing the required information. This part of web which is reachable only through search interface is deep web. Deep web pages are generated through dynamic query on the web databases. As compared to surface web or static web, contains larger amount of data with higher quality as well.

In order to get the information from the deep web, customized search engines provided by the individual websites are used. Such search engines submit the user query to the backend database and retrieve the structured records matching the query.

2. Literature Review

Various web data extractions have been listed in the literature review. In this section, previous work done on deep web data extraction are being reviewed briefly. Detailed survey about information extraction is given in [1-A survey of web Information Extraction System], where three parameters:

Task Difficulties, used techniques, Degree of Automation were considered.

TSIMMIS by Hammer et al. [2]: The target of extraction is specified manually using a specification file written using a declarative language. The command consists of three parts, namely, variables, source and pattern. The variables are used to store the result of extraction. The pattern specifies how to determine the content to be extracted and the source specifies the input text to be extracted. The output of extraction is represented using Object Exchange Model (OEM). It contains the actual data as well as structure of the data.

W4F by Sahuguet et al. [3] (World Wide Web Wrapper Factory): It is a Java Tool Kit used to generate extraction rules. It consists of three steps, namely, retrieval, extraction and mapping layers. The first layer is the retrieval layer which involves cleaning the input HTML document and converting it to the DOM tree. The second layer, namely, the extraction layer involves applying the extraction rules to the DOM tree in order to extract the target data. The output target data is represented as Nested String List structure.

Supervised Extraction Techniques

It requires less human intercession than the past procedure. This method doesn't include manual formation of covering. All things considered, a bunch of marked pages is given as contribution to the framework which creates the covering consequently. Some of generally utilized apparatuses dependent on this strategy are:

WIEN by Kushmerick et al. (Covering Induction for data ExtractioN) [4]: In this paper, the creators presented covering acceptance, which can develop coverings naturally by summing up from model question reactions. The authors have proposed six types of wrappers, namely, LR (Left-Right), HLRT (Head-Left-Right-Tail), OCLR (Open (Left token Right)* Close), HOCLR (Head Open (Left token Right)* Close), N-LR (Nested-Left-Right) and N-HLRT (Nested-Head-Left-Right-Tail). LR (Left-Right) wrappers are simple class of wrappers which scans the input document to determine left hand delimiter and right-hand delimiter for each and every attribute, until all the columns in the table are extracted. HLRT (Head-Left-Right-Tail) wrappers are more sophisticated than LR wrappers which can be used to extract content from country/code resource. OCLR uses O and C to identify tuples and LR is repeated for individual elements. HOCLR merges the functionalities of HLRT and OCLR. N-LR and N-HLRT searches for the delimiters in a different way and they are extension of LR and HLRT wrappers which can perform extraction of nested data records. Soft Mealy by Hsu et al. [5]: Hsu et al. introduced FST (Finite State Transducer) in order to handle different types of extraction structures. It consists of two different modules namely body transducer and tuple transducer. Body transducer is used to extract the data rich region containing the tuples and the tuple transducer is used to extract the tuples from the data rich region.

STALKER by Muslea et al. (Learning Extraction Rules for Semi-structured, Web based Information Sources) [6]: The authors have introduced a concept called Embedded Catalog (EC) formalism which is a tree-like structure in which the leaves are the attributes to be extracted and the non-leaf nodes represent the data records. It uses multiple passes to determine missing attributes and differently arranged attributes.

- *Semi-Supervised Extraction Techniques*

Unlike the supervised technique in which labelled examples are needed, these tools require post-effort (i.e. manual labelling after extraction). These techniques are appropriate for record-level extraction. The user's effort is needed to specify extraction targets using GUI. Unlike the supervised techniques which require comprehensive set of labelled training samples, these techniques require rough example for wrapper creation.

In [7], Chang et al. proposed a semi-supervised technique in which part of input data set is used for inferring schema. A Finite State Machine is built based on inferred schema and the inferred template is used for extraction of data. This technique is greatly affected by structural changes of the web pages as it might require changes to the inferred template.

➢ *Unsupervised Extraction Techniques*

These techniques require neither labelled examples nor user interaction for wrapper induction. Some of the matured unsupervised page-level extraction techniques are:
RoadRunner by Crescenzi et al. [8], EXALG by Arasu et al. [9], FivaTech by Kayed et al. [10] are page-level extraction techniques whereas certain others like DeLa by Wang et al. [11] and DEPTA by Zhai et al. [12] are record-level extraction techniques.

DeLa (Data Extraction and Label Assignment) by Wang et al. [13]: It consists of four major modules, namely, *form crawler, wrapper generator, aligner* and *label assigner*. The form crawler is used to submit queries and to obtain the response pages pertaining to the query. The wrapper generator determines the template in which the data records are embedded. The aligner is responsible for grouping the attribute values so that all the attribute values in a group belong to the same concept. The label assigner assigns meaningful labels to individual data units. Hierarchical Web Crawler is used to perform form submission and to get the Search response pages. DeLa uses wrapper generator derived from IEPAD [14] system. Data-rich Section Extraction (DSE) [15] technique has been employed to strip off non-informative sections of the web page. DSE [16] has been identified by comparing the HTML source code of multiple web pages generated using same server-side templates. In order to identify C-repeated pattern (Continuous Repeated Pattern), suffix-tree has been constructed. Pattern Tree is constructed by starting from an empty root node. Whenever a pattern is determined then it is added as a child to the root node in the Pattern Tree. Longest occurring pattern in the Pattern Tree (PT) represents the general template used for generating the multiple input pages. The shortcoming of this technique is that it won't work with disjunctions in the structure of the web page. DeLa [17] not only handles extraction but also annotation. The next step post extraction is aligning the attribute values. It computes similarity between data units using features such as tag path, presentation styles, etc., and groups the attribute values into disjoint sets. Annotation phase uses four different heuristics such as form keywords, common prefix/suffix, values matching conventional formats and keywords in the query/response page to determine appropriate label to each attribute value group.

RoadRunner by Crescenzi et al. [18]: This tool attempts to perform unsupervised extraction by comparing two documents generated using the same template. It finds the similarities and differences between the documents. It generates a union-free regular expression which matches both the input documents. It assumes one of the input pages as the wrapper and compares it with the other page. Whenever a mismatch is encountered, the wrapper is generalized. The mismatches are of two types, namely, tag mismatches and string mismatches. Tag mismatches are resolved by including optional or iterator in the wrapper. Whenever a tag mismatch is encountered, it finds the initial tag and the final tag of the mismatched block and denotes it by a square. Then it matches this block to its upper portion of the source code to find whether it represents a repeated pattern. If it represents a repeated pattern, then it is enclosed inside (…) + to represent repetition. If no such pattern can be found then, it tries to search the wrapper to find whether the block can be skipped. In that case, it represents an optional block which is denoted by enclosing the pattern shows result of applying Road Runner technique [19] for two sample input pages. It shows how repetition of block is determined and how the wrapper is generalized. Whenever string mismatch is encountered, it represents data units and therefore it is replaced by #PCDATA which is used to denote any sequence of characters.

Limitations:

i. Time complexity is exponential in the number of tokens of the input documents. In order to reduce the complexity the authors

have proposed certain strategies such as reducing the number of alternatives to be considered, the number of backtracking to be performed to identify iterations, and removal of some regular sub-expressions.
i. Biases introduced have negative impact on the effectiveness of the technique.
ii. It requires the input documents to be well-formed.
iii. It does not perform automatic annotation.
OMINI[20] also performs unsupervised extraction similar to Road Runner[21]. It consists of three phases, namely, preparing the input web page, identifying target of extraction, and extracting target the data records. The input web page is pre-processed. Pre-processing involves cleaning the web page in order to ensure that it is well-formed and then transforming it to tag tree representation. Identifying target of extraction requires two steps, namely, object rich sub-tree discovery and object separator extraction. The portion of sub-tree containing all data objects of interest is identified by considering features such as number of child nodes of a given node (fan out), the number of bytes of data corresponding to each node and the count of sub-trees. Object separator extraction involves finding the tags used as separators of data records. Five different heuristics are applied for this purpose.

Authors have made assumptions such as most of the tokens should be associated with type constructor, tokens belonging to the template must have unique roles, and the data must be enclosed between tags. Most of the real world websites do not satisfy all these assumptions which affect the effectiveness of the technique.

MDR by Liu et al. (Mining Data Records) [21]: MDR uses two observations: i. a group of data records representing similar objects are placed in contiguous section in the web page and formatted using similar HTML elements, ii. a set of similar records appear as child sub-trees of a common parent node in the tag tree representation of the HTML page. In DEPTA [22], tag tree is constructed based on nested rectangles since each HTML element is rendered using a rectangular box by the browser. The co-ordinates of all the boxes are determined and the co-ordinates are compared in order to know the containment relationship among boxes and then the tag tree is constructed accordingly. Figure 1 shows the construction of tag tree from the visual clues. In order to find the data records region, a measure called string edit distance as in OLERA by Chang et al. [23] is used for the comparison of tag strings of individual nodes and combination of multiple adjacent nodes.
FiVaTech by Kayed et al. [24] (Page-level web data extraction from template pages): It is an unsupervised approach for carrying out page-level extraction task. It consists of two phases: the first phase involves transforming the input documents into collection of DOM trees. The DOM trees are converted into pattern trees which are used to determine the server-side templates involved in generating the input documents; and the second phase involves producing schema from the pattern tree and extracting the embedded data records. But limitations are 1) FiVaTech requires well-formed input documents and error during parsing affects the effectiveness of the approach. 2) It uses pattern matching algorithm which is computationally expensive. 3) It uses a threshold for identifying peer nodes which impacts accuracy if proper threshold is not used.
Almost all methods discussed above methods can extract web data in the mentioned conditions.

3. PROPOSED ARCHITECTURE

From Literature Review it is clear that most of the existing systems have made assumptions such as the availability of multiple input pages; occurrence of missing or optional attributes is rare and so on. Also, these techniques are based on string pattern matching or DOM tree matching and therefore, change in structure of the web page requires occurrence of repeated patterns in the wrappers generated. This problem is addressed in the proposed technique by considering semantic information as the basis of extraction rather than structure or template of the web page. The steps proposed AWDES is as shown in the Figure 1:

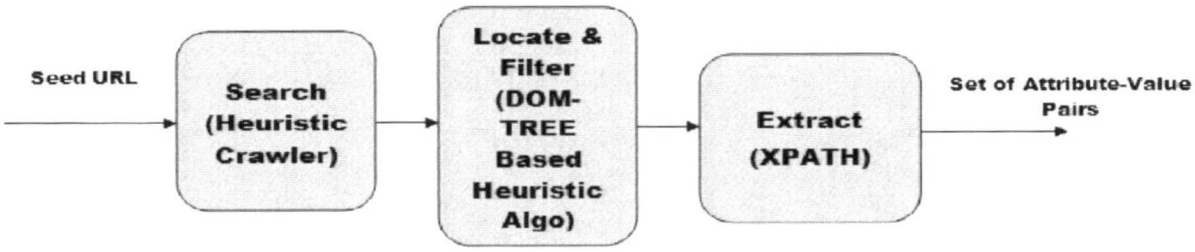

Figure 1: 1 Steps in Web Data Extraction

It involves four steps, namely, Search, Locate, Filter and Extract. The search step involves finding the target web pages given the URL of the website as input. The locate step refers to determining data rich regions in the target web pages and filter step involves the process of extracting the data rich nodes after eliminating non-informative sections such as navigation links, pop-up menus, advertisements, etc. The last step of web data extraction involves extracting the attribute-value pairs from the target web pages, which represents the description of the object.

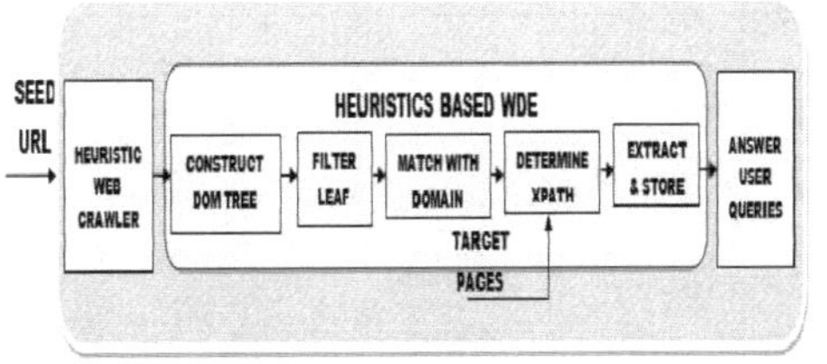

Figure 2: Overall Architecture of the Journal Information Extraction System

3.1 Mathemaical Model

Heuristic Algorithm for Automatic Navigation

Given the URL of the logical distributer's site urlp, explore to the site and concentrate the arrangement of URLs comparing to the diary home pages orchestrated in sequential request (i.e.) A-Z. In the wake of exploring to diary's landing page, it's anything but a bunch of diary particular addressing data like title, ISSN, URL of the diary's landing page, and so on URLs for every one of the diaries are requested by A-Z. The arrangement of target URLs is meant by H = {hi, 1 ≤ I ≤ n}. Every diary landing page hello is parsed to gather the required properties like title, ISSN, Impact

3.2 Heuristic Technique for Data Extraction

The proposed approach is based on the observation that the journal home pages linked to publishers' website are well formatted. The uniformity in presentation across several web pages of the website is achieved since they are generated using the same server-side template. If the location of target data is identified for a single web page (XPATH for attribute name-value pairs) from the scientific publisher's website, then the same XPATH can be used for extraction from similarly formatted journal specification pages. Overall architecture is shown in Figure 2:

The point of this procedure is to recognize the area of the objective ascribes, separate the quality qualities, explain and store the removed information records containing property estimations into a social data set. The procedure depends on the heuristics that the objective data to be separated is available as apparent substance which is shown in the substance space of the program. In the DOM tree addressing the site page, all the content hubs will be available as leaf hubs. The method includes visiting every URL and getting the DOM tree of the page containing the diary determination. The calculation getLeafNodes is utilized to acquire the leaf hubs and furthermore, to check whether it's anything but a content hub. On the off chance that they match with the area catchphrases like SJR, SNIP, Impact Factor, and so on, then, at that point its XPath is resolved utilizing the system getFullXPath. The calculation convert_XPath_to_Selector is utilized for changing the XPath over to a selector string. A similar selector string can be utilized to do extraction from correspondingly organized website pages. Following are the algorithm used in proposed approach:

```
Algorithm Extract (Seed_URL)
//Input: Seed URL(URL of Publishers' site)
//Output: Structured records
begin
  target_urls=extractURLs(Seed_URL);
  for each url in target_urls do
  impact_factor, issn = display_journal_metrics(url)
  create database connectivity
  store impact_factor, issn
  end for
end
```

Figure 3: Algorithm Extract

```
Algorithm displayJournalMetrics(target_url)
//Input: target url
//Output: impact_factor,issn
begin
   useragent.visit(target_url);
   issn_pattern = "\\d\\d\\d\\d-\\d\\d\\d\\d|\\d\\d\\d\\d-\\d\\d\\dX"
   impact_pattern = "\\d.\\d{1,3}|\\d\\d.\\d{1,3}"
if (issn_xpath is NULL) then
begin
  doc = construct DOM tree
   root = node corresponding to Body element
   leaf = getLeafNodes(root)
for each node in leaf do
if node.text matches issn_pattern then
   issn_xpath = getFullXPath(node)
   issn = node.text.substring(start, end)
end if
if node.text matches impact_pattern then
   impact_xpath = getFullXPath(node)
   impact_factor = node.text.substring(start,end)
end if
else
  impact_factor_selector = convert_xpath_to_selector(impact_xpath)
  issn_selector = convert_xpath_to_selector(issn_xpath)
  impact_node = doc.select(impact_factor_selector)
  impact_factor = impact_node.text.substring(start,end)
  issn_node = doc.select(issn_selector)
  issn = issn_node.text.substring(start, end)
end if
  return impact_factor,issn
end
```

Figure 4: Algorithm displayJournalMetrics

```
Algorithm getLeafNodes (Node root)
//Input: root
//Output: leaf_nodes
begin
if (root.getType() equals Node.ELEMENT_TYPE) then
for each node in root.getChildNodes()
begin
  if node.getType() equals Node.TEXT_TYPE then
    leaf_nodes.add(node)
  else
    getLeafNodes(node)
  end if
return leaf_nodes
end
```

Figure 5: Algorithm getLeafNodes

4. Result & Analysis

The experiment was conducted on a machine having Intel Core 2 Duo T5870 @ 2.0 GHz with 3 GB of RAM. This machine was running with Windows 7 OS. Tests were performed using WAMP server equipped with php v. 5.2.6 and mysql v. 5.0.51b., Microsoft Visual Basic 2008 .net 3.5 and Net beans IDE. All the tests were performed on Firefox 3.5.4. and were performed multiple times to minimize measurement

Simulation of proposed architecture of deep web crawler and their result are analysed with the help of performance metrics like Precision, Recall and F-measure. Three domains have used i.e. book, auto, job for our analysis. Detailed analysis regarding the book domain is presented in this section. Six commercial implementation of book domain i.e. www.borders.com, www.bookshare.org, www.textbooks.com, www.ecampus.com, www.textbooksrus.com, ww.bookfinder.com for extracting the content of deep web by querying through search interface have been used. QIIIEP server is used for fetching query words related to domain for level value assignment. The GET and POST method is analysed through action tag and stored in query words to the url relation manager. The method implemented at local host is named as fetch query interface and the query argument used as sdd, stands for search deep domain. http:\\localhost\fetchqueryinterface?sdd=borders.com

To measure the performance of proposed deep web crawler "DWC", three metrics are defined [32] are as follows:

i) Precision, which is defined as *PR = CDWP/ (CDWP + WDWP)*

ii) Recall is *RE = CDWP/ (CDWP + NCDWP) and*

iii) F-measure is *FM = 2PR.RE/ (PR + RE)*

Where Precision (PR) and Recall (RE) are weighted equally.

For Book domain, about 1400 query words received & fired and response pages were analysed. It may be noted from Fig. 8 that as the number of received queries word increases for Site1:www.borders.com, the value of Precision, Recall and F-measure increases. When about 200 received query words were fired, the value of the Precision, Recall and F-measure were 73%, 75%, and 74.03% respectively. As the number of query word reached to 1400, the values of the Precision, Recall and F-measure reached at 97%, 89% and 92% respectively.

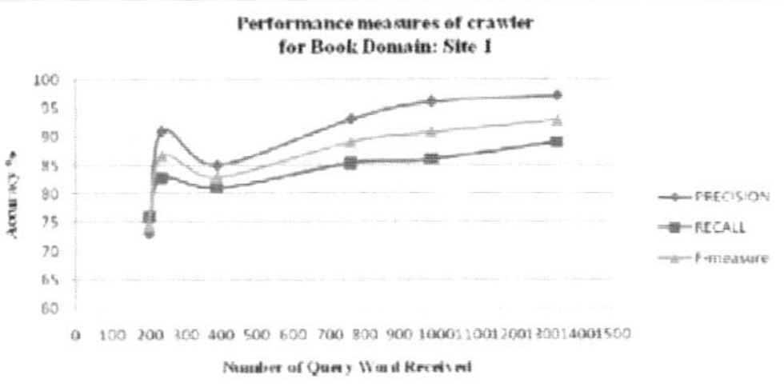

Figure 6: Performance Measure

The number of received queries word increases for Site2: www.bookshare.org, the value of Precision, Recall and F-measure increases. When about 200 received query words were fired, the value of the Precision, Recall and F- measure were 67%, 70%, and 68% respectively. As the number of query word reached to 1400, the values of the Precision, Recall and F-measure reached at 93%, 85% and 88% respectively.

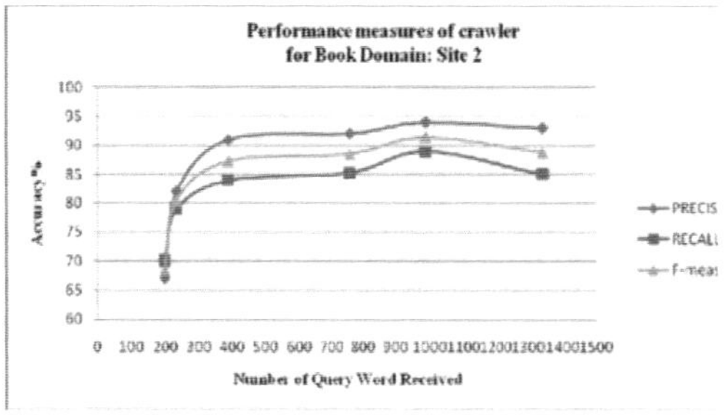

Figure 7: Performance Measure

Figure 6 and 7 shows the performance measures of proposed deep web crawler "DWC" at different quantity of query words received by QIIIEP server that are used to fetch the content from the deep web sites. From the analysis of this graph we can easily visualize that the number of query words received for specific domain, the precision, recall and f-measure of information extraction have continuously increased. So it can be concluded that when enough knowledge base is collected at QIIEP server, the result is obviously improved.

5. Conclusion

The presence of high quality information in Deep Web Pages which are dynamically generated by embedding structured data records in server-side templates has made many researchers design techniques for automatic extraction of such data. The limitation of the existing systems includes their dependency on the structure of HTML pages for inducing wrappers. Our proposed research addressed this challenge.

This system enables integration of journal information from multiple publishers' websites and it enables the user to get the complete information about journals in a single interaction with the system. The system can be used to extract data records from websites where the attribute label is present explicitly. It is not always true since some websites like social discussion forums are not having attribute labels mentioned explicitly.

6. References

[1] Luciano Barbosa and Juliana Freire, "Searching for Hidden-Web Databases", In Proc. of Web Database, 2015.

[2] J. Cope, N. Craswell, and D. Hawking, "Automated Discovery of Search Interfaceson the web", In Proceedings of the Fourteenth Australasian Database Conference (ADC2019), Adelaide, Australia, 2019.

[3] Z. Zhang, B. He, and K. Chang, "Understanding Web Query Interfaces: Best-Effort Parsing with Hidden Syntax", In Proceedings of ACM International Conference on Management of Data, pp.107-118, 2019.

[4] L. Barbosa and J. Freirel, "Siphoning Hidden-Web Data through Keyword-Based Interface" , In Proceedings of SBBD, 2004.

[5] X. B. Deng, Y. M. Ye, H. B. Li, and J. Z. Huang, "An Improved Random Forest Approach For Detection Of Hidden Web Search Interfaces", In Proceedings of the 7th International Conference on Machine Learning and Cybernetics, China. IEEE, 2008.

[6] Y. Ye et al., "Feature Weighting Random Forest for Detection of Hidden Web Search Interfaces", In Computational Linguistics and Chinese Language Processing , Vol.13, No. 4, pp.387-404, 2019.

[7] P. Bai, and J. Li, "The Improved Naive Bayesian WEB Text Classification Algorithm", In International Symposium on Computer Network and Multimedia Technology, IEEE Explorer, 2019.

[8] Anuradha and A.K. Sharma, "A Novel Approach For Automatic Detection and Unification of Web Search Query Interfaces Using Domain Ontology", In International Journal of Information Technology and Knowledge Management , Vol. 2, No. 2, 2010.

[9] J. Madhavan, P. A. Bernstein, and E. Rahm, "Generic Schema Matching with Cupid", In the 27th VLBB Conference, Rome, 2009.

[10] H. H. Do, and E. Rahm, "COMA-a System for Flexible Combination of Schema Matching Approaches", In Proceedings of the 28th Intl. Conference on Very Large

Databases (VLDB), Hong Kong, 2012.
[11] A. H. Doan, P. Domingos, and A. Levy, "Learning source descriptions for data integration", In Proceedings of the WebDB Workshop, pp. 81-92, 2017.
[12] A. Doan, P. Domingos, and A. Halevy, "Reconciling schemas of disparate data sources: A machine learning approach", In Proceedings of the International Conference on Management of Data (SIGMOD), Santa Barbara, CA, New York: ACM Press, 2011.
[13] S. Melnik, H. Garcia-Molina, and E. Rahm, "Similarity Flooding: A Versatile Graph Matching Algorithm", In Proceedings of the 18th International Conference on Data Engineering (ICDE), San Jose, CA,2019.
[14] O. Kaljuvee, O. Buyukkokten, H. G. Molina, and A. Paepcke, "Efficient Web form entry on PDAs", In Proceedings of the 10th International Conference on World Wide Web, pp. 663- 672, 2018.
[15] H. He, W. Meng, , C. Yu, and Z. Wu, "Constructing interface schemas for search interfaces of Web databases", In Proceedings of the 6th International Conference on Web Information Systems Engineering (WISE'05),pp. 29-42, 2015.
[16] W. Wu, C. Yu, A. Doan, and W. Meng, "An interactive clustering-based approachto integrating source query interfaces on the Deep Web", In Proceedings of the ACM SIGMOD International Conference on Management of Data, pp. 95-106, 2014.
[17] B. He, and K. C. C. Chang, "Automatic complex schema matching across web query interfaces: A correlation mining approach", In Proceedings of the ACM Transactionon Database Systems, Vol.31, pp.1-45, 2016.
[18] P. Lage et al., "Collecting Hidden Web Pages for Data Extraction" In Proceedingsof the 4th International workshop on Web Information and Data Management, pp. 69-75, 2012.
[19] B. He, and K. C. C. Chang, "Statistical schema matching across web queryinterfaces", In the SIGMOD Conference, 2013.
[20] X. Zhong, et al., "A Holistic Approach on Deep Web Schema Matching", In Proceedings of the International Conference on Convergence Information Technology, 2017.
[21] S. Raghavan and H. Garcia-Molina, "Crawling the Hidden Web", In Proceedings of the 27th International Conference on Very Large Data Bases, Roma, Italy, pp. 129–138, 2016.
[22] B. Qiang, J. Xi, B. Qiang, and L. Zhang, "An Effective Schema Extraction Algorithm on the Deep Web. Washington, DC: IEEE, 2018.
[23] M. Niepert, C. Buckner, and C. Allen, "A Dynamic Ontology for a Dynamic Reference Work", In Proceedings of the (JCDL'17), Vancouver, Canada, 2017.

[24] A. Deitel, C. Faron, and R. Dieng, "Learning ontologies from rdf annotations", Inthe IJCAI Workshop in Ontology Learning, 2001.

A Parallel Apriori Algorithm and FP- Growth Based on SPARK

Priyanka Gupta[1], *Dr.Vinaya Sawant*[2]

[1]Dwarkadas J Sanghvi College of Engineering, Mumbai
[2]Dwarkadas J Sanghvi College of Engineering, Mumbai

Abstract. Frequent Itemset Mining is an important data mining task in real-world applications. Distributed parallel Apriori and FP-Growth algorithm is the most important algorithm that works on data mining for finding the frequent itemsets. Originally, Map-Reduce mining algorithm-based frequent itemsets on Hadoop were resolved. For handling the big data, Hadoop comes into the picture but the implementation of Hadoop does not reach the expectations for the parallel algorithm of distributed data mining because of its high I/O results in the transactional disk. According to research, Spark has an in-memory computation technique that gives faster results than Hadoop. It was mainly acceptable for parallel algorithms for handling the data. The algorithm working on multiple datasets for finding the frequent itemset to get accurate results for computation time. In this paper, we propose on parallel apriori and FP-growth algorithm to finding the frequent itemset on multiple datasets to get the mining itemsets using the Apache SPARK framework. Our experiment results depend on the support value to get accurate results.

Keywords:-Data Mining, Association Rule Mining, Apriori Algorithm, FP- Growth Algorithm, RDD, Apache Spark.

1 INTRODUCTION

The Association Rule Mining (ARM) methodology is used to extract relevant associations from a transactional dataset. Even though association rules are produced from them, the development of frequent itemsets is the most computationally intensive element of the association rule mining process. For the regularity of the used itemsets, various adaptions have been evolved by various experts for the procedures to speed the range of the items. These algorithms were unable to process large data sets due to their sequential computing structure. For larger datasets, parallel and distributed types of frequent itemsets mining algorithms have been evolved to build the frequent itemsets. To handle the massive amount of data, the Hadoop platform is designed. During execution, it moves data processing to the nodes rather than sending data to the nodes. HDFS (Hadoop Distributed File System) and MapReduce, Hadoop's two primary components, manage data storage and computing, respectively. MapReduce is a scalable and fault-tolerant parallel programming technology. HDFS is a file system that separates enormous amounts of data into blocks and replicates them across all nodes in a cluster. Throughout the cluster, an application is stated as a MapReduce job, that runs on separates tasks which are divided into various tasks and run on different blocks of data simultaneously.

1.1 Apache Spark Framework

Apache SPARK is a cluster computing framework that is free and open source. It was first developed in 2009 at UC Berkeley's AMPLab. In Java, it provides a high level of APIs. Due to the in-memory computing technique, Spark provides the implementation 10 times faster than the Hadoop framework. It provides batch and real-time iterative processing of data. Spark provides a programming interface for "Resilient Distributed Dataset", a type of data structure. In Apache Spark, RDD is an abstraction. Spark is platform-independent it can run on any platform.

1.2 Resilient Distributed Dataset (RDD)

In Apache Spark, RDD is the basic abstraction of data structure. The main feature of RDD is that it maintains a partitioned, immutable collection of elements that can be managed simultaneously. RDD can improve the execution and makes Spark acceptable for interacting with datasets. Dataset will divide into logical patterns which can be measured on the individual node. There are two types of RDD operations: Transformation and Action. Transformation is used to create a new RDD from the existing whereas action is used to perform the computation and return the result to the main program.

2 LITERATURE REVIEW

Sanjay Rathee et.al [1] have proposed an adaptive-miner algorithm for efficient use of distributed association rule mining algorithm using spark. They make use of a reduced approach when several frequent itemsets are large else basic Apriori is used. For each iteration the execution plan with the minimum cost. The approach they have used for every iteration will give fast and efficient results.

Ravi Ranjan et.al [2] they perform the Apriori and FP-growth algorithm on Hadoop-MapReduce on the various dataset. They stated that big data can be used to make accurate predictions and enhance the association rule mining algorithm. Also, they raised some research questions based on which technique is used and how they will execute on various datasets and more questions are there.

Samba N et.al [3] represented a parallel version of FP-Growth algorithm to work as efficiently for processing large datasets. They demonstrate the proposed algorithm can work efficiently to process large datasets.

Shaozang C et.al [4] used distributed FP-Growth algorithm to mining the frequent itemset in three steps: Step 1 is to calculate 1-itemsets, step 2 i.e for conditional pattern base, step 3 is for mining frequent itemset. Their implementation results state that distributed FP- growth algorithm works better than Parallel FP- Growth algorithm.

Zhaowei Q, et.al [5] present an approach for imputation algorithm based on Apriori they aimed at the issue of incomplete data. They used the distributed approach to decrease the execution time of datasets. They also stated that they need more approaches to find the solution of how the problems will be resolved.

Shaosong Y, et.al [6] have used the improved kind of parallel Apriori algorithm based on Spark to mine the frequent itemsets. They stated that using the IAP algorithm, they worked on the parallelization and verified by the various comparative datasets.

Ravi R, et.al [7] have used the Apriori algorithm on Hadoop MapReduce to get accurate results using various datasets to mine the frequent itemsets. They have worked on a single node to get the results. They also stated that they were used to imitate the performance of the distributed environment.

Feng G, et.al [8] have used the matrix-based pruning method to mine the frequent itemsets under the MapReduce framework. They have used the novel pruning technique to get accurate results. Also, implemented the algorithm under the Spark framework to stimulate the efficiency of the results.

Sanjay R, et.al [9] presented a new algorithm based on Apriori i.e R-apriori used to mine the frequent pattern from large datasets having various attributes. They have used the reduced approach to mine the frequent itemsets. They stated that the conventional Apriori algorithm takes too much time than the parallel apriori algorithm.

Hongjian Q, et.al [10] frequent itemset mining algorithm i.e Apriori algorithm used to determine the fast and efficient algorithm that can handle large datasets using spark. They proposed the YAFIM algorithm for the use of the fast and efficient result of frequent itemset mining. They used the concept of RDD of the algorithm. They also mentioned the application which can be used for the association rule mining algorithm.

3 PROPOSED SYSTEM

Basic Association Rule Mining algorithms are used the serial approach but they were not beneficial for the larger data size. As the size of datasets become larger, their accuracy began to degrade. As a result, to manage the larger datasets parallel algorithm comes into the picture. There are many cluster-based algorithms that are there to handle the larger datasets but they produced several problems such as data replication, synchronization, etc. Therefore, the parallel approach is replaced by the MapReduce approach. An algorithm like Apriori and FP-Growth is the best for finding the frequent itemsets. In Apriori, key-value pairs are produced easily. Using Map-Reduce approach, all of which display a significant performance enhancement over the traditional algorithm. For implementing the Apriori algorithm, MapReduce Approach is the best platform. But using Hadoop also has several limitations. For each iteration, results are saved to HDFS on the Hadoop platform and for the next iteration, input is taken from HDFS which shows the reduction in productivity because of input-output time. So, RDD (Resilient Distributed Dataset) in Spark helps to resolve these problems. The main point of using these features is that it stores the result at the end of the iteration in a local cache and it's available for subsequent iterations. Both Apriori and FP-Growth algorithms are implemented on the Spark platform with multiple datasets to mine the frequent itemsets and generates the association rules.

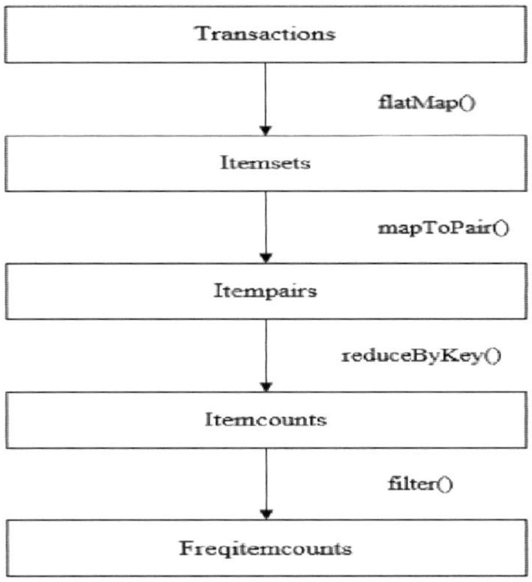

Fig 1:- RDD of Apriori Phase 1

Phase 1 generates the frequent items only.
It returns an RDD of objects that separates the transaction with the flatMap() transformation.
mapToPair() transformation generates the (item, 1) pairs for each item.
It adds all the values of the similar key together by the reduceByKey() function.
Lastly, the items are filtered out the pairs whose value is less than the minimal support with the filter() transformation and leaves those pairs which are frequent items then it took to evaluate them.
The action operation collect() returns the driver's frequently used objects for usage in the next phase.

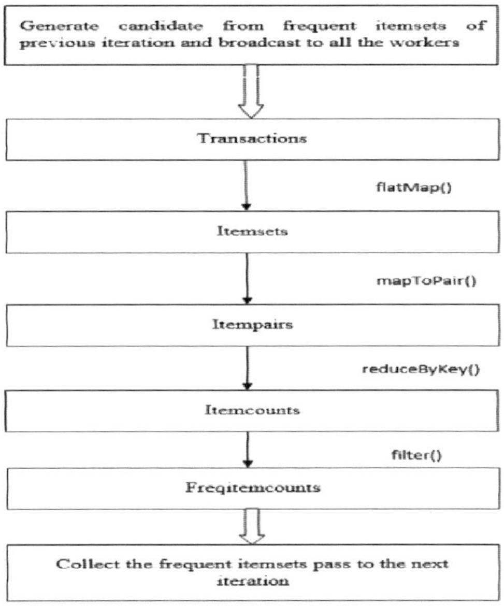

Fig 2:- RDD of Apriori Phase 2

The output of the previous iteration is the input of phase 2.
In previous iterations, (k-1) frequent are created i.e used for k^{th} iteration.
It returns a list of all candidate k-itemsets with the flatMap() transformation that the subset() operation has generated. Each itemset is converted into (itemset, 1) pairs, or key-value pairs, by the transformation mapToPair(). By summing all values with the same key, reduceByKey() reduces these pairs. Finally, the filter() method filters out infrequent itemsets, returning just frequent k- itemsets

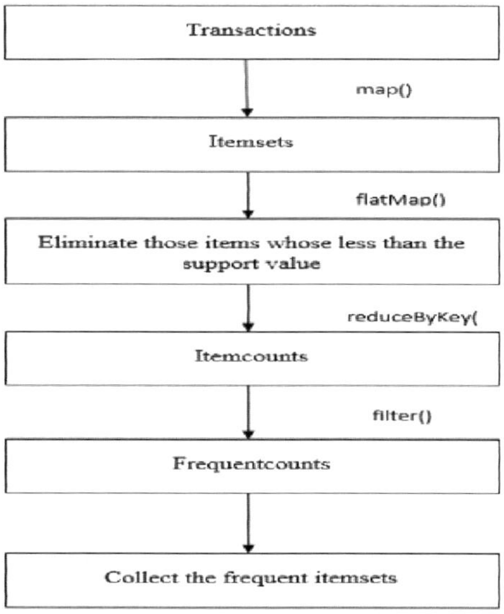

Fig 3:- RDD of FP-Growth

It involves the execution of two scans on the transaction database.
The first scan is to build up the support of every item and then select those items with the minimum support. It produces the frequent itemsets which store in frequency in descending order. The second scan builds the FP-Tree.
From the diagram, first, a transaction t is a scan from the database. It checks the prefix of t maps to the path in the FP- tree. The support count of the corresponding node is incremented in a tree.
New nodes are created if there is no overlapped path, and then it will eliminate whose items which is less than the support value by using reduceByKey() transformation.
Lastly, the items are filtered out the pairs which are frequent items using filter() transformation.
Then, action operation collect() returns the frequent items pairs.

4 EXPERIMENT RESULT

The experiment analysis was based on the support value provided by the user at the time of execution. It was performed on a single workstation. We have used four

https://doi.org/10.1051/itmconf/20214003046

datasets with the sample dataset to examine the results. We present the performance evaluation of both algorithms based on execution time.

4.1 Experimental Setup

The setup of implementation was built on a workstation with an Intel i7 processor, 16 GB RAM, 2 TB of the hard disk. Hadoop – 2.2.0 and python language is used for coding which is running on windows 10.

All source codes are written in python. On a single workstation, algorithms are tested.

4.2 Evaluation Measures

In this paper, different evaluation measure has been used to evaluate the quality of proposed methodology. The figure shows the implementation results of each dataset.

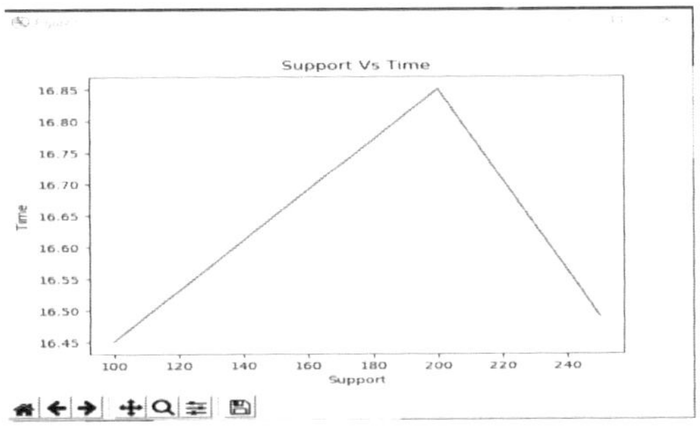

Fig 4:- Apriori Of Instacart Dataset

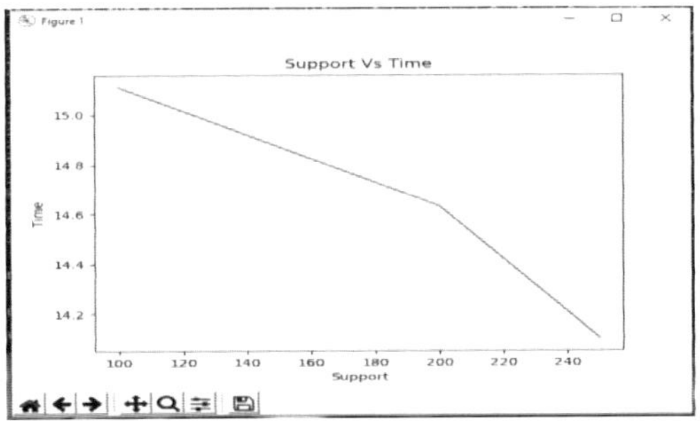

Fig 5:- Apriori Of Online Retail I Dataset

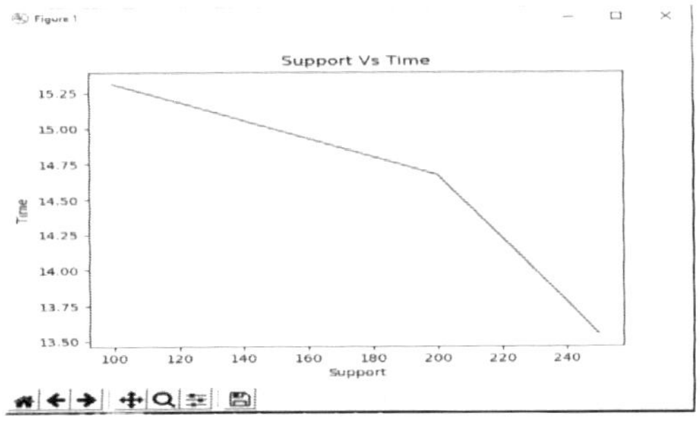

Fig 6:- Apriori Of Online Retail II Dataset

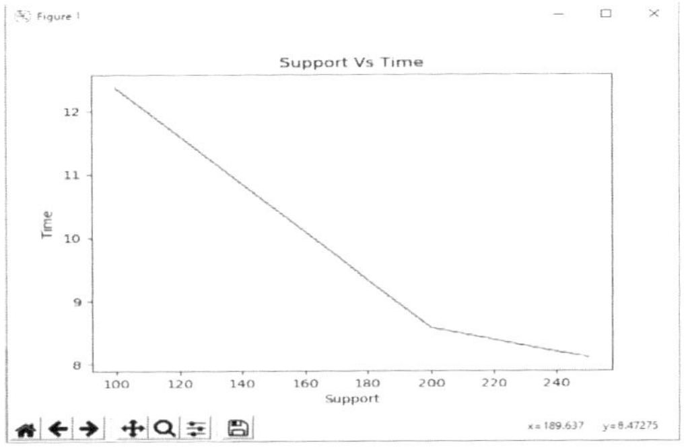

Fig 7:- Apriori Of Movie Rating Dataset

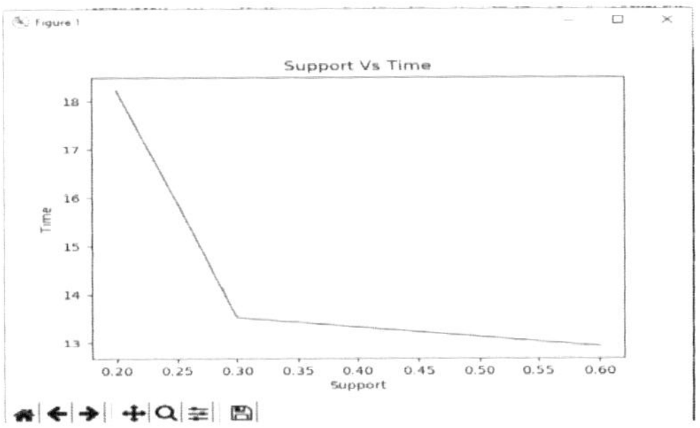

Fig 8:- FP-Growth of Instacart Dataset

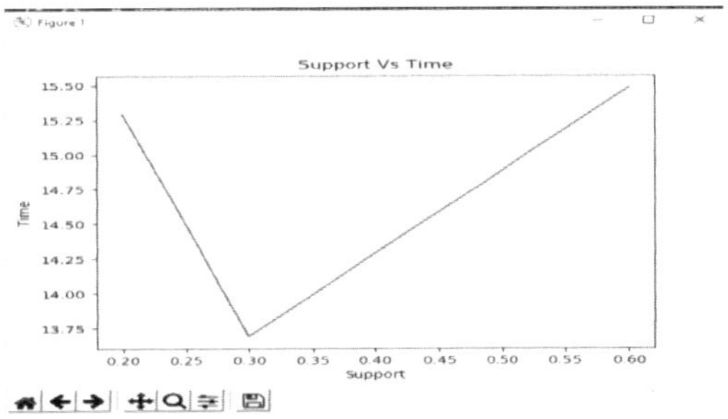

Fig 9:- FP-Growth of Online Retail I Dataset

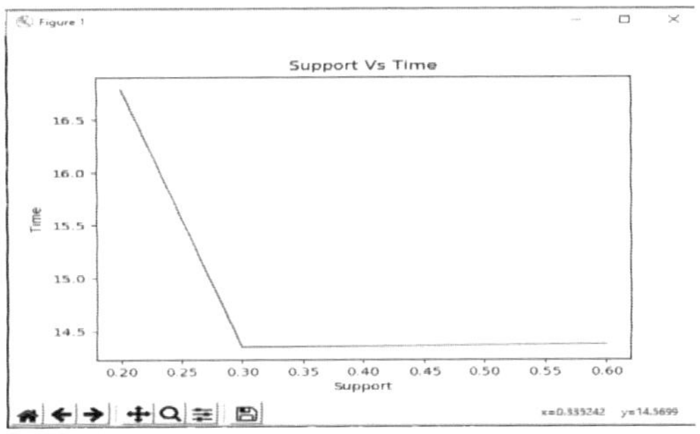

Fig 10:- FP-Growth of Online Retail II Dataset

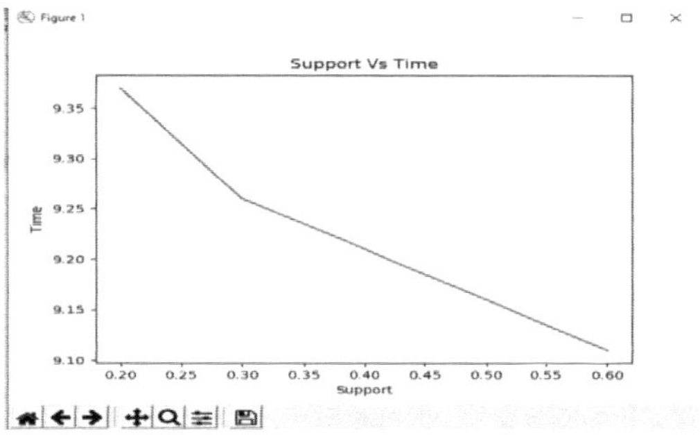

Fig 11:- FP-Growth Of Movie Rating Dataset

Acknowledgment

The authors would like to express their gratitude towards their guide, Dr. Vinaya Sawant, The Head of Information Technology Department, and the Principal, Dr. Hari Vasudevan for the opportunity provided to them to carry out implementation and research on this topic.

CONCLUSION

In the existing system, we performed the parallel Apriori algorithm based on MapReduce on the Hadoop platform, which results in the lack of implementation cost and communication cost. Using Hadoop, finding the association rules need a sufficient computation cost and memory. Thus, we execute the Parallel Apriori algorithm and FP-Growth algorithm on the SPARK platform with multiple datasets to get the results. Using, Apache Spark the performance of the results provides the fast and efficient than the Hadoop platform. During execution, this dynamic method makes very fast and efficient results for every iteration. Apache Spark architecture allows for highly parallel and distributed computing techniques. Due to its in-memory distributed computation, Apache Spark is well suited for both the algorithm for finding the frequent itemsets from the distributed data mining. Overall, both the algorithm performs admirably in the Apache Spark environment, although sequential Apriori performed poorly.

References

[1] S, Evaluation of Frequent Itemset Mining Platforms using Apriori and FP Growth algorithm, ICCM, vol. 2, pp. 1-6, (2019)

[2] Sanjay R, Arti K, Adaptive-Miner: an efficient distributed association rule mining algorithm on Spark, Springer, pp. 1-17, (2018)

[3] Singh S, Garg R, Performance optimization of MapReduce based apriori algorithm on Hadoop cluster, CEE, pp. 348-364, (2018)

[4] Diaby D, Fode C, S-FPG: A Parallel version of FP-growth algorithm under Apache Spark, IEEE, pp. 98-101, (2017)

[5] Xiujin S, Shazong C, DFPS: Distributed FP-Growth algorithm based on Spark, IEEE, pp. 1725-1731, (2017)

[6] Singh S, Garg R, Mishra P, Review of Apriori based algorithm on MapReduce framework, ICC, pp. 593-604, (2017)

[7] Zhaowei, Jianru Y, Association Rule based data imputations with Spark, CCIS, pp. 1725-1731, (2016)

[8] Xun Y, Zhang J Fidoop: Parallel mining of frequent itemsets using MapReduce, IEEE, pp. 313-325, (2016)

[9] Shaosong Y, Guoyan X, Zhijian W, The parallel Improved Apriori Algorithm research based on Spark, ICFCT, pp. 354-359, (2015)

[10] Ravi R, Performance Analysis of Apriori and FP Growth on different Mapreduce frameworks, DTU, pp. 1-7, (2015)

[11] Feng G, Yunlogo M, A Distributed frequent itemset mining algorithm based on Spark, IEEE, pp. 271-275, (2015)

[12] Sanjay R, Manohar K, R-Apriori: an efficient Apriori based algorithm on Spark, PIKM, pp. 1-8, (2015)

[13] Zhang F, Lin M, A distributed frequent itemset mining using Spark for big data analytics, CC, pp. 1493-1501, (2015)

[14] Singh S, Garg R, Performance analysis of Apriori algorithm with different data structure on Hadoop cluster, IJCA, pp. 45-51, (2015)

[15] Hongjian Q, Yihua H, YAFIM: A parallel frequent itemset mining algorithm with Spark, IEEE, pp. 1664-1671, (2014)

[16] Aavdh S, Ajeet K, Ashish M, An empirical analysis and comparison of Apriori and FP-Growth algorithm for frequent pattern mining, IEEE, pp. 1599-1602, (2014)

[17] Moens S, Aksehirli E, Frequent Itemset Mining for Big Data, IEEE, pp. 111-118, (2013)

[18] Frequent Itemset Mining Dataset Repository: http://fimi.ua.ac.be/data. Accessed 15 April 2021.

[19] Online Data Repository: https://www.kaggle.com/coldperformer/online-retail-data-v3.

[20] Online Data Repository: https://www.kaggle.com/rounakbanik/the-movies-dataset?select=ratings.csv.

https://doi.org/10.1051/itmconf/20214003047

Real Time Gender Classification using Convolutional Neural Network

Pooja Mishra[1,*], *Neha* Singh[2,**], and *Pallavi* Chavan[3,***]

[1]Information Technology, Ramrao Adik Institute of Technology, Nerul, Navi Mumbai, MH, India.
[2]Information Technology, Ramrao Adik Institute of Technology, Nerul, Navi Mumbai, MH, India.
[3]Information Technology, Ramrao Adik Institute of Technology, Nerul, Navi Mumbai, MH, India.

Abstract. This paper presents real time gender classification using Convolutional Neural Network. Automatic classification of gender has become important to an growing array of applications particularly with the emergence of web networks and social media. Slavery was a significant moral problem in the nineteenth century. It was a struggle toward fascism in the modern period. The fight for gender equality across the world, as well as the need to divide gender for meaningful purposes, would, we conclude, be the most critical moral issue of this century. Differences are needed at different places, such as restrooms for men and restrooms for women; attire for men and attire for women; and so on, in order to plan and advance further in the technological sector. To decrease crime rates, to place the advertisements in malls precisely attracting more people based on gender, to keep track of genders in respective toilets or in trains, for personal services, etc. The authors propose the gender classification dilemma for real-time applications, in which a tool decides if the faces within the exposure belong to a female or a male. The primary fundamental region of experimentation in this venture is adjusting a few already distributed, successful designs utilized for gender orientation classification. Generally, facial structure variations have an effect on gender classification accuracy considerably, as a result of facial form and skin texture modification as they become old. This requires re- examination on the sexual orientation classification framework. By learning representations through the utilization of deep-convolutional neural networks (CNN), a major increase in performance is obtained on these tasks.

1 Introduction

Gender plays a central role in social experiences. The number of picture uploads to the Internet has risen at an almost unprecedented pace over the last decade. This new discovered abundance of data has allowed computer scientists to solve previously either irrelevant or unsolvable machine vision issues. Recognition of gender is an inevitably challenging problem, far more than most other computer-vision activities.The main explanation for this complexity difference lies in the design of the data used to train these types of systems.According to a record , out of 7.8 billion people (current stats of year 2020) in total population of world, there are 4.48 billion internet users (statistics of October 2019) which has increased in this COVID pandemic situation. So there is a growth in the usage of automatic gender recognition because of online website shopping, online transactions, online social networking websites and social media. However, recognition of gender is an inevitably challenging problem, far more than most other computer-vision activities.The main explanation for this complexity difference lies in the design of the data used to train these types of systems.And for many such digital progress we should be ready with a better gender classification system which can process every person with a high accuracy even of a low quality light surroundings. Creating efficient, precise frameworks for tasks of gender distinction, and seeking to expand their solutions to boost performance and have a very good accuracy. Nevertheless, there is still a significant lack of performance of current methods on real-world photographs, particularly as compared to the enormous performance strides recently recorded for the related face recognition mission. In this modern world where everything is getting automated and online, starting from shopping for phones to jewellery. It is very important for websites to be efficient in giving right choice for customers as per their gender, taste and interest.

*e-mail: mishra.pooja.17it1082@gmail.com
**e-mail: singh.neha.17it2012@gmail.com
***e-mail: pallavi.chavan@rait.ac.in

2 Gender Detection Data Analysis

Thanks to the increasing usage of face recognition technologies in the field of safety, a slew of profile databases have sprung up in recent years. The authors have developed the data set of faces to get the right idea of how our data enhances the accuracy for Indian faces' gender recognition in real-time. Gender classification schemes are in a peculiar situation since they only describe 2 groups (male and female). As a result, classification and authentication methods are used in this situation without incurring any penalties.

https://doi.org/10.1051/itmconf/20214003047

2.1 Data collection

The authors collected data from the public domain, Social media like Instagram and Facebook (with the user's permission). The authors uses a web scraping tool to scrap the photos from their profile pictures and crop them around the frame. These pictures are improved in terms of quality by reducing the noise. The authors collected photos in one folder which was divided into 2 folders namely: Male and Female. These photos were then subdivided into 2 folders in each folder: Modified and Raw. Raw folder consist of photos which are collected from the Web scraping tool and through friends and families using WhatsApp and google drive link and it contains folders of Instagram, Group, Whats app. Modified folder consisted of photos which are improved versions of the raw photos by cropping and data improving tools which are manually segregated in male and female photos. As the difficulties increased in improving accuracies authors added more photos to the Raw and Modified photos folder.

2.2 Facial Image Databases

For improved accuracy, authors have used different types of faces and the number of times we have trained the model with each type of face to generate an accurate report of the database which was most useful to predict better results. While the images in real-time tend to recalculate the gender continuously so it requires every point to distinguish between the genders. Authors have created a mixture of each database type and trained the model with one database one at a time.

2.2.1 ORL database

The first widely used database created is of type YALE database which was first originated in the Olivetti Research Laboratory database. These datasets contain pictures shot from a frontal or nearly frontal face posture shown in Fig. 1.

Figure 1. Examples from collected ORL database

2.2.2 YALE database

The YALE database, on the other side, displays people in all sorts of combinations, including front/sides of face, high brightness/low brightness, with/without spectacles, and joyful/surprised/tired/sluggish/sleepy/laughing/crying/unhappy or wink emotions. As shown in Fig. 2 .

Figure 2. Examples from collected YALE database

2.2.3 YALEb database

Large security measures placed at airport terminals or public spaces are used in more strong and complicated scenarios. These systems must contend with uncontrollable lightning, – anti individuals, and huge stores of identity. In order to assess systems against these kind of scenarios, the dataset contains almost all the features.As shown in Fig. 3 .

Figure 3. Examples from collected YALEb database

2.2.4 FERET database

The FERET dataset includes pictures taken in a moderately controlled setting from 200 people in various face postures. As shown in Fig. 4. This dataset has the fascinating virtue of providing substantial statistics describing positions of face features, ethnicity, sexual orientation, and face traits such as moustache, beards, and spectacles details.

Figure 4. Examples from collected FERET database

2.2.5 FRGC database

Another comprehensive resource is the FRGC dataset. The FRGC dataset, like FERET, is made up of high-resolution pictures from a group of over 200 individuals, as well as comprehensive statistical information. This dataset, on the other hand, contains pictures of the entire body captured in various settings, implying significant

https://doi.org/10.1051/itmconf/20214003047

variations in backdrop and lighting conditions. As shown in Fig. 5. In summary, while evaluating a program, it is critical to consider the type of dataset being utilised. Using different databases creates diverse circumstances, allowing us to verify it against variety of challenges.

Figure 5. Examples from collected FRGC database

3 Proposed Architecture

In this work authors are using Convolutional Neural Network to train the model for Gender Detection. It is a Deep Learning Algorithm which specialises in taking input as image and learning various aspects from it. It processes images as per the data given to train and the output expected out of it. It is able to perform this classification because the data given to train the model are assigned among two class. Female and Male are the only two classes it has to classify between. As per the label assigned and using back propagation it assigns various learnable weights and biases to the objects present in the image. The preprocessing required are less in ConvNet when compared with other classification algorithms. Convolutional Network has ability to teach itself various aspects, characteristics and filters. The architecture of ConvNet is similar to how neurons are connected in Human Brains. Individual neurons respond to specific region in front of the visual field known as receptive field. ConvNet's primary aim is to reduce pictures in a way that makes it easier to analyse while preserving essential classification information.

3.1 Methodology

Following is the methodology:

1. C = {$Instagram, Whatsapp, GroupPhotos, Family$}

2. Data = D-A(Data) - Noise
   where Data $\rightarrow$ Enhanced Data,
   D-A $\rightarrow$ Data Augmentation,
   Noise $\rightarrow$ Noise removal,

3. CrImg = mtcnn.faceDetect(Data)
   Mtcnn uses face detection model to crop face, which is stored in Male and Female Folders

4. TrainSet , TestSet = sklearn.split(Folders),
   where Folder $\rightarrow$ Female Folder, Male Folder,

5. Model = CNNModel.add(TrainSet) ,

6. RealTime Video = Opencv.videoCapture(Camera),

7. Gender = Model.Predict(RealTime Video)
   Gender is predicted and displayed on Realtime Video

As mentioned above the flow of work started with data collection by sources mentioned in first step. Collected Raw data is then preprocessed to enhance for better classification and faces are cropped from the fed data which is followed by segregating the data into male and female folders respectively. Two different image folders are then divided into train and test set which trains the model and tests model accuracy on test set. The generated model is used to classify the gender on realtime video.

4 System Architecture

The Figure 6. shows the flow of System that is divided into different stages. It starts with data collection and ending Gender Detection by the features that got extracted using the CNN Model. The middle stages contain stages on data preprocessing, manual segregation to provide for classification, training the model and testing its accuracy.

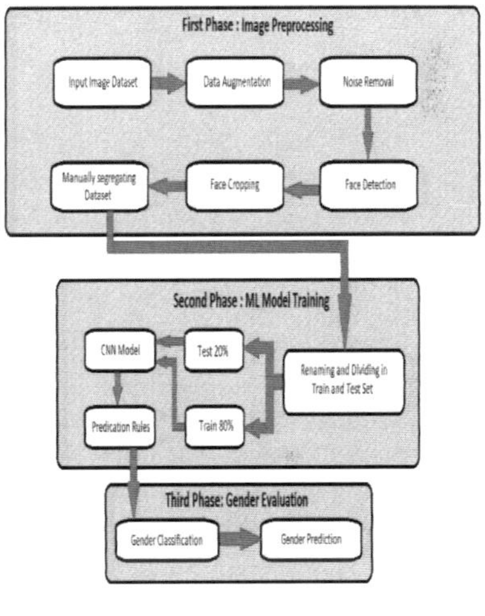

Figure 6. Real Time Gender Classification Architecture

4.1 CNN Algorithm

In this work, Keras Tensorflow library is used. Different layers are added to the sequential Model starting from conv2D. This layer marks the start of taking 2d images as input. Various layers that are included in the model are explained below

The Layers used are as follow:

1. **Conv2D:** The first layer of the Convolutional Neural Network remains the layer of the Convolutional Layer. Conversion layers apply the convolution effect to the input, transferring the result to the following layer. The answer converts all pixels in

their reception field to a single value. The foremost common kind of convolution used could be a 2D convolution layer and is commonly abbreviated as conv2D. The filter or kernel within the conv2D layer "slides" over 2D input file, enabling element-wise duplication. As a result, it summarizes the results into one single pixel output. The kernel will perform the identical function everywhere it goes, converting a 2D feature matrix into a separate 2D feature matrix.

2. **BatchNormalization:** To avoid overfitting, familiarity with a pre-data processing tool used to bring numerical data to a standard scale without distorting its structure. Generally, when authors enter data into a machine or in-depth reading algorithm authors tend to change values on an average scale. The reason authors are accustomed in part ensures that the model can adapt properly. Now back to the Batch normalization, it is a process of making neural networks faster and more stable by adding additional layers to a deeper neural network. The new layer performs the measurement and normalization functions of the layer input from the previous layer.

3. **Max Pooling:** Pooling layers are worn to reduce the scale of the feature maps. Therefore, reduce the quantity of learning parameters and also the number of computers created within the network. The integration layer summarizes the features present within the feature map region formed by the convolution layer. Therefore, some activities are performed on abbreviated elements instead of precisely set elements produced by the convolution layer. This makes the model more powerful in changing the form of the elements within the input image. Integration function includes sliding a two-sided filter at each feature map channel and summarizing features within a filter-covered region. Max Pooling maybe a merging function that selects the highest element from the feature map region covered by the filter. Therefore, the output after the max-pooling layer are a feature map that contains the foremost prominent features of the previous feature map.

4. **Flatten:**Classifying the 'Gender.' We created a Classification model, which means that the data used should be a good fit for the model. It needs to be in the form of a 1-sided vector. Rectangular or cubic shapes cannot be a straight input. And that's why we need flatten layers and fully connected. Flattening converts data into a 1-dimensional array to insert it into the next layer. We consider the removal of convolutional layers to create a single vector of long feature. It is also connected to a final separation model, called a fully connected layer. In other words, we put all the pixel data in a single line and make connections to the last layer. And again. 'Gender' segregation.

5. **Dense:**A dense layer is a layer of deeply connected neural network, which means that each neuron in a dense layer receives input from all the neurons of its previous layer. The dense layer is found as the most widely used layer in models. In the background, a dense layer creates a matrix-vector multiplication. The values used in the matrix are actually in parameters that can be trained and updated with the help of back propagation.The output produced by the dense layer is in the 'm' dimensional vector. Therefore, a dense layer is used to change the vector size. The dense layers work and function as rotation, scale, vector translation.

6. **Sigmoid:**Sigmoid activation function, also called activation function, is traditionally the most active opening function of neural networks. Input input is converted to a value of between 0.0 and 1.0. Inputs greater than 1.0 are converted to a value of 1.0, similarly, values greater than 0.0 are reduced to 0.0.

5 Experimental Results and Performance Analysis

5.1 Accuracy and Loss

Figure 7 shows the accuracy and loss comparison between train and test set when only celeb-A dataset is used.

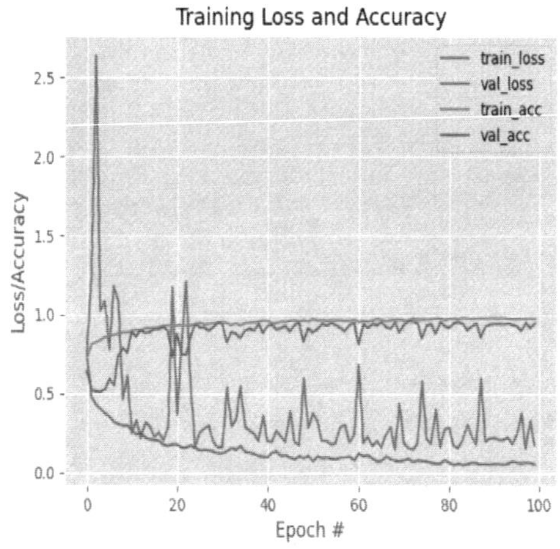

Figure 7. Loss Accuracy Graph on Celeb-A

Figure 8 shows the accuracy and loss comparison between train and test set when combined data of celeb-A and our collected data was fed to the Model
It is easily clear seeing both the graphs that the accuracy of Model when trained with only Celeb-A dataset which is readily available on Internet has a poor performance on

https://doi.org/10.1051/itmconf/20214003047

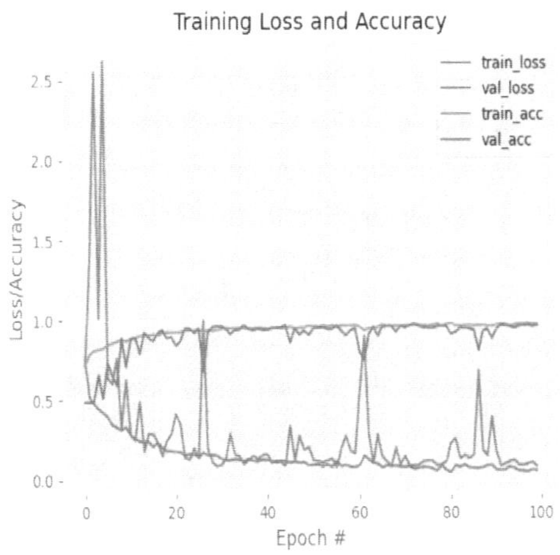

Figure 8. Loss Accuracy Overall

Unseen data whereas when it is trained with combined data yields a better performance.

5.2 Result Images

Test on realtime images are shown in the figures below. Different angles and look have been tried to check whether it was able to distinguish clearly or not.

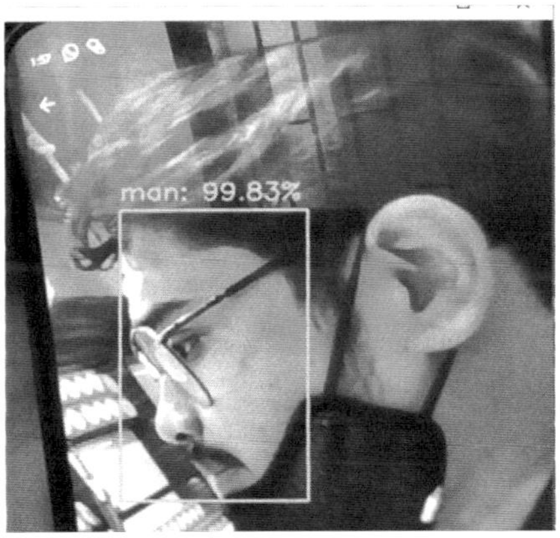

Figure 9. Male Face with side face and one-third face covered with mask

5.3 Performance Analysis

1. Earlier the model were trained with less raw images and that caused decrease in accuracy when classifying the human in realtime with different surrounding conditions. This problem was overcame by creating our own dataset which resembled realtime images to train the Model.

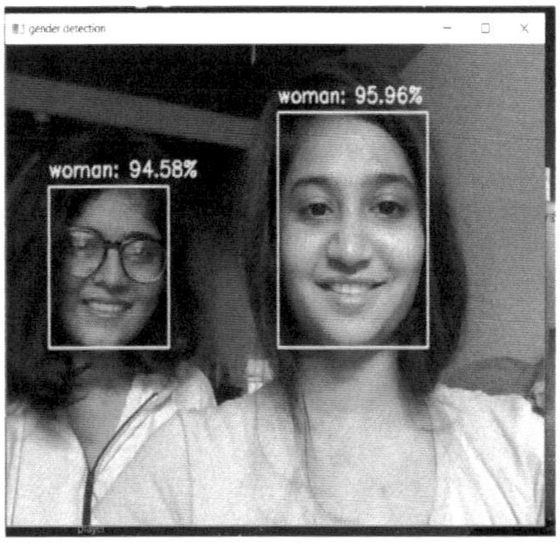

Figure 10. Two Females

Figure 11. Unmasked and Masked Male-Female

2. Second major problem was that many of the literature surveys tried gender classification but did not perform on realtime, while had slow performance in classifying the real-time feed instantly. This was solved by giving the model the required data and adding the right layers on the right places such as dropout and batch normalization in order to prevent overfitting on the training data.

6 Conclusion and Future Scope

The live gender classification and prediction dilemma, from which a machine automatically decides whether a source face relates to a female or a male, was developed in this chapter. While several previous approaches have tackled the problem of gender classification of images, this paper lays out a benchmark for the challenge based on state-of-the-art network architectures and shows that chaining age prediction with gender prediction will increase overall precision. Through this work, authors can draw two significant conclusions. First, CNN should be used to produce greater outcomes in gender labeling, even with the much smaller scale of contemporary unconstrained picture sets that are classified for gender. A gender classified properly can help cut down half of the choices for clothing, jewellery etc, which helps to give better experience to users surfing on the website. Not only for shopping but gender classification can also help for marketing. With the upcoming new social changes in the future there is one more gender which needs to be added Gender classification models that is for transgenders. To expand their methods so that outcomes can be strengthened. An end-to-end multi-task learning scheme for categorizing gender.

References

[1] Amit D. Ranjit K. Vijay B. Gender Recognition Through Face Using Deep Learning.Procedia Computer Science 132 (2018) 2–10

[2] Sajid Ali Khan, Muhammad Nazir, Naveed Riaz(2013)"Gender Classification using Multi-Level Wavelets on Real World Face Images", Springer Acta Polytechnica Hungarica Vol. 10, No. 4.

[3] S. Baluja and H. A. Rowley. Boosting sex identification performance. Int. J. Comput. Vision, 71(1):111–119, 2007.

[4] Jridi, M. Napoléon, T. Alfalou, A. One lens optical correlation: Application to face recognition. Appl. Opt. 2018.

[5] Tivive FHC, A.Bouzerdoum(Sep 2016) "A gender recognition system using shunting inhibitory convolutional neural networks" In: International Joint Conference on Neural Networks; Vancouver, Canada. New York, NY, USA: IEEE.

[6] Vaishnavi Y. Mali. Human Gender Classification using Machine Learning. y (IJERT) Vol. 8 Issue 12, December-2019

[7] Ma, Z. Ding, Y. Li, B. Yuan, X. Deep CNNs with Robust LBP Guiding Pooling for Face Recognition. Sensors 2018.

[8] Van de Wolfshaar, J. Karaaba, M.F. Wiering, M.A. Deep convolutional neural networks and support vector machines for gender recognition. In Proceedings of the 2015 IEEE Symposium Series on Computational Intelligence, Cape Town, South Africa, 7–10 December 2015.

[9] Duan, M. Li, K. Yang, C. Li, K. A hybrid deep learning CNN–ELM for age and gender classification Neurocomputing 2018.

[10] Levi G., Hassner T. Age and gender classification using convolutional neural networks. Boston, MA, USA, 7–12 June 2015.

[11] Rajeev, R. Vishal, M.P. Rama, C. HyperFace. A deep multi-task learning framework for face detection, landmark localization, pose estimation, and gender recognition. IEEE Trans. Pattern. Anal. Mach. Intell. 2019

[12] Yassin Kortli, Maher Jridi, Ayman Al Falou and Mohamed Atri. Face Recognition Systems: A Survey. Sensors January 2020

[13] Khalil Khan, Muhammad Attique , Ikram Syed and Asma Gul. Automatic Gender Classification through Face Segmentation. Symmetry 2019

[14] Alessandra Lumini, Loris Nanni, and Sheryl Brahnam. 2016. Ensemble of texture descriptors and classifiers for face recognition. Appl. Comput. Info. (2016).

[15] Shubham Patil Bhagyashree Patil, Ganesh Tatkare.Gender Recognition and Age Approximation using Deep Learning Techniques.y (IJERT) Vol. 9 Issue 04, April-2020

Movie Posters' Classification into Multiple Genres

Vaibhav Narawade
Ramrao Adik Institute of Technology
Nerul, Navi Mumbai
vnarawade@gmail.com

Aneesh Potnis
Ramrao Adik Institute of Technology
Nerul, Navi Mumbai
apotnis07@gmail.com

Vishwaroop Ray
Ramrao Adik Institute of Technology
Nerul, Navi Mumbai
vishwaroopray98@gmail.com

Pratik Rathor
Ramrao Adik Institute of Technology
Nerul, Navi Mumbai
pratikrathor2298@gmail.com

Abstract—**Our project intends to classify movies into the three most probable genres that they belong to, from a predefined set of 25 genres, based on only one image i.e the movie poster. We have made use of Convolutional Neural Networks (CNN) to realize this project as we believe it would be of help to extract the features and visual information from the image. Instead of a multi-class classification problem in which the input is classified into any one class, this project would be more correctly described as a multi-label classification problem as a movie belongs to more than one genre. In this project we see a comparative study of different architectures and tune them to yield the best result based on the metric of accuracy. We have applied various techniques such as data augmentation and L2 regularization to comparatively deduce the model that performs best from all the tested models.**

Index Terms—**Multi-label classification, movie genre classification, neural networks**

I. Introduction

In this problem we have used Multi-label classification and Convolutional Neural Networks for Genres based classification of Movie Posters. Multi-label classification originated from the investigation of text categorisation problem, where each document may belong to many pre-defined topics simultaneously. Multi-label classification of textual data is a crucial problem. Examples range from news articles to emails. In multi-label classification, the training set consists of instances that are labelled from a group of labels and the label for a unseen instance needs to be rightly predicted by analyzing the training instances that the model has been trained on with known label sets. In machine learning, Convolutional Neural Networks (CNN or ConvNet) are complex feed forward neural networks. CNNs are used for image classification and recognition due to its high accuracy. It was proposed by scientist Yann LeCun within the late 90s, when he was inspired from the human beholding of recognizing things. The CNN follows a hierarchical model which works on building a network, sort of a funnel, and eventually gives out a fully- connected layer where all the neurons are connected to each other and the output is processed.

A. Objective

The Objectives of this project is to categorize a movie according to the given poster into predetermined genres with the use of CNN. This classification would able the user to make informed decisions by providing the genres of the movies based on the features present in the poster. The proper classification of movies will enable seamless process of categorizing the movies according to the previous preferences of the user. Humans are able to recognize certain elements in a movie to classify it into genres and to automate this process CNN can be used to compare the posters present in the dataset and accordingly assign the relevant genres to the movie.

B. Motivation

The motivation behind this problem statement is to let the user identify if they would be interested in a particular movie by classifying the movie into multiple genres by identifying certain features from the poster of the said movie. This would be helpful to the user as the poster of the movie gets released up-to an year in advance, as opposed to the trailers which are released much later. The model would look for specific features in a movie poster that are predominantly present in a poster belonging to that specific genre, this would serve as a tool to notify the user about an upcoming movie that they might be interested in.

II. Literature Survey

1) Binary classification and Multi-Class classification have been highly discussed topics but a lot of newer technology and research has seen the development of Multi-Label classification. This has also highlighted the problems which come along with this classification which were previously not known in binary or multi-class classification. Purvi Prajapati, Amit Thakkar, Amit Ganatra (2012) shine a light on these challenges in their work in 2012. Their paper proposed the two main methods to approach for the solution of multi label problems, namely Problem Transformation Method and Algorithm Adaption Method. [1]
2) There have been a few attempts of classifying movie into different genres based on it's poster, the earliest being the work done by Marina Ivasic-Kos, Miran Pobar

and Ivo Ipsic (2014) which focused on extracting low-level features like dominant colours and colour layout. They also made the use of GIST image descriptor to extract the dominant spatial structures of an image characterized by it's boundary properties. Their classification techniques included Naïve Bayes classifier, a multi-label problem adaptation method called RAndom k-labELsets (RAKEL) and ML-kNN, a variation of the traditional k-Nearest Neighbour algorithm. This was applied on a dataset containing 6,739 movie posters and up-to 18 different genres each poster was classified into. [2]

3) This research served as groundwork for a later work done by Wei-Ta Chu and Hung-Jui Guo (2017) which applied deep learning for a modern approach to classification. Their research included a combination of fully connected neural network, convolutional neural network and a pre-trained YOLO v2 network for object detection. They also did a comparative study of the Ivasic-Kos, Pobar and Ipsic model and version versions of their own model by dividing the entire model into three parts and disabling or replacing one or more parts at a time including switching neural networks with a Support Vector Machine (SVM) classifier. Their dataset consisted of 8,191 movie posters belonging to 23 different genres. [3]
4) Jonatas Wehrmann and Rodrigo C. Barros (2017) proposed a neural network architecture on the basis of convolutional neural networks for accomplishing the task of genre classification. But what they did differently than the previous works is that they made use of movie trailers (audio and visual) to train their model. They developed a an approach which they called Convolution-Through-Time or CTT, which is an architecture that makes use of trailer frames through time to extract features for the model. [4]
5) Speaking of convolutional neural networks, the study by Rahul Chokshi and Samuel Sung (2018) expands on the trend of using deep neural networks to solve image classification problems. Samuel Sung and Rahul Chokshi utilised customized versions of three new and promising deep learning architectures namely: ResNet-50, VGG-16, and DenseNet-169. Their work included a dataset of 36898 posters classifying into 7 genres. [5]
6) A recent study on predicting movie genres from movie posters was done by Gabriel Barney and Kris Kaya (2019). Their work acknowledges both the previous attempts at the topic and even cites Ivasic-Kos, Pobar and Ipsic's work as the groundwork they attempted to expand upon. Of all the studies mentioned here, Barney and Kaya used the largest dataset, viz the Full MovieLens Dataset from Kaggle, which consists of meta-data collected from TMDB and GroupLens containing entries for 45,466 movies. After pre-processing their dataset, it yielded a total of 35,000 movies belonging to 20 different genres. They too did a comparative study in their research by comparing different techniques namely, ML-kNN, One vs. Many multi-label strategy (OvR), a pre-trained version of Residual Network (ResNet34) for transfer learning purposes and a custom neural network architecture. [6]
7) Jeong A. Wi, Soojin Jang and Youngbin Kim (2020) proposed a method to extract the optimal information and characteristics from movie posters to aid the classification of movies into genres calling it the Gram layer in a convolutional neural network (CNN). It initially extracts style features by first applying Gram matrix onto the feature map of the movie poster. This is used as style weight, the existing feature map is used along with style information for performing inference. For their study, a total of 20,764 movie posters were collected and 12 genres were defined. [7]
8) An extensive survey on modern architectures of Deep Convolutional Neural Networks was done by Asifullah Khan, Anabia Sohail, Umme Zahoora I& Aqsa Saeed Qureshi (2020). While it has become common knowledge in the deep learning community that deeper neural networks result in higher accuracy, this has also given rise to problems like disappearing gradient problems which occur in which more layers cause the loss function gradient to approach zero, making it difficult to train the network. [8]
9) The latest approach to solving multi-label classification problems was brought up by the work of Emanuel Ben-Baruch, Tal Ridnik, Nadav Zamir, Asaf Noy, Itamar Friedman, Matan Protter and Lihi Zelnik-Manor (2020). In their study, they have discussed the most common problem associated with multi label classification, the high imbalance in the positive vs negative labels for a single input(picture). This problem, dubbed a the Positive-Negative imbalance, takes up most of the time in optimization process and if not taken care of, can result in under-emphasizing gradients from positive labels which in turn will result in lower accuracy. [9]

A. Limitations of Existing System

1) The paper by Purvi Prajapati, Amit Thakkar, Amit Ganatra (2012) [1] presented a research of problem transformation vs algorithm adaption methods to solve multi label classification problems. One of the challenges stated by them in this field is the preprocessing of data that is required before the machine learning algorithms are applied. These techniques include but are not limited to pruning, feature selection and handling missing values to increase the performance of such problems.
2) According to Ivasic-Kos, Pobar and Ipsic's [2] comparative study, the Naïve Bayes classifier yielded the best result when considering the F1 score as the parameter of judgement, in which Naïve Bayes' F1 score was 0.38. Their application of GIST image descriptor along with classic machine learning algorithms can be considered

outdated as per modern standards as neural network architectures have proved to outperform machine learning algorithms when it comes to large datasets especially the ones which involve image classification.

3) Chu and Guo's [3] study showcases the effectiveness of incorporating object information in neural networks using the YOLO v2 pre-trained object detection module which works quite well in detecting animals, which often appear in animation movies. Although their improvement over previous works has been promising, the neural network architecture they used (AlexNet) is an early one (2012) which is not in the trend nowadays.
4) Although the work done by Jonatas Wehrmann and Rodrigo C. Barros (2017) [4] showed very promising results especially when compared to low level features and improving accuracy on certain genres, their usage of movie trailers as their data set is not only computationally heavy but also redundant because by the time the trailers are released, the genres of the movies are already made very clear.
5) Sung and Chokshi's [5] study was an approach to find the optimum model for identifying genres and led to the comparative study of the THEN promising new models among which:DenseNet-169 was able to generate the best results. They can be still improved. They propose that the improvement can come through various techniques They forecasted that this improvement could come through various techniques, some being, the use of an object identifier algorithm such as YOLO, collection of better data, and application of other convolutional neural network (CNN) architectures to the problem.
6) Barney and Kaya (2019) [6] used the largest dataset and after preprocessing, it yielded a total of 35,000 movies belonging to 20 different genres. Possible improvement is using more balanced data. As the dataset that is used for the model is imbalanced, data augmentation can be used to increase the data the model can train from. A mixture of undersampling and oversampling can be applied respectively to the classes that have less data and the classes that have greater number of entries.

III. Goals

- The project aims at classifying the movies into appropriate genre based only on the movie poster of the said movie by performing Multi-Label Classification and Deep Learning to provide an automated way of categorizing the movies into specific genres in order to accommodate the viewing preferences of users consuming the content.
- The Convolutional Neural Network has been identified as the Deep learning model for processing the images of movie posters.
- The dataset of 7242 movies has been used for this problem where these movies are spread across 25 different genres viz. Action, Adventure, Animation, Biography, Comedy, Crime, Documentary, Drama, Family, Fantasy, Horror, Music, Musical, Mystery, News, Reality-TV, Romance, Sci-Fi, Short, Sport, Thriller, War, Western.
- We propose to classify the movie poster into specific categories of genres from the 25 genres that are pre-defined. The movie would be categorized into different top three genres based on the probability of them belonging to the said genres. This classification would enable the user to make an informed decision regarding their interest in watching the movie when it releases. These classifications of movie posters into genres would be useful as the posters for various movies are released months before the trailers are released.
- The method incorporated for this project is deep learning. We have used the CNN which is widely used for processing images and recognizing patterns from these images. The first component of CNN called the convolution layer is for feature extraction, specifies the number of filters or feature detectors that make a feature map based on the feature that match with the filters and add a higher number in the feature map in such a place. The second component is the Max Pool where the size of the feature map is reduced while preserving the information about the features. The third component called Flattening transforms a two-dimensional matrix into a vector that can be fed into a fully connected layer. The fourth and final component before the output layer is the fully connected layer. In this layer as the name suggests every neuron is connected to every other neuron from the previous layer. As the genres can be classified into from a total of 25 the output layer has 25 nodes.

A. Scope

- Movie Poster Classification is a multi-label classification problem with a large movie dataset the aim is to achieve multiple movie genre classification based only on movie posters images.
- For movie-goers, movie posters are among the very first impressions used to get an idea about the content of the movie and its what genres it could possibly be. Humans can get ideas based on things like color, objects, expressions on the faces of objects etc to quickly determine the genre (horror,comedy,animation,etc).
- We can assume that the poster possesses some characteristics which might be utilized in deep learning algorithms to predict its genres.

IV. Design of the System

The system has been developed using TensorFlow. The Convolutional Neural Network has been identified as the Deep learning model for processing the images of movie posters. The dataset of 7242 movies has been used for this problem where these movies are spread across 25 different genres viz. Action, Adventure, Animation, Biography, Comedy, Crime,

Documentary, Drama, Family, Fantasy, Horror, Music, Musical, Mystery, News, Reality-TV, Romance, Sci-Fi, Short, Sport, Thriller, War, Western.

The system takes an input in the form of an image of a movie poster and this image is passed through a CNN model to make the appropriate inference of the genre of the movie. [Figure 1]

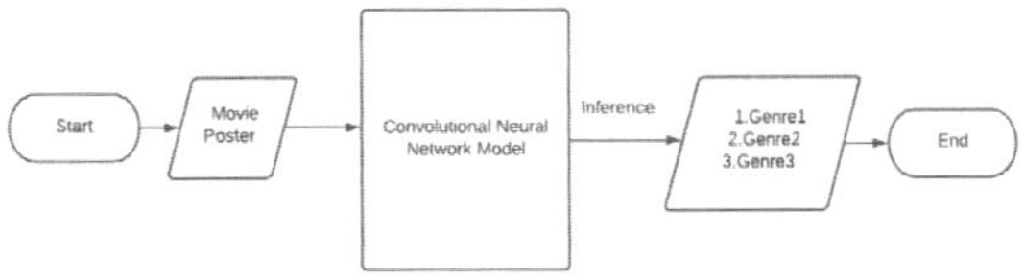

Fig. 1. Design of the system

V. Implementation Details

A. Models

We have created variation of models to make the inference on the movie poster and the models are described below:

a) Model 1: The baseline model is a 4 layer CNN model with batch normalization and dropout regularization after each convolutional layer. The baseline model that was created for performing inference on the movie posters.

b) Model 2: VGG inspired model i.e block based model with each block containing 2 convolutional layers having 3x3 filters. We have 3 blocks with 32, 64 and 128 filters respectively. We have also applied Dropout regularisation after each block. The VGG16 inspired model used has the following layers.[Figure 2][Figure 3]

c) Model 3: Same as previous models but with data augmentation. In this iteration of the model we have used data augmentation which randomly performs horizontal and vertical flips with rotations upto 90 degrees.

d) Model 4: Same model as Model 2 but with L2 regularization in 2 blocks of the model. (each block has 2 CNNs).

e) Model 5: Same model as Model 2 but with L2 regularization AND data augmentation.

f) Model 6: VGG16 architecture was used (using transfer learning).

g) Model 7: VGG16 with data augmentation.

Layer (type)	Output Shape	Param #
conv2d_12 (Conv2D)	(None, 128, 128, 32)	896
conv2d_13 (Conv2D)	(None, 128, 128, 32)	9248
max_pooling2d_6 (MaxPooling2	(None, 64, 64, 32)	0
dropout_8 (Dropout)	(None, 64, 64, 32)	0
conv2d_14 (Conv2D)	(None, 64, 64, 64)	18496
conv2d_15 (Conv2D)	(None, 64, 64, 64)	36928
max_pooling2d_7 (MaxPooling2	(None, 32, 32, 64)	0
dropout_9 (Dropout)	(None, 32, 32, 64)	0
conv2d_16 (Conv2D)	(None, 32, 32, 128)	73856
conv2d_17 (Conv2D)	(None, 32, 32, 128)	147584
max_pooling2d_8 (MaxPooling2	(None, 16, 16, 128)	0
dropout_10 (Dropout)	(None, 16, 16, 128)	0
flatten_2 (Flatten)	(None, 32768)	0
dense_4 (Dense)	(None, 128)	4194432
dropout_11 (Dropout)	(None, 128)	0
dense_5 (Dense)	(None, 25)	3225

Total params: 4,484,665
Trainable params: 4,484,665
Non-trainable params: 0

Fig. 2. VGG16 inspired Model

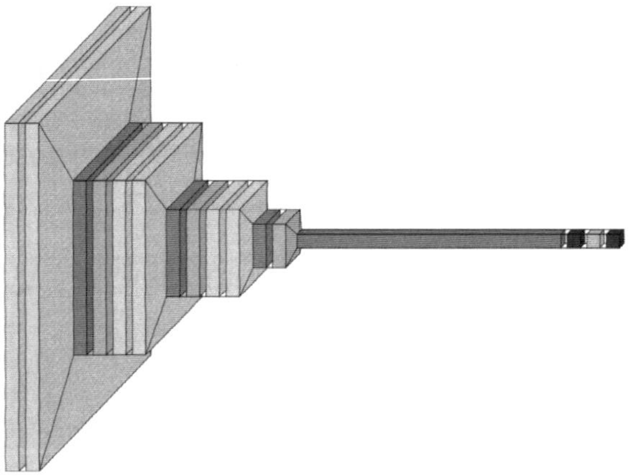

Fig. 3. Visual representation of the layers of Model 2

Yellow = Convolutional layer; Red = Max Pooling layer; Green = Dropout layer; Blue = Flatten layer; Dark blue= Dense layer

 https://doi.org/10.1051/itmconf/20214003048

VI. Results

A. Evaluation metric

Accuracy is the fraction of number of correctly predicted outcomes over the total number of predicted outcomes. Informal formula:

Accuracy = Correct Predictions/Total Predictions

In classification problems, the formula of Accuracy:

Accuracy = (TP + TN)/(TP + FP + TN + FN)

True Positives(TP): These are the outcomes in which the model predicts positive class and is correct. True Negatives(TN): These are the outcomes in which the model predicts negative class and is correct. False Positives(FP): These are the outcomes in which the model predicts positive class and is incorrect. False Negatives(FN): These are the outcomes in which the model predicts negative class and is incorrect. The models are compared on the basis of their accuracy and the cross entropy loss in the table[Table I] below.

TABLE I
Accuracy comparison of the models

Sr. No.	Model Details		
	Name	*Accuracy*	*Loss*
1	Model 1	0.279	0.245
2	**Model 2**	**0.364**	**0.236**
3	Model 3	0.352	0.239
4	Model 4	0.323	0.232
5	Model 5	0.341	0.237
6	Model 6	0.338	0.261
7	Model 7	0.329	0.242

The baseline model, Model 1[Figure 4], and the best performing model, Model 2[Figure 5], are compared through graphs for cross entropy loss and accuracy.

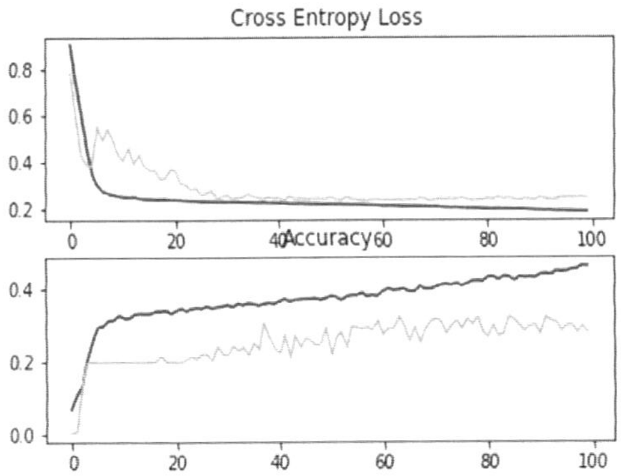

Fig. 4. Baseline model loss and accuracy

The inference from the model is given below where the most probable three genres are predicted. These genres are compared with actual listed genres of the movie.

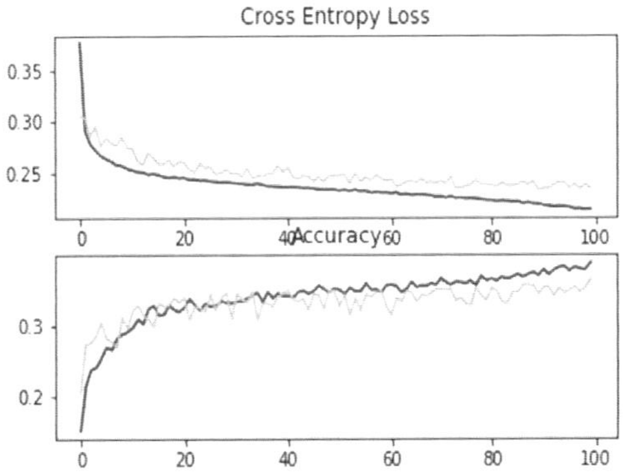

Fig. 5. VGG16 inspired model loss and accuracy

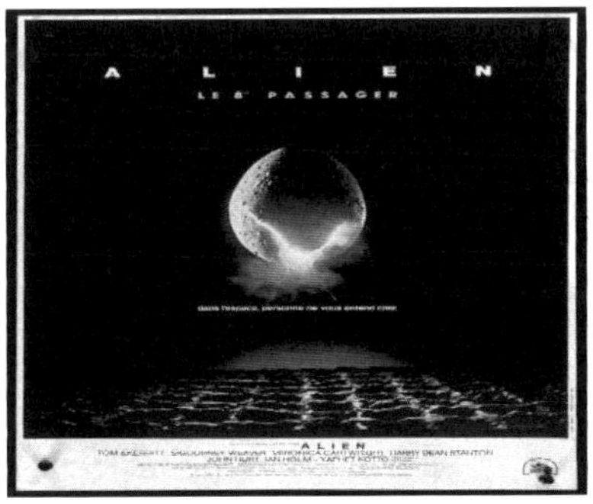

Fig. 6. Alien (1979)

Output: Drama, Thriller, Horror
Ground Truth: Sci-Fi, Thriller, Horror

Fig. 7. To All the Boys I've Loved Before (2018)

Output: Comedy, Drama, Romance
Ground Truth: Comedy, Drama, Romance

Fig. 8. Green Book (2018)

Output: Drama, Comedy, Adventure
Ground Truth: Drama, Comedy, Biography

Fig. 9. The Hitman's Bodyguard (2017)

Output: Comedy, Drama, Action
Ground Truth: Comedy, Adventure, Action

Fig. 10. Results

VII. Conclusion

With the results shown and on comparing the accuracies of different models, we can see that deeper and more complex neural network architectures tend to give significantly better results as compared to simpler architectures. After a certain point, the accuracies become stagnant and even different regularization techniques do not give improved results in terms of the metric of accuracy, although the loss is diminished by them. The regularization techniques also helped prevent overfitting of the model. The dataset too, as of now, is not very robust and is heavily partial towards the Drama, Comedy and Romance genres. Especially the genres, Comedy (2900 entries) and Drama (3619 entries) in a dataset of 7242. As is evident by the previous works, the imbalanced dataset problem has been common. Since our problem statement is a multi-label classification problem, classic methods of over-sampling or under-sampling of dataset, which works well for multi-class problems, will not provide us with desired results here. So in future, a well balanced dataset is sure to give much better results on our model.

References

[1] P. Prajapati, A. Thakkar, and A. Ganatra, "A survey and current research challenges in multi-label classification methods," *International Journal of Soft Computing and Engineering (IJSCE)*, vol. 2, no. 1, pp. 248–252, 2012.
[2] M. Ivasic-Kos, M. Pobar, and I. Ipsic, "Automatic movie posters classification into genres," in *International Conference on ICT Innovations*, pp. 319–328, Springer, 2014.
[3] W.-T. Chu and H.-J. Guo, "Movie genre classification based on poster images with deep neural networks," in *Proceedings of the Workshop on Multimodal Understanding of Social, Affective and Subjective Attributes*, pp. 39–45, 2017.
[4] J. Wehrmann and R. C. Barros, "Movie genre classification: A multi-label approach based on convolutions through time," *Applied Soft Computing*, vol. 61, pp. 973–982, 2017.
[5] S. Sung and R. Chokshi, "Classification of movie posters to movie genres,"
[6] G. Barney and K. Kaya, "Predicting genre from movie posters," 2019.
[7] J. A. Wi, S. Jang, and Y. Kim, "Poster-based multiple movie genre classification using inter-channel features," *IEEE Access*, vol. 8, pp. 66615–66624, 2020.
[8] A. Khan, A. Sohail, U. Zahoora, and A. S. Qureshi, "A survey of the recent architectures of deep convolutional neural networks," *Artificial Intelligence Review*, vol. 53, no. 8, pp. 5455–5516, 2020.
[9] E. Ben-Baruch, T. Ridnik, N. Zamir, A. Noy, I. Friedman, M. Protter, and L. Zelnik-Manor, "Asymmetric loss for multi-label classification," *arXiv preprint arXiv:2009.14119*, 2020.

ITM Web of Conferences **40**, 03049 (2021) https://doi.org/10.1051/itmconf/20214003049
ICACC-2021

Ensuring social distancing using machine learning

Gaurav Mathurkar[1], *Chinmay* Parkhi[1], *Manish* Utekar[1] and *Ms. Pallavi* H. Chitte[1]

[1]Department of Computer Engineering, Ramrao Adik Institute of Technology, India

Abstract—COVID-19 (Coronavirus) has affected our daily life and is slackening the global economy. In this fight against the COVID-19, social-distancing is proving to be an effective measure to slow down the transmission of coronavirus. In order to control the spread of transmission of COVID-19, social distancingis the best measure which aims to avoid close contact of people. In order to control the spread of transmission of COVID-19, social distancing is the best measure which aims to avoid close contact of people. With the aim of ensuring social- distancing norms in workplace and public places, we can develop a social-distancing detection tool that monitors if people are staying at a safe distance from each other. This can be done by examining real-time video streams from the cameras. In order to provide an effective solution, we are developing a model which has mainly two phases - firstly object detection phase where people will be detected from the video streams. Secondly showing the statistical analysis in the form of dashboards andsending alerts to concerned authorities to take necessary actions.

Keywords— COVID-19, object detection, YOLO V3, alerts

I.INTRODUCTION

Social distancing is an important step to avoid spreading COVID-19 virus which involves increasing the physical space between people. Implementing social distancing during the ongoing global pandemic is an important step. Maintaining a particular distance from other people lessens your chances of contracting COVID-19. Businesses need to ensure that social distancing policies are followed to keep customers, workers safe. Determining the exact distance between people in any workplace or location can be challenging and also resource consuming. Social-distancing is an effective practice to help reduce the contacts with the people affected with COVID-19 outside the home. Anyone can get and spread COVID-19, the risk of serious illness may or may not be different for everyone. So it's a duty of each person in the society to follow social distancing norms asper mandated by Government officials which results in protecting the entire community. To support government efforts and the given public health guidelines, a Social distancing detector that helps the concerned institutions to determine whether people are following social distancing guidelines can be developed. The system provides a detailed analysis of the situation through the dashboard, thus helping the relevant authorities to obtain an accurate analysis of the situation in the Region of Interest (ROI). While work from home for office work is advisable practice to scale back the transmission of coronavirus, some people don't have that option and that they may need to run bent the grocery, pharmacies or the other public areas. In these situations, it'll be important to take care of social distancing norms so as to take care of a healthy society and perform the activities productively.

The main functions include:

- Generate statistical analysis for the required region
- Sending alerts (if social distancing norms are violated above a particular threshold) to the respective authorities to take actions as per requirement.

We can use OpenCV and YOLO to monitor whether people are maintaining social distancing or not.

II.BACKGROUND STUDY AND RELATED WORK

In order to provide a solution for effective social distancing management, several approaches have been proposed. Among noteworthy systems some of the systems are analyzed and considered for our system development. Using Faster R-CNN and yolo v4 a system was proposed to detect people in a region of interest and measure the interpersonal distances in real time without data recording to protect the privacy of individuals which can be misused intentionally or unintentionally [3]. Using Yolo v3 and deep sort techniques a system was developed using three object detection models and a performance comparison was done for the three object detection models like Faster R-CNN, SSD and Yolo v3 [2]. Refer the following figure 2.1 for performance comparison of different object detection models as discussed in the paper [3].

Model	TT(in sec)	mAP	NoI	TL	FPS
SSD	9651	0.969	12135	0.02	3
Faster RCNN	2124	0.691	1200	0.22	10
YOLO v3	5659	0.846	7560	0.87	23

Table 1: Comparison of different object detection models

From the Table.1 we can notice that YOLOv3 can process at higher number of frames per second (fps) as compared to SSD and Faster RCNN. Using architecture of 19 layers a model named SocialdistancingNet-19 was developed and provided the solution for maintenance of social distancing by showing the number of violations in the area. The calculated accuracy of the model was that the maximum accuracy was estimated as 93.3 and the minimum accuracy was estimated as 74.3% [1]. Using Passive Infrared Sensors (PIR) which are used to detect human movement when it comes in a particular range, actual design was implemented using raspberry pie microcontroller and the proposed device created a alert on the mobile device using ML algorithm to determine the possibility of covid patients around a person

[4]. In the paper [6] a solution is proposed for developing a mobile application for monitoring with the help of mobile camera with 85% accuracy using Single Shot Detection model.

III.PROPOSED APPROACH

The proposed work includes usage of yolo v3 which is a fully convolutional neural network algorithm to detect the people in the video frames. OpenCV which is a real-time computer vision library is used to feed the input framed from the images or videos captured by the cameras to the yolo v3 neural network. The system will then provide the results for the particular region and provide a statistical analysis of the region captured by the cameras.

Yolo v3 which is capable of recognizing 80 different objects in given input images and videos is used for the object detection phase in the algorithm. Firstly, Yolo v3 is having 53 layer network trained on Imagenet. It uses a variant of darknet. But for the purpose of detecting object, more 53 layers are used on to the previous set of layers. Thus yolo v3 has total 106 layers in its network. The basic design of Yolo v3 is given in figure 3.1.

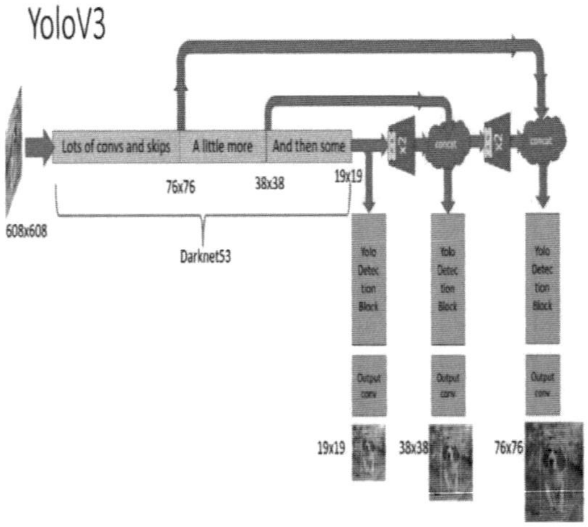

Fig 3.1:YOLO V3 architecture

The yolo v3 algorithm breaks the given input image into an S × S grid. A grid cell will be accountable for detection of object if the center of the object is present within the grid cell. To predict multiple labels for an object a threshold is used and for predicting each class score logistic regression is used by the algorithm. Classes that have scores higher than 50% which is considered as a confidence parameter to filter out the object that belongs to "people" class in the image. The interpersonal distance between the people is calculated using the distance between centroids i.e. the distance between the bounding boxes within the image. If the distance is found to be greater than a particular threshold then it is considered as a social distancing norms violation and displayed as unsafe with red bounding box in the image. If the social distancing norms are followed then the safe people are displayed with green bounding box.

Mathematical equations required for the system:
The yolo v3 network predicts four coordinates for each bounding box.

$z_x = \sigma(s_x) + m_x$ ....(i)

$z_y = \sigma \sum(s_y) + m_y$ ....(ii)

$z_h = q_h e^{s_h}$ ....(iii)

$z_w = q_w e^{s_w}$ ....(iv)

where z_x, z_y, z_h, z_w are the x and y center co-ordinates, height and width of our prediction and s_x, s_y, s_h, s_w, are the grid outputs. m_x and m_y are the upper left coordinates of the network. q_h and q_w are anchors dimensions of the box [14].

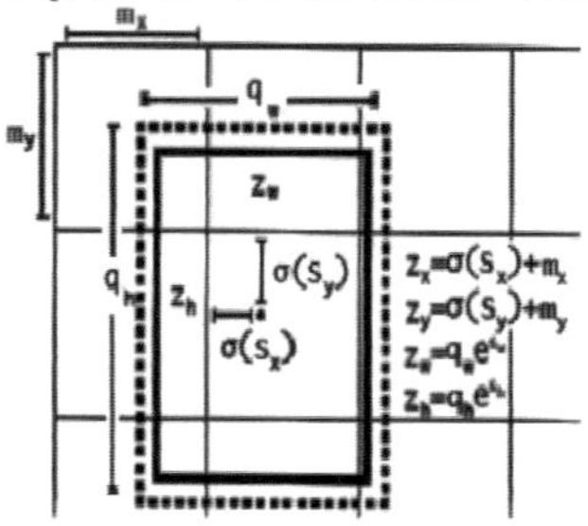

Fig 3.2:- Location prediction with coordinates in Bounding Boxes.

To find the distance between the centers of each bounding boxes which represent a person we use following formula,

$$\sqrt{(X_{\max} - X_{\min})^2 + (Y_{\max} - Y_{\min})^2}$$

The distance between the two people that is represented by the centers of the bounding boxes is given by the above equation.

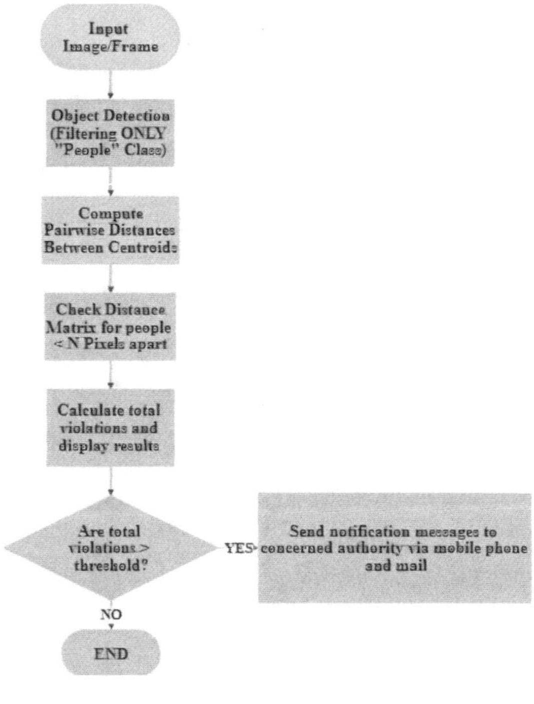

Fig 3.3 : System Design

https://doi.org/10.1051/itmconf/20214003049

IV.IMPLEMENTATION DETAILS

In order to process the frames and detect "people" class from the processed frames, some steps are to be followed that can be illustrated in the following flowchart. The flowchart gives a brief overview about each step in the system.

STEP 1:
The system takes an image or a frame from a real time video as an input using OpenCV. Since it wouldn't be very productive the system don't run the process for every frame. The video frame then will be processed to create a blob which is required for preprocessing our frame. The blob from Image can be used to process the frame which performs mean subtraction, scaling and swapping functions on the given image. After preprocessing the image will be provided to the Yolo neural network which then detects the objects within the image. For detecting the objects that belongs to the "People" class, a confidence of 50 % is set to filter out the objects with "People class" whose confidence is greater than 50 % as detected by our Yolo neural network.

STEP 2:
After detection of persons as an object, the system will treat this detections as a bird's eye view image as shown in figure. Further for transformation the system would select four points on the resultant image that are going to be the corner points of the image. With the two opposite sides being parallel, a rectangle will be formed with the help of those points.

STEP 3:
For each person detected, the system returns two points that are required to build a bounding box. These two points are nothing but the upper left corner and lower right corner of that box. Then from these the centroid of the box will be calculated by first calculating the midpoint of the line joining these two opposite sides. Then the distance between two centroids (i.e. Distance between two persons) is calculated with the threshold set as 2 feet or approximately 120 pixels.

STEP 4:
Using Matplotlib which is a plotting library we can showour results in the form of pie charts with detailed analysis of number of violations in the particular region as captured by the cameras on a dashboard adjacent to the video stream. We can provide total safe and unsafe number of violations which can provide statistical analysis for the particular region.

STEP 5:
The system is designed in such a way that if total number of violations are exceeding a particular predefined threshold then an alert email will be sent to the respective authority with total active violations at a particular point of time, which allows the concerned authority to keep a track for the particular area easily. The smtplib module in python which defines the SMTP client object is used to send the email.

STEP 6:
At the same time of sending email, a bulk SMS message will be sent to the registered mobile numbers of the respective authorities when the total number of violations are exceeding a particular predefined threshold. The SMS Gateway API (Fast2SMS) is used to send messages to mobile phones using python.

V.ALGORITHM

1. Taking video file as input using cv2.VideoCapture()
2. Setting up the YOLO v3 network
3. Detecting the "person" class in the current video frame and using cv2.dnn.blobFromImage() for performing mean subtraction, scaling and swapping functions.
   3.1. If confidence is greater than 50% then proceed to step 4.
4. Calculating the co-ordinates of bounding boxes.
5. Using non maximal suppression for removing overlapping boxes
6. Calculating distance between two people using the formula
   6.1. If distance<120 pixels:
   Increase the unsafe people count
   6.2. If distance>120 pixels:
   Increase the safe people count
   Show the results on the dashboard
7. If unsafe people count > predefined threshold:
   Send a mail using smtplib module and text message using Fast2SMS API to authorities.

https://doi.org/10.1051/itmconf/20214003049

VI.EXPERIMENTS AND RESULTS

The yolo v3 algorithm used in this system detects "people" class from its set of 80 different classes from the frames as seen in the figure 5.1 and figure 5.2. The system processes with frame rate of 1.39 fps. The system is tested using video streams and images. Each frame captured from the video are labelled as safe and unsafe according to the distance between the people. A dashboard is shown with total number of safe and unsafe people. A pie chart is shown on the dashboard which helps in getting a statistical view for a particular area. The pie chart shows safe percentage of people in an area using blue colour and unsafe percentage of people in an area using orange colour. Currently the threshold value (i.e. maximum number of violations allowed in captured area) is set to 7.

The application detects people who are close by and not adhering to the distancing norms and marks them in RED bounding boxes, signifying risk. Others are marked in GREEN.

Fig 5.1: Social distancing violations < Threshold value

Fig 5.2: Social distancing violations > Threshold value

If total number of violations are exceeding a particular predefined threshold then an alert email is sent to the higher authority in-charge with total active violations at a particular point of time.

Also if the total number of violations are exceeding a particular predefined threshold, then an alert text message is sent to the concerned sub-authorities.

VII.CONCLUSION

The proposed system provides a systematic approach for tracking social distancing violations. It is possible to focus on areas where social distancing is not maintained and get statistical analysis for the region of interest (ROI) in the form of a visual representation through the dashboard. In order to alert the concerned authorities about the violations in a particular area, the system focuses on sending alert messages to the respective authorities after the particular number of violations(threshold) is exceeded in particular area. This will help the concerned authorities to track the social distancing norms violations easily and take actions as per the requirement. This proposed model is beneficial rather than physical monitoring which reduces the human efforts required and giving a vast coverage of areas for detection purposes. The system can be used in most of the public places such as railway stations, religious places, marketplaces and hospitals to minimize the spread of any contagious disease. Thus the proposed model is useful for giving alerts to maintain social distancing in this pandemic.

REFERENCES

[1] Ninad Mehendale, Rinkal Keniya, Real-time social distancingdetector using Social-distancing,Net-19 deep learning network,Scribd, (2020), https://www.scribd.com/document/481289110/social-distance- detector.

[2] N. Singh Punn, S. Agarwal, S.K. Sonbhadra, Monitoring Covid-19 social-distancing with person-detection and tracking via fine-tuned YOLOv3 and Deep-sort techniques, arXiv, (2020), https ://arxiv.org/abs/2005.01385

[3] V. Renganathan, D. Yang, E. Yurtsever, U. Ozguner, K.A. Redmil1, A vision based social-distancing and critical density detection-system for COVID19, Researchgate, (2020), https://www.researchgate.net/publication/342763051_A_Vision_based_Social_Distancing_and_Critical_Density_Detection_System_for_COVID-19

[4] P. Whig, S.M. Mohammad, R.R. Nadikattu, Novel economical social-distancing smart device for COVID19, International Journal of Electrical Engineering and Technology(IJEET,) (2020), https://papers.Ssrn.com/sol3/papers.cfm?abstract_id =364 0230

[5] S. Gawde, D. Kalbande, A. Ghorai, Digital solution for enforcingsocial-distancing, SSRN 3614898, International Conference onInnovative Computing & Communications (ICICC), (2020), https://papers.ssrn.com/sol3/papers.cfm?abstract_id=3614898

[6] Dr. Neelavathy Pari S , Geetha A V , Jeevitha V K, Balaji Vasu,Monitoring Social-distancing by Smartphone App in the Effect of COVID19, IJERTV9IS090469, (2020), https://www.researchgate.net/publication/344536998_Monitoring_Social_Distancing_by_Smart_Phone_App_in_the_Effect_of_COVID-19

[7] R. Zust, M. Bielecki, Z.G. Stanga, D. Siegrist, D. Meyerhofer,A. Stettbacher, T.W. Buehrer, G.A.G.Crameri, J.W. Deuel, Social-distancing alters the clinical course of COVID19 in young adults: A comparative cohort study, Clinical Infectious Diseases, PubMed, (2020), https://pubmed.ncbi.nlm.nih.gov/32594121/

[8] Pratibha Rathi, Pranav Adarsh, Manoj Kumar, YOLO v3-Tiny: Object Detection and Recognition using one stage improved model, IEEE, (2020), https://ieeexplore.ieee.org/document/9074315

[9] S Yana, Dedi Satria, S. Syahreza, Rizal Munadi, Implementati on of SMS-Gateway in the Flood Notification System,Researc hgate, (2020), https://www.researchgate.net/publication/340490994_Implementatin_of_SMS_Gateway_in_the_Flood_Notificatin System_ using_Raspberry_Pi

[10] Shuaiyang Li, Liquan Zhao, Object Detection Algorithm Based onImproved YOLOv3, Researchgate, (2020), https://www.researchgate.net/publication/340145010_Object_Detecti on_Algorithm_Based_on_Improved_YOLOv3

[11] Dr. T. Uma Devi, Sakshi Gupta, YOLOv2 based Real Time ObjectDetection, IJCST, (2020), http://www.ijcstjournal.org/volume-8/issue-3/IJCST-V8I3P4.pdf

[12] Shuyuan Yang, Fan Zhang, Licheng Jiao, Fang Liu, Rong Qu, Lingling Li, Zhixi Feng, A Survey of Deep Learning-based ObjectDetection, arxiv, (2019), https://arxiv.org/pdf/1907.09 408.pdf

[13] Chulhee Lee, Youhak Lee, Jin-Sung Kim, Hyuk-Jae Lee; Fast Detection of Objects Using a YOLOv3 Network for a VendingMachine, IEEE, (2019), https://ieeexplore.ieee.org/document/8771517

[14] Farhadi, Ali., Redmon Joseph, YOLOv3: An IncrementalImprovement, arXiv:1804.02767,(2018), https://arxiv.org/abs/18 04.02767

[15] Tareq Abed Mohammed; Saad Albawi; Saad Al-Zawi, Understanding of a convolutional neural network, IEEE,(2018), https://ieeexplore.ieee.org/document/8308186

[16] Varsha Pathak, Manish Joshi, A survey of SMS based InformationSystems, Researchgate,(2015), https://www.researchgate.net/publication/277334889_A _survey_of_SMS_based_Inf ormation_Systems.

[17] Deva Ramanan, Zitnick Piotr Dollar, Lin Michael, Tsung-Yi, Belongie Lubomir, Maire Serge, Bourdev Ross, James Hays, Girshick, Pietro Perona, C. Lawrence, Microsoft COCO (Common Objects in Context),arxiv,(2015),https://arxiv.org/pdf/1405.0312.p df

[18] M. Gordon, "An introduction to network-programming the Pythonway", IEEE, (2005), https://ieeexplore.ieee.org/document/ 1541900/authors#authors

Early Stage Detection of Parkinson Disease

Megha Navada[1], Deepshikha Mishra[2], Saloni Parkar[3],Parag Patil[4],Chaitanya Jage[5]
Department of Electronics Engineering,
Ramrao Adik Institute of Technology,
Nerul, Navi Mumbai.
Email : meghanavada@gmail.com[1], deepshikhamishra777@gmail.com [2], salonipatkar1999@gmail.com[3], parag.patil@rait.ac.in[4], chaitanya.jage@rait.ac.in[5]

Abstract—**Parkinson's disease is a chronic neurodegenerative condition that demonstrate the progressive loss of the ability to correlate movements mainly occurs in the elderly. For the purpose of monitoring tremors in Parkinson's disease, a system has to be designed and developed. For coordination of movements, people with Parkinson's, deprive of a chemical called dopamine which behaves as the messenger between the brain parts and the nervous system .Detecting Parkinson's disease is a very arduous task as there is no evidence currently present to do this. Therefore, the main intention of our work is the designing of a system for recognizing Parkinson's disease at an initial stage. An Android application is being designed that allows the status of PD patients to be assessed based on the tests found on the Unified Parkinson's Disease Rating Scale approved by the Movement Disorders Society (MDS-UPDRS).**

Keywords- Neurodegenrative, Tremor, Dopamine, Android application.

I. Introduction

In Parkinson's disease the central nervous system of the patients brain is effected. the central nervous system of the patients. Because of this, there is a direct effect on the movements of the body. The motor neurons gets damaged. It slowly starts with tremors. At the initial stage, it occurs only in one hand. It gets worse with the ongoing time. Although, there is no cure for Parkinson's, medications might improve the conditions slightly.

The method used for detection of Parkinson's disease is performing motor tasks manually as mentioned in MDS-UPDRS scale which is the Movement Disorder Society (MDS)- sponsored new version of the UPDRS. . The Task Force for Rating scales in this disease, formulated this. The MDS-UPDRS scale has a disadvantage which is the requirement of a trained clinician while performing the tasks [9]. According to the paper "Using wearables to assess bradykinesia and rigidity in patients with Parkinson's disease" , a focused, narrative review of the literature, the reliability of these methods lie on the final statement given by the clinician [14]. This may lead to intra and inter-rater changeability. This also reduces the accuracy of assessments in both clinical and research settings. Next is that according to the website "Parkinson's Foundation", the most used medication for Parkinson's disease is Levodopa. Common Side Effects of this medicine is lowering of blood pressure, sickness, confused behaviour, dyskinesia. Hence there is a need of system which discards this disadvantages [8].

There are no specific tests to detect PD. Along with tests like MRI, ultrasound, etc. Medication gives levodopia carbidopia, a Parkinson's drug. A dose sufficient to show the benefit. The confirmation of the presence of this disease is recognised by the slow improvements with the usage of this drug. Sometimes , in order to diagnose this disease , it takes lots of time [3]. The recommendation of doctors is then, a regular follow-up appointment with a neurologist who has training in movement disorders so that he can assess the condition and symptoms for this purpose.

Due to these problems that exist, our goal is to develop a system that helps detect PD mainly focusing on symptoms Tremor and bradykinesia (slowness of movement) at first stage without the use of drugs that ultimately has a lot of side effects. Also, making our system profitable is one of the main goal. The system that will be designed by us can be used by a common man thus reducing the regular check-ups of the patient to hospitals.

So, the goal is to develop a system to detect the existence of Parkinson's disease at an early stage. This system must be profitable. Everyone should be able to use this system, which reduces the number of visits to the doctor.

The rest of the paper is organized as follows: In section 2, the literature survey performed while Parkinson;s disease research was done. In section 3, the software system design used during the process of system used to detect PD is explained. Section 4 represents the implementation details of the designed system. Finally all the results are given in section 5 and the conclusions drawn from this work are presented in section 6.

II. Literature Survey

Parkinson's Disease (PD) is a disease that causes neurodegeneration. Within the substantia nigra there is a reduction of dopaminergic cells which leads to the cause of this disease. Mainly, Parkinson's disease affects the functions which causes motor movements.The major of the PD are bradykinesia, rigidity, resting tremor and loss of balance. the bradykinesia causes slowness of movements. Rigidity causes stiffness or inflexibility of muscles [10]. Resting tremor is a most common symptom while early stage detection of PD. The frequency of occurrence of these tremors is 3 Hz to 7 Hz. Overall 75 percent of PD patients has this symptom. The moment of thumb and index fingers with each other is significant feature of resting tremor. This tremor may extend to fore arm, elbow and arm. The neurologist subjectively evaluates and diagnoses the patients who are being suffered from this disease by checking the movement disorders, based on the Unified Parkinson's Disease Rating Scale (UPDRS).

There is a revised version of the UPDRS scale which is called as MDS-UPDRS scale. The license and ownership of this is with the International Parkinson and Movement Disorder Society (MDS) .For the usage of this scale, there is requirement of permission . On submitting a permissions request form, this can be done . The submission of the request has to be done on the MDS website.

The MDS-UPDRS consists of four parts:

- Part I includes non-motor experiences of daily living
  1) It is concerned with the behavioural changes that are assessed by the investigators with the help of information provided by patients or caregivers.
  2) The completion of this part is done by the patient itself independent of investigator
- Part II includes motor experiences of daily living
- Part III includes motor examination
- Part IV includes motor complications

However, for ensuring whether all the questions are answered clearly, this sections can be reviewed by the rater. Also any perceived ambiguities can be helped by rater . The design of part II is just like part IB, a self-administered questionnaire. Inspite of this, for ensuring completeness and clarity, review of investigator can be done. There are no "ON" or "OFF" rating to the official versions of PartIA, Part IB and Part II of the MDS-UPDRS. but, if "ON" and "OFF" and ratings are required to the individual programs, the same questions can be used separately. The instructions to the rater for completing and demonstrating to the patient is given in Part III. It is completed by the rater. The instructions which are to be read to the patient and also the one for rater are mentioned in Part IV. The completion of this is done by integrating the information derived by patients and the clinical observations by rater [9]. The Bradykinesia Akinesia Incoordination (BRAIN) test has been shown to be a reliable software method for identifying symptoms of neurological illness, such as Parkinson's disease and cerebellar dysfunction [1]. The Unified Parkinson's Disease Rating Scale (UPDRS) and other PD severity scales have also been compared using the alternative finger tapping measure [2]. The routine neurological examination includes sequential finger tapping to detect bradykinesia, which is described as "slowness of initiation of voluntary movement with gradual reduction in speed and amplitude of repetitive motion" [16]. The authors created a system consisting of tremor sensing device for detecting parkinson's disease which works with android application that tracks the tremors in upper limbs. The tremors at back of patients hand is the area where most of the tremors can be sensed [10]. These therapies include pharmacological regimens (the most critical of which is levodopa) and neurosurgical methods for later disease (e.g., deep brain stimulation; Dams et al. 2011, 2013; Tomaszewski and Holloway 2001) [17]. Chronic disorders with a high prevalence and high medical costs, such as multiple neurodegenerative diseases, including Parkinson's disease, have piqued the attention of health care professionals.

III. System Design

System software design is done using Android software. An application has been developed for detecting Parkinson's disease. Android app the programs are written in Java. This application will consist of several tasks. Tasks developed focuses on detecting symptoms of tremor and bradykinesia. Some of them the tasks included are the following:

- Hand Tremor
- Foot Tremor
- Fimger Tapping
- Spiral Drawing

All the gyroscope and accelerometer values obtained by performing the above tasks will be stored in the database. These values as well as the values obtained by the hardware part of the system will then be evaluated. The results thus obtained after evaluation helps to detect the Parkinson's disease.

TASK 1: HAND TREMOR

Parkinson's tremors are rhythmic, at rest, asymmetrical. The tremors of Parkinson's disease often start in the fingers or hands with what is called a pill-rolling motion. This test focuses on hand tremors.

Directions:

- In this Balance test you stand still for 30 seconds.
- Posture of this test asks you to hold the smartphone with stretched hands for 30 seconds.
- There is a build in counter that increases when any movement is detected.

Detection :

The value will be compared with the default value i.e the counter value of non PD person. If the value is greater than the default value then the patient may have this disease.

TASK 2: FOOT TREMOR

For people with Parkinson's Disease, walking may be difficult (PD). Since the disorder affects a part of the brain that regulates movement, this is the case. As the disease progresses, all activities, including walking, may become slower and smaller. People with PD can develop a stiff appearance as they walk slowly with their chest bent forward, with short quick "shuffling" steps, and with less arm and body movement. People with Parkinson's disease also experience "freezing" episodes, in which they become "stuck" and unable to take a step or continue walking [18].

Direction :

- Walk 30 yards in front
- The examination It can be done at your preferred walking pace or at the highest possible speed.

- To avoid influencing the patient's pace, the test administrator will walk at least a half step behind the patient.
- The gait of the patient is observed during this part of the examination. The patient should walk back and forth several times, preferably in a corridor at least 10 feet long, to maximise the likelihood of picking up any anomalies that might be present.

Detection :

The value will be compared with the default value i.e the counter value of non PD person. If the value is greater than the default value then the patient may have this disease.

TASK 3: FINGER TAPPING TEST

The finger-tapping test (FTT) is a neuropsychological assessment of motor function, especially motor speed and lateralized coordination. This exercise assesses the pace at which you tap each hand. Slowing of movement, lack of amplitude, and delays in finger taps are all signs of 'bradykinesia,' a slowing of movement that is one of Parkinson's disease's most common symptoms [19].

Directions:

- The patient is advised to tap the index finger on the thumb as quickly and as large as possible in finger tapping..
- This means that before tapping the two fingers, the patient can try to differentiate them as much as possible.
- Make sure to test both the right and the left side.

Detection : The counter value will be stored in the database. After comparing all the result of the activites, final result will be displayed

TASK 4: SPIRAL DRAWING Patients with Parkinson's typically write and draw at a slow pace. Patient with PD may notice shakiness when they're holding their hands up. As a result, handwriting and spirals tend to be smaller and tightly bunched. The slower speed and smaller size of the writing movements signal a motor symptom associated with Parkinson's. Spiral tests are typically a better diagnostic tool than using handwriting. Test also allows the neurologist or physician to better determine the severity of the tremor.

Directions :

- A test that involves drawing a spiral on a piece of paper may be used to detect Parkinson's disease in its early stages.
- Both are helpful in identifying the disease, which induces muscle rigidity and trembling.
- In the detection process, patient is told to draw spiral drawing coinciding with the template which is already present in the application.

Detection : After the drawing, the clinician with whom the patient is present, must check whether the drawing overlaps with the template. Accordingly, points will be provided which will be useful for the detection.

IV. System implementation

In the last section, hardware and software design of the system was discussed. By using these designs, the system is implemented. A survey is conducted, in which people from different age groups are told to perfrom the tasks. Based on the results of their tasks, a threshold value ranging from 0 to 5 is decided. This range is decided using MDS-UPDRS scale.

Algorithm of the tasks in the android application is mentioned below

1) Application starts with the basic information of the disease.

2)TASK 1: HAND TREMOR

- Declare all the necessary variables such as sensor, textviews, sensor services required
- Declare database variable
- Initialize all the sensors, textviews, buttons.
- Writing OnClickListeners for the buttons Start, Stop, Reset and Submit.
- Under OnClickListener for start button, describe a function onClick which will start the functioning of the accelerometer sensor on clicking the button.
- Under OnClickListener for stop button, describe a function onClick which will stop the functioning of the accelerometer sensor on clicking the button.
- Under onClickListener for reset button, describe a function onClick. In this function, write a code for resetting the counter value.
- Under onClickListener for submit button, describe function onClick which will add the counter value obtained on performing the task to the database.

3) The above algorithm is same for both hand tremor and foot tremor task.

4) TASK 3: FINGER TAPPING TEST

- Declare all the necessary variables such as sensor, textviews, sensor services required.
- Declare database variable.
- Initialize all the sensors, textviews, buttons.
- Writing OnClickListeners for the buttons Start, Stop, Reset button for timer, tap1 and tap2 button for counter as well as reset button for counter and Submit button.
- Under OnClickListener for start button, describe a function onClick which will start the functioning of the accelerometer sensor.

5) TASK 4: SPIRAL DRAWING TEST

- Declare all the necessary variables such as sensor, textviews and buttons required.
- Declare database variable.

- Initialize all the sensors, textviews, buttons for color selection, increasing pen size in order to draw
- Initialize Paint view variable
- In the Paint view java class,
    1) Declare variables for Path which will give the path in which pen is moving and Paint which will give the pen color.
    2) Under canvas object, using canvas.drawPath() command, drawing can be done.
- Write onClickListener for buttons increment, decrement of pen size, selecting the pen color and reset button.

V. Results and Analysis

All the counter values calculated using various sensors in the application, are stored in the database. Patient needs to click the submit button, at the end of every task. Because of this, the result of individual task will be displayed at the end of the application.

For finalizing the range as mentioned in MDS UPDRS, a survey was taken of people having age more than 50. Results of the survey helped us to analyze and set the range. The range is between 0 to 4, each having its own meaning as mentioned below: Here,

0 : No problem

1 : Slight

2 : Mild

3 : Moderate

4 : Severe

Lastly, based on the rating of individual tasks, final rating for the disease is provided. The following figure represents the result page in the application.

Figure 1: RESULTS OF THE TASK

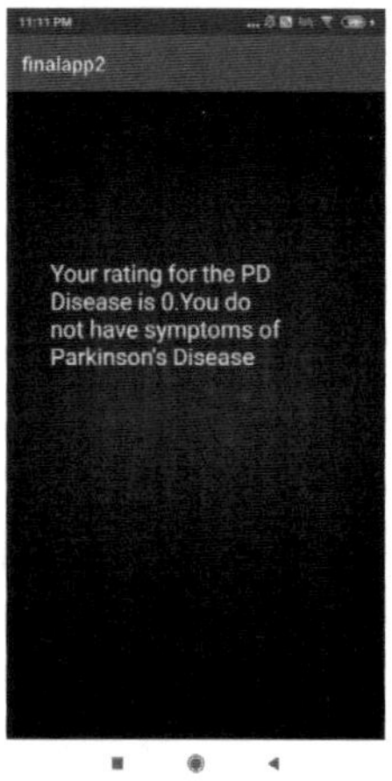

Figure 2: FINAL RESULT OF THE TASK

VI. Conclusion

The system designed has following advantages:

1) Since any medications are not used while detecting using our system, all the side effects of the medications like carbidopia – levodopia are thus discarded
2) Cost effective
3) Less complications involved
4) Time required for detection is less as compared to other systems

Thus using our proposed system, early stage detection of Parkison's disease is possible. This will result in cost effectiveness thus making it an efficient system.

References

[1] " Development of digital biomarkers for resting tremor and bradykinesia using a wrist-worn wearable device " by Nikhil Mahadevan1, Charmaine Demanuele1, Hao Zhang1, Dmitri Volfson , Bryan Ho2 , Michael Kelley Erb 1 and Shyamal Patel1*

[2] Journal of Neuroscience Methods. A palm-worn device to quantify rigidity in Parkinson's disease Thushara Pereraa, Wee-Lih Leea, Mary Jonesa,c, Joy L. Tana,b, Elizabeth L. Proudd,Angus Begga, Nicholas C. Sinclaira,b, Richard Pepparda, Hugh J. McDermotta,b

[3] Mayo Clinic, "MayoClinic.org"

[4] "Home Monitoring of Patients with Parkinson's Disease via Wearable Technology and a Web-based Application" by Shyamal Patel, Bor-rong Chen, Thomas Buckley, Ramona Rednic, Doug McClure, Daniel Tarsy, Ludy Shih, Jennifer Dy, Matt Welsh, Paolo Bonato

[5] "Quantitative Measurement of Rigidity in Parkinsons Disease: A Systematic Review" by María del Rosario Ferreira-Sánchez 1,*, Marcos Moreno-Verdú 1,2 and Roberto Cano-de-la-Cuerda 3

[6] "Quantitative Analysis of Motor Status in Parkinson's Disease Using Wearable Devices: From Methodological Considerations to Problems in Clinical Applications" by Masahiko Suzuki,1 Hiroshi Mitoma,2 and Mitsuru Yoneyama3

[7] "System to Monitor Tremors in Patients with Parkinson's Disease" by A. Bermeo, Student Member, IEEE, M. Bravo, Student Member, IEEE, M. Huerta, Senior Member, IEEE and A. Soto, Member, IEEE

[8] Parkinson's Foundation "www.parkinson.org"

[9] The International Parkinson and Movement Disorder Society "www.movementdisorders.org"

[10] "A System for Finger Tremor Quantification in Patients with Parkinson's Disease" by M. Bravo, Student Member, IEEE, A. Bermeo, Student Member, IEEE, M. Huerta, Senior Member, IEEE, C. Llumiguano, J. Bermeo, Member, IEEE, R. Clotet, Student Member, IEEE, and A. Soto, Member, IEEE.

[11] T. Lennon, et al. "Multi-sensory system for monitoring dyskinesia in movement disorders," in Proc. Biomedical Engineering Conference NEBEC, 41st Annual Northeast. IEEE, NY, USA, April 2015, pp. 1-2.

[12] G. Figueras, V. Parra, M. Huerta, A. Marzinotto, et al. "Smartphone application for quantitative measurement of Parkinson tremors," in Proc. VI Latin American Congress on Biomedical Engineering CLAIB 2014, Buenos Aires, Argentina 29, October 2014, pp. 785-788.

[13] "Evaluating the cost-effectiveness of an early detection of Parkinson's disease through innovative technology" by David A. Muñoza,b, Mehmet Serdar Kilinca,b, Harriet B. Nembharda,b, Conrad Tucker a,c, and Xuemei Huangd

[14] "Using wearables to assess bradykinesia and rigidity in patients with Parkinson's disease: a focused, narrative review of the literature" by Itay Teshuva1 · Inbar Hillel1 · Eran Gazit1 · Nir Giladi1,2,3 · Anat Mirelman1,2,3 · Jeffrey M. Hausdorff1,2,4,5,6

[15] MDS-UPDRS, The MDS-sponsored Revision of the Unified Parkinson's Disease Rating Scale.

[16] "Bradykinesia-Akinesia Incoordination Test: Validating an Online Keyboard Test of Upper Limb Function" by Alastair J. Noyce1,2*, Anna Nagy2, Shami Acharya2, Shahrzad Hadavi2, Jonathan P. Bestwick3, Julian Fearnley4, Andrew J. Lees1, Gavin Giovannoni2

[17] Taylor and Francis Online : Peer reviewed Journals, "www.tandfonline.com "

[18] Academy of Neurological physical therapy, "https://www.neuropt.org/"

[19] ClearSky medical diagnostics, "www.clearskymd.com"

AUTHOR INDEX

AUTHOR INDEX

AUTHOR INDEX